Springer Series in
Computational
Mathematics

20

Frank Stenger

Numerical Methods Based on Sinc and Analytic Functions

Springer-Verlag
New York Berlin Heidelberg London Paris
Tokyo Hong Kong Barcelona Budapest

Frank Stenger
Department of Computer Science
The University of Utah
Salt Lake City, UT 84112
USA

Mathematics Subject Classification: 30-00; 65-00

With 8 Illustrations.

Library of Congress Cataloging-in-Publication Data
Stenger, Frank.
Numerical methods based on Sinc and analytic functions/Frank Stenger.
p. cm. — (Springer series in computational mathematics; 20)
Includes bibliographical references and index.
ISBN 0-387-94008-1
1. Galerkin methods. 2. Differential equations—Numerical solutions. I. Title. II. Series.
QA372.S82 1993
515′.352—dc20 93-9403

Printed on acid-free paper.

Production managed by Natalie Johnson; manufacturing supervised by Vincent Scelta.
Photocomposed copy prepared from the author's LaTEX file.
Printed and bound by Edwards Brothers, Inc., Ann Arbor, MI.
Printed in the United States of America.

9 8 7 6 5 4 3 2 1

ISBN 0-387-94008-1 Springer-Verlag New York Berlin Heidelberg
ISBN 3-540-94008-1 Springer-Verlag Berlin Heidelberg New York

Preface

The primary purpose of this text is to familiarize the student and computational scientist with the use of *Sinc* methods. These methods fill a void in computational mathematics in a manner similar to the way in which polynomials, splines, and Fourier polynomials fill a void; although we can get by without the methods based on these functions, our environment is made richer when the methods are available. Indeed, once one gets accustomed to using the methods in any one of these classes, it is difficult to get along without them. Each existing method excels in some particular class of problems. For example, polynomials excel in the approximation of analytic functions without singularities, splines are particularly good for approximating measured, or noisy data, while Fourier polynomials excel in the approximation of functions that are both smooth and periodic over the whole real line. Sinc methods excel for problems with singularities, for boundary-layer problems and for problems over infinite or semi-infinite ranges.

The error of rational function approximation is the same as the error of Sinc approximation, in spaces where Sinc methods excel. On the other hand, Sinc methods have other features which make them more attractive than rational function methods. Sinc methods are replete with many properties and identities, which make them amenable for computation. The methods depend on a positive parameter h; when h is halved, every second evaluation point in the $h/2$ formula is the same as that in the h formula, and the $h/2$ formula has twice as many correct significant figures of accuracy as the h formula. Thus, convergence is very rapid, and this feature enables the computer to easily decide when a desired accuracy has been achieved. Furthermore, the computer programs based on Sinc methods are usually considerably shorter than the corresponding ones based on classical methods of approximation. Sinc methods are particularly powerful when the problem has singularities, in which case the error of an n-point approximation converges at an incredible $\mathcal{O}(e^{-cn^{1/2}})$ rate, whereas in such circumstances polynomial methods, at best, converge at the $\mathcal{O}(n^{-a})$ rate, where c is some positive number. Thus Sinc methods are particularly suited for solving partial differential and integral equations; for such problems the resulting system of algebraic equations can always be written down explicitly, and is relatively small in size, the determination of the close to optimal mesh spacing is automatic, curvilinear boundaries or corners do not pose problems, and the use of the fast Fourier transform (FFT—which is based on one of the Sinc methods—the DFT approximation!) can be conveniently applied to yield a highly efficient code.

Sinc methods are based on the *Cardinal function.* We use the notation

$$\mathrm{S}(k,h)(x) = \frac{\sin[\pi(x-kh)/h]}{\pi(x-kh)/h},$$

to denote the *Sinc function*, with h a positive number, and k an integer. Given a function f defined and bounded for all x on $(-\infty,\infty)$, the Cardinal function of the function f is defined by

$$\mathrm{C}(f,h)(x) = \sum_{k=-\infty}^{\infty} f(kh)\,\mathrm{S}(k,h)(x).$$

The Cardinal function plays the same role for Sinc methods as polynomials play for the more classical numerical methods, in that operations such as differentiation, integration, etc. on Sinc approximations of functions produce approximations of these operations on the functions themselves. Just as for polynomials, many operations on $C(f,h)$ can be carried out explicitly, and each of these yields an explicit methods of approximation. The function $\mathrm{C}(f,h)$ connects many approximation formulas such as the trapezoidal rule, the FFT method, and many techniques from the area of signal processing, all of which are easy to derive from $\mathrm{C}(f,h)$.

The Cardinal function dates back to the works of Bohr, de la Vallée Poussin, and E.T. Whittaker. It occupies an important place in the theory of analytic functions. My mentor, J.J. McNamee, rightly called this beautiful function $C(f,h)$ "a function of royal blood, whose distinguished properties separate it from its bourgeois brethren".

Research involving $\mathrm{C}(f,h)$ slowed somewhat after J.M. Whittaker's book appeared. The function $\mathrm{C}(f,h)$ reappeared later in the literature, this time on engineering applications, in the important papers of Hartley, Nyquist and Shannon, who illustrated the role of $\mathrm{C}(f,h)$ in communication theory. The term "sinc function"—different from our definition of "Sinc function" above—was first defined and used in these papers, where $\mathrm{sinc}(x)$ is defined by

$$\mathrm{sinc}(x) = \frac{\sin(\pi x)}{\pi x}.$$

This notation is currently well known to engineers. We have chosen not to use this notation, since the above form $\mathrm{S}(k,h)(x)$ is more convenient and powerful for our purposes.

The theory of analytic functions presented in Chapter 1 provides the majority of mathematical tools required for this text. While the derivation of the Sinc approximations requires relatively little of this theory, the more extensive coverage of the theory of complex variables makes it possible to establish the rapid $\mathcal{O}(e^{-cn^{1/2}})$ convergence rate for the many different applications of Sinc methods described in the text, and also it helps us to understand the role of Sinc methods relative to other classical methods of approximation. This chapter thus presents a reasonably complete coverage of elementary analytic function theory, with special emphasis on certain topics such as the Paley–Wiener theorem, the Cardinal function, Fourier, Laplace and Hilbert transforms, and Riemann–Hilbert problems, which are required for the presentation of this text. The presentations emphasize applications. For example, Taylor's and Laurent's theorem are given in truncated polynomial form, along with bounds on the error of approximation by these polynomials, these bounds being close to the best possible in the case of the Laurent polynomial. The connection is illustrated between Laurent's theorem and Fourier series. The transforms of Hilbert, Fourier, and Laplace are presented from an analytic function viewpoint. Some conformal and other special transformations are presented, as well as a discussion of some important spaces of analytic functions, the Paley–Wiener theorem, and the derivation of the cardinal series. The material is covered in an elementary fashion, so that this part of the text could be used in an advanced undergraduate or beginning graduate course on applied complex variable theory. On the other hand, if this text book is used in an advanced course in numerical analysis, or applied approximation theory, and all the students understand complex variables, most of this material can be skipped, or referred to, as the need arises.

Chapter 2 involves polynomial approximation. The discrete Fourier transform (DFT) is carefully developed here, since it provides vital links between the Cardinal series, FFT, and Fourier series, as well as with the important special case of Chebyshev polynomial approximation. The use of Laurent's theorem enables us to get excellent bounds on the error of polynomial approximation. One also finds, here, an analytic function derivation of the general Lagrange interpolation polynomial, and a derivation of the Faber polynomial.

Chapters 3 to 7 attempt to summarize research on Sinc numerical methods, which my students and I, as well as some other mathematicians have carried out over the past thirty years. Besides aiming at a unified presentation, it is also desirable to achieve a certain amount of completeness. Thus, many of the results presented in this text are new, resulting from my "straying" to fill gaps in existing research.

Chapter 3 deals with various methods of approximation via the direct use of the Cardinal function. Let f be a function analytic in an infinite strip $\mathcal{D}_d$ containing the real axis. Derivations are given here for the finite and infinite-term interpolation of f, and various norms of the error are derived for these interpolations, not only on the real line, but within an arbitrary sub-strip of $\mathcal{D}_d$. The trapezoidal rule may be obtained by termwise integration of the Cardinal function, and the quadrature error of f by the trapezoidal rule is also studied. When taking the Fourier transform of the Cardinal series, one obtains a simple explicit Fourier series representation of the Fourier transform, and an accurate bound on the error is given by the approximation of the Fourier transform of f by this Fourier series. We also obtain a Fourier series representation of the same Fourier transform via the DFT procedure of Section 2.2, along with an accurate error bound. Similar error bounds are obtained on the approximations of the Hilbert transform, the derivatives, and the indefinite integral of f by the explicit expressions provided by the Hilbert transform, the derivatives, and the indefinite integral of the Cardinal function of f.

In Chapter 4, results of the previous chapter involving approximation over the real line are extended to approximation over a contour Γ. Specifically, an explicit conformal mapping, ϕ, is required, of a domain $\mathcal{D}$ of analyticity of a function F onto the infinite strip $\mathcal{D}_d$ containing the real line. This function ϕ also maps the contour Γ contained in $\mathcal{D}$ to the real line. The new basis functions, $\mathrm{S}(k,h) \circ \phi(x)$ for Γ, now replace the previous basis functions, $\mathrm{S}(k,h)(x)$, for the real line. New interpolation points $\{z_k\}_{k=-\infty}^{\infty} = \{\phi^{-1}(kh)\}_{k=-\infty}^{\infty}$ for F now replace the previous points $\{kh\}_{k=-\infty}^{\infty}$ which were used for interpolating f on the real line. The important spaces $\mathbf{L}_\alpha(\mathcal{D})$ and $\mathbf{M}_\alpha(\mathcal{D})$ are defined, and some properties of these spaces are derived. In the case of a bounded region, the space $\mathbf{M}_\alpha(\mathcal{D})$ is equivalent to the space of all functions that are analytic and of class $\mathbf{Lip}_\alpha$ in the region. These spaces house solutions of differential and integral equations arising in applications, and they are also the spaces in which Sinc approximations excel, i.e., in which we can carry out n-point Sinc approximation to within an error $\mathcal{O}(e^{-cn^{1/2}})$. For finite, or semi-infinite regions these spaces include functions that have singularities at endpoints of the interval. Derivations are then given of finite and infinite-point formulas for a general contour Γ, as well as for the important specific finite intervals $\Gamma = (0,1)$, (a,b), and for the semi-infinite interval $\Gamma = (0,\infty)$. We discuss interpolation and quadrature in Section 4.2, Hilbert and related transforms in Section 4.3, approximation of derivatives in Section 4.4, approximation of indefinite integrals in Section 4.5, and new formulas for the approximation of indefinite integral convolutions in Section 4.6. All of these formulas converge at the above cited exponential and close to optimal rate.

Chapter 5 discusses methods that are related to Sinc methods, which use the same Sinc points z_k as evaluation points of the function F, although the basis functions are not necessarily the functions $\mathrm{S}(k,h) \circ \phi(x)$. Elliptic function approximants are derived, which have similar properties and power of approximation as the previously derived Sinc approximants. These elliptic function approximants are then used to generate a family of rational function approximants which also share many similarities with previously derived Sinc formulas, including the rapid rate of convergence. These rational approximants make it possible to deduce *a priori* when we can expect rational approximation to work, as well as conditions under which we can expect extrapolation to the limit (right up to the singularity) based on rational functions to be accurate. Other topics covered in this chapter include a method to invert the Laplace transform based on using only the values of the transform on the real line, several non-interpolatory methods to approximate a function via the use of exponential, rational, and *sech* functions, as well as some rational function methods to approximate the characteristic, the Heaviside, and the delta function. This chapter is in a sense not fully developed; while it contains some practically useful results, it merely demonstrates without extensive study, that the formulas with the approximating power of Sinc formulas can be derived in other ways, and that there exist families of other approximation formulas which use the same function values as the Sinc formulas.

Chapter 6 illustrates the application of Sinc methods to the approximate solution of integral equations. We present here a brief Banach-space development of the theory of Volterra and (mainly linear) Fredholm integral equations, the contraction mapping principle, convergence and error of approximation, the Fredholm alternative, convergence and error of approximate eigenvalues and eigenspaces, regularization, the use of Galerkin or Nyström's method to get a system of algebraic equations, the generalized inverse solution of a system of linear algebraic equations, and errors in the solution of these equations. We then illustrate the use of Sinc methods to obtain approximate solutions of non-linear Volterra integral equations, of a class of three-dimensional potential theory problems, along with an example, of the solution of the three-dimensional Lippmann–Schwinger integral equation on a half space, of a Cauchy singular integral equation over an arc in the complex plane, of one-dimensional convolution and Wiener–Hopf integral equations, and of the inversion of the Laplace transform. Special care is taken in all of these cases to establish the presence of the analyticity in the solutions which enables us to establish the rapid $\mathcal{O}(e^{-cn^{1/2}})$ convergence rate made possible by Sinc methods.

The last chapter of the text, Chapter 7, illustrates the use of Sinc methods to obtain approximate solutions of ordinary (ODE) and partial differential (PDE) equations for both initial and boundary value problems. Initially we discuss analyticity of solutions of ordinary differential equations, and the type of analyticity which we require for effective Sinc approximation of solutions for such equations. In Section 7.1, we present a new method to solve ordinary differential equation initial value problems, and since stiffness is a boundary-layer phenomena, it is almost as easy via this method to solve stiff as non-stiff problems, regardless of whether the interval on which the solution is defined is finite or infinite. We then illustrate the application of the methods on some explicit examples. In Section 7.2 we present derivations of slightly differing Sinc–Galerkin methods, one (a new version) by collocation, and one (the original version) by integration by parts. By proving the convergence of these methods, we thus show that

Galerkin, finite element, spectral, and collocation methods are essentially the same for the Sinc methods, since they all yield nearly the same system of algebraic equations, whose solutions have the same order of accuracy. We also derive the symmetric Sinc–Galerkin method, and its nearly equivalent collocation version. Our illustrative applications include the use of Sinc methods to obtain approximate solutions of ODE boundary value problems, including eigenvalue problems, and problems whose solutions have algebraic-type singularities at the endpoints of the interval, singular perturbation ODE boundary value problems, and an ODE formulation of a Klein–Gordon equation. In Section 7.3, we discuss analyticity of solutions of partial differential equations, spaces of solutions which lend themselves to effective Sinc approximation, and transformations of curvilinear regions to rectangular ones. These transformations are such that they preserve the accuracy of Sinc approximation and, hence, they provide the key for domain decomposition and parallel computation for Sinc methods. We also present a version of Fichera's theorem, which for our purposes provides a sufficiently accurate description of singularities due to corners in the domain. In Section 7.4 we illustrate the use of Sinc methods for solving elliptic partial differential equations. Examples include the Poisson problem $\nabla^2 u = F$ on a square and over curvilinear regions. In the latter case, we compare the complexity of various popular methods of solution for such a problem. We also illustrate the use of a special technique based on previously developed Sinc methods to obtain an approximate solution of a planar Dirichlet problem whose boundary consists of a finite number of smooth arcs. In Section 7.5 we briefly illustrate the use of Sinc methods for solving simple linear hyperbolic partial differential equation problems. In Section 7.6 we illustrate the use of Sinc methods for solving parabolic problems. We include a discussion of the solutions of the heat equation problem $u_t = \nabla^2 u$ for both the real line and for a finite interval, as well as of a conservation-law (shock) problem in n dimensions.

Each section of the text ends with a set of problems, with a "$*$" beside the more difficult ones. Each chapter closes with historical remarks that indicate the source of the material.

I am very grateful for criticisms and corrections of this manuscript, which were given to me by S. Å. Gustafson, A. Morlet, A. Naghsh–Nilchi, M. O'Reilly, M. Sugihara, and G.Y. Yin. I also wish to acknowledge that I was supported by an IBM grant on distributed computation during part of the duration of the writing of this text. I discovered some of the novel results of the text during this time, and with the help of my students, developed these into effective computer algorithms.

Frank Stenger February 13, 1993

Contents

1

Mathematical Preliminaries

In this chapter we introduce some mathematical concepts which we shall invoke at least once in the remainder of this monograph. Section 1.1 is a self-contained presentation of concepts of analytic function theory which are both useful and powerful in applications. Sections 1.2 through 1.9 introduce some other mathematical concepts which are important to this text, such as Hilbert, Fourier and Laplace transforms, Fourier series, conformal mapping, spaces of analytic functions and representation of classes of functions via the cardinal function.

1.1 Properties of Analytic Functions

This section provides a simple and self-contained presentation of basic concepts of analytic function theory which are used in presenting the results of the remainder of this text. These concepts have provided important insights to the author, in choosing or developing methods to obtain approximate solutions to problems in applications.

Let $\mathbb{R}$ denote the *real line*, i.e. $\mathbb{R} = (-\infty, \infty)$, let $\mathbb{C}$ denote the *complex plane*, $\mathbb{C} = \{z = x + iy; x \in \mathbb{R}, y \in \mathbb{R}\}$, and let $\mathbb{Z}$ denote the set of all integers. Let $\mathcal{D}(a; R)$ denote the *open disc* with center $a \in \mathbb{C}$ and radius $R > 0$, i.e.,

$$\mathcal{D}(a; R) = \{z \in \mathbb{C} : |z - a| < R\}. \tag{1.1.1}$$

Let S denote a subset of $\mathbb{C}$. A point $a \in \mathbb{C}$ is called an *interior point* of S if there exists a disc $\mathcal{D}(a; R)$ such that all points of $\mathcal{D}(a; R)$ belong to S. Point a is a *boundary point* of S if every disc $\mathcal{D}(a; R)$ contains both points in S and points which do not belong to S. The *boundary* of S, denoted by ∂S, is equal to the union of all boundary points of S. We will call S *open* if no boundary point of S belongs to S, and we denote $\overline{S} = S \cup \partial S$ to be the *closure* of S. The set S is *closed* if $\overline{S} = S$. The set S is said to be *bounded* if there exist numbers a in $\mathbb{C}$ and $R > 0$ such that the disc $\mathcal{D}(a; R)$ contains the closure $\overline{S}$ of S.

The sequence $\{z_n\}_{n=1}^{\infty}$ with $z_n \in \mathbb{C}$ (or more simply written $\{z_n\}$ if there is no ambiguity about the limits on n) is said to *converge to the point*, a, in $\mathbb{C}$, and we write $\lim_{(n \to \infty)} z_n = a$, if given any $\varepsilon > 0$ there exists an integer $N > 0$ such that whenever $n > N$, then $z_n \in \mathcal{D}(a; \varepsilon)$. The sequence $\{z_n\}$ is called a *Cauchy sequence* if given any $\varepsilon > 0$ there exists an integer $N > 0$ such that whenever $m > N$, $n > N$, then $|z_m - z_n| < \varepsilon$. A set S in $\mathbb{C}$ is said to have a *limit point* a if there exists a sequence $\{z_n\}$ of distinct points in $\mathbb{C}$, such that $\lim_{(n\to\infty)} z_n = a$. A limit point is also called an *accumulation point.*

Theorem 1.1.1 (Bolzano–Weiestrass Theorem): *Every bounded infinite subset S of $\mathbb{C}$ has at least one limit point.*

Proof. Let $\overline{S}$ belong to $\mathcal{D}(a;R)$, and define the closed square σ_0 by

$$\sigma_0 = \{z \in \mathbb{C} : \max\left[|\Re(z-a)|, |\Im(z-a)|\right] \leqq R\} \tag{1.1.2}$$

then σ_0 contains $\mathcal{D}(a;R)$, and hence $\overline{S}$. We also define four other closed squares $\sigma_1^k, k = 0,1,2,3$, whose union is σ_0, by

$$\sigma_1^k = \Big\{z \in \mathbb{C} : \max\Big[|\Re(z - a - i^{k+1/2}R/2^{1/2})|, \\ |\Im(z - a - i^{k+1/2}R/2^{1/2})|\Big] \leqq R/2\Big\} \tag{1.1.3}$$

Since σ_0 contains an infinite set (S), at least one of the four squares σ_1^k also contains an infinite set. Denote the σ_1^k which does contain such an infinite set by σ_1. We then subdivide σ_1 into four equal squares σ_2^k, $(k = 0,1,2,3)$ and identify one of these, σ_2, which contains an infinite subset of S, and so on. Proceeding in this way, we obtain a sequence of closed squares, σ_0, σ_1, $\sigma_2, \cdots$, $\sigma_m, \cdots$, each of which contains an infinite set of points, such that the squares are ordered by containment, $\sigma_0 \supset \sigma_1 \supset \sigma_2 \supset \cdots \supset \sigma_m \supset \cdots$ and such that each side of the square σ_m has length $2R/2^m$. As $m \to \infty$, σ_m converges to a single point, say, ζ, and since each σ_m is closed, ζ belongs to every σ_m. That is, we can pick an arbitrary point z_0 belonging to σ_0 and S, then pick a point z_1, belonging to σ_1 and S, such that $z_1 \neq z_0$, next, pick z_2 such that z_2 belongs to both σ_2 and S, and such that $z_2 \neq z_0$, $z_2 \neq z_1$, and so on, to obtain a Cauchy sequence $\{z_n\}$, with $\lim_{(n\to\infty)} z_n = \zeta$. ■

Corollary 1.1.2 *Every Cauchy sequence in $\mathbb{C}$ has a limit.*

Proof. See Problem 1.1.1. ■

Some Definitions

An *arc* Γ is defined by a mapping $z(t) = x(t) + iy(t)$, with x and y real-valued functions that are continuous on $(0,1)$, such that the derivative $z'(t) = x'(t) + iy'(t)$ exists and is sectionally continuous on $(0,1)$ and such that $x'(t)^2 + y'(t)^2 > 0$ at each point of continuity of $z'(t)$. A *contour* is an arc $\mathcal{L} = \{z(t) : 0 \leqq t \leqq 1\}$ such that $z(t)$ is continuous on the closed interval $[0,1]$, and such that $z(0) = z(1)$.

We shall next discuss the concept of a *domain.* Let $\mathcal{L} = \{z(t) : 0 \leqq t \leqq 1\}$ be a contour, such that whenever $t_1 \in (0,1)$, $t_2 \in (0,1)$, and $t_1 \neq t_2$, then $z(t_1) \neq z(t_2)$. It is known that such a contour always divides $\mathbb{C}$ into two regions, B_1 and B_2 having $\mathcal{L}$ as a common boundary, such that if $z_1 \in B_1$, and $z_2 \in B_2$, and such that if $\mathcal{P}$ is an arc connecting z_1 and z_2, then $\mathcal{P}$ cuts $\mathcal{L}$ at least once. All of the points in one region, say B_1, have a bounded modulus, whereas the region B_2 contains points of arbitrary large modulus.

Let $\mathcal{L}$ be the boundary of a region B, i.e., $\mathcal{L} = \partial B$. We say that $\mathcal{L}$ is *positively oriented* if upon traversing $\mathcal{L}$ with increasing t, the region B is always to the left of $\mathcal{L}$.

Definition 1.1.3 *(Definition of a k-Connected Domain): Let k be a positive integer, and for $m = 1, 2, \cdots, k$, let us consider k sequences of contours $\{\mathcal{L}_{mj}\}_{j=1}^{\infty} = \{z_{mj}(t) : 0 \leqq t \leqq 1\}_{j=1}^{\infty}$, having the following properties:*

(a) *$\mathcal{L}_{mj} = \partial\mathcal{B}_{mj}$, with $\mathcal{B}_{mj}$ a bounded open set in $\mathbb{C}$, and with $\mathcal{L}_{mj}$ positively oriented with respect to $\mathcal{B}_{mj}$;*

(b) *For each fixed m, the regions $\mathcal{B}_{mj}$ are nested, i.e., $\mathcal{B}_{mj} \subset \mathcal{B}_{m,j+1}, j = 1, 2, \cdots$, and furthermore $\mathcal{L}_{mj}$ and $\mathcal{L}_{m,j+1}$ have no common points;*

(c) *For each fixed j, the k regions $\mathcal{B}_{mj}$ $(m = 1, 2, \cdots, k)$ have no common points;*

(d) *Let regions $\mathcal{D}_j$ and $\mathcal{D}$ be defined by:*

$$\begin{aligned} \mathcal{D}_j &= \bigcup_{m=1}^{k} \mathcal{B}_{mj} \quad , \\ \mathcal{D} &= \lim_{j \to \infty} \mathcal{D}_j \quad . \end{aligned} \tag{1.1.4}$$

The regions $\mathcal{D}_j$ and $\mathcal{D}$ defined in this manner are called *k-connected domains.* For example, if $a \in \mathbb{C}$ and $R > 0$, the $\mathcal{D}(a;R)$ is a simply-connected domain. If $R > 1$, then the region $\{z \in \mathbb{C} : 1/R < |z| < R\}$ is a doubly-connected domain.

All of the domains $\mathcal{D}$ which we shall consider will be assumed to be k-connected, for some positive (finite) integer k. *If $k = 1$, then $\mathcal{D}$ is called simply-connected.*

A function f defined on a domain $\mathcal{D}$ is said to be a *mapping* of $\mathcal{D}$ into $\mathbb{C}$, and we write $f : \mathcal{D} \to \mathbb{C}$. The function $f : \mathcal{D} \to \mathbb{C}$ is said to have *limit* b at the point a in $\mathcal{D}$ if given any $\varepsilon > 0$ there exists an $R > 0$ such that the disc $\mathcal{D}(a;R)$ is in $\mathcal{D}$, and such that whenever $z \in \mathcal{D}(a;R), z \neq a$, then

$f(z) \in \mathcal{D}(b;\varepsilon)$. We can then write $\lim_{(z\to a)} f(z) = b$. The function $f : \mathcal{D} \to \mathbb{C}$ is said to be *continuous at the point* a *in* $\mathcal{D}$ if $\lim_{(z\to a)} f(z) = f(a)$, and f is said to be continuous in $\mathcal{D}$ if f is continuous at each point of $\mathcal{D}$. The continuous function $f : \mathcal{D} \to \mathbb{C}$ is said to be *differentiable at a point* $a \in \mathcal{D}$ if there exists a number $c \in \mathbb{C}$ such that

$$\lim_{z\to a} \frac{f(z) - f(a)}{z - a} = c. \tag{1.1.5}$$

In this case c is the *derivative* of f at a, and we write $f'(a) = c$. The same function f is said to be *analytic* or *holomorphic* in $\mathcal{D}$, *if* f *is differentiable at each point of* $\mathcal{D}$. *The family of all functions* f *that are analytic in a domain* $\mathcal{D}$ *will be denoted by* $\mathbf{Hol}(\mathcal{D})$.

Analytic functions are rather special when compared with real-valued differentiable functions, since, as we shall see, the existence of f' in $\mathcal{D}$ implies the existence of $f^{(n)}$ in $\mathcal{D}$ for every positive integer n.

Let $f \in \mathbf{Hol}(\mathcal{D})$, and let us set

$$\begin{aligned} u(x,y) &= \Re f(x+iy) \\ v(x,y) &= \Im f(x+iy) \end{aligned} \tag{1.1.6}$$

with $x \in \mathbb{R}, y \in \mathbb{R}$ and $z = x + iy \in \mathcal{D}$. The functions u and v are called *harmonic functions*. Then, with z fixed in $\mathcal{D}$, we have

$$\begin{aligned} f'(z) &= \lim_{h\to 0, h\in \mathbf{R}} \frac{f(z+h) - f(z)}{h} \\ &= u_x(x,y) + iv_x(x,y). \end{aligned} \tag{1.1.7}$$

Repeating the process, we also have

$$\begin{aligned} f'(z) &= \lim_{h\to 0, h\in \mathbf{R}} \frac{f(z+ih) - f(z)}{ih} \\ &= v_y(x,y) - iu_y(x,y). \end{aligned} \tag{1.1.8}$$

In (1.1.7) and (1.1.8) we have used the notation $u_x = \partial u/\partial x$, etc. Upon equating (1.1.7) and (1.1.8), we get

Theorem 1.1.4 *Let* f *be in* $\mathbf{Hol}(\mathcal{D})$, *and let* u *and* v *be defined by (1.1.6). Then the Cauchy–Riemann equations, i.e.*

$$\begin{aligned} u_x(x,y) &= v_y(x,y) \\ u_y(x,y) &= -v_x(x,y) \end{aligned} \tag{1.1.9}$$

hold at each point of $\mathcal{D}$.

The integral of a function f defined in $\mathcal{D}$ over an arc (or contour) $\mathcal{L}$ in $\mathcal{D}$ is defined by

$$\begin{aligned} \int_{\mathcal{L}} f(z)\,dz &= \int_0^1 f(z(t))\,z'(t)\,dt \\ &= \int_0^1 \{[u(x(t),y(t))\,x'(t) - v(x(t),y(t))\,y'(t)] \\ &\quad +i\,[u(x(t),y(t))\,y'(t) + v(x(t),y(t))\,x'(t)]\}\,dt, \end{aligned} \tag{1.1.10}$$

which is said to exist, provided that each of the four integrals on the right-hand side of (1.1.10) exist in the real variable sense.

While the construction of the sequence $\{\mathcal{D}_j\}_{j=1}^{\infty}$ in Definition 1.1.3 serves many purposes, the set of all such sequences $\{\mathcal{D}_j\}$ that satisfy Definition 1.1.3 for a given domain $\mathcal{D}$ is too large to define integrals over the boundary of $\mathcal{D}$, since, for example, when performing integrals of absolute values of functions we may get different answers for different choices of sequences $\{\mathcal{D}_j\}$. For our purposes, we thus adopt the following notion for the definition of integrals over the boundary of a domain $\mathcal{D}$.

Definition 1.1.5 *Let $\mathcal{D}$ be a k-connected domain, and let $\mathbf{D}(\mathcal{D})$ denote the family of all sequences $\{\mathcal{D}_j\}$ defined as in Assumption 1.1.3. Let $f \in$ $\mathbf{Hol}(\mathcal{D})$, except possibly at a countable set of isolated points in $\mathcal{D}$, where f' does not exist. Let p be a fixed number such that $1 \leqq p \leqq \infty$. Let $\mathbf{D}_p(f,\mathcal{D})$ denote the set of all sequences $\{\mathcal{D}_j\} \in \mathbf{D}(\mathcal{D})$ such that if $\{\mathcal{D}_j\} \in \mathbf{D}_p(f,\mathcal{D})$ then $N_p(f,\mathcal{D}) < \infty$, with*

$$\begin{aligned} N_p(f,\mathcal{D}) &= \lim_{j\to\infty} \left(\int_{\partial\mathcal{D}_j} |f(z)|^p\,|dz| \right)^{1/p} \\ &\equiv \inf_{\mathcal{D}_j \in \mathbf{D}_p(f,\mathcal{D})} \sup_{j\to\infty} \left(\int_{\partial\mathcal{B}_j} |f(z)|^p\,|dz| \right)^{1/p}, \quad 1 \leqq p < \infty, \qquad (1.1.11) \\ N_\infty(f,\mathcal{D}) &= \lim_{j\to\infty} \sup_{z\in\partial\mathcal{D}_j} |f(z)| \\ &\equiv \inf_{\mathcal{D}_j \in \mathbf{D}_\infty(f,\mathcal{D})} \sup_{j\to\infty} \sup_{z\in\partial\mathcal{D}_j} |f(z)|. \end{aligned}$$

The case $p = 1$ is particularly important in this text, and we shall write $N(f,\mathcal{D})$ instead of $N_1(f,\mathcal{D})$.

Given any f, with $N_p(f, \partial\mathcal{D}) < \infty$, we shall always define $\int_{\partial\mathcal{D}} f(z)\,dz$ by

$$\int_{\partial\mathcal{D}} f(z)\,dz = \lim_{j\to\infty} \int_{\partial\mathcal{D}_j} f(z)\,dz, \tag{1.1.12}$$

with $\{\mathcal{D}_j\} \in \mathbf{D}_p(f, \mathcal{D})$. Furthermore, we denote the special set of all $f \in$ $\mathbf{Hol}(\mathcal{D}) \cap \mathbf{D}_p(f, \mathcal{D})$ by $\mathbf{H}^p(\mathcal{D})$.

Theorem 1.1.6 (Cauchy's Theorem): *If $\mathcal{D}$ is a k-connected domain, and if $f \in \mathbf{H}^p(\mathcal{D})$, for some $p \in [1, \infty]$, then*

$$\int_{\partial\mathcal{D}} f(z)\,dz = 0. \tag{1.1.13}$$

Proof. We set $f = u + iv$, $z = x + iy$ with u and v defined, as in (1.1.6), and with x and y are real. Then, with $\{\mathcal{D}_j\} \in \mathbf{D}_p(f, \mathcal{D})$ defined, as in Definition 1.1.5, we have

$$\int_{\partial\mathcal{D}_j} f(z)\,dz = \int_{\partial\mathcal{D}_j} [u\,dx - v\,dy + i(u\,dy + v\,dx)]. \tag{1.1.14}$$

Now applying Green's theorem to the integral on the right-hand side of (1.1.14), we get

$$\int_{\partial\mathcal{D}_j} f(z)\,dz = \int\int_{\mathcal{D}_j} [-u_y - v_x + i(u_x - v_y)]\,dx\,dy = 0 \tag{1.1.15}$$

since the Cauchy–Riemann equations are satisfied on $\mathcal{D}_{j+1} \supset \mathcal{D}_j$. Equation (1.1.13) thus follows by taking the limit as $j \to \infty$ in (1.1.15). ■

Theorem 1.1.7 (Cauchy's Formula): *If $\mathcal{D}$ is a k-connected domain, if $f \in \mathbf{H}^p(\mathcal{D})$, for some $p \in [1, \infty]$, and if a is a point in $\mathcal{D}$, then*

$$\frac{1}{2\pi i} \int_{\partial\mathcal{D}} \frac{f(z)}{z - a}\,dz = f(a). \tag{1.1.16}$$

Proof. See Problem 1.1.2. ■

Cauchy's formula has many applications. For example, if f is analytic in $\mathcal{D}$, then for arbitrary $\mathcal{D}(a; R)$ in $\mathcal{D}$, it immediately yields the result

$$f(a) = \frac{1}{2\pi} \int_{-\pi}^{\pi} f(a + Re^{i\theta})\,d\theta, \tag{1.1.17}$$

where, to get (1.1.17), we have set $z = a + Re^{i\theta}$ in (1.1.16). Upon setting $u = \Re f$, $v = \Im f$, then we also have

$$u(a) = \frac{1}{2\pi} \int_{-\pi}^{\pi} u(a + Re^{i\theta})\, d\theta,$$
$$v(a) = \frac{1}{2\pi} \int_{-\pi}^{\pi} v(a + Re^{i\theta})\, d\theta. \tag{1.1.18}$$

Thus $f(a), u(a)$ and $v(a)$ are respectively the average values of f, u and v on $\partial\mathcal{D}(a;R)$. This proves

Theorem 1.1.8 (Maximum Principle): *If f is analytic in $\mathcal{D}$,*

$$\sup_{z \in \mathcal{D}} |f(z)| = \sup_{z \in \partial\mathcal{D}} |f(z)|. \tag{1.1.19}$$

Moreover, for all z in $\mathcal{D}$,

$$\inf_{w \in \partial\mathcal{D}} u(w) \leqq u(z) \leqq \sup_{w \in \partial\mathcal{D}} u(w),$$
$$\inf_{w \in \partial\mathcal{D}} v(w) \leqq v(z) \leqq \sup_{w \in \partial\mathcal{D}} v(w). \tag{1.1.20}$$

Theorem 1.1.9 (Taylor's Theorem): *Let $1 \leqq p \leqq \infty$, let $f \in \mathbf{H}^p(\mathcal{D}(a;R))$, and let a_n $(n = 0, 1, 2, \cdots)$ be defined by*

$$a_n = \frac{1}{2\pi i} \int_{\partial\mathcal{D}(a;R)} f(z)(z-a)^{-n-1}\, dz. \tag{1.1.21}$$

Then, $a_n = f^{(n)}(a)/n!$, and

$$f(z) = \sum_{n=0}^{\infty} a_n (z-a)^n, \tag{1.1.22}$$

where the series on the right-hand side of (1.1.22) converges to $f(z)$ for all z in $\mathcal{D}(a;R)$. Furthermore, if f is bounded by M on $\mathcal{D}(a;R)$, then

$$|a_n| \leqq \frac{M}{R^n}, \quad n = 0, 1, 2, \cdots, \tag{1.1.23}$$

and moreover, it then follows for every positive integer m, that

$$\left| f(z) - \sum_{n=0}^{m-1} a_n (z-a)^n \right| \leqq \frac{MR}{R - |z-a|} \left(\frac{|z-a|}{R} \right)^m. \tag{1.1.24}$$

Proof. We defer the proof of this important theorem to Problem 1.1.3.

However we shall prove the more complicated Laurent's Theorem, which follows. In that theorem we shall use the notations

$$\begin{aligned} \mathcal{A}(a;b) &= \{z \in \mathbb{C} : a < |z| < b\} \\ \mathcal{A}_R &= \{z \in \mathbb{C} : 1/R < |z| < R\} \end{aligned} \tag{1.1.25}$$

where we assume for the case of the annulus $\mathcal{A}(a;b)$ that $0 \leqq a < b \leqq \infty$, and in the case of $\mathcal{A}_R$, that $R > 1$. ■

Theorem 1.1.10 (Laurent's Theorem): *Let $1 \leqq p \leqq \infty$, let f be analytic in $\mathcal{A}(a;b)$, and for $n = 0, \pm 1, \pm 2, \cdots$, define c_n by*

$$c_n = \frac{1}{2\pi i} \int_{|z|=\rho} f(z) z^{-n-1}\, dz, \tag{1.1.26}$$

with ρ an arbitrary number on the interval $a < \rho < b$. Then for all z in $\mathcal{A}(a;b)$,

$$f(z) = \sum_{n=-\infty}^{\infty} c_n z^n, \tag{1.1.27}$$

where the series on the right-hand side of (1.1.27) converges to $f(z)$ for all z in $\mathcal{A}(a;b)$. Moreover, if f is bounded by M in $\mathcal{A}(a;b)$, then

$$|c_n| \leqq \begin{cases} M b^{-n}, & n = 0, 1, 2, \cdots \\ M a^{-n}, & n = -1, -2, \cdots, \end{cases} \tag{1.1.28}$$

and in this case, for any z in $\mathcal{A}(a;b)$, and m a positive integer,

$$\begin{aligned} &\left| f(z) - \sum_{n=-(m-1)}^{m-1} c_n z^n \right| \\ &\leqq \frac{Mb}{b-|z|} (|z|/b)^m + \frac{M/a}{1/a - 1/|z|} (a/|z|)^m . \end{aligned} \tag{1.1.29}$$

Proof. Define a_j, b_j and $\mathcal{D}_j$ by

$$\begin{aligned} a_j = a + \frac{b-a}{2^{j+1}}, \quad b_j = b - \frac{b-a}{2^{j+1}}, \\ \mathcal{D}_j = \mathcal{A}(a_j; b_j). \end{aligned} \tag{1.1.30}$$

Given z in $\mathcal{A}(a;b)$, we can choose j sufficiently large, so that $z \in \mathcal{A}(a_j; b_j) = \mathcal{D}_j$. Then, by Theorem 1.1.7 we have

$$\begin{aligned} f(z) &= \frac{1}{2\pi i}\int_{\partial\mathcal{D}_j} \frac{f(\zeta)}{\zeta - z}\,d\zeta \\ &= \frac{1}{2\pi i}\int_{|\zeta|=b_j} \frac{f(\zeta)}{\zeta - z}\,d\zeta - \frac{1}{2\pi i}\int_{|\zeta|=a_j} \frac{f(\zeta)}{\zeta - z}\,d\zeta. \end{aligned} \tag{1.1.31}$$

Substituting c_n defined by (1.1.26) into the left-hand side of (1.1.29) and using the representation (1.1.31) for $f(z)$, we get

$$\begin{aligned} f(z) &- \sum_{n=-(m-1)}^{m-1} c_n z^n \\ &= \frac{1}{2\pi i}\int_{|\zeta|=b_j} \frac{f(\zeta)}{\zeta - z}\left(\frac{z}{\zeta}\right)^m d\zeta \\ &-\frac{1}{2\pi i}\int_{|\zeta|=a_j} \frac{f(\zeta)}{\zeta - z}\left(\frac{\zeta}{z}\right)^{m-1} d\zeta. \end{aligned} \tag{1.1.32}$$

Now, if $\max_{(z\in D_j)} |f(z)| = M_j$ then (1.1.26) yields

$$|c_n| \leqq \frac{M_j}{2\pi}\int_{|\zeta|=\rho} \frac{d\zeta}{|\zeta|^{n+1}} = \frac{M_j}{\rho^n}, \tag{1.1.33}$$

with ρ an arbitrary number, such that $a_j \leqq \rho \leqq b_j$. Hence (1.1.28) follows from (1.1.32). Next, letting $\max_{(z\in\mathcal{D}_j)} |f(z)| = M_j$ in (1.1.32), we get

$$\begin{aligned} &\left| f(z) - \sum_{j=-(m-1)}^{m-1} c_n z^n \right| \\ &\leqq \frac{b_j M_j}{b_j - |z|}\left(\frac{|z|}{b_j}\right)^m + \frac{a_j M_j}{|z| - a_j}\left(\frac{a_j}{|z|}\right)^{m-1} . \end{aligned} \tag{1.1.34}$$

Since both $|z|/b_j$ and $a_j/|z|$ are less than 1 in magnitude, Equation (1.1.29) follows upon letting $m \to \infty$ in (1.1.34). Equation (1.1.29) also follows from (1.1.34). ∎

An important special case results if we take $\mathcal{A}(a;b) = \mathcal{A}_R$.

Corollary 1.1.11 *Let $R > 1$, let f be analytic and bounded by M in $\mathcal{A}_R$, and set*

$$c_n = \frac{1}{2\pi}\int_0^{2\pi} f(e^{i\theta})e^{-in\theta}\,d\theta, n = 0, \pm 1, \pm 2, \cdots. \tag{1.1.35}$$

Then

$$|c_n| \leqq \frac{M}{R^{|n|}}, \tag{1.1.36}$$

and for all $\theta \in [0, 2\pi]$ and arbitrary positive integer m,

$$\left| f(e^{i\theta}) - \sum_{n=-(m-1)}^{m-1} c_n e^{in\theta} \right| \leqq \frac{2MR}{R-1} R^{-m}. \tag{1.1.37}$$

In applications it is sometimes convenient to invoke Morera's theorem, which provides a converse of Cauchy's theorem.

Theorem 1.1.12 (Morera's Theorem): *If f is continuous in a simply-connected domain $\mathcal{D}$, and if $\int_{\mathcal{L}} f(z)\, dz = 0$ for every closed contour $\mathcal{L}$ lying in $\mathcal{D}$, then f is analytic in $\mathcal{D}$.*

Proof. The proof is left to Problem 1.1.12. ■

We define a *singularity* of f to be a point $z_0 \in \mathbb{C}$ such that $f'(z_0)$ does not exist. Of course a singularity of f can occur only on the boundary, or on the exterior of the domain of analyticity of f. The point z_0 is called an isolated singularity of f if $f'(z)$ is known to exist for all z in the annulus $\{z \in \mathbb{C} : 0 < |z - z_0| < R\}$ for some $R > 0$. In this case, f has a *Laurent expansion*

$$f(z) = \sum_{n=-\infty}^{\infty} c_n (z - z_0)^n, \quad 0 < |z - z_0| < R, \tag{1.1.38}$$

The point z_0 in this expansion is called:

(a) *a removable singularity* if $c_n = 0$ for all $n < 0$;

(b) *a pole of order* $m > 0$ if $c_n = 0$ for all $n < -m$ but $c_{-m} \neq 0$; or

(c) *an essential singularity* if the number of non-zero c_n with $n < 0$ is infinite.

The coefficient c_{-1} is called the *residue* of f at z_0.

We remark that if $P(z)$ is a power series,

$$P(z) = \sum_{n=0}^{\infty} a_n z^n \tag{1.1.39}$$

which converges for all z in $\mathcal{D}(0; R)$ then $P(z)$ is an analytic function on $\mathcal{D}(0; R)$. The radius of convergence R of any power series of the form (1.1.39) is given by

$$R = \left(\lim_{N\to\infty} \sup_{j>N} |a_j|^{1/j} \right)^{-1}. \tag{1.1.40}$$

For example, if the series (1.1.39) converges for $z = z_0$, then $|a_j z_0^j| < 1$ for all j sufficiently large, which implies by (1.1.40) that $|z_0| \leqq R$. The mth derivative of the series (1.1.39) is

$$P^{(m)}(z) = \sum_{n=0}^{\infty} a_{n+m}(m+n)(m+n-1)\cdots(n+1)z^n \tag{1.1.41}$$

and it is readily shown that if R is defined by (1.1.40), then we also have

$$R = \left[\lim_{N\to\infty} \sup_{j>N} |a_{j+n}(j+n)(j+n-1)\cdots(j+1)|^{1/j} \right]^{-1}. \tag{1.1.42}$$

As noted earlier, it is interesting to compare Taylor's Theorem with its real valued version found in calculus texts. Taylor's theorem shows that if f is analytic in $\mathcal{D}$, then $f^{(n)}$ exists for every positive integer n, a situation which generally does not occur in real variables.

Corollary 1.1.13 *If for some $p \in [1,\infty]$, we have $f \in \mathbf{H}^p(\mathcal{D})$, and if a is a point in $\mathcal{D}$, then for $n = 0, 1, 2, \cdots$,*

$$\frac{f^{(n)}(a)}{n!} = \frac{1}{2\pi i} \int_{\partial\mathcal{D}} \frac{f(z)}{(z-a)^{n+1}}\, dz. \tag{1.1.43}$$

Theorem 1.1.14 *Let f be in $\mathbf{Hol}(\mathcal{D})$, let $\{z_j\}_{j=1}^{\infty}$ be a Cauchy sequence of distinct points which converges to the point a in $\mathcal{D}$, and set $\phi_{m,n}(z) = \prod_{j=0}^{n}(z - z_{m+j})$. Then, for each fixed non-negative integer n,*

$$\lim_{m\to\infty} \sum_{j=0}^{n} \frac{f(z_{m+j})}{\phi'_{m,n}(z_{m+j})} = \frac{f^{(n)}(a)}{n!}. \tag{1.1.44}$$

Proof. Let $R > 0$ be such that the closed disc $\mathcal{D}[a;R]$ is contained in $\mathcal{D}$. Let f be bounded by M in $\mathcal{D}(a;R)$. Since the sequence $\{z_j\}$ converges to a, there exists a positive integer N such that $z_j \in \mathcal{D}(a;R)$ for all $j > N$. Hence, if $m > N$, then the identity

$$\frac{1}{\phi_{m,n}(z)} = \sum_{j=0}^{n} \frac{1}{(z - z_{m+j})\phi'_{m,n}(z_{m+j})} \tag{1.1.45}$$

combined with Cauchy's formula (1.1.16) yields the result

$$\frac{1}{2\pi i}\int_{\partial\mathcal{D}(a:R)}\frac{f(z)}{\phi_{m,n}(z)}\,dz=\sum_{j=0}^{n}\frac{f(z_{m+j})}{\phi'_{m,n}(z_{m+j})}. \tag{1.1.46}$$

Now let an arbitrary positive number ε be selected, such that $\varepsilon < R$, and let N be selected so that $|z_j - a| < \varepsilon$ for all $j > N$. Then, for $|z-a| = R$,

$$\begin{aligned}&\left|\frac{1}{(z-a)^{n+1}}-\frac{1}{\phi_{m,n}(z)}\right|\\&=\left|\frac{1}{(z-a)^{n+1}}\right|\left|\left[1-\prod_{j=0}^{n}\left(1-\frac{z_{m+j}-a}{z-a}\right)^{-1}\right]\right|\\&\leqq R^{-n-1}\left[(1-\varepsilon/R)^{-n-1}-1\right]\\&\leqq\frac{(n+1)\varepsilon}{R^{n+2}}(1-\varepsilon/R)^{-n-2}.\end{aligned} \tag{1.1.47}$$

Hence, using (1.1.43) and the assumption that $|f(z)| \leqq M$ in $\mathcal{D}(a;R)$, we get

$$\begin{aligned}&\left|\frac{f^{(n)}(a)}{n!}-\sum_{j=0}^{n}\frac{f(z_{m+j})}{\phi'_{m,n}(z_{m+j})}\right|\\&\leqq M\frac{(n+1)\varepsilon}{R^{n+1}}(1-\varepsilon/R)^{-n-2}.\end{aligned} \tag{1.1.48}$$

Equation (1.1.44) now follows, since ε in (1.1.48) may be taken to be as small as we please. ■

Corollary 1.1.15 *If f is in* $\mathbf{Hol}(\mathcal{D})$, *and if f vanishes at a sequence of distinct points z_n having a limit point a in $\mathcal{D}$, then $f(z) \equiv 0$ in $\mathcal{D}$.*

Theorem 1.1.16 (Montel's Theorem): *Let F be an infinite family of functions such that each $f \in F$ is analytic and bounded by M in a k-connected domain $\mathcal{D}$. Then there exists a sequence of functions $\{f_n\}_{n=1}^{\infty}$, such that each f_n belongs to F, and such that $\lim_{(n\to\infty)} f_n(z) = f(z)$, with f analytic and bounded by M in $\mathcal{D}$.*

Proof. Let a sequence of domains $\{\mathcal{D}_j\}_{j=1}^{\infty}$ be defined corresponding to $\mathcal{D}$, by Assumption 1.1.3. Let the integer j be arbitrary, and let $\partial\mathcal{D}_j = \bigcup_{m=1}^{k}\{z_{mj}(t) : 0 \leqq t \leqq 1\}$. Let $t_1, t_2, \cdots$ denote a countable enumberation of the rational numbers on $[0, 1)$. For every positive integer n, we can choose a positive integer m in the range $1 \leqq m \leqq k$ and a positive integer ℓ such that $n = (\ell - 1)k + m$, and then define ζ_n by $\zeta_n = z_{mj}(t_\ell)$. Since the points $\{t_\ell\}$ are dense on $[0, 1)$ and since the z_{mj} are continuous on $[0, 1]$, it follows that the ζ_n are uniformly bounded and dense on $\partial\mathcal{D}_j$.

Corresponding to the sequence $\{\zeta_j\}_{j=1}^{\infty}$ set $S_1 = \{f(\zeta_1) : f \in F\}$. Since the set S_1 is an infinite set of points in $\mathbb{C}$ that are bounded by M, by the Bolzano–Weierstrass theorem we can select from S_1 a Cauchy sequence, $S_{11} = \{f_{1n}(\zeta_1)\}_{n=1}^{\infty}$ which converges to a number b_1 in $\mathbb{C}$, with $|b_1| \leqq M$. Next, from the infinite set of numbers $S_{12} = \{f_{1n}(\zeta_2)\}_{n=1}^{\infty}$, we select a Cauchy sequence $\{f_{2n}(\zeta_2)\}_{n=1}^{\infty}$, such that $\{f_{2n}(\zeta_2)\}_{n=1}^{\infty}$ converges to point b_2 in $\mathbb{C}$, with $|b_2| \leqq M$. By repeating this process, we can derive sequences of functions $\{f_{mn}\}_{n=1}^{\infty}, m = 1, 2, \cdots$, such that $\{f_{mn}\}_{n=1}^{\infty} \subseteq \{f_{m-1,n}\}_{n=1}^{\infty}$ $(m = 2, 3, 4, \cdots)$, and such that $\{f_{mn}(\zeta_\ell)\}_{n=1}^{\infty}$ is a Cauchy sequence which converges to b_ℓ for $\ell = 1, 2, \cdots, m$, with $b_\ell \in \mathbb{C}$ and $|b_\ell| \leqq M$. Hence, the "diagonal sequence" $\{f_{nn}\}_{n=1}^{\infty}$, which is obtained taking f_{11} from $\{f_{1n}\}$, f_{22} from $\{f_{2n}\}$, and so on, is a sequence of functions belonging to F, and such that each of the sequences $\{f_{nn}(\zeta_\ell)\}_{n=1}^{\infty}$ is a Cauchy sequence, where $f_{nn}(\zeta_\ell) \to b_\ell \in \mathbb{C}$ as $n \to \infty$, and $|b_\ell| \leqq M$.

We shall next show that $\{f_{nn}(\zeta)\}_{n=1}^{\infty}$ is also a Cauchy sequence, for each $\zeta \in \mathcal{D}$. To this end let us first show that the family F is equicontinuous on $\partial\mathcal{D}_j$. If m is a fixed positive integer, set $d = \inf\{|\zeta - z| : \zeta \in \partial\mathcal{D}_j, z \in \partial\mathcal{D}_{j+m}\}$. The number d is positive, by Assumption 1.1.3. Also, let L denote the length of the boundary $\partial\mathcal{D}_{j+m}$, which is finite, by Assumption 1.1.3. Then, if ζ' and ζ'' are arbitrary points of $\partial\mathcal{D}_j$, we have by Cauchy's formula, that for any $f \in F$,

$$f(\zeta') - f(\zeta'') = \frac{1}{2\pi i} \int_{\partial\mathcal{D}_{j+m}} f(z) \left\{ \frac{1}{z - \zeta'} - \frac{1}{z - \zeta''} \right\} dz, \qquad (1.1.49)$$

i.e.

$$|f(\zeta') - f(\zeta'')| \leqq \frac{LM}{2\pi d^2} |\zeta' - \zeta''|. \qquad (1.1.50)$$

This inequality shows that the family F is equicontinuous on $\partial\mathcal{D}_j$.

Let us now consider the sequence $\{f_{nn}(\zeta)\}$ with $\{f_{nn}\}$ selected as above, and ζ an arbitrary point of $\partial\mathcal{D}_j$. Given an arbitrary positive number ε we can:

(a) Select ζ_s from the set $\{\zeta_\ell\}_{\ell=1}^{\infty}$, which is dense in $\partial\mathcal{D}_j$, such that $|f_{nn}(\zeta) - f_{nn}(\zeta_s)| < \varepsilon/3$ for every f_{nn} in $\{f_{nn}\}$, since the sequence $\{f_{nn}\}$ is equicontinuous on $\partial\mathcal{D}_j$;

(b) Choose $N > 0$, such that whenever $m > N$ and $n > N$, then we have $|f_{mm}(\zeta_s) - f_{nn}(\zeta_s)| < \varepsilon/3$, this being possible, since $\{f_{nn}(\zeta_s)\}_{n=1}^{\infty}$ is a Cauchy sequence.

Then, we have for such m, n and ζ_s, that

$$\begin{gathered} |f_{mm}(\zeta) - f_{nn}(\zeta)| \leqq \\ |f_{mm}(\zeta) - f_{mm}(\zeta_s)| + |f_{mm}(\zeta_s) - f_{nn}(\zeta_s)| + |f_{nn}(\zeta) - f_{nn}(\zeta_s)| < \varepsilon. \end{gathered} \qquad (1.1.51)$$

That is, the sequence $\{f_{nn}(\zeta)\}_{n=1}^{\infty}$ is a Cauchy sequence. Thus $\{f_{nn}(\zeta)\}_{n=1}^{\infty}$ converges for every $\zeta \in \partial\mathcal{D}_j$.
In order to show continuity of the limit function $f(\zeta) = \lim_{(n\to\infty)} f_{nn}(\zeta)$, on $\partial\mathcal{D}_j$ we use the identity

$$f(\zeta') - f(\zeta'') = \tag{1.1.52}$$
$$f(\zeta') - f_{nn}(\zeta') + f_{nn}(\zeta') - f_{nn}(\zeta'') + f_{nn}(\zeta'') - f(\zeta''),$$

with ζ' and ζ'' are selected on $\partial\mathcal{D}_j$ such that $|f_{nn}(\zeta') - f_{nn}(\zeta'')| < \varepsilon/3$ for all $n = 1, 2, 3, \cdots$, with n sufficiently large so that both $|f_{nn}(\zeta') - f(\zeta')|$ and $|f_{mm}(\zeta'') - f(\zeta'')|$ are less than $\varepsilon/3$. Then $|f(\zeta') - f(\zeta'')| < \varepsilon$.
Finally, if z is an arbitrary point in $\mathcal{D}_j$, we have for $n = 1, 2, 3, \cdots$ that

$$f_{nn}(z) = \frac{1}{2\pi i}\int_{\partial\mathcal{D}_j} \frac{f_{nn}(\zeta)}{\zeta - z}\, d\zeta, \tag{1.1.53}$$

$$f'_{nn}(z) = \frac{1}{2\pi i}\int_{\partial\mathcal{D}_j} \frac{f_{nn}(\zeta)}{(\zeta - z)^2}\, d\zeta. \tag{1.1.54}$$

The convergence of $f_{nn}(z)$ to $f(z)$ at each interior point z of $\mathcal{D}_j$ is now readily shown using (1.1.53). Clearly $|f(z)| \leqq M$ in $\mathcal{D}_j$, the analyticity of $f(z)$ in $\mathcal{D}_j$ is then readily established via (1.1.53) and (1.1.54). The details of the proof are left to the problems.

The proof of Theorem 1.1.16 is now completed by letting $j \to \infty$. ■

The family F in Theorem 1.1.16 is called a *compact family* of *analytic functions.*

Corollary 1.1.17 (Vitali's Theorem): *Let $\{f_n\}_{n=1}^{\infty}$ denote a sequence of functions, each of which is analytic and bounded by M in a k-connected domain $\mathcal{D}$. Let $\{z_j\}_{j=1}^{\infty}$ denote a Cauchy sequence of distinct points z_j which has a limit point ζ in $\mathcal{D}$. If each of the sequences $\{f_n(z_j)\}_{n=1}^{\infty}$ is a Cauchy sequence, then $\lim_{(n\to\infty)} f_n(z) = f(z)$, for all points z in $\mathcal{D}$, with f analytic and bounded by M in $\mathcal{D}$.*

Proof. By Montel's Theorem, we can construct a subsequence $\{f_{nn}\}_{n=1}^{\infty}$ of the sequence $\{f_n\}_{n=1}^{\infty}$, such that f_{nn} converges to a function f that is analytic and bounded by M in $\mathcal{D}$. Furthermore, it is clear then, that $\{f(z_n)\}_{n=1}^{\infty}$ is a Cauchy sequence. Now consider the set $\{g_n\}_{n=1}^{\infty} = \{f_n\}_{n=1}^{\infty} \setminus \{f_{nn}\}_{n=1}^{\infty}$. If $\{g_n\}_{n=1}^{\infty}$ is a finite sequence, we can include it in $\{f_{nn}\}_{n=1}^{\infty}$ without changing the properties of $\{f_{nn}\}$. If $\{g_n\}_{n=1}^{\infty}$ is an infinite sequence, then by Montel's theorem we can construct a subsequence $\{g_{nn}\}$ of the sequence $\{g_n\}$, such that $g_{nn} \to g$ as $n \to \infty$, with g analytic and bounded by M

in $\mathcal{D}$. Moreover, we have $f(z_j) = g(z_j)$ at the Cauchy sequence of distinct points $z_j, j = 1, 2, \cdots$, which has a limit point ζ in $\mathcal{D}$. Hence, by Corollary 1.1.15, $f(z) - g(z) \equiv 0$ in $\mathcal{D}$. That is $\{f_n\}$ converges to a unique function f that is analytic and bounded by M in $\mathcal{D}$. ∎

We conclude this section with some remarks on *analytic continuation.* At the outset, let us extend the definition of a domain to that of a *Riemann surface.* Let us consider an example.

The function

$$f'(z) = \frac{1}{z} \tag{1.1.55}$$

is analytic in $\mathbb{C}$, except at $z = 0$, which is a pole of f'. If we define

$$f(z) = \log z = \int_1^z \frac{d\zeta}{\zeta} \tag{1.1.56}$$

with

$$\begin{aligned} \int_1^z \frac{d\zeta}{\zeta} &= \int_1^{|z|} \frac{dt}{t} + \int_0^{\arg z} \frac{i|z|e^{i\theta}}{|z|e^{i\theta}}\, d\theta \\ &= \log|z| + i \arg z, \end{aligned} \tag{1.1.57}$$

we see by taking $z = 2e^{\pi i}$, and then $z = 2e^{-\pi i}$ (whose values are both -2) that we get two different answers:

$$\begin{aligned} \log(2e^{\pi i}) &= \log(-2) = \log 2 + \pi i \\ \log(2e^{-\pi i}) &= \log(-2) = \log 2 - \pi i. \end{aligned} \tag{1.1.58}$$

That is, whereas the analyticity of f' in $\mathcal{D}$ implies the analyticity of $f^{(2)}$, $f^{(3)}, \cdots$ in $\mathcal{D}$, this is not the case for $f(z) = \int_1^z f'(\zeta)\, d\zeta$, since $f(z)$ may not be single-valued in $\mathcal{D}$. Hence, in the case of $f(z) = \log z$, we can preserve the single-valuedness either by introducing a *cut*, i.e. a continuous curve which starts at $z = 0$, ends up at ∞, yet never crosses itself, to be part of the boundary of the domain of $\log z$, or else by defining the domain of $\log z$ to be the *Riemann surface* on which $\log z$ is single valued. The first case is accomplished, e.g. by the straight line cut which emanates from 0, passes through -1, and ends up at $-\infty$, so that the domain of $\log z$ is $\{z \in \mathcal{D} : -\pi < \arg z < \pi\}$. From the need for representation of inverse functions, it is often desirable to allow the range of $\arg z$ to be unbounded. To this end, we can think of the points $z = |z|e^{i(\theta + 2n\pi)}$ to be distinct, for distinct integers n. Thus, to each integer n the domain of $\log z$ is a different *"sheet"*, and the union of all these sheets is the Riemann surface, i.e., the complete domain of definition of the function $f(z) = \log z$.

Whereas the Riemann surface of the function $\log z$ has an infinite number of sheets, some functions require only a finite number of sheets to completely define the Riemann surface which is their domain. For example, the function f defined by $f(z) = z^{1/2}$ has a two-sheeted Riemann surface

$$\{z = |z|e^{i\theta} : 0 < |z| < \infty, 0 \leqq \theta < 4\pi\}, \tag{1.1.59}$$

as its domain. Note that

$$z^{1/2} = |z|^{1/2} e^{i\theta/2}, \tag{1.1.60}$$

so that the points $|z|e^{i0}$ and $|z|e^{i4\pi}$ yield the same values of $f(z)$. Similarly, it may be shown that if $a_1, a_2, \cdots, a_n$ are arbitrary complex numbers, then the solution w, as a function of z, of the equation

$$w^n + a_1 w^{n-1} + a_2 w^{n-2} + \cdots + a_n = z, \tag{1.1.61}$$

is analytic on an n-sheeted Riemann surface.

In general, the complete extended domain of analyticity of a function f is any minimally-sheeted Riemann surface on which f is analytic and single-valued. An arbitrary domain of analyticity of f is any open connected subset of the extended domain of analyticity of f.

Theorem 1.1.18 (Principle of Analytic Continuation): *Let $\mathcal{D}_1$ and $\mathcal{D}_2$ be two domains, and let $\mathcal{D}_3$ be a non-empty domain belonging to both $\mathcal{D}_1$ and $\mathcal{D}_2$. Let f be analytic in $\mathcal{D}_1$, let g be analytic in $\mathcal{D}_2$, and let $f \equiv g$ in $\mathcal{D}_3$. Then there exists an analytic function h defined on $\mathcal{D}_1 \cup \mathcal{D}_2$, such that $h \equiv f$ on $\mathcal{D}_1$, and $h \equiv g$ on $\mathcal{D}_2$.*

It is assumed in Theorem 1.1.18 that, if necessary, $\mathcal{D}_1 \cup \mathcal{D}_2$ is defined as part of a Riemann surface, so that the function f remains single-valued on its new domain.

The use of *power series* is a frequent tool for carrying out analytic continuation, and we shall later employ such a device. Sometimes a *recurrence relation* provides a tool for carrying out analytic continuation, as for example in the case of the Gamma function. The *Schwarz Reflection Principle*, which we now state, provides yet another tool.

Theorem 1.1.19 (Schwarz's Reflection Principle): *Let $\mathcal{D}$ be a domain in Ω_+, where*

$$\Omega_+ = \{z = x + iy : x \in \mathbb{R}, y > 0\}, \tag{1.1.62}$$

and such that an interval $(a, b) \subset \mathbb{R}$ belongs to the boundary of $\mathcal{D}$. Let f be analytic in $\mathcal{D}$, and let the limit $w(x) = \lim_{(y \to 0^+)} f(x + iy)$ be a real-valued function that is continuous on (a, b). Define $\tilde{\mathcal{D}}$ by $\tilde{\mathcal{D}} = \{\overline{z} = x - iy : x \in \mathbb{R}, y \in \mathbb{R},\ z = x + iy \in \mathcal{D}\}$. Then g defined for $z \in \mathcal{D}$ by $g(z) = f(z)$, and for $x + iy$ in $\tilde{\mathcal{D}}$ by $g(x + iy) = f(x - iy)$ provides the analytic continuation of f to $\mathcal{D} \cup \tilde{\mathcal{D}} \cup \{(a, b)\}$.

Proof. See Problem 1.1.13. ■

Another useful theorem, whose proof is similar to that of the Schwarz Reflection Principle, and which we shall simply refer to as the *Theorem on Analytic Continuation*, is the following.

Theorem 1.1.20 (Theorem on Analytic Continuation): *Let $\mathcal{D}_1$ and $\mathcal{D}_2$ be two domains without common points, but such that their boundaries, $\partial\mathcal{D}_1$ and $\partial\mathcal{D}_2$ have in common an open arc Γ. Let f be analytic in $\mathcal{D}_1$, let g be analytic in $\mathcal{D}_2$, and let both functions have the same limiting value as z tends to an arbitrary point on Γ, from within their respective domains of definition. Moreover, let this limiting function be continuous on Γ. Then there exists a function h which is analytic in $\mathcal{D}_1 \cup \mathcal{D}_2 \cup \Gamma$, such that $h = f$ in $\mathcal{D}_1$, and $h = g$ in $\mathcal{D}_2$.*

Proof. See Problem 1.1.14. ■

Analytic continuation is a very powerful tool in applications, especially when combined with Liouville's Theorem. If f is analytic and uniformly bounded by M in $\mathbb{C}$, then the inequality (1.1.23) holds for all $R > 0$, and it thus shows that $a_n = 0$ for $n = 1, 2, 3, \cdots$. This yields the following result.

Lemma 1.1.21 *If f is analytic and uniformly bounded in $\mathbb{C}$, then f is a constant.*

Although this lemma is the original version of Liouville's theorem, the following more general version is more conveniently employed in applications.

Theorem 1.1.22 (Liouville's Theorem): *Let m denote a positive integer, and let:*

$\mu_0, \mu_1, \cdots, \mu_m$ denote $m+1$ positive integers;
$z_0, z_1, \cdots, z_m$ denote $m+1$ distinct points in $\overline{\mathbb{C}}$, with $z_0 = \infty$;
$g_0(z) = \sum_{n=1}^{\mu_0} a_{0n} z^n$ with the a_{0n} constants;
$g_j(z) = \sum_{n=1}^{\mu_j} a_{jn}/(z - z_j)^n$, $j = 1, 2, \cdots, m$, with the a_{jn} constants;
$R_0 = max_{(j=1,2,\cdots,m)} |z_j|$;
$R_j = min_{(i=1,2,\cdots,m;i\neq j)} |z_j - z_i|$;
$\mathcal{D}_0 = \{z \in \mathbb{C} : |z| > R_0\}$; and
$\mathcal{D}_j = \{z \in \mathbb{C} : |z - z_j| < R_j\}$, $j = 1, 2, \cdots, m$.
Let F be analytic in $\mathbb{C}$, except at the isolated points $z_0, z_1, \cdots, z_m$, and let $F - g_j \in \mathbf{Hol}(\mathcal{D}_j)$, $j = 0, 1, 2, \cdots, m$. Then

$$F = a_{00} + \sum_{j=0}^{m} g_j, \tag{1.1.63}$$

with a_{00} a constant.

Proof. This follows directly, upon noting that $F - \sum_{j=0}^{m} g_j$ satisfies the conditions of Lemma 1.1.21. ∎

•

Problems for Section 1.1

1.1.1 Prove that every Cauchy sequence in $\mathbb{R}$ and $\mathbb{C}$ has a limit.

1.1.2 Prove Theorem 1.1.7. [Hint: Set $z = a + re^{i\theta}$, with r sufficiently small so that $\mathcal{D}(a;r) \subset \mathcal{D}$, and show that

$$\int_{\partial\mathcal{D}(a;r)} \frac{f(z)}{z-a}\,dz$$

$$= i\int_0^{2\pi} f(a + re^{i\theta})\,d\theta$$

$$\to 2\pi i\, f(a) \quad \text{as } r \to 0.$$

Then note that $F(z) = f(z)/(z-a)$ satisfies the conditions of Cauchy's Theorem on $\mathcal{D} \setminus \mathcal{D}(a;r)$.]

1.1.3 Prove Taylor's Theorem.

1.1.4 Show that if x is an arbitrary finite complex number, then the function $\exp\{x(z - z^{-1})/2\}$ is an analytic function of z in $\mathcal{A}(0;\infty)$ for all $x \in \mathbb{C}$, and hence it has a Laurent expansion of the form

$$\exp\left\{\frac{x}{2}(z - z^{-1})\right\} = \sum_{n=-\infty}^{\infty} J_n(x) z^n, \quad z \in \mathcal{A}(0;\infty).$$

(a) Obtain the estimate

$$|J_n(x)| \leqq R^{-|n|} \exp\left\{\frac{|x|}{2}(R + R^{-1})\right\}.$$

(b) By minimizing $R^{-|n|}\exp\left[\frac{|x|}{2}R\right]$ as a function of $R > 0$, show that

$$|J_n(x)| \leqq \left(\frac{e|x|}{2|n|}\right)^{|n|} \exp\left(\frac{|x|^2}{4|n|}\right).$$

1.1.5 Let $\Gamma(z)$ be defined for $\Re z > 0$ by

$$(i) \qquad \Gamma(z) = \int_0^\infty e^{-t} t^{z-1}\, dt.$$

(a) Prove that $\Gamma(z)$ is analytic in the domain $\mathcal{D}_0$, defined by $\mathcal{D}_0 = \{z \in \mathbb{C} : \Re z > 0\}$.

(b) Prove that the relation

$$\Gamma(z+1) = z\Gamma(z),$$

is valid for all $z \in \mathcal{D}_0$. Conclude that if n is a positive integer, then $\Gamma(n+1) = n!$.

(c) In view of the fact that $\Gamma(z+1)$ is analytic in the domain $\mathcal{D}_{-1} = \{z \in \mathbb{C} : \Re z > -1\}$ (why?) and the result $\Gamma(1) = 1$, prove that $\Gamma(z)$ is analytic in $\mathcal{D}_{-1}$ except for a simple pole at $z = 0$.

(d) Repeat the above process to carry out the analytic continuation of $\Gamma(z)$ to the entire complex plane, with the exception of $z = -n$, $n = 0, 1, 2, \cdots$.

(e) What is the residue of $\Gamma(z)$ at $z = -n$, $n = 0, 1, 2, \cdots$?

(f) Prove that if a is a fixed complex number then

$$(ii) \qquad (2\pi)^{-1/2} z^{-z-a} e^z \Gamma(z + a + 1/2) \to 1$$

as $z \to \infty$ along the line

$$\mathcal{L} = \{z \in \mathbb{C} : z = te^{i\alpha},\ \alpha \text{ fixed in } (-\pi, \pi),\ 0 \leqq t \leqq \infty\}.$$

(g) Use formula (i) above to deduce that if $\Re z > 0$, $\Re \zeta > 0$, then

$$(iii) \qquad \Gamma(z)\,\Gamma(\zeta) = 4 \int_0^\infty \int_0^\infty e^{-u^2-v^2} u^{2z-1}\, v^{2\zeta-1}\, du\, dv$$

$$= \Gamma(z+\zeta) \left\{ 2 \int_0^{\pi/2} \cos^{2z-1}(\theta)\, \sin^{2\zeta-1}(\theta)\, d\theta \right\},$$

where the last identity is obtained via the use of polar coordinates in (iii). Deduce the formulas for the Beta function

$$(iv) \qquad \begin{aligned} B(z,\zeta) &= 2 \int_0^{\pi/2} \cos^{2z-1}(\theta)\, \sin^{2\zeta-1}(\theta)\, d\theta \\ &= \int_0^1 t^{z-1}\,(1-t)^{\zeta-1}\, dt \\ &= \frac{\Gamma(z)\,\Gamma(\zeta)}{\Gamma(z+\zeta)}. \end{aligned}$$

(h) Let $B(z, \zeta)$ be defined as in (g) above. Prove that

$$B(z,\zeta) = \int_0^\infty t^{z-1}\,(1+t)^{-z-\zeta}\,dt = 2\int_0^\infty t^{2z-1}\,(1+t^2)^{-z-\zeta}\,dt.$$

(i) Prove that if $x \in \mathbb{R}$, $y \in \mathbb{R}$, then

$$\left|\frac{\Gamma(x)}{\Gamma(x+iy)}\right|^2 = \prod_{n=0}^{\infty}\left[1+\frac{y^2}{(x+n)^2}\right].$$

Prove that the right-hand side of this equation reduces to $\sinh(\pi y)/(\pi y)$ when $x = 1$, and conclude that if $0 \leqq x \leqq 1$, then

$$\frac{x^2}{x^2+y^2} \leqq \left|\frac{\Gamma(x+iy)}{\Gamma(x)}\right|^2 \frac{\sinh(\pi y)}{\pi y} \leqq 1.$$

1.1.6 Prove that if R is defined by (1.1.40) then (1.1.42) holds for every fixed integer $m > 0$.

1.1.7 Use (1.1.51) to prove that $\{f_{nn}(z)\}_{n=1}^{\infty}$ is a Cauchy sequence, for each z in $\mathcal{D}_j$.

1.1.8 Use (1.1.54) to prove that

$$\begin{aligned} f'(z) &= \frac{d}{dz}\lim_{n\to\infty} f_{nn}(z) \\ &= \lim_{n\to\infty} f'_{nn}(z). \end{aligned}$$

1.1.9 (a) Prove that if f is analytic in a simply-connected domain $\mathcal{D}$ and if $\mathcal{C}$ is a closed contour in $\mathcal{D}$ such that $f \neq 0$ on $\mathcal{C}$, then $\kappa(\mathcal{C}) = (2\pi i)^{-1}\int_{\mathcal{C}}\{f'(z)/f(z)\}dz$ is the number of zeros of f in $\mathcal{C}$ counted according to multiplicity.

(b) Prove **Hurwitz's Theorem:** Let $\{f_n\}$ be a sequence of functions that are analytic in a simply-connected domain $\mathcal{D}$, let $f_n \to f$ uniformly on every closed and bounded subset of $\mathcal{D}$ and let $\mathcal{C}$ denote a closed contour in $\mathcal{D}$, such that $f \neq 0$ on $\mathcal{C}$. If f has $\kappa(\mathcal{C})$ zeros in the interior of $\mathcal{C}$, counted according to multiplicity, then for all n sufficiently large f_n also has exactly $\kappa(\mathcal{C})$ zeros in the interior of $\mathcal{C}$.

1.1.10 What is wrong with the following argument:

$$e^z = (e^{2\pi i})^{z/(2\pi i)} = 1^{z/(2\pi i)} = 1?$$

1.1.11 Prove that if R defined by (1.1.40) is finite, then the function $P(z)$ in (1.1.39) has at least one singularity on $\partial\mathcal{D}(0;R)$.

1.1.12 Prove Morera's Theorem (Theorem 1.1.12). [Hint: (a) Let F be defined by $F(z) = \int_a^z f(t)\,dt$, with $a \in \mathcal{D}$, $z \in \mathcal{D}$, and with the integral taken along an arc $\Gamma = \Gamma(a,z)$ lying in $\mathcal{D}$, and prove that the value of F is independent of Γ; (b) Prove that $F'(z) = f(z)$ at all points z in $\mathcal{D}$; and (c) conclude that since F is analytic in $\mathcal{D}$, then f is analytic in $\mathcal{D}$.]

1.1.13 Prove Theorem 1.1.19. [Hint: (a) Prove that g is analytic in $\mathcal{D}$; (b) prove that if $\mathcal{C}$ is any closed contour in $\mathcal{D} \cup \tilde{\mathcal{D}}$, then $\int_{\mathcal{C}} g(z)\,dz = 0$; (c) now invoke Morera's Theorem, proved in Problem 1.1.12.]

1.1.14 Prove Theorem 1.1.20. [Hint: Use the method of proof for the previous problem].

1.1.15 Let Ω_+ denote the upper half plane, so that $\partial\Omega_+ = \mathbb{R}$, and set $f(z) = (z+i)^{-2}$, $g(z) = z^{-2}\exp\{-z^{-2}\}$. Explain that, although $f \in \mathbf{Hol}(\Omega_+)$, $g \in \mathbf{Hol}(\Omega_+)$, $\int_{\mathbb{R}} |f(x)|\,dx = \pi < \infty$, $\int_{\mathbb{R}} |g(x)|\,dx = \sqrt{\pi} < \infty$, we have $f \in \mathbf{H}^1(\Omega_+)$, $g \notin \mathbf{H}^1(\Omega_+)$.

1.1.16 Let $0 < b < a$, let $\mathcal{D} = \{z \in \mathbb{C} : (\Re z)^2/a^2 + (\Im z)^2/b^2 < 1\}$, and set $c = \sqrt{a^2 - b^2}$. Prove that

$$\frac{2ab}{c^2}\int_{-c}^{c} \sqrt{c^2 - x^2}\, f(x)\,dx = \int\!\!\int_{\mathcal{D}} f(x+iy)\,dx\,dy$$

for all $f \in \mathbf{Hol}(\mathcal{D}) \cap \mathbf{L}^1(\mathcal{D})$. ([Davis, Da3]).

1.1.17 Let $R = (5^{1/2}+2)^{1/2}$, and let $\mathcal{E}_R$ denote the ellipse with foci at ± 1 and sum of semi-axes equal to R. Prove that

$$\int_{-1}^{1} \frac{f(x)}{\sqrt{1-x^2}}\,dx = \int\!\!\int_{\mathcal{E}_R} f(x+iy)\,dx\,dy$$

for all $f \in \mathbf{Hol}(\mathcal{E}_R) \cap \mathbf{L}^1(\mathcal{E}_R)$. (New).

1.1.18 The transformation $\phi(w) = (1/\pi)\log\{(1+w\alpha)/(1-w\alpha)\}$ maps the interior of the unit disc in $\mathbb{C}$ onto a certain region $\mathcal{B}$ in the complex plane. Determine α such that

$$\int_{-1}^{1} f(x)\,dx = \int\!\!\int_{\mathcal{B}} f(x+iy)\,dx\,dy$$

for all $f \in \mathbf{Hol}(\mathcal{B}) \cap \mathbf{L}^1(\mathcal{B})$ ([Davis, Da1]).

1.1.19 Let $a > 0, b > 0, c = \sqrt{a^2+b^2}, \theta = \arctan(b/a), \omega = e^{i\theta}$, and set $\mathcal{B} = \{z = x+iy : -a < x < a, -b < y < b\}$. Prove that

$$\frac{2ab}{c^2}\int_{-c}^{c}[c-|t|][f(\omega t)+f(\overline{\omega}t)]\,dt = \int_{-a}^{a}\int_{-b}^{b} f(x+iy)\,dx\,dy$$

for all $f \in \mathbf{Hol}(\mathcal{B}) \cap \mathbf{L}^1(\mathcal{B})$. ([Stenger, S9]).

1.1.20 Let $a > 0, b > 0, c = \sqrt{a^2+b^2}, \theta = \arctan(b/a), C = \cos(\theta),\ S = \sin(\theta)$, and set $\mathcal{B} = \{z = x+iy : -a < x < a, -b < y < b\}$. Prove that

$$\frac{2ab}{c^2}\int_{-c}^{c}\frac{\sin\{k[c-|t|]\}}{k}[f(tC,tS)+f(tC,-tS)]\,dt$$
$$= \int_{-a}^{a}\int_{-b}^{b} f(x+iy)\,dx\,dy$$

for all $f \in \mathbf{L}^1(\mathcal{B})$ which satisfy the equation $\nabla^2 f + k^2 f = 0$ in $\mathcal{B}$. ([Kratz, Kz3]).

1.1.21 * (Research problem) Let μ be a sigma-finite positive measure with support on the Borel sets of $\mathbb{R}$, and let μ have compact support on $\mathbb{R}$. Prove there exists a unique bounded region $\mathcal{B} \subset \mathbb{C}$ such that $\mathcal{B}$ contains the support of μ, and such that

$$\int_{\mathbb{R}} f(x)\,d\mu(x) = \int\int_{\mathcal{B}} f(x+iy)\,dx\,dy$$

for all $f \in \mathbf{Hol}(\mathcal{B}) \cap \mathbf{L}^1(\mathcal{B})$.

1.2 Hilbert Transforms

Let $1 \leqq p < \infty$ (resp. $p = \infty$) and let $\mathbf{L}^p(a,b)$ denote the family of all measurable functions f defined on the interval (a,b), such that $\|f\|_p < \infty$, with

$$\|f\|_p = \begin{cases} \left(\int_a^b |f(x)|^p\,dx\right)^{1/p}, & \text{if } 1 \leqq p < \infty, \\ \sup\limits_{x\in(a,b)} f(x), & \text{if } p = \infty. \end{cases} \tag{1.2.1}$$

Given $f \in \mathbf{L}^p(a,b)$, we define the Hilbert transform of f, by

$$(\mathcal{S}f)(x) = \frac{P.V.}{\pi i} \int_a^b \frac{f(t)}{t-x} dt \tag{1.2.2}$$

with the *principal value integral* (1.2.2) defined for $a < x < b$ by

$$\begin{aligned} P.V. \int_a^b \frac{f(t)}{t-x} dt \\ = \lim_{\varepsilon \to 0^+} \left\{ \int_a^{x-\varepsilon} \frac{f(t)}{t-x} dt + \int_{x+\varepsilon}^b \frac{f(t)}{t-x} dt \right\} . \end{aligned} \tag{1.2.3}$$

It will be seen that if $1 < p < \infty$, then the limit (1.2.3) exists a.e. on (a, b).

Theorem 1.2.1 (Pichorides' Theorem): *Let $1 < p < \infty$, and let f be a real-valued function in $\mathbf{L}^p(a, b)$. Then $\mathcal{S}f \in \mathbf{L}^p(a, b)$, and moreover,*

$$\|\mathcal{S}f\|_p \leqq B_p \|f\|_p; \quad B_p \leqq \begin{cases} \tan\left(\frac{\pi}{2p}\right) & \text{if } 1 < p \leqq 2; \\ \cot\left(\frac{\pi}{2p}\right) & \text{if } 2 \leqq p < \infty. \end{cases} \tag{1.2.4}$$

Furthermore, the bounds on the right-hand side of (1.2.4) are sharp.

Proof. The proof is somewhat lengthy, and hence it is outlined in Problems 1.2.8-1.2.14 at the end of this section. ■

Let $(a, b) \subseteq \mathbb{R}$, let $z \in \mathbb{C}$, let $1 < p < \infty$, and let w be defined for a given $f \in \mathbf{L}^p(a, b)$ by

$$w(z) = \frac{1}{2\pi i} \int_a^b \frac{f(t)}{t-z} dt. \tag{1.2.5}$$

Clearly, w is analytic for $z \notin [a, b]$.

Theorem 1.2.2 *Let w be defined as in (1.2.5), and for $x \in \mathbb{R}$, $y > 0$, set*

$$\begin{aligned} (\mathcal{P}f)(x, y) &= w(x + iy), \\ [(I - \mathcal{P})f](x, -y) &= -w(x - iy), \end{aligned} \tag{1.2.6}$$

$$\begin{aligned} u(x, y) = (\mathcal{P}f)(x, y) + [(I - \mathcal{P})f](x, -y), \\ v(x, y) = (\mathcal{P}f)(x, y) - [(I - \mathcal{P})f](x, -y). \end{aligned} \tag{1.2.7}$$

Then

$$u(x,y) = \frac{y}{\pi}\int_a^b \frac{f(t)}{(t-x)^2+y^2}dt, \qquad (1.2.8)$$

$$v(x,y) = \frac{1}{\pi i}\int_a^b f(t)\frac{t-x}{(t-x)^2+y^2}dt. \qquad (1.2.9)$$

Both u and v are harmonic functions in the upper half plane, $\Omega_+ = \{z \in \mathbb{C} : \Im z > 0\}$, and they satisfy the Plemelj formulas

$$\lim_{y\to 0+} u(x,y) = f(x), \qquad (1.2.10)$$

$$\lim_{y\to 0+} v(x,y) = (\mathcal{S}f)(x). \qquad (1.2.11)$$

a.e. on (a,b).

Proof. The fact that u and v are harmonic in Ω_+ is a consequence of the analyticity of w in Ω_+. The identities (1.2.8) and (1.2.9) follow by taking $z = x \pm iy$ in (1.2.5) and then substituting into (1.2.7), using the definitions (1.2.6).

We shall prove the relations (1.2.10) and (1.2.11) only in the practically important case of $f \in \mathbf{L}^\infty(a,b) \cap \mathbf{Lip}_\alpha(a,b)$, i.e., in the case of f uniformly bounded on (a,b), and there are positive constants C and α such that for all t_1 and t_2 on (a,b), $|f(t_1) - f(t_2)| \leqq C|t_1 - t_2|^\alpha$.

To prove (1.2.10), let us assume that $x \in (a,b)$; in this case

$$\frac{y}{\pi}\int_a^b \frac{dt}{(t-x)^2+y^2} = \frac{1}{\pi}\left\{\tan^{-1}\left(\frac{b-x}{y}\right) - \tan^{-1}\left(\frac{a-x}{y}\right)\right\} \qquad (1.2.12)$$
$$\to 1 \text{ as } y \to 0.$$

Hence,

$$u(x,y) - f(x) = \frac{y}{\pi}\int_a^b \frac{f(t)-f(x)}{(t-x)^2+y^2}dt$$
$$+\left[\frac{1}{\pi}\left\{\tan^{-1}\left(\frac{b-x}{y}\right) + \tan^{-1}\left(\frac{x-a}{y}\right)\right\} - 1\right] f(x). \qquad (1.2.13)$$

Clearly, the last term, "[ ]", on the right-hand side of (1.2.13) approaches zero as $y \to 0$. For the first term, we take $\varepsilon = y^{3/4}$, so that the interval $(x-\varepsilon, x+\varepsilon) \subset (a,b)$ for all y sufficiently small, and therefore

$$\frac{y}{\pi}\int_a^b \frac{f(t)-f(x)}{(t-x)^2+y^2}dt = \frac{y}{\pi}\int_a^{x-\varepsilon} \frac{f(t)-f(x)}{(x-t)^2+y^2}dt$$
$$+\frac{y}{\pi}\int_{x-\varepsilon}^{x+\varepsilon} \frac{f(t)-f(x)}{(t-x)^2+y^2}dt + \frac{y}{\pi}\int_{x+\varepsilon}^{b} \frac{f(t)-f(x)}{(t-x)^2+y^2}dt. \tag{1.2.14}$$

Let

$$M = \sup_{x\in(a,b)} |f(x)|. \tag{1.2.15}$$

Then, for the first integral on the right-hand side of (1.2.14), we have

$$\left|\frac{y}{\pi}\int_a^{x-\varepsilon} \frac{[f(t)-f(x)]}{(t-x)^2+y^2}dt\right| \leqq \frac{2My}{\pi}\int_{-\infty}^{x-\varepsilon} \frac{dt}{(t-x)^2+y^2}$$
$$= \frac{2M}{\pi}\tan^{-1}\left(\frac{y}{\varepsilon}\right) = \frac{2M}{\pi}\tan^{-1} y^{1/4} \to 0 \tag{1.2.16}$$

as $y \to 0$. Similarly, the third integral on the right-hand side of (1.2.14) is also bounded by $(2M/\pi)\tan^{-1} y^{1/4}$. For the second integral, it follows by our $\mathbf{Lip}_\alpha$ assumption on f, that $|f(t)-f(x)| \leqq C|t-x|^\alpha \leqq Cy^{3\alpha/4}$, since $|t-x| \leqq \varepsilon = y^{3/4}$. Hence,

$$\left|\frac{y}{\pi}\int_{x-\varepsilon}^{x+\varepsilon} \frac{f(t)-f(x)}{(t-x)^2+y^2}dt\right|$$
$$\leqq \frac{Cy^{1+3\alpha/4}}{\pi}\int_{\mathbb{R}} \frac{dt}{(t-x)^2+y^2} = Cy^{3\alpha/4} \to 0 \tag{1.2.17}$$

as $y \to 0$.

To prove (1.2.11), let $\varepsilon > 0$ be fixed, independent of $y > 0$, and sufficiently small, so that interval $(x-\varepsilon, x+\varepsilon) \subset (a,b)$. Then, setting $I(\varepsilon) = \{t \in (a,b) : |x-t| > \varepsilon\}$, $q = p/(p-1)$, we have

$$\left|\frac{1}{\pi i}\int_{I(\varepsilon)} f(t)\left\{\frac{1}{t-x} - \frac{t-x}{(t-x)^2+y^2}\right\}dt\right|$$
$$= \left|\frac{y^2}{\pi i}\int_{I(\varepsilon)} \frac{f(t)\,dt}{(t-x)[(t-x)^2+y^2]}\right|$$

$$\leqq \frac{y^2}{\pi}\|f\|_p \left(\int_{I(\varepsilon)} |t-x|^{-q}[(t-x)^2+y^2]^{-q}dt\right)^{1/q} \tag{1.2.18}$$
$$\leqq \frac{y^2}{\pi}\|f\|_p \left(\int_{I(\varepsilon)} |t-x|^{-3q}dt\right)^{1/q}$$
$$\leqq \frac{y^2}{\pi}\|f\|_p \varepsilon^{-3}(b-a)^{1/q} \quad \to 0$$

as $y \to 0$.

On the other hand,

$$\begin{aligned}
&\left|\frac{1}{\pi i}\int_{x-\varepsilon}^{x+\varepsilon} \frac{(t-x)f(t)}{(t-x)^2+y^2}dt\right| \\
&= \left|\frac{1}{\pi i}\int_{x-\varepsilon}^{x+\varepsilon} \frac{(t-x)f(x)\,dt}{(t-x)^2+y^2} + \frac{1}{\pi i}\int_{x-\varepsilon}^{x+\varepsilon} \frac{[f(t)-f(x)](t-x)\,dt}{(t-x)^2+y^2}\right| \\
&= \left|\frac{1}{\pi}\int_{x-\varepsilon}^{x+\varepsilon} \frac{[f(t)-f(x)](t-x)}{(t-x)^2+y^2}dt\right| \\
&\leqq \frac{1}{\pi}\int_{x-\varepsilon}^{x+\varepsilon} \frac{C|t-x|^{1+\alpha}}{(t-x)^2+y^2}dt \\
&\leqq \frac{C}{\pi}\int_{x-\varepsilon}^{x+\varepsilon} |t-x|^{\alpha-1}dt = \frac{2C\varepsilon^\alpha}{\alpha\pi},
\end{aligned} \tag{1.2.19}$$

where the first integral on the second line is zero. Hence, (1.2.11) follows, since $\varepsilon > 0$ was arbitrary. ∎

Remark 1.2.3 Equations (1.2.10) and (1.2.11) imply yet another form of the Plemelj formulas, namely, if w is defined by (1.2.5) and if $x \in (a,b)$, then

$$\lim_{y\to 0+} w(x+iy) = \frac{1}{2}\left\{f(x) + (\mathcal{S}f)(x)\right\}, \tag{1.2.20}$$

$$\lim_{y\to 0+} w(x-iy) = \frac{1}{2}\left\{-f(x) + (\mathcal{S}f)(x)\right\}. \tag{1.2.21}$$

Remark 1.2.4 It is readily seen, via Hölder's inequality, that if $f \in \mathbf{L}^p(\mathbb{R})$, $1 < p < \infty$, and if u and v are respectively defined by (1.2.8) and (1.2.9), then

$$\|u(\cdot,y)\|_p = \left(\int_{\mathbb{R}} |u(x,y)|^p dx\right)^{1/p} < \infty, \tag{1.2.22}$$

$$\|v(\cdot,y)\|_p = \left(\int_{\mathbb{R}} |v(x,y)|^p dx\right)^{1/p} < \infty. \tag{1.2.23}$$

•

Problems for Section 1.2

1.2.1 Show that if $x \in \mathbb{R}, y > 0$, then

$$\frac{y}{\pi} \int_{\mathbb{R}} \frac{dt}{(x-t)^2 + y^2} = 1.$$

1.2.2 Set $\varepsilon = y^{3/4}$, and prove that

$$\frac{y}{\pi} \int_{\mathbb{R} \setminus [x-\varepsilon, x+\varepsilon]} \frac{dt}{(x-t)^2 + y^2} \to 0 \quad as \quad y \to 0^+.$$

1.2.3 Use the results of Problems 1.2.1 and 1.2.2 to show that if f is continuous and uniformly bounded by M on $\mathbb{R}$, then

$$\lim_{y \to 0+} \frac{y}{\pi} \int_{\mathbb{R}} \frac{f(t)\, dt}{(x-t)^2 + y^2} = f(x).$$

1.2.4 Prove that if $\mathcal{S}f$ is defined as in (1.2.2), and if $(a, b) = \mathbb{R}$, then

$$\mathcal{S}\{\mathcal{S}f\} = f.$$

1.2.5 Prove that if F is defined for $x \in \mathbb{R}$, $y > 0$, and $t \in \mathbb{R}$, by

$$F(x) = \frac{y}{\pi} \frac{1}{(x-t)^2 + y^2},$$

then

$$(\mathcal{S}F)(x) = \frac{1}{\pi i} \frac{t-x}{(t-x)^2 + y^2}.$$

1.2.6 Prove that if u and v are defined as in (1.2.7), and if $\|u(\cdot, y)\|_p$ and $\|v(\cdot, y)\|_p$ are defined as in (1.2.23), then both $\|u(\cdot, y)\|_p$ and $\|v(\cdot, y)\|_p$ are monotonically decreasing functions of y on the interval $[0, \infty]$.

1.2.7 Let $n > 0$, and let f in $\mathbf{L}^p(\mathbb{R})$, with $1 < p < \infty$. Set

$$(\mathcal{S}_n f)(x) = \frac{P.V.}{2n\pi i} \int_{-\pi n}^{\pi n} f(t) \cot(\frac{t-x}{2n})\, dt.$$

Prove that

$$\lim_{n \to \infty} (\mathcal{S}_n f)(x) = (\mathcal{S}f)(x).$$

The next five problems involve the proof of Theorem 1.2.1 (see [Pichorides, Pi1]).

1.2.8 Let $x \in \mathbb{R}, |x| \leqq \pi/2$, $0 < \gamma < \pi/2$, and for $1 < p \leqq 2$, set

$$A(p,\gamma) = \frac{\tan^{p-1}(\gamma)}{\tan[(p-1)\gamma]},$$

$$B(p,\gamma) = \frac{\sin^{p-1}(\gamma)}{\sin[(p-1)\gamma]},$$

$$C(p) = -\sec(p\pi/2).$$

Prove that

$$[\cos^p(x) - \cos(px)]C(p) \leqq |\sin(x)|^p$$

$$\leqq A(p,\gamma)\cos^p(x) - B(p,\gamma)\cos(px).$$

1.2.9 For given $f \in \mathbf{L}^p(-\pi,\pi)$, with $1 < p \leqq 2$, define $\tilde{f}$ on $(-\pi,\pi)$, by

$$\tilde{f}(x) \equiv \mathcal{S}f(x) = \frac{P.V.}{\pi i}\int_{-\pi}^{\pi} f(t)\cot(\frac{t-x}{2})\,dt.$$

Use the results of Problem 1.2.9 to show that if $f \geqq 0$, then for $0 < \gamma < \pi/2$,

$$C(p)(\|f\|_p^p - \|f\|_1^p) \leqq \|\tilde{f}\|_p^p \leqq A(p,\gamma)\|f\|_p^p - B(p,\gamma)\|f\|_1^p.$$

1.2.10 If f satisfies the conditions in Problem 1.2.10, prove that

$$\|\tilde{f}\|_p = A(p,\gamma)\|f\|_p^p - B(p,\gamma)\|f\|_1^p$$

if and only if $|\tilde{f}| = (\tan\gamma)|f|$ a.e.

1.2.11 Prove that if $g(z)$ and $\tilde{g}(z)$ are defined for $|z| < 1$ by

$$g(z) = \Re(\frac{1+z}{1-z})^{2\gamma/\pi}, \quad \tilde{g}(z) = \Im(\frac{1+z}{1-z})^{2\gamma/\pi}$$

then

$$f(t) = \lim_{r\to 1-} g(re^{it}), \; \tilde{f}(t) = \lim_{r\to 1-} \tilde{g}(re^{it})$$

satisfy

$$|\tilde{f}| = (\tan\gamma)|f|.$$

1.2.12 Under the conditions of Problem 1.2.10, prove that

(a) $A(p,\gamma)$ is the smallest value of A such that

$$\|\tilde{f}\|_p^p \leqq A\|f\|_p^p - B(p,\gamma)\|f\|_1^p,$$

(b) $C(p)$ is the largest value of C such that

$$C|\ \|f\|_p^p - \|f\|_1^p\ | \leqq \|\tilde{f}\|_p^p.$$

1.2.13 Prove that the least value B_p of the number A such that $\|\tilde{f}\|_p \leqq A\|f\|_p$, for every real and 2π-periodic $f \in \mathbf{L}^p(-\pi,\pi)$ is $\tan(\pi/2p)$ if $1 < p \leqq 2$ and $\cot(\pi/2p)$ if $2 \leqq p < \infty$.

1.2.14 Use the results of Problem 1.2.8 to complete the proof of Theorem 1.2.1.

1.2.15 Let (a,b) be a finite interval, let $f \in \mathbf{Lip}_\alpha(a,b)$, and let $f(a) = f(b) = 0$. Let $\mathcal{S}f$ be defined as in (1.2.2). Prove that $\mathcal{S}f \in \mathbf{Lip}_\alpha(a,b)$.

1.2.16 Let $\mathcal{D}$ be a simply-connected domain, let $f \in \mathbf{Hol}(\mathcal{D})$. Assume that $\mathcal{D}$ and f are such that to two fixed boundary points a and b of $\mathcal{D}$, there is a line segment Γ joining a and b, which lies totally in $\mathcal{D}$, and there are constants C and α such that $|f(z)| \leqq C|z-a|^\alpha|b-z|^\alpha$ for all $z \in \mathcal{D}$. Let $\mathcal{S}f$ be defined on Γ.

(i) Prove that $\mathcal{S}f$ may be continued analytically into $\mathcal{D}$.

(ii) Prove that there exists a constant C_1, such that

$$\left|(\mathcal{S}f)(z) - \frac{(b-z)\mathcal{S}(a) - (z-b)\mathcal{S}(b)}{b-a}\right| \leqq C_1|z-a|^\alpha|b-z|^\alpha$$

for all $z \in \mathcal{D}$.

1.3 Riemann–Hilbert Problems

We consider here the solution of the homogeneous and nonhomogeneous Riemann–Hilbert problems in the case of a smooth arc Γ, or of a smooth contour Γ. We then apply these results to solve the dominant equation of a Cauchy singular integral equation.

At the outset, we introduce some basic concepts and assumptions, such as the domain, the class of functions, the Hilbert transform, and the Plemelj formulas. These concepts are simple extensions of the corresponding ones of the previous section. We derive a new formula for computing the index, and we then solve the homogeneous Riemann–Hilbert problem in the case of a contour, and of an open arc. These solutions are then used to solve the non-homogeneous Riemann–Hilbert problem in the case of a contour, and also of an open arc. These concepts are applied in Chapter 6 to obtain the solution of a Cauchy singular integral equation.

1.3.1 Some Definitions

Let $\Gamma = \Gamma(a,b) = \{z(s) : 0 < s < 1\}$ denote either a bounded smooth open arc oriented in direction from $a = z(0+)$ to $b = z(1-)$, or a bounded smooth oriented closed contour (for which $a = b$). If Γ is an open arc, the points a and b do not belong to Γ. We denote $\overline{\Gamma} = \Gamma \cup \{a, b\}$ to be the closure of Γ. $\mathcal{D}^+$ will denote the domain immediately to the left of Γ, and $\mathcal{D}^-$ will denote the domain immediately to the right of Γ, as Γ is traversed for increasing s, and we set $\mathcal{D} = \mathbb{C} \setminus \overline{\Gamma}$. We will assume furthermore, that $\mathcal{D}^+$ and $\mathcal{D}^-$ have been suitably defined so that $\mathcal{D} = \mathcal{D}^+ \cup \mathcal{D}^-$.

A function $w : \Gamma \to \mathbb{C}$ is said to satisfy a Lipschitz condition with exponent α $(0 < \alpha \leqq 1)$ at the point $t \in \Gamma$, if for all points $\tau \in \Gamma$ we have

$$|w(\tau) - w(t)| \leqq C|\tau - t|^{\alpha}, \tag{1.3.1}$$

with C a positive constant depending only on t. The function w is said to satisfy a Lipschitz condition with exponent α on Γ if Equation (1.3.1) holds for all t and τ on Γ, with C a positive constant independent of t and τ. In this case, we shall write $w \in \mathbf{Lip}_{\alpha}(\Gamma)$. If $w \in \mathbf{Lip}_{\alpha}(\Gamma)$, then w can be uniquely defined at the ends a and b, so that $w \in \mathbf{Lip}_{\alpha}(\overline{\Gamma})$.

Let $w : \Gamma \to \mathbb{C}$, with $w \in \mathbf{Lip}_{\alpha}(\Gamma)$, and with Γ either a finite open arc, or a finite contour, and for $z \in \mathbb{C} \setminus \overline{\Gamma}$, let us set

$$W(z) = \frac{1}{2\pi i} \int_{\Gamma} \frac{w(\tau)}{\tau - z} d\tau. \tag{1.3.2}$$

Let $t \in \Gamma$, and set

$$\begin{aligned} W^+(t) &= \lim_{z \to t, z \in \mathcal{D}^+} W(z) \\ W^-(t) &= \lim_{z \to t, z \in \mathcal{D}^-} W(z). \end{aligned} \tag{1.3.3}$$

Then it may be shown (Problem 1.3.1) proceeding as in the proof of Theorem 1.2.2, that the limits (1.3.3) exist for every $t \in \Gamma$, provided that $z \to t$ along an arc which is not tangent to Γ, and moreover, each of the functions W^+ and W^- then satisfy a Lipschitz condition at t, with exponent

α. Hence, if Γ is a contour, then $W^{\pm}$ belong to $\mathbf{Lip}_\alpha(\Gamma)$. If Γ is an open arc, the functions W^+ and W^-, belong to the class $\mathbf{H}^*$ of functions which satisfy a Lipschitz condition with exponent α everywhere on Γ, except at the endpoints, where W^+ and W^- have integrable singularities.

If $w \in \mathbf{Lip}_\alpha(\Gamma)$, and if $t \in \Gamma$, then the Plemelj formulas are valid (Problem 1.3.2):

$$\begin{aligned} W^+(t) &= \frac{1}{2}\left[w(t) + (\mathcal{S}w)(t)\right] \\ W^-(t) &= \frac{1}{2}\left[-w(t) + (\mathcal{S}w)(t)\right]. \end{aligned} \tag{1.3.4}$$

Here W^+ and W^- are defined as in (1.3.3), and the Hilbert transform $\mathcal{S}w$ is given by

$$(\mathcal{S}w)(t) = \frac{P.V.}{\pi i} \int_\Gamma \frac{w(\tau)}{\tau - t} d\tau. \tag{1.3.5}$$

The principal value integral in (1.3.5) is defined for $t \in \Gamma$ by

$$P.V. \int_\Gamma \frac{w(\tau)}{\tau - t} d\tau = \lim_{\varepsilon \to 0} \int_{\Gamma_\varepsilon} \frac{w(\tau)}{\tau - t} d\tau \tag{1.3.6}$$

with Γ_ε the portion of Γ remaining after deleting from Γ all those points that are within a distance ε of t.

1.3.2 The Index of a Function

We shall here define the index of a function in the case of a closed contour, and we shall describe a simple formula which is convenient for computing the index.

Definition 1.3.1 *Let Γ denote a closed contour, and let $G = u + iv$ be a continuous and nonvanishing function defined on Γ, with u and v real-valued functions. The index κ of G with respect to Γ is then defined by the formula*

$$\begin{aligned} \kappa &= \operatorname{ind}(G) \\ &= \frac{1}{2\pi} \int_\Gamma \frac{u(t)\, dv(t) - v(t)\, du(t)}{u^2(t) + v^2(t)}. \end{aligned} \tag{1.3.7}$$

We next describe a simple formula to compute the index κ defined in Definition 1.3.1. To this end, for a real number c, we define $\operatorname{sgn} c$ by

$$\operatorname{sgn} c = \begin{cases} 1 & \text{if} \quad c > 0 \\ 0 & \text{if} \quad c = 0 \\ -1 & \text{if} \quad c < 0. \end{cases} \tag{1.3.8}$$

Theorem 1.3.2 (Computation of Index): *Let $G = u + iv$ and κ be defined as in Definition 1.3.1. Let $z_0, z_1, \ldots, z_n$ denote points on Γ which one meets consecutively as one traverses Γ in a counterclockwise manner, such that $z_n = z_0$, and such that uv changes sign at most once on each closed sub-arc $\overline{z_{j-1} z_j}$ of Γ, $j = 1, 2, \ldots, n$. Then*

$$\kappa = \frac{1}{8} \sum_{j=1}^{n} \begin{vmatrix} \operatorname{sgn} u(t_{j-1}) & \operatorname{sgn} v(t_{j-1}) \\ \operatorname{sgn} u(t_j) & \operatorname{sgn} v(t_j) \end{vmatrix}. \tag{1.3.9}$$

Proof. Let us define a vector function (U, V) on Γ by $(U, V) = (\operatorname{sgn} u, \operatorname{sgn} v)$. Let us then simultaneously examine the two integrals

$$I = \frac{1}{2\pi} \int_\Gamma \frac{u\,dv - v\,du}{u^2 + v^2} \tag{1.3.10}$$

and

$$J = \frac{1}{8} \int_\Gamma \begin{vmatrix} U & V \\ dU & dV \end{vmatrix}. \tag{1.3.11}$$

Setting $(U_j, V_j) = (\operatorname{sgn} u(t_j), \operatorname{sgn} v(t_j))$, Equation (1.3.11) yields

$$\begin{aligned} J_j &\equiv \frac{1}{8} \int_{z_{j-1}}^{z_j} \begin{vmatrix} U & V \\ dU & dV \end{vmatrix} \\ &= \frac{1}{8} [U_{j-1}(V_j - V_{j-1}) - V_{j-1}(U_j - U_{j-1})] \\ &= \frac{1}{8} \begin{vmatrix} U_{j-1} & V_{j-1} \\ U_j & V_j \end{vmatrix} \end{aligned} \tag{1.3.12}$$

with the integral defining J_j taken along the segment $\overline{z_{j-1}, z_j}$ of Γ, and with the identities in Equation (1.3.12) valid, provided UV changes sign at most once on the segment $\overline{z_{j-1}, z_j}$. Hence, we have

$$J = \sum_{j=1}^{n} J_j \tag{1.3.13}$$

$$= \frac{1}{8}\sum_{j=1}^{n}\begin{vmatrix} \text{sgn } u(t_{j-1}) & \text{sgn } v(t_{j-1}) \\ \text{sgn } u(t_j) & \text{sgn } v(t_j) \end{vmatrix},$$

and this is just the value of the sum in (1.3.9). If uv does not change sign on Γ, then the vector $u(t) + iv(t)$ does not circle the origin as t traverses Γ, and so by (1.3.10), we have $I = 0$. In this case $(U_{j-1}, V_{j-1}) = (U_j, V_j)$ for $j = 1, 2, \ldots, n$, and hence by (1.3.13), we have $J = 0$. Hence if uv does not change sign on Γ, then $I = J$.

Next, let $\zeta_0, \zeta_1, \ldots, \zeta_m$ denote the distinct consecutive points located on Γ in a counterclockwise manner, at which uv changes sign, and such that $\zeta_m = \zeta_0$. Let us set

$$\begin{aligned} \mathcal{I}_j &= \frac{1}{2\pi}\int_{\zeta_{j-1}}^{\zeta_j} d\,\arctan[u(t)/v(t)] \\ \mathcal{J}_j &= \frac{1}{8}\int_{\zeta_{j-1}}^{\zeta_j}\begin{vmatrix} U & V \\ dU & dV \end{vmatrix} \end{aligned} \tag{1.3.14}$$

for $j = 1, 2, \ldots, m$, with the integration taken along Γ. As t traverses Γ from ζ_{j-1} to ζ_j, the vector (u, v) changes according to one of the following:

$$\begin{array}{llll} (i) & (0,a) & \rightarrow & (0,b) \\ (ii) & (0,a) & \rightarrow & (b,0) \\ (iii) & (a,0) & \rightarrow & (b,0) \\ (iv) & (a,0) & \rightarrow & (0,b), \end{array}$$

with a and b non-zero, since $(u, v) \neq (0, 0)$ on Γ.

In case (i), a and b must be of the same sign, for otherwise the function v which is continuous on Γ would change sign on Γ between ζ_{j-1} and ζ_j. Hence the vector (u, v) has the same direction at ζ_{j-1} as it does at ζ_j, and therefore $\mathcal{I}_j = 0$. Now, if t is an arbitrary point on the segment $\overline{\zeta_{j-1}, \zeta_j}$ belonging to Γ, we have

$$\begin{aligned} \mathcal{J}_j &= \frac{1}{8}\int_{\zeta_{j-1}}^{t}\begin{vmatrix} U & V \\ dU & dV \end{vmatrix} + \frac{1}{8}\int_{t}^{\zeta_j}\begin{vmatrix} U & V \\ dU & dV \end{vmatrix} \\ &= \frac{1}{8}\begin{vmatrix} 0 & \text{sgn } a \\ \text{sgn } u(t) & \text{sgn } v(t) \end{vmatrix} + \frac{1}{8}\begin{vmatrix} \text{sgn } u(t) & \text{sgn } v(t) \\ 0 & \text{sgn } b \end{vmatrix} \\ &= \frac{1}{8}\text{sgn } u(t)[\text{sgn } b - \text{sgn } a] \\ &= 0 \end{aligned} \tag{1.3.15}$$

since $\text{sgn } b = \text{sgn } a$. In the case (i) we therefore have $\mathcal{I}_j = \mathcal{J}_j = 0$.

In the case (ii), neither u nor v vanish on the segment $\overline{\zeta_{j-1}, \zeta_j}$ of Γ. Thus, as t traverses the segment $\overline{\zeta_{j-1}, \zeta_j}$, the vector (u, v) rotates through an angle $-(\pi/2)\mathrm{sgn}(ab)$. That is, $\mathcal{I}_j = -(1/4)\mathrm{sgn}(ab)$, with $a \neq 0$, $b \neq 0$. On the other hand,

$$\begin{aligned}
\mathcal{J}_j &= \frac{1}{8}\int_{\zeta_{j-1}}^{t} \begin{vmatrix} U & V \\ dU & dV \end{vmatrix} + \frac{1}{8}\int_{t}^{\zeta_j} \begin{vmatrix} U & V \\ dU & dV \end{vmatrix} \\
&= \frac{1}{8}\begin{vmatrix} 0 & \mathrm{sgn}\ a \\ \mathrm{sgn}\ b & \mathrm{sgn}\ a \end{vmatrix} + \frac{1}{8}\begin{vmatrix} \mathrm{sgn}\ b & \mathrm{sgn}\ a \\ \mathrm{sgn}\ b & 0 \end{vmatrix} \\
&= -\frac{1}{4}(\mathrm{sgn}\ a)(\mathrm{sgn}\ b) \\
&= -\frac{1}{4}\mathrm{sgn}(ab),
\end{aligned} \tag{1.3.16}$$

since $\mathrm{sgn}\ u(t) = \mathrm{sgn}\ b$, and $\mathrm{sgn}\ v(t) = \mathrm{sgn}\ a$. Hence, in case (ii), $\mathcal{I}_j = \mathcal{J}_j = -(1/4)\mathrm{sgn}(ab)$.

The argument in the case (iii) is similar to the one in the case (i), and the argument in the case (iv) is similar to the one in the case (ii), and these arguments show that in all cases we have $\mathcal{I}_j = \mathcal{J}_j$. Hence

$$\begin{aligned}
\kappa &= \frac{1}{2\pi}\int_{\Gamma} d\ \arctan(v/u) \\
&= \sum_{j=1}^{m} \mathcal{I}_j \\
&= \sum_{j=1}^{m} \mathcal{J}_j \\
&= \frac{1}{8}\int_{\Gamma} \begin{vmatrix} U & V \\ dU & dV \end{vmatrix} \\
&= \frac{1}{8}\sum_{j=1}^{n} \begin{vmatrix} \mathrm{sgn}\ u(t_{j-1}) & \mathrm{sgn}\ v(t_{j-1}) \\ \mathrm{sgn}\ u(t_j) & \mathrm{sgn}\ v(t_j) \end{vmatrix}.
\end{aligned} \tag{1.3.17}$$

This completes the Proof of Theorem 1.3.2. ∎

Remark 1.3.3 We remark that if u and v are continuous and have at most a finite number of zeros on Γ, then by subdividing Γ into smaller and smaller segments, we eventually arrive at the stage for which the conditions of Theorem 1.3.2 are satisfied. Therefore, the addition of new points $t_j \in \Gamma$ no longer changes the value of the sum on the right hand side of (1.3.9).

1.3.3 Homogeneous Problem

In this section we shall solve the homogeneous Riemann–Hilbert problem, first in the case of a closed contour, then in the case of an open arc.

Riemann–Hilbert Problem: *Let the simple, closed smooth contour* Γ *be given, let* $G \in \mathbf{Lip}_\alpha(\Gamma)$, *and let* $G \neq 0$ *on* Γ. *Let* $\mathcal{D}^+$ *denote the part of* $\mathbb{C}$ *interior to* Γ, *and set* $\mathcal{D}^- = \mathbb{C} \setminus \mathcal{D}^+ \setminus \overline{\Gamma}$. *We attempt to determine two functions:* W^+, *analytic in* $\mathcal{D}^+$, *and* W^-, *analytic in* $\mathcal{D}^-$, *including* $z = \infty$, *such that for every* $t \in \Gamma$ *we have*

$$W^+(t) = G(t)W^-(t). \tag{1.3.18}$$

We seek a solution having one of the forms

$$W(z) = \frac{1}{2\pi} \int_\Gamma \frac{w(\tau)}{\tau - z} d\tau, \tag{1.3.19}$$

with $w \in \mathbf{Lip}_\alpha(\Gamma)$ *in the case of a closed contour, or* $w \in \mathbf{H}^*$ *for the case of an open arc, so that* W *satisfies the condition* $W(\infty) = 0$, *or else the equation*

$$W(z) = \frac{1}{2\pi} \int_\Gamma \frac{w(\tau)}{\tau - z} d\tau + c \tag{1.3.20}$$

with w *defined as in (1.3.19), and* c *a constant, so that* $W(\infty) = c$.

i. The Case $\kappa = 0$.

It is most important to consider first the case $\kappa = \text{ind}(G) = 0$. In this case $\log(G)$ is continuous on Γ, and furthermore, since it is readily seen that if $G \neq 0$ on Γ, we have $\log(G) \in \mathbf{Lip}_\alpha(\Gamma)$. We thus see from Equation (1.3.18) that for every $t \in \Gamma$ we have

$$\log W^+(t) - \log W^-(t) = \log G(t). \tag{1.3.21}$$

Comparison with the Plemelj formulas (1.3.4) now yields the simple solution

$$\log W(z) = \frac{1}{2\pi i} \int_\Gamma \frac{\log G(\tau)}{\tau - z} d\tau. \tag{1.3.22}$$

Notice that the function W defined by (1.3.22) satisfies the relation $W(\infty) = 1$, and that W^+ (resp. W^-) is uniformly bounded and non-vanishing on $\overline{\mathcal{D}^+}$ (resp. $\overline{\mathcal{D}^-}$). In this case we define the *canonical function* Z by $Z = W$. This canonical function will be useful for purposes of solving the non-homogeneous Riemann–Hilbert problem.

ii. The Case $\kappa \neq 0$.

Next, we consider the case of a closed contour, when the integer (why integer?) $\kappa = \text{ind}(G) \neq 0$. In this case we assume, without loss of generality, that the point $z = 0 \in \mathcal{D}^+$. In this case $\text{ind}(t^\kappa) = \kappa$ (why?), and by Problem 1.3.10(b) it therefore follows that $\text{ind}(G_1) = 0$, with $G_1(t) = t^{-\kappa}G(t)$. We may thus factor G_1 in the same manner as we factored the function G above, to get $G_1 = V^+/V^-$ with $V^+ = \exp(W_1^+)$, and $V^- = \exp(W_1^-)$. Upon substituting this factorization into (1.3.18), we get

$$\frac{W^+(t)}{V^+(t)} = t^\kappa \frac{W^-(t)}{V^-(t)}. \tag{1.3.23}$$

Let us note that, the function V^+ (resp. V^-) is non-zero on $\Gamma \cup \mathcal{D}^+$ (resp. $\Gamma \cup \mathcal{D}^-$) (Why?). The canonical function is defined in the case $\kappa \neq 0$ by the equations

$$Z^+(z) = V^+(z), \quad Z^-(z) = z^{-\kappa}V^-(z). \tag{1.3.24}$$

Now, consider first *the case when* $\kappa > 0$. In this case, the function on the left-hand side of (1.3.23) is analytic in $\mathcal{D}^+$, and the function on the right-hand side of (1.3.23) is analytic in $\mathcal{D}^-$, except at ∞, where it can have a pole of order at most κ. These two functions are the analytic continuations of one another through the contour Γ. It thus follows from Liouville's Theorem, that these two functions may be identified with a function which is analytic in all of $\mathbb{C}$, except at ∞, where it can have a pole of order at most κ, i.e., it is a polynomial of degree at most κ. Hence, in the case $\kappa > 0$, the function W is given in terms of the canonical function by the formula

$$W(z) = Z(z)P_\kappa(z), \tag{1.3.25}$$

with Z defined as in (1.3.24), and $P_\kappa(z)$ an arbitrary polynomial in z of at most degree κ.

The above procedure still applies in *the case* $\kappa < 0$, except now we must conclude by Liouville's Theorem that the two functions on the right-hand side of (1.3.24) may be identified with one function which is analytic in all of $\mathbb{C}$ except at $z = \infty$, where it has a pole of order at most κ. That is, W^- is no longer analytic in $\mathcal{D}^-$, so that our factorization problem, as stated originally, does not have a solution. However, we still define the canonical function in this case by (1.3.24), and we shall be able to use this canonical function to solve the non-homogeneous Riemann–Hilbert problem.

iii. The Case Γ *is an Arc.*

Let us recall how easy it was to obtain the factorization (1.3.18) in the case $\kappa = 0$ above, and how effectively we used this result to obtain the factorization in the case $\kappa \neq 0$. We now extend this procedure to obtain a factorization (1.3.18) in the case where Γ is an arc. This case may be considered as a special case of Γ being a contour, on which the function G

has two discontinuities at the points a and b. We must thus give up our assumption that $G \in \mathbf{Lip}_\alpha(\Gamma)$.

Let us assume that $G \in \mathbf{Lip}_\alpha$ at all points of a contour Γ_1 except at the endpoints a and b of Γ_1, where G has simple discontinuities. Let $G \equiv 1$ on the arc $\Gamma_1 \setminus \Gamma$, with Γ being the subarc $\overline{a,b}$ of Γ_1. In this case, we set

$$Y(z) = \frac{1}{2\pi i} \int_\Gamma \frac{\log G(\tau)}{\tau - z} d\tau. \tag{1.3.26}$$

It is readily seen from this equation (see Problem 1.3.7) that for z near an end $c = a$ or $c = b$, we have

$$Y(z) = \pm \frac{\log G(c)}{2\pi i} \log(z - c) + U_c(z), \tag{1.3.27}$$

with the plus sign taken for $c = a$ and the minus sign for $c = b$, and with $U_c(z)$ bounded in a neighborhood of c and $\lim_{z \to c} U_c(z)$ exists and is finite. Let α_j and β_j denote real numbers, such that $\gamma_j = \alpha_j + i\,\beta_j$ is defined by the equation

$$\gamma_j = (-1)^{j-1} \frac{\log G(c)}{2\pi i}, \tag{1.3.28}$$

with $j = 1$ corresponding to $c = a$, and $j = 2$ to $c = b$. It follows then (see Problem 1.3.9) that $\exp(Y)$ will be integrable over Γ provided $\alpha_j > -1$. Since this will in general not be the case, in order to construct a canonical function integrable over Γ, we determine integers κ_j by the requirement

$$-1 < \alpha_j - \kappa_j \leqq 0, \; j = 1, 2. \tag{1.3.29}$$

The sum

$$\kappa = \sum_{j=1}^{2} \kappa_j \tag{1.3.30}$$

will then be called the index of G with respect to Γ. Setting

$$\Pi(z) = (z - a)^{-\kappa_1} (b - z)^{-\kappa_2}, \tag{1.3.31}$$

we define the canonical function Z by

$$Z(z) = \Pi(z)\, e^{Y(z)}. \tag{1.3.32}$$

It is then readily seen (Problem 1.3.12) that the function Z defined by Equation (1.3.32) is integrable over Γ, and moreover, it satisfies Equation (1.3.18), with Γ an open arc. Also, Z is non-vanishing on $\overline{\Gamma}$. Furthermore, the function Z defined in (1.3.32) satisfies the relation $\lim_{(z \to \infty)} z^\kappa\, Z(z) = 1$, i.e., Z has a zero of order κ at ∞ if $\kappa > 0$, a pole of order κ if $\kappa < 0$, and $Z(\infty) = 1$ if $\kappa = 0$. We remark also that if we multiply Z as defined by

Equation (1.3.32) by any polynomial it again satisfies Equation (1.3.18); for $\kappa > 0$, we may multiply Z by an arbitrary polynomial of degree $\kappa - 1$ to get a function W which is representable in the form (1.3.19), whereas, if the degree of the polynomial is κ, the resulting function W is representable in the form (1.3.20).

1.3.4 Solution of the Non-Homogeneous Problem

In the previous subsection we constructed a fundamental function Z, which played a role in the solution of the homogeneous Riemann–Hilbert problem. We now use this same fundamental function to solve the non-homogeneous Riemann–Hilbert problem in the cases of either a contour or an open arc, via the same simple technique. Let us now define the non-homogeneous problem.

Non-Homogeneous Problem. *Let Γ denote either a smooth contour or a smooth open arc, and let G and g be some given functions belonging to $\mathbf{Lip}_\alpha(\Gamma)$, with $G \neq 0$ on $\overline{\Gamma}$. We want to determine a function W representable in either of the form (1.3.19) or (1.3.20), such that the following equation is satisfied for all t on Γ:*

$$W^+(t) = G(t)\, W^-(t) + g(t). \tag{1.3.33}$$

Let us now solve the non-homogeneous Riemann–Hilbert problem. To this end, we recall that in all cases the function G is representable on Γ by means of the equation

$$G(t) = \frac{Z^+(t)}{Z^-(t)}; \tag{1.3.34}$$

substitution of this expression into (1.3.34), we may replace (1.3.33) by the equation

$$\frac{W^+(t)}{Z^+(t)} - \frac{W^-(t)}{Z^-(t)} = \frac{g(t)}{Z^+(t)}. \tag{1.3.35}$$

Now we recall that Z is non-vanishing on $\overline{\Gamma}$, and hence, by assumption, the function W/Z is thus holomorphic on $\mathcal{C} \setminus \overline{\Gamma}$, and moreover W/Z is integrable over Γ. By considering the index κ of G, it thus follows from the Plemelj formulas that W is representable in the form

$$W(z) = \frac{Z(z)}{2\pi i} \int_\Gamma \frac{g(t)}{Z^+(t)(t-z)} dt + Z(z) P_k(z), \tag{1.3.36}$$

with P_k any polynomial of degree k, with $k = \kappa - 1$ if W vanishes at ∞, or with $k = \kappa$ if $W(\infty) = c$, with c is a constant, and with $P_k = 0$ if $k < 0$.

If $\kappa < 0$, there is an additional difficulty stemming from the fact that Z then has a pole of order $-\kappa$ at ∞. In this case, the polynomial P_k appearing in (1.3.36) will of course be zero; on the other hand, if we want a solution having the representation (1.3.19), then W must vanish at ∞, and in order to neutralize the effect of the pole of Z at ∞, and to have $W(\infty) = 0$, we require the additional conditions

$$\int_\Gamma \frac{t^j\, g(t)}{Z(t)}\, dt = 0, \quad j = 0, 1, \cdots, -\kappa - 1. \tag{1.3.37}$$

It is now readily verified, upon setting

$$X(t) = \frac{1}{2}\mathcal{S}(\log\, G)(t) \tag{1.3.38}$$

with $\mathcal{S}$ defined as in (1.3.5), then for all $t \in \Gamma$,

$$Z^\pm(t) = G(t)^{\pm 1/2} e^{X(t)}. \tag{1.3.39}$$

•

Problems for Section 1.3

1.3.1 Prove that if $w \in \mathbf{Lip}_\alpha(\Gamma)$, then each of the limits (1.3.3) exist for every $t \in \Gamma$.

1.3.2 Let $w \in \mathbf{Lip}_\alpha(\Gamma)$, and let the functions $W^\pm$ be defined as in (1.3.3).

(a) Prove that each function $W^\pm$ satisfies a Lipschitz condition with exponent α at every point $t \in \Gamma$.

(b) Prove that if $\Gamma(a, b)$ is an open arc, and if $w(a) = w(b) = 0$, then $W^\pm \in \mathbf{Lip}_\alpha(\Gamma)$.

(c) Prove that the Equations (1.3.4) hold.

1.3.3 Let W be defined by (1.3.22). Determine the explicit form of W as a function of G on Γ.

1.3.4 Let W be defined by (1.3.25). Determine the explicit form of W on Γ.

1.3.5 Let Γ be a simple, closed contour, and let w and W be related as in (1.3.2).

(a) Prove that if w is analytic and bounded in $\mathcal{D}^+$, then for all $t \in \Gamma$,

$$-\frac{1}{2}w(t) + \frac{P.V.}{2\pi i}\int_\Gamma \frac{w(\tau)}{\tau - t}\, d\tau = 0;$$

(b) Prove that if w is analytic and bounded in $\mathcal{D}^-$, then for all $t \in \Gamma$,

$$\frac{1}{2}w(t) + \frac{P.V.}{2\pi i}\int_\Gamma \frac{w(\tau)}{\tau - t}\, d\tau = w(\infty);$$

(c) Prove that if $w = w_0 + P$, with w_0 analytic in $\mathcal{D}^-$ and vanishing at ∞, and P a polynomial in t, then

$$\frac{1}{2}w(t) + \frac{P.V.}{2\pi i}\int_\Gamma \frac{w(\tau)}{\tau - t}d\tau = P(t).$$

1.3.6 Let Γ be a closed contour, let $u \in \mathbf{Lip}_\alpha(\Gamma)$, and set

$$v(t) = \frac{P.V.}{\pi i}\int_\Gamma \frac{u(\tau)}{\tau - t}d\tau.$$

Prove that

$$u(t) = \frac{P.V.}{\pi i}\int_\Gamma \frac{v(\tau)}{\tau - t}d\tau.$$

1.3.7 Let $\Gamma = \Gamma(a,b)$ be an open arc, and let $u \in \mathbf{Lip}_\alpha(\Gamma)$. Let U be defined for $z \in \mathbb{C} \setminus \overline{\Gamma}$, by

$$U(z) = \frac{P.V.}{2\pi i}\int_\Gamma \frac{u(\tau)}{\tau - z}d\tau.$$

Prove that if z is near an endpoint c of Γ, then

$$U(z) = \frac{\pm u(c)\log(z-c)}{2\pi i} + U_c(z),$$

with the plus sign taken for $c = a$ and the minus one for $c = b$, the function $U_c(z)$ is bounded in a neighborhood of c, and tends to a finite limit as $z \to c$.

1.3.8 Let Γ_1 and Γ_2 denote open arcs, where Γ_1 has endpoints a and c, Γ_2 has endpoints c and b, and where $\Gamma = \Gamma_1 \cup \Gamma_2$ is a smooth arc. Let $w \in \mathbf{Lip}_\alpha(\Gamma_1) \cap \mathbf{Lip}_\alpha(\Gamma_2)$, and let W be defined in terms of w by (1.3.2).

$$W(z) = \frac{w(c-0) - w(c+0)}{2\pi i}\log(z-c) + W_c(z),$$

with $W_c(z)$ bounded in a neighborhood of c and such that $\lim_{z\to c} W_c(z)$ exists and is finite.

1.3.9 Let $\gamma = \alpha + i\beta$ be a constant, with α and β real, and $0 \leqq \alpha < 1$. Let $t \in \overline{\Gamma}$, and set

$$W(z) = \frac{1}{2\pi i}\int_\Gamma \frac{d\tau}{(\tau - t)^\gamma(\tau - z)}.$$

If $\Gamma(a,b)$ is an open arc, then

$$W(z) = \pm \frac{e^{\pm i\pi\gamma}}{2i\sin(\pi\gamma)}(z-c)^{-\gamma} + W_c(z),$$

with the plus sign taken for $c = a$, and the minus sign for $c = b$, and with $W_c(z)$ analytic in a neighborhood of $z = c$.

1.3.10 (a) Prove using (1.3.7) that if $G = u + iv$ is continuous and nonvanishing on the closed contour Γ, then

$$\kappa = \operatorname{ind}(G) = \frac{1}{2\pi i}\int_\Gamma d \arg G(t) = \frac{1}{2\pi i}\int_\Gamma d \log G(t).$$

(b) Conclude from part (a) of this problem that if G_1 and G_2 are continuous and nonvanishing on the closed contour Γ, then

$$\operatorname{ind}(G_1\ G_2) = \operatorname{ind}(G_1) + \operatorname{ind}(G_2).$$

1.3.11 Let $\Gamma = \Gamma(a,b)$ be an open arc, such that Γ is a subset of a closed arc Γ_1. Let G be continuous and nonvanishing on Γ, and extend the definition of G to Γ_1, such that $G = 1$ on $\Gamma_1 \setminus \Gamma$, as in Section 1.3.3 (iii). Next, modify G on small neighborhoods of a and b, so that the resulting function is continuous on all of Γ_1. Then, discuss the use of the method of Theorem 1.3.2 to compute the index of G on Γ_1, and show how this result compares with that given by Equation (1.3.30).

1.3.12 Prove that the function Z defined in Equation (1.3.32) is integrable over Γ.

1.3.13 Derive Equation (1.3.39).

1.3.14 Let Z be defined by (1.3.32). Determine the explicit form of Z on Γ.

1.3.15 Let W be defined by (1.3.36). Determine the explicit form of W on Γ.

1.3.16 Let the canonical function Z be defined as in (1.3.32). Prove that $Z \in \mathbf{Hol}(\mathcal{D}^+ \cup \mathcal{D}^-)$, and that there exist positive constants C_1 and C_2, such that for all $t \in \Gamma$,

$$C_1 \leqq \frac{|Z(t)|}{|t-a|^{\alpha_1-\kappa_1}|b-t|^{\alpha_2-\kappa_2}} \leqq C_2.$$

1.3.17 Let $R > 1$, let $f \in \mathcal{A}_R$, with $\mathcal{A}_R$ defined as in (1.1.25), so that for all $w \in \mathcal{A}_R$, we have

$$f(w) = \sum_{n=-\infty}^{\infty} c_n \, w^n.$$

(a) Define $\mathcal{S}f$ for $|w| = \rho \in (R^{-1}, R)$, by

$$\mathcal{S}f(w) = \frac{P.V.}{\pi i} \int_{|t|=\rho} \frac{f(t)}{t-w} dt,$$

and use the results of Problem 1.3.5 to prove that

$$\mathcal{S}f(w) = -\sum_{n=-\infty}^{-1} c_n \, w^n + \sum_{n=0}^{\infty} c_n \, w^n.$$

(b) Conclude that the series representation for $\mathcal{S}f$ holds for all $w \in \mathcal{A}_R$.

1.4 Fourier Transforms

In this section we discuss the Fourier exponential, the Fourier sine, and the Fourier cosine integral, with emphasis on results which are important in applications. Among these particular results are the usual inversion formulas, results about convolutions, the Riemann–Lebesgue lemma, Parseval's Theorem, and connections with Hilbert transforms and functions analytic in the upper half plane. We also discuss being able to recognize whether a function is a Fourier transform or not, such as, e.g., if it is in $\mathbf{L}^p(\mathbb{R})$, with $1 < p \leqq 2$, and the powerful but less well-known Wiener–Levy Theorem. Another not so well-known result of this nature is discussed in Problem 1.4.17.

Let $1 \leqq p < \infty$, and let the *Fourier transform* F of f be denoted by

$$F(x) = \int_{\mathbb{R}} e^{ixt} f(t) \, dt. \tag{1.4.1}$$

Although the integral in (1.4.1) does not always exist, it may be defined a.e. on $\mathbb{R}$ as the limit

$$F(x) = \lim_{c \to 0+} F(x, c), \tag{1.4.2}$$

with

$$F(x, c) = \int_{\mathbb{R}} e^{-c|t|} e^{ixt} f(t) \, dt. \tag{1.4.3}$$

Let $1 < p < \infty$, $q = p/(p-1)$, let $f \in \mathbf{L}^p(\mathbb{R})$, $g \in \mathbf{L}^q(\mathbb{R})$ and let F and G denote the Fourier transforms of respectively f and g. If $\delta > 0$, then

$$\begin{aligned}
&\frac{1}{2\pi}\int_{\mathbb{R}} e^{-\delta|x|} F(x)\overline{G(x)}\,dx \\
&\quad = \frac{1}{2\pi}\int_{\mathbb{R}} e^{-\delta|x|}\left\{\int_{\mathbb{R}} e^{ixt} f(t)\,dt\right\}\left\{\int_{\mathbb{R}} e^{-ix\tau}\overline{g(\tau)}\,d\tau\right\}dx \\
&\quad = \int_{\mathbb{R}} f(t)\left\{\int_{\mathbb{R}}\overline{g(\tau)}\left[\int_{\mathbb{R}}\frac{e^{-\delta|x|+ix(t-\tau)}}{2\pi}dx\right]d\tau\right\}dt \\
&\quad = \int_{\mathbb{R}} f(t)\left\{\int_{\mathbb{R}}\overline{g(\tau)}\frac{\delta}{\pi[(\tau-t)^2+\delta^2]}d\tau\right\}dt.
\end{aligned} \tag{1.4.4}$$

It now follows from the previous section that the function $g(t;\delta)$ defined by

$$g(t;\delta) = \frac{\delta}{\pi}\int_{\mathbb{R}}\frac{g(\tau)}{(\tau-t)^2+\delta^2}d\tau, \tag{1.4.5}$$

is in $\mathbf{L}^q(\mathbb{R})$ for every $\delta > 0$, that

$$\|g(\cdot;\delta)\|_q \leqq \|g(\cdot;0)\|_q = \|g\|_q,$$

and that

$$\lim_{\delta\to 0} g(t;\delta) = g(t),$$

a.e. on $\mathbb{R}$. Hence, we have proved

Theorem 1.4.1 (Parseval's Theorem): *If $1 < p < \infty$, $q = p/(p-1)$, if $f \in \mathbf{L}^p(\mathbb{R})$, $g \in \mathbf{L}^q(\mathbb{R})$, and if F and G denote the Fourier transforms of f and g respectively, then*

$$\int_{\mathbb{R}} f(t)\overline{g(t)}\,dt = \frac{1}{2\pi}\int_{\mathbb{R}} F(x)\overline{G(x)}\,dx. \tag{1.4.6}$$

Now, by taking $G(x) = e^{-\delta|x|}e^{i\zeta x}$ in (1.4.6), so that $g(t) = \overline{g(t)} = (\delta/\pi)/[(t-\zeta)^2+\delta^2]$, we get

Corollary 1.4.2 *Given $F \in \mathbf{L}^p(\mathbb{R})$, let F be defined as in Equations (1.4.1)–(1.4.3). Then we have*

$$f(t) = \lim_{\delta\to 0}\frac{1}{2\pi}\int_{\mathbb{R}} e^{-\delta|x|-ixt}F(x)\,dx, \tag{1.4.7}$$

a.e. on $\mathbb{R}$.

In place of (1.4.1) and (1.4.7), we shall, henceforth, use the simpler notations for the transform pair,

$$F(x) = \int_{\mathbb{R}} e^{ixt} f(t)\, dt,$$
$$f(t) = \frac{1}{2\pi} \int_{\mathbb{R}} e^{-ixt} F(x)\, dx, \tag{1.4.8}$$

where it is understood that it may in fact be necessary to define the integrals on the right-hand sides by the limits (1.4.2) or (1.4.7).

Corollary 1.4.3 *Let $1 \leqq p < \infty$, let $f \in \mathbf{L}^p(\mathbb{R})$, and let F denote the Fourier transform of f. Then $F \in \mathbf{L}^q(\mathbb{R})$, where $1/p + 1/q = 1$.*

Proof. (a) In (1.4.6), take $0 < \varepsilon < 1$,

$$g(t) = \begin{cases} 0 \text{ if } t < 0 \\ e^{-t} t^{-1/(q+\varepsilon)}, \text{ if } t > 0. \end{cases} \Rightarrow G(x) = \frac{\Gamma\left(\frac{q+\varepsilon-1}{q+\varepsilon}\right)}{(1-ix)^{\frac{q+\varepsilon-1}{q+\varepsilon}}}, \tag{1.4.9}$$

Then $g \in \mathbf{L}^{q'}(\mathbb{R})$ for all $q' \in [1, q]$, and $G \in \mathbf{L}^{p'}(R)$ for all $p' \in [p, \infty]$. Hence, from (1.4.6) it follows that $F \in \mathbf{L}^{q'}$, for some $q' \in [1, q]$.

(b) Similarly, we take for $0 < \varepsilon < 1$,

$$g(t) = \frac{\Gamma\left(\frac{1}{q-\varepsilon}\right)}{2\pi(1-it)^{1/(q-\varepsilon)}} \Rightarrow G(x) = \begin{cases} x^{-\frac{q-\varepsilon-1}{q-\varepsilon}} e^{-x}, & x > 0, \\ 0, & x < 0. \end{cases} \tag{1.4.10}$$

which yields $F \in \mathbf{L}^{q'}(\mathbb{R})$ for some $q' \in [q, \infty]$. The conclusion of the corollary now follows, by letting $\varepsilon \to 0$. ∎

Theorem 1.4.4 *Let $1 \leqq p < \infty$, let $f \in \mathbf{L}^p(\mathbb{R})$, $g \in \mathbf{L}^q(\mathbb{R})$, with $1/p + 1/q = 1$, and let F and G denote their respective Fourier transforms. Then*

$$\frac{1}{2\pi} \int_{\mathbb{R}} e^{ixt} F(x) G(x)\, dx = \int_{\mathbb{R}} f(\tau) g(t-\tau)\, d\tau. \tag{1.4.11}$$

Proof. Let $\delta > 0$. Then, by Fubini's theorem, we have

$$\begin{aligned}
&\int_{\mathbb{R}} e^{-\delta|\tau|} f(\tau) g(t-\tau) d\tau \\
&\quad = \int_{\mathbb{R}} \frac{e^{-\delta|\tau|}}{(2\pi)^2} \int_{\mathbb{R}} \int_{\mathbb{R}} F(u) e^{-iu\tau} G(v) e^{-iv(t-\tau)} du\, dv\, d\tau \qquad (1.4.12) \\
&\quad = \frac{1}{2\pi} \int_{\mathbb{R}} \left\{ \frac{\delta}{\pi} \int_{\mathbb{R}} \frac{F(u)}{(u-v)^2+\delta^2} du \right\} e^{-ivt} G(v)\, dv,
\end{aligned}$$

which is just the left-hand side of (1.4.11), after letting $\delta \to 0$, and applying Theorem 1.2.2. ∎

Theorem 1.4.5 *Let $f \in \mathbf{L}^1(\mathbb{R})$, $g \in \mathbf{L}^2(\mathbb{R})$, and define h by*

$$h(t) = \int_{\mathbb{R}} f(t-\tau) g(\tau)\, d\tau. \tag{1.4.13}$$

Then $h \in \mathbf{L}^2(\mathbb{R})$, and moreover

$$\int_{\mathbb{R}} |h(t)|^2 dt \leqq \left(\int_{\mathbb{R}} |f(t)|\, dt \right)^2 \int_{\mathbb{R}} |g(t)|^2 dt. \tag{1.4.14}$$

Proof. Let F, G, and H denote the Fourier transform of respectively f, g, and h, so that from (1.4.11) and (1.4.8)

$$H(x) = F(x) G(x). \tag{1.4.15}$$

Hence

$$\begin{aligned}
\int_{\mathbb{R}} |H(x)|^2 dx &= \int_{\mathbb{R}} |F(x)|^2 |G(x)|^2\, dx \\
&\leqq \left(\sup_{y \in \mathbb{R}} |F(y)| \right)^2 \int_{\mathbb{R}} |G(x)|^2\, dx \qquad (1.4.16) \\
&\leqq \left(\int_{\mathbb{R}} |f(t)| dt \right)^2 \int_{\mathbb{R}} |G(x)|^2\, dx
\end{aligned}$$

and (1.4.14) now follows from Parseval's theorem. ∎

Given $f \in \mathbf{L}^p(\mathbb{R})$, with $1 \leqq p < \infty$, we can define the three functions

$$
\begin{aligned}
F(x) &= \int_{\mathbb{R}} e^{ixt} f(t)\,dt, \\
F_+(x) &= \int_0^\infty e^{ixt} f(t)\,dt, \qquad (1.4.17)\\
F_-(x) &= \int_{-\infty}^0 e^{ixt} f(t)\,dt.
\end{aligned}
$$

Then we have

Theorem 1.4.6 *Let F, F_+, and F_- be defined by Equations (1.4.17), with $f \in \mathbf{L}^p(\mathbb{R})$ for some fixed $p, 1 \leqq p < \infty$. Then F_+ and F_- are also related to F by*

$$
\begin{aligned}
F_+(x) = (\mathcal{P}F)(x) &= \lim_{y \to 0+} \frac{1}{2\pi i} \int_{\mathbb{R}} \frac{F(t)}{t - x - iy} dt, \\
F_-(x) = (I - \mathcal{P})F(x) &= \lim_{y \to 0+} \frac{-1}{2\pi i} \int_{\mathbb{R}} \frac{F(t)}{t - x + iy} dt \quad .
\end{aligned} \qquad (1.4.18)
$$

Furthermore

$$
\begin{aligned}
F_+ + F_- &= F \\
F_+ - F_- &= \mathcal{S}F
\end{aligned} \qquad (1.4.19)
$$

with for all $x \in \mathbb{R}$,

$$
(\mathcal{S}F)(x) = \frac{P.V.}{\pi i} \int_{\mathbb{R}} \frac{F(t)}{t - x} dt, \qquad (1.4.20)
$$

$$
(\mathcal{S}F)(x) = \int_{\mathbb{R}} \operatorname{sgn}(t) e^{ixt} f(t)\,dt. \qquad (1.4.21)
$$

Proof. The proof is left to Problem 1.4.16. ■

Henceforth, let Ω_+ and Ω_- respectively denote the upper and lower half planes, defined by

$$
\begin{aligned}
\Omega_+ &= \{z \in \mathbb{C} : \Im z > 0\}, \\
\Omega_- &= \{z \in \mathbb{C} : \Im z < 0\}.
\end{aligned} \qquad (1.4.22)
$$

Corollary 1.4.7 *Define F_+ and F_- by*

$$F_+(z) = \int_0^\infty e^{izt} f(t)\,dt, \quad z \in \Omega_+,$$
$$F_-(z) = \int_{-\infty}^0 e^{izt} f(t)\,dt, \quad z \in \Omega_-, \tag{1.4.23}$$

with $f \in \mathbf{L}^p(\mathbb{R})$, $1 \leqq p < \infty$. Thus F_+ is analytic on Ω_+, F_- is analytic on Ω_- and for $x \in \mathbb{R}$

$$\lim_{y \to 0+} F_+(x+iy) = F_+(x),$$
$$\lim_{y \to 0+} F_-(x-iy) = F_-(x), \tag{1.4.24}$$

a.e. on $\mathbb{R}$, with F_+ and F_- are defined by (1.4.18).

Aside from their importance in applications, the Fourier transforms of functions in $\mathbf{L}^1(\mathbb{R})$ have beautiful mathematical properties, so we give special consideration to them.

Theorem 1.4.8 (Riemann–Lebesgue Lemma): *If $f \in \mathbf{L}^1(\mathbb{R})$ and F is the Fourier transform of f, then*

$$\lim_{x \to \pm\infty} F(x) = 0. \tag{1.4.25}$$

Proof. Let $f \in \mathbf{L}^1(\mathbb{R})$. Then given any $\varepsilon > 0$, we can determine $M > 0$ such that

$$\int_{-\infty}^{-M} |f(t)|dt < \varepsilon, \quad \int_M^\infty |f(t)|dt < \varepsilon. \tag{1.4.26}$$

On the interval $[-M, M]$, there exists a polynomial p such that

$$\int_{-M}^{M} |f(t) - p(t)|dt < \varepsilon \tag{1.4.27}$$

(see Problem 1.4.6). Hence

$$F(x) = \int_{\mathbb{R}} e^{ixt} f(t)\,dt$$
$$= \int_{-M}^{M} [f(t) - p(t)]e^{ixt}dt + \int_{-M}^{M} p(t)e^{ixt}dt + \int_{|x|>M} f(t)e^{ixt}dt. \tag{1.4.28}$$

Integrating by parts,

$$\begin{aligned}
&\int_{-M}^{M} e^{ixt} p(t)\, dt \\
&= \frac{1}{ix}\left\{ e^{ixM} p(M) - e^{-ixM} p(-M) - \int_{-M}^{M} e^{ixt} p'(t)\, dt \right\} \qquad (1.4.29) \\
&\to 0 \quad as \quad x \to \pm\infty.
\end{aligned}$$

Thus, from (1.4.27), the first term on the extreme right-hand side of (1.4.28) is bounded by ε; the second term by ε for all $x \in \mathbb{R}$ such that $|x|$ is sufficiently large; and the third term by 2ε, from (1.4.26). Hence, given any $\varepsilon > 0$, we can determine an $N > 0$, such that, if $x \in \mathbb{R}$ and $|x| > N$, then $|F(x)| < 4\varepsilon$. ∎

Let $\mathbf{L}$ denote $\mathbf{L}^1(\mathbb{R})$, let $\mathbf{L}_+$(resp. $\mathbf{L}_-$) denote the subset of those functions f in $\mathbf{L}$ such that, $f(t) = 0$ if $t < 0$ (resp. $f(t) = 0$ if $t > 0$). Let W, W_+ and W_- denote respectively the set of all Fourier transforms of functions in $\mathbf{L}$, $\mathbf{L}_+$ and $\mathbf{L}_-$ (see Equation (1.4.17)).

Theorem 1.4.9 *Let $\mathbf{V}$ denote any one of the three classes of functions $\mathbf{W}$, $\mathbf{W}_+$ or $\mathbf{W}_-$. Then every $F \in \mathbf{V}$ is continuous and uniformly bounded on $\mathbb{R}$. If $F \in \mathbf{V}$ and $G \in \mathbf{V}$, then $FG \in \mathbf{V}$, and for arbitrary complex number α and β, $\alpha F + \beta G \in \mathbf{V}$.*

Proof. Let us introduce a norm on $\mathbf{L}$ by

$$\|f\| = \int_{\mathbb{R}} |f(t)| dt. \qquad (1.4.30)$$

If $F \in \mathbf{W}$ is defined in terms of f as in (1.4.17), then we also define

$$\|F\| = \int_{\mathbb{R}} |f(t)| dt = \|f\|. \qquad (1.4.31)$$

It is readily seen that both $\mathbf{L}$ and $\mathbf{W}$ become Banach spaces (see Chapter 6) under these norms.
It is clear from (1.4.17), that $\sup_{(x\in\mathbb{R})} |F(x)| \leqq \|F\|$, which proves that each F is uniformly bounded on $\mathbb{R}$.
We next prove the continuity of F on $\mathbb{R}$, by means of the identity

$$F(x+h) - F(x) = \int_{\mathbb{R}} (e^{iht} - 1) e^{ixt} f(t)\, dt. \qquad (1.4.32)$$

If $\|f\| = 0$, then $F \equiv 0$, and there is nothing to prove. Hence, let $\|f\| > 0$, and corresponding to a given $\varepsilon > 0$, select $M > 0$, so that $\int_{|t|>M} |f(t)| dt <$

$\varepsilon/4$. For $t \in [-M, M]$, we have $|e^{iht} - 1| \leqq h|t| \leqq hM$. We thus select h by $h = \varepsilon/[2M\|f\|]$. Then, for any $x \in \mathbb{R}$, we have

$$\begin{aligned} &|F(x+h) - F(x)| \\ &\leqq \int_{|t|>M} |(e^{iht} - 1)e^{ixt} f(t)| dt + \int_{-M}^{M} |(e^{iht} - 1)e^{ixt} f(t)| dt \\ &\leqq 2 \int_{|t|>M} |f(t)| dt + hM \int_{\mathbb{R}} |f(t)| dt < \varepsilon. \end{aligned} \tag{1.4.33}$$

Hence F is continuous on $\mathbb{R}$.
Let $F \in \mathbf{V}$ and $G \in \mathbf{V}$ denote the respective Fourier transforms of $f \in \mathbf{L}$ and $g \in \mathbf{L}$. Then $K = FG$ is the Fourier transform of k, with

$$k(t) = \int_{\mathbb{R}} f(\tau) g(t-\tau)\, d\tau. \tag{1.4.34}$$

Thus

$$\begin{aligned} \|k\| &= \int_{\mathbb{R}} |k(t)| dt \\ &\leqq \int_{\mathbb{R}} |f(\tau)| d\tau \int_{\mathbb{R}} |g(t-\tau)| dt \\ &= \|f\| \ \|g\|, \end{aligned} \tag{1.4.35}$$

so that $k \in \mathbf{L}$, i.e., $FG \in \mathbf{W}$. The representation (1.4.34) shows moreover, that if both $f \in \mathbf{L}_+$, $g \in \mathbf{L}_+$, then $k(t) = 0$ if $t < 0$, that is, $k \in \mathbf{L}_+$. That is, if $F \in \mathbf{W}_+$, $G \in \mathbf{W}_+$, then $FG \in \mathbf{W}_+$. Similarly, if $F \in \mathbf{W}_-$, $G \in \mathbf{W}_-$, then $FG \in \mathbf{W}_-$.

This completes the proof of Theorem 1.4.9. ∎

Theorem 1.4.10 (Wiener–Levy Theorem): *Let* $\mathbf{V}$ *denote any one of the three classes of functions* $\mathbf{W}$, $\mathbf{W}_+$ *or* $\mathbf{W}_-$. *Let* w *be an arbitrary analytic and single-valued function in a domain (or Riemann surface)* $\mathcal{D}$. *If* $F \in \mathbf{V}$ *and if* $\mathcal{C} \equiv \{F(x) : x \in \mathbb{R}\} \in \mathcal{D}$, *then* $w \circ F - w(0) \in \mathbf{V}$.

Proof. The proof is somewhat lengthy, and we thus leave it to Problems 1.4.10 to 1.4.15 which are at the end of this section. ∎

We next consider *Fourier sine and cosine transforms* which also play an important role in applications.

Let $f \in \mathbf{L}^p(\mathbb{R})$ be given, with $1 \leqq p < \infty$, and let us define f_e and f_o by

$$\begin{aligned} f_e(t) &= \frac{1}{2}\{f(t)+f(-t)\}, \\ f_o(t) &= \frac{1}{2}\{f(t)-f(-t)\}. \end{aligned} \tag{1.4.36}$$

We see that f_e is an *even* function of t $(f_e(-t) = f_e(t))$ and f_o is an *odd* function of t $(f_o(-t) = -f_o(t))$, and for all $t \in \mathbb{R}$,

$$f(t) = f_e(t) + f_o(t). \tag{1.4.37}$$

Clearly, if $f \in \mathbf{L}^p(\mathbb{R})$, then f_e and f_o are both in $\mathbf{L}^p(\mathbb{R})$.

Let F be defined for a given $f \in \mathbf{L}^p(\mathbb{R})$, by (1.4.1). Then we have

$$F = F_c + iF_s, \tag{1.4.38}$$

with

$$\begin{aligned} F_c(x) &= \int_{\mathbb{R}} \cos(xt) f_e(t)\, dt, \\ F_s(x) &= \int_{\mathbb{R}} \sin(xt) f_o(t)\, dt. \end{aligned} \tag{1.4.39}$$

Inspection of (1.4.39) shows that F_c is an even function of x, and F_s is an odd function.

Using the second expression of (1.4.8), we have

$$\begin{aligned} f(t) &= \frac{1}{2\pi}\int_{\mathbb{R}} e^{-ixt}\{F_c(x) + iF_s(x)\}dx \\ &= \frac{1}{2\pi}\int_{\mathbb{R}} \cos(xt)F_c(x)\, dx + \frac{1}{2\pi}\int_{\mathbb{R}} \sin(xt)F_s(x)\, dx \\ &= f_e(t) + f_o(t). \end{aligned} \tag{1.4.40}$$

That is,

$$\begin{aligned} f_e(t) &= \frac{1}{2\pi}\int_{\mathbb{R}} F_c(x)\cos(xt)\, dx \\ f_o(t) &= \frac{1}{2\pi}\int_{\mathbb{R}} F_s(x)\sin(xt)\, dx. \end{aligned} \tag{1.4.41}$$

Theorem 1.4.11 *Let g be an arbitrary function belonging to $\mathbf{L}^p(\mathbb{R}_+)$, with $\mathbb{R}_+ = (0,\infty)$, and with $1 < p < \infty$, and define G_c and G_s by*

$$G_c(x) = \int_{\mathbb{R}_+} \cos(xt) g(t)\, dt,$$
$$G_s(x) = \int_{\mathbb{R}_+} \sin(xt) g(t)\, dt. \tag{1.4.42}$$

Then both G_c and G_s belong to $\mathbf{L}^q(\mathbb{R}_+)$, with $1/p + 1/q = 1$. Moreover, if G_c and G_s are defined for some g in $\mathbf{L}^p(\mathbb{R}_+)$ by (1.4.42), then a.e. on $\mathbb{R}_+$,

$$g(t) = \frac{2}{\pi} \int_{\mathbb{R}_+} \cos(xt) G_c(x)\, dx,$$
$$g(t) = \frac{2}{\pi} \int_{\mathbb{R}_+} \sin(xt) G_s(x)\, dx. \tag{1.4.43}$$

Proof. Let us verify the first equations in (1.4.42) and (1.4.43). The verification of the second identities in these equations is similar, and we therefore defer this to the problems.

If $g \in \mathbf{L}^p(\mathbb{R}_+)$ we may define a function $f \in \mathbf{L}^p(\mathbb{R})$ by

$$f(t) = \begin{cases} g(t), & t > 0 \\ g(-t), & t < 0. \end{cases} \tag{1.4.44}$$

Then $f = f_e$, with f_e defined as in (1.4.36), and hence its Fourier transform, F, is

$$\begin{aligned} F(x) &= \int_{\mathbb{R}} e^{ixt} f(t)\, dt \\ &= \int_{\mathbb{R}} \cos(xt) f(t)\, dt \\ &= 2\int_{\mathbb{R}_+} \cos(xt) f(t)\, dt \\ &= 2\int_{\mathbb{R}_+} \cos(xt) g(t)\, dt = 2G_c(x), \end{aligned} \tag{1.4.45}$$

with G_c defined as in (1.4.42). Now, noting that G_e is an even function of x, and taking $t > 0$, Equation (1.4.8) yields

$$\begin{aligned}
g(t) &= f(t) = \frac{1}{2\pi}\int_{\mathbb{R}} e^{-ixt}F(x)\,dx \\
&= \frac{1}{2\pi}\int_{\mathbb{R}} e^{-ixt}\{2G_c(x)\}dx \\
&= \frac{2}{2\pi}\int_{\mathbb{R}} \cos(xt)G_c(x)\,dx \\
&= \frac{2}{\pi}\int_0^{\infty} \cos(xt)G_c(x)\,dx.
\end{aligned} \tag{1.4.46}$$

This completes the proof. ■

•

Problems for Section 1.4

1.4.1 Evaluate

$$\int_{\mathbb{R}} e^{-y|t|}e^{ixt}dt$$

where $y > 0$.

1.4.2 Show that if $t \in \mathbb{R}$, $x \in \mathbb{R}$, then

$$\int_{\mathbb{R}} \log\left|\frac{u+t}{u-t}\right| e^{ixu}du = \frac{2\pi i\sin(xt)}{|x|}.$$

1.4.3 Show that if $x \in \mathbb{R}$, $z \in \mathbb{R}$, $z \neq 0$, then

$$\int_{\mathbb{R}} \log\left|\frac{e^u+z}{e^u-z}\right| e^{ixu}du = \frac{\pi^2}{2}\frac{\tan(\pi x/2)}{\pi x/2}\cdot e^{ix\log|z|}.$$

1.4.4 Show that if $x \in \mathbb{R}$, $z = |z|e^{i\theta}$, $|z| \neq 0$, and $0 \leqq |\theta| \leqq \pi$, then

$$\int_{\mathbb{R}} \log\left|\frac{e^u+z}{e^u-z}\right| e^{ixu}du$$

$$= \frac{\pi\sinh[(\pi/2-|\theta|)x]}{x\cosh[\pi x/2]}e^{ix\log|z|}.$$

1.4.5 Prove that

$$\int_{\mathbb{R}} \frac{e^{ixt}dt}{\cosh(t-y)} = \frac{\pi e^{iy}}{\cosh(\pi x/2)}.$$

1.4.6 How does the inequality (1.4.27) follow from Weierstrass' theorem?

1.4.7 Prove that every $F \in \mathbf{L}^p(\mathbb{R})$ is the Fourier transform of some $f \in \mathbf{L}^q(\mathbb{R})$ whenever $1 < p \leqq 2$, with $1/p + 1/q = 1$.

1.4.8 Prove that if $2 < p$ and if $1/p + 1/q = 1$, then not every $F \in \mathbf{L}^p(\mathbb{R})$ is the Fourier transform of some $f \in \mathbf{L}^q(\mathbb{R})$.

1.4.9 Prove the second equations in (1.4.42) and (1.4.43).

Problems 1.4.10 to 1.4.15 involve the proof of the Wiener–Levy Theorem.

1.4.10 (a) Using the notation of Theorem 1.4.10, prove that since $\mathcal{C} \subset \mathcal{D}$ and $F \in \mathbf{W}$, we have $\rho \equiv \inf_{(x \in \mathcal{C}, y \in \partial\mathcal{D})} |x - y| > 0$.

(b) Conclude that if ζ is an arbitrary point on $\mathcal{C}$, and $|z - \zeta| < \rho$, then

$$w(z) = \sum_{n=0}^{\infty} \frac{w^{(n)}(\zeta)}{n!}(z - \zeta)^n.$$

1.4.11 Let $x \in \mathbb{R}$, and define $S(x)$ by

$$S(x) = \begin{cases} 0 & \text{if } |x| > 1, \\ 1 - |x| & \text{if } |x| \leqq 1. \end{cases}$$

Prove that if $h > 0$, and $\zeta \in \mathbb{R}$, then

$$\frac{1}{2\pi} \int_{\mathbb{R}} S\left(\frac{x - \zeta}{h}\right) e^{-ixt} dx = h\, e^{i\zeta t} s(ht),$$

with

$$s(t) = \frac{\sin^2(t/2)}{2\pi(t/2)^2}.$$

1.4.12 Let $h > 0$, j an integer, $x \in \mathbb{R}$, and set

$$K_j(x) = [F(x) - F(jh)] \sum_{\ell=j-1}^{j+1} S\left(\frac{x - \ell h}{h}\right).$$

(a) Let $\|K_j\|$ be defined as in (1.4.31), and prove that if h is sufficiently small, then $\|K_j\| < \rho/2$, with ρ defined as in Problem

1.4.10. [Hint: Show that if K_j, F and $\sum_{\ell=-1}^{1} S(x-\ell)$ are respectively the Fourier transforms of k_j, f and σ, then

$$k_j(t) = \int_{\mathbf{R}} e^{ijhy} f(y) \left\{[h\sigma(ht-hy) - h\sigma(ht)]e^{ijht}\right\} dy,$$

and conclude that

$$\|K_j\| = \|k_j\| \leqq \int_{\mathbf{R}} \|\sigma(t-hy) - \sigma(t)\| |f(y)| dy].$$

(b) Conclude that $K_j \in \mathbf{W}$, $K_j^m \in \mathbf{W}$ for $m = 1, 2, \cdots$, that

$$\left[\sum_{m=0}^{\infty} \frac{w^{(m)}(F(jh))}{m!} K_j^m(x)\right] S\left(\frac{x-jh}{h}\right) \in \mathbf{W},$$

and that for arbitrary positive integer N,

$$\sum_{j=-N}^{N} S\left(\frac{x-jh}{h}\right) [w \circ F(x) - w(0)] \in \mathbf{W}.$$

1.4.13 (a) Prove that if $h > 0$, and N is an arbitrary positive integer, then $G_N \in \mathbf{W}$, with

$$G_N(x) = \left[1 - \sum_{j=-N}^{N} S\left(\frac{x-jh}{h}\right)\right] F(x);$$

(b) Prove that $\|G_N\| \to 0$ as $N \to \infty$.

1.4.14 Use the results of Problems 1.4.12 and 1.4.13 to conclude that $w \circ F - w(0) \in \mathbf{W}$.

1.4.15 Let $\mathcal{C}$ be defined as in Problem 1.4.10, let $F \in \mathbf{W}_+$, so that $\mathcal{C} = \partial\{w \circ F(z) : z \in \Omega_+\}$. Prove that if $z = x + iy \in \Omega_- = \mathbb{C} \setminus \overline{\Omega}_+$, then

$$\frac{1}{2\pi i} \int_{\mathbf{R}} \frac{(w \circ F)(t) - w(0)}{t - x - iy} dt = 0.$$

Conclude that if $F \in \mathbf{W}_+$ (resp. $F \in \mathbf{W}_-$) then $w \circ F - w(0) \in \mathbf{W}_+$ (resp. $w \circ F - w(0) \in \mathbf{W}_-$).

1.4.16 Give a detailed proof of Theorem 1.4.6.

1.4.17 * Let $\mathbf{M}$ denote the set of all sigma finite complex measures μ defined on the Borel sets of $\mathbb{R}$, such that $\int_{\mathbb{R}} |d\mu| < \infty$, and let $\widehat{\mathbf{M}}$ denote the set of all functions of the form $\hat{\mu}(x) = \int_{\mathbb{R}} e^{ixt}\, d\mu(t)$, where $\mu \in \mathbf{M}$.

(a) Prove that if $\hat{\mu} \in \widehat{\mathbf{M}}$, $\hat{\nu} \in \widehat{\mathbf{M}}$, then $\hat{\mu} \pm \hat{\nu} \in \widehat{\mathbf{M}}$, and $\hat{\mu}\,\hat{\nu} \in \widehat{\mathbf{M}}$.

(b) Prove that if $\hat{\mu} \in \mathbf{C}^2(\mathbb{R})$, and if $\hat{\mu}$ has compact support on $\mathbb{R}$, then $\hat{\mu} \in \widehat{\mathbf{M}}$.

(c) Prove that if $\hat{\mu} \in \widehat{\mathbf{M}}$, $\hat{\nu} \in \widehat{\mathbf{M}}$, and if for some compact subset $\mathcal{B}$ of $\mathbb{R}$ we have $\hat{\mu} \neq 0$ on $\mathcal{B}$, $\hat{\nu} = 0$ on $\mathbb{R} \setminus \mathcal{B}$, then $\hat{\nu}\,/\hat{\mu} \in \widehat{\mathbf{M}}$.

1.5 Laplace Transforms

Let $\mathbb{R}_+ = (0, \infty)$, let f be defined on $\mathbb{R}_+$, and let g be defined by the integral

$$g(s) = (\mathcal{L}F)(s) \equiv \int_0^\infty e^{-st} f(t)\, dt. \tag{1.5.1}$$

Upon setting $s = \xi + i\eta$ in (1.5.1), we get

$$g(\xi + i\eta) = \int_0^\infty \left[e^{-\xi t} f(t)\right] e^{-i\eta t} dt. \tag{1.5.2}$$

We may thus think of the *Laplace transform* (1.5.1) as a Fourier transform; this analogy immediately leads to the inversion formula

$$f(t) = \frac{1}{2\pi} \int_{-\infty}^\infty g(\xi + i\eta) e^{(\xi + i\eta)t} d\eta. \tag{1.5.3}$$

That is, we get the *Bromwich inversion integral*,

$$f(t) = \frac{1}{2\pi i} \int_{\xi - i\infty}^{\xi + i\infty} g(s) e^{st} ds. \tag{1.5.4}$$

For example, if $f \in \mathbf{L}^2(\mathbb{R}_+)$ then g defined by (1.5.1) is also an $\mathbf{L}^2$ function, i.e., by Theorem 1.4.1, we have the identity

$$\frac{1}{2\pi} \lim_{\xi \to 0+} \int_{-\infty}^\infty |g(\xi + i\eta)|^2 d\eta = \int_0^\infty |f(t)|^2 dt. \tag{1.5.5}$$

Note in this case that g is analytic in Ω^+, with Ω^+ the right-half plane defined by

$$\Omega^+ = \{z \in \mathbb{C} : \Re z > 0\}. \tag{1.5.6}$$

Note here, the difference in notation between the right-half plane, Ω^+, and the upper-half plane, Ω_+, with the latter defined as in (1.4.22).

•

Problems for Section 1.5

1.5.1 Prove the following identities, valid for $\Re s > 0$:

(a) $\displaystyle\int_0^\infty e^{-st}\sin(t)\,dt = \frac{1}{s^2+1}$;

(b) $\displaystyle\int_0^\infty e^{-st}t^{\alpha-1}dt = \Gamma(\alpha)/s^\alpha, \Re\alpha > 0$;

(c) $\displaystyle\int_0^\infty e^{-st}t^{-1/2}e^{-1/t}dt = e^{-2s^{1/2}}(\pi/s)^{1/2}$;

(d) $\displaystyle\int_0^\infty e^{-st}t^\alpha J_\alpha(t)\,dt = 2^\alpha\Gamma(\alpha+1/2)\pi^{-1/2}(1+s^2)^{-\alpha-1/2}, \Re\alpha >$ $-1/2$;

(e) $\displaystyle\int_0^\infty e^{-st}\left[1+2\sum_{n=1}^\infty e^{-n^2\pi^2 t}\right]dt = \frac{1}{s^{1/2}\tanh(s^{1/2})}$;

(f) $\displaystyle\int_0^\infty e^{-st}J_\alpha(t)\,dt = \frac{[(1+s^2)^{1/2}-s]^\alpha}{(1+s^2)^{1/2}}$;

(g) $\displaystyle\int_0^\infty e^{-st}e^{-t}dt = \frac{1}{s+1}$;

(h) $\displaystyle\int_0^\infty e^{-st}\cos(t)\,dt = \frac{s}{s^2+1}$;

(i) $\displaystyle\int_0^\infty e^{-st}e^{-t^2}dt = e^{s^2/4}\int_{s/2}^\infty e^{-t^2}dt.$

1.5.2 * Let $g(s) = (\mathcal{L}F)(s)$, for some F, such that $\int_0^\infty \exp(-at)\,|F(t)|dt < \infty$, for all $a > a_0$. Prove that we cannot have $g(s) \equiv 0$ for all s on an open interval $(c,d) \subset (0,\infty)$.

1.6 Fourier Series

Historically, the discovery of *Fourier series* predated the discovery of complex variables. In the present monograph we have already demonstrated (see Corollary 1.1.11) that Laurent's theorem provides considerable insight to the power and accuracy of Fourier series approximation. In the present section we develop some well-known and important practical results about Fourier series, their convergence, and the Poisson summation formula.

Let $h > 0$, and let f be defined on the interval $(-\pi/h, \pi/h]$, such that $\left(\int_{-\pi/h}^{\pi/h} |f(x)|^p dx\right)^{1/p} < \infty$, $1 \leqq p \leqq \infty$. We may then extend the definition of f to a periodic function on $\mathbb{R}$ by means of the formula

$$f(x) = f(x + 2\pi/h). \tag{1.6.1}$$

Let us consider the representation

$$f(x) = \sum_{n=-\infty}^{\infty} c_n e^{inhx}, \tag{1.6.2}$$

with

$$c_n = \frac{h}{2\pi} \int_{-\pi/h}^{\pi/h} f(x) e^{-inhx} dx. \tag{1.6.3}$$

If (1.6.1) is valid, then it follows that if c_n is defined by (1.6.3), we also have

$$c_n = \frac{h}{2\pi} \int_{a-\pi/h}^{a+\pi/h} f(x) e^{-inhx} dx \tag{1.6.4}$$

with a an arbitrary real number.

Lemma 1.6.1 *If N is a positive integer, and θ is arbitrary, then*

$$\frac{\sin N\theta}{\tan \frac{1}{2}\theta} = \frac{1}{2} e^{-iN\theta} + \sum_{k=-(N-1)}^{N-1} e^{ik\theta} + \frac{1}{2} e^{iN\theta}. \tag{1.6.5}$$

Moreover, for any $a \in \mathbb{R}$, and $h > 0$, we have

$$\frac{h}{2\pi} \int_{a-\pi/h}^{a+\pi/h} \frac{\sin(Nh\theta)}{\tan(h\theta/2)} d\theta = 1. \tag{1.6.6}$$

Proof. See Problem 1.6.2. ∎

In view of Equations (1.6.1) through (1.6.6), it now follows that

$$\begin{aligned} &\frac{1}{2} c_{-N} e^{-iNhx} + \sum_{k=-(N-1)}^{N-1} c_k e^{ikhx} + \frac{1}{2} c_N e^{iNhx} - f(x) \\ &= \frac{h}{2\pi} \int_{x-\pi/h}^{x+\pi/h} \frac{f(t) - f(x)}{\tan[h(t-x)/2]} \sin[Nh(t-x)]\, dt \\ &= \frac{h}{2\pi} \int_{-\pi/h}^{\pi/h} \frac{f(x+t) - f(x)}{\tan(ht/2)} \sin(Nht)\, dt. \end{aligned} \tag{1.6.7}$$

It may be shown now that under our assumptions on f, the right-hand side of (1.6.7) approaches zero in $\mathbf{L}^p(-\pi/h, \pi/h)$, as $N \to \infty$. However, it is more suitable for our aims, and also for practical applications, to somewhat restrict the class of functions f.

Definition 1.6.2 (Recall Equation (1.3.1).) *A function f defined on an interval (a,b) is said to belong to $\mathbf{Lip}_\alpha(a,b)$, with α a positive number, if there exists a constant C such that whenever x and t belong to (a,b), then $|f(x) - f(t)| \leqq C|x-t|^\alpha$. The function f is said to be piecewise $\mathbf{Lip}_\alpha$ on (a,b), it there exist a finite number of subintervals (a_i,b_i) whose union is (a,b), such that f belongs to $\mathbf{Lip}_\alpha(a_i,b_i)$ on each subinterval (a_i,b_i) of $(a,b)\setminus\{a_i\}\setminus\{b_i\}$.*

Theorem 1.6.3 *Let f be defined on $(-\pi/h, \pi/h)$, let f be piecewise $\mathbf{Lip}_\alpha$ on this interval, and let f be defined on $\mathbb{R}$ via (1.6.1). Furthermore, assume that for each point x on $\mathbb{R}$,*

$$f(x) = \lim_{h\to 0+} \frac{1}{2}\{f(x+h) + f(x-h)\}. \tag{1.6.8}$$

Let c_n be defined by (1.6.3). Then

$$\frac{1}{2}c_{-N}e^{-iNhx} + \sum_{k=-(N-1)}^{N-1} c_k e^{ikhx} + \frac{1}{2}c_N e^{iNhx} \to f(x) \tag{1.6.9}$$

for every $x \in \mathbb{R}$.

Proof. Let us denote the left-hand side of (1.6.7) by $\varepsilon_N(f,x)$. Upon replacing t by $-t$ on the right-hand side of (1.6.7) we see that

$$\varepsilon_N(f,x) = \frac{h}{4\pi}\int_{-\pi/h}^{\pi/h} \frac{f(x+t) - 2f(x) + f(x-t)}{\tan(ht/2)} \sin(Nht)\,dt. \tag{1.6.10}$$

Now by our $\mathbf{Lip}_\alpha$ assumption on f, and by (1.6.8), it follows that there exists a constant C such that

$$|f(x+t) - 2f(x) + f(x-t)| \leqq 2C|t|^\alpha,$$

uniformly as a function of x, $x \in \mathbb{R}$. Since $|\tan(ht/2)| \geqq h|t|/2$ on $(-\pi/h, \pi/h)$, it thus follows that

$$\left|\frac{f(x+t) - 2f(x) + f(x-t)}{\tan(ht/2)}\right| \leqq \frac{4C|t|^{\alpha-1}}{h}. \tag{1.6.11}$$

That is, the function $\{(f(x+t) - 2f(x) + f(x-t)\}/\tan(ht/2)$ belongs to $\mathbf{L}^1(-\pi/h, \pi/h)$ as a function of t, uniformly as a function of x, $x \in \mathbb{R}$. Hence $\varepsilon_N(f,x) \to 0$ as $N \to \infty$, by the Riemann–Lebesgue lemma.

As for the Fourier transforms, we can also represent f via either sines or cosines. ∎

Theorem 1.6.4 *Let both f_e and f_O satisfy the conditions of Theorem 1.6.3, and in addition, let f_e be an even function, and f_o an odd function, i.e., for all $x \in \mathbb{R}$, $f_e(x) = f_e(-x)$, $f_o(-x) = -f_o(x)$. Define $a_n (n = 0, 1, 2, \cdots)$ and $b_n (n = 1, 2, 3, \cdots)$ by*

$$a_n = \frac{2h}{\pi} \int_0^{\pi/h} f_e(t) \cos(nht)\, dt \tag{1.6.12}$$

$$b_n = \frac{2h}{\pi} \int_0^{\pi/h} f_o(t) \sin(nht)\, dt \tag{1.6.13}$$

then

$$f_e(x) - \frac{1}{2} a_0 - \sum_{k=1}^{N-1} a_k \cos(khx) \to 0 \tag{1.6.14}$$

$$f_o(x) - \sum_{k=1}^{N-1} b_k \sin(khx) \to 0 \tag{1.6.15}$$

as $N \to \infty$, uniformly in x, $x \in \mathbb{R}$.

Proof. The proof of Theorem 1.6.4. uses the results of Theorem 1.6.3, and is left to Problem 1.6.3. ∎

Theorem 1.6.5 *Let $f \in \mathbf{L}^2(-\pi/h, \pi/h)$, and let c_n be defined as in (1.6.3). Then*

$$\frac{h}{2\pi} \int_{-\pi/h}^{\pi/h} |f(x)|^2 dx = \sum_{m=-\infty}^{\infty} |c_m|^2. \tag{1.6.16}$$

Proof. The proof may be carried by forming $f(x)\overline{f(x)}$ using the representation (1.6.2) of f, and then carrying out term-by-term integration over $(-\pi/h, \pi/h)$. ∎

We now state and prove the discrete form of Theorem 1.4.5.

Theorem 1.6.6 *Let $\{u_k\}_{k=-\infty}^{\infty}$ and $\{v_k\}_{k=-\infty}^{\infty}$ be two sequences of complex numbers such that*

$$\sum_{k=-\infty}^{\infty} |u_k|^2 < \infty, \quad \sum_{k=-\infty}^{\infty} |v_k| < \infty. \tag{1.6.17}$$

Set

$$w_k = \sum_{m=-\infty}^{\infty} u_{k-m} v_m. \tag{1.6.18}$$

Then

$$\sum_{k=-\infty}^{\infty} |w_k|^2 \leqq \sum_{k=-\infty}^{\infty} |u_k|^2 \left(\sum_{m=-\infty}^{\infty} |v_m| \right)^2. \tag{1.6.19}$$

Proof. Define u, v, and w on $(-\pi, \pi)$ by

$$\begin{aligned} u &= \sum_{k=-\infty}^{\infty} u_k e^{ikx}, & v &= \sum_{m=-\infty}^{\infty} v_m e^{imx}, \\ w &= uv = \sum_{k=-\infty}^{\infty} w_m e^{ikx}, & w_k &= \sum_{m=-\infty}^{\infty} u_{k-m} v_m. \end{aligned} \tag{1.6.20}$$

Then by the previous theorem,

$$\begin{aligned} \sum_{k=-\infty}^{\infty} |w_k|^2 &= \frac{1}{2\pi} \int_{-\pi}^{\pi} |w(x)|^2 dx = \frac{1}{2\pi} \int_{-\pi}^{\pi} |u(x) v(x)|^2 dx \\ &\leqq \frac{1}{2\pi} \int_{-\pi}^{\pi} |u(x)|^2 dx \left(\sup_{-\pi < x < \pi} |v(x)| \right)^2 \\ &\leqq \sum_{k=-\infty}^{\infty} |u_k|^2 \left(\sum_{m=-\infty}^{\infty} |v_m| \right)^2. \end{aligned} \tag{1.6.21}$$

■

Let us now give a straightforward proof of the *Poisson Summation Formula.* To this end, let $1 \leqq p < \infty$, and set

$$W^p(\mathbb{R}) = \left\{ F : F(x) = \lim_{\delta \to 0+} \int_{\mathbb{R}} e^{ixt - \delta|t|} f(t)\, dt, f \in \mathbf{L}^p(\mathbb{R}) \right\}; \tag{1.6.22}$$

$$\mathcal{L}^p(\mathbb{R}) = \left\{ f : f(t) = \frac{1}{2\pi} \lim_{\delta \to 0+} \int_{\mathbb{R}} e^{-ixt - \delta|x|} F(x)\, dx, F \in W^p(\mathbb{R}) \right\}. \tag{1.6.23}$$

The following theorem contains two equivalent formulations of the *Poisson Summation formula.*

Theorem 1.6.7 (Poisson Summation Formula): *Let $1 \leqq p < \infty$, let $f \in \mathcal{L}^p(\mathbb{R})$, let $F \in W^p(\mathbb{R})$ denote the Fourier transform of f, and let $h > 0$. Then, for all $x \in \mathbb{R}$,*

$$h \sum_{m=-\infty}^{\infty} f(mh - x) = \sum_{n=-\infty}^{\infty} F\left(\frac{2n\pi}{h}\right) e^{i2n\pi x/h}; \tag{1.6.24}$$

and

$$h \sum_{m=-\infty}^{\infty} f(mh) e^{imhx} = \sum_{n=-\infty}^{\infty} F\left(\frac{2n\pi}{h} + x\right). \tag{1.6.25}$$

Proof. We shall only prove (1.6.25), as the proof of (1.6.24) is similar, and we shall therefore leave it to the problems at the end of this section.

Let us for the moment assume that the series on the right-hand side of (1.6.25) converges absolutely to $\mathcal{F}$, i.e.,

$$\mathcal{F}(x) = \sum_{n=-\infty}^{\infty} F\left(\frac{2n\pi}{h} + x\right). \tag{1.6.26}$$

Then clearly $\mathcal{F}$ is periodic, with period $2\pi/h$, so that we may set

$$\mathcal{F}(x) = \sum_{m=-\infty}^{\infty} c_m e^{imhx}, \tag{1.6.27}$$

where by Equation (1.6.3),

$$c_m = \frac{h}{2\pi} \int_{-\pi/h}^{\pi/h} \mathcal{F}(x) e^{-imhx} dx = h\, f(mh), \tag{1.6.28}$$

and where the last identity following by (1.4.8). Hence (1.6.25) follows if we combine (1.6.26), (1.6.27) and (1.6.28), and keep in mind that as a consequence of $F \in W^p(\mathbb{R})$, (1.6.27) is an identity for all $x \in \mathbb{R}$.
In the case when the series on the right-hand side of (1.6.26) does not converge absolutely to $\mathcal{F}$, we consider instead of $\mathcal{F}$, the function $\mathcal{F}_\delta$, with $\delta > 0$, and with

$$\mathcal{F}_\delta(x) = \sum_{n=-\infty}^{\infty} F\left(\frac{2n\pi}{h} + x\right) \exp\left[-\delta \left|\frac{2n\pi}{h} + x\right|\right]. \tag{1.6.29}$$

In this case all of the steps (1.6.26) through (1.6.28) may be carried out as before, for arbitrary fixed $\delta > 0$. We then define both sides of (1.6.25) by passing to the limit as $\delta \to 0$. ∎

Finally, let us prove a result on the Hilbert transform of a Fourier series.

Theorem 1.6.8 *Let f be periodic, with period $2\pi/h$, let f be of class $\mathbf{Lip}_\alpha$ on $\mathbb{R}$, with $0 < \alpha < 1$, let the conditions of Theorem 1.6.3 be satisfied, and let the numbers c_n be defined as in (1.6.3). Then F defined by*

$$F(x) = \frac{P.V.}{2\pi i} \int_{-\pi/h}^{\pi/h} \frac{h\, f(t)}{\tan[(t-x)h/2]}\, dt \tag{1.6.30}$$

is periodic and of class $\mathbf{Lip}_\alpha$ on $\mathbb{R}$, and moreover, for all $x \in \mathbb{R}$, we have

$$F(x) = \sum_{n=-\infty}^{\infty} c_n \operatorname{sgn}(n) e^{inhx}, \tag{1.6.31}$$

with sgn(n) *defined as in (1.3.8).*

Proof. By Problem 1.3.5 (a) and (b), we have for any $\alpha \in \mathbb{R}$, and any integer n, that

$$\frac{P.V.}{\pi i} \int_{\alpha-\pi}^{\alpha+\pi} \frac{e^{in\theta}}{e^{i\theta} - e^{i\alpha}} i e^{i\theta}\, d\theta = e^{in\alpha} \operatorname{sgn}(n). \tag{1.6.32}$$

We also note the identity

$$\frac{ie^{i\theta}}{e^{i\theta} - e^{i\alpha}} = \frac{1}{2\tan[(\theta-\alpha)/2]} + \frac{i}{2}. \tag{1.6.33}$$

The proof thus follows by taking the Hilbert transform as defined by (1.3.5) of the polynomial identity (1.6.5), using (1.6.32) and (1.6.33), then proceeding as in the proof of Theorem 1.6.3, and recalling the fact that if Γ is a closed contour, and if $f \in \mathbf{Lip}_\alpha(\Gamma)$, then $Sf \in \mathbf{Lip}_\alpha(\Gamma)$. ■

•

Problems for Section 1.6

1.6.1 Let z be a complex number. By finding the explicit Fourier series of e^{ixht} on the interval $-\pi/h < t < \pi/h$, conclude that if m is an integer

$$\sum_{n=-\infty}^{\infty} \frac{(-1)^n e^{inht}}{z-n}$$

$$= \begin{cases} \dfrac{\pi e^{izh(t-2m\pi/h)}}{\sin(\pi z)} & \text{if } (2m-1)\frac{\pi}{h} < t < (2m+1)\frac{\pi}{h} \\ \pi \cot(\pi z) & \text{if } t = (2m \pm 1)\frac{\pi}{h}. \end{cases}$$

1.6.2 Prove Lemma 1.6.1.

1.6.3 Prove Theorem 1.6.4.

1.6.4 Prove (1.6.18).

1.6.5 (a) Show that if $x \in \mathbb{R}$ and $h > 0$, then

$$h \sum_{m=-\infty}^{\infty} e^{-|mh-x|} = \sum_{n=-\infty}^{\infty} \frac{e^{i2n\pi x/h}}{1+(n\pi x/h)^2}.$$

(b) Prove that if $a \in \mathbb{R}$, if $m = 0, \pm 1, \pm 2, \cdots$, and if $2m\pi/h \leqq t < (2m+2)\pi/h$, and if $a \neq 0$, then

$$\frac{ah}{\pi} \sum_{n=-\infty}^{\infty} \frac{e^{-inht}}{a^2+n^2h^2} = 2\pi \frac{\cosh\{a[\pi/h - |t-2m\pi/h|]\}}{\sinh(a\pi/h)}.$$

1.6.6 Prove that for all $x \in \mathbb{R}, h > 0$ and $d > 0$,

$$\frac{h}{2d} \sum_{k=-\infty}^{\infty} \frac{1}{\cosh[\pi(x-kh)/(2d)]} = \sum_{k=-\infty}^{\infty} \frac{e^{i2k\pi x/h}}{\cosh[2\pi dk/h]}$$

$$= 1 + 2\sum_{k=0}^{\infty} (-1)^k e^{-2\pi d(2k+1)/h} \cdot$$

$$\cdot \left\{ \frac{1}{e^{2\pi z/h} - e^{-2\pi d(2k+1)/h}} + \frac{1}{e^{-2\pi z/h} - e^{-2\pi d(2k+1)/h}} \right\}.$$

1.6.7 Prove that if $x \in \mathbb{R}$, $t \in \mathbb{R}$, $v \in \mathbb{C}$, then

$$2(1-2t\cos x + t^2)^{-v}$$
$$= \sum_{n=-\infty}^{\infty} \frac{t^{|n|}}{|n|!} \frac{\Gamma(|n|+v)}{\Gamma(v)} F(v, |n|+v; |n|+1; t^2) \cdot e^{inx}$$

with $F(a, b; c; z)$ the usual hypergeometric function.

1.6.8 Prove that if $a \in \mathbb{R}$, $x \in \mathbb{R}$ $y \in \mathbb{R}$, $0 \leqq |x| < |y|$, then

$$\frac{\sinh(ax)}{\sinh(ay)} = 2\pi \sum_{n=1}^{\infty} \frac{(-1)^{n-1} n \sin(n\pi x/y)}{n^2\pi^2 + a^2y^2}.$$

1.6.9 (a) Prove that if $a \in \mathbb{R}$, $x \in \mathbb{R}$, $y \in \mathbb{R}$, $0 \leqq |x| < |y|$, then

$$2\frac{\cosh(ax)}{\sinh(ay)} = ay \sum_{n=-\infty}^{\infty} \frac{(-1)^n e^{in\pi x/y}}{n^2\pi^2 + a^2y^2}.$$

(b) Prove that if $m = 0, \pm 1, \pm 2, \cdots$, if $(2m-1)\pi/h \leqq t < (2m+1)\pi/h$, and if $a \in \mathbb{R}$, $a \neq 0$, then

$$\frac{ah}{\pi} \sum_{n=-\infty}^{\infty} \frac{(-1)^n e^{-inht}}{a^2 + n^2h^2} = \frac{\cosh\{a[t - 2m\pi/h]\}}{\sinh(a\pi/h)}.$$

1.6.10 Prove that if $a \in \mathbb{R}$, $x \in \mathbb{R}$, $y \in \mathbb{R}$, $0 \leqq |x| < |y|$, then

$$\frac{\sinh(ax)}{\cosh(ay)} = ay \sum_{n=-\infty}^{\infty} \frac{(-1)^n e^{i(n+1/2)\pi x/y}}{(n+\frac{1}{2})^2\pi^2 + a^2y^2}.$$

1.6.11 Prove that if $a \in \mathbb{R}$, $x \in \mathbb{R}$, $y \in \mathbb{R}$, $0 \leqq |x| < |y|$, then

$$\frac{\cosh(ax)}{\cosh(ay)} = 2\pi \sum_{n=0}^{\infty} \frac{(-1)^n (n+1/2)\cos[(n+1/2)\pi x/y]}{(n+1/2)^2\pi^2 + a^2y^2}.$$

1.6.12 Prove that if $x \in \mathbb{R}$, $y \in \mathbb{C}$, $xy \neq 0$ then

$$\frac{1}{2x} + \frac{\pi}{2(xy)^{1/2}\tanh[\pi(x/y)^{1/2}]} = \sum_{n=0}^{\infty} \frac{1}{x+n^2y}.$$

1.6.13 (a) Prove that if $x, y, z \in \mathbb{C}$, then

$$S(x,y,z) \equiv \frac{xy}{\pi} \sum_{n=-\infty}^{\infty} \frac{1}{(z+ny)^2 + x^2}$$
$$= \frac{\sinh(2\pi x/y)}{\cosh(2\pi x/y) - \cos(2\pi z/y)}.$$

(b) Prove that if $S(x,y,z)$ is defined in part (a) of this problem, and if $x, y, z \in \mathbb{R}$, then

$$|S(x,y,z)| \leqq \frac{1}{|\tanh(\pi x/y)|}.$$

1.6.14 Prove that the identities in Problems 1.6.8 through 1.6.13, may all be obtained from the identities in Problem 1.6.1.

1.6.15 Let $\theta = |\arg(z)|$, such that $0 < \theta < \pi$. Verify the formula

$$\sum_{n=-\infty}^{\infty} \log\left|\frac{e^{nh}+z}{e^{nh}-z}\right| = \frac{2\pi}{h}\left(\frac{\pi}{2} - \theta\right)$$
$$+2\sum_{m=1}^{\infty} \frac{\sinh\left[2m\pi\left(\frac{\pi}{2}-\theta\right)/h\right]\cos\left[(2m\pi/h)\log|z|\right]}{m\cosh\left(m\pi^2/h\right)}, \quad h > 0.$$

1.7 Transformations of Functions

Some results on transformation of analytic functions, which are useful to our purpose, will be presented in this section. At the outset we discuss conformal transformations. We shall also present some transformations which are practically useful, although they are not conformal.

Let $\mathcal{D}_1$ and $\mathcal{D}_2$ denote two domains. A function ϕ mapping $\mathcal{D}_1$ onto $\mathcal{D}_2$ is called one-to-one if to each distinct pair of points z_1 and z_2 in $\mathcal{D}_1$, the points $\zeta_1 = \phi(z_1)$ and $\zeta_2 = \phi(z_2)$ in $\mathcal{D}_2$ are also distinct. It is called one-to-one and onto if it is one-to-one, and if given any point ζ in $\mathcal{D}_2$, there is a point z in $\mathcal{D}_1$ such that $\zeta = \phi(z)$. The mapping $\phi(z)$ is called a conformal transformation of $\mathcal{D}_1$ onto $\mathcal{D}_2$, if ϕ is one-to-one, onto and analytic in $\mathcal{D}_1$.

Theorem 1.7.1 (Riemann Mapping Theorem): *Let $\mathcal{D}$ be a simply-connected domain, and let a be an arbitrary point in $\mathcal{D}$. Then there is a unique conformal transformation ϕ mapping $\mathcal{D}$ onto the unit disk $\mathcal{D}(0;1)$ such that $\phi(a) = 0$, $\phi'(a) > 0$.*

Proof. We omit the proof of this theorem due to its length and because most proofs of this theorem, while constructive, do not provide a useful construction of the conformal map ϕ. We thus leave the proof as a problem (Problem 1.7.4), for the student who enjoys a challenge. ∎

Circles and straight lines play an important role in the applications of complex variables. It is readily shown that if a, b, and c are arbitrary complex numbers such that

$$b\overline{b} > (a + \overline{a})(c + \overline{c}), \tag{1.7.1}$$

where e.g. $\overline{b}$ denotes the complex conjugate of b, then the set

$$\{z \in \mathbb{C} : (a + \overline{a})z\overline{z} + bz + \overline{bz} + c + \overline{c} = 0\} \tag{1.7.2}$$

is either a circle or a straight line. Furthermore, if α, β, γ, and δ are arbitrary complex numbers such that $\alpha\delta - \beta\gamma \neq 0$, then setting

$$z = \frac{\alpha\zeta + \beta}{\gamma\zeta + \delta} \tag{1.7.3}$$

in (1.7.2) yields an equation of the same form, i.e., we get the equation

$$(a' + \overline{a}')\zeta\overline{\zeta} + b'\zeta + \overline{b'\zeta} + c' + \overline{c'} = 0,$$

where if (1.7.1) holds, then we again have $b'\overline{b'} > (a' + \overline{a'})(c' + \overline{c'})$. That is, the transformation (1.7.3), with $\alpha\delta - \beta\gamma \neq 0$, always transforms a circle or a straight line into another circle or a straight line.

Example 1.7.2 Let w and a be arbitrary complex numbers such that $|w| = 1$ and $|a| < 1$. Then the mapping

$$\zeta = w\frac{(z-a)}{1-\overline{a}z}, \tag{1.7.4}$$

is a conformal transformation of the unit disc $\mathcal{D}(0;1)$ in the z-plane onto the unit disc in the ζ-plane.

Example 1.7.3 Let w and a be arbitrary complex numbers such that $|w| = 1$ and $\Im a > 0$. Then the mapping

$$\zeta = \frac{w(z-a)}{z-\overline{a}} \tag{1.7.5}$$

is a conformal transformation of the region $\Omega_+ = \{z : \Im z > 0\}$ onto the region $\mathcal{D}(0;1) = \{\zeta : |\zeta| < 1\}$.

The most important applications of conformal transformations are those for which it is possible to explicitly express the function $w = \phi(z)$, as well as its inverse transformation, $z = \psi(w) = \phi^{-1}(w)$. A large family of such transformations can be found in textbooks on conformal transformations. In the remainder of this section we shall present several transformations (not all of which are conformal) which we shall later find particularly useful.

Example 1.7.4 Let $R > 1$, and let $\mathcal{E}_R$ denote the ellipse

$$\mathcal{E}_R = \left\{z = x + iy \in \mathbb{C} : x \in \mathbb{R},\ y \in \mathbb{R},\ \left(\frac{2x}{R+R^{-1}}\right)^2 + \left(\frac{2y}{R-R^{-1}}\right)^2 < 1\right\}. \tag{1.7.6}$$

Let $\mathcal{A}_R$ be defined as in (1.1.25), and set

$$\begin{aligned} \mathcal{A}_R^+ &= \{\zeta \in \mathcal{A}_R : |\zeta| > 1\} \\ \mathcal{A}_R^- &= \{\zeta \in \mathcal{A}_R : |\zeta| < 1\}. \end{aligned} \tag{1.7.7}$$

The mapping

$$z = \frac{1}{2}(\zeta + \zeta^{-1}) \tag{1.7.8}$$

is a conformal transformation of $\mathcal{A}_R^+$ onto $\mathcal{E}_R$, and furthermore, (1.7.8) is also a conformal transformation of $\mathcal{A}_R^-$ onto $\mathcal{E}_R$. However, whereas (1.7.8) is *not* a conformal transformation of $\mathcal{A}_R$ onto $\mathcal{E}_R$, it is, nevertheless, a useful transformation for purposes of studying functions that are defined on $\mathcal{A}_R$, and which take the same value at each point ζ in $\mathcal{A}_R$ as at the point $1/\zeta$.

Example 1.7.5 Let $d > 0$, and let D_d denote the domain

$$D_d = \{w \in \mathbb{C} : |\Im w| < d\}. \tag{1.7.9}$$

Let us define some other regions:

$$D_d^1 = \{z \in \mathbb{C} : |\arg z| < d\} \tag{1.7.10}$$

$$D_d^2 = \left\{z \in \mathbb{C} : |\arg\left(\frac{z-a}{b-z}\right)| < d\right\}, -\infty < a < b < \infty; \tag{1.7.11}$$

$$D_d^3 = \{z \in \mathbb{C} : |\arg(\sinh z)| < d\}; \tag{1.7.12}$$

$$D_d^4 = \{z \in \mathbb{C} : |\arg(z + \sqrt{1+z^2})| < d\}. \tag{1.7.13}$$

Then

$$w = \log z (\Leftrightarrow z = e^w) \tag{1.7.14}$$

is a conformal map of D_d^1 onto D_d;

$$w = \log\left(\frac{z-a}{b-z}\right) \quad \left(\Leftrightarrow z = \frac{a+be^w}{1+e^w}\right) \tag{1.7.15}$$

is a conformal map of D_d^2 onto D_d;

$$w = \log(\sinh z) \quad \left(\Leftrightarrow z = \log\left[e^w + \sqrt{1+e^{2w}}\right]\right) \tag{1.7.16}$$

is a conformal map of D_d^3 onto D_d; and

$$w = \log\left(z + \sqrt{1+z^2}\right) \quad (\Leftrightarrow z = \sinh w) \tag{1.7.17}$$

is a conformal map of D_d^4 onto D_d.

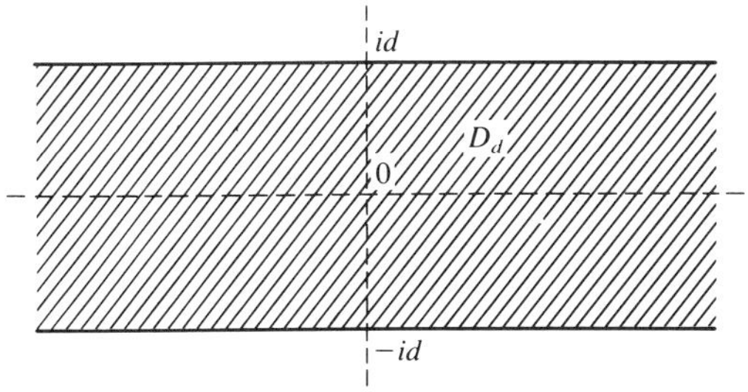

Figure 1.7.4a: The Region D_d of Equation (1.7.9)

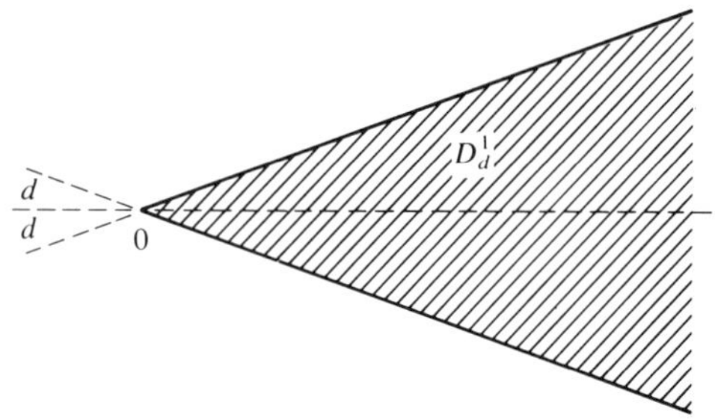

Figure 1.7.4b: The Region D_d^1 of Equation (1.7.10)

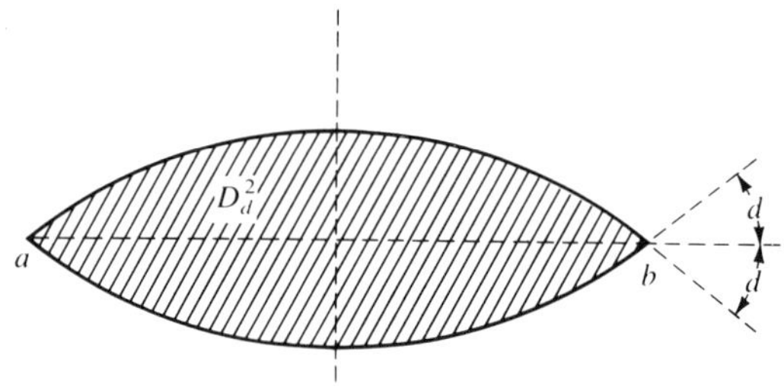

Figure1.7.4c: The Region D_d^2 of Equation (1.7.11)

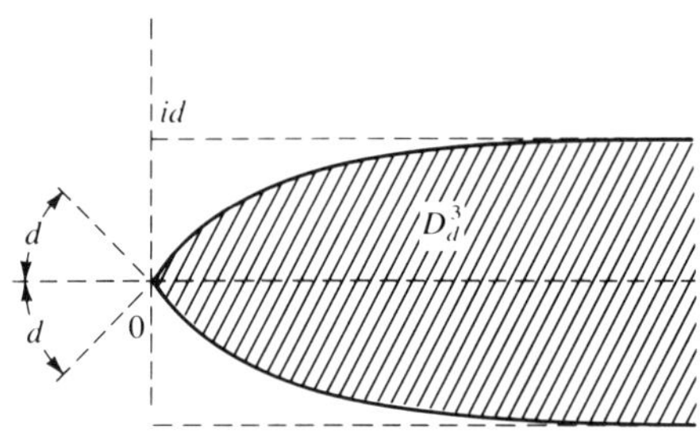

Figure 1.7.4d: The Region D_d^3 of Equation (1.7.12)

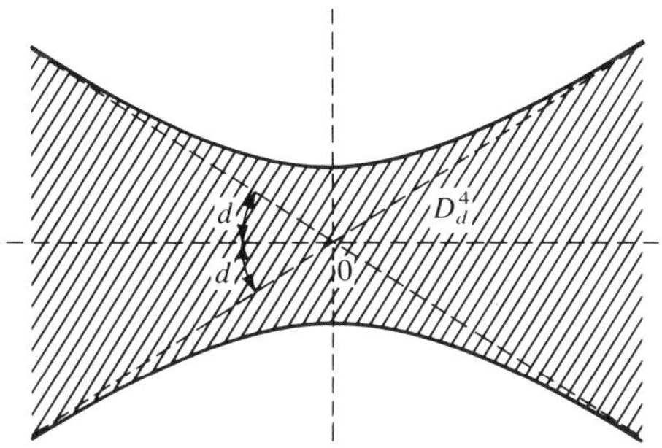

Figure 1.7.4e: The Region D_d^4 of Equation (1.7.13)

Example 1.7.6 Let $\rho > 0$ and $R > 1$. The function

$$w(z) = \exp\left\{\frac{2\log R}{\pi i}\log\left[\rho\left(\frac{z-1}{z+1}\right)\right] - \log\ R\right\} \tag{1.7.18}$$

is analytic in Ω_+ (see Equation (1.4.22)) and satisfies

$$R^{-1} < |w(z)| < R \tag{1.7.19}$$

for all z in Ω_+. Moreover, w maps the interval $(-1, 1)$ onto the circle $|w| = R$, and each of the intervals $(-\infty, -1)$ and $(1, \infty)$ onto $|w| = R^{-1}$.

Example 1.7.7 Let $0 < k < 1$, and set

$$\left\{\begin{array}{l} k_1 = (1-k^2)^{1/2} \\ u = u(z) = \int_0^z \{(1-t^2)(1-k^2t^2)\}^{-1/2}\,dt \Leftrightarrow z = \mathrm{sn}(\mathrm{u};\mathrm{k}) \\ K = K(k) = u(1),\ K_1 = K(k_1),\ q = q(k) = \exp(-\pi K_1/K), \\ q_1 = q(k_1),\ R = q_1^{-1}. \end{array}\right. \tag{1.7.20}$$

The function

$$\zeta = \zeta(w) = \mathrm{sn}\left[\frac{K_1}{\pi}\log w; k\right] \tag{1.7.21}$$

maps the annulus $\mathcal{A}_R$ (see Equation (1.1.25)) conformally onto the ζ-plane minus the slits

$$\left\{\begin{array}{l} S_1 = \{\zeta \in \mathbb{C} : \Im\zeta = 0, -k^{-1} \leqq \Re\zeta \leqq -1\} \\ \\ S_2 = \{\zeta \in \mathbb{C} : \Im\zeta = 0, 1 \leqq \Re\zeta \leqq k^{-1}\}. \end{array}\right. \tag{1.7.22}$$

Moreover, $\zeta(\cdot)$ maps the circle $\{w \in \mathbb{C} : |w| = R^{-1}\}$ onto S_1, and the circle $\{w \in \mathbb{C} : |w| = R\}$ onto S_2. Furthermore, ζ is analytic in $\mathcal{A}_R$, except for a simple pole at $\zeta = -1$. The proof of this statement is left to Problem 1.7.5. •

Problems for Section 1.7

1.7.1 Prove that the transformation (1.7.3), with $\alpha\delta - \beta\gamma \neq 0$, transforms (1.7.2) into $(A + \overline{A})\zeta\overline{\zeta} + B\zeta + \overline{B\zeta} + C + \overline{C} = 0$ with A, B and C complex numbers satisfying $B\overline{B} > (A + \overline{A})(C + \overline{C})$.

1.7.2 Verify in detail the results of Example 1.7.6.

1.7.3 Let $0 < \gamma < 2\pi$, and consider the transformation $w = \varphi(z) = \log\left\{\dfrac{e^{i\gamma/2}(z-1)}{e^{i\gamma} - z}\right\}$ which transforms the circular arc

$$\Gamma = \left\{z = e^{i\theta} : 0 < \theta < \gamma\right\}$$

onto $\mathbb{R}$. What region does φ map onto $\mathcal{D}_d$?

1.7.4 * Prove the Riemann Mapping Theorem.

1.7.5 Verify in detail the results of Example 1.7.7.

1.7.6 Let $k_1, k_2, w(z)$ and its inverse function $\mathrm{sn}(w;k)$ as well as K, K_1, q and q_1 be defined as in (1.7.20). Let Ω_+ denote the upper half plane, and let $\mathcal{R}$ denote the rectangular region $\mathcal{R} = \{u + iv : -K < u < K,\ 0 < v < K_1\}$.

(a) Verify the integral identities

$$K_1 = \int_1^{1/k} \frac{dt}{\sqrt{(t^2-1)(1-k^2t^2)}},$$

$$K = \int_{1/k}^{\infty} \frac{dt}{\sqrt{(t^2-1)(k^2t^2-1)}},$$

$$K_1 = \int_0^{\infty} \frac{dt}{\sqrt{(t^2+1)(1+k^2t^2)}};$$

(b) Prove that as z increases from 0 to 1, w increases from 0 to K;

(c) Consider $w(z)$ for $z = 1 + \rho\, e^{i\theta}$, with ρ fixed on the interval $0 < \rho < \min(1, 1/k - 1)$, and with $0 \leqq \theta \leqq \pi$. Conclude that $w(1+\rho) = K + iv$, with $v = \int_1^{1+\rho} \{(t^2+1)(1-k^2t^2)\}^{-1/2}\, dt$;

(d) Consider $w(z)$ for $z = 1/k + \rho\, e^{i(\pi-\theta)}$, with ρ fixed on the interval $0 < \rho < \min(1, 1/k - 1)$, and with $0 \leqq \theta \leqq \pi$. Conclude that $w = K + iK' - u$, with $u = \int_{1/k}^{1/k+\rho} \{(t^2-1)(k^2t^2-1)\}^{-1/2}\, dt$;

(e) Prove that if $z \in \Omega_+$, then $w'(z) \neq 0$, and also, if z_1 and z_2 belong to Ω_+, with $z_1 \neq z_2$, and if $\mathcal{L}$ is the straight line path connecting z_1 and z_2, then $w(z_2) - w(z_1) = \int_{\mathcal{L}} \{(1-t^2)(1-k^2t^2)\}^{-1/2}\, dt \neq 0$. Conclude that $w(z)$ is a conformal map of Ω_+ onto $\mathcal{R}$;

(f) Let m and n be integers. Prove that if $w \in \mathbb{C}$, then

$$\begin{aligned} \mathrm{sn}(0;k) = 0, \quad \mathrm{sn}(K;k) = 1, \\ \mathrm{sn}(iK_1/2;k) = \mathrm{sn}(3iK_1/2;k) = ik^{-1/2}, \\ \mathrm{sn}(2mK + 2niK_1 + w;k) = (-1)^m\, \mathrm{sn}(w;k), \end{aligned}$$

and that $\mathrm{sn}(w;k)$ has simple poles at $w = 2mK + (2n+1)iK_1$, with residues $(-1)^m\, k^{-1}$.

(g) Prove that if $\zeta(w)$ is defined as in (1.7.21), then for arbitrary integer m,

$$\begin{aligned} \zeta(w) = 0 \Rightarrow w = q_1^{2m}, \\ \zeta(w) = (-1)^m \Rightarrow w = q_1^{2m+1}, \\ \zeta(w) = \infty \Rightarrow w = -q_1^{2m}. \end{aligned}$$

1.7.7 Let $\mathrm{sn}(w;k)$ be defined as in (1.7.20).

(a) Use Liouville's Theorem to derive the formula

$$\frac{d}{dw}\log\left\{\mathrm{sn}\left[\frac{K_1}{\pi}\log\, w;k\right]\right\}$$

$$= 2\left\{\frac{1}{w^2-1} + \sum_{j=1}^{\infty} q_1^{2j}\left[\frac{1}{w^2 - q_1^{4j}} - \frac{1}{1 - w^2 q_1^{4j}}\right]\right\}.$$

(b) Deduce that for some constant c,

$$\mathrm{sn}\left[\frac{K_1}{\pi}\log\, w;k\right] = c\frac{w-1}{w+1}\prod_{j=1}^{\infty}\frac{(w - q_1^{2j})(1/w - q_1^{2j})}{(w + q_1^{2j})(1/w + q_1^{2j})}.$$

(c) Using the identities $\mathrm{sn}(K;k) = 1$ and $\mathrm{sn}(K + iK_1;k) = k^{-1}$, deduce that the constant c in part (b) of this problem is given by $c = k^{-1/2}$.

(d) Use the result in (c) and the fact that $\mathrm{sn}(K;k) = 1$ to deduce, upon setting $q_1 = e^{-h/2}$, that that

$$\log k^{1/2} = \sum_{n=-\infty}^{\infty} \log \left| \frac{e^{h/2} - e^{nh}}{e^{h/2} + e^{nh}} \right|.$$

(e) Use the result of Problem 1.6.15 to deduce that

$$\log k^{1/2} = -\frac{\pi^2}{h} - 2 \sum_{m=1}^{\infty} \frac{(-1)^m}{m} \frac{\sinh(m\pi^2/h)}{\cosh(m\pi^2/h)}.$$

(f) Prove that

$$\left(\frac{2K_1 k^{1/2}}{\pi}\right)^{1/2} = \prod_{j=1}^{\infty} \frac{1 - q_1^{2j}}{1 + q_1^{2j}} = \prod_{j=1}^{\infty} \frac{1 - e^{-jh}}{1 + e^{-jh}},$$

and conclude, using part (d) of this problem, that

$$\left(\frac{2K}{\pi}\right)^{1/2} = \sum_{j=-\infty}^{\infty} e^{-j^2 h/2}.$$

(g) Now use the Poisson summation formula to derive the result

$$\left(\frac{2K}{\pi}\right)^{1/2} = \left(\frac{2\pi}{h}\right)^{1/2} \sum_{j=-\infty}^{\infty} e^{-2\pi^2 j^2/h}.$$

1.7.8 In (1.7.18), take $R = q_1^{-1}$, and $\rho = q^{1/2}$. Set $v(z) = \zeta(w(z))$, with ζ defined as in (1.7.21), set $\chi(z) = (1+k)^{-1}[1+kv(z)]$, and for $z \in \mathbb{R}$, define χ^* by $\chi^*(z) = 0$ if $|z| > 1$, and $\chi^*(z) = 1$ if $|z| < 1$. Also, define a and b by $a = (1-k)/(1+k)$, and $b = 4E/(1+k) - 2$, with $E = \int_0^1 (1 - k^2t^2)^{1/2}(1-t^2)^{-1/2}\, dt$.

(i) Prove that if $z \in \mathbb{R}$, then $0 \leqq \chi(z) - \chi^*(z) \leqq a$;

(ii) Prove that

$$\int_{\mathbb{R}} |\chi(z) - \chi^*(z)|\, dz = \int_{\mathbb{R}} [\chi(z) - \chi^*(z)]\, dz = b;$$

(iii) Prove that if p and p' are numbers on $[1, \infty]$ such that $1/p + 1/p' = 1$, then

$$\left(\int_{\mathbb{R}} |\chi(z) - \chi^*(z)|^{p'}\, dz\right)^{1/p'} = a^{1/p}\, b^{1/p'}.$$

(iv) Prove that for $z \in \mathbb{C}$, $\chi(z)$ has the convergent sum representation

$$\chi(z) = \sum_{m=-\infty}^{\infty} \left\{ \frac{r_m}{z - \alpha_m} + \frac{\overline{r}_m}{z - \overline{\alpha}_m} \right\},$$

with

$$\alpha_m = \frac{1 + iq^m}{1 - iq^m},$$

$$r_m = - \frac{\pi}{(1+k)K} \frac{q^m}{(1 - iq^m)^2}.$$

(v) Prove that $\chi(z) = \mathcal{O}(z^{-2})$ as $z \to \infty$.

(vi) Prove by integration of $f\,\chi$ over $\mathbb{R}$, that if $f \in \mathbf{H}^p(\Omega_+)$, with Ω_+ the upper half plane, then

$$\int_{-1}^{1} f(x + i0^+)\, dx - \sum_{m=-\infty}^{\infty} r_m\, f(\alpha_m) = \varepsilon,$$

with

$$|\varepsilon| = \left| \int_{\mathbb{R}} [\chi^*(x) - \chi(x)]\, f(x + i0^+)\, dx \right| \leq a^{1/p}\, b^{1/p'}\, N_p(f, \Omega_+),$$

with $N_p(f, \Omega_+)$ defined as in (1.8.21).

(vii) Prove that if $q = e^{-h}$, then as $h \to 0$, we have

$$\begin{aligned}
&\frac{\pi^2}{(1+k)K} = h\,[1 + \mathcal{O}(e^{-2\pi^2/h})],\\
&a = 4\,e^{-\pi^2/h}[1 + \mathcal{O}(e^{-2\pi^2/h})],\\
&b = 8\,e^{-\pi^2/h}[1 + \pi^2/h][1 + \mathcal{O}(e^{-2\pi^2/h})].
\end{aligned}$$

1.7.9 Prove that

$$\operatorname{sn}\left[\frac{K_1}{\pi} \log\, w; k\right] = \frac{\pi}{i\,kK} \sum_{n=-\infty}^{\infty} \frac{w^{i(n+1/2)K_1/K}}{q^{-n-1/2} - q^{n+1/2}}.$$

1.8 Spaces of Analytic Functions

In this section we discuss some spaces of analytic functions, which we shall encounter again throughout this monograph. More specifically, we present some explicit formulas for representing analytic functions in terms of their boundary values; and we obtain estimates on the bounds of such functions. These results are useful for deriving Sinc methods, and also for studying error in Sinc methods. At the outset we discuss some important spaces of functions that are analytic on the unit disc. We then derive analogous results for functions that are analytic in the upper half plane, and we also obtain some results for functions that are analytic on the strip $\mathcal{D}_d = \{z \in \mathbb{C} : |\Im z| < d\}$.

1.8.1 Functions Analytic on the Unit Disc

Let $\mathcal{U}$ denote the unit disc, $\mathcal{D}(0;1)$, and let $\mathbf{H}^p(\mathcal{U})$ denote the family of all functions f that are analytic in $\mathcal{U}$, such that $N_p(f,\mathcal{U}) < \infty$, where

$$N_p(f,\mathcal{U}) = \begin{cases} \lim\limits_{r\to 1^-} \left(\displaystyle\int_0^{2\pi} |f(re^{i\theta})|^p \, d\theta \right)^{1/p} & \text{if } 1 \leqq p < \infty, \\ \sup\limits_{z\in\mathcal{U}} |f(z)| & \text{if } p = \infty. \end{cases} \tag{1.8.1}$$

We remark that this definition of $N_p(f,\mathcal{U})$ is equivalent to the one given in Definition 1.1.5. Given $g \in \mathbf{L}^p(\partial\mathcal{U})$ for some $p \in [1,\infty]$, we introduce the operators

$$(\mathcal{T}g)(z) = \frac{1}{2\pi} \int_{|\zeta|=1} \frac{\zeta + z}{\zeta - z} g(\zeta) |d\zeta|, \quad z \in \mathcal{U}; \tag{1.8.2}$$

$$(\mathcal{R}g)(z) = \frac{1-|z|^2}{2\pi} \int_{|\zeta|=1} \frac{g(\zeta)}{|\zeta - z|^2} |d\zeta|, \quad z \in \mathcal{U}; \tag{1.8.3}$$

$$(\mathcal{I}g)(z) = \frac{1}{\pi i} \int_{|\zeta|=1} \frac{\Im(\bar{\zeta} z)}{|\zeta - z|^2} g(\zeta) |d\zeta|, \quad z \in \mathcal{U}; \tag{1.8.4}$$

and

$$(\mathcal{S}g)(z) = \frac{P.V.}{\pi i} \int_{|\zeta|=1} \frac{g(\zeta)}{\zeta - z} d\zeta, \quad z \in \partial\mathcal{U}. \tag{1.8.5}$$

We can then prove

Theorem 1.8.1 *Let the operators $\mathcal{T}, \mathcal{R}, \mathcal{I}$ and $\mathcal{S}$ be defined as in (1.8.2) through (1.8.5), with $g \in \mathbf{L}^p(\partial\mathcal{U})$ for some $p \in [1,\infty]$. Then:*
(a) *$\mathcal{R}g$ and $\mathcal{I}g$ are harmonic in $\mathcal{U}$, and for almost every $w \in \partial\mathcal{U}$,*

$$\lim_{r \to 1^-} (\mathcal{R}g)(rw) = g(w). \tag{1.8.6}$$

Moreover, if $1 < p < \infty$, *then we also have*

$$\lim_{r \to 1^-} (\mathcal{I}g)(rw) = (\mathcal{S}g)(w). \tag{1.8.7}$$

(b) $\mathcal{T}g$ *is analytic in* $\mathcal{U}$. *Moreover*

$$\mathcal{T}g = \mathcal{R}g + \mathcal{I}g, \tag{1.8.8}$$

and

$$\mathcal{I}g = \mathcal{R}\mathcal{S}g. \tag{1.8.9}$$

(c) *Every* $f \in \mathbf{H}^p(\mathcal{U})$ *can be represented either in the form* $f = \mathcal{T}g + iC$ *with* g *defined for almost every* $w \in \partial\mathcal{U}$ *by* $g(w) = \lim_{(r \to 1^-)} \Re f(rw)$, *and* C *a constant, or else in the form* $f = \mathcal{R}g_1$, *with* g_1 *defined for almost every* $w \in \partial\mathcal{U}$ *by* $g_1(w) = \lim_{(r \to 1^-)} f(rw)$.
(d) *Every* $f \in \mathbf{H}^p(\mathcal{U})$ *satisfies the inequality*

$$\sup_{|z|=r} |f(z)| \leqq \left[2\pi(1-r^2)\right]^{-1/p} \left[\frac{\Gamma(2q-1)}{\Gamma(q)\Gamma(q)}\right]^{1/q} N_p(f,\mathcal{U}), \tag{1.8.10}$$

with $r \in (0,1)$, *and with* q *defined by* $1/p + 1/q = 1$; *and*
(e) *Every* f *in* $\mathbf{H}^p(\mathcal{U})$ *such that* f *is real-valued on* $(-1,1)$ *can be represented on* $(-1,1)$ *in the form*

$$f(x) = \frac{1-x^2}{\pi} \int_0^\pi \frac{g(e^{i\theta})}{|e^{i\theta} - x|^2} d\theta \tag{1.8.11}$$

where for almost every $\theta \in (0,\pi)$, $g(e^{i\theta}) = \lim_{(r \to 1^-)} \Re f(re^{i\theta})$.

Proof. Let us first describe a simple construction for $\mathcal{R}g$. If $f \in \mathbf{H}^p(\mathcal{U})$ with $1 \leqq p \leqq \infty$, then Cauchy's formula (see Equation (1.1.17)) yields

$$\begin{aligned} f(0) &= \frac{1}{2\pi} \int_0^{2\pi} f(e^{i\theta})\, d\theta \\ &= \frac{1}{2\pi} \int_{|\zeta|=1} f(\zeta) |\, d\zeta| \end{aligned} \tag{1.8.12}$$

with the second identity in (1.8.12) obtained after setting $\zeta = e^{i\theta}$ in the first.

Now, if z and w denote complex numbers such that $|z| < 1$ and $w \in \mathcal{U}$, then the mapping

$$\zeta = \zeta(w) = \frac{w - z}{1 - \overline{z}w} \tag{1.8.13}$$

provides a conformal map of $\mathcal{U}$ onto itself, and hence (1.8.12) yields

$$f(0) = \frac{1 - |z|^2}{2\pi} \int_{|w|=1} f\left(\frac{w - z}{1 - \overline{z}w}\right) \frac{|dw|}{|1 - \overline{z}w|^2} \tag{1.8.14}$$

that is, upon setting

$$F(w) = f\left(\frac{w - z}{1 - \overline{z}w}\right), \tag{1.8.15}$$

we get

$$F(z) = \frac{1 - |z|^2}{2\pi} \int_{|w|=1} \frac{F(w)}{|1 - \overline{z}w|^2} |dw|. \tag{1.8.16}$$

This equation provides a constructive derivation of the operator $\mathcal{R}$.
Part (a) of Proof: The identities (1.8.6) and (1.8.7) are left to the problems, (Problem 1.8.6) inasmuch as we have already established analogous results in the proof of Theorem 1.2.2.
Part (b): The identity (1.8.8) follows directly from the definitions of $\mathcal{T}g, \mathcal{R}g$ and $\mathcal{I}g$, while (1.8.9) is a consequence of (1.8.7) and the fact that $\mathcal{I}g$ is harmonic in $\mathcal{U}$. We recall here, by Theorem 1.2.1, that $\mathcal{S}f \in \mathbf{L}^p(\mathbb{R})$ whenever $f \in \mathbf{L}^p(\mathbb{R})$, provided that $1 < p < \infty$.
Part (c): The proof of part (c) of Theorem 1.8.1 is a simple consequence of the relations in (a) and (b), and the fact that both the real and imaginary parts of an analytic function are harmonic.
Part (d): Let f be an arbitrary function in $\mathbf{H}^p(\mathcal{U})$ so that $f = \mathcal{R}g_1$, with g_1 defined as in part (c) of Theorem 1.8.1, so that the $\mathbf{H}^p(\mathcal{U})$-norm of f and the $\mathbf{L}^p(\partial\mathcal{U})$-norm of g_1 are the same. Applying Hölder's inequality to the equation $f = \mathcal{R}g_1$, we obtain

$$|f(re^{i\theta})| \leqq \frac{1 - r^2}{2\pi} \left(\int_0^{2\pi} |1 - re^{i\theta}|^{-2q} d\theta\right)^{1/q} N_p(f, \mathcal{U}). \tag{1.8.17}$$

But by using the notation $(q)_m = \Gamma(m + q)/\Gamma(q)$, with Γ defined as in Problem 1.1.5, we have

$$\left(\int_0^{2\pi} |1-re^{i\theta}|^{-2q} d\theta\right)^{1/q} = \left(\int_0^{2\pi} (1-re^{i\theta})^{-q}(1-re^{-i\theta})^{-q} d\theta\right)^{1/q}$$
$$= \left(\int_0^{2\pi} \sum_{m=0}^{\infty}\sum_{n=0}^{\infty} \frac{(q)_m}{m!}\frac{(q)_n}{n!} r^{m+n} e^{i(m-n)\theta} d\theta\right)^{1/q} \qquad (1.8.18)$$
$$= \left(2\pi \sum_{m=0}^{\infty} \left[\frac{(q)_m}{m!}\right]^2 r^{2m}\right)^{1/q}.$$

Now, by Problem 1.8.4, we have for $0 < r < 1$,

$$\sum_{m=0}^{\infty} \left[\frac{(q)_m}{m!}\right]^2 r^{2m} = (1-r^2)^{1-2q} \sum_{m=0}^{\infty} \left[\frac{(1-q)_m}{m!}\right]^2 r^{2m}, \qquad (1.8.19)$$

and also,

$$\sum_{m=0}^{\infty} \left[\frac{(1-q)_m}{m!}\right]^2 r^{2m} \leqq \sum_{m=0}^{\infty} \left[\frac{(1-q)_m}{m!}\right]^2 = \frac{\Gamma(2q-1)}{\Gamma(q)\Gamma(q)}, \qquad (1.8.20)$$

where $\Gamma(x)$ denotes Euler's Gamma function (see Problem 1.1.5), and with the identity on the extreme right-hand side of (1.8.20) derived in Problem 1.8.5.

Part (e): Finally if $f \in \mathbf{H}^p(\mathcal{U})$ and if f restricted to $(-1,1)$ is real, then by the Schwarz Reflection Principle we have for $z \in \mathcal{U}$ and $\Im z < 0$, that $f(z) = \overline{f(\overline{z})}$. That is, if $f = \mathcal{R} g_1$, with g_1 defined as in the (c)-part of Theorem 1.8.1, then for all $\theta \in (0,\pi)$, $g_1(e^{-i\theta}) = \overline{g_1(e^{i\theta})}$. Substituting this result into Equation (1.8.3) and taking $z = x$, we get (1.8.11).

This completes the proof of Theorem 1.8.1. ■

1.8.2 Functions Analytic on Ω_+

Let Ω_+ denote the upper half of the complex plane, as defined in Equation (1.4.22). Let $1 \leqq p \leqq \infty$, and let $\mathbf{H}^p(\Omega_+)$ denote the family of all functions f that are analytic on Ω_+, such that $N_p(f,\Omega_+) < \infty$, with

$$N_p(f,\Omega_+) = \begin{cases} \lim\limits_{y\to 0+} \left(\int_{\mathbb{R}} |f(x+iy)|^p dx\right)^{1/p} & \text{if } 1 \leqq p < \infty, \\ \sup\limits_{z\in\Omega_+} |f(z)| & \text{if } p = \infty. \end{cases} \qquad (1.8.21)$$

Given $g \in \mathbf{L}^p(\mathbb{R})$, let us set

$$(\mathcal{T}g)(z) = \frac{1}{\pi i}\int_{\mathbb{R}} \frac{(1+zt)g(t)}{(t-z)(1+t^2)}dt, z \in \Omega_+; \quad (1.8.22)$$

$$(\mathcal{R}g)(z) = \frac{\Im z}{\pi}\int_{\mathbb{R}} \frac{g(t)}{|t-z|^2}dt, z \in \Omega_+; \quad (1.8.23)$$

$$(\mathcal{I}g)(z) = (\mathcal{T}g)(z) - (\mathcal{R}g)(z), z \in \Omega_+ \quad (1.8.24)$$

$$(\mathcal{S}g)(x) = \frac{P.V.}{\pi i}\int_{\mathbb{R}} \frac{g(t)}{t-x}dt, x \in \mathbb{R}. \quad (1.8.25)$$

We shall then establish

Theorem 1.8.2 *Let the operators $\mathcal{T}, \mathcal{R}, \mathcal{I}$ and $\mathcal{S}$ be defined as in (1.8.22) through (1.8.25), with $g \in \mathbf{L}^p(\mathbb{R})$, for some $p \in [1, \infty]$. Then*

(a) *$\mathcal{R}g$ and $\mathcal{I}g$ are harmonic on Ω_+, and $\mathcal{T}g = \mathcal{R}g + \mathcal{I}g$ is analytic on Ω_+. Moreover*

$$\lim_{y \to 0+} (\mathcal{R}g)(x+iy) = g(x) \quad (1.8.26)$$

a.e. on $\mathbb{R}$ and if $1 < p < \infty$, then for all $z \in \Omega_+$,

$$(\mathcal{I}g)(z) = (\mathcal{R}\mathcal{S}g)(z) - iC_1 \quad (1.8.27)$$

with

$$C_1 = \frac{1}{\pi}\int_{\mathbb{R}} \frac{tg(t)}{1+t^2}dt; \quad (1.8.28)$$

(b) *Every $f \in \mathbf{H}^p(\Omega_+)$ can be represented in the form $f = \mathcal{R}g_1$, with g_1 defined a.e. on $\mathbb{R}$ by $g_1(x) = \lim_{(y \to 0+)} f(x+iy)$. If $1 < p < \infty$, then every $f \in \mathbf{H}^p(\Omega_+)$ can also be represented in the form $f = \mathcal{T}g + iC_1$, with C_1 defined as in (1.8.28), and $g(x) = \lim_{(y \to 0+)} \mathcal{R}f(x+iy)$ a.e. on $\mathbb{R}$;*
(c) *Every $f \in \mathbf{H}^p(\Omega_+)$ satisfies the inequality*

$$\sup_{x \in \mathbb{R}} |f(x+iy)| \leqq (\pi y)^{-1/p} \left[\frac{\Gamma(q-1/2)}{\Gamma(q)}\right]^{1/q} N_p(f, \Omega_+), \quad (1.8.29)$$

on Ω_+, with $1/p + 1/q = 1$; and
(d) *Every $f \in \mathbf{H}^p(\Omega_+)$ such that f is real-valued on the line $x = 0$ in the upper-half plane may be represented for $y > 0$ by the formula*

$$f(iy) = \frac{2y}{\pi}\int_0^{\infty} \frac{g(t)\,dt}{t^2+y^2}, \quad (1.8.30)$$

with $g(t) = \lim_{(y \to 0+)} f(t+iy)$ a.e. on $\mathbb{R}$.

Proof. Let us briefly outline a procedure for constructing the formulas. Setting

$$z = \frac{i\zeta + 1}{i\zeta - 1}, \quad w = \frac{it + 1}{it - 1} \tag{1.8.31}$$

in (1.8.16), with the first equation in (1.8.31) a conformal transformation of Ω_+ onto $\mathcal{U}$, we get the representation

$$f(\zeta) = \frac{\Im\zeta}{\pi} \int_{\mathbb{R}} \frac{f(t)}{|t - \zeta|^2} dt. \tag{1.8.32}$$

This representation may be readily extended to all $f \in \mathbf{H}^p(\Omega_+)$, using Theorem 1.2.2.

Part (a): Clearly $\mathcal{T}g$ is analytic and hence harmonic on Ω_+. Also $\mathcal{R}g$ is harmonic on Ω_+, as we saw earlier in Theorem 1.2.2, and moreover $\mathcal{R}g$ satisfies (1.8.26). Thus $\mathcal{I}g = \mathcal{T}g - \mathcal{R}g$ is also harmonic on Ω_+.

Let us now take $x \in \mathbb{R}$, $y > 0$, $g \in \mathbf{L}^p(\mathbb{R})$, and let us note the identity

$$\begin{aligned} (\mathcal{T}g)(x + iy) &= \frac{1}{\pi i} \int_{\mathbb{R}} \frac{[1 + (x + iy)t]g(t)}{[(t - x) - iy](t^2 + 1)} dt \\ &= \frac{y}{\pi} \int_{\mathbb{R}} \frac{g(t)}{(t - x)^2 + y^2} dt + \frac{1}{\pi i} \int_{\mathbb{R}} \frac{(t - x)g(t)}{(t - x)^2 + y^2} dt - iC_1, \end{aligned} \tag{1.8.33}$$

with C_1 defined in Equation (1.8.28). Both integral terms on the extreme right-hand side of (1.8.33) are harmonic on Ω_+. The first term, $(\mathcal{R}g)(x+iy)$ converges to $g(x)$ a.e. on $\mathbb{R}$ as $y \to 0^+$, while the second integral term converges to $(\mathcal{S}g)(x)$ a.e. on $\mathbb{R}$ as $y \to 0^+$. If $g \in \mathbf{L}^p(\mathbb{R})$ for $1 < p < \infty$, then so is $\mathcal{S}g$, by Theorem 1.2.1. Hence, in this case, the second integral on the extreme right-hand side of (1.8.33) is just $\mathcal{R}\mathcal{S}g(x + iy)$. Hence (1.8.27) follows from (1.8.33) and (1.8.24).

Part (b): The representation $f = \mathcal{R}g_1$, with g_1 defined as in Theorem 1.8.2, is again a consequence of Theorem 1.2.2. From (1.8.33) it follows that if $g \in \mathbf{L}^p(\mathbb{R})$, where $1 < p < \infty$, then $f_1 = \mathcal{T}g + iC_1 \in \mathbf{H}^p(\Omega_+)$ and that $\Re f(x + i0^+) = g(x) = \Re f_1(x + i0^+)$ a.e. on $\mathbb{R}$. From the Cauchy–Riemann equations it thus follows that f and f_1 can differ only by an imaginary constant. However, this constant is zero, for otherwise we could not have $f \in \mathbf{H}^p(\Omega_+)$.

Part (c): The estimate (1.8.29) comes from Hölder's inequality applied to the representation $f = \mathcal{R}g_1$, with g_1 defined as in part (b) of Theorem 1.8.2, so that the $\mathbf{H}^p(\mathcal{U})$-norm of f is the same as the $\mathbf{L}^p(\mathbb{R})$ norm of g_1. Thus, we have

$$\begin{aligned} |f(x+iy)| &\leqq \frac{y}{\pi}\left(\int_{\mathbb{R}} \frac{dt}{[y^2+t^2]^q}\right)^{1/q} \|f\|_p \\ &= (\pi y)^{-1/p}\left[\frac{\Gamma(q-1/2)}{\Gamma(q)}\right]^{1/q} N_p(f,\Omega_+). \end{aligned} \tag{1.8.34}$$

Part (d): If $f(iy)$ is real for $y > 0$, then by the Schwarz Reflection Principle, we have for $t \in \mathbb{R}$,

$$f(-t+i0^+) = \overline{f(t+i0^+)}. \tag{1.8.35}$$

Setting $f(t+i0^+) = g(t) + ih(t)$ where g and h are real-valued functions in $\mathbf{L}^p(\mathbb{R})$, it follows from (1.8.35), that

$$g(-t) = g(t), \quad h(-t) = -h(t), \tag{1.8.36}$$

thus starting with the representation

$$f(x+iy) = \frac{y}{\pi}\int_{\mathbb{R}} \frac{g(t)+ih(t)}{(t-x)^2+y^2}\,dt, \tag{1.8.37}$$

and setting $x = 0$, we get

$$\begin{aligned} f(iy) &= \frac{y}{\pi}\int_{\mathbb{R}} \frac{g(t)}{t^2+y^2}\,dt \\ &= \frac{2y}{\pi}\int_0^\infty \frac{g(t)}{t^2+y^2}\,dt, \end{aligned} \tag{1.8.38}$$

where the first equation in (1.8.38) results since h is an odd function, and the second, since g is an even function. This proves the representation (1.8.30). ∎

Clearly, the representation (1.8.38) has an analytic continuation to all of Ω_+. An analytic continuation of $f(iy), y > 0$ to Ω_+ is given by $\mathcal{T}g$, in (1.8.22). Another is given in Problem 1.8.16.

1.8.3 Functions Analytic in the Strip $\mathcal{D}_d$

Let $\mathcal{D}_d$ be defined as in (1.7.9). Let $1 \leqq p \leqq \infty$, and let $\mathbf{H}^p(\mathcal{D}_d)$ denote the family of all functions f that are analytic on $\mathcal{D}_d$, and such if $\mathcal{D}_d(\varepsilon)$ is defined for $0 < \varepsilon < 1$ by

$$\mathcal{D}_d(\varepsilon) = \{z \in \mathbb{C} : |\Re z| < 1/\varepsilon, \ |\Im z| < d(1-\varepsilon)\}, \tag{1.8.39}$$

then $N_p(f, \mathcal{D}_d) < \infty$, with

$$N_p(f, \mathcal{D}_d) = \begin{cases} \lim_{\varepsilon \to 0} \left(\int_{\partial \mathcal{D}_d(\varepsilon)} |f(z)|^p |dz| \right)^{1/p} & \text{if } 1 \leqq p < \infty \\ \lim_{\varepsilon \to 0} \sup_{z \in \mathcal{D}_d(\varepsilon)} |f(z)| & \text{if } p = \infty. \end{cases} \tag{1.8.40}$$

We remark, that this definition of $N_p(f, \mathcal{D}_d)$ is equivalent to the one given in Definition 1.1.5.
We also let $\mathbf{L}^p(\partial \mathcal{D}_d)$ to the family of all measurable functions g defined on $\partial \mathcal{D}_d$, such that $\|g\|_{\mathbf{L}^p(\partial \mathcal{D}_d)} < \infty$, with

$$\|g\|_{\mathbf{L}^p(\partial \mathcal{D}_d)} \equiv \begin{cases} \left(\int_{\mathbb{R}} [|g(t - id)|^p + |g(t + id)|^p] \, dt \right)^{1/p} & \text{if } 1 \leqq p < \infty \\ \sup_{\Im z = \pm d} |g(z)| < \infty & \text{if } p = \infty. \end{cases} \tag{1.8.41}$$

For a given $g \in \mathbf{L}^p(\partial \mathcal{D}_d)$, let us set

$$\begin{aligned} (\mathcal{T} g)(z) &= \frac{1}{4id} \int_{\mathbb{R}} \frac{\left\{1 + i \exp\left[\frac{i\pi}{2d}(t + z)\right]\right\} g(t - id)}{\left[\exp\left(\frac{\pi}{2d} t\right) - i \exp\left(\frac{\pi}{2d} z\right)\right] \cosh\left(\frac{\pi}{2d} t\right)} \, dt \\ &\quad - \frac{1}{4id} \int_{\mathbb{R}} \frac{\left\{1 + i \exp\left[\frac{i\pi}{2d}(t + z)\right]\right\} g(t + id)}{\left[\exp\left(\frac{\pi}{2d} t\right) + i \exp\left(\frac{\pi}{2d} z\right)\right] \cosh\left(\frac{\pi}{2d} t\right)} \, dt \end{aligned} \tag{1.8.42}$$

$$\begin{aligned} (\mathcal{R} g)(z) &= \frac{\cos\left(\frac{\pi}{2d} y\right)}{4d} \int_{\mathbb{R}} \left\{ \frac{g(t - id)}{\cosh\left[\frac{\pi}{2d}(x - t)\right] + \sin\left(\frac{\pi}{2d} y\right)} \right. \\ &\quad \left. + \frac{g(t + id)}{\cosh\left[\frac{\pi}{2d}(x - t)\right] - \sin\left(\frac{\pi}{2d} y\right)} \right\} dt. \end{aligned} \tag{1.8.43}$$

Theorem 1.8.3 *Let the operators $\mathcal{T}$ and $\mathcal{R}$ be defined as in (1.8.42) and (1.8.43), with $g \in \mathbf{L}^p(\partial \mathcal{D}_d)$, for some $p \in [1, \infty]$. Then*

(a) *$\mathcal{T} g$ is analytic in $\mathcal{D}_d$, and $\mathcal{R} g$ is harmonic in $\mathcal{D}_d$, and moreover for almost every $x \in \mathbb{R}$,*

$$\lim_{y \to \pm d} (\mathcal{R} g)(x + iy) = g(x \pm id). \tag{1.8.44}$$

Furthermore, if g is real-valued then $\mathcal{R}(\mathcal{T} g)(z) = (\mathcal{R} g)(z)$;

(b) *Every* $f \in \mathbf{H}^p(\mathcal{D}_d)$ *can be represented in the form* $f = \mathcal{R} g_1$ *with* g_1 *defined for almost every* $x \in \mathbb{R}$ *by* $g_1(x \pm id) = \lim_{(y \to \pm d)} f(x + iy)$. *Every* $f \in \mathbf{H}^p(\mathcal{D}_d)$ *such that* $\Re f$ *is uniformly bounded in* $\mathcal{D}_d$ *can be represented in the form* $f = \mathcal{T} g + i C_2$ *with* g *real-valued on* $\partial \mathcal{D}$ *and with* C_2 *constant and real;*
(c) *Every* $f \in \mathbf{H}^p(\mathcal{D}_d)$ *satisfies the inequality*

$$|f(x+iy)| \leqq \left[d \cos\left(\frac{\pi y}{2d}\right)\right]^{-1/p} \left[\frac{(q+1)\Gamma(q-1/2)}{2\Gamma(q)}\right]^{1/q} N_p(f, \mathcal{D}_d); \tag{1.8.45}$$

(d) *Every* $f \in \mathbf{H}^p(\mathcal{D}_d)$ *such that* f *is real-valued on* $\mathbb{R}$ *may also be represented in* $\mathcal{D}_d$ *by the formula*

$$f(x) = \frac{1}{2d} \int_{\mathbb{R}} \frac{g_1(t)\, dt}{\cosh\left[\frac{\pi}{2d}(x-t)\right]}, \tag{1.8.46}$$

with $g_1(t) = \lim_{(y \to -id)} \Re f(t + iy)$ *a.e. on* $\mathbb{R}$.

Proof. The formulas (1.8.42) and (1.8.43) may be obtained from (1.8.2) and (1.8.3) respectively, upon setting $z = (e^w - 1)/(e^w + 1)$, $\zeta = (e^t - 1)/(e^t + 1)$, and then replacing w by z. Parts (a) and (b) of Theorem 1.8.3 are thus direct consequences of Theorem 1.8.1.
Part (c): Setting $c = \cos[\pi y/(2d)]$, $s = \sin[\pi y/(2d)]$ in (1.8.43) and noting from the definition of g, that $N_p(f, \mathcal{D}_d) = \|g\|_{\mathbf{L}^p(\partial \mathcal{D}_d)}$ we get

$$|f(x+iy)| \leqq \frac{c}{2d} \left(\int_{\mathbb{R}} \frac{dt}{\{\cosh\left[\frac{\pi}{2d}(x-t)\right] - |s|\}^q} \right)^{1/q} N_p(f, \mathcal{D}_d). \tag{1.8.47}$$

Now the integral on the right-hand side of (1.8.47) may be evaluated directly via the formula (see Problem 1.8.9)

$$\begin{aligned} J(\sigma) &\equiv \int_{\mathbb{R}} [\cosh u - \sigma]^{-q} du \\ &= \sqrt{\pi}(1-\sigma^2)^{1/2-q} \left[\frac{\Gamma\left(\frac{q}{2}\right)}{\Gamma\left(\frac{q+1}{2}\right)} F\left(\frac{1-q}{2}, \frac{1-q}{2}; \frac{1}{2}; \sigma^2\right) \right. \\ &\quad \left. + 2\sigma \frac{\Gamma\left(\frac{q+1}{2}\right)}{\Gamma\left(\frac{q}{2}\right)} F\left(1 - \frac{q}{2}, \frac{1-q}{2}; \frac{3}{2}; \sigma^2\right)\right], \end{aligned} \tag{1.8.48}$$

with $F(a, b; c; \sigma^2)$ the standard hypergeometric function. Each of these hypergeometric functions may be bounded explicitly by setting $\sigma^2 = 1$, so that we get (see also Problem 1.8.11)

$$J(\sigma) \leqq \frac{\pi\Gamma(q-1/2)}{\Gamma(q/2)}\Gamma\left(\frac{q+1}{2}\right)(1-\sigma^2)^{1/2-q}. \tag{1.8.49}$$

Next, taking $\sigma = |s|$, we get (1.8.45).
Part (d): This is left to Problem 1.8.13. ∎

The next theorem will be useful for purposes of establishing analyticity of some operators in later sections of this text.

Theorem 1.8.4 *(a) Let $1 \leqq p < \infty$, and let q be defined by $1/p + 1/q = 1$. A function f belongs to $\mathbf{H}^p(\mathcal{D}_d)$, if and only if its Fourier transform F defined by (1.4.1) is expressible on $\mathbb{R}$ in the form $F(x) = G(x)/\cosh(dx)$ for some function $G \in \mathbf{L}^q(\mathbb{R})$, the Fourier transform of a function $g \in \mathbf{L}^p(\mathbb{R})$.*
(b) A function f satisfies $f \in \mathbf{H}^2(\mathcal{D}_d)$, and $|f(t)| \leqq C\exp(-\alpha|t|)$, for all $t \in \mathbb{R}$, with C, d and α positive constants, if and only if its Fourier transform F satisfies $F \in \mathbf{H}^2(\mathcal{D}_\alpha)$, and $|F(x)| \leqq C_1\exp(-d|x|)$ for all $x \in \mathbb{R}$, with C_1 a positive constant.

Proof. Part (a): If f can be represented in the form (1.8.46), then the Fourier transform of f has the form $(*) F(x) = G(x)/\cosh(dx)$. On the other hand, if F does indeed have the representation $(*)$, then by taking the inverse Fourier transform of F we end up with a function having the form (1.8.46).
Part (b): This follows from part (a), and Parseval's Theorem. ∎

Theorem 1.8.5 *Let $d' > 0$, $\alpha > 0$, let $f \in \mathbf{Hol}(\mathcal{D}_{d'})$, and for all $z \in \mathcal{D}_{d'}$, let $f(z) \leqq C\exp\{-\alpha|\Re z|\}$, with C a constant. Let $d \in (0, d')$. Then*

$$|f^{(n)}(z)| \leqq C_n \exp\{-\alpha|\Re z|\} \tag{1.8.50}$$

for $n = 0, 1, 2, \cdots$, for all $z \in \mathcal{D}_d$, with C_n a constant depending only on n.

Proof. Using (1.1.21) with $R = d' - d$, we find that for any $z \in \mathcal{D}_d$, we have

$$|f^{(n)}(z)| \leqq C n! \exp\{-\alpha[|\Re z| + (d' - d)]\}/(d' - d)^n. \tag{1.8.51}$$

Hence (1.8.50) follows, with $C_n = \exp\{\alpha(d' - d)\}/(d' - d)^n$. ∎

Theorem 1.8.6 *Let f satisfy the conditions of Theorem 1.8.5, let $c \in \mathcal{D}_d$ be fixed, let $z \in \mathcal{D}_d$ be arbitrary, and define g by $g(z) = \int_c^z f(t)\,dt$. Then $g \in \mathbf{H}^\infty(\mathcal{D}_d)$. Moreover, there exists a constant C_1, such that*

$$\begin{aligned} |g(z) - g(\infty)| &\leqq C_1 \exp\{-\alpha\Re z\}, \quad \Re z > 0, \\ |g(z) - g(-\infty)| &\leqq C_1 \exp\{\alpha\Re z\}, \quad \Re z < 0. \end{aligned} \tag{1.8.52}$$

Proof. If $\Re z > 0$, we have with $z = x + iy$, that

$$\begin{aligned} |g(z) - g(\infty)| &= \left| \int_z^\infty f(t)\, dt \right| \\ &\leqq C \int_x^\infty e^{-\alpha t} dt \qquad (1.8.53) \\ &= \frac{C}{\alpha} e^{-\alpha \Re z}. \end{aligned}$$

The proof for the case of $\Re z < 0$ is similar, and we omit it. ■

•

Problems for Section 1.8

1.8.1 Prove that if $1 \leqq p \leqq \infty$, $f \in \mathbf{H}^p(\mathcal{U})$ and if $0 \leqq r_1 \leqq r_2 \leqq 1$, then

$$\left(\int_0^{2\pi} |f(r_1 e^{i\theta})|^p d\theta \right)^{1/p} \leqq \left(\int_0^{2\pi} |f(r_2 e^{i\theta})|^p d\theta \right)^{1/p} \leqq N_p(f, \mathcal{U}).$$

1.8.2 Prove that if $f = \mathcal{R}g$ with g a real-valued function in $\mathbf{L}^p(\partial\mathcal{U})$, then for all $z \in \mathcal{U}$,

$$\inf_{\zeta \in \partial\mathcal{U}} g(\zeta) \leqq (\Re f)(z) \leqq \sup_{\zeta \in \partial\mathcal{U}} g(\zeta).$$

1.8.3 Given the transformation (1.8.13) with $z \in \mathcal{U}$ fixed, and with $|\zeta| = |w| = 1$. Show that $|d\zeta| = [1 - |z|^2]|1 - \overline{z}w|^{-2}|dw|$.

1.8.4 Establish the identity (1.8.19). [Hint: Note first that if $0 \leqq r < 1$, then the left-hand side of (1.8.19) is an entire function of q. If $0 < q < 1$, then the left-hand side of (1.8.19) has the representation

$$\frac{1}{\Gamma(q)\Gamma(1-q)} \int_0^1 t^{q-1}(1-t)^{-q}(1-r^2 t)^{-q} dt.$$

Now substitute $t = (1-\tau)/(1-r^2\tau)$ in this integral, expand in powers of τ, and carry out termwise integration, to arrive at (1.8.19), for $0 < q < 1$. Now appeal to analytic continuation, to conclude that (1.8.19) must in fact hold for all q.]

1.8.5 Establish the relations in (1.8.20). [Here $0 < r < 1$; the identity on the extreme right-hand side may be obtained evaluating the standard hypergeometric function at "1".]

1.8.6 Use the validity of (1.8.26) to prove (1.8.6) and (1.8.7).

1.8.7 Prove that if $1 \leqq p \leqq \infty$, if $f \in \mathbf{H}^p(\Omega_+)$, and if $0 \leqq y_1 \leqq y_2$, then

$$\left(\int_{\mathbb{R}} |f(x+iy_2)|^p dx\right)^{1/p} \leqq \left(\int_{\mathbb{R}} |f(x+iy_1)|^p dx\right)^{1/p}$$

$$\leqq N_p(f, \Omega_+).$$

1.8.8 Prove that if $1 < p < \infty$, if $g \in \mathbf{L}^p(\mathbb{R})$, and if C_1 is defined as in (1.8.28), then for all $y > 0$

$$\int_{\mathbb{R}} |(\mathcal{T}g)(x+iy) + iC_1|^p dx < \infty.$$

1.8.9 Evaluate $\left(\int_{\mathbb{R}} \left[y^2 + t^2\right]^{-q} dt\right)^{1/q}$, for arbitrary $q \in [1, \infty]$.

1.8.10 Prove the formula (1.8.48). [Hint: Use the series

$$(\cosh u - \sigma)^{-q} = \sum_{n=0}^{\infty} \frac{(q)_n}{n!} \frac{\sigma^n}{\cosh^{n+q} u}$$

and carry out termwise integration to get

$$J = \sqrt{\pi} \sum_{n=0}^{\infty} \frac{(q)_n}{n!} \frac{\Gamma\left(\frac{q+n}{2}\right)}{\Gamma\left(\frac{q+n+1}{2}\right)} \sigma^n.$$

Both the odd and even sums of this series may be expressed explicitly via hypergeometric function notation. We may similarly extract the factor $(1-s^2)^{1/2-q}$ from each of these series by proceeding as in (1.8.19)].

1.8.11 Derive (1.8.49) from (1.8.48).

1.8.12 Let ϕ denote a conformal map of a simply-connected domain $\mathcal{D}$ onto $\mathcal{D}_{\pi/2}$, with $\mathcal{D}_d$ as in (1.7.9), and set $\rho = e^\phi$.

(a) Prove that every f analytic in $\mathcal{D}$ and such that $\sup_{(z \in \mathcal{D})} |\Re f(z)| < \infty$, is representable in the form

$$f(z) = \frac{1}{\pi i} \int_{\partial \mathcal{D}} \frac{\rho(z)\rho(\zeta) + 1}{\rho(\zeta) - \rho(z)} \left[\frac{|\rho(\zeta)\, \phi'(\zeta)\, g(\zeta)\, d\zeta|}{|\rho(\zeta)|^2 + 1}\right]$$

where a.e. on $\partial\mathcal{D}$, $g(\zeta) = \lim_{(z \to \zeta)} \Re f(z)$.

(b) Prove that if f satisfies the conditions of (a), and if g_1 is defined a.e. on $\partial\mathcal{D}$ by $g_1(\zeta) = \lim_{(z\to\zeta)} f(z)$, then for all $z \in \mathcal{D}$,

$$f(z) = \frac{\Im[\rho(z)]}{\pi} \int_{\partial\mathcal{D}} \frac{|\rho(\zeta)\,\phi'(\zeta)\,g_1(\zeta)\,d\zeta|}{|\rho(z)-\rho(\zeta)|^2}.$$

1.8.13 Prove Part (d) of Theorem 1.8.3. [Hint: Apply the Schwarz Reflection Principle.]

1.8.14 Let $g \in \mathbf{L}^\infty(\mathbb{R})$, let $g(t+T) = g(t)$ for all $t \in \mathbb{R}$, with $T > 0$, and let $\mathcal{R}g$ be defined as in (1.8.23). Prove that if $z = x + iy$, where $x \in \mathbb{R}, y > 0$, then

$$(\mathcal{R}g)(z) = \frac{1}{T}\int_0^T \frac{\sinh\left(\frac{2\pi y}{T}\right) g(t)\,dt}{\cosh\left(\frac{2\pi y}{T}\right) - \cos\left(\frac{2\pi(x-t)}{T}\right)}.$$

1.8.15 Prove that if $g_1 \in \mathbf{L}^\infty(\mathbb{R})$, if $g_1(t+T) = g_1(t)$ for all $t \in \mathbb{R}$, where $T > 0$, and if f is defined in $\mathcal{D}_d$ as in (1.8.46), then f is also given in $\mathcal{D}_d$ by the formulas

$$f(x) = \frac{1}{T}\int_0^T g_1(t) \sum_{n=-\infty}^{\infty} \frac{e^{i2n\pi(x-t)/T}}{\cosh\left(\frac{2n\pi d}{T}\right)} dt,$$

and

$$f(x) = \frac{1}{T}\int_0^T g_1(t) \left\{1 + 2\sum_{n=1}^{\infty} \frac{\cos\left[\frac{2n\pi(x-t)}{T}\right]}{\cosh\left(\frac{2n\pi d}{T}\right)}\right\} dt.$$

1.8.16 Prove that if g satisfies the conditions of Theorem 1.8.2 (d), and if $f(x+iy)$ is the analytic continuation to Ω_+ of $f(iy)$, then for all $z = x + iy \in \Omega_+$,

$$f(x+iy) = \frac{2y}{\pi}\int_0^\infty \frac{(t^2+x^2+y^2)g(t) + 2ixt\,h(t)}{[(t-x)^2+y^2][(t+x)^2+y^2]} dt,$$

with

$$h(t) = \lim_{y\to 0+} \Im f(t+iy).$$

1.8.17 Let Ω_+ denote the upper half plane, defined as in (1.4.22), let $f \in \mathbf{L}^2(\Omega_+)$, and let $g \in \mathbf{L}^2(\mathbb{R})$. Prove that the problem to determine $u = u(z)$, with $z = x + iy$, such that

$$\nabla^2 u = f, \quad z \in \Omega_+,$$
$$\lim_{y \to 0} u(x + iy) = g(x),$$

has the following solution, for $\zeta \in \Omega_+$:

$$\begin{aligned} u(\zeta) &= \frac{1}{2\pi} \int\int_{\Omega_+} \log \left| \frac{z - \zeta}{z + \overline{\zeta}} \right| f(z)\, dx dy \\ &\quad + \frac{\Im\zeta}{\pi} \int_{\mathbb{R}} \frac{g(x)}{|z - \zeta|^2} dx. \end{aligned}$$

1.9 The Paley–Wiener Theorem

In this section we shall discuss some properties of functions that are analytic in $\mathbb{C}$.

Definition 1.9.1 *An entire function is a function which is analytic in $\mathcal{D}(0; R)$ for all $R > 0$, with $\mathcal{D}(0; R)$ defined as in (1.1.1).*

Let f be an entire function, so that for every $z \in \mathbb{C}$ we have

$$f(z) = \sum_{n=0}^{\infty} a_n\, z^n, \tag{1.9.1}$$

with a_n defined by (1.1.21), with $a = 0$. The entire function f is said to have (finite) *order* ρ if

$$\rho = \limsup_{n \to \infty} \frac{\log n}{\log |a_n|^{-1/n}} < \infty. \tag{1.9.2}$$

The entire function of finite order ρ is said to have *type* σ if

$$\sigma = \limsup_{n \to \infty} \rho^{-1} (n!\, |a_n|)^{1/n}. \tag{1.9.3}$$

It may be shown (Problem 1.9.1) that if for a given entire function f we set

$$M(R) = \max_{|z|=R} |f(z)|, \tag{1.9.4}$$

then the same values ρ and σ may also be defined by

$$\rho = \limsup_{R \to \infty} \frac{\log\{\log M(R)\}}{\log R}, \tag{1.9.5}$$

and

$$\sigma = \limsup_{R \to \infty} \frac{\log M(R)}{R^\rho}. \tag{1.9.6}$$

For example, the function $f(z) = \exp(\sigma z^\rho)$ is of order ρ, and type σ. In what follows, we shall analyze an important class of functions of order 1 and type π/h.

Definition 1.9.2 *Let $h > 0$, and let $\mathbf{W}(\pi/h)$ denote the family of all functions f that are analytic in $\mathbb{C}$, such that*

$$\int_{\mathbb{R}} |f(t)|^2 dt < \infty, \tag{1.9.7}$$

and such that for all z in $\mathbb{C}$

$$|f(z)| \leqq Ce^{\pi|z|/h}, \tag{1.9.8}$$

with C a positive constant.

Theorem 1.9.3 (Paley–Wiener Theorem): *To every f in $\mathbf{W}(\pi/h)$ there exists a function g in $\mathbf{L}^2(-\pi/h, \pi/h)$, such that for all z in $\mathbb{C}$,*

$$f(z) = \frac{1}{2\pi} \int_{-\pi/h}^{\pi/h} g(t) e^{-izt} dt. \tag{1.9.9}$$

Proof. We split the proof of Theorem 1.9.3 into the proof of 3 lemmas.

Lemma 1.9.4 *If $f \in \mathbf{W}(\pi/h)$, then for all z in $\mathbb{C}$,*

$$f(z) = \sum_{n=0}^{\infty} \frac{a_n}{n!} z^n \tag{1.9.10}$$

with

$$\lim_{N \to \infty} \sup_{n > N} |a_n|^{1/n} \leqq \frac{\pi}{h}. \tag{1.9.11}$$

Proof of Lemma 1.9.4: Since f is an entire function, it follows from (1.1.43) that

$$a_n = \frac{n!}{2\pi i} \int_0^{2\pi} \frac{f(Re^{i\theta})}{R^n} e^{-in\theta} d\theta \tag{1.9.12}$$

where R is an arbitrary positive number in Equation (1.9.12). Bounding the integral in (1.9.12) and using (1.9.8), we get

$$|a_n| \leqq C \frac{n!}{R^n} \exp(\pi R/h). \tag{1.9.13}$$

Since the bound (1.9.13) is valid for *all* $R > 0$, we may minimize the right-hand side of (1.9.13) as a function of R, to get

$$|a_n| \leqq C\, n! \left(\frac{e\pi}{nh}\right)^n . \tag{1.9.14}$$

Now, using Stirling's formula for $n!$ (see Problem 1.1.5), i.e.,

$$n! = (2\pi n)^{1/2} (n/e)^n \left[1 + \mathcal{O}(1/n)\right], \quad n \to \infty, \tag{1.9.15}$$

on the right-hand side of (1.9.14), taking the n'th root of each side of the resulting expression, and then letting $n \to \infty$, we get (1.9.11). ∎

Let Ω_+ and Ω_- be defined as in (1.4.22).

Lemma 1.9.5 *Let $f \in \mathbf{W}(\pi/h)$. The function $F_+(z)$ defined on Ω_+ by*

$$F_+(z) = \int_0^\infty e^{izt} f(t)\, dt \tag{1.9.16}$$

is analytic on Ω_+, and its boundary function F_+ defined a.e. on $\mathbb{R}$ by

$$F_+(x) = \lim_{y \to 0+} F_+(x + iy) \tag{1.9.17}$$

belongs to $\mathbf{L}^2(\mathbb{R})$.

Proof of Lemma 1.9.5: This is a consequence of Corollary 1.4.7, the formulas (1.4.19) and Parseval's Theorem. ∎

Lemma 1.9.6 *Let F_+ be defined on Ω_+ by (1.9.16). Then F_+ may be continued analytically to the exterior of the disc of radius π/h, as a single-valued, analytic function.*

Proof of Lemma 1.9.6: If $\Re(iz) = -y < -\pi/h$, we may substitute the series (1.9.10) for f into the right-hand side of (1.9.16) and carry out termwise integration to get

$$F_+(z) = \sum_{n=0}^{\infty} \frac{a_n}{(iz)^{n+1}}, \tag{1.9.18}$$

since the series on the right-hand side of (1.9.18) converges, in view of Lemma 1.9.4 for all $|z| > \pi/h$. Hence, the desired analytic continuation may be carried out via the series (1.9.18), (see Theorem 1.1.18). ∎

Let us now complete the proof of Theorem 1.9.3. To this end, set

$$F_-(z) = \int_{-\infty}^0 e^{izt} f(t)\, dt. \tag{1.9.19}$$

Then we see that

(a) F_- is analytic in Ω_-;

(b) $\lim_{(y \to 0+)} F_-(x - iy) = G_-(x)$, with G_- belonging to $\mathbf{L}^2(\mathbb{R})$;

(c) F_- can be continued analytically to a function which is single valued and analytic on the exterior of the disc of radius π/h; indeed we have here, that

$$F_-(z) = -F_+(z), \tag{1.9.20}$$

with F_+ the function in (1.9.18).

Consider now, the function w defined by

$$w(x, y) = F_+(x + iy) + F_-(x - iy). \tag{1.9.21}$$

Then

(a) The limit $\lim_{(y \to 0+)} w(x, y) = F(x)$ exists a.e. on $\mathbb{R}$, and F is in $\mathbf{L}^2(\mathbb{R})$;

(b) $F(x) = 0$ if $x \in \mathbb{R}, |x| > \pi/h$;

(c) The transform

$$F(x) = \int_{\mathbb{R}} e^{ixt} f(t)\, dt \tag{1.9.22}$$

exists a.e. on $\mathbb{R}$, and from the Fourier transform pair (1.4.8),

$$f(t) = \frac{1}{2\pi} \int_{-\pi/h}^{\pi/h} F(x) e^{-ixt} dx. \tag{1.9.23}$$

This completes the proof of Theorem 1.9.3. ■

•

Problems for Section 1.9

1.9.1 Let f be an entire function. Prove that the Equations (1.9.2) through (1.9.3) and (1.9.4) through (1.9.5) are equivalent.

1.9.2 Let ϕ denote a conformal map of a simply-connected domain $\mathcal{D}$ into $\mathcal{D}_d = \{z \in \mathbb{C} : |\Im z| < d\}$, let $\psi = \phi^{-1} : \mathcal{D}_d \to \mathcal{D}$ denote the inverse map, and set $\Gamma = \psi(\mathbb{R})$. Let $f \in \mathbf{Hol}(\mathcal{D})$, such that $g = f \circ \phi$ has an an analytic continuation to all of $\mathbb{C}$, such that $|g(\zeta)| \leqq C \exp\{\pi|\zeta|/h\}$ for all $\zeta \in \mathbb{C}$, with C and h positive numbers, and such that $\int_\Gamma |f(x)|^2 \, |\phi'(x)\, dx| < \infty$. Prove that there exists a function $F \in \mathbf{L}^2(-\pi/h, \pi/h)$ such that for all $z \in \mathcal{D}$,

$$f(z) = \frac{1}{2\pi} \int_{-\pi/h}^{\pi/h} e^{-it\phi(z)}\, F(t)\, dt.$$

1.10 The Cardinal Function

The cardinal function is truly "a function of royal blood, whose distinguished properties separate it from its bourgeois brethren". In this section we shall present the cardinal function and some of its many beautiful identities, thus giving justification to this quote.

Although the term $\mathrm{sinc}(x)$ which is defined by

$$\mathrm{sinc}(x) = \frac{\sin(\pi x)}{\pi x} \tag{1.10.1}$$

is very well-known in engineering literature, we shall adopt a more powerful notation for the *sinc* function. Namely, if $h > 0$ and k is an integer, we define $\mathrm{S}(k,h)$ by

$$\mathrm{S}(k,h)(x) = \frac{\sin\left[\pi(x-kh)/h\right]}{\pi(x-kh)/h}. \tag{1.10.2}$$

We shall thus refer to $\mathrm{S}(k,h)(x)$ as the k'th *Sinc* function, with step size h, evaluated at x.

Probably the most important properties of the Cardinal function are summarized in Theorem 1.10.1 below. Other beautiful properties and identities are presented in Corollaries 1.10.2 through 1.10.6.

Theorem 1.10.1 *Let $h > 0$, and let $\mathbf{W}(\pi/h)$ be defined as in Definition 1.9.2. The sequence $\{h^{-1/2}\mathrm{S}(k,h)\}_{k=-\infty}^{\infty}$ is a complete orthonormal sequence in $\mathbf{W}(\pi/h)$. Every f in $\mathbf{W}(\pi/h)$ has the cardinal series representation*

$$f(x) = C(f,h)(x) \tag{1.10.3}$$

with

$$C(f,h) = \sum_{k=-\infty}^{\infty} f(kh)\, \mathrm{S}(k,h). \tag{1.10.4}$$

Moreover,

$$\lim_{N\to\infty} \int_{-N}^{N} f(x)\, dx = \lim_{N\to\infty} h \sum_{k=-N}^{N} f(kh). \tag{1.10.5}$$

Proof. By the Paley–Wiener theorem, given $f \in \mathbf{W}(\pi/h)$, there exists $F \in \mathbf{L}^2(-\pi/h, \pi/h)$, such that

$$f(t) = \frac{1}{2\pi} \int_{-\pi/h}^{\pi/h} e^{-ixt} F(x)\, dx. \tag{1.10.6}$$

The function F in (1.10.6) may be represented on the interval $(-\pi/h, \pi/h)$ by its Fourier series representation, i.e.

$$F(x) = \sum_{k=-\infty}^{\infty} c_k e^{ikhx}, \quad -\pi/h < x < \pi/h \tag{1.10.7}$$

with

$$c_k = \frac{h}{2\pi} \int_{-\pi/h}^{\pi/h} F(x) e^{-ikhx} dx = h f(kh) \tag{1.10.8}$$

and where the last identity follows from (1.10.6). That is, for $x \in \mathbb{R}$, we have the identity

$$F(x) = \int_{\mathbb{R}} f(t) e^{ixt}\, dt = \begin{cases} h \displaystyle\sum_{k=-\infty}^{\infty} f(kh) e^{ikhx}, & \text{if } |x| < \pi/h; \\ 0, & \text{if } |x| > \pi/h. \end{cases} \tag{1.10.9}$$

Substituting the extreme right-hand side of (1.10.9) into (1.10.6) and performing termwise integration yields (1.10.3) through (1.10.4), since

$$\frac{1}{2\pi} \int_{-\pi/h}^{\pi/h} h e^{ikht - ixt} dt = \mathrm{S}(k,h)(x). \tag{1.10.10}$$

Now, if k and ℓ are integers, we have by Equation (1.10.10) and Parseval's theorem (see (1.4.6)), that

$$\int_{\mathbb{R}} \mathrm{S}(k,h)(x) \mathrm{S}(\ell,h)(x)\, dx = \frac{h^2}{2\pi} \int_{-\pi/h}^{\pi/h} e^{i(k-\ell)ht}\, dt = h \delta_{k-\ell}, \tag{1.10.11}$$

with $\delta_{k-\ell}$ denoting the Kronecker delta, which proves the orthonormality assertion of Theorem 1.10.1. Finally, the identity (1.10.5) follows by taking $x = 0$ in (1.10.9). ∎

Corollary 1.10.2 *If $f \in \mathbf{W}(\pi/h)$, then for $x \in \mathbb{R}$,*

$$\int_{\mathbb{R}} f(t)\, e^{ixt}\, dt = \begin{cases} h \displaystyle\sum_{n=-\infty}^{\infty} f(nh) e^{inhx} & \text{if } |x| < \pi/h \\ 0 & \text{if } |x| > \pi/h. \end{cases} \tag{1.10.12}$$

Moreover,

$$\int_{\mathbb{R}} |f(x)|^2\, dx = h \sum_{n=-\infty}^{\infty} |f(nh)|^2. \tag{1.10.13}$$

Proof. The first identity in (1.10.12) was already obtained in (1.10.9). The identity (1.10.13) follows from Parseval's theorem, using Equations (1.10.3), (1.10.4), and (1.10.11). ■

Corollary 1.10.3 *Let Ω_+ be defined as in (1.4.22).*
(a) *For $x \in \mathbb{R}$, $y \in \mathbb{R}$, $z = x + iy \in \Omega_+$, we have*

$$\frac{1}{2\pi i} \int_{\mathbb{R}} \frac{\mathrm{S}(k,h)(t)}{t-z} dt = \frac{e^{ix(z-kh)/h} - 1}{2i\pi(z-kh)/h}. \tag{1.10.14}$$

In particular, as $y \to 0^+$,

$$\begin{aligned} \Re \frac{1}{\pi i} \int_{\mathbb{R}} \frac{\mathrm{S}(k,h)(t)}{t-x-iy} dt &= \Re \left\{ \frac{e^{i\pi(z-kh)/h} - 1}{i\pi(z-kh)/h} \right\} \\ &= \frac{y}{\pi} \int_{\mathbb{R}} \frac{\mathrm{S}(k,h)(t)}{|t-z|^2} dt \to \mathrm{S}(k,h)(x); \end{aligned} \tag{1.10.15}$$

$$\begin{aligned} \Im \frac{1}{\pi i} \int_{\mathbb{R}} \frac{\mathrm{S}(k,h)(t)}{t-x-iy} dt &= \Im \left\{ \frac{e^{i\pi(x-kh)/h} - 1}{i\pi(z-kh)/h} \right\} \\ \to \frac{P.V.}{\pi} \int_{\mathbb{R}} \frac{\mathrm{S}(k,h)(t)}{t-x} dt &= \frac{\cos[\pi(x-kh)/h] - 1}{\pi(x-kh)/h} \quad . \end{aligned} \tag{1.10.16}$$

(b) *Let $u_{xx} + u_{yy} = 0$ in Ω_+, $u(x, 0^+) = f(x)$, with $f \in \mathbf{W}(\pi/h)$. Then, with $z = x + iy \in \Omega_+$,*

$$u(x,y) = \sum_{k=-\infty}^{\infty} f(kh) \Re \left\{ \frac{e^{i\pi(z-kh)/h} - 1}{i\pi(z-kh)/h} \right\}. \tag{1.10.17}$$

(c) *Let $u_{xx} + u_{yy} = 0$ in Ω_+, $u(x, 0^+) = (\mathcal{S}f)(x)$ with $f \in \mathbf{W}(\pi/h)$ and with $\mathcal{S}f$ defined as in (1.4.20). Then, with $z = x + iy \in \Omega_+$,*

$$u(x,y) = \sum_{k=-\infty}^{\infty} f(kh) \Im \left\{ \frac{e^{i\pi(z-kh)/h} - 1}{i\pi(z-kh)/h} \right\}. \tag{1.10.18}$$

Proof. The proof of Corollary 1.10.3 is left to the problems at the end of this section. ■

Corollary 1.10.4 *If* $f \in \mathbf{W}(\pi/h)$, *then* $f' \in \mathbf{W}(\pi/h)$. *If* $m \in \mathbb{Z}$, $n \in \mathbb{Z}$, *and if* $\delta_{m-n}^{(k)}$ *is defined by*

$$\delta_{m-n}^{(k)} = \left(\frac{d}{dx}\right)^k \mathrm{S}(m,1)(x)|_{x=n}, \quad k = 0, 1, 2, \cdots, \tag{1.10.19}$$

then

$$f^{(k)}(nh) = h^{-k} \sum_{m=-\infty}^{\infty} f(mh)\delta_{m-n}^{(k)}. \tag{1.10.20}$$

In particular,

$$\delta_{m-n}^{(0)} = \begin{cases} 1 & \text{if } m = n \\ 0 & \text{if } m \neq n, \end{cases} \quad \delta_{m-n}^{(1)} = \begin{cases} 0 & \text{if } m = n \\ \dfrac{(-1)^{n-m}}{n-m} & \text{if } m \neq n, \end{cases} \tag{1.10.21}$$

$$\delta_{m-n}^{(2)} = \begin{cases} -\pi^2/3 & if\ m = n \\ \dfrac{-2(-1)^{n-m}}{(n-m)^2} & if\ m \neq n, \end{cases}$$

and for f *in* $\mathbf{W}(\pi/h)$ *and* $k \geqq 0$ *an integer,*

$$f^{(k)}(x) = h^{-k} \sum_{n=-\infty}^{\infty} \left\{ \sum_{m=-\infty}^{\infty} f(mh)\delta_{m-n}^{(k)} \right\} \mathrm{S}(n,h)(x). \tag{1.10.22}$$

Proof. Differentiating each side of the identity (1.10.6), we get

$$f^{(k)}(t) = \frac{1}{2\pi} \int_{-\pi/h}^{\pi/h} e^{-ixt}(-ix)^k F(x)\, dx. \tag{1.10.23}$$

Since $F \in \mathbf{L}^2(-\pi/h, \pi/h)$, $G \in \mathbf{L}^2(-\pi/h, \pi/h)$, with G defined by $G(x) = (-ix)^k F(x)$, and since by Schwarz's inequality, we have

$$\int_{-\pi/h}^{\pi/h} |G(x)|^2 dx \leqq \left(\int_{-\pi/h}^{\pi/h} |x|^{2k} dx\right) \left(\int_{-\pi/h}^{\pi/h} |F(x)|^2 dx\right). \tag{1.10.24}$$

That is, $f \in \mathbf{W}(\pi/h)$ implies that $f^{(k)} \in \mathbf{W}(\pi/h)$ for every non-negative integer k. The relations 1.10.21 follow directly from (1.10.10) and (1.10.19).

The identity (1.10.20) follows from (1.10.4), (1.10.19) and that $f^{(k)} \in \mathbf{W}(\pi/h)$. The relation (1.10.22) follows from (1.10.20), the fact that $f^{(k)} \in \mathbf{W}(\pi/h)$, and the representation (1.10.4). ∎

Corollary 1.10.5 *If $f \in \mathbf{W}(\pi/h)$, then*

$$f(x) = \frac{1}{h} \int_{\mathbb{R}} \frac{\sin[\pi(x-t)/h]}{\pi(x-t)/h} f(t)\, dt. \tag{1.10.25}$$

Proof. From the representation (1.10.6), we have for $x \in \mathbb{R}$,

$$\begin{aligned} &\int_{\mathbb{R}} f(t) \frac{\sin[\pi(x-t)/h]}{\pi(x-t)/h} dt \\ &= \frac{1}{2\pi} \int_{-\pi/h}^{\pi/h} F(\zeta) \left\{ \int_{\mathbb{R}} \frac{\sin[\pi(x-t)/h]}{\pi(x-t)/h} e^{-i\zeta t} dt \right\} d\zeta. \end{aligned} \tag{1.10.26}$$

But

$$\begin{aligned} \int_{\mathbb{R}} \frac{\sin[\pi(x-t)/h]}{\pi(x-t)/h} e^{-i\zeta t}\, dt &= e^{i\zeta x} \int_{\mathbb{R}} \frac{\sin(\pi u/h)}{\pi u/h} e^{-i\zeta u} du \\ &= e^{i\zeta x} \int_{\mathbb{R}} \frac{\sin(\pi u/h)\cos(\zeta u)}{\pi u/h} du \\ = e^{-i\zeta x} \int_{\mathbb{R}} &\left\{ \frac{\sin[(\pi/h - \zeta)u]}{2\pi u/h} + \frac{\sin[(\pi/h+\zeta)u]}{2\pi u/h} \right\} du \\ &= e^{i\zeta x} \begin{cases} h & \text{if} \quad -\pi/h < \zeta < \pi/h \\ h/2 & \text{if} \quad \zeta = \pm\pi/h \\ 0 & \text{if} \quad \zeta \in \mathbb{R}, |\zeta| > \pi/h \end{cases} \end{aligned} \tag{1.10.27}$$

where the last three identities were obtained from the fact that if $k \in \mathbb{R}$, then

$$\int_{\mathbb{R}} \frac{\sin(ku)}{u} du = \begin{cases} \pi & \text{if} \quad k > 0 \\ 0 & \text{if} \quad k = 0 \\ -\pi & \text{if} \quad k < 0. \end{cases} \tag{1.10.28}$$

The relation (1.10.25) now follows as a consequence of replacing the extreme left-hand side of (1.10.27) by the extreme right-hand side, in (1.10.26). ∎

Corollary 1.10.6 *If $f \in \mathbf{W}(\pi/h)$, and if $g(x) = \int_{-\infty}^{x} f(t)\,dt$ exists for all $x \in \mathbb{R}$, then for all $k = 0, \pm 1, \pm 2, \cdots$,*

$$\int_{-\infty}^{kh} f(t)\,dt = h \sum_{\ell=-\infty}^{\infty} \delta_{k-\ell}^{(-1)} f(\ell h) \tag{1.10.29}$$

with

$$\begin{aligned} \delta_k^{(-1)} &= \frac{1}{2} + \sigma_k, \\ \sigma_k &= \int_0^k \frac{\sin(\pi t)}{\pi t}\,dt. \end{aligned} \tag{1.10.30}$$

Moreover, if $g \in \mathbf{L}^2(\mathbb{R})$, then for all $x \in \mathbb{C}$,

$$\int_{-\infty}^{x} f(t)\,dt = h \sum_{k=-\infty}^{\infty} \left\{ \sum_{\ell=-\infty}^{\infty} \sigma_{k-\ell} f(\ell h) \right\} \mathrm{S}(k,h)(x). \tag{1.10.31}$$

Proof. By our assumptions on f, and from the definition of $\delta_k^{(-1)}$, it follows that

$$\int_{-\infty}^{kh} f(t)\,dt = \sum_{\ell=-\infty}^{\infty} f(\ell h) \int_{-\infty}^{kh} \mathrm{S}(l,h)(t)dt \tag{1.10.32}$$

and since by (1.10.30), $\int_{-\infty}^{kh} \mathrm{S}(\ell,h)(t)\,dt = \delta_{k-\ell}^{(-1)}$, this yields (1.10.29). Furthermore, it follows that

$$\int_{-\infty}^{x} f(t)\,dt = \frac{1}{2\pi} \int_{-\pi/h}^{\pi/h} e^{-ix\zeta} \frac{F(\zeta)}{-i\zeta} d\zeta. \tag{1.10.33}$$

Since the left-hand side of (1.10.31) is in $\mathbf{L}^2(\mathbb{R})$, it follows by Parseval's theorem, that $\int_{-\pi/h}^{\pi/h} |F(\zeta)/\zeta|^2 d\zeta < \infty$. That is $g \in \mathbf{W}(\pi/h)$, and hence, by Theorem 1.10.1,

$$\int_{-\infty}^{x} f(t)\,dt = \sum_{k=-\infty}^{\infty} \left(\int_{-\infty}^{kh} f(t)\,dt \right) \mathrm{S}(k,h)(x). \tag{1.10.34}$$

Termwise integration of (1.10.4) from $-\infty$ to x, the definition (1.10.30) of σ_k and the fact that $\int_{\mathbb{R}} f(t)\,dt = h \sum_{k=-\infty}^{\infty} f(kh) = 0$ now yield (1.10.29). ∎

Some of the numbers σ_k are tabulated in Table 1.10.1.

n	SIGMA(n)	n	SIGMA(n)
0	0.0000000000000000	1	0.5894898722360839
2	0.4514116667901405	3	0.5330932376182722
4	0.4749696698836553	5	0.5201071641913087
6	0.4832052174977473	7	0.5144159971233055
8	0.4873742250578202	9	0.5112301526369977
10	0.4898881711538789	11	0.5091957420082168
12	0.4915683516686011	13	0.5077846578125663
14	0.4927702093748033	15	0.5067486944720118
16	0.4936724151782212	17	0.5059559079171770
18	0.4943745528336628	19	0.5053297104401591
20	0.4949364995706957	21	0.5048226073041721
22	0.4953964150848574	23	0.5044035851980506
24	0.4957797661366434	25	0.5040515358439449
26	0.4961041974885161	27	0.5037515950319094
28	0.4963823201653523	29	0.5034929932779406
30	0.4966233866311803	31	0.5032677369477720
32	0.4968343388552230	33	0.5030697682021877
34	0.4970204870276673	35	0.5028944125557015
36	0.4971859623359650	37	0.5027380053824386
38	0.4973340269269068	39	0.5025976332159279
40	0.4974672909775609	41	0.5024709506908217
42	0.4975878678043933	43	0.5023560485249016
44	0.4976974867059375	45	0.5022513566773183
46	0.4977975763910226	47	0.5021555722141233
48	0.4978893275646686	49	0.5020676048274522
50	0.4979737405030820	51	0.5019865351657797
52	0.4980516616563340	53	0.5019115825934454
54	0.4981238121215731	55	0.5018420799807533
56	0.4981908100179338	57	0.5017774537988254
58	0.4982531882343348	59	0.5017172082611159
60	0.4983114086292758	61	0.5016609125833092

Table 1.10.1: The Numbers σ_k

•

Problems for Section 1.10

1.10.1 Use the results of Problem 1.6.1 to prove that if $h > 0$, $z \in \mathbb{C}$, then for arbitrary integer m

$$\sum_{n=-\infty}^{\infty} e^{inht} \mathrm{S}(n,h)(z)$$

$$= \begin{cases} e^{iz(t-2m\pi/h)}, & \text{if } (2m-1)\pi/h < t < (2m+1)\pi/h; \\ \cos(\pi z/h), & \text{if } t = (2m \pm 1)\pi/h. \end{cases}$$

1.10.2 Use the result of Problem 1.10.1 to derive an alternate proof of Equation (1.10.3).

1.10.3 Verify the validity of the identity

$$\sum_{n=0}^{\infty} \left\{ \binom{n+1-x/h}{2n+1} \delta^{2n} g(0) + \binom{n+x/h}{2n+1} \delta^{2n} g(h) \right\}$$

$$= \sum_{n=-\infty}^{\infty} g(nh)\mathrm{S}(n,h)(x)$$

in the case

$$|g(0)| + |g(h)| + |g(-h)| + \sum_{n=2}^{\infty} [|g(nh)| + |g(-nh)|] \cdot \frac{\log n}{n} < \infty.$$

Here

$$\delta g(x) = g(x+h/2) - g(x-h/2)$$

$$\delta^2 g(x) = \delta\{\delta g(x)\}, \quad etc.$$

1.10.4 Prove part (a) of Corollary 1.10.3.

1.10.5 Prove part (b) of Corollary 1.10.3.

1.10.6 Prove part (c) of Corollary 1.10.3.

1.10.7 Prove that if $a \in \mathbb{C}$, $z \in \mathbb{C}$, then for $h > 0$,

$$\sum_{n=-\infty}^{\infty} \frac{1}{a^2 + n^2 h^2} \mathrm{S}(n,h)(z) = \frac{1}{a^2 + z^2} \left[1 + \frac{z \sin(\pi/h)}{a \sinh(\pi a/h)} \right].$$

1.10.8 Let Γ denote the Gamma function, which is defined as in Problem (1.1.5), and which may be continued analytically to all of $\mathbb{C}$ via the formula $\Gamma(z+1) = z\Gamma(z)$. Use the identity $1/\Gamma(z) = \Gamma(1-z)\sin(\pi z)/\pi$ to deduce the formula

$$(*) \qquad \frac{1}{\Gamma(z)} = \sum_{n=1}^{\infty} \frac{\mathrm{S}(n,1)(z)}{(n-1)!} + \frac{\sin(\pi z)}{\pi} \int_1^{\infty} e^{-t} t^{-z} dt,$$

which is valid for all z in $\mathbb{C}$. Discuss the error in approximating $1/\Gamma(z)$ by the sum ($\sum_{n=1}^{\infty}$) on the right-hand side of (*).

1.10.9 Show that if $\theta \in \mathbb{R}$, $\alpha \in \mathbb{R}$, then

$$\sum_{n=-\infty}^{\infty} \left[\frac{\sin(n\alpha + \theta)}{n\alpha + \theta}\right]^2 = \frac{\pi}{\alpha}.$$

1.10.10 Show that if $h > 0$, and

$$f(z) = \frac{1}{2\pi} \int_{-\pi/h}^{\pi/h} g(t) e^{-izt} dt, \quad z \in \mathbb{C},$$

with $g \in \mathbf{C}(\mathbb{R})$, $g(t) = 0$ for $|t| > \pi/h$, then $\sum_{n=-\infty}^{\infty} (-1)^n f(nh) = 0$.

1.10.11 Let $F \in \mathbf{W}(\pi/h)$, where for some entire function f

$$F(z) = \frac{f(z) - f(0)}{z}.$$

Show that

$$f(z) = f(0) + \frac{z \sin(\pi z/h)}{\pi/h} f'(0)$$

$$+ \frac{\sin(\pi z/h)}{\pi/h} \sum_{n \neq 0} (-1)^n \left\{\frac{1}{nh} + \frac{1}{z - nh}\right\} f(nh).$$

1.10.12 Show that if $f \in \mathbf{W}(\pi/h)$, then

$$f(z) = \sum_{n=-\infty}^{\infty} \{f(2nh) + (z - 2nh) f'(2nh)\} [\mathrm{S}(n, 2h)(z)]^2.$$

1.10.13 Show that if $f,\, g \in \mathbf{W}(\pi/h)$, then

$$\int_{\mathbb{R}} f(t) g(z-t)\, dt = h \sum_{n=-\infty}^{\infty} f(nh) g(z-nh).$$

1.10.14 * Let $K(\cdot,t) \in \mathbf{L}^2(I)$ for each $t \in \mathbb{R}$, where $I = [a,b]$ denotes a finite closed interval, and let $\{t_k\}_{k=-\infty}^{\infty}$ be a countable sequence of real numbers such that $\{K(\cdot,t_k)\}_{k=-\infty}^{\infty}$ forms a complete orthogonal set of functions in $\mathbf{L}^2(I)$. If

$$f(t) = \int_I K(x,t)\, g(x)\, dx \qquad (t \in \mathbb{R})$$

for some $g \in \mathbf{L}^2(I)$, then f admits the sampling representation

$$f(t) = \sum_{k=-\infty}^{\infty} f(t_k)\, S_k^*(t),$$

with

$$S_k^*(t) \equiv \frac{\int_I K(x,t)\, K(x,t_k)\, dx}{\int_I |K(x,t_k)|^2 dx}.$$

(Kramer [Kra1]).

1.10.15 * Let $[a,b]$ denote a finite interval on $\mathbb{R}$, let q be continuous and real-valued on $[a,b]$, and let $y = \Phi = \Phi(x,z)$ denote the solution to the problem

$$\frac{d^2y}{dx^2} + q(x)\, y = -z^2 y,$$
$$y(a)\, \cos(\alpha) + y'(a)\, \sin(\alpha) = 0,$$

for some $\alpha \in \mathbb{R}$. Let $\{\Phi(x,z_k)\}_{k=1}^{\infty}$ be a discrete sequence of such functions which also satisfy the condition

$$y(b)\, \cos(\beta) + y'(b)\, \sin(\beta) = 0$$

for some $\beta \in \mathbb{R}$. Prove that:

(a) Each of the numbers $z_k, k = 1, 2, \cdots$ is real and these numbers are distinct from one another, and

$$z_k = \frac{k\,\pi}{b-a} + \mathcal{O}(\frac{1}{k}), \qquad k \to \infty.$$

(b) Each of the functions $\Phi(x, z_k)$, $k = 1, 2, \cdots$, is a real-valued function of x defined on $[a, b]$, and moreover,

$$\int_a^b \Phi(x, z_k)\, \Phi(x, z_\ell)\, dx = c_k \delta_{k-\ell},$$

with

$$c_k = \int_a^b |\Phi(x, z_k)|^2\, dx > 0.$$

(c) For each fixed $x \in [a, b]$, $\Phi(x, z)$ is an entire, even function of z, of order 1 and type $\leqq (b - a)$.

(d) Let f be defined corresponding to some $g \in \mathbf{L}^2(a, b)$ and $z \in \mathbb{C}$ by

$$f(z) = \int_a^b g(x)\, \Phi(x, z)\, dx.$$

Then f is an entire function of order 1 and type $\leqq (b - a)$, which may also be represented in the form

$$f(z) = \sum_{k=1}^{\infty} f(z_k) \frac{2\, z_k\, G(z)}{(z^2 - z_k^2) G'(z_k)},$$

provided that no $z_k = 0$, where

$$G(z) = \prod_{k=1}^{\infty} \left(1 - \frac{z^2}{z_k^2}\right).$$

On the other hand, if e.g. $z_1 = 0$, then

$$f(z) = \frac{f(0)\, G(z)}{z^2} + \sum_{k=2}^{\infty} f(z_k) \frac{2\, z_k\, G(z)}{(z^2 - z_k^2)\, G'(z_k)},$$

with

$$G(z) = z^2 \prod_{k=2}^{\infty} \left(1 - \frac{z^2}{z_k^2}\right).$$

([Zayed (Z1)]).

1.10.16 Let $\cot^{-1} : \mathbb{R} \to (0, \pi)$ (as usual), let $a \geqq 0$, and set $\alpha = \cot^{-1}(a)$. Prove that if u is defined by $u = (a - t)/\sqrt{1 + a^2}$, if $|u| < 1$, and if *sinc* is defined as in (1.10.1), then

$$\cot^{-1}(t) = \alpha \sum_{m=0}^{\infty} \{\mathrm{sinc}(m\alpha)\} \, u^m.$$

[Hint: Show first that if $|u| < 1$, then

$$\frac{\mathrm{sinc}(\alpha)}{1 - 2\, u \, \cos(\alpha) + u^2} = \sum_{m=1}^{\infty} \{\mathrm{sinc}(m\alpha)\} \, m \, u^{m-1}.]$$

[M. Marsden, private communication, (1990)].

Historical Remarks on Chapter 1

The elementary method of presentation of the material of this chapter is new. For example, although the connection of the Laurent series, Fourier series, Chebyshev polynomials and the error of approximation via truncation of these series may be encountered in part in several texts on approximation theory, these subjects have never before appeared altogether in a text on computational mathematics, where they do, in fact, belong.

I have included *Pichorides'* Theorem [Pi1] (Theorem 1.2.1) since I believe it to be important, and since it is not well known.

Although the Riemann–Hilbert procedure of Section 1.3 is very powerful for solving problems from many areas of applications, it is not very well known, since its complex variable presentation has appeared previously only in textbooks that were wholly devoted to this subject; hopefully, the brief yet elementary complex variable presentation given here will make it more accessible to scientists. The Computation of Index Theorem (Theorem 1.3.2) is new, although I published a related result, for computing the topological degree via a different proof in [S18].

All of the Sinc approximation methods of this text could in fact be derived by means of the beautiful function χ, which I derived in 1970 along the lines of Examples 1.7.6 and 1.7.7, and Problems 1.7.5 and 1.7.8, although the paper first appeared in 1973 [S16], since it was originally rejected by a different journal. In [S16], I derived various Sinc-related quadrature procedures via this function; however, my mentor, *J. McNamee*, later talked me into using the cardinal series instead of χ, for purposes of developing and studying Sinc-numerical methods, since he believed that the latter derivations and studies would be simpler to understand.

The simple derivation of the Poisson kernel in the proof of Theorem 1.8.1 is believed to be new. The Tauberian-like Theorem 1.8.4 is believed to be new. It, as well as the other theorems of Section 1.8, are very useful for proving the convergence of Sinc methods as well as for deriving new methods.

The Cardinal function was first discovered by *Borel* [Bor1,Bor2], although *Whittaker* [We1] appears to be the first to have recognized its significance in the theory of analytic functions. Many have since contributed to the enrichment of this theory. An excellent account of the history of the Cardinal function may be found in *Higgins* [Hs1]. An excellent paper of the application of the Cardinal function to signal processing is due to *Butzer* [Bz1]. Yet another excellent paper on the application of the Cardinal function to statistics is due to *Jerry* [Je1]. I have not been able to find the original sources of the many identities associated with the Cardinal function, which are given in Section 1.10, and which I derived via preceding methods of this text. Some of these, such as (1.10.14) through (1.10.23) and (1.10.29) may, in fact, be new.

The powerful notation $\mathrm{S}(k, h)(x)$ which is so suitable for this text is originally due to one of my Ph. D. students, *Lundin* [Lu1].

2

Polynomial Approximation

In our presentation, we saw Fourier series emerge in Corollary 1.1.11, by way of Laurent's Theorem. In Section 1.6 we studied Fourier exponential, sine and cosine series in greater detail. In Section 2.1, we again turn to complex variables, using Laurent's Theorem and a transformation of Section 1.7.4 to obtain a simple derivation of the Chebyshev polynomials. We thus witness the explicit orthogonality over $[0, 2\pi]$ of integer powers of e^{ix} transcend, upon taking real and imaginary parts to sines, to cosines, and to Chebyshev polynomials. These results have enriched science by making possible a vast number of explicit expressions in analysis, as well as a wide range of applications. That such orthogonality exists in such a discrete form further enriches science, and especially applications, with many beautiful formulas. The powerful *Fast Fourier Transform* (FFT) algorithm is made possible by the existence of this discrete orthogonality; this algorithm has had a tremendous impact in computing and applications. This *discrete Fourier transform* (DFT) is also developed in the setting of Laurent's Theorem, since this is the space of functions where it best illustrates its true power and accuracy.

The continuous Fourier polynomial which is presented in Section 2.1 is, of course, the Fourier transform of a Sinc approximation. The second section of this chapter thus presents the mathematics for the algorithms which make approximate methods based on Sinc approximations very efficient, especially for problems, such as three-dimensional, partial differential, and integral equation problems, for which the matrices resulting from Sinc approximation are Toeplitz matrices.

In the third section of this chapter we briefly present the more general Lagrange polynomial, in order to enable us to demonstrate, in the problems, both the power and shortcomings of the popular method of polynomial approximation. In the final section of the Chapter we present *Faber polynomials*, which are a practically important generalization of Chebyshev polynomials.

2.1 Chebyshev Polynomials

We have already seen, by Taylor's Theorem, that the Taylor polynomial (see Equation (1.1.24)) can provide a method to approximate an analytic function f in a neighborhood of a point of analyticity, provided that we know the values of a few derivatives of f at the point, i.e., provided that we know the values of the fixed few numbers a_n defined by (1.1.21). We have also studied the approximation of f by its Fourier polynomial, in Corollary 1.1.11. The statement of Corollary 1.1.11 may in fact be used to derive an effective method of approximating a function on an interval by a polynomial, as well as provide an error bound for such an approximation.

Let $\mathbf{M}_R$ denote the family of all functions f that are analytic and bounded by M in $\mathcal{A}_R$ (see Equation (1.1.25)). By Theorem 1.1.10, each member f of this family has a Laurent expansion of the form

$$f(z) = \sum_{n=-\infty}^{\infty} c_n z^n, \quad z \in \mathcal{A}_R \tag{2.1.1}$$

with c_n defined, for example by Equation (1.1.26). Let $\mathbf{M}_{R,e}$ denote that subclass of functions f in $\mathbf{M}_R$ for which $c_{-n} = c_n$. Then, every f in $\mathbf{M}_{R,e}$ has a convergent series representation of the form

$$f(z) = c_0 + \sum_{n=1}^{\infty} c_n (z^n + z^{-n}), \quad z \in \mathcal{A}_R. \tag{2.1.2}$$

Furthermore, it is clear, from (2.1.2), that if $f \in \mathbf{M}_{R,e}$ and if $z \in \mathcal{A}_R$, then

$$f(1/z) = f(z). \tag{2.1.3}$$

Let us now invoke the mapping $w = (z + 1/z)/2$ which was discussed in Example 1.7.4. This mapping satisfies the equation $w(1/z) = w(z)$, and although it is not a conformal map from $\mathcal{A}_R$ onto $\mathcal{E}_R$ (see Equation (1.7.6)), it provides a one to one correspondence of functions F that are analytic and bounded by M in $\mathcal{E}_R$ with the functions f in $\mathbf{M}_{R,e}$. Indeed, under this mapping we have

$$\frac{1}{2}(z^n + z^{-n}) = T_n(w) \tag{2.1.4}$$

with $T_n(w)$ the Chebyshev polynomial, which may also be defined recursively by means of the equations

$$\begin{cases} T_0(w) = 1, \quad T_1(w) = w \\ \\ T_{n+1}(w) = 2wT_n(w) - T_{n-1}(w), \quad n = 1, 2, \cdots. \end{cases} \tag{2.1.5}$$

Hence, it follows that if f defined by (2.1.2) is in $\mathbf{M}_{R,e}$, then the corresponding function F that is analytic and bounded by M in $\mathcal{E}_R$ is given by

$$F(w) = c_0 + 2\sum_{n=1}^{\infty} c_n T_n(w), \quad w \in \mathcal{E}_R. \tag{2.1.6}$$

Then, by combining Theorem 1.1.10 and Corollary 1.1.11, we get

Theorem 2.1.1 *Let $R > 1$, and let F be analytic and bounded by M in $\mathcal{E}_R$. Then the expansion (2.1.6) holds for all $w \in \mathcal{E}_R$, with T_n the Chebyshev polynomial of order n, defined as in (2.1.5), and with*

$$c_n = \frac{1}{\pi} \int_{-1}^{1} \frac{F(w) T_n(w) dw}{\sqrt{1-w^2}}. \tag{2.1.7}$$

Moreover,

$$|c_n| \leqq \frac{M}{R^n}, \tag{2.1.8}$$

and if for any given w in $\mathcal{E}_R$, the numbers $\rho \in [1, R)$ and $\theta \in [-\pi, \pi]$ are determined by $w = \frac{1}{2}(\rho e^{i\theta} + \rho^{-1} e^{-i\theta})$, then for $m = 1,\ 2,\ 3, \cdots$,

$$\left| F(w) - c_0 - 2 \sum_{n=1}^{m-1} c_n T_n(w) \right| \leqq \frac{2MR}{R-\rho} \left(\frac{\rho}{R} \right)^m. \tag{2.1.9}$$

Note the important special case of Theorem 2.1.1, which occurs when $\rho = 1$; in this case w is on the interval $[-1, 1]$.

Let us now consider, under conditions similar to those of Theorem 2.1.1, an important kind of Hilbert transform of the function F.

Theorem 2.1.2 *Let $R > 1$, let F be analytic and bounded by M in $\mathcal{E}_R$, and let c_n be defined by (2.1.7). Let g be defined on the interval $(-1, 1)$ by*

$$g(x) = \frac{P.V.}{\pi i} \int_{-1}^{1} \frac{F(t)\, dt}{\sqrt{1-t^2}(t-x)}. \tag{2.1.10}$$

Then g has an analytic continuation G to all of $\mathcal{E}_R$. Moreover, for all $w \in \mathcal{E}_R$, we have

$$G(w) = -2 \sum_{n=1}^{\infty} c_n\, U_{n-1}(w) \tag{2.1.11}$$

with U_{n-1} defined for $n = 1, 2, \cdots$, by the formula

$$U_{n-1}(w) = \frac{1}{n} \frac{d\, T_n(w)}{dw}. \tag{2.1.12}$$

Moreover, for all $w \in \mathcal{E}_R$, such that $w = \frac{1}{2}(\rho e^{i\theta} + \rho^{-1} e^{-i\theta})$, with $\rho \in [1, R)$ and $\theta \in [-\pi, \pi]$, we have for all $m > 1$,

$$\left| G(w) + 2 \sum_{n=1}^{m-1} c_n U_{n-1}(w) \right| \leqq \frac{2M\,R}{R-\rho} \left(\frac{\rho}{R} \right)^m. \tag{2.1.13}$$

Proof. It is convenient to carry out the proof on $\mathcal{A}_R$, recalling that the transformation (*) $w = \frac{1}{2}(z+z^{-1})$ maps $\mathcal{A}_R$ onto $\mathcal{E}_R$. The result being that if in (2.1.10) we take $F(t) = T_n(t)$, then $g(x) = -U_{n-1}(x)$ follows from Problem 2.1.5. Under the transformation (*), we have $T_n(w) = \frac{1}{2}(z^n + z^{-n})$, and $U_{n-1}(w) = n^{-1}(z^n - z^{-n})/(z - z^{-1})$. Hence, both $T_n(w)$ and $U_{n-1}(w)$ are bounded by ρ^n for $z = \rho e^{i\theta}$. Using (2.1.8), we find that the sum $\sum_{n=m}^{\infty} 2M(\rho/R)^n$ is bounded by the right-hand side of (2.1.13). ■

We also note the important special case of Theorem 2.1.2, which occurs when $\rho = 1$, i.e., when w is on the interval $[-1, 1]$.

•

Problems for Section 2.1

2.1.1 Prove that if $w = \frac{1}{2}(z + z^{-1})$, then for any integer n, $\frac{1}{2}(z^n + z^{-n})$ is, in fact, a polynomial of degree $|n|$ in w. [Hint: Use Eqs. (2.1.5)].

2.1.2 Give a detailed proof of Theorem 2.1.1.

2.1.3 Let $0 < \alpha < 1$, and let $S(\mathbf{x}_n, x)$ denote the piecewise linear interpolant of x^α at the *knots* $\mathbf{x}_N = \{x_i\}_{i=0}^n$, with $0 \leqq x_0 < x_1 < \cdots < x_n \leqq 1$.

(a) Prove that if $x_i = i/n$, $i = 0, 1, \cdots, n$, then there exists a positive constant c, independent of n, such that

$$\max_{0 \leqq x \leqq 1} |x^\alpha - S(\mathbf{x}_n, x)| \geqq \frac{c}{n^\alpha}.$$

(b) Prove that there exists a constant c_1 which is independent of n, such that

$$\inf_{\mathbf{x}_n} \max_{0 \leqq x \leqq 1} |x^\alpha - S(\mathbf{x}_n, x)| \leqq \frac{c_1}{n^2}.$$

[Hint: Determine β such that a $x_i = (i/n)^\beta$ will produce the desired inequality.]

2.1.4 * Let $\mathbf{P}_n$ denote the set of all polynomials of degree at most $n - 1$. Given $f \in \mathbf{C}[0, 1]$, let $E_n(f)$ denote the error,

$$E_n(f) = \inf_{q \in \mathbf{P}_n} \sup_{0<x<1} |f(x) - q(x)|.$$

(a) Let $0 < \alpha < 1$. Prove there exists a positive constant c, independent of n, such that

$$E_n(x^\alpha) \geqq cn^{-2\alpha}.$$

(b) Given $\mathbf{x} = \{x_i\}_0^m$, with $0 = x_m < x_{m-1} < \cdots < x_0 = 1$, and integers $\mathbf{n} = \{n_j\}_0^{m-1}$, $n_0, \cdots, n_{m-1}$, let $S_{\mathbf{n},\mathbf{x}}$ denote the space of piecewise polynomials with knots $\mathbf{x}$ such that if $S \in S_{\mathbf{n},\mathbf{x}}$ then S is a polynomial of at most degree n_i on $[x_{i+1}, x_i]$, $i = 0, 1, \cdots, m-1$. If N is a positive integer, set

$$S_N = \cup_{\mathbf{n},\mathbf{x},\sum n_i = N} S_{\mathbf{n},\mathbf{x}}$$

and

$$E_N^*(f) = \inf_{S \in S_N} \sup_{0<x<1} |f(x) - S(x)|.$$

Prove that if $0 < \alpha < 1$, then there exist positive constants C_1 and C_2, independent of N, such that

$$C_2 N^{-2\alpha-1} \delta^{(\alpha N)^{1/2}} \leqq E_N^*(x^\alpha) \leqq C_1 \delta^{(\alpha N)^{1/2}}$$

with $\delta = (2^{1/2} - 1)^2 \approx e^{-1.7627}$. (DeVore and Scherer, [dVSr1]).

2.1.5 Prove that if in (2.1.10) we take $F(t) = T_n(t)$, then $g(x) = -U_{n-1}(x)$. [Hint: Use the result of Problem 1.3.17.]

2.1.6 Let $w \in \{\mathbb{C} : |w| = 1\}$. Use the results of Problems 1.3.5, or 1.3.17, to prove the identity

$$\frac{P.V.}{\pi i} \int_{|t|=1} \frac{t^n + t^{-n}}{t - w} dt = w^n - w^{-n}, \; n = 1, 2, \cdots.$$

Then use this identity to prove Theorem 2.1.2.

2.1.7 * Let V_n denote the Vandermonde matrix of order n, with (i,j)th element x_i^{j-1}, with $x_i = \cos[(2i-1)\pi/(2n)]$, $i = 1, 2, \cdots, n$. Prove that, if $\|V_n\| = \sup_{c^T \in \mathbb{C}^n} \|V_n c\| / \|c\|$, with $c = (c_1, \cdots, c_n)^T$, $\|c\| = \max_i |c_i|$, then

$$\|V_n^{-1}\| \|V_n\| \sim 3^{3/4} 4^{-1} (1 + \sqrt{2})^n, \quad n \to \infty.$$

(Gautschi [Ga1])

2.1.8 * Let $\xi > 1$. Find a polynomial $p(x)$ of degree n in x, such that

$$\delta(n) = \max_{-1 \leqq x \leqq 1} |\frac{1}{x - \xi} - p(x)|$$

has the smallest possible value $\{(\xi^2 - 1)(\xi + \sqrt{\xi^2 - 1})^n\}^{-1}$.

2.2 Discrete Fourier Polynomials

The trapezoidal and midordinate integration methods are among the simplest numerical integration methods, and yet they are also very powerful. Given an interval $[a, b]$ on $\mathbb{R}$, and given a positive integer N, we set $h = (b-a)/N$. We can thus define the *trapezoidal*-$\mathcal{T}_N$ and the *midordinate*-$\mathcal{M}_N$ rules to approximate the integral of a function F defined on an interval $[a, b]$ by

$$\begin{aligned} \mathcal{T}_N F &= h[\tfrac{1}{2}F(a) + \textstyle\sum_{j=1}^{N-1} F(a+jh) + \tfrac{1}{2}F(b)] \\ \mathcal{M}_N F &= h \textstyle\sum_{j=1}^{N} F(a + (j-\tfrac{1}{2})h) \end{aligned} \tag{2.2.1}$$

with the sum $\sum_{j=1}^{N-1}$ on the right-hand side of (2.2.1) to be omitted if $N = 1$. It is important for computational purposes to note that

$$\mathcal{T}_{2N} F = \frac{1}{2}(\mathcal{T}_N F + \mathcal{M}_N F). \tag{2.2.2}$$

This formula enables us to circumvent repeated evaluation of F at the same points. Furthermore, since the errors of approximation of $\int_a^b F(x)\,dx$ by $\mathcal{T}_N$ and $\mathcal{M}_N$ usually have a different sign, i.e., if $F \in C^2[a,b]$, then,

$$\begin{aligned} \int_a^b F(x)\,dx - \mathcal{T}_N F &= -\frac{F''(\xi)}{12N^2}(b-a)^2 \\ \int_a^b F(x)\,dx - \mathcal{M}_N F &= \frac{F''(\eta)}{24N^2}(b-a)^2 \end{aligned} \tag{2.2.3}$$

with ξ and η depending on F and N, and located on (a, b). Hence the difference between $\mathcal{T}_N F$ and $\mathcal{M}_N F$ usually gives us a bound on the error of each method.

Henceforth, we shall confine our attention to the important case of $(a, b) = (-\pi, \pi)$. In this case, to approximate $\frac{1}{2\pi}\int_0^{2\pi} f(\theta)\,d\theta$ we use the formulas:

$$\begin{aligned} \mathcal{T}_N f &\equiv \frac{1}{N}\left(\frac{1}{2}f(-\pi) + \sum_{j=1}^{N-1} f\left\{(2j-N)\frac{\pi}{N}\right\} + \frac{1}{2}f(\pi)\right) \\ \mathcal{M}_N F &\equiv \frac{1}{N}\sum_{j=1}^{N} f\left(\frac{2j-N-1}{N}\pi\right). \end{aligned} \tag{2.2.4}$$

Lemma 2.2.1 *Let $\mathcal{T}_N$ and $\mathcal{M}_N$ be defined as in (2.2.4). Then, for any integer k,*

$$\mathcal{T}_N(e^{ikx}) = \begin{cases} (-1)^{\sigma N} & \text{if } k = \sigma N, \ \sigma \text{ aninteger} \\ 0 & \text{otherwise} \end{cases} \tag{2.2.5}$$

$$\mathcal{M}_N(e^{ikx}) = \begin{cases} (-1)^{\sigma(1+N)} & \text{if } k = \sigma N, \ \sigma \text{ aninteger} \\ 0 & \text{otherwise.} \end{cases} \tag{2.2.6}$$

Proof. This is left to Problem 2.1.4. ■

The error of quadrature by $\mathcal{T}_N$ or $\mathcal{M}_N$ decreases much more rapidly than indicated by the $\mathcal{O}(N^{-2})$ rates in (2.2.3), if the function F is a smooth function of period 2π. The function F defined by $F(\theta) = f(e^{i\theta})$, with f analytic in $\mathcal{A}_R$, is of this type.

Theorem 2.2.2 *Let f be analytic in $\mathcal{A}_R$ (Equation (1.1.2)) and let c_m be defined as in Equation (1.1.26). Define F by $F(\theta) = f(e^{i\theta})$. Then*

$$\mathcal{T}_N F = \sum_{\sigma=-\infty}^{\infty} (-1)^{\sigma N} c_{\sigma N} \tag{2.2.7}$$

$$\mathcal{M}_N F = \sum_{\sigma=-\infty}^{\infty} (-1)^{\sigma(1+N)} c_{\sigma N}. \tag{2.2.8}$$

Proof. The results (2.2.7) and (2.2.8) follow by substituting the series expansion (2.1.1) with $z = e^{i\theta}$ into (2.2.4), then we use the identities (2.2.5) and (2.2.6). ■

Theorem 2.2.3 *Let k be an integer, and let $\tilde{c}_k$ be defined by either of the formulas*

$$\begin{aligned} \tilde{c}_k &= \mathcal{T}_N\{e^{-ik\theta} f(e^{i\theta})\} \\ \tilde{c}_k &= \mathcal{M}_N\{e^{-ik\theta} f(e^{i\theta})\} \end{aligned} \tag{2.2.9}$$

with f analytic and bounded by M in $\mathcal{A}_R$. If m denotes an arbitrary positive integer, then

$$\max_{-\pi \leqq \theta \leqq \pi} \left| f(e^{i\theta}) - \sum_{k=-(m-1)}^{m-1} \tilde{c}_k e^{ik\theta} \right| \leqq \frac{2MR}{R-1} \left\{ \frac{1}{R^m} + \frac{R^{m-1}}{R^N - 1} \right\}. \tag{2.2.10}$$

Proof. We have

$$f(e^{i\theta}) - \sum_{k=-(m-1)}^{m-1} \tilde{c}_k e^{ik\theta} = \sum_{k=-(m-1)}^{m-1} (c_k - \tilde{c}_k) e^{ik\theta} + \sum_{|k| \geqq m} c_k e^{ik\theta}. \tag{2.2.11}$$

As was shown in (1.1.37), the sum on the extreme right-hand side of (2.2.11) is bounded by $[2MR/(R-1)]R^{-m}$. In order to bound the first sum on the right-hand side of (2.2.11), we note that

$$\begin{aligned} \mathcal{T}_N\{e^{-ik\theta} f(e^{i\theta})\} &= \sum_{\sigma=-\infty}^{\infty} (-1)^{\sigma N} c_{k+\sigma N} \\ \mathcal{M}_N\{e^{ik\theta} f(e^{i\theta})\} &= \sum_{\sigma=-\infty}^{\infty} (-1)^{\sigma(1+N)} c_{k+\sigma N} \end{aligned} \tag{2.2.12}$$

so that, using (1.1.36) and (2.2.9), we get

$$|c_k - \tilde{c}_k| \leqq M(R^k + R^{-k}) \sum_{\sigma=1}^{\infty} R^{-N\sigma}. \tag{2.2.13}$$

This inequality yields

$$\sum_{k=-m+1}^{m-1} |c_k - \tilde{c}_k| \leqq \frac{2M}{R-1} \frac{R^m}{R^N - 1} \tag{2.2.14}$$

and the result (2.2.10) follows. ■

Remark 2.2.4 The identities (2.2.9), and (2.2.14), show that if $F(e^{i\theta})$ is identically equal to the Fourier polynomial $\sum_{k=-m+1}^{m-1} c_k e^{ik\theta}$, then in order to have $\tilde{c}_k = c_k$ for $k = -(m-1), \cdots, m-1$, we must have $N \geqq m$.

The above results may be applied immediately to achieve Chebyshev polynomial approximation on $[-1, 1]$.

Theorem 2.2.5 *Let g be analytic and bounded by M in $\mathcal{E}_R$ (Equation (1.7.6)), and let $\tilde{c}_k$ $(k = 0, 1, \cdots, m-1)$ be determined by*

$$\tilde{c}_k = \frac{1}{N} \sum_{j=1}^{N} g(x_{N,j}) T_k(x_{N,j}) \tag{2.2.15}$$

with

$$x_{N,j} = \cos[\frac{2j-1}{2N}\pi] \tag{2.2.16}$$

and with T_k the Chebyshev polynomial, as defined in (2.1.5). Let U_n be defined as in (2.1.12). Then

$$\left.\begin{array}{l} \max_{-1 \leqq x \leqq 1} \left| g(x) - \tilde{c}_0 - 2 \sum_{k=1}^{m-1} \tilde{c}_k T_k(x) \right| \\ \\ \max_{-1 \leqq x \leqq 1} \left| \frac{P.V.}{\pi i} \int_{-1}^{1} \frac{g(t)\,dt}{\sqrt{1-t^2}(t-x)} + 2 \sum_{k=1}^{m-1} \tilde{c}_k U_{k-1}(x) \right| \end{array} \right\} \leqq \delta \tag{2.2.17}$$

with

$$\delta = \frac{2MR}{R-1} \left\{ \frac{1}{R^m} + \frac{R^{m-1}}{R^{2N}-1} \right\}. \tag{2.2.18}$$

Proof. Let f be analytic in $\mathcal{A}_R$, such that f is defined by $f(w) = g(\frac{1}{2}w + \frac{1}{2}w^{-1})$. Then, for all w in $\mathcal{A}_R$, $f(w) = f(w^{-1})$, and moreover

$$f(w) = c_0 + \sum_{k=1}^{\infty} c_k (w^k + w^{-k}). \tag{2.2.19}$$

Thus, if $\tilde{c}_k$ is defined by

$$\begin{aligned} \tilde{c}_k &= \mathcal{M}_{2N} \left\{ e^{ik\theta} f\left(e^{i\theta}\right) \right\} \\ &= \frac{1}{2N} \sum_{j=1}^{2N} e^{ik(\frac{2j-2N-1}{2N})\pi} f\left(e^{i(\frac{2j-2N-1}{2N})\pi}\right), \end{aligned} \tag{2.2.20}$$

we see from (2.2.12) that $\tilde{c}_k = \tilde{c}_{-k}$. Thus by taking the average of the right-hand sides of (2.2.20) with k and $-k$, we get

$$\tilde{c}_k = \frac{1}{2N} \sum_{j=1}^{2N} \cos\left\{k\left(\frac{2j-2N-1}{2N}\right)\pi\right\} f\left(e^{i(\frac{2j-2N-1}{2N})\pi}\right). \qquad (2.2.21)$$

Finally, since both $\cos(k\theta)$ and $f(e^{i\theta})$ are even functions of θ on $(-\pi, \pi)$ we can replace (2.2.21) by

$$\tilde{c}_k = \frac{1}{N} \sum_{j=1}^{N} \cos\left\{k\frac{2j-1}{2N}\pi\right\} f\left(e^{i\frac{(2j-1)}{2N}\pi}\right) \qquad (2.2.22)$$

which is just (2.2.15), in view of the definitions of T_k (Equation (2.1.4)), and since $f(1/w) = f(w)$.

We omit the proof of the second inequality in (2.2.17), which is similar to the proof of Theorem 2.1.2. ■

We also omit the proof of the following result, which is based on the identity (1.6.7), on the identities of Lemma 2.2.1, and on the inequality (2.1.8).

Theorem 2.2.6 *Let N be a positive integer, and let θ_k be defined by*

$$\theta_k = \begin{cases} (k-\frac{1}{2})\pi/N, \ k = 1, 2, \cdots, N \\ \\ (k+\frac{1}{2})\pi/N, \ k = -1, -2, \cdots, -N. \end{cases} \qquad (2.2.23)$$

Let f be a complex-valued continuous and periodic function of period 2π defined on $\mathbb{R}$, and define $\mathcal{P}_N$, $\mathcal{S}$ and $\mathcal{S}_N$ by

$$\begin{aligned} \mathcal{P}_N f(\theta) &= \frac{1}{2N} \sum_{k=-N}^{N} f(\theta_k) \frac{\sin[N(\theta_k - \theta)]}{\tan[(\theta_k - \theta)/2]} \\ \mathcal{S} f(\theta) &= \frac{P.V.}{2\pi i} \int_{-\pi}^{\pi} \frac{f(\omega)\, d\omega}{\tan[(\omega - \theta)/2]}, \\ \mathcal{S}_N f(\theta) &= \frac{1}{iN} \sum_{k=-N}^{N} f(\theta_k) \frac{\sin^2[N(\theta_k - \theta)]}{\tan[(\theta_k - \theta)/2]}. \end{aligned} \qquad (2.2.24)$$

Then $\mathcal{P}_N f = f$ and $\mathcal{S}_N f = \mathcal{S} f$ for all functions f of the form

$$f(\theta) = \sum_{k=-(N-1)}^{N-1} c_k e^{ik\theta}. \qquad (2.2.25)$$

Moreover, if F defined by $F(e^{i\theta}) = f(\theta)$ is holomorphic and bounded by M in $\mathcal{A}_R$, then we have

$$\sup_{-\pi<\theta<\pi} |f(\theta) - (\mathcal{P}_N f)(\theta)| \leqq \frac{2MR}{R-1} R^{-N}$$

$$\sup_{-\pi<\theta<\pi} |(\mathcal{S}f)(\theta) - (\mathcal{S}_N f)(\theta)| \leqq \frac{2MR}{R-1} R^{-N}. \tag{2.2.26}$$

■

•

Problems for Section 2.2

2.2.1 Prove that if $\mathcal{T}_N$ and $\mathcal{M}_N$ are defined as in (2.2.1), then Equations (2.2.3) are satisfied for $F \in \mathbf{C}^2[a,b]$.

2.2.2 Verify (2.2.2), with $\mathcal{T}_N$ and $\mathcal{M}_N$ defined as in (2.2.1).

2.2.3 Prove Lemma 2.2.1.

2.2.4 What is the least value of N in Theorem 2.2.5, so that if $g(x) = c_0 + 2\sum_{k=1}^{m-1} c_k T_k(x)$, then $\tilde{c}_k$ defined in (2.2.15) is equal to c_k, for $k = 0, 1, \cdots, m-1$?

2.2.5 Provide a detailed proof of the second inequality in (2.2.17).

2.2.6 Prove Theorem 2.2.6.

2.2.7 Let $\mathcal{M}_N$ be defined as in (2.2.1), for $(a,b) = (0,1)$. Prove that if α is a positive constant, and if $F(x) = x^{\alpha-1}$, then

$$\int_0^1 F(x)\,dx - \mathcal{M}_N F \geqq [\frac{1}{\alpha} - 2^{1-\alpha}]\, N^{-\alpha}.$$

2.2.8 Let $h > 0$, let N be a positive integer, define x_m and a_m for $m = 0, \pm 1, \pm 2, \cdots, \pm N$, by $x_m = \pi m/(Nh)$, $a_{\pm N} = 1/2$, and $a_m = 1$ if $|m| < N$. Let f and g be functions defined on the interval $[-\pi/h, \pi/h]$, and define the inner product

$$< f, g >_N = \frac{1}{2N} \sum_{j=-N}^{N} a_j f(x_j)\overline{g(x_j)}.$$

(i) Set $\psi_m(t) = e^{imht}$, and prove that

$$< \psi_m, \psi_n >_N = \begin{cases} 1 & \text{if} \quad m = n + 2sN \\ 0 & \text{if} \quad m \neq n + 2sN \end{cases}$$

with s some integer.

(ii) Define K_N by

$$K_N(x) = \sum_{m=-N}^{N} \psi_m(x)$$

and prove that

$$K_N(x) = \frac{\sin[(N+1/2)hx]}{\sin(hx/2)}.$$

(iii) Let f be sectionally continuous on $[-\pi/h, \pi/h]$, i.e., let f be continuous at each point of $[-\pi/h, \pi/h]$, except at a finite number of points, with f allowed to be arbitrarily finite-valued, and discontinuous. Set

$$\begin{aligned} c_{N,m} &= \langle f, \psi_m \rangle_N \\ f_N(x) &= \sum_{m=-N}^{N} c_{N,m}\psi_m(x). \end{aligned}$$

Prove that for all $x \in [-\pi/h, \pi/h]$

$$f_N(x) = \langle f, K_N(x - \cdot) \rangle_N .$$

(iv) Prove that if h is fixed and independent of N, then

$$\int_{-\pi/h}^{\pi/h} |f(x) - f_N(x)|^2 dx \to 0$$

as $N \to \infty$. [Hint: It helps to evaluate $\int_{-\pi/h}^{\pi/h} K_N(x - x_m)\, K_N(x - x_n)\, dx$.]

(v) Prove that if γ is a positive constant independent of N, if $h = \gamma/N^{1/2}$, and if f is equicontinuous on $\mathbb{R}$, such that $\int_{\mathbb{R}} |f(x)|^2 dx < \infty$, then

$$\int_{-\pi/h}^{\pi/h} |f(x) - f_N(x)|^2 dx \to 0$$

as $N \to \infty$.

2.2.9 Let h, N, ψ_m, $K_N(x)$, and the inner product $< f, g >_N$ be defined as in the previous problem. Let f be defined on the interval $[-\pi/h, \pi/h]$ by the series

$$f(x) = h \sum_{m=-\infty}^{\infty} c_m e^{imhx}$$

where, corresponding to fixed positive numbers C and α independent of N and h, we have

$$|c_m| \leqq C e^{-|m|\alpha h}$$

for all integers m.
(a) Define $\tilde{f}_N$ on $[-\pi/h, \pi/h]$ by

$$\tilde{f}_N(x) = h \sum_{m=-N}^{N} c_m e^{imhx}$$

and prove that

$$\max_{-\pi/h \leqq x \leqq \pi/h} |f(x) - \tilde{f}_N(x)| \leqq \frac{2C}{\alpha} e^{-\alpha N h}.$$

Define f_N by

$$f_N(x) =< f, K_N(x - \cdot) >_N .$$

and prove that there exists a positive number C_1, independent of N and h, such that

$$\max_{-\pi/h \leqq x \leqq \pi/h} |f(x) - f_N(x)| \leqq C_1 e^{-\alpha N h}.$$

[Hint: Establish the identity

$$< f, \psi_m >_N = h \sum_{s=-\infty}^{\infty} c_{m+2sN}$$

and then use it, along with the above given bound on $|c_m|$ to bound $c_m - < f, \psi_m >_N$, and then to bound $f - f_N$, by means of the inequality

$$|f(x) - f_N(x)| \leqq h \sum_{|m|>N} |c_m| + h \sum_{m=-N}^{N} |c_m - < f, \psi_m >_N |.]$$

2.2.10 Use Theorem 2.2.3 and Parseval's Theorem to prove that if V_n denotes the Vandermonde matrix of order n, with (k,ℓ)th element $x_k^{\ell-1}$, with $x_k = \exp[i(k-1)2\pi/n]$, $k = 1, 2, \cdots, n$, if $\|V_n\| = \sup_{c \in \mathbb{C}^n} \|V_n c\|/\|c\|$, with $c = (c_1, \cdots, c_n)^T$, $\|c\| = (\sum_{k=1}^n |c_k|^2)^{1/2}$, then

$$\|V_n^{-1}\| \|V_n\| = 1.$$

2.2.11 * Let f denote the even periodic function of period 2π defined on the interval $(0, \pi)$ by

$$f(x) = [x(\pi - x)]^\alpha$$

with $0 < \alpha < 1$. Let f_N be defined as in Problem 2.2.8, above, with $h = 1$. Prove that there exist positive constants c_1 and c_2 which are independent of N, such that

$$\frac{c_1}{N^\alpha} \leqq \max_{0 \leqq x \leqq \pi} |f(x) - f_N(x)| \leqq \frac{c_2}{N^\alpha}.$$

2.3 The Lagrange Polynomial

Let us start by describing a general procedure to obtain a formula which may be used to interpolate a function at the zeros of another function, via the use of contour integration. Let f and G be functions belonging to $\mathbf{Hol}(\mathcal{D})$, let Y be a subset of $\mathcal{D}$, and let us set $X = \{z \in \mathcal{D} : G(z) = 0\}$. Furthermore, let us assume that every $z \in X$ is a simple zero of G. Under suitable conditions for f and G we may then deduce that

$$\frac{G(\xi)}{2\pi i} \int_{\partial \mathcal{D}} \frac{f(z)}{(z - \xi)\, G(z)} dz = f(\xi) - \sum_{x \in X} \frac{f(x)\, G(\xi)}{(\xi - x)\, G'(x)}. \tag{2.3.1}$$

The left-hand side of (2.3.1) may then be considered to be the interpolation error of f by the sum $\sum_{x \in X} \{f(x)\, G(\xi)/[(\xi - x)\, G'(x)]\}$. Furthermore, if $f \in H_1(\mathcal{D})$, so that $N_1(f, \mathcal{D}) < \infty$, (recall Definition 1.1.5) then

$$\sup_{\xi \in Y} \left| f(\xi) - \sum_{x \in X} \frac{f(x)\, G(\xi)}{(\xi - x)\, G'(x)} \right| \leqq \frac{\sup_{\xi \in Y} |G(\xi)|}{\inf_{z \in \partial \mathcal{D}} |G(z)|} N_1(f, \mathcal{D}). \tag{2.3.2}$$

The interpolation of f by the sum $\sum_{x \in X} \{f(x)\, G(\xi)/[(\xi - x)\, G'(x)]\}$ may then be considered accurate if the factor multiplying $N_1(f, \mathcal{D})$ in (2.3.2) is small.

Let us now apply this concept to polynomial interpolation. Let f be a given analytic function in $\mathcal{D}$, and let $z_1, z_2, \cdots, z_m$ denote m distinct points on $\mathcal{D}$. Let us set

$$\varphi(z) = \prod_{j=1}^{m} (z - z_j) \tag{2.3.3}$$

and let us examine the contour integral

$$\varepsilon(\xi) = \frac{\varphi(\xi)}{2\pi i} \int_{\partial\mathcal{D}} \frac{f(z)\, dz}{(z - \xi)\varphi(z)} \tag{2.3.4}$$

with ξ a point in $\mathcal{D}$.

Theorem 2.3.1 *Under the above conditions we have the error of Lagrange interpolation given by*

$$\varepsilon(\xi) = f(\xi) - \sum_{j=1}^{m} \frac{f(z_j)\phi(\xi)}{(z - z_j)\varphi'(z_j)}. \tag{2.3.5}$$

Moreover, $\varepsilon(z_j) = 0$, $j = 1, 2, \cdots, m$, *and if* $f(z)$ *is a polynomial of at most degree* $m - 1$ *in* z, *then* $\varepsilon(\xi) \equiv 0$ *for all* ξ.

Proof. We may carry a proof similar to that of Theorem 1.1.14, using the identity

$$\frac{1}{\varphi(z)} = \sum_{j=1}^{m} \frac{1}{(z - z_j)\varphi'(z_j)}. \tag{2.3.6}$$

The details are left to Problems 2.3.2. ■

An important case of Theorem 2.2.1 is that of interpolating at the zeros of the Chebyshev polynomials, for purposes of approximation on the interval $[-1, 1]$.

Corollary 2.3.2 *Let* m *be a positive integer, let* z_j *be defined by* $z_j = \cos[(2j-1)\pi/(2m)]$, $j = 1, 2, \cdots, m$. *If* f *is analytic and bounded by* M *in* $\mathcal{E}_R$ *(Equation (1.7.6)) then*

$$\max_{-1 \leqq x \leqq 1} \left| f(x) - \frac{T_m(x)}{m} \sum_{j=1}^{m} \frac{(-1)^{j-1}\sqrt{1 - z_j^2}\, f(z_j)}{x - z_j} \right| \tag{2.3.7}$$

$$\leq \frac{2M(R^2+1)}{(R-1)^2(R^m - R^{-m})}.$$

Proof. We consider, instead of (2.3.4), the integral

$$\varepsilon(x) = \frac{T_m(x)}{2\pi i} \int_{\partial \mathcal{E}_R} \frac{f(z)\, dz}{(z-x)T_m(z)}. \tag{2.3.8}$$

Now, using (2.1.4), it is easily shown that $T_m'(z_j) = (-1)^{j-1} m / \sqrt{1-z_j^2}$, so that

$$\frac{1}{T_m(z)} = \frac{1}{m} \sum_{j=1}^{m} \frac{(-1)^{j-1}\sqrt{1-z_j^2}}{z-z_j}. \tag{2.3.9}$$

The bound on the right-hand side of (2.3.7) thus results from the estimates:

$$\begin{aligned} |T_m(x)| &\leqq 1 \text{ on } [-1,1]; \\ |T_m(z)| &\geqq \frac{1}{2}(R^m - R^{-m}) \text{ on } \partial \mathcal{E}_R; \\ |f(z)| &\leqq M \text{ on } \mathcal{E}_R; \\ |z-x| &\geqq \frac{1}{2}(R+R^{-1}) - 1 \text{ for } x \in [-1,1],\ z \in \partial \mathcal{E}_R; \\ \int_{\partial \mathcal{E}_R} |dz| &\leqq \pi(R+R^{-1}). \end{aligned} \tag{2.3.10}$$

This completes the proof. ■

Now let $\mathbf{P}_m$ denote the class of all polynomials of degree $m-1$, let f be an arbitrary continuous function on $[-1,1]$, and let α_m and β_m be defined by

$$\alpha_m = \inf_{p \in \mathbf{P}_m} \sup_{-1<x<1} |f(x) - p(x)| \tag{2.3.11}$$

$$\beta_m = \sup_{-1<x<1} \left| f(x) - \frac{T_m(x)}{m} \sum_{j=1}^{m} \frac{(-1)^{j-1}\sqrt{1-z_j^2}\, f(z_j)}{x - z_j} \right| \tag{2.3.12}$$

with the z_j and T_m defined as in Corollary 2.3.2. It may then be shown (Problem 2.3.6) that

$$\begin{aligned} \beta_m &\leqq 4\alpha_m \quad \text{if } m \leqq 20 \\ \beta_m &\leqq 5\alpha_m \quad \text{if } m \leqq 100. \end{aligned} \tag{2.3.13}$$

The $\mathcal{O}(R^{-m})$ bounds on the right-hand sides of (2.1.9), (2.1.13), and (2.3.7) are optimal in the sense that for example, in the case of (2.3.7), there exists an f which is uniformly bounded by M in $\mathcal{E}_R$, such that

$\alpha_m \geqq (C/m)R^{-m}$ with C a positive number. That is, the R in the above $\mathcal{O}(R^{-m})$ bounds cannot be replaced by a larger number.

We defer the proof of the following generalization of Theorem 2.2.1 to the problems at the end of this section.

Theorem 2.3.3 *Let f be analytic in a domain $\mathcal{D}$, and let $z_1, z_2, \cdots, z_k$ denote k distinct points in $\mathcal{D}$. Let $\mu_1, \mu_2, \cdots, \mu_k$, be positive integers, and set*

$$\begin{aligned} \varphi(z) &= \prod_{j=1}^{k} (z - z_j)^{\mu_j} \\ \ell_n(z) &= \frac{\varphi(z)}{(z - z_n)^{\mu_n}}, \quad n = 1, 2, \cdots, k. \end{aligned} \tag{2.3.14}$$

If $\varepsilon(\xi)$ is defined in $\mathcal{D}$ by

$$\varepsilon(\xi) = \frac{\varphi(\xi)}{2\pi i} \int_{\partial\mathcal{D}} \frac{f(z)\, dz}{(z - \xi)\varphi(z)} \tag{2.3.15}$$

then

$$\varepsilon(\xi) = f(\xi) - \sum_{n=1}^{k} \ell_n(\xi) \sum_{j=0}^{\mu_n - 1} \left\{ \frac{f(z)}{\ell_n(z)} \right\}^{(j)}_{z=z_n} \frac{(\xi - z_n)^j}{j!}. \tag{2.3.16}$$

Moreover, $\varepsilon^{(j)}(z_n) = 0$ for $j = 0, 1, \cdots, \mu_{n-1}$, $n = 1, 2, \cdots, k$, and if $\sum_{j=1}^{k} \mu_j = m$, then $\varepsilon(\xi) \equiv 0$ whenever f is a polynomial of degree $< m$.

An important special case of Theorem 2.3.3 is that of cubic spline interpolation.

Example 2.3.4 If we take $k = 2, z_1 = a$, $z_2 = b$ and $\mu_1 = \mu_2 = 2$ in Theorem 2.3.3, we get a cubic spline approximation of f on (a, b), i.e.,

$$\begin{aligned} f(x) &\approx (\frac{b-x}{b-a})^2 \left\{ [1 + \frac{2(x-a)}{b-a}] f(a) + (x-a) f'(a) \right\} \\ &+ (\frac{x-a}{b-a})^2 \left\{ [1 + \frac{2(b-x)}{b-a}] f(b) - (b-x) f'(b) \right\}. \end{aligned} \tag{2.3.17}$$

The right-hand side of (2.3.17) interpolates f at a and b, and the derivative of the right-hand side of (2.3.17) interpolates f' at a and b.

Example 2.3.5 *Aitken–Neville Algorithm.* Let $n+1$ pairs of complex numbers (a_j, b_j), $j = 0, 1, \cdots, n$, be given, with distinct a_j. Consider the evaluation at z of the polynomial p_n of degree at most n, defined by the equations $p(a_j) = b_j, j = 0, 1, \cdots, n$, by means of the following algorithm:

$$\begin{aligned} p_{jn} &= b_j, \quad j = 0, 1, \cdots, n, \\ p_{jk} &= \frac{\begin{vmatrix} p_{k+1,k+1} & a_{k+1} - z \\ p_{j,k+1} & a_j - z \end{vmatrix}}{a_j - a_{k+1}}, \quad \begin{array}{l} j = 0, 1, \cdots, k; \\ k = n-1, n-2, \cdots, 0. \end{array} \end{aligned} \tag{2.3.18}$$

Let us now show that $p_{00} = p(z)$. To this end, we make the induction hypothesis that for $k = n-1, n-2, ..., 0$, p_{k+1} is a polynomial of degree at most $k+1$, which satisfies $p_{k+1}(z) = p_n(z)$. The induction hypothesis is certainly true for $k = n-1$. Assuming the hypothesis to be true for k, we observe that if we define a polynomial π_{k+1} by the equation

$$\pi_{k+1}(\zeta) = \begin{vmatrix} p_{k+1,k+1} & a_{k+1} - z \\ p_{k+1}(\zeta) & \zeta - z \end{vmatrix} \tag{2.3.19}$$

then $\pi_{k+1}(\zeta)$ is of degree at most $k+1$ in ζ, and that $\pi(a_{k+1}) = 0$. Therefore, $p_k(\zeta) = \pi_{k+1}(\zeta)/(\zeta - a_{k+1})$ is a polynomial of degree at most k in ζ, which may be defined by the $k+1$ values $p_{jk}, j = 0, 1, \cdots, k$. Moreover, we see by inspection of (2.3.19) that $p_k(z) = \pi_{k+1}(z)/(z - a_{k+1}) = p_{k+1}(z)$. Hence, it follows that $p_n(z) = p_{n-1}(z) = \cdots = p_0(z)$. Since, however, p_0 is a constant, it takes on the same value for all z, and therefore $p_{00} = p_0(z) = p_n(z)$.

•

Problems for Section 2.3

2.3.1 Let φ be defined as in Equation (2.3.3), and set

$$\frac{\varphi(x)}{(x - z_j)\varphi'(z_j)} = \sum_{\ell=1}^{m} c_{\ell j} x^{\ell-1}$$

and let V denote the square matrix whose (i,j) th entry is z_j^{i-1}. Prove that $V^{-1} = C$ with C a square matrix whose (i,j)th entry is c_{ij}.

2.3.2 (a) Prove the identity (2.3.5), using (2.3.6).

(b) Show that if f is a polynomial then ε defined by the right-hand side of (2.3.5) is also given by

$$\varepsilon = \frac{\varphi(\xi)}{2\pi i} \int_{|z|=R} \frac{f(z)\, dz}{(z - \xi)\varphi(z)}$$

with R sufficiently large so that z_j $(j = 1, 2, \cdots, m)$ and ξ are both in $\mathcal{D}(0; R)$, but R is otherwise arbitrary.

(c) Let f be an arbitrary polynomial of degree at less than m. Show there exists a number $R_0 > 0$, such that whenever $z = Re^{i\theta}$, with $R > R_0$ and ξ uniformly bounded, then

$$|f(z)| < M_0 R^{m-1},$$
$$|(z-\xi)\varphi(z)| > M_1 R^{m+1}$$

with M_0 and M_1 constants. Use these results to conclude from the equation in part (b) of this problem that $\varepsilon \equiv 0$.

2.3.3 Let μ be a positive integer, and let ξ, z and a be arbitrary numbers in $\mathbb{C}$. Prove that

$$\left(\frac{\xi - a}{z - a}\right)^{\mu} \frac{1}{z-\xi} = \frac{1}{z-\xi} - \sum_{n=0}^{\mu-1} \frac{(\xi-a)^n}{(z-a)^{n+1}}.$$

2.3.4 (a) Let r be sufficiently small so that if $\xi \in \mathcal{D}$, $\xi \neq z$, $\xi \neq z_1$, $\cdots, \xi \neq z_k$, then $\mathcal{D}(\xi; r)$ lies in $\mathcal{D}$ but contains none of the points z_j $(j = 1, \cdots, k)$. Prove that if φ is defined by (2.3.3), then

$$\frac{\varphi(\xi)}{2\pi i} \int_{\partial\mathcal{D}(\xi;r)} \frac{f(z)\,dz}{(z-\xi)\varphi(z)} = f(\xi).$$

(b) Let $\xi, z_1, \cdots, z_k$ be distinct points in $\mathcal{D}$. Let $r > 0$ be chosen such that the closure of each disc $\mathcal{D}(z_n; r)$, $n = 1, 2, \cdots, k$ contains z_n but neither ξ nor z_j, $j \neq n$. Prove using the identity in Problem 2.3.3 that

$$\frac{\varphi(\xi)}{2\pi i} \int_{\partial\mathcal{D}(z_n;r)} \frac{f(z)\,dz}{(z-\xi)\varphi(z)}$$
$$= -\ell_n(\xi) \sum_{s=0}^{\mu_n - 1} \frac{(\xi - z_n)^s}{s!} \left[\frac{wf(z)}{\ell_n(z)}\right]^{(s)} (z_n).$$

(c) Prove that

$$\int_{\partial\mathcal{D}} \frac{f(z)\,dz}{(z-\xi)\varphi(z)} = \int_{\partial\mathcal{D}(\xi;r)} \frac{f(z)\,dz}{(z-\xi)\varphi(z)}$$
$$+ \sum_{n=1}^{k} \int_{\partial\mathcal{D}(z_n;r)} \frac{f(z)\,dz}{(z-\xi)\varphi(z)}.$$

(d) Use the procedure of Problem 2.3.2 (c) to deduce that if $m = \sum_{j=1}^{k} \mu_j$ then ε defined by (2.3.15) is identically zero whenever f is a polynomial of degree $< m$.

2.3.5 Prove that if $N = m$ and $g = f$ in Theorem 2.2.5, then the polynomials

$$\tilde{c}_0 + 2 \sum_{k=1}^{m-1} \tilde{c}_k T_k(x)$$

with $\tilde{c}_k$ defined as in Theorem 2.2.5, and

$$\frac{T_m(x)}{m} \sum_{j=1}^{m} \frac{(-1)^{j-1} \sqrt{1 - z_j^2} f(z_j)}{x - z_j}$$

defined in Corollary 2.3.2, are identical.

2.3.6 * Prove (3.3.13). [Hint: Let $\varphi(x)$ be defined as in (2.3.3), and set

$$\mu_m = \max_{x \in [-1,1]} \sum_{j=0}^{m} \left| \frac{\varphi(x)}{(x - z_j)\varphi'(z_j)} \right| .$$

Then show that $\mu_m \leqq \mu_m^*$, with μ_m^* the μ_m in the case of the z_j the $m+1$ zeros of $T_{m+1}(x)$, and that $\mu_m^* \sim (2/\pi) \log m$, as $m \to \infty$. Then numerically evaluate the first 100 μ_m^*].

2.3.7 Let $w \in \mathbf{L}^1(-1,1)$ and let $w > 0$ on $(-1,1)$. Let $\{p_n\}_{n=0}^{\infty}$ be a sequence of polynomials, such that $p_n(x)$ is of degree n in x, and such that $\int_{-1}^{1} w(x)\, p_m(x)\, p_n(x)\, dx = \delta_{mn}$, with δ_{mn} the Kronecker delta. Let $z_1, z_2, \cdots, z_m$ denote the m zeros of $p_m(x)$.

(a) Prove that the z_j are distinct, and that they are all located on the interval (-1,1).

(b) Set

$$w_j = \int_{-1}^{1} w(x) \frac{p_m(x)}{(x - z_j)\, p_m'(z_j)} dx$$

then prove that $w_j > 0$ for $j = 1, 2, \cdots, m$.

(c) Let $q_m(z)$ be defined for $z \notin [-1,1]$ by

$$q_m(z) = \frac{-1}{2\pi i} \int_{-1}^{1} w(x) \frac{p_m(x)}{x - z} dx.$$

Prove that $q_m \in \mathbf{Hol}(\mathbb{C} \setminus [-1,1])$, and that $q_m(z) = \mathcal{O}(z^{-m-1})$, as $z \to \infty$, and hence, that $q_m(z)/p_m(z) = \mathcal{O}(z^{-2m-1})$ as $z \to \infty$.

(d)* Prove that if $f \in \mathbf{Hol}(\mathcal{E}_R)$, with $\mathcal{E}_R$ defined for $R > 1$ as in Example 1.7.4, and if f is uniformly bounded by M in $\mathcal{E}_R$, then setting

$$\delta \equiv \int_{-1}^{1} w(x)\, f(x)\, dx - \sum_{j=1}^{m} w_j\, f(z_j)$$

we have

$$\delta = \int_{\partial \mathcal{E}_R} f(z) \frac{q_m(z)}{p_m(z)} dz$$

and

$$|\delta| \leqq \frac{8\,\mu_0}{\pi} \frac{M}{R^{2m}}$$

with $\mu_0 = \int_{-1}^{1} w(x)\, dx$.

(e) Use the result in (c) or (d) of this problem to prove that if f is any polynomial of degree at most $2m-1$, then $\delta = 0$, with δ defined as in (d) above.

2.3.8 * Let α be a fixed number between 0 and 1, and define f and g by $f(x) = x\, e^x$, and $g(x) = x^\alpha\, e^x$, with $0 \leqq x \leqq 1$. Let $\mathbf{P}$ denote the class of all polynomials of degree at most m, and for a given continuous function h defined on the interval [0,1], set

$$E_m(h) = \inf_{p \in \mathbf{P}} \max_{-1 \leqq x \leqq 1} |h(x) - p(x)|.$$

Prove that $E_m(f) \leqq [e/(4m)]^m$, and that there exists a positive constant c independent of m, such that $E_m(g) \geqq c\, m^{-2\alpha}$.

2.3.9 Let f and g be defined as in Problem 2.3.8. Apply the Aitken–Neville algorithm to values of f and g at equi-spaced points on the interval $1/2 \leqq x \leqq 1$ to estimate the values $f(0)$ and $g(0)$. Discuss your results in light of the conclusions of Problem 2.3.8.

2.3.10 * Let $\mathbf{P}$ denote the set of all polynomials of degree at most $m-1$, let $\{z_j\}_{j=0}^m$ and $\{w_j\}_{j=0}^m$ be any two sets of complex numbers, such that the z_j are distinct, and set $\phi(z) = \prod_{j=0}^m (z - z_j)$. Let p^* and q^* be polynomials in $\mathbf{P}$ such that

$$\inf_{p\in\mathbf{P}}\left(\max_{j=0,1,\cdots,m}|w_j-p(z_j)|\right)=\max_{j=0,1,\ldots,m}|w_j-p^*(z_j)|$$

and

$$\inf_{q\in\mathbf{P}}\sum_{j=0}^{m}\frac{|w_j-q(z_j)|^2}{|\phi'(z_j)|}=\sum_{j=0}^{m}\frac{|w_j-q^*(z_j)|^2}{|\phi'(z_j)|}.$$

Prove that $p^*=q^*$. (New).

2.3.11 Let V_n denote the *Vandermonde matrix* of order n, with the (k,ℓ)th element $v_{k\ell}=p_{\ell-1}(x_k)$, with p_m defined as in Problem 2.3.7, and with $p_n(x_k)=0, k=1,\cdots,n$. Let w_j denote the Gaussian quadrature weights as defined in Problem 2.3.7.

(a) Prove that, if $\|V_n\|=\left(\sum_{k,\ell=1}^n|v_{k\ell}|^2\right)^{1/2}$ then

$$\|V_n^{-1}\|\|V_n\|=\left(\sum_{k=1}^n w_k\sum_{\ell=1}^n\frac{1}{w_\ell}\right)^{1/2}.$$

(Gautschi [Ga2])

(b) Prove that if $\|V_n\|$ is the standard Euclidean norm corresponding to the vector norm $\|(c_1,\cdots,c_n)^T\|=\left(\sum_{k=1}^n|c_k|^2\right)^{1/2}$ then

$$\|V_n^{-1}\|\|V_n\|=\left(\frac{\sup_{k=1,\cdots,n}w_k}{\inf_{k=1,\cdots,n}w_k}\right)^{1/2}.$$

(J. Lund, private communication, 1991)

2.3.12 Let V_n denote the Vandermonde matrix of order n, with the (k,ℓ)th element $v_{k\ell}=x_k^{\ell-1}$. Prove that, $\|V_n\|=\sup_{c^T\in\mathbf{C}^n}\|V_nc\|/\|c\|$, $c=(c_1,\cdots,c_n)^T$, $\|c\|=\max_i|c_i|$, then

$$\max_j\prod_{i\neq j}\frac{\max(1,|x_i|)}{|x_j-x_i|}<\|V_n^{-1}\|\leqq\max_j\prod_{i\neq j}\frac{1+|x_i|}{|x_j-x_i|}.$$

(Gautschi [Ga2]).

2.4 Faber Polynomials

It is possible to achieve polynomial approximation of analytic functions over very complicated subsets $\mathcal{E}$ of the complex plane $\mathbb{C}$, as long as $\mathbb{C} \setminus \mathcal{E}$ is connected. This condition is violated, for example, in the case of $\mathcal{E} = \{z \in \mathbb{C} : |z| = 1\}$. Although $f(z) = 1/z$ is analytic on $\mathcal{E}$, one cannot accurately approximate f by polynomials on $\mathcal{E}$, since for any polynomial p of degree $m \geqq 0$, we have

$$1 = \left| \frac{1}{2\pi i} \int_{|z|=1} \left[\frac{1}{z} - p(z) \right] dz \right| \leqq \sup_{z \in \mathcal{E}} \left| \frac{1}{z} - p(z) \right|. \tag{2.4.1}$$

Theorem 2.4.1 *Let $\mathcal{D}$ be a simply-connected domain, let $f \in$ **Hol**$(\mathcal{D})$, and let $\mathcal{E}$ be a compact subset of $\mathcal{D}$ such that $\mathcal{E}$ consists of more than one point, and such that $\mathbb{C} \setminus \mathcal{E}$ is connected. Then there exists a sequence of polynomials which converges uniformly to f on $\mathcal{E}$.*

Proof. We shall prove Theorem 2.4.1 only in the case of $\overline{\mathbb{C}} \setminus \mathcal{E}$ simply-connected. The Riemann mapping theorem asserts that there exists a function ϕ^*, which provides a conformal map of the domain $\overline{\mathbb{C}} \setminus \mathcal{E}$ onto the unit disc $\mathcal{U}$. Then the function $w = \phi = 1/\phi^*$ provides a conformal map of the domain $\overline{\mathbb{C}} \setminus \mathcal{E}$ in the z-plane onto the domain $\overline{\mathbb{C}} \setminus \mathcal{U}$ in the w-plane. We may assume, moreover, that $\phi(\infty) = \infty$, and $\phi'(\infty) > 0$, so that, in a neighborhood of $z = \infty$,

$$\phi(z) = \frac{z}{c} + b_0 + \frac{b_1}{z} + \frac{b_2}{z^2} + \cdots \qquad c > 0. \tag{2.4.2}$$

Under the transformation $z = \psi(w)$, with $\psi = \phi^{-1}$ denoting the inverse of ϕ, F defined by $F(w) = f \circ \psi(w)$ now has an analytic continuation into a bounded domain $\mathcal{D}'$ containing $\overline{\mathcal{U}}$, and, hence, by Taylor's Theorem, it follows that for all $w \in \overline{\mathcal{U}}$,

$$F(w) = \sum_{n=0}^{\infty} a_n \, w^n. \tag{2.4.3}$$

Let us now choose a positive number $R > 1$ such that the disc $\mathcal{U}_R = \{w \in \mathbb{C} : |w| < R\}$ is in $\mathcal{D}'$, and let us define a closed contour Γ_R by $\Gamma_R = \psi(\partial \mathcal{U}_R)$. Let $\mathcal{B}_R$ denote the bounded domain such that $\Gamma_R = \partial \mathcal{B}_R$. The series (2.4.3) then clearly converges to F for all $w \in \overline{\mathcal{U}}_R$, so that by (1.1.23), we have $|a_n| = \mathcal{O}(R^{-n})$ as $n \to \infty$.

At this point we take note that

$$\begin{aligned} \{\phi(z)\}^n &= \left\{\frac{z^n}{c^n} + \cdots\right\} + \frac{1}{z} G_n(z) \\ &= F_n(z) + \frac{1}{z} G_n(z) \end{aligned} \tag{2.4.4}$$

with $F_n(z)$ a polynomial of degree n in z, and with G_n analytic in a neighborhood of $z = \infty$. The polynomials F_n is called the *n th degree Faber polynomial for the set* $\mathcal{E}$. We may also note, in view of (2.4.4) that if $z \in \mathcal{B}_R$, then

$$\begin{aligned} F_n(z) &= \frac{1}{2\pi i} \int_{\Gamma_R} \frac{\{\phi(t)\}^n dt}{t - z} \\ &= \frac{1}{2\pi i} \int_{\partial \mathcal{U}_R} \frac{w^n \psi'(w)\, dw}{\psi(w) - z}. \end{aligned} \tag{2.4.5}$$

We may thus deduce by Cauchy's formula, that if $z \in \mathcal{B}_R$, then

$$\begin{aligned} f(z) &= \frac{1}{2\pi i} \int_{\Gamma_R} \frac{f(t)\, dt}{t - z} \\ &= \frac{1}{2\pi i} \int_{\partial \mathcal{U}_R} \frac{F(w)\, \psi'(w)\, dw}{\psi(w) - z} \\ &= \sum_{n=0}^{\infty} \frac{a_n}{2\pi i} \int_{\partial \mathcal{U}_R} \frac{w^n \psi'(w)\, dw}{\psi(w) - z} \\ &= \sum_{n=0}^{\infty} a_n\, F_n(z) \end{aligned} \tag{2.4.6}$$

with a_n defined as in (2.4.3) and F_n as in (2.4.5).

This completes the Proof of Theorem 2.4.1. ∎

In Chapter 4 we shall demonstrate a more constructive procedure than the above theorem and proof, to produce a sequence of functions that uniformly approximate a given analytic function f on an arbitrary compact subset of its domain of analyticity, in the case (as is required for the practical implementation of Theorem 2.4.1) of the existence of an explicit conformal map of the domain of analyticity of f to $\mathcal{U}$.

•

Problems for Section 2.4

2.4.1 (a) In the notation of this section, show that if $\mathcal{E} = [-1, 1]$, then $\phi(z) = z + \sqrt{1 - z^2}$, and $\psi(w) = (w + w^{-1})/2$.

(b) Prove that $F_n(z) = T_n(z)$, with T_n denoting the Chebyshev polynomial, defined as in (2.1.5).

2.4.2 Verify the first identity in (2.4.5).

2.4.3 (a) Use (2.4.5) to prove that if $z \in \mathcal{E}$ then

$$|F_n(z)| \leqq \frac{C}{R^n},$$

with C a constant independent of N.

(b) Use the result in part (a) of this problem to prove that there exists a constant C' independent of n, such that

$$\sup_{z \in \mathcal{E}} \left| f(z) - \sum_{m=0}^{n} a_m F_m(z) \right| \leqq \frac{C'}{R^n}.$$

2.4.4 Let $n+1$ pairs of complex numbers (a_j, b_j), $j = 0, 1, \cdots, n$, be given, where the a_j are distinct and are located in a simply-connected domain $\mathcal{D}$. Let ω denote a conformal map of $\mathcal{D}$ onto the unit disc $\mathcal{U}$. Consider the evaluation at z of the polynomial $p_n \circ \omega(z)$ of degree at most n in $\omega(z)$, defined by the equations $p_n \circ \omega(a_j) = b_j$, $j = 0, 1, \cdots, n$, by means of the following algorithm:

$$\begin{aligned} p_{jn} &= b_j, \quad j = 0, 1, \cdots, n, \\ p_{jk} &= \frac{\begin{vmatrix} p_{k+1,k+1} & \omega(a_{k+1}) - \omega(z) \\ p_{j,k+1} & \omega(a_j) - \omega(z) \end{vmatrix}}{\omega(a_j) - \omega(a_{k+1})}, \quad \begin{aligned} & j = 0, 1, \cdots, k; \\ & k = n-1, n-2, \cdots, 0. \end{aligned} \end{aligned}$$

(a) Show that $p_{00} = p \circ \omega(z)$.

(b) Let f be analytic in $\mathcal{D}$, let $f(a_j) = b_j$, $j = 0, 1, \cdots$, with the a_j all distinct, let $|f(z)|$ be uniformly bounded in $\mathcal{D}$, and let $|\omega(z)| = r$, with $r < 1$. Prove that if a sequence of polynomials $\{p_n \circ \omega(z)\}$, are constructed as above, with one for each integer n, then

$$f(z) - p_n \circ \omega(z) = \mathcal{O}(r^n), \quad n \to \infty.$$

[Hint: For part (a) of this problem, use the argument given in Example 2.3.5. For part (b), use the inequalities of Taylor's Theorem.]

2.4.5 Let F_n be defined as in (2.4.5), and set

$$\psi(w) = \phi^{-1}(w) = c\left(w + \sum_{n=0}^{\infty} c_n w^{-n}\right).$$

Prove that

$$F_1(z) = \frac{z}{c} - c_0$$

$$F_{n+1}(z) = F_1(z)\,F_n(z) - \sum_{k=1}^{n-1} c_k F_{n-k}(z) - (n+1)\,c_n.$$

Historical Remarks on Chapter 2

The majority of the results of this chapter are well-known. The earliest complex variable-type bounds on the error of trigonometric polynomial approximation are due to *Davis* [Da1,Da2]. Other references to complex variable bounds on the error of trigonometric polynomial approximation, the trapezoidal rule (2.2.4) and the approximation of the Hilbert transform $\mathcal{S}f$ by $\mathcal{S}_N f$ as defined in (2.2.24) are *Kress* [Ks2–Ks6], *Chawla and Kaul* [ChKu1], *Forst* [Fo1], *Knauff* [Kn1,Kn2], *Knauff and Kress* [KnKs1], and *Schönhage* [Sh1].

The identities (2.2.24) are due to *Gabdulhaev* [Gb1,Gb2].

The procedure of relating trigonometric and ordinary polynomial approximation that we have followed in this chapter was known to *Bernstein*—see *Achieser* [Ac1]. In [GaV1], *Gautschi and Varga* obtain a sharp complex variable bound on the error of Gaussian quadrature. A historical summary of other such bounds may be found in *Gautschi* [Ga2]. See also *Davis and Rabinowitz* [DaRz1], and *Engels* [En1], for excellent summaries of complex variable quadrature bounds. An excellent historical account of polynomial approximation in the complex domain is due to *Saff* [Sa1]. See also [SaSn1] for a discussion of complex variable bounds for quadrature.

The results of Problems 2.3.10 and 2.4.4 are new.

3

Sinc Approximation in Strip

In this chapter we develop methods to approximate analytic functions via Sinc series. Initially we obtain bounds on the error of approximation of functions that are analytic in the strip $\mathcal{D}_d$ of Equation (1.7.9), via the Sinc series derived in Section 1.9. The region $\mathcal{D}_d$ is a natural choice, since the modulus of the function $\sin(\pi z/h)$ which we employ to get the error of Sinc approximation is both large and nearly constant on the boundary of $\mathcal{D}_d$. We also make an assumption on the growth of f in $\mathcal{D}_d$, which has far-reaching consequences, and which enables us to replace the infinite Sinc series by a finite one.

The approximations are then extended to other intervals and in the following chapter, to contours. Our notation is such that the approximations derived earlier for $\mathbb{R}$ and $\mathcal{D}_d$ extend directly, with minimal additional analysis. Some new approximations which are peculiar to the new special situations are also derived. Examples are given, illustrating the approximations in important special cases.

3.1 Sinc Approximation in $\mathcal{D}_d$

Let d be a positive number d, and let the domain $\mathcal{D}_d$ be defined as in Equation (1.7.9).

Definition 3.1.1 *Let $1 \leqq p \leqq \infty$, and let $\mathbf{H}^p(\mathcal{D}_d)$ denote the family of all functions analytic in $\mathcal{D}_d$, such that if $\mathcal{D}_d(\varepsilon)$ is defined for $0 < \varepsilon < 1$ by*

$$\mathcal{D}_d(\varepsilon) = \{z \in \mathbb{C} : |\Re z| < 1/\varepsilon, \quad |\Im z| < d(1-\varepsilon)\} \tag{3.1.1}$$

then $N_p(f, \mathcal{D}_d) < \infty$, with

$$N_p(f, \mathcal{D}_d) = \begin{cases} \lim\limits_{\varepsilon \to 0} \left(\displaystyle\int_{\partial \mathcal{D}_d(\varepsilon)} |f(z)|^p |dz| \right)^{1/p} & \text{if } 1 \leqq p < \infty \\ \lim\limits_{\varepsilon \to 0} \sup\limits_{z \in \mathcal{D}(\varepsilon)} |f(z)| \ \text{ if } p = \infty. & \end{cases} \tag{3.1.2}$$

Let $S(k,h)(x)$ be defined as in (1.10.2), let $f \in \mathbf{H}^p(\mathcal{D}_d)$, $h > 0$, let N be a positive integer, and let us set

$$\begin{cases} C(f,h) = \sum_{k=-\infty}^{\infty} f(kh)S(k,h) \\ C_N(f,h) = \sum_{k=-N}^{N} f(kh)S(k,h) \\ E(f,h) = f - C(f,h) \\ E_N(f,h) = f - C_N(f,h). \end{cases} \tag{3.1.3}$$

It was shown in Section 1.10, that if $f \in \mathbf{W}(\pi/h)$ (see Def. 1.9.2), then $C(f,h) = f$. While this is no longer true for $f \in \mathbf{H}^p(\mathcal{D}_d)$, the function $C(f,h)$ nevertheless provides a very accurate approximation of $f \in \mathbf{H}^p(\mathcal{D}_d)$.

Theorem 3.1.2 *Let $f \in \mathbf{H}^p(\mathcal{D}_d)$, with $1 \leqq p \leqq \infty$. Then for z in $\mathcal{D}_d$,*

$$\begin{aligned} E(f,h)(z) &= f(z) - C(f,h)(z) \\ &= \frac{\sin(\pi z/h)}{2\pi i} \int_{\mathbb{R}} \left\{ \frac{f(t - id^-)}{(t - z - id)\sin[\pi(t - id)/h]} \right. \\ &\qquad \left. - \frac{f(t + id^-)}{(t - z + id)\sin[\pi(t + id)/h]} \right\} dt. \end{aligned} \tag{3.1.4}$$

Proof. Let $0 < \delta < d$, let n denote a positive integer, let $\mathcal{D}(n,\delta)$ denote the region

$$\mathcal{D}(n,\delta) = \left\{ z \in \mathbb{C} : |\Re\, z| < (n + \frac{1}{2})h, \quad |\Im\, z| < \delta \right\}, \tag{3.1.5}$$

and consider the integral

$$E(n,\delta,f)(z) = \frac{\sin(\pi z/h)}{2\pi i} \int_{\partial\mathcal{D}(n,\delta)} \frac{f(\zeta)d\zeta}{(\zeta - z)\sin[\pi\zeta/h]}. \tag{3.1.6}$$

Let $z = x+iy$ be fixed in $\mathcal{D}_d$, where x and y are real, and set $\zeta = \xi+i\eta$, with ξ and η real. It thus follows that if n is sufficiently large and δ is sufficiently close to d, then $z \in \mathcal{D}(n,\delta)$. In this case $|z - \zeta| \geqq \min\{(n + \frac{1}{2})h - |x|, \delta - |y|\} > 0$, which implies, by our assumptions on f, that the integral in (3.1.6) is bounded. Along the vertical segments of $\partial\mathcal{D}(n,\delta)$, $z = (n+\frac{1}{2})h+iy$, and $|\sin(\pi z/h)| = |(-1)^n \cosh(\pi y/h)| \geqq 1$; thus the contribution to $E(n,\delta,f)$ of the integral along these segments is bounded by

$$\begin{aligned}
&\frac{|\sin(\pi z/h)|}{2\pi}\int_{-\delta}^{\delta}\left\{\frac{|f((n+\frac{1}{2})h+iy)|}{|(n+\frac{1}{2})h-x|}+\frac{|f((-n-\frac{1}{2})h+iy)|}{|(-n-\frac{1}{2})h-x|}\right\}dy \\
&\leq \frac{|\sin(\pi z/h)|}{2\pi}(2\delta)^{1/q}N_p(f,\mathcal{D}_d)\left\{\frac{1}{|(n+\frac{1}{2})h-x|}+\frac{1}{|(n+\frac{1}{2})h+x|}\right\} \\
&\to 0 \quad \text{as} \quad n\to\infty.
\end{aligned} \tag{3.1.7}$$

In (3.1.7), q is defined by the expression $1/p+1/q=1$. Along the horizontal segments, we have $|z-\zeta| = [(x-\xi)^2+(|y|-\delta)^2]^{1/2}$. The upper and lower estimates which follow for the function $\sin[\pi z/h]$, for z on the horizontal portions of $\partial\mathcal{D}(n,\delta)$ will often be used throughout the remainder of this text:

$$\begin{aligned}
\sinh(\pi\delta/h) &\leq [\cosh^2(\pi\delta/h)-\cos^2(\pi t/h)]^{1/2} \\
&= |\sin[\pi(t\pm i\delta)/h]| \\
&\leq \cosh(\pi\delta/h).
\end{aligned} \tag{3.1.8}$$

These inequalities enable us to conclude that the integral (3.1.6) is bounded by

$$\frac{|\sin(\pi z/h)|}{2\pi\sinh(\pi\delta/h)}N_p(f,\mathcal{D}_d)\left(\int_{\mathbb{R}}\{(x-\xi)^2+(|y|-\delta)^2\}^{-q/2}d\xi\right)^{1/q} < \infty. \tag{3.1.9}$$

Hence, we may let $n\to\infty$ and $\delta\to d^-$ in (3.1.6) to conclude that this limit exists, and is equal to the extreme right-hand side of (3.1.4).
On the other hand, it is readily seen (Problem 3.1.1) that

$$\begin{aligned}
&E(n,\delta,f)(z) \\
&= f(z)-\sin(\pi z/h)\sum_{k=-n}^{n}\frac{(-1)^k f(kh)}{\pi(z-kh)/h}.
\end{aligned} \tag{3.1.10}$$

We may also let $n\to\infty$ and $\delta\to d^-$ in this expression, since we have already shown that this limit exists.
This completes the proof of Theorem 3.1.2. ∎

Let $z=x+iy$, with $x\in\mathbb{R}$, $y\in\mathbb{R}$, and $|y|<d$, and let $E(f,h)$ be defined as in (3.1.3). For given $\sigma\in(1,\infty]$, we use the notation

$$\|E(f,h)(\cdot+iy)\|_\sigma = \left(\int_{\mathbb{R}}|E(f,h)(x+iy)|^\sigma dx\right)^{1/\sigma}. \tag{3.1.11}$$

Theorem 3.1.3 *Let $f \in \mathbf{H}^p(\mathcal{D}_d)$, and let $z = x + iy$, with $x \in \mathbb{R}$, $y \in \mathbb{R}$, and $|y| < d$.*
(a) *If $f \in \mathbf{H}^1(\mathcal{D}_d)$, then*

$$\left.\begin{array}{r} 2\pi(d-|y|)\|E(f,h)(\cdot+iy)\|_\infty, \\ 2\pi^{1/2}(d-|y|)^{1/2}\|E(f,h)(\cdot+iy)\|_2 \end{array}\right\} \leqq \frac{\cosh(\pi y/h)}{\sinh(\pi d/h)} N_1(f,\mathcal{D}_d). \quad (3.1.12)$$

(b) *If $f \in \mathbf{H}^2(\mathcal{D}_d)$, then*

$$\begin{aligned} &2\pi^{1/2}(d-|y|)^{1/2}\|E(f,h)(\cdot+iy)\|_\infty, \quad \|E(f,h)(\cdot+iy)\|_2 \\ &\qquad \leqq \frac{\cosh(\pi y/h)}{\sinh(\pi d/h)} N_2(f,\mathcal{D}_d). \end{aligned} \quad (3.1.13)$$

Proof. We give a separate proof for each part of the inequalities (3.1.12) and (3.1.13).

(3.1.12)-(i): Using (3.1.8) in (3.1.4), we get

$$\begin{aligned} |E(f,h)(x+iy)| &\leqq \frac{\cosh(\pi y/h)}{2\pi\sinh(\pi d/h)} \cdot \\ &\cdot \left\{ \frac{1}{d+y}\int_{\mathbb{R}} |f(t-id^-)|dt + \frac{1}{d-y}\int_{\mathbb{R}} |f(t+id^-)|dt \right\}, \end{aligned} \quad (3.1.14)$$

which yields (3.1.12)-(i).

(3.1.12)-(ii): In this case we apply Theorem 1.4.5, to get

$$\begin{aligned} &\left(\int_{\mathbb{R}} |E(f,h)(x+iy)|^2 dx\right)^{1/2} \\ &\qquad \leqq \frac{\cosh(\pi y/h)}{2\pi\sinh(\pi d/h)} \left\{ \left(\frac{\pi}{d+y}\right)^{1/2} \int_{\mathbb{R}} |f(t-id^-)|\, dt \right. \quad (3.1.15) \\ &\qquad \left. + \left(\frac{\pi}{d-y}\right)^{1/2} \int_{\mathbb{R}} |f(t+id^-)|\, dt \right\}. \end{aligned}$$

The inequality (3.1.12)-(ii) now follows.

(3.1.13)-(i): Both $f(\cdot \pm id^-)$, and $f(\cdot \pm id^-)/\sin[\pi(\cdot \pm id)/h]$ are functions in $\mathbf{L}^2(\mathbb{R})$, and, therefore, by applying Schwarz's inequality to the terms in (3.1.4), we get

$$|2\pi i\, E(f,h)(z)/\sin(\pi z/h)|$$

$$\leqq \left(\frac{\pi}{d+y}\right)^{1/2} \left(\int_{\mathbb{R}} \left|\frac{f(t-id^-)}{\sin[\pi(t-id)/h]}\right|^2 dt\right)^{1/2} \tag{3.1.16}$$

$$+ \left(\frac{\pi}{d-y}\right)^{1/2} \left(\int_{\mathbb{R}} \left|\frac{f(t+id^-)}{\sin[\pi(t+id)/h]}\right|^2 dt\right)^{1/2},$$

and after applying (3.1.8) to this last inequality we get (3.1.13)-(i).

(3.1.13)-(ii): Since $f(\cdot \pm id) \in \mathbf{L}^2(\mathbb{R})$, it follows that $F_\pm(\cdot) \equiv f(\cdot \pm id)/\sin[\pi(\cdot \pm id)/h] \in \mathbf{L}^2(\mathbb{R})$. Hence

$$F_\pm(t) = \int_{\mathbb{R}} \varphi_\pm(\lambda) e^{i\lambda t}\, d\lambda, \tag{3.1.17}$$

where $\varphi_\pm \in \mathbf{L}^2(\mathbb{R})$ and where, from (3.1.8) and Parseval's theorem,

$$\frac{\|f(\cdot \pm id)\|_2^2}{\sinh^2(\pi d/h)} \geqq \|F_\pm\|_2^2 = 2\pi \|\varphi_\pm\|_2^2. \tag{3.1.18}$$

In view of (1.4.17) and (1.4.18), it thus follows from (3.1.4), that

$$E(f,h)(x+iy)/\sin[\pi(x+iy)/h]$$

$$= \int_0^\infty \varphi_-(\lambda) e^{i\lambda(z+id)}\, d\lambda + \int_{-\infty}^0 \varphi_+(\lambda) e^{i\lambda(z-id)}\, d\lambda \tag{3.1.19}$$

$$\equiv \int_{\mathbb{R}} \Phi(\lambda) e^{i\lambda x}\, d\lambda,$$

where

$$\Phi(\lambda) = \begin{cases} \varphi_-(\lambda) e^{-\lambda(d+y)} & \text{if} \lambda > 0 \\ \varphi_+(\lambda) e^{\lambda(d-y)} & \text{if} \lambda < 0. \end{cases} \tag{3.1.20}$$

Hence, by Parseval's theorem, and (3.1.18),

$$\begin{aligned} \|E(f,h)(\cdot + iy)\|_2^2 &\leqq 2\pi \cosh^2(\pi y/h) \|\Phi\|_2^2 \\ &\leqq 2\pi \cosh^2(\pi y/h) [\int_0^\infty |\varphi_+(\lambda)|^2 d\lambda + \int_{-\infty}^0 |\varphi_-(\lambda)|^2 d\lambda] \\ &= \cosh^2(\pi y/h) \int_{\mathbb{R}} \left\{ |F_+(t)|^2 + |F_-(t)|^2 \right\} dt \\ &\leqq \frac{\cosh^2(\pi y/h)}{\sinh^2(\pi d/h)} (N_2(f, \mathcal{D}_d))^2, \end{aligned} \tag{3.1.21}$$

which is just (3.1.13)-(ii). ∎

Remark 3.1.4 We emphasize that Theorem 3.1.3 tells us that if $f \in \mathbf{H}^p(\mathcal{D}_d)$, $p = 1$ *or* 2, and if $-d < y < d$, then both $\|E(f,h)(\cdot + iy)\|_\infty$ and $\|E(f,h)(\cdot + iy)\|_2$ are $\mathcal{O}(e^{-\pi(d-|y|)/h})$ as $h \to 0^-$. That is, $C(f,h)(z) \to f(z)$ as $h \to 0$, for all $z \in \mathcal{D}_d$.

It is important for the purpose of practical application to identify those functions that are analytic in $\mathcal{D}_d$, for which we can achieve accurate approximation of f via $C_N(f,h)$ for relatively small N.

Definition 3.1.5 *Let α and β denote positive numbers, and let $\mathbf{L}_{\alpha,\beta}(\mathcal{D}_d)$ denote the family of functions f analytic in $\mathcal{D}_d$, and such that for some constant $c > 0$, and all $z \in \mathcal{D}_d$, we have*

$$|f(z)| \leqq c\frac{|e^{\alpha z}|}{(1+|e^z|)^{\alpha+\beta}}. \tag{3.1.22}$$

Let $0 < \alpha \leqq 1$, $0 < \beta \leqq 1$, let $0 < d < \pi$, and let $\mathbf{M}_{\alpha,\beta}(\mathcal{D}_d)$ denote the family of functions $g \in \mathbf{H}^\infty(\mathcal{D}_d)$ such that $f \in \mathbf{L}_{\alpha,\beta}(\mathcal{D}_d)$, where $f(x) = g(x) - [g(-\infty) + e^x g(\infty)]/[1+e^x]$. We simply write $\mathbf{L}_\alpha(\mathcal{D}_d)$ for $\mathbf{L}_{\alpha,\alpha}(\mathcal{D}_d)$, and $\mathbf{M}_\alpha(\mathcal{D}_d)$ for $\mathbf{M}_{\alpha,\alpha}(\mathcal{D}_d)$.

For the sake of simplicity of presentation of the material, we shall nearly always present our approximating formulas for the spaces $\mathbf{L}_\alpha(\mathcal{D}_d)$ and $\mathbf{M}_\alpha(\mathcal{D}_d)$ instead of for the slightly more complicated spaces $\mathbf{L}_{\alpha,\beta}(\mathcal{D}_d)$ and $\mathbf{M}_{\alpha,\beta}(\mathcal{D}_d)$. Several results of approximation in $\mathbf{L}_{\alpha,\beta}(\mathcal{D}_d)$ and in $\mathbf{M}_{\alpha,\beta}(\mathcal{D}_d)$ will, however, also be considered in this, as well as in the next chapter, in order to illustrate the similar treatment of the approximations and of bounding the error of approximation in these spaces.

Let us note, by way of simplification of (3.1.22), that for all $x \in \mathbb{R}$, we have

$$\frac{1}{2^{2\alpha}}e^{-\alpha|x|} \leqq \frac{e^{\alpha x}}{(1+e^x)^{2\alpha}} \leqq e^{-\alpha|x|}. \tag{3.1.23}$$

Lemma 3.1.6 *If $f \in \mathbf{L}_\alpha(\mathcal{D}_d)$, then $f \in \mathbf{H}^p(\mathcal{D}_d)$, $1 \leqq p \leqq \infty$, and moreover,*

$$N_p(f,\mathcal{D}_d) = \left(\int_{\partial\mathcal{D}_d} |f(z)|^p\,|dz|\right)^{1/p} \leqq \begin{cases} c\left(\frac{4}{\alpha p}\right)^{1/p}, & \text{if } 1 \leqq p < \infty, \\ c, & \text{if } p = \infty. \end{cases} \tag{3.1.24}$$

Proof. Clearly, $f \in \mathbf{Hol}(\mathcal{D}_d)$. Using (3.1.22), the definition of $N_p(f,\mathcal{D}_d)$, and (3.1.23), we have for the case of $1 \leqq p < \infty$, that

$$\begin{aligned} N_p(f, \mathcal{D}_d) &= \left(\int_{\partial \mathcal{D}_d} |f(z)|^p |dz| \right)^{1/p} \\ &\leqq c \left(\int_{\partial \mathcal{D}_d} \left(\frac{e^{\alpha x}}{(1+e^x)^{2\alpha}} \right)^p dx \right)^{1/p} \\ &\leqq c \left(\frac{4}{\alpha p} \right)^{1/p} . \end{aligned} \tag{3.1.25}$$

The case $p = \infty$ is similar, using the right-hand side of (3.1.23). This completes the proof of Lemma 3.1.6. ■

Theorem 3.1.7 *Let either:*

(a) $f \in \mathbf{L}_\alpha(\mathcal{D}_d)$; *or else*

(b) f *satisfies one of the conditions of Theorem 3.1.3, and furthermore, for all* $x \in \mathbb{R}$, *let*

$$|f(x)| < ce^{-\alpha|x|} \tag{3.1.26}$$

with c *a positive number. Then, taking*

$$h = \left(\frac{\pi d}{\alpha N} \right)^{1/2} \tag{3.1.27}$$

there exists a positive number c_1*, depending only on* f*,* d*,* α*, and* y*, such that for* $s = 2$ *or* $s = \infty$*,*

$$\begin{aligned} &\|E_N(f,h)(\cdot + iy)\|_s \\ &\leqq c_1 N^{(1-1/s)/2} \exp\left\{ -\left(\frac{\pi\alpha}{d} \right)^{1/2} (d - |y|) N^{1/2} \right\}. \end{aligned} \tag{3.1.28}$$

Proof. Part (a): If $f \in \mathbf{L}_\alpha(\mathcal{D}_d)$, then by Lemma 3.1.6, $f \in \mathbf{H}^p(\mathcal{D}_d)$ for $p = 1$ and $p = 2$, and so the conditions of Theorem 3.1.3 are satisfied. From this point on, the proofs of parts (a) and (b) of Theorem 3.1.7 are the same.

Parts (a) and (b): The definition (3.1.3) of $E_N(f,h)$ implies that

$$E_N(f,h) = E(f,h) - \sum_{|k|>N} f(kh)S(k,h). \tag{3.1.29}$$

Hence, if $s \geqq 2$, then

$$\begin{aligned} &\|E_N(f,h)(\cdot + iy)\|_s \\ &\leqq \|E(f,h)(\cdot + iy)\|_s + \left\| \sum_{|k|>N} f(kh)S(k,h)(\cdot + iy) \right\|_s . \end{aligned} \tag{3.1.30}$$

Bounds of the first term on the right-hand side of (3.1.30) have already been obtained in Theorem 3.1.3. Let us then get bounds for the second term on the right-hand side of (3.1.30).

Now by (1.10.10),

$$S(m,h)(x+iy) = \frac{1}{2\pi} \int_{-\pi/h}^{\pi/h} h e^{yt + i(mh-x)t} dt \tag{3.1.31}$$

and hence

$$\begin{aligned} \sup_{x \in \mathbf{R}} |S(m,h)(x+iy)| &\leqq \frac{h}{2\pi} \int_{-\pi/h}^{\pi/h} e^{yt} dt \\ &= \frac{\sinh(\pi y/h)}{\pi y/h} \leqq \cosh(\pi y/h), \end{aligned} \tag{3.1.32}$$

with the last inequality obtained by comparing the power series expansions about $x = 0$ of $x^{-1}\sinh(x)$ and $\cosh x$.

Hence, using (3.1.32) and either (3.1.22) or (3.1.26), we get

$$\begin{aligned} &\| \sum_{|k|>N} f(kh)\, S(k,h)(\cdot + iy)\|_\infty \\ &\leqq \;\; 2c\cosh(\pi y/h) \sum_{k=N+1}^{\infty} e^{-\alpha kh} \\ &\leqq \;\; 2c\cosh(\pi y/h)\, e^{-\alpha(N+1)h}/(1 - e^{-\alpha h}) \\ &\leqq \;\; 2c\cosh(\pi y/h) e^{-\alpha N h}/\alpha h, \end{aligned} \tag{3.1.33}$$

with c the constant in (3.1.22) or (3.1.26).

The inequality (3.1.28) now follows in the case of $s = \infty$, taking h according to (3.1.27).
If $s = 2$, we use the inequality

$$\begin{aligned} &\| \sum_{|k|>N} f(kh)S(k,h)(\cdot + iy)\|_2^2 \\ &= \sum_{|m|>N, |n|>N} f(mh)\overline{f(nh)} I_{m,n} \end{aligned} \tag{3.1.34}$$

where, by Parseval's theorem and (3.1.31),

$$\begin{aligned} I_{mn} &= \int_{\mathbb{R}} S(m,h)(x+iy)\overline{S(n,h)(x+iy)}dx \\ &= \frac{1}{2\pi}\int_{-\pi/h}^{\pi/h} (he^{yt+imht})(he^{yt-inht})dt \\ &= \frac{h^2(-1)^{m-n}}{\pi[2y+i(m-n)h]}\sinh(\frac{2\pi y}{h}). \end{aligned} \tag{3.1.35}$$

Now estimating the right-hand side of (3.1.35), we get

$$\begin{aligned} |I_{mn}| &\leqq h^2 \frac{\sinh(2\pi y/h)}{2\pi y} \\ &\leqq h\cosh^2(\pi y/h), \end{aligned} \tag{3.1.36}$$

with the last inequality obtained by comparing the power series expansions about $x = 0$ of $\sinh(2x)/(2x)$ and $\cosh^2(x)$.
Substituting (3.1.36) into (3.1.34) and using either (3.1.22) or (3.1.26), we get

$$\begin{aligned} &\| \sum_{|k|>N} f(kh)S(k,h)(\cdot + iy)\|_2^2 \\ &\leqq h\cosh^2(\pi y/h) \sum_{|m|>N, |n|>N} |f(mh)| \; |f(nh)| \\ &\leqq (C')^2 e^{-2\alpha N h} \cosh^2(\pi y/h)/(\alpha^2 h), \end{aligned} \tag{3.1.37}$$

with C' a constant.
Combining (3.1.37) with the bounds in Theorem 3.1.3, and using (3.1.27), we get (3.1.28) in the case $s = 2$.
This completes the proof of Theorem 3.1.7. ∎

The determination of h by (3.1.27) has the desirable feature that in the case of uniform Sinc approximation, the error $E_N(f,h)(x)$ defined as in (3.1.3) has roughly the same magnitude near the center and far from the center of the interval of approximation. In addition the extreme points of the error seem to occur close to the Sinc points $\{kh\}$. Choosing an h which is larger than the optimal value predicted by (3.1.27) results in a less accurate approximation near the origin and a more accurate one farther away from the origin, while choosing an h which is smaller than the optimal value predicted by (3.1.27) results in a less accurate approximation far from the origin, and a more accurate approximation near the origin.

Let us conclude with some remarks about the interpolation of f at $x = \pm\infty$. To this end, let us assume that f is analytic in $\mathcal{D}_d$, and that the limits

$$f(\pm\infty) = \lim_{x \to \pm\infty} f(x) \tag{3.1.38}$$

exist and are finite, but that at least one of the values $f(\pm\infty)$ is non-zero. Although f cannot be accurately interpolated via Sinc functions alone, it may nevertheless be possible to accurately interpolate g, with

$$g(x) = f(x) - \frac{f(-\infty)e^{-\beta x} + f(\infty)e^{\beta x}}{e^{-\beta x} + e^{\beta x}} \tag{3.1.39}$$

The function g in (3.1.39) may satisfy an inequality of the type

$$|g(x)| \leqq Ce^{-\alpha|x|}, \quad x \in \mathbb{R}, \tag{3.1.40}$$

with C and α independent of β, for all β sufficiently large. In this case, the error bound for $2N+1$-point Sinc interpolation of g on $\mathbb{R}$ is minimized as a function of β, by taking $\beta = \alpha/2$.

Let us now also state a result on the construction of f given its values on $\mathbb{R}$, which, along with its extension in the next section, will be useful for purposes of extrapolating the limit, in Chapter 5.

Theorem 3.1.8 *Given a sequence of numbers $\{c_k\}_{k\in\mathbb{Z}}$, such that for all $k \in \mathbb{Z}$, we have $|c_k| \leqq C \exp(-\alpha|k|h)$, with C and α positive numbers, there exists an entire function f of order 1 and type π/h, such that $f(kh) = c_k$ for all $k \in \mathbb{Z}$. Moreover, for all $z \in \mathbb{C}$, we have*

$$|f(z)| \leqq \frac{C \cosh(\pi|\Im z|/h)}{\tanh(\alpha h/2)}, \tag{3.1.41}$$

f belongs to $\mathbf{L}_\alpha(\mathcal{D}_d)$, and for any positive integer N, we have

$$\left| f(z) - \sum_{k=-N}^{N} c_k S(k,h)(z) \right| \leqq \frac{2C}{\alpha h} \cosh(\pi|\Im z|/h)\, e^{-\alpha N h}. \tag{3.1.42}$$

Proof. The proof is straightforward, taking $f = \sum_{k\in\mathbb{Z}} c_k S(k,h)$, then using the inequality (3.1.32) and the method of proof of Theorem 3.1.7. ∎

It is also possible, and indeed, at times convenient for interpolation, quadrature, and Hilbert transforms to be able to include simple poles within $\mathcal{D}_d$. To this end, we have

Theorem 3.1.9 (Bialecki): *Let $f \in \mathbf{H}^1(\mathcal{D}_d)$, except for simple poles $b_k \in \mathcal{D}_d \setminus \mathbb{R}$, and such that $g \in \mathbf{Hol}(\mathcal{D}_d)$, with*

$$g(z) = f(z) - \sum_{k=1}^{M} \sum_{j=1}^{m_k} B_{jk}(z - b_k)^{-j}. \tag{3.1.43}$$

Set

$$\gamma_j(h; b) = \frac{1}{j!} \left(\frac{d}{dz} \right)^j \frac{1}{\sin(\pi z/h)} \Bigg|_{z=b}. \tag{3.1.44}$$

Then for any $x \in \mathbb{R}$,

$$\begin{aligned} f(x) - \sum_{k=-\infty}^{\infty} f(kh)\, S(k,h)(x) - \sum_{k=1}^{M} \sum_{j=1}^{m_k} B_{kj} \sum_{\ell=1}^{j} \frac{\sin(\pi x/h)}{(x - b_k)^\ell} \gamma_{j-\ell}(h; b_k) \\ = \frac{\sin(\pi z/h)}{2\pi i} \int_{\mathbb{R}} \left\{ \frac{f(t - id^-)}{(t - z - id) \sin[\pi(t - id)/h]} \right. \\ \left. - \frac{f(t + id^-)}{(t - z + id) \sin[\pi(t + id)/h]} \right\} dt. \end{aligned} \tag{3.1.45}$$

Proof. The proof is left to Problem 3.1.9. ∎

•

Problems for Section 3.1

3.1.1 Let $E(n, \delta, f)$ be defined by Equation (3.1.6). Prove that if $\zeta \in \mathcal{D}_d$, then

$$E(n, \delta, f)(\zeta)$$
$$= f(\zeta) - \sin(\pi\zeta/h) \sum_{k=-n}^{n} \frac{(-1)^k f(kh)}{\pi(\zeta - kh)/h}.$$

3.1.2 Give a detailed proof of part (a) of Theorem 3.1.7.

3.1.3 Let f be analytic in $\mathcal{D}_d$, and assume that the limits (3.1.38) exist and are non-zero. Furthermore, assume that (3.1.40) holds for all β sufficiently large, where C and α are independent of β, and where β is defined as in (3.1.39). Prove that the error bound for $2N+1$-point Sinc interpolation of g on $\mathbb{R}$ is minimized by taking $\beta = \alpha/2$.

3.1.4 Find the coefficient of x^{2n} in $(2x)^{-1}\sinh(2x)$, $\cosh(2x)$, and $\cosh^2(x)$ and prove that if $x \in \mathbb{R}$, then $(2x)^{-1}\sinh(2x) \leqq \cosh^2(x) \leqq \cosh(2x)$.

3.1.5 (a) Prove that

$$(*)\|C_N\|_\infty \equiv \sup_{x\in\mathbf{R}} \sum_{k=-N}^{N} |S(k,h)(x)| \leqq \frac{2}{\pi}[3/2+\gamma+\log(N+1)],$$

$$\leqq \frac{2}{\pi}[3+\log(N)],$$

with γ Euler's constant,

$$\gamma = \lim_{N\to\infty} 1+\frac{1}{2}+\frac{1}{3}+\cdots+\frac{1}{N-1}-\log(N) = .577\cdots.$$

[Hint: Prove first that the sum $(*)$ is even in x; next, in the case of $0 \leqq x \leqq h$, prove that

$$\max_{0\leqq x\leqq h} \frac{\sin(\pi x/h)}{\pi}\left\{\frac{1}{x/h}+\frac{1}{1-x/h}\right\} = \frac{4}{\pi}.$$

On the interval $0 \leqq x \leqq h$, prove that for the remaining sum,

$$b_N(x) \equiv$$

$$\frac{\sin(\pi x/h)}{\pi}\left\{\left[\frac{1}{1+x/h}+\frac{1}{2-x/h}\right]\right.$$
$$+\left[\frac{1}{2+x/h}+\frac{1}{3-x/h}\right]+\cdots$$
$$\left.+\left[\frac{1}{N-1+x/h}+\frac{1}{N-x/h}\right]+\frac{1}{N+x/h}\right\},$$

each of the terms in square brackets has a maximum value at $x=0$ and at $x=h$, and hence

$$|b_N(x)| \leqq b_N \equiv \frac{2}{\pi}\left\{-\frac{1}{2}+\sum_{k=1}^{N}\frac{1}{k}\right\}$$
$$\leqq \frac{2}{\pi}\{\gamma+\log(N+1)-\frac{1}{2}\}.$$

Finally, show that if $k>0$, and if $kh \leqq x \leqq (k+1)h$, then the sum in $(*)$ is less than it is for $0 \leqq x \leqq h$.]

(b) Prove that if $\|C_N\|_\infty$ is defined as in part (a) of this problem, then

$$\|C_N\|_\infty = \frac{2}{\pi}\log(N)[1+o(1)]$$

as $N \to \infty$. [Hint: Take $x = h/2$ in the sum in $(*)$, then carefully estimate the result using the relation $\sum_{k=0}^{N}\{1/(2k+1)\} = \gamma + (1/2)\{1+\log(4N)+o(1)\}$ as $N \to \infty$].

3.1.6 Explain that, whereas the bound $(*)$ given in Problem 3.1.5 may be used in the form $\|C_N f\|_\infty \leqq \|C_N\|_\infty \|f\|_\infty$ for f an arbitrary bounded function defined on $\mathbb{R}$, if $f \in \mathbf{H}^p(\mathcal{D}_d)$, $d > 0$, and $1 \leqq p \leqq \infty$, we do in fact have $\|C_N f\|_\infty \leqq [1+\varepsilon(h,N)]\|f\|_\infty$, with $\varepsilon(h,N)$ uniformly bounded for all positive integers N, i.e., for $f \in \mathbf{H}^p(\mathcal{D}_d)$ we have $\|C_N\|_\infty \leqq [1+\varepsilon_N]$. [Hint: Use the result of Theorem 3.1.3 in the inequality $\|C_N f\|_\infty \leqq \|f\|_\infty + \|f - C_N f\|_\infty$.]

3.1.7 Consider the inequality (3.1.30) in the form

$$\|E_N(f,h)\|_s \leqq \|E(f,h)\|_s + \|\sum_{|k|>N} f(kh)S(k,h)\|_s,$$

with the s-norm taken over $\mathbb{R}$. Prove that if $|f(kh| \leqq C/(|k|h)^{1+\alpha}$, where $\alpha > 0$, then taking $N = \exp\{\pi d/(\alpha h)\}/h^{1+1/\alpha}$ we get $\|E_N(f,h)\|_s \leqq c_1 \exp\{-\pi d/h\}$, with c_1 a constant independent of h.

3.1.8 Give a detailed proof of Theorem 3.1.8.

3.1.9 Prove Theorem 3.1.9. [Hint: Find the coefficient of $(z-b_k)^{-1}$ in the Laurent expansion of $f(z)/[(z-x)\sin(\pi z/h)]$ about $z = b_k$.]

3.1.10 Prove that if $x \in \mathbb{R}$, then

$$\sum_{k=-\infty}^{\infty} \{S(k,h)(x)\}^2 = 1.$$

[Hint: Take $t = \pi/h$ in the identity of Problem 1.6.1, then differentiate both sides with respect to z, or else use Parseval's theorem on the identity in Problem 1.10.1.]

3.2 Sinc Quadrature on $(-\infty, \infty)$

In this section we shall bound the error of the trapezoidal rule (Equation (1.10.5)) and also the error when the identity (1.10.25) is applied to a function f in $\mathbf{H}^1(\mathcal{D}_d)$.

Let us use the notations

$$\left\{\begin{array}{l} I(f) = \int_{\mathbb{R}} f(x)dx \\ T_N(f,h) = h \sum_{k=-N}^{N} f(kh) \\ T(f,h) = h \sum_{k=-\infty}^{\infty} f(kh) \\ \eta_N(f,h) = I(f) - T_N(f) \\ \eta(f,h) = I(f) - T(f) \end{array}\right. \tag{3.2.1}$$

and let us recall Definitions 3.1.1 and 3.1.5. We prove

Theorem 3.2.1 *Let $f \in \mathbf{H}^1(\mathcal{D}_d)$, and let $\eta(f,h)$ be defined as in (3.2.1). Then*

$$\eta(f,h) = \frac{i}{2} \int_{\mathbb{R}} \left\{ \frac{f(t - id^-)e^{-\pi(d+it)/h}}{\sin[\pi(t - id)/h]} - \frac{f(t + id^-)e^{-\pi(d-it)/h}}{\sin[\pi(t + id)/h]} \right\} dt. \tag{3.2.2}$$

Moreover,

$$|\eta(f,h)| \leqq \frac{e^{-\pi d/h}}{2\sinh(\pi d/h)} N_1(f, \mathcal{D}_d). \tag{3.2.3}$$

If in addition, f satisfies (3.1.26) on $\mathbb{R}$, then, taking

$$h = \left(\frac{2\pi d}{\alpha N}\right)^{1/2}, \tag{3.2.4}$$

we obtain

$$|\eta_N(f,h)| \leqq c_1 e^{-(2\pi d \alpha N)^{1/2}} \tag{3.2.5}$$

with c_1 depending only on f, d, and α.

Proof. Let $E(f,h)$ be defined as in (3.1.4). Upon replacing z by x in (3.1.4), we have

$$\eta(f,h) = \int_{\mathbb{R}} E(f,h)(x)dx. \tag{3.2.6}$$

We now interchange the order of integration in the resulting double integral, and use the identities

$$\frac{1}{2\pi i}\int_{\mathbb{R}} \frac{\sin(\pi x/h)}{t-x-id}\,dx = \frac{i}{2}e^{-\pi(d+it)/h}$$

$$\frac{1}{2\pi i}\int_{\mathbb{R}} \frac{\sin(\pi x/h)}{x-t-id}\,dx = \frac{i}{2}e^{-\pi(d-it)/h}, \tag{3.2.7}$$

to get (3.2.2). Inequality (3.2.3) now follows from (3.2.2), using (3.1.8) and the fact that $f \in \mathbf{H}^1(\mathcal{D}_d)$. Finally, if f satisfies (3.1.26), then it follows by proceeding similarly as in (3.1.33) that

$$\begin{aligned} |h \sum_{|k|>N} f(kh)| &\leqq \frac{2Che^{-\alpha(N+1)h}}{1-e^{-\alpha h}} \\ &< \frac{2C}{\alpha}e^{-\alpha N h}. \end{aligned} \tag{3.2.8}$$

Hence, since

$$|\eta_N(f,h)| \leqq |\eta(f,h)| + h \sum_{|k|>N} |f(kh)|, \tag{3.2.9}$$

the inequality (3.2.5) follows by taking h as in (3.2.4), and then using the bounds (3.2.3) and (3.2.8) in (3.2.9). ■

Corollary 3.2.2 *Let $f \in \mathbf{H}^1(\mathcal{D}_d)$. Then*

$$\begin{aligned} \eta^*(f,h) &\equiv \int_{\mathbb{R}} f(t)\,S(k,h)(t)dt - hf(kh) \\ &= \frac{ih(-1)^k}{2\pi}\int_{\mathbb{R}} \left\{ \frac{f(t-id^-)e^{-i\pi(t-id)/h}}{t-id-kh} - \frac{f(t+id)e^{i\pi(t+id)/h}}{t+id-kh} \right\} dt. \end{aligned} \tag{3.2.10}$$

Moreover,

$$|\eta^*(f,h)| \leqq \frac{h\,e^{-\pi d/h}}{2\pi d} N_1(f,\mathcal{D}_d). \tag{3.2.11}$$

Proof. This follows directly from Theorem 3.2.1, after replacing f in Theorem 3.2.1 by $f\,S(k,h)$. ■

•

PROBLEMS FOR SECTION 3.2

3.2.1 Prove the identities (3.2.7).

3.2.2 (a) Instead of starting with Equation (3.2.2), let $\mathcal{D}(n, \delta)$ be defined as in Equation (3.1.5) and for $f \in \mathbf{H}^1(\mathcal{D}_d)$, consider the integral

$$\frac{1}{2\pi i} \int_{\partial \mathcal{D}(n,\delta)} f(z) \pi \cot(\pi z / h)\, dz$$

and use it to derive the quadrature error (3.2.2).

(b) Prove that if f satisfies the conditions of Theorem 3.2.1, then

$$\int_{\mathbb{R}} f(x)\, dx - h \sum_{n=-\infty}^{\infty} f((n + 1/2)h)$$
$$= \int_{\mathbb{R}} \left(\frac{e^{-2\pi(d+it)/h} f(t + id^-)}{1 + e^{-2\pi(d+it)/h}} + \frac{e^{-2\pi(d-it)/h} f(t - id^-)}{1 + e^{-2\pi(d-it)/h}} \right) dt$$

$$\int_{\mathbb{R}} f(x)\, dx - h \sum_{n=-\infty}^{\infty} f(nh)$$
$$= - \int_{\mathbb{R}} \left(\frac{e^{-2\pi(d+it)/h} f(t + id^-)}{1 - e^{-2\pi(d+it)/h}} + \frac{e^{-2\pi(d-it)/h} f(t - id^-)}{1 - e^{-2\pi(d-it)/h}} \right) dt.$$

3.2.3 Obtain an accurate bound on the error when approximating $I(f)$ by $T_N(f, h)$, with these quantities defined as in (3.2.1), in the case

(a) $f(x) = e^{-x^2}$;
(b) $f(x) = e^{-x^2}/(1 + x^2)^{1/3}$;
(c) $f(x) = e^{-\sqrt{1+(x-1)^2}}$; and
(d) $f(x) = e^{-1/2\sqrt{(x-2)^2+1}} \mathrm{sn}(x; k)$.

3.2.4 Why would $T_N(f, h)$ not yield an accurate approximation of $\int_{\mathbb{R}} f(x)\, dx$ in the case $f(x) = [(x - 2)^2 + 4]^{-3/4}$?

3.2.5 Let $T(f, h)$ be defined as in (3.2.1), and set

$$M(f, h) = h \sum_{k=-\infty}^{\infty} f((k - 1/2)h).$$

Verify that $T(f, h/2) = (1/2)[T(f, h) + M(f, h)]$, and construct an efficient algorithm to approximate $\int_{\mathbb{R}} f(x)\, dx$ based on this identity.

3.2.6 Let the conditions of Theorem 3.1.9 be satisfied. Prove that

$$\left| \int_{\mathbb{R}} f(t)\,dt - h \sum_{n=-\infty}^{\infty} f(nh) \right.$$
$$\left. - \pi \sum_{k=1}^{M} e^{i\pi s_k b_k/h} \sum_{j=1}^{m_k} B_{kj} \sum_{\ell=1}^{j} \frac{(i\pi s_k/h)^{\ell-1}}{(\ell-1)!} \gamma_{j-\ell}(h; b_k) \right|$$
$$\leqq \frac{e^{-2\pi d/h}}{1 - e^{-2\pi d/h}} N(f, \mathcal{D}_d).$$

[Hint: Integrate both sides of (3.1.45) with respect to x over $\mathbb{R}$, interchange the order of integration in the contour integral term, and for $b_k \in \mathcal{D}_d$, $\Im b_k \neq 0$, use the identity

$$\int_{\mathbb{R}} \frac{\sin(\pi x/h)}{(x - b_k)^{\ell}} dx = \frac{(i\pi s_k/h)^{\ell-1}}{(\ell-1)!} e^{i\pi s_k b_k/h},$$

which is valid for $\ell = 1, 2, \cdots$, and with $s_k = \operatorname{sgn}\{\Im b_k\}$.]

3.2.7 Prove with reference to Corollary 3.2.2 that if $f \in \mathbf{H}^p(\mathcal{D}_d)$ for some $p \in [1, \infty]$, then there exists a constant c which is independent of h, k and p, such that

$$\left| \int_{\mathbb{R}} f(t)\, S(k,h)(t) dt - h f(kh) \right| \leqq c\, h\, e^{-\pi d/h}.$$

3.3 Discrete Fourier Transforms on $(-\infty, \infty)$

In this section we shall present two methods to approximate the Fourier transform

$$F(x) = \int_{\mathbb{R}} e^{ixt} f(t)\, dt \tag{3.3.1}$$

via a Fourier series. The first approximation of F uses values of f—it is just the sum obtained approximating of the integral (3.3.1) via the trapezoidal rule, while the second method uses the DFT procedure described in Section 2.2, and it thus requires values of F.

We shall use the notations and definitions of Sections 1.9 and 3.1. If $E_N(f,h)$ and $E(f,h)$ are defined as in (3.1.3), take $x \in \mathbb{R}$, and set

$$\left\{ \begin{array}{lcl} \delta_N(f,h)(x) & = & \displaystyle\int_{\mathbb{R}} e^{ixt} E_N(f,h)(t) dt, \\ \delta(f,h)(x) & = & \displaystyle\lim_{N \to \infty} \delta_N(f,h)(x), \\ & = & \displaystyle\int_{\mathbb{R}} e^{ixt} E(f,h)(t) dt. \end{array} \right. \tag{3.3.2}$$

Then, by Equation (1.10.9), we have

$$\delta_N(f,h)(x) = \begin{cases} \displaystyle\int_{\mathbb{R}} e^{ixt} f(t)dt - h \sum_{k=-N}^{N} f(kh) e^{ikhx}, & \text{if } |x| < \pi/h; \\ \displaystyle\int_{\mathbb{R}} e^{ixt} f(t)dt, \text{ if } |x| > \pi/h. \end{cases} \tag{3.3.3}$$

Theorem 3.3.1 *Let $f \in \mathbf{H}^p(\mathcal{D}_d)$, with p either 1 or 2. Let $x \in \mathbb{R}$, and let $\delta(f,h)$ and $\delta_N(f,h)$ be defined by Equations (3.3.2). Then*

$$\begin{aligned} &\delta(f,h)(x) \\ &\quad = e^{xd} \int_{\mathbb{R}} e^{ixt} f(t - id^-)\, dt \quad \text{if } x < -\pi/h \\ &\quad = \frac{i}{2} \int_{\mathbb{R}} \left\{ \frac{e^{(x-\pi/h)(d+it)} f(t-id^-)}{\sin[\pi(t-id)/h]} + \frac{e^{-(x+\pi/h)(d-it)} f(t+id^-)}{\sin[(\pi+id)/h]} \right\} dt \\ &\qquad \text{if } \quad -\pi/h < x < \pi/h \\ &\quad = e^{-xd} \int_{\mathbb{R}} e^{ixt} f(t+id^-)dt \quad \text{if } x > \pi/h. \end{aligned} \tag{3.3.4}$$

Moreover, if p is either 1 or 2, then

$$\|\delta(f,h)\|_2 = (2\pi)^{1/2} \|E(f,h)(\cdot + i0)\|_2 \tag{3.3.5}$$

with $\|E(f,h)(\cdot + iy)\|$ bounded as in Theorem 3.1.3. If $p = 1$, then we also have

$$\|\delta(f,h)\|_\infty \leqq \frac{N_1(f, \mathcal{D}_d)}{2\sinh(\pi d/h)} \tag{3.3.6}$$

If, in addition to the above assumption, f satisfies (3.1.26) for all $x \in \mathbb{R}$, or else if $f \in \mathbf{L}_\alpha(\mathcal{D}_d)$, then taking h as in (3.1.27), there exists a constant c_1 depending only on f, d, and α, such that

$$\|\delta_N(f,h)\|_s \leqq c_1 e^{-(\pi d \alpha N)^{1/2}} \tag{3.3.7}$$

for all $s \in [2, \infty]$.

Proof. Let $\tau \in \mathbb{R}$. Then the identities

$$\frac{1}{2\pi i}\int_{\mathbb{R}} \frac{\sin(\pi t/h)e^{ixt}}{\tau - t - id}\,dt$$

$$= \begin{cases} \sin[\pi(\tau - id)/h]e^{ix(\tau-id)} & \text{if } x < -\pi/h \\ \frac{i}{2}e^{-i(\pi/h - x)(\tau - id)} & \text{if } -\pi/h < x < \pi/h \\ 0 & \text{if } x > \pi/h \end{cases} \tag{3.3.8}$$

and

$$-\frac{1}{2\pi i}\int_{\mathbb{R}} \frac{\sin(\pi t/h)e^{ixt}}{\tau - t + id}\,dt$$

$$= \begin{cases} 0 & \text{if } x < -\pi/h \\ -\frac{i}{2}e^{i(\pi/h + x)(\tau + id)} & \text{if } -\pi/h < x < \pi/h \\ \sin[\pi(\tau + id)/h]e^{ix(\tau+id)} & \text{if } x > \pi/h \end{cases} \tag{3.3.9}$$

immediately yield (3.3.4), after substitution of (3.1.4) into (3.3.2).
We mention that if $f \in \mathbf{L}_\alpha(\mathcal{D}_d)$, then f is in both $\mathbf{H}^1(\mathcal{D}_d)$ and $\mathbf{H}^2(\mathcal{D}_d)$.
The last identity in (3.3.3) shows that the bound (3.3.5) is a consequence of Parseval's theorem.

In order to bound (3.3.6), we note that each of the factors

$$\begin{cases} e^{x(d+it)} & \text{for } x < -\pi/h \\ \dfrac{e^{(x-\pi/h)(d+it)}}{\sin[\pi(t-id)/h]} & \text{for } -\pi/h < x < \pi/h \\ \dfrac{e^{-(x+\pi/h)(d-it)}}{\sin[\pi(t+id)/h]} & \text{for } -\pi/h < x < \pi/h \\ e^{x(-d+it)} & \text{for } x > \pi/h \end{cases} \tag{3.3.10}$$

is bounded by $1/\sinh(\pi d/h)$, and also that if $f \in \mathbf{H}^1(\mathcal{D})$, then $N_1(f, \mathcal{D}_d)$ is an upper bound of each of the factors $\int_{\mathbb{R}} |f(t \pm id^-)|dt$ and $\int_{\mathbb{R}}[|f(t - id^-)| + |f(t + id^-)|]dt$.
The bound (3.3.7) follows; we note that

$$|\eta_N(f,h)(x)| \leqq |\eta(f,h)(x)| + h \sum_{|k|>N} |f(kh)| \tag{3.3.11}$$

and proceeding as in the proof of (3.2.5). ■

It is worthwhile to recall that the numbers $\{h\, f(kh)\}$ were the coefficients of the Fourier series approximation of F in (3.3.3). We next state and prove an analogous result for the approximation of F via a Fourier series whose coefficients involve values of F, and are obtained via the DFT procedure of Section 2.2. To this end, we take $h > 0$, N a positive integer, and we introduce the notations

$$\begin{aligned} x_j &= \pi j/(Nh); \\ a_{-N} = a_N &= \frac{1}{2}, \quad a_j = 1 \text{ if } |j| < N; \\ \omega_j(x) &= e^{ijhx}; \\ K_N(x) \equiv \sum_{j=-N}^{N} \omega_j(x) &= \frac{\sin\{(2N+1)hx/2\}}{\sin(hx/2)}. \end{aligned} \tag{3.3.12}$$

We also define the inner product

$$< F, G >_{2N} = \frac{1}{2N} \sum_{j=-N}^{N} a_j F(x_j) \overline{G(x_j)}. \tag{3.3.13}$$

Letting F be defined as in (3.3.1), we set

$$h c_j = < F, \omega_j >_{2N} . \tag{3.3.14}$$

It is then easily verified that

$$\mathcal{T}_{2N} F(x) \equiv \begin{cases} < F, K_N(x - \cdot) >_{2N} = h \displaystyle\sum_{j=-N}^{N} c_j \omega_j(x) \text{ if } |x| < \pi/h, \\ \\ 0 \text{ if } |x| > \pi/h. \end{cases} \tag{3.3.15}$$

Now using a notation analogous to (3.3.3), we consider the error

$$\hat{\delta}_N(F,h)(x) = \begin{cases} F(x) - h \displaystyle\sum_{k=-N}^{N} c_j e^{ijhx}, & \text{if } |x| < \pi/h; \\ F(x), & \text{if } |x| > \pi/h. \end{cases} \tag{3.3.16}$$

The following theorem is, in a certain sense, a dual to Theorem 3.3.1, with respect to Theorem 1.8.4. Note, in particular, the interchange of the roles of d and α.

Theorem 3.3.2 *Either let $F \in \mathbf{H}^2(\mathcal{D}_\alpha)$ and $|F(x)| \leqq c_3 \exp(-d|x|)$ for all $x \in \mathbb{R}$, or else let $F \in \mathbf{L}_d(\mathcal{D}_\alpha)$, with d, α and c_3 positive numbers. Let $\hat{\delta}_N(F,h)$ be defined as in (3.3.16) above. Then there exist constants c_1 and c_2, independent of h and N, such that*

$$\|\hat{\delta}_N(F,h)\|_2 \leqq c_1 e^{-\pi d/h} + c_2 e^{-\alpha N h}. \tag{3.3.17}$$

Moreover, if $h = [\pi d/(\alpha N)]^{1/2}$, then there exists a constant, c, independent of N, such that

$$\|\hat{\delta}_N(F,h)\|_2 \leqq c e^{-(\pi d \alpha N)^{1/2}}. \tag{3.3.18}$$

Proof. It follows from Theorem 1.1.8, that there exists $f \in \mathbf{H}_2(\mathcal{D}_d)$, such that F is the Fourier transform of f as defined in (3.3.1); and moreover, for all $t \in \mathbb{R}$, we have $|f(t)| \leqq c_4 \exp\{-\alpha|t|\}$, with c_4 a constant. Hence, defining $F_h(x)$ for $x \in \mathbb{R}$ by

$$F_h(x) = \begin{cases} h \sum_{j=-\infty}^{\infty} f(jh)\omega_j, & \text{if} \quad |x| < \pi/h \\ 0 & \text{if} \quad |x| > \pi/h, \end{cases} \tag{3.3.19}$$

it follows from Theorem 3.3.1, that

$$\|F - F_h\|_2 \leqq c_5 e^{-\pi d/h}, \tag{3.3.20}$$

with c_5 a constant. Next, the term $F_h - \mathcal{T}_{2N} F_h$ is bounded as in Problem 2.2.9, yielding

$$\|F_h - \mathcal{T}_{2N} F_h\|_2 \leqq c_6 e^{-\alpha N h}, \tag{3.3.21}$$

with c_6 independent of N and h. Finally, by Problem 2.2.9(iv), we have

$$\|\mathcal{T}_{2N}(F_h - F)\|_2 \leqq (1 + \varepsilon_N)\|F_h - F\|_2, \tag{3.3.22}$$

with $\varepsilon_N \to 0$ as $N \to \infty$. Taking norms of each of the terms in braces in the identity,

$$\{F - \mathcal{T}_{2N} F\} = \{F - F_h\} + \{F_h - \mathcal{T}_{2N} F_h\} + \{\mathcal{T}_{2N}(F_h - F)\}, \tag{3.3.23}$$

and then combining with the inequalities (3.3.20)–(3.3.22), we get inequality (3.3.17). ■

•

Problems for Section 3.3

3.3.1 Derive in detail the identity (3.3.4).

3.3.2 Prove (3.3.5) and (3.3.7).

3.3.3 Rephrase Theorem 3.3.1 so that it becomes an analogous theorem to the approximation of the Fourier sine transform.

3.3.4 Rephrase Theorem 3.3.1 so that it becomes an analogous theorem to the approximation of the Fourier cosine transform.

3.4 Cauchy-Like Transforms on $(-\infty, \infty)$

Given $f \in \mathbf{H}^p(\mathcal{D}_d)$, where $1 \leqq p < \infty$, let $\mathcal{S}f$ be defined as in (1.2.2), with $(a, b) = \mathbb{R}$, and define $\mathcal{P}$ by $\mathcal{P}f = \frac{1}{2}(f + \mathcal{S}f)$. Then, clearly $\mathcal{P}f$ defined on $\mathbb{R}$ may be analytically continued to $\{z \in \mathbb{C} : \Im z > -d\}$, and $\mathcal{S}f$ to $\mathcal{D}_d$.

Theorem 3.4.1 *Let $f \in \mathbf{H}^p(\mathcal{D}_d)$, with p either 1 or 2, and let $E(f,h)$ be defined as in (3.1.4). Let $z = x + iy$, with $x \in \mathbb{R}$, $y > -d$. Then*

$$[\mathcal{P}E(f,h)](z) \tag{3.4.1}$$

$$= (\mathcal{P}f)(z) - \sum_{j=-\infty}^{\infty} f(jh) \frac{e^{i\pi(z-jh)/h} - 1}{2\pi i(z - jh)/h}$$

$$= \frac{1}{4\pi} \int_{\mathbb{R}} \left\{ \frac{[e^{-i\pi(t-id)/h} - e^{i\pi z/h}] f(t - id^-)}{(t - id - z) \sin[\pi(t - id)/h]} \right. \tag{3.4.2}$$

$$\left. - \frac{[e^{i\pi(t+id)/h} - e^{i\pi z/h}] f(t + id)}{(t + id - z) \sin[\pi(t + id)/h]} \right\} dt.$$

Define δ by $\delta = \min(y, d)$.

(a) *Let $f \in \mathbf{H}^1(\mathcal{D}_d)$, and define μ and ν by*

$$\mu = \max\left\{\frac{1}{d+y}, 1\right\}, \quad \nu = \max\left\{\left(\frac{4\pi}{d+y}\right)^{1/2}, \left(\frac{2\pi^2}{h}\right)^{1/2}\right\}.$$

Then

$$\mu^{-1}\|\mathcal{P}E(f,h)(\cdot + iy)\|_\infty, \quad \nu^{-1}\|\mathcal{P}E(f,h)(\cdot + iy)\|_2 \tag{3.4.3}$$

$$\leqq \frac{e^{-\pi\delta/h} N_1(f, \mathcal{D}_d)}{4\pi \sinh(\pi d/h)}.$$

(b) *Let $f \in \mathbf{H}^2(\mathcal{D}_d)$, and define μ and ν by*

$$\mu = \left(\frac{4\pi}{y+d} + \frac{2\pi^2}{h}\right)^{1/2}, \quad \nu = 2\sqrt{2}$$

then

$$\mu^{-1}\|\mathcal{P}E(f,h)(\cdot+iy)\|_\infty, \quad \nu^{-1}\|\mathcal{P}E(f,h)(\cdot+iy)\|_2$$
$$\leqq \frac{e^{-\pi\delta/h}N_2(f,\mathcal{D}_d)}{4\pi\sinh(\pi d/h)}. \tag{3.4.4}$$

Proof. Let $\xi \in \mathbb{R}$, so that $E(f,h)(\xi)$ is given by (3.1.4), z replaced by ξ. We now interchange the order of integration and use the identity

$$\frac{1}{2\pi i}\int_{\mathbb{R}} \frac{\sin(\pi\xi/h)}{(\xi-w)(\xi-z)}d\xi =$$
$$= \frac{\exp[i(\pi/h)w\, sgn\, \Im w] - \exp[i(\pi/h)z \operatorname{sgn} \Im z]}{2i(w-z)} \tag{3.4.5}$$

to evaluate $(2\pi i)^{-1}\int_{\mathbb{R}}\{E(f,h)(\xi)/(\xi-x-iy)\}d\xi$, with $y>0$, thus getting (3.4.2). However, by analytic continuation, (3.4.2) holds for all z such that $\Im z > -d$.

The $\|\mathcal{P}E(f,h)\|_\infty$ bound in (3.4.3) is a consequence of inequalities (3.1.8), as well as

$$\begin{gathered}|e^{-i\pi(t-id)/h} - e^{i\pi z/h}| \leqq 2e^{-\pi\delta/h}\\ |t-id-z| \geqq d+y \\ \left|\frac{e^{i\pi(t+id)/h} - e^{i\pi z/h}}{t+id-z}\right| \leqq e^{-\pi\delta/h}.\end{gathered} \tag{3.4.6}$$

The $\|\mathcal{P}E(f,h)(\cdot+iy)\|_2$ bound is based on Theorem 1.4.5, as well as on the relations[1]

$$\int_{\mathbb{R}}\left|\frac{e^{i\pi(x+iy)/h} - e^{-i\pi(t-id)/h}}{x+iy-(t-id)}\right|^2 dx \tag{3.4.7}$$

$$\left(= \int_{\mathbb{R}}\left|\frac{e^{i\pi(x+iy)/h} - e^{-i\pi(t-id)/h}}{x+iy-(t-id)}\right|^2 dt\right) \tag{3.4.8}$$

$$= \frac{\pi}{y+d}[e^{-2\pi y/h} + e^{-2\pi d/h} - 2e^{-2\pi(y+d)/h)}\cos(2\pi t/h)]$$
$$\leqq \frac{4\pi}{y+d}, \tag{3.4.9}$$

[1]The value of the integral (3.4.8) is the same as that of (3.4.7), after interchanging t and x in (3.4.9).

and [2]

$$\int_{\mathbb{R}} \left| \frac{e^{i\pi(x+iy)/h} - e^{i\pi(t+id)/h}}{x+iy-t-id} \right|^2 dx \tag{3.4.10}$$

$$\left(= \int_{\mathbb{R}} \left| \frac{e^{i\pi(x+iy)/h)} - e^{i\pi(t+id)/h}}{x+iy-t-id} \right|^2 dt \right) \tag{3.4.11}$$

$$= \frac{\pi}{|d-y|}[e^{-2\pi y/h} + e^{-2\pi d/h} - 2e^{-\pi(y+d+|d-y|)}] \tag{3.4.12}$$

$$\leqq \frac{2\pi^2}{h} e^{-2\pi\delta/h}.$$

We mention in passing that we have included (3.4.8) and (3.4.11) for purposes of obtaining the $\|\mathcal{P}E(f,h)\|_\infty$ bound in (3.4.4). The $\|\mathcal{P}E(f,h)\|_2$ bound in (3.4.4) is a consequence of the first inequality in (3.4.6), the inequality $|e^{i\pi(x+iy)/h} - e^{i\pi(t+id)/h}| \leqq 2e^{-\pi\delta/h}$, and the fact that if $G \in \mathbf{L}^2(\mathbb{R})$, then $\|\mathcal{P}G\|_2 \leqq \|G\|_2$. Hence

$$\|\mathcal{P}E(f,h)(\cdot + iy)\|_2 \leqq \frac{2e^{-\pi\delta/h}}{4\pi \sinh(\pi d/h)}\cdot$$

$$\cdot \left[\left(\int_{\mathbb{R}} |f(t-id^-)|^2 dt \right)^{1/2} + \left(\int_{\mathbb{R}} |f(t+id^-)|^2 dt \right)^{1/2} \right] \tag{3.4.13}$$

$$\leqq \frac{2\sqrt{2}e^{-\pi\delta/h}}{4\pi \sinh(\pi d/h)} N_2(f, \mathcal{D}_d).$$

■

Corollary 3.4.2 *Let $f \in \mathbf{L}_\alpha(\mathcal{D}_d)$, let $x \in \mathbb{R}$, let y be fixed in the range $-d < y < \infty$, and let δ be defined as in Theorem 3.4.1. Let N be a positive integer, let h be defined by (3.1.27), and set*

$$\mathcal{P}E_N(f,h)(x+iy) \tag{3.4.14}$$

$$= (\mathcal{P}f)(x+iy) - \sum_{j=-N}^{N} f(jh) \frac{e^{i\pi(x+iy-jh)/h} - 1}{2\pi i(x+iy-jh)/h}.$$

---

[2]The value of the integral (3.4.11) is the same as that of (3.4.10).

Then there exists a constant c_1, depending only on f, d, y, and α, such that for all $s \in [2, \infty]$,

$$\|\mathcal{P}E_N(f_1 h)(\cdot + iy)\|_s \leqq c_1 N^{1/2} e^{-\gamma N^{1/2}} \tag{3.4.15}$$

with

$$\gamma = \begin{cases} (\pi d\alpha)^{1/2} & \text{if } y \geqq 0 \\ (\frac{\pi\alpha}{d})^{1/2}(d+\delta) & \text{if } y < 0. \end{cases} \tag{3.4.16}$$

Proof. The proof is based on the inequality

$$\begin{aligned} &\|\mathcal{P}E_N(f,h)(\cdot + iy)\|_s \\ &\leqq \|\mathcal{P}E(f,h)(\cdot + iy)\|_s + \left\| \sum_{|j|>N} f(jh) \frac{e^{i\pi(z-jh)/h} - 1}{2\pi i(z-jh)/h} \right\|_s . \end{aligned} \tag{3.4.17}$$

The first term on the right-hand side of (3.4.17) has been bounded in Theorem 3.4.1. For the second term, we have by Parseval's theorem (for the case $s = 2$) that

$$\begin{aligned} &\left| \int_{\mathbb{R}} \left[\frac{e^{i\pi(z-jh)/h} - 1}{2\pi i(z-jh)/h} \right] \left[\overline{\frac{e^{i\pi(z-\ell h)/h} - 1}{2\pi i(z-\ell h)/h}} \right] dx \right| \\ &= \left| \frac{h^2}{4\pi^2} \int_0^{\pi/h} e^{-2yt} e^{-i(j-\ell)ht} dt \right| \\ &= \left| \frac{h^2}{4\pi^2} \frac{e^{-2\pi y/h}(-1)^{j-\ell} - 1}{2y + i(j-\ell)h} \right| \\ &\leqq \frac{h}{2\pi} \cdot \begin{cases} e^{-2\pi y/h} & \text{if } y \leqq 0, \\ 1 & \text{if } y > 0. \end{cases} \end{aligned} \tag{3.4.18}$$

Also, we have

$$\left| \frac{e^{i\pi(z-jh)/h} - 1}{2\pi i(z-jh)/h} \right| \leqq \begin{cases} \frac{1}{2} e^{-\pi y/h} & \text{if } y \leqq 0 \\ \frac{1}{2} & \text{if } y > 0. \end{cases} \tag{3.4.19}$$

Hence, using (3.4.17), we have

$$\left\| \sum_{|j|>N} f(jh) \frac{e^{i\pi(\cdot+iy)/h} - 1}{2\pi i(\cdot + iy - jh)/h} \right\|_2^2$$

$$\leqq \frac{h}{2\pi} \sum_{|j|,|\ell|>N} |f(jh)|\ |f(\ell h)| \cdot \begin{cases} e^{-2\pi y/h} & \text{if} \quad y \leqq 0 \\ 1 & \text{if} \quad y > 0 \end{cases} \tag{3.4.20}$$

$$= \frac{4C^2}{2\pi\alpha^2 h} e^{-2\alpha N h} \cdot \begin{cases} e^{-2\pi y/h} & \text{if } y < 0 \\ 1 & \text{if } y \geqq 0, \end{cases}$$

using the estimate $|f(jh)| \leqq C e^{-\alpha|j|h}$ which we obtain from (3.1.22). In the case of $s = 2$ norm, the second term on the right-hand side of (3.4.17) is therefore bounded by $(C/\alpha)[2/(\pi h)]^{1/2} C\ e^{-\alpha N h - \pi y/h}$ if $y \leqq 0$, and by $(C/\alpha)[2/(\pi h)]^{1/2} C\ e^{-\alpha N h}$ if $y > 0$. Hence, by combining it with (3.4.3) or (3.4.4), we get (3.4.15), h as in (3.1.27).
Next, combining (3.4.19) and (3.1.22) and taking $s = \infty$ in (3.4.17), we obtain the estimate

$$\left\| \sum_{|j|>N} f(jh) \frac{e^{i\pi(\cdot+iy-jh)/h} - 1}{2\pi i(\cdot + iy - jh)/h} \right\|_\infty$$

$$\leqq \frac{C}{\alpha h} e^{-\alpha N h} \cdot \begin{cases} e^{-\pi y/h} & \text{if} \quad y \leqq 0 \\ 1 & \text{if} \quad y > 0. \end{cases} \tag{3.4.21}$$

Then, once again using either (3.4.3) or (3.4.4), we get (3.4.15). Since (3.4.15) holds for $s = 2$ and $s = \infty$, it follows from Hölder's inequality that it holds for all $s \in [2, \infty]$. This completes the proof. ∎

Let us briefly note a connection with Dirichlet problems, which are encountered occasionally in applications. To this end, we recall that $\mathcal{P}f$ is analytic in the upper half plane, and indeed, has an analytic continuation to $\{z \in \mathbb{C} : \Im z > -d\}$. Furthermore, $\mathcal{P}f = u + iv$, with both u and v harmonic in this region, and for $1 \leqq p \leqq \infty$, both $\|u(\cdot, y)\|_p$ and $\|v(\cdot, y)\|_p$ are bounded by $\|\mathcal{P}f(\cdot, y)\|_p$. Hence we have

Corollary 3.4.3 *Let $f \in \mathbf{H}^p(\mathcal{D}_d)$, with p either 1 or 2, and let u and v be solutions to the partial differential equation problems (3.4.22) and (3.4.23) respectively, i.e.,*

$$\begin{cases} \dfrac{\partial^2 u}{\partial x^2} + \dfrac{\partial^2 u}{\partial y^2} = 0, & x + iy \in \Omega_+ \\ u(x, 0^+) = f(x), & x \in \mathbb{R} \end{cases} \tag{3.4.22}$$

$$\begin{cases} \dfrac{\partial^2 v}{\partial x^2} + \dfrac{\partial^2 v}{\partial y^2} = 0, & x + iy \in \Omega_+ \\ v(x, 0^+) = (\mathcal{S}f)(x), & x \in \mathbb{R}, \end{cases} \tag{3.4.23}$$

with Ω_+ defined in Equation (1.4.22). Then, for all $y > -d$ and $s = 2$ or ∞,

$$\left\| u(\cdot, y) - \sum_{k=-\infty}^{\infty} f(kh) \Re \left\{ \frac{e^{i\pi(\cdot + iy - kh)/h} - 1}{2\pi i(\cdot + iy - kh)/h} \right\} \right\|_s , \\ \leqq \|\mathcal{P}E(f,h)(\cdot + iy)\|_s \tag{3.4.24}$$

$$\left\| v(\cdot, y) - \sum_{k=-\infty}^{\infty} f(kh) \Im \left\{ \frac{e^{i\pi(\cdot + iy - kh)/h} - 1}{2\pi i(\cdot + iy - kh)/h} \right\} \right\|_s \\ \leqq \|\mathcal{P}E(f,h)(\cdot + iy)\|_s , \tag{3.4.25}$$

with $\|\mathcal{P}E(f,h)(\cdot + iy)\|_s$ bounded in Theorem 3.4.1. Moreover, if $f \in \mathbf{L}_\alpha(\mathcal{D}_d)$, if N is a positive integer, and if h is selected as in Equation (3.1.27), then for all $s \in [2, \infty]$

$$\left\| u(\cdot, y) - \sum_{k=-N}^{N} f(kh) \Re \left\{ \frac{e^{i\pi(\cdot + iy - kh)/h} - 1}{2\pi i(\cdot + iy - kh)/h} \right\} \right\|_s \\ \leqq \|\mathcal{P}E_N(f,h)(\cdot + iy)\|_s \tag{3.4.26}$$

$$\left\| v(\cdot, y) - \sum_{k=-N}^{N} f(kh) \Im \left\{ \frac{e^{i\pi(\cdot + iy - kh)/h} - 1}{2\pi i(\cdot + iy - kh)/h} \right\} \right\|_s \\ \leqq \|\mathcal{P}E_N(f,h)(\cdot + iy)\|_s \tag{3.4.27}$$

with $\|\mathcal{P}E_N(f,h)(\cdot + iy)\|_s$ bounded in Corollary 3.4.2.

Let us next turn to the approximation of the Hilbert transform, defined as in Equation (1.4.20).

Theorem 3.4.4 *Let $f \in \mathbf{H}^p(\mathcal{D}_d)$, where p is either 1 or 2, and let $E(f,h)$ be defined as in Equation (3.1.4). Let $z = x + iy$, where $x \in \mathbb{R}$, and where $-d < y < d$. Then*

$$[\mathcal{S}E(f,h)](z)$$

$$= (\mathcal{S}f)(z) - \sum_{j=-\infty}^{\infty} f(jh) \frac{\cos[\pi(z-jh)/h] - 1}{\pi i(z-jh)/h} \tag{3.4.28}$$

$$= \frac{1}{2\pi} \int_{\mathbb{R}} \left\{ \frac{[e^{-i\pi(t-id)/h} - \cos(\pi z/h)] f(t - id^-)}{(t - id - z) \sin[\pi(t - id)/h]} \right.$$

$$\left. \frac{-[e^{i\pi(t+id)/h} - \cos(\pi z/h)] f(t + id^-)}{(z + id - z) \sin[\pi(t + id)/h]} \right\} dt. \tag{3.4.29}$$

(a) *If $f \in \mathbf{H}^1(\mathcal{D}_d)$, then*

$$\|\mathcal{S}E(f,h)(\cdot + iy)\|_\infty \leqq \frac{e^{-\pi d/h} + \cosh(\pi y/h)}{2\pi d \sinh(\pi d/h)} N_1(f, \mathcal{D}_d), \tag{3.4.30}$$

and

$$\|\mathcal{S}E(f,h)(\cdot + iy)\|_2$$

$$\leqq \frac{\cosh(\pi y/h)}{2\pi^{1/2}[d - |y|]^{1/2} \sinh(\pi d/h)} N_1(f, \mathcal{D}_d). \tag{3.4.31}$$

(b) *If $f \in \mathbf{H}^2(\mathcal{D}_d)$, then*

$$\|\mathcal{S}E(f,h)(\cdot + iy)\|_\infty$$

$$\leqq \frac{[e^{-\pi d/h} + \cosh(\pi y/h)]\, d^{1/2}}{[2\pi(d^2 - y^2)]^{1/2} \sinh(\pi d/h)} N_2(f, \mathcal{D}_d) \tag{3.4.32}$$

and

$$\|\mathcal{S}E(f,h)(\cdot + iy)\|_2 \leqq \frac{\cosh(\pi y/h)}{\sinh(\pi d/h)} N_2(f, \mathcal{D}_d). \tag{3.4.33}$$

Proof. The inequality (3.4.30) is a consequence of the fact that both coefficients of f in (3.4.29) are bounded by the coefficient in front of $N_1(f, \mathcal{D}_d)$ on the right-hand side of (3.4.30). To obtain (3.4.32) we apply Schwarz's inequality to each term on the right-hand side of (3.4.29), to get

$$\begin{aligned}
&\|\mathcal{S}E(f,h)(\cdot + iy)\|_\infty \\
&\qquad \leqq \frac{e^{-\pi d/h} + \cosh(\pi y/h)}{2\pi \sinh(\pi d/h)} \cdot \qquad (3.4.34) \\
&\cdot \sup_{x \in \mathbb{R}} \left[\left(\int_{\mathbb{R}} \frac{dt}{(t-x)^2 + (d+y)^2} \right)^{1/2} \left(\int_{\mathbb{R}} |f(t - id^-)|^2 dt \right)^{1/2} \right. \\
&\left. + \left(\int_{\mathbb{R}} \frac{dt}{(t-x)^2 + (d-y)^2} \right)^{1/2} \left(\int_{\mathbb{R}} |f(t + id^-)|^2 dt \right)^{1/2} \right] \\
&\leqq \frac{e^{-\pi d/h} + \cosh(\pi y/h)}{2\pi \sinh(\pi d/h)} \left(\frac{2\pi d}{d^2 - y^2} \right)^{1/2} N_2(f, \mathcal{D}_d).
\end{aligned}$$

Bounds for $\|\mathcal{S}E(f,h)(\cdot + iy)\|_2$ are much simpler to obtain. In this case, it follows from (1.4.21) and Parseval's theorem, that if $F \in \mathbf{L}^2(\mathbb{R})$, then $\|\mathcal{S}F\|_2 = \|F\|_2$. Hence $\|\mathcal{S}E(f,h)(\cdot+iy)\|_2 = \|E(f,h)(\cdot+iy)\|_2$, we can thus simply use the bounds obtained for $\|E(f,h)(\cdot + iy)\|_2$ in Theorem 3.1.3. ∎

Corollary 3.4.5 *Let $f \in \mathbf{L}_\alpha(\mathcal{D}_d)$, let $x \in \mathbb{R}$, and let y be fixed in the range $-d < y < d$. Let N, be a positive integer, let h be defined by (3.1.27), and set*

$$\begin{aligned}
&\mathcal{S}E_N(f,h)(x+iy) \qquad (3.4.35) \\
&= (\mathcal{S}f)(x+iy) - \sum_{j=-N}^{N} f(jh) \frac{\cos[\pi(x+iy-jh)/h] - 1}{\pi i(x+iy-jh)/h} \quad .
\end{aligned}$$

Then there exists a constant c_1, depending only on f, d, y, and α, such that for all $s \in [2, \infty]$,

$$\|\mathcal{S}E_N(f,h)(\cdot + iy)\|_s \leqq c_1 \, N^{1/2} e^{-(\frac{\pi\alpha}{d})^{1/2}(d-|y|)N^{1/2}}. \qquad (3.4.36)$$

Proof. From the formula

$$S(k,h)(x) = \int_{-\pi/h}^{\pi/h} e^{ixt} \left[\frac{h}{2\pi} e^{-ikht} \right] dt \qquad (3.4.37)$$

and Equation (1.4.21), we get

$$\begin{aligned} \mathcal{S}S(k,h)(x) &= \int_{-\pi/h}^{\pi/h} e^{ixt}[\frac{h}{2\pi} e^{-ikht}] \operatorname{sgn}(t) dt \\ &= \frac{\cos[\pi(x-kh)/h]-1}{\pi i(x-kh)/h}. \end{aligned} \tag{3.4.38}$$

Hence, by (3.4.38), we get

$$\begin{aligned} \sup_{x\in\mathbb{R}} |\mathcal{S}S(k,h)(x+iy)| &\leqq \frac{h}{2\pi} \int_{-\pi/h}^{\pi/h} e^{-yt} dt \\ &= \frac{e^{\pi y/h} - e^{-\pi y/h}}{2\pi y/h} \leqq \cosh(\pi y/h). \end{aligned} \tag{3.4.39}$$

Also using (3.4.38) and Parseval's theorem, and proceeding as in (3.1.35) through (3.1.36), we get

$$\begin{aligned} \int_{\mathbb{R}} &\{\mathcal{S}S(k,h)(x+iy)\}\{\overline{\mathcal{S}S(\ell,h)(x+iy)}\}\, dx \\ &\leqq h \cosh^2(\pi y/h). \end{aligned} \tag{3.4.40}$$

Combining the bounds (3.4.39) and (3.4.40) with the results of Theorem 3.4.4 and proceeding as in the proof of Corollary 3.4.2, we get the inequality (3.4.36). ■

•

Problems for Section 3.4

3.4.1 Verify formula (3.4.5).

3.4.2 Verify inequalities (3.4.6).

3.4.3 Evaluate (3.4.7).

3.4.4 Evaluate (3.4.8).

3.4.5 Evaluate (3.4.10).

3.4.6 Evaluate (3.4.11).

3.4.7 Prove that

$$\|\mathcal{S}_N\| \equiv \sup_{x\in\mathbb{R}} \sum_{k=-N}^{N} \left| \frac{\cos\left\{\frac{\pi}{h}(x-kh)\right\} - 1}{\frac{\pi}{h}(x-kh)} \right|$$
$$\leqq \frac{2}{\pi}\{1+\gamma+\log(N+1)\}.$$

[Hint: See the hint given in Problem 3.1.5.]

3.4.8 Prove that

$$\|\mathcal{P}_N\| \equiv \sup_{x\in\mathbb{R}} \sum_{k=-N}^{N} \left| \frac{\exp\left\{\frac{\pi i}{h}(x-kh)\right\} - 1}{\frac{2\pi i}{h}(x-kh)} \right|$$
$$\leqq \frac{\sqrt{2}}{\pi}\{2+\log(N+1)\}.$$

[Hint: See the hint given in Problem 3.1.5.]

3.4.9 Explain that, whereas the bound given in Problem 3.1.7 may be used in the form $\|\mathcal{S}_N f\|_\infty \leqq \|\mathcal{S}_N\|_\infty \|f\|_\infty$ for f an arbitrary bounded function defined on $\mathbb{R}$, in the case $f \in \mathbf{H}^p(\mathcal{D}_d)$, with $d > 0$ and $1 \leqq p \leqq \infty$, we do in fact have $\|C_N f\|_\infty \leqq [1+\varepsilon(h,N)]\|f\|_\infty$, with $\varepsilon(h,N)$ uniformly bounded for all positive integers N, i.e., for $f \in \mathbf{H}^p(\mathcal{D}_d)$ we have $\|C_N\|_\infty \leqq [1+\varepsilon(h,N)]$.

3.4.10 Do Problem 3.4.9, with reference to Problem 3.4.8 instead of to Problem 3.4.7, and with $\mathcal{S}_N$ replaced with $\mathcal{P}_N$.

3.4.11 (a) Replace $f(t)$ in Problem 3.2.6 by $F(t)/(t-z)$, with $F \in \mathbf{Hol}(\mathcal{D}_d)$, with $z \in \mathcal{D}_d$, and with $\Im z > 0$, to deduce the formula

$$\left| \frac{1}{2\pi i}\int_{\mathbb{R}} \frac{F(t)}{t-z}\,dt - \frac{h}{2\pi i}\sum_{n=-\infty}^{\infty} \frac{F(nh)}{nh-z} - \frac{F(z)}{1-e^{-2\pi i z/h}} \right|$$
$$\leqq \frac{e^{-2\pi d/h}}{1-e^{-2\pi d/h}} \int_{\mathbb{R}} \left\{ \left| \frac{F(t-id^-)}{t-z-id} \right| + \left| \frac{F(t+id^-)}{t-z+id} \right| \right\} dt.$$

(b) Use the Plemelj formulas to deduce what happens in the above formula, as $z \to x$, with $x \in \mathbb{R}$. Distinguish two cases: $x = mh$, and $x \neq mh$, with m is an integer. (Bialecki, [Bi3]).

3.5 Approximation of Derivatives in $\mathcal{D}_d$

In applications it is often desirable to find approximations Φ of f, such that $(d/dx)^n\Phi(x) = \Phi^{(n)}(x)$ accurately approximates $f^{(n)}(x)$ on an interval, for several values of n. Such approximations are readily given by $C_N(f,h)$.

Theorem 3.5.1 *Let $f \in \mathbf{H}^p(\mathcal{D}_d)$, where $1 \leqq p < \infty$. Then for arbitrary $x \in \mathbb{R}$, $y \in \mathbb{R}$, so that $z = x + iy \in \mathcal{D}_d$,*

$$E(f,h)^{(n)}(z)$$

$$= \frac{n!}{2\pi i}\int_{\mathbb{R}}\left\{\left[\sum_{j=0}^{n}\frac{\sin^{(n-j)}(\pi z/h)(\pi/h)^{n-j}}{(n-j)!(t-z-id)^{j+1}}\right]\frac{f(t-id^-)}{\sin[\pi(t-id)/h]}\right. \tag{3.5.1}$$

$$\left. - \left[\sum_{j=0}^{n}\frac{\sin^{(n-j)}(\pi z/h)(\pi/h)^{n-j}}{(n-j)!(t-z+id)^{j+1}}\right]\frac{f(t+id^-)}{\sin[\pi(t+id)/h]}\right\} dt.$$

Let $\tau > 0$, and define $a(n,h,\tau)$ and $b(n,h,\tau)$ by

$$a(n,h,\tau) = \frac{(\pi/h)^n n!}{2\pi\tau}\sum_{j=0}^{n}\left(\frac{h}{\pi\tau}\right)^j\frac{1}{(n-j)!} \tag{3.5.2}$$

$$b(n,h,\tau) = \tag{3.5.3}$$

$$\frac{(\pi/h)^n n!}{2\pi\tau^{1/2}}\left(\sum_{j=0}^{n}\left[\frac{1}{(n-j)!}\right]^2\right)^{1/2}\left(\sum_{j=0}^{n}\frac{(1/2)_j}{j!}\left(\frac{h}{\pi\tau}\right)^{2j}\right)^{1/2}.$$

(i) *If $f \in \mathbf{H}^1(\mathcal{D}_d)$, then*

$$\frac{\|E^{(n)}(f,h)(\cdot+iy)\|_\infty}{a(n,h,d-|y|)}, \quad \frac{\|E^{(n)}(f,h)(\cdot+iy)\|_2}{b(n,h,d-|y|)} \tag{3.5.4}$$

$$\leqq \frac{\cosh(\pi y/h)}{\sinh(\pi d/h)}N_1(f,\mathcal{D}_d).$$

(ii) *If $f \in \mathbf{H}^2(\mathcal{D}_d)$, then*

$$\frac{\|E^{(n)}(f,h)(\cdot+iy)\|_\infty}{b(n,h,d-|y|)},\quad \frac{\|E^{(n)}(f,h)(\cdot+iy)\|_2}{a(n,h,d-|y|)}$$

$$\leqq \frac{\sqrt{2}\cosh(\pi y/h)}{\sinh(\pi d/h)} N_2(f,\mathcal{D}_d). \tag{3.5.5}$$

Proof. The formula (3.5.1) is obtained upon differentiating (3.1.4) n times with respect to z.

(i) Let $f \in \mathbf{H}^1(\mathcal{D}_d)$. Since $\sin^{(n-j)}(\pi z/h)$ is either $\pm\sin(\pi z/h)$ or $\pm\cos(\pi z/h)$, and since both of these functions are bounded by $\cosh(\pi y/h)$ on the horizontal line $\mathcal{L} = \{z = x+iy : y \text{ is fixed on } (-d,d),\ x \in \mathbb{R}\}$, (see (3.1.8)), we have

$$\begin{aligned}\|E^{(n)}(f,h)(\cdot+iy)\|_\infty &\leqq \frac{\cosh(\pi y/h)}{\sinh(\pi d/h)}\cdot \\ &\cdot\left\{\frac{(\pi/h)^n n!}{2\pi}\sum_{j=0}^{n}\left[\frac{h}{\pi(d+y)}\right]^j \frac{1}{(n-j)!}\int_{\mathbb{R}}|f(t-id^-)|\,dt\right. \\ &\left.+\frac{(\pi/h)^n n!}{2\pi}\sum_{j=0}^{n}\left[\frac{h}{\pi(d-y)}\right]^j \frac{1}{(n-j)!}\int_{\mathbb{R}}|f(t+id^-)|\,dt\right\} \\ &\leqq \frac{\cosh(\pi y/h)}{\sinh(\pi d/h)} a(n,h,d-|y|)N_1(f,\mathcal{D}_d),\end{aligned} \tag{3.5.6}$$

with $a(n,h,\tau)$ as defined in (3.5.2). This establishes the first inequality in (3.5.4).

In order to bound $\|E^{(n)}(f,h)(\cdot+iy)\|_2$, it is convenient to set

$$F_\pm(z) = \frac{(\pi/h)^n n!}{2\pi i}\int_{\mathbb{R}}\frac{f(t\pm id)}{\sin[\pi(t\pm id)/h]}\sum_{j=0}^{n}\frac{(h/\pi)^j\sin^{(n-j)}(\pi z/h)}{(n-j)!(t-z\pm id)^{j+1}}dt. \tag{3.5.7}$$

Then, it follows that

$$\|E^{(n)}(f,h)(\cdot+iy)\|_2 \leqq \|F_+(\cdot+iy)\|_2 + \|F_-(\cdot+iy)\|_2. \tag{3.5.8}$$

Proceeding as above and applying Theorem 1.4.5,

$$\|F_\pm(\cdot + iy)\|_2 \leqq \frac{\cosh(\pi y/h)}{\sinh(\pi d/h)} \cdot \frac{(\pi/h)^n n!}{2\pi} \cdot$$
$$\cdot \left\| \int_{\mathbb{R}} |f(t \pm id)^-| \sum_{j=0}^{n} \frac{(h/\pi)^j}{(n-j)![(t-x)^2 + (d \mp y)^2]^{(j+1)/2}} dt \right\|_2$$
$$\leqq \frac{\cosh(\pi y/h)}{\sinh(\pi d/h)} \frac{(\pi/h)^n n!}{2\pi} \int_{\mathbb{R}} |f(t \pm id^-)|\, dt \cdot \tag{3.5.9}$$
$$\cdot \left(\sum_{j=0}^{n} \frac{1}{[(n-j)!]^2} \right)^{1/2} \left(\sum_{j=0}^{n} \left(\frac{h}{\pi}\right)^{2j} \int_{\mathbb{R}} \frac{dt}{[(t-x)^2 + (d \mp y)^2]^{j+1}} \right)^{1/2}.$$

The identity

$$\int_{\mathbb{R}} \frac{dt}{[t^2 + \tau^2]^{j+1}} = \frac{(1/2)_j}{j!\tau^{2j+1}}, \quad \tau > 0, \tag{3.5.10}$$

now yields

$$\|F_\pm(\cdot + iy)\|_2 \leqq \frac{\cosh(\pi y/h)}{\sinh(\pi d/h)} b(n, h, d \mp y) \int_{\mathbb{R}} |f(t \pm id^-)| dt, \tag{3.5.11}$$

and substituting (3.5.11) into (3.5.8), we get the second inequality in (3.5.4).

(ii) Now let $f \in \mathbf{H}^2(\mathcal{D}_d)$. If $F_\pm$ is defined as in (3.5.7), we have

$$\|F_\pm(\cdot + iy)\|_\infty \leqq \frac{\cosh(\pi y/h)}{\sinh(\pi d/h)} \left(\int_{\mathbb{R}} |f(t \pm id)|^2 dt \right)^{1/2}$$
$$\cdot \sup_{x \in \mathbb{R}} \frac{(\pi/h)^n n!}{2\pi} \left\{ \int_{\mathbb{R}} \sum_{j=0}^{n} \frac{(h/\pi)^j}{(n-j)![(t-x)^2 + (d \mp y)^2]^{\frac{j+1}{2}}} dt \right\}^{1/2}$$
$$\leqq \frac{\cosh(\pi y/h)}{\sinh(\pi d/h)} b(n, h, d \mp y) \left(\int_{\mathbb{R}} |f(t \pm id)|^2 dt \right)^{1/2}. \tag{3.5.12}$$

Hence

$$\begin{aligned}\|E^{(n)}(f,h)(\cdot+iy)\|_\infty &\leqq \|F_+(\cdot+iy)\|_\infty + \|F_-(\cdot+iy)\|_\infty \\ &\leqq \frac{\cosh(\pi y/h)}{\sinh(\pi d/h)}\left[b(n,h,d+y)\left(\int_{\mathbb{R}}|f(t-id^-)|^2dt\right)^{1/2}\right. \\ &\qquad \left. + b(n,h,d-y)\left(\int_{\mathbb{R}}|f(t+id^-)|^2dt\right)^{1/2}\right] \\ &\leqq \frac{\sqrt{2}\cosh(\pi y/h)}{\sinh(\pi d/h)} b(n,h,d-|y|)N_2(f,\mathcal{D}_d).\end{aligned} \tag{3.5.13}$$

This proves the first inequality in (3.5.5).

For the two-norm of $E^{(n)}(f,h)(\cdot+iy)$, we recall by the definition (1.4.18) of $\mathcal{P}G$, that if $G \in \mathbf{L}^2(\mathbb{R})$, then $\|\mathcal{P}G\|_2 \leqq \|G\|_2$. Inspection of (3.5.7) shows, after factoring out $(t-z\pm id)^{-1}$, that F_+ and F_- are in fact projections, i.e. $F_+ = \mathcal{P}G_+$, and $F_- = (I-\mathcal{P})G_-$, with

$$\begin{aligned}G_\pm(z) &= \frac{\pm(\pi/h)^n n!}{2\pi i}\int_{\mathbb{R}}\frac{f(t\pm id)}{\sin[\pi(t\pm id)/h]} \cdot \\ &\qquad \cdot \sum_{j=0}^{n}\frac{(h/\pi)^j \sin^{(n-j)}(\pi z/h)}{(n-j)!(t-z\pm id)^j}dt.\end{aligned} \tag{3.5.14}$$

Hence, taking the two-norm, and proceeding as above,

$$\begin{aligned}\|E^{(n)}(f,h)(\cdot+iy)\|_2 &\leqq \|G_+(\cdot+iy)\|_2 + \|G_-(\cdot+iy)\|_2 \\ &\leqq \frac{\cosh(\pi y/h)}{\sinh(\pi d/h)}\left[(d+y)a(n,h,d+y)\left(\int_{\mathbb{R}}|f(t-id^-)|^2dt\right)^{1/2}\right. \\ &\qquad \left. +(d-y)a(n,h,d-y)\left(\int_{\mathbb{R}}|f(t+id^-)|^2dt\right)^{1/2}\right] \\ &\leqq \frac{\sqrt{2}\cosh(\pi y/h)}{\sinh(\pi d/h)}(d-|y|)a(n,h,d-|y|)N_2(f,\mathcal{D}_d).\end{aligned} \tag{3.5.15}$$

This completes the proof of Theorem 3.5.1. ∎

Corollary 3.5.2 *Let either one of the following conditions be satisfied:*

(i) *f belongs either $\mathbf{H}^1(\mathcal{D}_d)$ or $\mathbf{H}^2(\mathcal{D}_d)$; and on $\mathbb{R}$, f satisfies the inequality (3.1.26);*

(ii) *f belongs to $\mathbf{L}_\alpha(\mathcal{D}_d)$.*

Let h be defined by (3.1.27), and let $-d < y < d$. Then there exists a constant c, depending only on f, d, α, n, and y such that for all $s \in [2, \infty]$,

$$\begin{aligned}\|E_N^{(n)}(f,h)(\cdot + iy)\|_s &= \|F^{(n)}(\cdot + iy) - C_N^{(n)}(f,h)(\cdot + iy)\|_s \\ &\leqq c\, N^{\frac{2n+1}{4}} \exp\left[-\left(\frac{\pi\alpha}{d}\right)^{1/2} (d - |y|) N^{1/2}\right].\end{aligned} \tag{3.5.16}$$

Proof. This is left to Problem 3.5.4. ■

•

Problems for Section 3.5

3.5.1 Prove the identity (3.5.11).

3.5.2 Prove that if L is defined as at the outset of the (i)-part of the proof of Theorem 3.5.1, then for $z = x + iy \in L$, with $x \in \mathbb{R}$, $y \in \mathbb{R}$, we have

$$|\sinh(\pi y/h)| \leqq |\sin(\pi z/h)| \leqq \cosh(\pi y/h)$$

$$|\sinh(\pi y/h)| \leqq |\cos(\pi z/h)| \leqq \cosh(\pi y/h).$$

3.5.3 (a) Use the representation

$$S(k,h)(z) = \int_{-\pi/h}^{\pi/h} \frac{h}{2\pi} e^{-ikht} e^{izt}\, dt$$

to prove that

$$S^{(n)}(k,h)(z) = \int_{-\pi/h}^{\pi/h} \frac{h}{2\pi} e^{-ikht} (it)^n e^{izt}\, dt;$$

(b) Prove that if $y \in \mathbb{R}$, then

$$\sup_{x \in \mathbb{R}} |S^{(n)}(k,h)(x+iy)| \leqq \frac{1}{n+1} \left(\frac{\pi}{h}\right)^n \cosh(\pi y/h);$$

(c) Prove that if $y \in \mathbb{R}$, then

$$\left| \int_{\mathbb{R}} S^{(n)}(k,h)(x+iy)\, S^{(n)}(\ell,h)(x-iy)\, dx \right|$$

$$\leq \frac{1}{2n+1} (\pi/h)^{2n-1} \cosh^2(\pi y/h).$$

3.5.4 Use the results of Problem 3.5.3 to prove Corollary 3.5.2.

3.6 The Indefinite Integral on $(-\infty, \infty)$

In this section we shall give a new proof for the approximation of $\int_{-\infty}^{x} f(t)dt$ along with an accurate error bound.

Let us first establish some notation. Let $\mathbb{Z}$ denote the set of all integers. Throughout this section, we shall assume that $f \in \mathbf{L}_{\alpha'}(\mathcal{D}_{d'})$, with α' and d' positive numbers. Let f be a given function, and define g and G by

$$g(z) = f(z) - \frac{\beta}{2\cosh^2(\beta z)} \int_{\mathbb{R}} f(t)dt \tag{3.6.1}$$

$$G(z) = \int_{-\infty}^{z} f(t)dt - \frac{e^{\beta z}}{2\cosh(\beta z)} \int_{\mathbb{R}} f(t)dt. \tag{3.6.2}$$

Here β is a positive constant, and $G(z) = \int_{-\infty}^{z} g(t)\, dt$. Let us also set

$$d = \min(d', \frac{\pi}{2\beta} - \delta), \quad \alpha = \min(\alpha', \beta) \tag{3.6.3}$$

with $\delta \in (0, \pi/(2\beta))$, but otherwise an arbitrary positive number. For $k \in \mathbb{Z}$, $\xi \in \mathbb{R}$, and $h > 0$, let us set

$$J(k,h)(\xi) = \int_{-\infty}^{\xi} S(k,h)(t)dt, \tag{3.6.4}$$

let σ_k be defined by Equation (1.10.30), let N be a positive number, and let us define h by Equation (3.1.27). Finally, in order to simplify our notation, we set

$$(\mathcal{J}g)(z) = \int_{-\infty}^{z} g(t)dt \tag{3.6.5}$$

$$(Cg)(z) = \sum_{k \in \mathbb{Z}} g(kh) S(k,h)(z) \tag{3.6.6}$$

$$(C_N g)(z) = \sum_{|k| \leq N} g(kh) S(k,h)(z). \tag{3.6.7}$$

We shall prove the following theorem.

Theorem 3.6.1 *Let $f \in \mathbf{L}_{\alpha'}(\mathcal{D}_{d'})$, with α' and d' positive numbers. Let β be a positive number, let d and α be defined by (3.6.3), where δ is an arbitrary positive number in the range $0 < \delta < \pi/(2\beta)$. Let σ_k be defined as in Equation (1.10.30), and let h be defined by (3.1.27), where N is a positive integer. Then there exists a positive constant K which is independent of N, such that*

$$\sup_{x \in \mathbb{R}} \left| \int_{-\infty}^{x} f(t)\, dt - \frac{e^{\beta x}}{2\cosh(\beta x)} h \sum_{|m| \leqq N} f(mh) - h \sum_{|k| \leqq N} \cdot \right.$$

$$\left. \cdot \left\{ \sum_{|\ell| \leqq N} \sigma_{k-\ell} \left[f(\ell h) - \frac{\beta}{2\cosh^2(\beta \ell h)} h \sum_{|j| \leqq N} f(jh) \right] \right\} S(k,h)(x) \right|$$

$$\leqq K\, N^{1/2}\, e^{-(\pi d \alpha N)^{1/2}}. \tag{3.6.8}$$

Proof. We split the proof of Theorem 3.6.1 into the proof of a series of lemmas. As a preliminary to the proof of the next two lemmas, let us note that since $0 < \alpha \leqq \alpha'$, we have for all $x \in \mathbb{R}$ that

$$\frac{e^{\alpha' x}}{(1+e^x)^{2\alpha'}} \leqq \frac{e^{\alpha x}}{(1+e^x)^{2\alpha}} \tag{3.6.9}$$

Lemma 3.6.2 *Let f satisfy the conditions of Theorem 3.6.1, let g be defined by (3.6.1), and let G be defined by (3.6.2). Then both g and G belong to $\mathbf{L}_\alpha(\mathcal{D}_d)$.*

Proof. Since $f \in \mathbf{L}_\alpha(\mathcal{D}_d)$, and since it readily follows from (3.6.3) that the second term on the right-hand side of (3.6.1) also belongs to $\mathbf{L}_\alpha(\mathcal{D}_d)$, we clearly have $g \in \mathbf{L}_\alpha(\mathcal{D}_d)$. Next, if G is defined by (3.6.2), then $G \in \mathbf{Hol}(\mathcal{D}_d)$. Also, since $f \in \mathbf{L}_\alpha(\mathcal{D}_d)$, $z = x + iy \in \mathcal{D}_d$, and $\Re z = x$, we immediately have by (3.1.23), that $f(x+iy) \leqq C\, e^{-\alpha|x|}$. In particular, it immediately follows that $|\int_{\mathbb{R}} f(t)\, dt| \leqq 2C/\alpha$.
Now, for $z \in \mathcal{D}_d$, let us assume that $x = \Re z \leqq 0$. Then, since $\int_{-\infty}^{z} f(t)\, dt$ is independent of the path for any path remaining in $\mathcal{D}_d$, we select a particular path depending on a positive number M, by choosing M such that $-M < x$. The initial portion of the path is along $\mathbb{R}$ from $-\infty$ to $-M$; the second, a vertical line segment from $-M$ to $-M + iy$; and the last, a horizontal line segment from $-M + iy$ to $x + iy$. Using (3.1.23) to estimate the integrals of $|f|$ along each of these three straight segments, we thus find that

$$\left| \int_{-\infty}^{z} f(t)\, dt \right| \leqq \frac{C}{\alpha} e^{-\alpha M} + C\,|y|\, e^{-\alpha M} + \frac{C}{\alpha}(e^{-\alpha|x|} - e^{-\alpha M}).$$

Letting $M \to \infty$, we get $|\int_{-\infty}^{x+iy} f(t)\,dt| \leqq C\,e^{-\alpha|x|}/\alpha$.
Let us now take $z = x + iy \in \mathcal{D}_d$, and $\Re z = x \leqq 0, \Im z = y$; in this case we have

$$\begin{aligned}\left|\frac{e^{\beta z}}{\cosh(\beta z)}\right| &= \frac{2\,e^{2\beta x}}{[(e^{2\beta x}-1)^2 + 4\,e^{2\beta x}\cos^2(\beta y)]^{1/2}} \\ &\leqq \frac{2\,e^{2\beta x}}{2\,e^{\beta x}\cos(\beta y)} = \frac{e^{\beta x}}{\cos(\beta y)} \\ &\leqq \frac{e^{\alpha x}}{\sin(\alpha\delta)},\end{aligned}$$

since $0 < \alpha \leqq \beta$, and $|y| < d \leqq \pi/(2\beta) - \delta < \pi/(2\beta)$. The above inequalities and the left hand side of (3.1.23) thus yield, for $z = x + iy \in \mathcal{D}_d$, and $x \leqq 0$, that

$$\begin{aligned}|G(x+iy)| &\leqq \frac{C}{\alpha}\left[1 + \frac{1}{\sin(\alpha\delta)}\right] e^{-\alpha|x|} \\ &\leqq \frac{2\,C}{\alpha}\left[1 + \frac{1}{\sin(\alpha\delta)}\right] \frac{e^{\alpha x}}{(1+e^x)^{2\alpha}}.\end{aligned} \tag{3.6.10}$$

Similarly, if $\Re z > 0$, $z \in \mathcal{D}_d$, we may replace (3.6.2) with the equivalent expression

$$G(z) = \frac{e^{-2\beta z}}{1 + e^{-2\beta z}} \int_{\mathbb{R}} f(t)\,dt - \int_z^{\infty} f(t)\,dt \tag{3.6.11}$$

to arrive at the same bound given on the extreme right-hand side of (3.6.10).

This completes the proof of Lemma 3.6.2. ■

Lemma 3.6.3 *If τ, t, $\xi \in \mathbb{R}$, $\tau \neq 0$, and if $w(h,t,\tau,\xi)$ is defined by*

$$w(h,t,\tau,\xi) = \frac{1}{2\pi i}\int_{-\infty}^{\xi} \frac{\sin(\pi x/h)dx}{(t - x - i\tau)}, \tag{3.6.12}$$

then

$$|w(h,t,\tau,\xi)| \leqq \frac{h}{4|\tau|}. \tag{3.6.13}$$

Proof. After splitting (3.6.12) into its real and imaginary parts, we find that

$$|w(h,t,\tau,\xi)| \leqq \frac{1}{2\pi}[\sup_{t,\xi\in\mathbb{R}} |u(t,\xi)|^2 + \sup_{t,\xi\in\mathbb{R}} |v(t,\xi)|^2]^{1/2} \tag{3.6.14}$$

with

$$u(t,\xi) = \int_{-\infty}^{\xi} \frac{x \sin[\pi(x+t)/h]dx}{x^2+\tau^2} \tag{3.6.15}$$

$$v(t,\xi) = \int_{-\infty}^{\xi} \frac{\tau \sin[\pi(x+t)/h]dx}{x^2+\tau^2}. \tag{3.6.16}$$

Let us first consider the function $u(t, \xi)$. Now the function

$$x/(x^2+\tau^2)$$

decreases monotonically from 0 to $-(2|\tau|)^{-1}$ as x increases from $-\infty$ to $-|\tau|$, and then it increases from $-(2|\tau|)^{-1}$ to 0 as x increases from $-|\tau|$ to 0. Hence, if $\xi \in (-\infty, -|\tau|)$, then setting $\xi + t = kh + \mu$, where $k \in \mathbb{Z}$ and $0 \leqq \mu < h$, we find that

$$\begin{aligned} u(t,\xi) &= \int_0^h \sin(\pi x/h) \sum_{\ell=0}^{\infty} \frac{(-1)^{k+\ell+1}(a-\ell h+x)}{(a-\ell h+x)^2+\tau^2} dx \\ &\quad +(-1)^k \int_0^{\mu} \frac{(kh-t+x)\sin(\pi x/h)}{(kh-t+x)^2+\tau^2} dx, \end{aligned} \tag{3.6.17}$$

with $a = kh - h - t$. We may rewrite the right-hand side of (3.6.17) in the form

$$u(t,\xi) = (-1)^k[\lambda a_0 - a_1 + a_2 - a_3 + \cdots] \tag{3.6.18}$$

with $0 \leqq \lambda < 1$, and with

$$a_j = \int_0^h \frac{(x-t+kh-jh)\sin(\pi x/h)dx}{(x-t+kh-jh)^2+\tau^2}. \tag{3.6.19}$$

Thus $a_j \to 0$, monotonically, as $j \to \infty$. Hence, if $\xi \in (-\infty, -|\tau|]$, then

$$\begin{aligned} |u(t,\xi)| \leqq |a_0| &\leqq \frac{1}{2|\tau|} \int_0^h \sin(\pi x/h)dx \\ &= \frac{h}{\pi|\tau|}. \end{aligned} \tag{3.6.20}$$

Next, if $\xi \in (-|\tau|, 0)$, then by a similar argument, we find that

$$\begin{aligned} |u(t,\xi)| &\leqq |u(t,-|\tau|)| + |u(t,-|\tau|) - u(t,\xi)| \\ &\leqq \frac{h}{\pi|\tau|} + \frac{h}{\pi|\tau|} \leqq \frac{2h}{\pi|\tau|}, \end{aligned} \tag{3.6.21}$$

and similarly for $\xi \in (0, |\tau|)$ we add another term $h/(\pi|\tau|)$ to the extreme right-hand side of (3.6.21), and finally, if $\xi \in (|\tau|, \infty)$ we add $2\,h/(\pi|\tau|)$. Hence, for all $\xi \in \mathbb{R}$, we have

$$|u(t, \xi)| \leqq \frac{4\,h}{\pi|\tau|}. \tag{3.6.22}$$

Let us next turn to $v(t, \xi)$. In this case we observe that the function $\tau/(x^2 + \tau^2)$ varies monotonically from 0 to $1/\tau$ as x increases from $-\infty$ to 0. Hence, if $-\infty \leqq \xi \leqq 0$, then repeating an argument similar to the one used for $u(t, \xi)$ in the case of $\xi \in (-\infty, -|\tau|)$, we find that $|v(t, \xi)| \leqq h/(\pi|\tau|)$, $-\infty \leqq \xi \leqq 0$. Hence, if $\xi \in \mathbb{R}$, it follows that

$$|v(t, \xi)| \leqq \frac{2\,h}{\pi|\tau|}. \tag{3.6.23}$$

Combining (3.6.22) and (3.6.23) we have, in view of (3.6.14), that if $\xi \in \mathbb{R}$, then

$$|w(h, t, \tau,\ \xi)| \leqq \frac{\sqrt{5}h}{\pi^2|\tau|} < \frac{h}{4\,|\tau|}. \tag{3.6.24}$$

This completes the proof of Lemma 3.6.3. ■

Lemma 3.6.4 *Let $w(h, t, \tau, \xi)$ be defined as in (3.6.12). Then, setting*

$$\delta(g, h, x) \tag{3.6.25}$$
$$= \int_{-\infty}^{x} \left\{ g(t) - \sum_{k \in \mathbf{Z}} g(kh) S(k, h)(t) \right\} dt$$

we have for $x \in \mathbb{R}$,

$$\delta(g, h, x) \tag{3.6.26}$$
$$= \int_{\mathbf{R}} \left\{ \frac{w(h, t, d, x) g(t - id)}{\sin[\pi(t - id)/h]} - \frac{w(h, t, -d, x) g(t + id)}{\sin[\pi(t + id)/h]} \right\} dt$$

and moreover, if $N_1(g, \mathcal{D}_d)$ is defined as in (1.1.11), then for $x \in \mathbb{R}$, we have

$$|\delta(g, h, x)| \leqq \frac{h\ N_1(g, \mathcal{D}_d)}{4\,d\,\sinh(\pi d/h)}. \tag{3.6.27}$$

Proof. By Lemma 3.6.2, we have $g \in \mathbf{L}_\alpha(\mathcal{D}_d)$, so that $g \in \mathbf{H}^1(\mathcal{D}_d)$. Let us replace f by g, and z by t in (3.1.4), and let us integrate with respect to t from $-\infty$ to x. Then, using the definition (3.6.12) of $w(h,t,\tau,\xi)$, we get (3.6.26). The bound (3.5.27) then follows, from the estimate (3.6.13). ■

Lemma 3.6.5 *Let $h > 0$, $k \in \mathbb{Z}$, and $x \in \mathbb{R}$, and let us set*

$$J(k,h)(x) = \int_{-\infty}^{x} S(k,h)(t)\,dt, \tag{3.6.28}$$

with $S(k,h)$ defined as in (1.10.2). Then for $x \in \mathbb{R}$,

$$|J(k,h)(x)| \leqq 1.1\ h. \tag{3.6.29}$$

Proof. Setting

$$I(\xi) = \frac{1}{\pi} \int_{-\infty}^{\xi} \frac{\sin\ t}{t} dt \tag{3.6.30}$$

Then, for $\xi = m \in \mathbb{R}$, we have

$$I(m) = \frac{1}{2} + \sigma_m, \tag{3.6.31}$$

with σ_m defined as in (1.10.30).

Next, setting

$$\gamma_m = \frac{1}{\pi} \int_0^{\pi} \frac{\sin\ t}{m\pi + t} dt, \tag{3.6.32}$$

with $m \geqq 0$, $m \in \mathbb{Z}$, we have, by inspection, that

$$\begin{cases} \sigma_{m+1} - \sigma_m = (-1)^m \gamma_m; \\ \sigma_0 = 0, \quad \sigma_1 = \gamma_0; \\ \gamma_0 > \gamma_1 > \gamma_2 > \cdots > \gamma_m > 0; \\ \gamma_m \to 0, \quad m \to \infty. \end{cases} \tag{3.6.33}$$

Moreover, the next identities follow readily:

$$\begin{cases} \sigma_m = \sum_{j=1}^{m} (\sigma_j - \sigma_{j-1}) = \sum_{j=0}^{m-1} (-1)^j \gamma_j \\ \sigma_\infty = \frac{1}{2} = \sum_{j=0}^{\infty} (-1)^j \gamma_j \\ \sigma_m = \frac{1}{2} - (-1)^m \sum_{j=0}^{\infty} (-1)^j \gamma_{m+j}. \end{cases} \tag{3.6.34}$$

Hence, by (3.6.34), we have

$$(-1)^m [\sigma_m - \frac{1}{2}] = -\gamma_m + \gamma_{m+1} - \cdots > -\gamma_m. \tag{3.6.35}$$

Now, from the definition (1.10.30) of σ_m and the numerical values in Table 1.10.1, we have

$$\begin{cases} \sigma_0 = 0 \\ 0.58 < \sigma_1 < 0.59 \Rightarrow 0.58 < \gamma_0 < 0.59 \\ 0.45 < \sigma_2 < 0.46 \Rightarrow 0.12 < \gamma_1 < 0.14 \\ 0.53 < \sigma_3 < 0.54 \Rightarrow 0.08 < \gamma_2 < 0.09 \end{cases} \tag{3.6.36}$$

and moreover, using (3.6.35), and (3.6.36), we get

$$\begin{cases} \sigma_0 = 0; \\ 0.41 < \frac{1}{2} - \gamma_{2\ell} < \sigma_{2\ell} < \frac{1}{2},\ \ell = 1, 2, \cdots; \\ \frac{1}{2} < \sigma_{2\ell+1} < \frac{1}{2} + \gamma_{2\ell+1}. \end{cases} \tag{3.6.37}$$

Let us now return to (3.6.30), with $\xi = m\pi + \eta$, $m \in \mathbb{Z}$, $0 \leqq \eta < \pi$. Then, by (3.6.32) and (3.6.37) we have

$$\begin{cases} I(\xi) = \dfrac{1}{2} + \dfrac{1}{\pi} \displaystyle\int_0^\eta \frac{\sin t}{t} dt \quad \text{if} \quad m = 0 \\ I(\xi) = \dfrac{1}{2} + \sigma_m + \dfrac{(-1)^m}{\pi} \displaystyle\int_0^\eta \frac{\sin t}{m\pi + t} dt, \quad m > 0. \end{cases} \tag{3.6.38}$$

That is, if $m \geqq 0$, we have

$$\begin{cases} \frac{1}{2} < I(\xi) < \frac{1}{2} + \sigma_1 < 1.1 \quad \text{if } m = 0 \\ \frac{1}{2} + \sigma_{2\ell} < I(\xi) < \frac{1}{2} + \sigma_{2\ell+1} \text{ if } m = 2\ell,\ \ell = 1, 2, 3, \cdots \\ \frac{1}{2} + \sigma_{2\ell+2} < I(\xi) < \frac{1}{2} + \sigma_{2\ell+1} \text{ if } m = 2\ell + 1,\ \ell = 0, 1, 2, \cdots \\ I(-\xi) = 1 - I(\xi), \xi \in \mathbb{R}. \end{cases} \tag{3.6.39}$$

It now follows from (3.6.39) and (3.6.36) that for all $\xi \in \mathbb{R}$, we have

$$|I(\xi)| < 1.1. \tag{3.6.40}$$

Finally, setting $t = kh + uh/\pi$ in (3.6.30), we get

$$J(k, h)(x) = hI(\pi[x - kh]/h), \tag{3.6.41}$$

so that (3.6.29) follows from (3.6.40) and (3.6.41). ∎

We are now in position to complete the proof of Theorem 3.6.1. To this end we shall first show, in the notation of (3.6.1) and (3.6.5) through (3.6.7), that the relation

$$|(\mathcal{J}g)(x) - (C_N \mathcal{J} C_N g)(x)| \leqq c_1 N^{1/2} e^{-(\pi d \alpha N)^{1/2}} \tag{3.6.42}$$

with c_1 a constant independent of N, holds uniformly for $x \in \mathbb{R}$. The passage from this relation to (3.6.8) is left as a problem (Problem 3.6.1).

To establish (3.6.42), we write

$$\mathcal{J}g - C_N\mathcal{J}C_Ng = \varepsilon_1 + \varepsilon_2 + \varepsilon_3 \tag{3.6.43}$$

where

$$\begin{cases} \varepsilon_1 = \mathcal{J}g - C_N\mathcal{J}g \\ \varepsilon_2 = C_N\mathcal{J}g - C_N\mathcal{J}Cg \\ \varepsilon_3 = C_N\mathcal{J}Cg - C_N\mathcal{J}C_Ng \end{cases} \tag{3.6.44}$$

Now $\mathcal{J}g = G$, and by Lemma 3.6.2, we have $G \in \mathbf{L}_\alpha(\mathcal{D}_d)$. Hence, by Theorem 3.1.7, we have

$$|\varepsilon_1(x)| \leqq c_2 N^{1/2} e^{-(\pi d\alpha N)^{1/2}} \tag{3.6.45}$$

with c_2 a constant which is independent of N.
Next, by Lemma 3.6.4, we have

$$|(\mathcal{J}g)(x) - (\mathcal{J}Cg)(x)| \leqq c_3 h e^{-\pi d/h} \tag{3.6.46}$$

with c_3 a constant independent of h (assuming, of course, that h is uniformly bounded). Hence by Problem 3.1.5,

$$\begin{aligned} |\varepsilon_2(x)| &= |(C_N\mathcal{J}g)(x) - (C_N\mathcal{J}Cg)(x)| \\ &\leqq \|C_N\| \sup_{x\in\mathbf{R}} |\mathcal{J}\,Cg(x) - \mathcal{J}\,C_Ng(x)| \\ &\leqq c_7 N^{-1/2} \log N\, e^{-(\pi d\alpha N)^{1/2}}, \end{aligned} \tag{3.6.47}$$

with c_7 is a constant which is independent of N.
Let us now consider ε_3. By Lemma 3.6.2, we have $g \in \mathbf{L}_\alpha(\mathcal{D}_d)$, and therefore there is a constant c_8 such that $|g(kh)| \leqq c_8 e^{-\alpha|k|h}$, for all $k \in \mathbb{Z}$. Using the results of Lemma 3.6.5 we therefore have

$$\begin{aligned} |\mathcal{J}Cg(x) - \mathcal{J}C_Ng(x)| &= \left|\int_{-\infty}^{x} \sum_{|k|>N} g(kh)S(k,h)(t)\,dt\right| \\ &= \left|\sum_{|k|>N} g(kh)J_k(x)\right| \leqq \sup_{x\in\mathbf{R},k\in\mathbf{Z}} |J_k(x)| \sum_{|k|>N} |g(kh)| \\ &\leqq 1.1\,h \sum_{|k|>N} |g(kh)| \leqq 1.1\,hc_8 \sum_{|k|>N} e^{-\alpha|k|h} \\ &= 2.2\,h\,c_8 \frac{e^{-\alpha h}}{1-e^{-\alpha h}} e^{-\alpha Nh} \leqq \frac{2.2\,c_8}{\alpha} e^{-\alpha Nh}. \end{aligned} \tag{3.6.48}$$

Hence, taking $h = [\pi d/(\alpha N)]^{1/2}$, and using the results of Problem 3.1.5, we get

$$|\varepsilon_3(x)| \leqq c_9 \log N \, e^{-(\pi d \alpha N)^{1/2}}, \tag{3.6.49}$$

with c_9 a constant.
Combining (3.6.45), (3.6.47), and (3.6.49), we get (3.6.42).
This completes the proof of Theorem 3.6.1. ■

We now also record some important consequential results of Theorem 3.6.1, which can be more readily applied to approximate indefinite integrals, to solve Volterra integral equations, and to solve initial value problems in differential equations.

Corollary 3.6.6 *Let $f \in \mathbf{L}_\alpha(\mathcal{D}_d)$, with $0 < \alpha \leqq 1$, and $0 < d < \pi$. Let σ_k be defined as in Equation (1.10.30), and let $h = [\pi d/(\alpha N)]^{1/2}$, where N is a positive integer. Then there exists a positive constant K which is independent of N and h, such that*

$$\begin{aligned}
&\sup_{x \in \mathbf{R}} \Bigg| (\mathcal{J} f)(x) - \frac{e^x}{1+e^x} h \sum_{|k| \leqq N} f(kh) \\
&-h \sum_{|k| \leqq N} \left\{ \sum_{|\ell| \leqq N} \sigma_{k-\ell} \left[f(\ell h) - \frac{e^{\ell h}}{1+e^{\ell h}} h \sum_{|j| \leqq N} f(jh) \right] \right\} \cdot S(k,h)(x) \Bigg| \\
&\leqq K \, N^{1/2} \, e^{-(\pi d \alpha N)^{1/2}}.
\end{aligned} \tag{3.6.50}$$

Proof. The proof is a simple consequence of the fact that if the conditions of Corollary 3.6.6 are satisfied, then so are the conditions of Theorem 3.6.1, with $\beta = 1/2$, $\alpha' = \alpha$, and $d' = d$. ■

Next, we also readily have

Corollary 3.6.7 *Let $f \in \mathbf{L}_\alpha(\mathcal{D}_d)$, with $\alpha > 0$, $d > 0$, let σ_k and $\delta_k^{(-1)}$ be defined as in (1.10.30), and let $h = [\pi d/(\alpha N)]^{1/2}$. Then there exists a constant K independent of N, such that*

$$\left| \int_{-\infty}^{kh} f(t)\, dt - h \sum_{\ell=-N}^{N} \delta_{k-\ell}^{(-1)} f(\ell h) \right| \leqq K e^{-(\pi d \alpha N)^{1/2}}. \tag{3.6.51}$$

Proof. Letting $E(f,h)$ be defined as in (3.1.4), we have

$$\int_{-\infty}^{kh} f(t)\,dt - h \sum_{\ell=-\infty}^{\infty} \delta_{k-\ell}^{(-1)} f(\ell h) = \int_{-\infty}^{kh} E(f,h)(t)\,dt. \tag{3.6.52}$$

Now, by Lemma 3.6.3, it follows that

$$\left| \int_{-\infty}^{kh} E(f,h)(t)\,dt \right| \leqq \frac{h}{4d\sinh(\pi d/h)} N_1(f, \mathcal{D}_d). \tag{3.6.53}$$

Also, by Lemma 3.6.5, we have $|\delta_k^{(-1)}| \leqq 1.1$. Hence, proceeding as in (3.6.48), we have

$$h \left| \sum_{|\ell|>N} \delta_{k-\ell} f(\ell h) \right| \leqq \frac{2.2C}{\alpha} e^{-\alpha N h}. \tag{3.6.54}$$

The inequality (3.6.51) now follows by combining (3.6.52)–(3.6.4), and then selecting h as in the statement of Corollary 3.6.7. ■

We also state the following result, leaving the proof to Problem 3.6.3.

Corollary 3.6.8 *Let $f \in \mathbf{L}_\alpha(\mathcal{D}_d)$, with $\alpha > 0$, and $d > 0$. Let $I^{(-1)}$ denote the matrix*

$$I^{(-1)} = \left[\delta_{k-\ell}^{(-1)}\right], \quad k, \ell = -N, \cdots, N, \tag{3.6.55}$$

and let h be selected as in Corollary 3.6.6. Set $F(x) = \int_{-\infty}^{x} f(t)\,dt$, $F_k = F(kh)$, $f_k = f(kh)$, $\hat{F} = (F_{-N}, \cdots, F_N)^T$, and $\hat{f} = (f_{-N}, \cdots, f_N)^T$, and $\hat{G} = (G_{-N}, \cdots, G_N)^T = h\,I^{(-1)}\hat{f}$, with the superscript "$T$" denoting the transpose. Then

$$\hat{F} = \hat{G} + \hat{\varepsilon}, \tag{3.6.56}$$

with $\hat{\varepsilon} = (\varepsilon_{-N}, \cdots, \varepsilon_N)^T$, and with ε_j bounded by the right-hand side of (3.6.51). Moreover, if $\alpha \leqq 1$ and if $d < \pi$, then there exists a constant K independent of N, such that for all $x \in \mathbb{R}$, we have

$$\left| F(x) - \frac{e^x}{1+e^x} G_N - \sum_{k=-N}^{N} \left(G_k - \frac{e^{kh}}{1+e^{kh}} G_N \right) S(k,h)(x) \right| \tag{3.6.57}$$
$$\leqq K_1 N^{1/2} e^{-(\pi d \alpha N)^{1/2}}.$$

•

Problems for Section 3.6

3.6.1 Derive (3.6.8) using (3.6.42).

3.6.2 Use Theorem 3.6.1 to find approximations of the integrals:

(a) $\int_0^x \exp\{-\sqrt{t^3+1}\}t^{-6/5} \sin t\, dt$;

(b) $\int_{-\infty}^x \exp\{-\sqrt{t^2+2}\} \cos t\, dt$;

(c) $\int_{-\infty}^{-x} t\, e^{-t^2}\, dt$.

3.6.3 Prove Corollary 3.6.8.

3.6.4 State and prove a theorem analogous to Theorem 3.6.1 to approximate $\int_x^\infty f(t)\, dt$.

3.6.5 In the notation of Corollary 3.6.7, prove that

$$\left| \int_{kh}^\infty f(t)\, dt - h \sum_{\ell=-N}^{N} \delta_{\ell-k}^{(-1)} f(\ell h) \right| \leqq K\, e^{-(\pi d \alpha N)^{1/2}}.$$

3.6.6 Let $f \in \mathbf{L}_\alpha(\mathcal{D}_d)$, with $\alpha > 0$ and $d > 0$. Let J denote the transpose of $I^{(-1)}$, and set $G_k = \int_{kh}^\infty f(t)\, dt$, $\hat{G} = (G_{-N}, \cdots, G_N)^T$, $f_k = f(kh)$, and $\hat{f} = (f_{-N}, \cdots, f_N)^T$.

(a) Prove that

$$\hat{G} = hJ\hat{f} + \hat{\varepsilon},$$

with $\hat{\varepsilon} = (\varepsilon_{-N}, \cdots, \varepsilon_N)^T$, and $|\varepsilon_j| \leqq K \exp[-(\pi d \alpha N)^{1/2}]$.

(b) Prove that if $G(x) = \int_x^\infty f(t)\, dt$, if $0 < \alpha \leqq 1$, and if $0 < d < \pi$, then there exists a constant K_1 independent of N, such that for all $x \in \mathbb{R}$,

$$\left| G(x) - \frac{G_{-N}}{1+\rho(x)} - \sum_{k=-N}^{N} \left(G_k - \frac{G_{-N}}{1+e^{kh}} \right) S(k,h)(x) \right| \leqq K_1 N^{1/2} e^{-(\pi d \alpha N)^{1/2}}.$$

3.6.7 Restate Corollary 3.6.8 for approximating $\int_x^\infty F(x)\, dx$.

3.6.8 Use Corollary 3.6.8 to derive a Sinc approximation of

$$\frac{1}{\sqrt{\pi}} \int_{-\infty}^x e^{-t^2} dt$$

accurate to 6 decimals for all $x \in \mathbb{R}$.

Historical Remarks on Chapter 3

Useful summaries involving the trapezoidal rule may be found in *Hardy* [Hr1] and *Lindelöf* [Ln1]. The rectangular contour integral method described in Problem 3.2.2 to derive the trapezoidal rule dates back to *Plana* [Pl1], *Abel* [Ab1], and *Cauchy* [Cu1]. *Goodwin* [Goo1] appears to be the first to have noted the incredible accuracy of the trapezoidal formula when applied to the integration of a certain class of functions over $\mathbb{R}$. It became clear from *McNamee's* paper [Mc1] (see also *Davis* [Da1]) that the appropriate class of functions was the one for which the bound on the contour integral could be made small.

Shortly thereafter, *McNamee* and my Ph.D. advisor, *Whitney*, submitted a paper about Whittaker's Cardinal function to SIAM Review, which was returned to them for modification. They were too proud to make the revisions, inasmuch as many of the referees' demands were unreasonable, and after some time, with their permission, I made the modifications and I also added a few novelties. They thus allowed me to become a joint author [McSW1], and to become familiar with the very beautiful cardinal function. In [McSW1] the rectangular contour integral method was used to bound both the interpolation error by the Cardinal function, as well as the error for the trapezoidal rule.

In [Ma1] *Martensen* wrote a pretty paper on the use of the contour integral method to derive and bound the error for the trapezoidal rule, and in [Ks1] (see also [Ks7]) *Kress* wrote interesting and pretty papers using the contour integral method to bound the error of interpolation, differentiation, and Hilbert transform, each approximating formula being a cardinal function identity (Section 1.10 of this text). I first derived $\mathbf{L}^p$ bounds for these approximations in [S12].

The assumption on the decrease rate of f as given in Definition 3.1.5, was first made in [S14]; this assumption yields the $\mathcal{O}(\exp(-c\,n^{1/2}))$ rate of convergence of all the methods, and is readily extended to transformations of the methods, as described in Chapter 4.

While these previous results are for approximation over the real line, the results in the present chapter establish convergence throughout $\mathcal{D}_{d'}$ for any $d' \in (0, d)$ and they are new. The results of Theorem 3.1.9, and Problems 3.2.6 and 3.4.11 were first obtained by Bialecki (see [*Bialecki*, Bi3]). Formulas and error analysis for the Fourier transform, the indefinite integral (see *Kearfott*, [K1], for the first published error analysis of the indefinite integral formula, and *Haber*, [H3] for a more recent version), the projection approximation (see Theorem 3.4.1 and Corollary 3.4.2), and the Dirichlet problem solutions of Corollary 3.4.3,—all of which use identities given in Section 1.10 for the cardinal function were first given in ([S24]).

4

Sinc Approximation on Γ

In this chapter we shall extend some of the results for approximation on $\mathbb{R}$, which were obtained in the previous section, to results for approximating on a certain arc Γ. At the outset we shall introduce some definitions and notations which will make it possible to deduce the correct approximations directly, from results obtained in the previous chapter. Many new formulas are thus possible, which are both simple to apply and very accurate. In addition, these formulas yield accurate methods for interpolating and integrating functions which may have singularities at the endpoints of Γ, a property which the polynomial methods shown in Chapter 2 do not possess.

4.1 Basic Definitions

Let us at once introduce a definition which will be very important throughout the remainder of this book.

Definition 4.1.1 *Let $\mathcal{D}$ be a simply-connected domain having boundary $\partial\mathcal{D}$. Let a and b denote two distinct points of $\partial\mathcal{D}$, and let ϕ denote a conformal map of $\mathcal{D}$ onto $\mathcal{D}_d$, where $\mathcal{D}_d$ is defined in Equation (1.7.9), such that $\phi(a) = -\infty$, and $\phi(b) = \infty$. Let $\psi = \phi^{-1}$ denote the inverse map, and let Γ be defined by*

$$\Gamma = \{z \in \mathbb{C} : z = \psi(u), u \in \mathbb{R}\}. \tag{4.1.1}$$

Given ϕ, ψ and a positive number h, let us set

$$z_k = z_k(h) = \psi(kh), \;\; k = 0, \pm 1, \pm 2, \cdots. \tag{4.1.2}$$

Let us also define ρ by

$$\rho(z) = e^{\phi(z)}. \tag{4.1.3}$$

Let $1 \leqq p \leqq \infty$, and let $\mathbf{H}^p(\mathcal{D})$ denote the family of all functions F that are analytic in $\mathcal{D}$, such that

$$N_p(F, \mathcal{D}) \equiv \begin{cases} \left(\int_{\partial\mathcal{D}} |F(z)|^p |dz| \right)^{1/p} < \infty \;\text{ if }\; 1 \leqq p < \infty, \\ \sup\limits_{z \in \mathcal{D}} |F(z)| < \infty \;\text{ if }\; p = \infty. \end{cases} \tag{4.1.4}$$

In the case $p = 1$ we shall simply write $N(F, \mathcal{D})$, instead of $N_1(F, \mathcal{D})$.

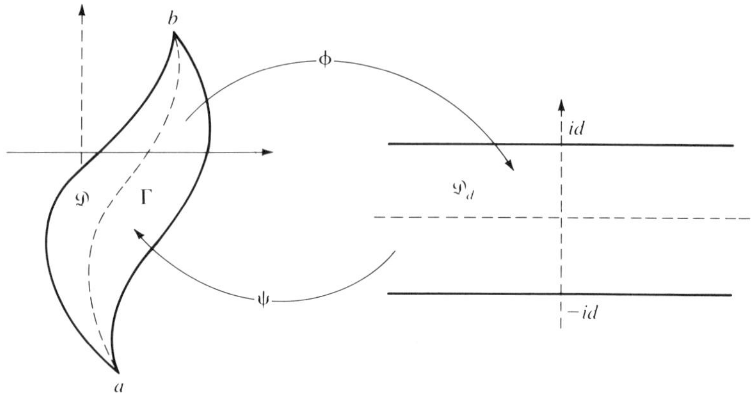

Figure 4.1.1: The General Picture

Corresponding to numbers α and β, let $\mathbf{L}_{\alpha,\beta}(\mathcal{D})$ denote the family of all functions $F \in \mathbf{Hol}(\mathcal{D})$ for which there exists a constant C such that

$$|F(z)| \leqq C \frac{|\rho(z)|^{\alpha}}{[1 + |\rho(z)|]^{\alpha+\beta}} \tag{4.1.5}$$

for all z in $\mathcal{D}$. Finally, let us define $\mathbf{L}_{\alpha}(\mathcal{D})$ by $\mathbf{L}_{\alpha}(\mathcal{D}) = \mathbf{L}_{\alpha,\alpha}(\mathcal{D})$.

It is convenient to define yet another important family of functions, $\mathbf{M}_{\alpha,\beta}(\mathcal{D})$, with $0 < \alpha \leqq 1$, $0 < \beta \leqq 1$, and $0 < d < \pi$. The family $\mathbf{M}_{\alpha,\beta}(\mathcal{D})$ consists of all those functions $F \in \mathbf{Hol}(\mathcal{D}) \cap \mathbf{C}(\overline{\mathcal{D}})$ such that $G \in \mathbf{L}_{\alpha,\beta}(\mathcal{D})$ where

$$G(z) = F(z) - \frac{F(a) + \rho(z) F(b)}{1 + \rho(z)}. \tag{4.1.6}$$

Just as for $\mathbf{L}_{\alpha,\beta}(\mathcal{D})$ and $\mathbf{L}_{\alpha}(\mathcal{D})$, we define $\mathbf{M}_{\alpha}(\mathcal{D})$ by $\mathbf{M}_{\alpha}(\mathcal{D}) = \mathbf{M}_{\alpha,\alpha}(\mathcal{D})$.

It is shown in Problem 4.2.5 that the class $\mathbf{M}_{\alpha}(\mathcal{D})$ contains the class $\mathbf{Lip}_{\alpha}(\overline{\mathcal{D}}) \cap \mathbf{Hol}(\mathcal{D})$ for the case of the bounded regions which we shall encounter in the examples of this chapter. We remark, also, that if F and G are related as in (4.1.6), then the term $L \equiv F - G = [F(a) + \rho\, F(b)]/[1+\rho]$ satisfies the relations $L \circ \psi \in \mathbf{Hol}(\mathcal{D}_{d'})$ for any $d' \in (0, \pi)$, $L(z) - L(a) = \mathcal{O}(|\rho(z)|)$ as $z \to a$ from within $\mathcal{D}$, and $L(z) - L(b) = \mathcal{O}(1/|\rho(z)|)$ as $z \to b$ from within $\mathcal{D}$, and this is why we require $0 < \alpha \leqq 1$ and $0 < d < \pi$ in the definition of the class $\mathbf{M}_{\alpha,\beta}(\mathcal{D})$.

The classes $\mathbf{L}_{\alpha}(\mathcal{D})$ and $\mathbf{M}_{\alpha}(\mathcal{D})$ are important to Sinc methods since they are easy to recognize in applications; and also, since they guarantee the rapid convergence of Sinc approximation described in the theorems of this chapter, as well as of the approximations used in later chapters of this text. This same rate of convergence can be achieved under more general conditions, but which may not be so easily checked. For example, as we shall see below, for purposes of accurate interpolation (respectively, quadrature) on Γ, all that is required is that $\phi' F$ (respectively, F) is in $\mathbf{H}^1(\mathcal{D})$, and that

$$|F(x)| \leqq Ce^{-\alpha|\phi(x)|} \left(\text{resp.,} \left|\frac{F(x)}{\phi'(x)}\right| \leqq Ce^{-\alpha|\phi(x)|}\right) \tag{4.1.7}$$

for all $x \in \Gamma$. If F (resp., F/ϕ') is in $\mathbf{L}_\alpha(\mathcal{D})$ then these conditions are satisfied.

Definition 4.1.1 enables us to easily extend directly the results of interpolation, quadrature, etc. on $\mathbb{R}$, obtained in the previous chapter, to corresponding results for interpolation, quadrature, etc. over Γ. The basis functions, $S(k,h) \circ (x)$, that were used in the previous chapter for approximating on $\mathbb{R}$, now become the basis functions $S(k,h) \circ \phi(x)$ for approximating on Γ. The new interpolation points $\{z_k\}_{k=-\infty}^{\infty}$ play the same role for approximating on Γ as the points $\{kh\}_{k=-\infty}^{\infty}$ played for approximating on $\mathbb{R}$.

Let us next state and prove some results which are simple consequences of these definitions.

Theorem 4.1.2 *Let $\alpha > 0$, $d' > 0$, let $\mathcal{D}_{d'}$ be defined as in (1.7.9), let $\mathcal{D}' = \psi(\mathcal{D}_{d'})$ and for some fixed $d \in (0, d')$, let $\mathcal{D} = \psi(\mathcal{D}_d)$. Let $F \in \mathbf{L}_\alpha(\mathcal{D}')$. Then:*

(a) $F'/\phi' \in \mathbf{L}_\alpha(\mathcal{D})$;

(b) $F''/(\phi')^2 \in \mathbf{L}_\alpha(\mathcal{D})$ whenever $(1/\phi')'$ is uniformly bounded in $\mathcal{D}$.

Proof. We shall use the result of Theorem 1.8.5.

Part (a): Since $F \in \mathbf{L}_\alpha(\mathcal{D}')$, $F \circ \psi \in \mathbf{L}_\alpha(\mathcal{D}_{d'})$, and therefore, by Theorem 1.8.5, $(F' \circ \psi)\psi' \in \mathbf{L}_\alpha(\mathcal{D}_d)$. That is, $F'/\phi' \in \mathbf{L}_\alpha(\mathcal{D})$.

Part (b): By Theorem 1.8.5 and part (a) of this theorem, we have $[(F' \circ \psi)\psi']' \in \mathbf{L}_\alpha(\mathcal{D}_d)$. But, since

$$(F'' \circ \psi)(\psi')^2 = [(F' \circ \psi)\psi']' - (F' \circ \psi)\psi'(\psi''/\psi'),$$

it now follows that each term on the right-hand side of this equation is in $\mathbf{L}_\alpha(\mathcal{D}_d)$, provided that ψ''/ψ' is uniformly bounded in $\mathcal{D}_d$. But since $\psi''/\psi' = -(1/\phi')'$, the result (b) follows. ■

Theorem 4.1.3 *Let $\alpha > 0$, $d > 0$, and let $F'/\phi' \in \mathbf{L}_\alpha(\mathcal{D})$. Set $F(z) = \int F'(z)\,dz$. Then $F \in \mathbf{H}^\infty(\mathcal{D})$. Moreover, if $0 < d < \pi$, and $0 < \alpha \leqq 1$, then $F \in \mathbf{M}_\alpha(\mathcal{D})$.*

Proof. Setting $z = \psi(\zeta)$, we see that $F'/\phi' \in \mathbf{L}_\alpha(\mathcal{D})$ if and only if $(F' \circ \psi)\psi' \in \mathbf{L}_\alpha(\mathcal{D}_d)$. Let G be defined in terms of F by (4.1.6). That $G \in \mathbf{L}_\alpha(\mathcal{D})$

is then a consequence of Lemma 3.6.2. We leave the details of the remainder of the proof to Problem 4.1.2. ■

Theorem 4.1.4 *Let $F \in \mathbf{L}_\alpha(\mathcal{D})$, and let $\mathcal{D}$ and $\mathcal{D}'$ be selected as in Theorem 4.1.2. Moreover, let $0 < d < \pi$, and let $0 \leqq \alpha \leqq 1$.*
(a) Define TF on Γ by

$$TF(x) = P.V. \int_\Gamma \frac{F(t)\,\phi'(t)}{\phi(t) - \phi(x)} dt. \tag{4.1.8}$$

Then TF has an analytic continuation into $\mathcal{D}'$ which belongs to $\mathbf{M}_\alpha(\mathcal{D})$.
(b) Let $\mathcal{D}$ be a bounded region, and let SF be defined by

$$SF(x) = P.V. \int_\Gamma \frac{F(t)}{t - x} dt. \tag{4.1.9}$$

Then SF has an analytic continuation into $\mathcal{D}'$ which belongs to $\mathbf{M}_\alpha(\mathcal{D})$.

Proof. We defer the proof of this theorem to Problem 4.1.5. ■

•

Problems for Section 4.1

4.1.1 Prove, in the notation of Example 1.7.5, that $(1/\phi')'$ is uniformly bounded in the region $\psi(\mathcal{D}_d)$ for each of the following cases:

(i) For all $d > 0$ if $\phi(z) = z$;

(ii) For all $d > 0$ if $\phi(z) = \log(z)$;

(iii) For all $d \in (0, \pi/2)$ if $\phi(z) = \log[\sinh(z)]$;

(iv) For all $d \in (0, \pi)$ if $\phi(z) = \log[(z-a)/(b-z)]$; and

(v) For all $d \in (0, \pi/2)$ if $\phi(z) = \log[z + \sqrt{1+z^2}]$.

4.1.2 Supply the details of the last part of the proof of Theorem 4.1.3.

4.1.3 Determine the maximum value of α and d, for which each of the following functions f is in $\mathbf{M}_\alpha(\mathcal{D})$, where the mapping $\phi : \mathcal{D} \to \mathcal{D}_d$ is given:

(i) $f(x) = e^{-x^2} + (2 - 3\,e^x)/(1 + e^x)$, $\phi(x) = x$;

(ii) $f(x) = (3 + 8\,x^2)^{1/3}/(1+x)^{2/3}$, $\phi(x) = \log(x)$;

(iii) $f(x) = (1+x)^{1/2} x^{-1} \log(1+x) e^{x/2}/(1+e^x)^{1/2}$, $\phi(x) = \log(\sinh(x))$;

(iv) $f(x) = x^{-1}(1+x^2)^{1/2} \log(1+x)$, $\phi(x) = \log[(1+x)/(1-x)]$; and

(v) $f(x) = (x + \sqrt{1+x^2})\, x\, e^{-x}/(4 + e^{-2x})^{1/2}$, $\phi(x) = \log\{\sinh(x + \sqrt{1+x^2})\}$.

4.1.4 Determine the explicit function g for each of the functions f in Problem 4.1.3, such that $g = f - [f(a) + \rho\, f(b)]/[1+\rho] \in \mathbf{L}_\alpha(\mathcal{D})$.

4.1.5 Give a detailed proof of Theorem 4.1.4. [Hint: Part (a) of the theorem follows directly from Theorem 3.4.4. Part (b) follows from the fact that if $\mathcal{C}$ is a closed contour, and if $F \in \mathbf{Lip}_\alpha(\mathcal{C})$ then $SF \in \mathbf{Lip}_\alpha(\mathcal{C})$, and also, since we have $F \in \mathbf{Lip}_\alpha(\overline{\mathcal{D}})$, under our assumptions.]

4.1.6 Let $R > 0$, and let $\mathcal{U}_R$ denote the interior of the disc with center at the origin, and radius R. Prove that if $f \in \mathbf{Hol}(\mathcal{U}_R)$, then $f \in \mathbf{M}_\alpha(\mathcal{U}_r)$ for every $r \in (0, R)$.

4.1.7 Let d, d', $\mathcal{D}$ and $\mathcal{D}'$ be defined as in Theorem 4.1.2. Prove that if $(1/\phi')'$ (or $1/\phi'$) is uniformly bounded in $\mathcal{D}'$, then $(1/\phi')^{(k)}/[\phi']^k$ is uniformly bounded in $\mathcal{D}$ for every positive integer k. [Hint: Define F by $F(w) = (1/\phi') \circ \psi(w)$, on $\mathcal{D}_{d'}$ and use contour integration formula (see Taylor's theorem) to express $F^{(m)}(v)$, for $v \in \mathcal{D}_d$ as an integral of $F(w)$ to conclude that $F^{(m)}$ is uniformly bounded in $\mathcal{D}_d$ for every integer $m \geqq 0$. What is $F'(w)$ as a function of z?]

4.1.8 Let d, d', $\mathcal{D}$ and $\mathcal{D}'$ be defined as in Theorem 4.1.2. Prove that if $F \in \mathbf{L}_\alpha(\mathcal{D})$, and if $(1/\phi')'$ is uniformly bounded in $\mathcal{D}'$, then $F^{(k)}/[\phi']^k \in \mathbf{L}_\alpha(\mathcal{D})$ for every positive integer k. [Hint: Use induction. Part of the proof comes from Theorem 4.1.2, and the remaining part by inspection of the identity

$$\frac{1}{\phi'}\left(\frac{F^{(k)}}{[\phi']^k}\right)' = \frac{F^{(k+1)}}{[\phi']^{k+1}} - k\left(\frac{1}{\phi'}\right)' \frac{F^{(k)}}{[\phi']^k},$$

in view of Theorem 4.1.2.]

4.1.9 (a) Prove that if ϕ_1 and ϕ_2 are conformal maps such that if $\phi_1 : \mathcal{D}_1 \to \mathcal{D}_d$, $\phi_2 : \mathcal{D}_2 \to \mathcal{D}_d$, and if $F \in \mathbf{L}_{\alpha,\beta}(\mathcal{D}_1)$, then $F \circ \phi_1 \circ \phi_2 \in \mathbf{L}_{\alpha,\beta}(\mathcal{D}_2)$. (b) Repeat the problem, replacing $\mathbf{L}$ with $\mathbf{M}$.

4.2 Interpolation and Quadrature on Γ

In this section we shall derive the basic formulas for interpolation and quadrature on Γ. These results are then illustrated with some examples.

The first theorem contains two very important identities.

Theorem 4.2.1 *Let $F \in \mathbf{H}^1(\mathcal{D})$.*
(a) *For all z in $\mathcal{D}$,*

$$\frac{F(z)}{\phi'(z)} - \sum_{k=-\infty}^{\infty} \frac{F(z_k)}{\phi'(z_k)} S(k,h) \circ \phi(z)$$
$$= \frac{\sin[\pi\phi(z)/h]}{2\pi i} \int_{\partial\mathcal{D}} \frac{F(\zeta) d\zeta}{[\phi(\zeta) - \phi(z)] \sin[\pi\phi(\zeta)/h]}. \tag{4.2.1}$$

(b) *Furthermore,*

$$\int_\Gamma F(z)\, dz - h \sum_{k=-\infty}^{\infty} \frac{F(z_k)}{\phi'(z_k)}$$
$$= \frac{i}{2} \int_{\partial\mathcal{D}} \frac{F(\zeta)\, \exp[i(\pi/h)\, \phi(\zeta)\, \mathrm{sgn}\, \Im\phi(\zeta)]\, d\zeta}{\sin[\pi\phi(\zeta)/h]}. \tag{4.2.2}$$

Proof. If $F \in \mathbf{H}^1(\mathcal{D})$, then it follows at once that $f = F \circ \psi\, \psi' \in \mathbf{H}^1(\mathcal{D}_d)$. Hence Equation (4.2.1) follows at once from Equation (3.1.4), after replacing $\phi(\zeta)$ by $t \pm id$, and $\phi(z)$ by z. Equation (4.2.2) is obtained from Equation (4.2.1) after multiplying both sides of (4.2.1) by $\phi'(z)$, integrating over Γ, interchanging the order of integration in the double integral, and using the identities

$$\int_\Gamma S(k,h) \circ \phi(z)\, \phi'(z)\, dz = h \tag{4.2.3}$$

and

$$\frac{1}{2\pi i} \int_\Gamma \frac{\sin[(\pi/h)\phi(z)]}{[\phi(\zeta) - \phi(z)]} \phi'(z)\, dz$$
$$= \frac{i}{2} \exp\left[i(\pi/h)\phi(\zeta)\ \mathrm{sgn}\{\Im\phi(\zeta)\}\right], \tag{4.2.4}$$

which follow from (1.10.28) and (3.2.7). ■

We remark that we readily obtain a method for interpolating F (rather than F/ϕ') if we assume that $\phi' F \in \mathbf{H}^1(\mathcal{D})$, and if we replace F by $\phi' F$ in (4.2.1). We then can easily bound the contour integrals on the right-hand sides of (4.2.1) and (4.2.2), to get

Theorem 4.2.2 (a) *Let $\phi' F \in \mathbf{H}^1(\mathcal{D})$. Then for $z \in \mathcal{D}$,*

$$\left| F(z) - \sum_{k=-\infty}^{\infty} F(z_k)\, S(k,h) \circ \phi(z) \right| \leqq \frac{\cosh\left[(\pi/h)\{\Im\phi(z)\}\right] N(\phi' F, \mathcal{D})}{2\pi\left[d - |\Im\phi(z)|\right]\sinh(\pi d/h)}. \tag{4.2.5}$$

(b) *If $F \in \mathbf{H}^1(\mathcal{D})$, then*

$$\left| \int_\Gamma F(z)dz - h \sum_{k=-\infty}^{\infty} \frac{F(z_k)}{\phi'(z_k)} \right| \leqq \frac{N(F,\mathcal{D})e^{-\pi d/h}}{2\sinh(\pi d/h)}. \tag{4.2.6}$$

Proof. Since $z \in \mathbb{R}$ and $\zeta \in \partial\mathcal{D}$, it follows that $\Im\phi(\zeta) = \pm d$, and $-d < \Im\phi(z) < d$. Hence (see (3.1.8))

$$|\sin\left[(\pi/h)\phi(\zeta)\right]| \geqq \sinh(\pi d/h) \tag{4.2.7}$$

$$|\sin\left[(\pi/h)\phi(z)\right]| \leqq \cosh\left[(\pi/h)\{\Im\phi(z)\}\right] \tag{4.2.8}$$

$$|\phi(\zeta) - \phi(z)| \geqq d - |\Im\phi(z)| \tag{4.2.9}$$

$$|\exp\left[i(\pi/h)\phi(\zeta)\operatorname{sgn}\{\Im\phi(\zeta)\}\right]| = e^{-\pi d/h}. \tag{4.2.10}$$

By applying these inequalities to (4.2.1) and (4.2.2), we readily obtain the bounds (4.2.5) and (4.2.6). ■

We remark that the p-norm results of Section 3.1 may be readily extended to the present case of interpolation on Γ. However, for purposes of avoiding excessive complexity in the statements and proofs of the results of this section, we have confined ourselves to the case of $p = \infty$.

Corollary 4.2.3 *Let $0 \leqq d' < d$, and let $\mathcal{D}'$ be defined by*

$$\mathcal{D}' = \{z \in \mathbb{C} : |\Im\phi(z)| \leqq d'\}. \tag{4.2.11}$$

If $\phi' F \in \mathbf{H}^1(\mathcal{D})$, then there exists a constant C which is independent of h, such that

$$\sup_{z\in\mathcal{D}'} \left| F(z) - \sum_{k=-\infty}^{\infty} F(z_k)\, S(k,h) \circ \phi(z) \right| \leqq C e^{-\pi(d-d')/h}. \tag{4.2.12}$$

We remark that the inequality (4.2.12) shows that the infinite sum $\sum_{k=-\infty}^{\infty} F(z_k)S(k,h) \circ \phi$ converges uniformly to F in $\mathcal{D}'$.

Lemma 4.2.4 *If $F \in \mathbf{L}_{\alpha,\beta}(\mathcal{D})$, then $\phi' F \in \mathbf{H}^1(\mathcal{D})$.*

Proof. Clearly $F \in \mathbf{Hol}(\mathcal{D})$. It will thus follow that $\phi' F \in \mathbf{H}^1(\mathcal{D})$, if $\int_{\partial D} |\phi'(z)F(z)\,dz| < \infty$. Now, if $z \in \partial\mathcal{D}$, then $\rho(z) = e^{\phi(z)} = e^{t+i\tau}$ where $t = \Re\phi(z)$, $\tau = \Im\phi(z) = \pm d$. Thus $d\rho(z) = \phi'(z)\rho(z)dz = e^{t+i\tau}dt = \rho(z)dt$, i.e., $dz = \{1/\phi'(z)\}dt$. Hence if $F \in \mathbf{L}_{\alpha,\beta}(\mathcal{D})$, then

$$\begin{aligned} \int_{\partial\mathcal{D}} |\phi'(z)F(z)\,dz| &\leqq C \int_{\partial\mathcal{D}} \frac{|\rho(z)|^{\alpha}|\phi'(z)\,dz|}{[1+|\rho(z)|]^{\alpha+\beta}} \\ &= 2C \int_{\mathbb{R}} \frac{e^{\alpha t}dt}{[1+e^t]^{\alpha+\beta}} = 2C\frac{\Gamma(\alpha)\,\Gamma(\beta)}{\Gamma(\alpha+\beta)} < 2\,C\left(\frac{1}{\alpha}+\frac{1}{\beta}\right) < \infty. \end{aligned} \tag{4.2.13}$$

■

Theorem 4.2.5 *Let $F \in \mathbf{L}_{\alpha,\beta}(\mathcal{D})$, let $\varepsilon = \min(\alpha,\beta)$, $\delta = \max(\alpha,\beta)$, let n be a positive integer, and let h be selected by the formula*

$$h = \left(\frac{\pi d}{\varepsilon n}\right)^{1/2}. \tag{4.2.14}$$

If $\varepsilon = \alpha$, let M and N be defined by

$$M = n, \quad N = [\alpha n/\beta], \tag{4.2.15}$$

while if $\varepsilon = \beta$, let M and N be defined by

$$M = [\beta n/\alpha], \quad N = n. \tag{4.2.16}$$

If $0 \leqq d' < d$ and if $\mathcal{D}'$ is defined by (4.2.11), then there exists a constant C which is independent of n, such that

$$\begin{aligned} &\sup_{z\in\mathcal{D}'} \left| F(z) - \sum_{k=-M}^{N} F(z_k)\,S(k,h)\circ\phi(z) \right| \\ &\leqq C n^{1/2} \exp\left\{-(d-d')\left(\frac{\pi\varepsilon n}{d}\right)^{1/2}\right\}. \end{aligned} \tag{4.2.17}$$

Proof. By Lemma 4.2.4 and Corollary 4.2.3 it follows that

$$\sup_{z \in \mathcal{D}'} \left| F(z) - \sum_{k=-M}^{N} F(z_k)\, S(k,h) \circ \phi(z) \right| \leqq S_0 + S_1 + S_2, \tag{4.2.18}$$

where

$$\begin{aligned} S_0 &= C_0 e^{-\pi(d-d')/h}, \\ S_1 &= \sup_{z \in \mathcal{D}'} \left| \sum_{k=-\infty}^{-M-1} F(z_k)\, S(k,h) \circ \phi(z) \right| \\ S_2 &= \sup_{z \in \mathcal{D}'} \left| \sum_{k=N+1}^{\infty} F(z_k)\, S(k,h) \circ \phi(z) \right| \end{aligned} \tag{4.2.19}$$

and where S_0 is the bound on the right-hand side of (4.2.12). But by (3.1.29), we have

$$\sup_{z \in \mathcal{D}'} |S(k,h) \circ \phi(z)| \leqq \cosh(\pi d'/h) < e^{\pi d'/h}. \tag{4.2.20}$$

Hence, since $F \in \mathbf{L}_{\alpha,\beta}(\mathcal{D})$, we use (4.1.5) to get, in the case when $\varepsilon = \alpha$, that

$$\begin{aligned} |S_1| &\leqq C_2 e^{\pi d'/h} \sum_{k=-\infty}^{-M-1} \frac{|\rho(z_k)|^{\alpha}}{[1 + |\rho(z_k)|]^{\alpha+\beta}} \\ &\leqq C_2 e^{\pi d'/h} \sum_{k=-\infty}^{-M-1} e^{\varepsilon k h} \\ &\leqq \frac{C_2}{\alpha h} e^{\pi d'/h} e^{-\varepsilon n h} \end{aligned} \tag{4.2.21}$$

where C_2 denotes the constant C on the right-hand side of (4.1.5), and where $M = n$. Similarly, if $M = [(\alpha/\beta)n]$, then S_2 is also bounded by the right-hand side of (4.2.21). These same bounds hold also for the case when $\varepsilon = \beta$. Thus, the sum of the three terms on the right-hand side of (4.2.18) satisfies the inequality

$$S_0 + S_1 + S_2 \leqq \frac{C_3}{h} e^{\pi d'/h} \left[h\, e^{-\pi d/h} + 2\, e^{-\varepsilon n h} \right] \tag{4.2.22}$$

where C_3 is a constant that is independent of n and h. Now, by taking h as in (4.2.14), we get (4.2.17). ■

The determination of h by (4.2.14) has the desirable feature that in the case of uniform Sinc approximation the error

$$F(z) - \sum_{k=-M}^{N} F(z_k) S(k,h) \circ \phi(z)$$

has roughly the same magnitude near the center as far from the center of Γ. In addition the extrema of the error seem to occur close to the Sinc points $\{z_k\}$. Choosing an h smaller than the optimal value predicted by (4.2.14) results in a more accurate approximation near the center (i.e., near the point z_0) of Γ and a less accurate one near the endpoints of Γ, while choosing an h larger than the optimal value predicted by (4.2.14) results in a more accurate approximation near the endpoints of Γ and a less accurate approximation near the center of Γ.

The following theorem involves bounding the error of $M + N + 1$-point *Sinc* quadrature of F on Γ. We omit proof of this theorem, which is similar, although simpler than the proof of Theorem 4.2.5.

Theorem 4.2.6 *Let $F/\phi' \in \mathbf{L}_{\alpha,\beta}(\mathcal{D})$, let $\varepsilon = \min(\alpha,\beta)$, $\delta = \max(\alpha,\beta)$, let n be a positive integer, and let h be selected by the formula*

$$h = \left(\frac{2\pi d}{\varepsilon n}\right)^{1/2}. \tag{4.2.23}$$

If $\varepsilon = \alpha$, let M and N defined by

$$M = n, \quad N = [\alpha n/\beta] \tag{4.2.24}$$

while if $\varepsilon = \beta$, let M and N be defined by

$$M = [\beta n/\alpha], \quad N = n. \tag{4.2.25}$$

Then there exists a constant C, independent of n, such that

$$\left| \int_\Gamma F(z)dz - h \sum_{k=-M}^{N} \frac{F(z_k)}{\phi'(z_k)} \right| \leq C e^{-(2\pi d\varepsilon n)^{1/2}}. \tag{4.2.26}$$

In applications, the constants d, α and β may be known to exist, yet their exact values may not be known. For this case, we take note of the following result.

Corollary 4.2.7 (a) *Let $F \in \mathbf{L}_\alpha(\mathcal{D})$, let N be a positive integer, and let h be selected by the formula*

$$h = c/N^{1/2} \tag{4.2.27}$$

with c a positive constant. Then there exist positive constants C and γ, independent of N, such that

$$\sup_{z\in\Gamma}\left|F(z)-\sum_{k=-N}^{N}F(z_k)S(k,h)\circ\phi(z)\right|\leq Ce^{-\gamma N^{1/2}}. \tag{4.2.28}$$

(b) *Let $F/\phi' \in \mathbf{L}_\alpha(\mathcal{D})$, let N be a positive integer, and let h be selected by the formula (4.2.27), with c a positive constant. Then there exist positive constants C and γ, independent of N, such that*

$$\left|\int_\Gamma F(z)dz - h\sum_{k=-N}^{N}\frac{F(z_k)}{\phi'(z_k)}\right|\leq Ce^{-\gamma N^{1/2}}. \tag{4.2.29}$$

Proof. The proof of part (a) of Corollary 4.2.7 follows at once from the inequality (4.2.22), by setting $d' = 0$, $n = N$ and then taking $h = c/N^{1/2}$. The proof of part (b) of the Corollary is similar, and we omit it. ∎

It may happen that $F \in \mathbf{H}^1(\mathcal{D})$, but $\phi' F \notin \mathbf{H}^1(\mathcal{D})$. If the limits

$$\lim_{x\to a}F(x)=F(a),\quad \lim_{x\to b}F(x)=F(b) \tag{4.2.30}$$

taken along Γ exist and are bounded, it may happen that $\phi' G \in \mathbf{H}^1(\mathcal{D})$, with

$$G(z)=F(z)-\frac{F(a)+\rho(z)F(b)}{1+\rho(z)}, \tag{4.2.31}$$

with ρ defined as in (4.1.3). These conditions are all satisfied, for example, if $F \in \mathbf{M}_\alpha(\mathcal{D})$. This interpolation device is often useful in applications.

Let us illustrate the above formulas with some examples.

Example 4.2.8 *The case of $\Gamma = (a, b)$, where $-\infty < a < b < \infty$.* In this case we take

$$\begin{aligned} w=\phi(z)&=\log\frac{z-a}{b-z}\leftrightarrow\psi(w)=\frac{a+be^w}{1+e^w} \\ \phi'(z)&=\frac{(b-a)}{(z-a)(b-z)},\quad z_k=\frac{a+be^{kh}}{1+e^{kh}}. \end{aligned} \tag{4.2.32}$$

The domain $\mathcal{D}$ is the eye-shaped region defined as in (1.7.11), illustrated in Figure 1.7.4c. If $0 < d < \pi$, the class $\mathbf{L}_{\alpha,\beta}(\mathcal{D})$ is the family of all functions F in $\mathbf{Hol}(\mathcal{D})$ which for some constant C satisfy the inequality

$$|F(z)|\leq C|z-a|^\alpha|b-z|^\beta \tag{4.2.33}$$

in $\mathcal{D}$. If $F \in \mathbf{M}_{\alpha,\beta}(\mathcal{D})$, $0 < \alpha \leqq 1$, $0 < \beta \leqq 1$, and $0 < d < \pi$, then $G \in \mathbf{L}_{\alpha,\beta}(\mathcal{D})$, where G defined as in (4.2.31) takes the form

$$G(z) = F(z) - \frac{(z-a)F(b) + (b-z)F(a)}{b-a}. \tag{4.2.34}$$

It may be shown (see Problem 4.2.5) that if $F \in \mathbf{Hol}(\mathcal{D}) \cap \mathbf{Lip}_{\alpha'}(\mathcal{D}')$, where $\alpha' \geqq \max(\alpha, \beta)$, then G is bounded throughout $\mathcal{D}$ by the right-hand side of (4.2.33).
If F is in $\mathbf{Hol}(\mathcal{D})$ and satisfies (4.2.33) in $\mathcal{D}$, then by taking ε, n, h, M and N as in Theorem 4.2.5, we get

$$\begin{aligned} &\sup_{a<x<b} \left| F(x) - \sum_{j=-M}^{N} F(z_j) S(j,h) \circ \log\left(\frac{x-a}{b-x}\right) \right| \\ &\qquad \leqq C_1 n^{1/2} e^{-(\pi d \varepsilon n)^{1/2}} \end{aligned} \tag{4.2.35}$$

with C_1 a positive constant. If F is in $\mathbf{Hol}(\mathcal{D})$ and if F satisfies

$$|F(z)| \leqq C_2 |z-a|^{\alpha-1} |b-z|^{\beta-1} \tag{4.2.36}$$

in $\mathcal{D}$, with C_2, α and β positive constants, then by taking ε, n, h, M and N as in Theorem 4.2.6, we get

$$\begin{aligned} &\left| \int_a^b F(x)\, dx - h(b-a) \sum_{k=-M}^{N} \frac{e^{kh}}{(1+e^{kh})^2} F(z_k) \right| \\ &\qquad \leqq C_3 e^{-(2\pi d \varepsilon n)^{1/2}} \end{aligned} \tag{4.2.37}$$

with C_3 a positive constant independent of N.

Example 4.2.9 *The special case* $\Gamma = (-1,1)$, *with* $\phi' F$ *or* F *in* $\mathbf{H}^p(\mathcal{U})$. In this case we have

$$\begin{aligned} w = \phi(z) &= \log[(1+z)/(1-z)] \\ \Leftrightarrow z = z(w) &= (e^w - 1)/(e^w + 1) \\ \phi'(z) = 2/\left(1-z^2\right), \quad z_k &= \left(e^{kh} - 1\right)/\left(e^{kh} + 1\right). \end{aligned} \tag{4.2.38}$$

The domain $\mathcal{D}$ is the unit disc $\mathcal{U}$, and the class of functions will consist of all those functions f such that $\phi' F$ is in $\mathbf{H}^p(\mathcal{U})$ in the case of interpolation on $(-1,1)$, and $F \in \mathbf{H}^p(\mathcal{U})$ in the case of quadrature on $(-1,1)$. Here we shall assume the notation of Section 1.2.1, and we shall also assume that $1 < p \leqq \infty$.

Now, if $g \in \mathbf{H}^p(\mathcal{U})$, then, with $q = p/(p-1)$,

$$\begin{aligned} N_1(g,\mathcal{U}) &= \lim_{r\to 1^-} \int_0^{2\pi} |g(re^{i\theta})| d\theta \\ &\leqq (2\pi)^{1/q} N_p(g,\mathcal{U}). \end{aligned} \tag{4.2.39}$$

(a) Let us consider first the case of interpolation of F on $(-1,1)$ when $\phi' F \in \mathbf{H}^p(\mathcal{U})$. By Theorem 4.2.2(a) and (4.2.39), we have, since $d = \pi/2$,

$$\begin{aligned} &\left| F(x) - \sum_{k=-\infty}^{\infty} F(z_k) S(k,h) \circ \phi(x) \right| \\ &\leqq \frac{(2\pi)^{1/q} N_p(\phi' F, \mathcal{U})}{\pi^2 \sinh[\pi^2/(2h)]}. \end{aligned} \tag{4.2.40}$$

We can now use (1.8.10); since $\phi' F \in \mathbf{H}^p(\mathcal{U})$, it follows that

$$\begin{aligned} &\phi'(x)|F(x)| \\ &\leqq (2\pi)^{-1/p}(1-x^2)^{-1/p} \left[\frac{\Gamma(2q-1)}{\Gamma(q)\Gamma(q)} \right]^{1/q} N_p(\phi' F, \mathcal{U}). \end{aligned} \tag{4.2.41}$$

That is, using ϕ' as given in (4.2.38),

$$|F(x)| \leqq 2^{-1}(2\pi)^{-1/p}(1-x^2)^{1/q} \left[\frac{\Gamma(2q-1)}{\Gamma(q)\Gamma(q)} \right]^{1/q} N_p(\phi' F, \mathcal{U}). \tag{4.2.42}$$

By taking z_k as in (4.2.38), we thus get

$$\begin{aligned} \sum_{|k|>N} \left(1 - z_k^2\right)^{1/q} &= \sum_{|k|>N} \left[\frac{4e^{kh}}{(1+e^{kh})^2} \right]^{1/q} \\ \leqq 4^{1/q} \sum_{|k|>N} e^{-kh/q} &= \frac{2^{1+2/q} e^{-(N+1)h/q}}{1 - e^{-h/q}} \\ &\leqq 2^{1+2/q}(q/h) e^{-Nh/q}. \end{aligned} \tag{4.2.43}$$

Hence, combining (4.2.43) and (4.2.42), we get

$$\sum_{|k|>N} |F(z_k) S(k,h) \circ \phi(x)| \leqq \sum_{|k|>N} |F(z_k)|$$

$$\leqq 2^{2/q} (2\pi)^{-1/p} \left[\frac{\Gamma(2q-1)}{\Gamma(q)\Gamma(q)} \right]^{1/q} (q/h) N_p(\phi' F, \mathcal{U}) \cdot e^{-Nh/q}. \tag{4.2.44}$$

That is, using (4.2.40) and (4.2.44), and taking $h = \pi[q/(2N)]^{1/2}$, and over-estimating $1/\sinh\left[\pi^2/(2h)\right]$ by $2e^{-\pi^2/(2h)}$, this estimate being valid, provided, e.g., that $e^{-\pi^2/h} < 1/3$, we get

$$\begin{aligned} &\sup_{-1 \leqq x \leqq 1} \left| F(x) - \sum_{k=-N}^{N} F(z_k) S(k,h) \circ \phi(x) \right| \\ &\leqq \frac{1}{\pi} (2\pi)^{-1/p} \left\{ 2 + 2^{2/q} (2Nq)^{1/2} \left[\frac{\Gamma(2q-1)}{\Gamma(q)\Gamma(q)} \right]^{1/q} \right\} \cdot \\ &\cdot N_p(\phi' F, \mathcal{U}) e^{-\pi[N/(2q)]^{1/2}}. \end{aligned} \tag{4.2.45}$$

(b) Similarly, if $F \in \mathbf{H}^p(\mathcal{U})$, then by (4.2.6),

$$\begin{aligned} &\left| \int_{-1}^{1} F(x) dx - h \sum_{k=-\infty}^{\infty} \frac{2e^{kh}}{(1+e^{kh})^2} F\left(\frac{e^{kh}-1}{e^{kh}+1} \right) \right| \\ &\leqq \frac{(2\pi)^{1/q} N_p(F, \mathcal{U}) e^{-\pi^2/(2h)}}{2 \sinh\left[\pi^2/(2h)\right]} \\ &\approx (2\pi)^{1/q} N_p(F, \mathcal{U}) e^{-\pi^2/h}. \end{aligned} \tag{4.2.46}$$

Also, using (1.8.10),

$$\left| h \sum_{|k|>N} \frac{2e^{kh}}{(1+e^{kh})^2} F\left(\frac{e^{kh}-1}{e^{kh}+1} \right) \right|$$

$$\leqq 2^{2/q} q (2\pi)^{-1/p} \left[\frac{\Gamma(2q-1)}{\Gamma(q)\Gamma(q)} \right]^{1/q} N_p(F, \mathcal{U}) e^{-Nh/q}. \tag{4.2.47}$$

By taking $h = \pi(q/N)^{1/2}$, and combining (4.2.46) and (4.2.47), we thus get

$$\int_{-1}^{1} F(x)dx - h \sum_{k=-N}^{N} \frac{2e^{kh}}{(1+e^{kh})^2} F\left(\frac{e^{kh}-1}{e^{kh}+1}\right)$$

$$\leqq \left\{(2\pi)^{1/q} + 2^{2/q} q (2\pi)^{-1/p} \left[\frac{\Gamma(2q-1)}{\Gamma(q)\Gamma(q)}\right]^{1/q}\right\} N_p(F, \mathcal{U}) \cdot e^{-\pi (N/q)^{1/2}}. \tag{4.2.48}$$

Example 4.2.10 *The case of* $\Gamma = [0, \infty]$*: the non-oscillatory case.* In this case we take

$$w = \phi(z) = \log z \leftrightarrow z = \psi(w) = e^w$$
$$\phi'(z) = 1/z, \quad z_k = e^{kh}, \quad \rho(z) = z. \tag{4.2.49}$$

The domain $\mathcal{D}$ is the sector defined as in (1.7.10) and illustrated in Figure 1.7.4b. The class of functions $\mathbf{L}_{\alpha,\beta}(\mathcal{D})$ is the family of all functions F in $\mathbf{Hol}(\mathcal{D})$, which for some constant C, satisfy the inequality

$$|F(z)| \leqq C|z|^{\alpha}[1+|z|]^{-\alpha-\beta} \tag{4.2.50}$$

in $\mathcal{D}$. If $F \in \mathbf{M}_{\alpha,\beta}(\mathcal{D})$, $0 < \alpha \leqq 1$, $0 < \beta \leqq 1$, and $0 < d < \pi$, then $G \in \mathbf{L}_{\alpha,\beta}(\mathcal{D})$, with

$$G(z) = F(z) - \frac{F(0) + zF(\infty)}{1+z}. \tag{4.2.51}$$

If F is in $\mathbf{Hol}(\mathcal{D})$ and satisfies (4.2.50) then taking ε, n, h, M, and N as in Theorem 4.2.5, we get

$$\sup_{0<x<\infty} \left| F(x) - \frac{h}{\pi} \sin\left\{\frac{\pi}{h} \log x\right\} \sum_{k=-M}^{N} \frac{(-1)^k F(e^{kh})}{\log x - kh} \right|$$
$$\leqq C_1 n^{1/2} e^{-(\pi d \varepsilon n)^{1/2}}, \tag{4.2.52}$$

with C_1 a positive constant, independent on n. If $F \in \mathbf{Hol}(\mathcal{D})$ and if F satisfies the inequality

$$|F(z) \leqq C_2 |z|^{\alpha-1}[1+|z|]^{-\alpha-\beta} \tag{4.2.53}$$

in $\mathcal{D}$, with C_2, α and β positive constants, then taking ε, n, h, M and N as in Theorem 4.2.6, we get

$$\left| \int_0^\infty F(x)dx - h \sum_{k=-M}^{N} e^{kh} F(e^{kh}) \right| \leqq C_3 e^{-(2\pi d \varepsilon n)^{1/2}} \tag{4.2.54}$$

with C_3 a constant independent of n.

Example 4.2.11 *The case of* $\Gamma = [0, \infty]$*: the oscillatory case.* In this case we take

$$w = \phi(z) = \log\{\sinh(z)\} \Leftrightarrow z = \psi(w) = \log\left[e^w + \sqrt{1 + e^{2w}}\right] \tag{4.2.55}$$

$$\rho(z) = \sinh(z), \ \phi'(z) = \frac{1}{\tanh(z)}, \ z_k = \log\left[e^{kh} + \sqrt{1 + e^{2kh}}\right].$$

The domain $\mathcal{D}$ is the "bullet-shaped" region defined by (1.7.12), illustrated in Figure 1.7.4d. The class of functions $\mathbf{L}_{\alpha,\beta}(\mathcal{D})$ is the family of all functions in $\mathbf{Hol}(\mathcal{D})$ which satisfy

$$|F(z)| \leqq C \left(\frac{|z|}{1 + |z|}\right)^\alpha e^{-\beta \Re z} \tag{4.2.56}$$

in $\mathcal{D}$. If $F \in \mathbf{M}_{\alpha,\beta}(\mathcal{D})$, with $0 < \alpha \leqq 1$, $0 < \beta \leqq 1$, and $0 < d < \pi$, then $G \in \mathbf{L}_{\alpha,\beta}(\mathcal{D})$, with

$$G(z) = F(z) - \frac{F(0) + \sinh(z) F(\infty)}{1 + \sinh(z)}. \tag{4.2.57}$$

If $F \in \mathbf{L}_{\alpha,\beta}(\mathcal{D})$, then taking ε, n, h, M and N as in Theorem 4.2.5, we get

$$\sup_{0<x<\infty} \left| F(x) - \sum_{k=-M}^{N} F(z_k) S(k,h) \circ \log\{\sinh(z)\} \right| \tag{4.2.58}$$

$$\leqq C_1 n^{1/2} e^{-(\pi d \varepsilon n)^{1/2}}$$

with C_1 a constant independent of n. If $F \in \mathbf{Hol}(\mathcal{D})$, and if F satisfies the inequality

$$|F(z)| \leqq C_2 \left(\frac{|z|}{1 + |z|}\right)^{\alpha - 1} e^{-\beta \Re z} \tag{4.2.59}$$

in $\mathcal{D}$, then taking ε, n, h, M and N as in Theorem 4.2.6, we get

$$\left| \int_0^\infty F(z)dz - h \sum_{k=-M}^{N} \left(1 + e^{-2kh}\right)^{-1/2} F(z_k) \right| \leq C_3 e^{-(2\pi d\varepsilon n)^{1/2}} \tag{4.2.60}$$

with C_3 a constant independent of n.

Example 4.2.12 *The case of* $\Gamma = \mathbb{R}$, *when* F *decays algebraically on* $\mathbb{R}$. In this case

$$\begin{aligned} w = \phi(z) = \log\left[z + \sqrt{1+z^2}\right] &\leftrightarrow z = \psi(w) = \sinh(w) \\ z_k = \sinh(kh), \quad \rho(z) &= z + \sqrt{1+z^2} \\ \phi'(z) &= \frac{1}{\sqrt{1+z^2}}. \end{aligned} \tag{4.2.61}$$

The domain $\mathcal{D}$ is defined in Equation (1.7.13). It is a region bounded by two hyperbolas, and it is illustrated in Figure 1.7.4e. The class of functions $\mathbf{L}_{\alpha,\beta}(\mathcal{D})$ consists of the family of all functions in $\mathbf{Hol}(\mathcal{D})$ which satisfy the inequalities

$$|F(z)| \leq \begin{cases} C\left(1+|z|\right)^{-\alpha}, z \in \mathcal{D},\ \Re z \leq 0 \\ C\left(1+|z|\right)^{-\beta}, z \in \mathcal{D},\ \Re z \geq 0. \end{cases} \tag{4.2.62}$$

If $F \in \mathbf{M}_{\alpha,\beta}(\mathcal{D})$, with $0 < \alpha \leq 1$, $0 < \beta \leq 1$, and $0 < d < \pi$, then $G \in \mathbf{L}_{\alpha,\beta}(\mathcal{D})$, with

$$G(z) = F(z) - \frac{F(-\infty) + \left[z + \sqrt{1+z^2}\right] F(\infty)}{1 + z + \sqrt{1+z^2}}. \tag{4.2.63}$$

If $F \in \mathbf{L}_{\alpha,\beta}(\mathcal{D})$, then taking ε, n, h, M and N as in Theorem 4.2.5, we get

$$\sup_{x\in\mathbb{R}} \left| F(x) - \sum_{k=-M}^{N} F(z_k) S(k,h) \circ \log\left[x + \sqrt{1+x^2}\right] \right| \leq C_1 n^{1/2} e^{-(\pi d\varepsilon n)^{1/2}} \tag{4.2.64}$$

with C_1 a constant independent of n. If $F \in \mathbf{Hol}(\mathcal{D})$ and if F satisfies the inequalities

$$|F(z)| \leq \begin{cases} C_2\left(1+|z|\right)^{-\alpha-1}, z \in \mathcal{D}, \Re z \leq 0 \\ C_2\left(1+|z|\right)^{-\beta-1}, z \in \mathcal{D}, \Re z \geq 0, \end{cases} \tag{4.2.65}$$

then taking ε, n, h, M and N as in Theorem 4.2.6, we get

$$\left| \int_{\mathbb{R}} F(x)dx - h \sum_{k=-M}^{N} \cosh(kh)F(\sinh(kh)) \right| \leqq C_3 e^{-(2\pi d\varepsilon n)^{1/2}} \tag{4.2.66}$$

with C_3 (as well as C_2) a constant independent of n.

Example 4.2.13 It may be advantageous for certain problems, to use more than one transformation, i.e., a *double exponential* transformation. A typical situation of this type occurs when F is holomorphic in an infinitely sheeted Riemann surface. For example, let us consider the integral

$$I = \int_0^1 t^{\alpha-1}(1-t)^{\beta-1}dt \tag{4.2.67}$$

where α and β are positive numbers. It then follows that the integrand is holomorphic not only in the whole plane $\mathbb{C} \setminus L_1 \setminus L_2$, where L_1 is the closed segment of the real axis joining $-\infty$ and 0, and L_2 is the closed segment of the real axis joining 1 and ∞, but in the larger region, $\mathcal{D}^* = \{t \in \mathbb{C} : |\arg[t/(1-t)]| < \infty\}$. After making the usual transformation (1.7.15) (in which $a = 0$ and $b = 1$), we get

$$I = \int_{\mathbb{R}} \frac{e^{\alpha u}}{(1+e^u)^{\alpha+\beta}} du. \tag{4.2.68}$$

If we now apply $T_N(f,h)$ as defined in (3.2.1) to approximate I, taking $h = [2\pi d/(\alpha N)]^{1/2}$, with $d = \pi$, we get

$$I - T_N(f,h) = \mathcal{O}(\exp[-\pi(2\alpha N)^{1/2}]). \tag{4.2.69}$$

One may note, on the other hand, that the integrand in (4.2.68) is analytic and bounded, not only in $\mathcal{D}_\pi$, the region defined in (1.7.9), but in the larger region $\mathcal{D}^3_{\pi/2}$ which was defined in (1.7.12). After making a second transformation, $u = \sinh(v)$, and then applying the rule $T_N(f,h)$ defined in (3.2.1), with $h = \log(N)/N$, the error of this latter quadrature approximation is $\mathcal{O}(\exp[-N/\log(N)])$.

A similar double exponential transformation procedure may be used to derive a $2N+1$-point Sinc approximation of $|x|^\alpha$ on the interval [-1,1], for which the error is $\mathcal{O}(\exp[-N/\log(N)])$ (see Problem 4.2.6).

Given a sequence of numbers $\{c_k\}$, the following result, which extends Theorem 3.1.8, enables us to deduce when there exists a function in $\mathbf{L}_\alpha(\mathcal{D})$ which interpolates this sequence at the Sinc points, z_k.

Theorem 4.2.14 *Let $\{c_k\}_{k \in \mathbb{Z}}$ denote a sequence of numbers, such that for all $k \in \mathbb{Z}$, we have $|c_k| \leqq C \exp(-\alpha|k|h)$, with C and α positive constants. Then there exists a function $f \in \mathbf{L}_\alpha(\mathcal{D})$ such that $f(z_k) = c_k$ for all $k \in \mathbb{Z}$. Moreover, for all $z \in \mathcal{D}$, we have*

$$|f(z)| \leqq \frac{C \cosh(\pi|\Im\phi(z)|/h)}{\tanh(\alpha h/2)}. \tag{4.2.70}$$

Moreover, letting N be a positive integer, we have

$$\left| f(z) - \sum_{k=-N}^{N} c_k S(k,h) \circ \phi(z) \right| \leqq \frac{2C}{\alpha h} \cosh(\pi|\Im\phi(z)|/h)\, e^{-\alpha N h}. \tag{4.2.71}$$

Proof. We leave the straightforward proof of this theorem to Problem 4.2.8. ■

Finally, we present a result which extends Corollary 3.2.2 to a contour Γ, one which is important when using a Sinc basis in Galerkin methods.

Corollary 4.2.15 *Let $F \in \mathbf{H}^1(\mathcal{D})$. Then*

$$\begin{aligned} \eta^*(F,h) &\equiv \int_\Gamma F(t)\, S(k,h) \circ \phi(t)\, dt - h\frac{F(z_k)}{\phi'(z_k)} \\ &= \frac{ih}{2\pi} \int_\Gamma \frac{F(\zeta) \exp[i(\pi/h)\phi(\zeta) \operatorname{sgn} \Im\phi(\zeta)]}{\phi(\zeta) - kh} d\zeta. \end{aligned} \tag{4.2.72}$$

Moreover,

$$|\eta^*(F,h)| \leqq h \frac{e^{-\pi d/h}}{2\pi d} N(F, \mathcal{D}). \tag{4.2.73}$$

Proof. We leave the proof to Problem 4.2.9. ■

•

Problems for Section 4.2

4.2.1 Derive an explicit formula for interpolation and quadrature along the arc Γ described in Problem 1.7.3.

4.2.2 Which of the formulas of this section would you use to carry out Sinc interpolation of the following functions:

(a) $(1+x)^{1/3}(1-x)^{1/4}x^{-1}\tan^{-1}x, -1 \leqq x \leqq 1;$

(b) $\left[(x-T)^2+1\right]^{-1/2}\left[(x+T)^2+1\right]^{1/2}e^{-(x^2+1)^{1/2}}, -\infty \leqq x \leqq \infty,$ $-5000 \leqq T \leqq 500;$

(c) $\left(x^2+2\right)^{-1/4}\left[(x+2)^2+4\right]^{-1/5}, -\infty \leqq x \leqq \infty;$

(d) $x^{1/3}\,(1+x)^{-1/2}\log(x), 0 < x < \infty;$

(e) $x^{3/4}\left[(x-T)^2+2\right]^{1/2}e^{-x}\cos(3x)\log(x), 0 \leqq x \leqq \infty,$ $0 \leqq T \leqq 200.$

Explain your answer in each case.

4.2.3 Which of the formulas of this section would you use to approximate each of the following integrals:

(a) $\displaystyle\int_{-1}^{1}(1+x)^{-1/2}(1-x)^{-2/3}x^{-1}\log(1+x)dx;$

(b) $\displaystyle\int_{\mathbb{R}} T^2 e^{-(x^2+1)^{1/2}}\left[(x+T)^2+1\right]^{-1/2}\left[(x+T)^2+1\right]^{-1/2}dx,$

with $-10^4 < T < 10^4$;

(c) $\displaystyle\int_{\mathbb{R}}\left(x^2+2\right)^{-1/3}\left[(x+2)^2+4\right]^{-1/4}dx;$

(d) $\displaystyle\int_{0}^{\infty} x^{-2/3}\,(1+x)^{-1/2}\log(x)dx;$

(e) $\displaystyle\int_{0}^{\infty} x^{-3/4}\left[(x-T)^2+2\right]^{1/2}e^{-2x}\cos 2x dx,$

with $-5 < T < 10^5$.

Explain your answer in each case.

4.2.4 Verify (4.2.13); i.e., prove that if $\alpha > 0$, $\beta > 0$, then

$$\int_{\mathbb{R}}\frac{e^{\alpha t}dt}{[1+e^t]^{\alpha+\beta}} = \frac{\Gamma(\alpha)\,\Gamma(\beta)}{\Gamma(\alpha+\beta)} < \frac{1}{\alpha}+\frac{1}{\beta} < \infty.$$

4.2.5 Let $F \in \mathbf{Hol}(\mathcal{D}) \cap \mathbf{Lip}_\alpha(\mathcal{D})$, with $\mathcal{D}$ defined as in (1.7.11), and with $\alpha \in (0,1]$. Let G be defined as in (4.2.34). Prove that there exists a constant C such that for all $z \in \mathcal{D}$ we have

$$|G(z)| \leqq C|z-a|^\alpha|b-z|^\alpha.$$

4.2.6 Let $R > 1$, and let $\alpha \in (1/R, R)$.

(a) Set $|x| = g(y) = (1 - y^2)^{1/2}$. The function $g \in \mathbf{L}_\alpha(\mathcal{D})$ with $\mathcal{D}$ defined in (1.7.11), with $(a,b) = (-1,1)$ and $d = \pi$. Prove, by taking $h = [\pi d/(\alpha N)]^{1/2}$, $u(x) = (1-x^2)^{1/2}\mathrm{sgn}(x)$, $v(x) = \log\{[1+u(x)]/[1-u(x)]\}$, and $\omega_k(x) = S(k,h) \circ v(x) + S(-k,h) \circ v(x)$, that there exists a constant C, independent of N, such that

$$\sup_{-1<x<1} \left| |x|^\alpha - \frac{1}{2}\omega_0(x) - \sum_{k=1}^{N} \frac{(2e^{kh})^\alpha}{(1+e^{kh})^\alpha} \omega_k(x) \right| \leqq C N^{1/2} e^{-(\pi d \alpha N)^{1/2}}.$$

(b) Set $G(u) = g(\tanh(u)) = [\cosh(u)]^{-\alpha}$, and note that $G \in \mathbf{Hol}(\mathcal{D})$, with $\mathcal{D}$ is defined as in (1.7.13), with $d = \pi/2$, and also, that $G(\sinh(kh)) = \mathcal{O}(\exp\{-(\alpha/2)e^{-|k|h}\})$, $k \to \pm\infty$. Conclude, by taking $h = \log(N)/N$, that there exists a constant C, which is independent of N, such that

$$\sup_{u \in \mathbb{R}} \left| G(u) - \sum_{k=-N}^{N} G(\sinh(kh)) S(k,h) \circ \log\{u + \sqrt{1+u^2}\} \right|$$
$$\leqq C \exp\{-\tfrac{N}{\log(N)}\}.$$

(c) Use the result of part (b) of this problem to construct an $N+1$ basis approximation to $|x|^\alpha$ on $[-1,1]$, which is accurate to within an error bounded by $C\exp\{-\frac{N}{\log(N)}\}$. (Keinert, [Ki1]).

4.2.7 Let $F \in \mathbf{Hol}(\mathcal{D})$, let $z \in \mathcal{D}$, such that z lies to the left of Γ, let m be a positive integer, and define G by

$$G(t) = \frac{F(t)}{(t-z)^m} - \sum_{j=0}^{m-1} \frac{F^{(j)}(z)}{(t-z)^{m-j}}.$$

Let $\gamma(h;b)$ be defined as in (3.1.44). Prove that if $G \in \mathbf{H}^1(\mathcal{D})$, then

$$\left| \int_\Gamma \frac{F(t)}{(t-z)^m} dt - h \sum_{k=-\infty}^{\infty} \frac{F(z_k)}{(z_k - z)^m \phi'(z_k)} \right.$$
$$\left. - \pi \rho(z)^{i\pi/h} \sum_{j=1}^{m} B_j(z) \sum_{\ell=1}^{j} \frac{(i\pi/h)^{\ell-1}}{(\ell-1)!} \gamma_{\ell-1}(h;\phi(z)) \right|$$
$$\leqq \frac{e^{-2\pi d/h}}{1 - e^{-2\pi d/h}} N(G, \mathcal{D}).$$

with

$$B_j(z) = \sum_{\ell=j}^{m} \eta_{j,\ell} \frac{F^{(\ell)}(z)}{\ell!},$$

and with

$$\eta_{j,\ell} = \frac{1}{(\ell-1)!} \left(\frac{d}{dt}\right)^{\ell-1} [\phi(t) - \phi(z)]^{j-1} \bigg|_{t=z}.$$

([Bialecki, Bi3, Bi4]).

4.2.8 Give a detailed proof of Theorem 4.2.14. [Hint: See Theorem 3.1.8, and the remark in its proof.]

4.2.9 Give a detailed proof of Corollary 4.2.15.

4.2.10 Let $f/\phi' \in \mathbf{L}_{\alpha,\beta}(\mathcal{D})$, with $\mathcal{D}$ such that $\sup_{z\in\mathcal{D}} |\Im z| \leqq \delta < \infty$, and let $\Gamma \subseteq \mathbb{R}$.

(a) Show that we may approximate

$$c_m = \int_\Gamma f(x)\, e^{imx}\, dx$$

within a relative error of $e^{-\beta N h}$ via the quadrature formula in (4.2.26), after selecting an integer c such that $c \geqq |m|$, then setting $M = [\beta/\alpha N]$, and taking

$$h = \frac{2\pi d}{\delta c + \sqrt{(\delta c)^2 + 2\pi d \beta N}}.$$

Show, moreover, that all coefficients c_m, $m = -c, \cdots, c$ may be computed to the same accuracy using the same h and N. [J. Lund, private communication, 1991.]

(b) Illustrate, by, for example, taking

$$f(x) = x^{-2/3}(\pi - x)^{-3/2} \log(\pi x),$$

$\mathcal{D} = \{z \in \mathbb{C} : |z - \pi/2| < \pi/2\}$, and $\Gamma = (0, \pi)$, that the method of part (a) of this problem can be used to accurately compute (say, to 6 decimals) several hundred of the coefficients c_j, whereas this endeavor may be quite troublesome using either Gaussian quadrature (regardless of which weight function is used) or any open Newton–Cotes formula.

4.2.11 Prove with reference to Corollary 4.2.15 that if $F \in \mathbf{H}^\infty(\mathcal{D})$, then there exists a constant c which is independent of h and k, such that

$$\left| \int_\Gamma F(t)\, S(k,h) \circ \phi(t) \phi'(t) dt - hF(z_k) \right| \leqq c\, h\, e^{-\pi d/h}.$$

4.2.12 Let $F \in \mathbf{H}^\infty(\mathcal{D})$, let $h = \gamma / N^{1/2}$, with N a positive integer and γ a constant, and set

$$\begin{aligned} \omega_j &= S(j,h) \circ \phi, \quad j = -N+1, \cdots, N-1 \\ \omega_{-N} &= \frac{1}{1+\rho} - \sum_{j=-(N-1)}^{N} \frac{S(j,h) \circ \phi}{1+e^{jh}} \\ \omega_N &= \frac{\rho}{1+\rho} - \sum_{j=-N}^{N-1} \frac{e^{jh} S(j,h) \circ \phi}{1+e^{jh}}. \end{aligned}$$

Prove that as $N \to \infty$,

$$\sup_{x \in \Gamma} \left| F(x) - \sum_{j=-N}^{N} F(z_j)\, \omega_j(x) \right| \to 0.$$

4.3 Hilbert and Related Transforms on Γ

In this section we shall extend the results of Section 3.4, which involve Hilbert and related transforms in $\mathcal{D}_d$ to analogous approximations over subregions of $\mathcal{D}$, where $\mathcal{D}$ is related to $\mathcal{D}_d$ as in Definition 4.1.1. It will be seen in Section 6.7 that some of the results of this section can be applied to the approximate solution of Cauchy singular integral equations, and in Section 7.7 we shall see how these results may be applied to obtain approximate solutions of partial differential equations. However, we shall be concerned here only with the *direct* extension of the results of Section 3.4, under the notation of Definition 4.1.1. Other variants of some of these formulas which may be more suited to applications are derived in Chapter 5.

We shall assume here that Γ (see Definition 4.1.1) is oriented, its positive orientation being in the direction from a to b as one traverses Γ. Let $\mathcal{D}^+$ denote the portion of $\mathcal{D}$ to the left of Γ. Given that $\phi' F \in \mathbf{H}^1(\mathcal{D})$, let us set

$$(\Pi F)(z) = \frac{1}{2\pi i} \int_\Gamma \frac{F(t)\phi'(t)}{\phi(t) - \phi(z)} dt \tag{4.3.1}$$

where z lies in $\mathcal{D}^+$. Let us also define two functions $\mathcal{R}F$ and $\mathcal{L}F$ which are closely related to ΠF, by

$$(\mathcal{R}F)(z) = \frac{\Im\phi(z)}{\pi}\int_\Gamma \frac{F(t)\phi'(t)}{|\phi(t)-\phi(z)|^2}dt \tag{4.3.2}$$

$$(\mathcal{L}F)(z) = \frac{1}{\pi}\int_\Gamma \frac{[\Re\phi(z)-\phi(t)]F(t)\phi'(t)dt}{|\phi(t)-\phi(z)|^2} \tag{4.3.3}$$

where z again lies in $\mathcal{D}^+$. Clearly, both $\mathcal{R}F$ and $\mathcal{L}F$ are harmonic in $\mathcal{D}^+$ where they are related by the formula

$$\Pi F = \frac{1}{2}(\mathcal{R}F + i\mathcal{L}F). \tag{4.3.4}$$

Next, letting ζ denote a point on Γ, and letting $z \to \zeta$ along an arc which is not tangent to Γ, we obtain the formulas

$$\lim_{z\to\zeta}(\mathcal{R}F)(z) = F(\zeta) \tag{4.3.5}$$

$$\lim_{z\to\zeta}(\mathcal{L}F)(z) = (\mathcal{S}F)(\zeta) \tag{4.3.6}$$

where $\mathcal{S}F$ is the principal value integral

$$(\mathcal{S}F)(\zeta) = \frac{P.V.}{\pi}\int_\Gamma \frac{F(t)\phi'(t)}{\phi(t)-\phi(\zeta)}dt,\ \zeta\in\Gamma. \tag{4.3.7}$$

As expected, the principal value integral $\mathcal{S}F$ may be defined either by letting $z \to \zeta$ in (4.3.3) or equivalently, by the formula

$$(\mathcal{S}F)(\zeta) = \lim_{\varepsilon\to 0}\frac{1}{\pi}\int_{\Gamma_\varepsilon} \frac{F(t)\phi'(t)}{\phi(t)-\phi(\zeta)}dt \tag{4.3.8}$$

where Γ_ε is the portion of Γ remaining after we delete the part in the disc with center ζ and radius ε.

Returning to (4.3.4), and using (4.3.5) and (4.3.6), we get the following generalization of the Plemelj formulas (see Equations (1.2.20) and (1.2.21)):

$$(\Pi F)(\zeta) = \frac{1}{2}[F(\zeta) + \mathcal{S}F(\zeta)],\ \zeta\in\Gamma. \tag{4.3.9}$$

Clearly, ΠF is analytic in $\mathcal{D}^+$. Since F is analytic in $\mathcal{D}$, and since $\mathcal{R}F$ is harmonic in $\mathcal{D}^+$, it follows from (4.3.2) and (4.3.5) that $\mathcal{R}F$ and $\mathcal{L}F$ are analytic in $\mathcal{D}^+$. Moreover, it is readily seen that all three of these functions which were defined via integration over Γ have an analytic continuation to $\mathcal{D}$. Furthermore, we also have the identity

$$\mathcal{L} = \frac{1}{2}\mathcal{R}\mathcal{S}. \tag{4.3.10}$$

Let us next consider the approximation of the above functions, either on a subset of $\mathcal{D}$, or on Γ. To this end we recall Corollary 1.10.3, which yields the identities

$$\Pi\{S(k,h)\cdot\phi\}(z) = \frac{\exp\{i\pi\,[\phi(z)-kh]\}-1}{\pi i\,[\phi(z)-kh]\,/h} \tag{4.3.11}$$

$$\begin{aligned} T(k,h)\circ\phi(z) &\equiv \mathcal{R}\{S(k,h)\circ\phi\}(z) \\ &= \Re\{\pi S(k,h)\circ\phi(z)\} \end{aligned} \tag{4.3.12}$$

$$\mathcal{S}\{S(k,h)\circ\phi\}(z) = \frac{\cos\{\pi\,[\phi(z)-kh]\,/h\}-1}{\pi\,[\phi(z)-kh]\,/h}. \tag{4.3.13}$$

We may now select a positive integer, N, and in terms of the above functions define the differences

$$(\Pi E_N F)(z) = (\Pi F)(z) - \sum_{|k|\leqq N} F(z_k)\frac{\exp\{i\pi[\phi(z)-kh]/h\}-1}{\pi i[\phi(z)-kh]/h} \tag{4.3.14}$$

$$(\mathcal{R}E_N F)(z) = (\mathcal{R}F)(z) - \sum_{|k|\leqq N} F(z_k)\Re\left\{\frac{\exp\{i\pi[\phi(z)-kh]/h\}-1}{\pi i[\phi(z)-kh]/h}\right\} \tag{4.3.15}$$

$$(\mathcal{S}E_N)F(z) = (\mathcal{S}F)(z) - \sum_{|k|\leqq N} F(z_k)\frac{\cos\{\pi[\phi(z)-kh]/h\}-1}{\pi[\phi(z)-kh]/h}, \tag{4.3.16}$$

as well as the limits

$$\begin{aligned} &((\Pi E)F, (\mathcal{R}E)F, (\mathcal{S}E)F) \\ &= \lim_{N\to\infty}((\Pi E_N)F, (\mathcal{R}E_N)F, (\mathcal{S}E_N)F)\,. \end{aligned} \tag{4.3.17}$$

By collecting results of Section 3.4, we then get

Theorem 4.3.1 *Let $\phi' F \in \mathbf{H}^1(\mathcal{D})$, and for given $\eta \in (-d,d)$, define $\mu, \Gamma(\eta)$ and $\mathcal{D}(\eta)$ by $\mu = \max(1,(d+\eta)^{-1})$*

$$\begin{aligned} \Gamma(\eta) &= \{z\in\mathcal{D} : \Im\phi(z)=\eta\} = \{z\in\mathcal{D} : z=\psi(\xi+i\eta),\quad \xi\in\mathbb{R}\} \\ \mathcal{D}(\eta) &= \{z\in\mathcal{D} : |\Im\phi(z)|<|\eta|\}. \end{aligned} \tag{4.3.18}$$

Then

$$\sup_{\eta \leqq \Im \phi(z) \leqq d} |(\Pi E F)(z)| \leqq \frac{e^{-\pi\eta/h} N(\phi' F, \mathcal{D})}{4\pi\mu \sinh(\pi d/h)}, \tag{4.3.19}$$

$$\sup_{z \in \mathcal{D}(\eta)} |(\mathcal{S} E F)(z)| \leqq \frac{e^{-\pi d/h} + \cosh(\pi\eta/h)}{2\pi d \sinh(\pi d/h)} N(\phi' F, \mathcal{D}). \tag{4.3.20}$$

Moreover, if $F \in \mathbf{L}_\alpha(\mathcal{D})$*, then* $\mathcal{R}F \in \mathbf{L}_\alpha(\mathcal{D}$*,* $\mathcal{S}F \in \mathbf{L}_\alpha(\mathcal{D}$*, and by taking* h *as in Equation (3.1.27) we get the inequalities*

$$\sup_{\eta \leqq \Im \phi(z) \leqq d} |(\Pi E_N F)(z)| \leqq C_1 N^{1/2} e^{-(d+\eta)(\pi\alpha N/d)^{1/2}} \tag{4.3.21}$$

$$\sup_{\eta \leqq \Im \phi(z) \leqq d} |(\mathcal{R} E_N F)(z)| \leqq C_2 N^{1/2} e^{-(d+\eta)(\pi\alpha N/d)^{1/2}} \tag{4.3.22}$$

$$\sup_{z \in \mathcal{D}(\eta)} |\mathcal{S}(E_N F)(z)| \leqq C_3 N^{1/2} e^{-(d-|\eta|)(\pi\alpha N/d)^{1/2}} \tag{4.3.23}$$

where C_1*,* C_2*, and* C_3 *are constants which are independent of* N*.*

Proof. These results follow directly from those obtained in Section 3.4 after transformation to $\mathcal{D}_d$ via replacement of z by $\psi(z)$. ■

•

Problems for Section 4.3

4.3.1 Prove that if for some $p \in [1, \infty]$, we set

$$\|F\|_p = \lim_{\eta \to -d} \left(\int_{\Gamma(\eta)} |F(t)|^p |\phi'(t) dt| \right)^{1/p}$$

where $\Gamma(\eta)$ is defined as in (4.3.18), then

(a)

$$\|\mathcal{R}F\|_p = \|F\|_p$$

and also

(b)

$$\|\mathcal{S}F\|_p \leqq B_p \|f\|_p ; B_p = \begin{cases} \tan[\pi/(2p)], 1 < p \leqq 2 \\ \cot[\pi/(2p)], 2 \leqq p < \infty. \end{cases}$$

(Compare (1.2.4).)

4.3.2 Why do ΠF, $\mathcal{R}F$ and $\mathcal{S}F$ have an analytic continuation to $\mathcal{D}$?

4.3.3 Describe in detail how the identity (4.3.9) is obtained.

4.3.4 Discuss the identity (4.3.10) showing how, why and where it is valid.

4.3.5 (a) Give an explicit statement of Theorem 4.3.1 in the case where Γ, ϕ and z_k are in Example 4.2.8.

(b) Give an explicit formula for approximating

$$I \equiv \frac{P.V.}{\pi} \int_a^b \frac{F(t)\, \phi'(t)}{\phi(t) - \phi(\zeta)} dt, \ \zeta \in (a, b)$$

with ϕ as in (a), but with $F \in \mathbf{L}_\alpha(\mathcal{D})$. [Hint: Compare Theorem 3.4.5.]

4.3.6 Give an explicit statement of Theorem 4.3.1 in the case where Γ, ϕ and z_k are defined as in Example 4.2.10.

4.3.7 Give an explicit statement of Theorem 4.3.1 in the case where Γ, ϕ and z_k are defined as in Example 4.2.11.

4.3.8 Give an explicit statement of Theorem 4.3.1 in the case where Γ, ϕ and z_k are defined as in Example 4.2.12.

4.3.9 Show that when $z_k = \psi\left((k + 1/2)h\right)$, k an integer, then for an integer, j,

$$|(\mathcal{S}EF)(jh)| \leqq \frac{e^{-\pi d/h} N(\phi' F, \mathcal{D})}{2\pi d \sinh(\pi d/h)}.$$

4.3.10 Let $z \in \mathcal{D}$, so that z lies to the left of Γ, let $F \in \mathbf{Hol}(\mathcal{D}_d)$, and let

$$N(G, \mathcal{D}) \equiv \int_{\partial \mathcal{D}} |G(t)\, dt| < \infty,$$

with $G(t) = F(t)/(t - z)$.

(a) Prove that

$$\frac{1}{2\pi i} \int_\Gamma \frac{F(t)}{t - z} dt$$

$$- \frac{h}{2\pi i} \sum_{k=-\infty}^{\infty} \frac{F(z_k)}{(z_k - z)\phi'(z_k)} + \frac{\rho(z)^{i\pi/h} F(z)}{\rho(z)^{i\pi/h} - \rho(z)^{-i\pi/h}} \equiv \varepsilon$$

with

$$\varepsilon = -\frac{1}{4\pi}\int_{\partial\mathcal{D}} G(t)\frac{\exp[(i\pi/h)\phi(t)\mathrm{sgn}\Im\phi(t)]}{\sin[\pi\phi(t)/h]}dt$$

so that

$$|\varepsilon| \leqq \frac{1}{2\pi}\frac{e^{-2\pi d/h}}{1-e^{-2\pi d/h}}N(G,\mathcal{D}).$$

(Bialecki [Bi3]).

(b) Let z and F be as in part (a) of this problem. Assuming that $\int_{\partial\mathcal{D}} |F(t)/(t-z)^m dt| < \infty$, obtain an approximation of

$$\frac{1}{2\pi i}\int_{\partial\mathcal{D}} \frac{F(t)}{(t-z)^m}dt$$

by differentiating with respect to z the result in (a). Derive explicit formulas for the case of $m = 2$.

(c) Use the Plemelj formulas to express

$$\lim_{z\to x\in\Gamma}\frac{1}{2\pi i}\int_\Gamma \frac{F(t)}{t-z}dt$$

and then to deduce an approximation of the Hilbert transform of F over Γ. Consider both the case $x = z_\ell$ for some integer ℓ, and also the case $x \neq z_\ell$.

(d) Repeat part (c) of this problem in the case $m = 2$, with m defined as in part (b).

4.3.11 Prove that if $\mathcal{U}$ denotes the unit disc in $\mathbb{C}$, if $x \in (-1,1)$, and if $F \in \mathbf{H}^2(\mathcal{U})$, then:

(a)

$$P.V.\int_{-1}^{1}\frac{F(t)}{t-x}dt = P.V.\int_{-1}^{1} F\left(\frac{x+t}{1+xt}\right)\frac{dt}{t(1+xt)}.$$

(b) Prove that if $h > 0$, and we set $u_k = \exp[-(k-1/2)h]$, and $y_k = (u_k - 1)/(u_k + 1)$, then there exists a finite positive number $C(x)$, which is independent of h, such that

$$\left|P.V.\int_{-1}^{1}\frac{F(t)}{t-x}dt - \frac{h}{2}\sum_{k=-\infty}^{\infty}\frac{1-y_k^2}{y_k(1+x\,y_k)}F\left(\frac{x+y_k}{1+xy_k}\right)\right|$$
$$\leqq \frac{\exp(-\pi^2/h)}{1-\exp(-\pi^2/h)}C(x).$$

(c) Obtain an accurate bound on $C(x)$. In particular, prove that if $F \in \mathbf{L}_\alpha(\mathcal{U})$, then $\sup_{x\in(-1,1)} C(x) < \infty$. (Stenger, [S19]).

4.3.12 In the notation of Definition 4.1, let $\mathcal{D}_\Gamma^+$ denote that part of $\mathcal{D}$ lying to the left of Γ, as one traverses Γ from a to b. Let $z \in \mathcal{D}_\Gamma^+$, and let $\zeta \in \Gamma$, and for $k \in \mathbb{Z}$ and $h > 0$, set

$$T(k,h)\circ\phi(z) = \frac{\Im\phi(z)}{\pi}\int_\Gamma \frac{S(k,h)\circ\phi(t)\,\phi'(t)}{|\phi(z)-\phi(t)|^2}dt.$$

(a) Show that

$$T(k,h)\circ\phi(z) = \Im\left\{\frac{1-e^{i\pi\{\phi(z)-kh\}/h}}{\pi\{\phi(z)-kh\}/h}\right\}$$

and that $T(k,h)\circ\phi$ is harmonic in $\mathcal{D}_\Gamma^+$.

(b) Show that

$$\lim_{z\to\zeta} T(k,h)\circ\phi(z) = S(k,h)\circ\phi(\zeta).$$

[Hint: See Corollary 1.10.3.]

4.4 Approximation of Derivatives on Γ

It is sometimes desirable to have a simple and accurate approximation F_h of F, such that the nth derivative, $F_h^{(n)}$ of F_h is also an accurate approximation of $F^{(n)}$ for $n = 0, 1, \cdots, m$. This feature is particularly useful in the approximate solution of some differential equations.

Except on $\mathbb{R}$, (4.2.5) may not be useful for approximating derivatives of F on Γ, since the terms $\phi' S'(k,h)\circ\phi$ may be unbounded on Γ. A formula for approximating $F^{(n)}$ may nevertheless be obtained by replacing the interpolating basis $\{S(k,h)\circ\phi\}$ by $\{gS(k,h)\circ\phi\}$, where g is a “nullifier function”, with the property that $(d/dx)^n[g(x)S(k,h)\circ\phi(x)]$ is bounded on Γ, for $n = 0, 1, \cdots, m$. The nullifier function $[\phi'(x)]^{-m}$ is usually satisfactory, in view of Theorem 4.1.2.

Upon replacing F in Theorem 4.2.1 by $\phi' F/g$, we get

Theorem 4.4.1 *Let $\phi' F/g \in \mathbf{H}^1(\mathcal{D})$. Then, for all x on Γ,*

$$\begin{aligned} &F(x) - \sum_{k=-\infty}^{\infty} \frac{F(z_k)}{g(z_k)} g(x) S(k,h)\circ\phi(x) \\ &= \frac{g(x)\sin[\pi\phi(x)/h]}{2\pi i}\int_{\partial\mathcal{D}} \frac{[\phi'(z)F(z)/g(z)]dz}{[\phi(z)-\phi(x)]\sin[\pi\phi(z)/h]}. \end{aligned} \tag{4.4.1}$$

Let us next prove the main Theorem of this section.

Theorem 4.4.2 *Let $\phi' F/g \in \mathbf{H}^1(\mathcal{D})$, and let*

$$\sup_{x\in\Gamma, -\pi/h \leqq t \leqq \pi/h} \left| \left(\frac{d}{dx}\right)^n g(x) e^{it\phi(x)} \right| \leqq C_2 h^{-n}, \tag{4.4.2}$$

for $n = 0, 1, , \cdots, m$, with C_2 a constant depending only on m, ϕ, and g. Then there exists a constant C_3, depending only on m, ϕ, g, d and F, such that for $n = 0, 1, \cdots, m$, we have

$$\sup_{x\in\Gamma} \left| F^{(n)}(x) - \left(\frac{d}{dx}\right)^n \sum_{k=-\infty}^{\infty} \frac{F(z_k)}{g(z_k)} g(x) S(k,h)\circ\phi(x) \right| \tag{4.4.3}$$
$$\leqq C_3 h^{-n} e^{-\pi d/h}.$$

If in addition we have $F/g \in \mathbf{L}_\alpha(\mathcal{D})$, α a positive constant, then taking

$$h = \left(\frac{\pi d}{\alpha N}\right)^{1/2} \tag{4.4.4}$$

it follows that

$$\sup_{x\in\Gamma} \left| F^{(n)}(x) - \left(\frac{d}{dx}\right)^n \sum_{k=-N}^{N} \frac{F(z_k)}{g(z_k)} g(x) S(k,h)\circ\phi(x) \right| \tag{4.4.5}$$
$$\leqq C_4 N^{(n+1)/2} e^{-(\pi d \alpha N)^{1/2}}$$

for $n = 0, 1, \cdots, m$, with C_4 a constant depending only on m, ϕ, g, d, α, and F.

Proof. If (4.4.2) holds, then we immediately get the inequalities

$$\left| \left(\frac{d}{dx}\right)^n g(x) S(k,h)\circ\phi(x) \right| \leqq C_2 h^{-n} \tag{4.4.6}$$

and

$$\left| \left(\frac{d}{dx}\right)^n g(x) \sin[\pi\phi(x)/h] \right| \leqq C_2 h^{-n} \tag{4.4.7}$$

with each of these valid for $n = 0, 1, \cdots, m$. Here (4.4.7) follows immediately from (4.4.2), whereas (4.4.6) follows by differentiating both sides of the identity

$$g(x)S(k,h)\circ\phi(x) = \frac{hg(x)}{2\pi}\int_{-\pi/h}^{\pi/h} e^{it[\phi(x)-kh]}dt \tag{4.4.8}$$

and then bounding the integral using (4.4.2). Furthermore, upon successively examining each derivative, $(d/dx)^n g(x)e^{it\phi(x)}$ for $n = 0, 1, \cdots$, we conclude from (4.4.2) that

$$g^{(j)}\phi^{(k)}[\phi']^{n-k} \tag{4.4.9}$$

is uniformly bounded on Γ for all non-negative integers j, k, and n such that $j + n \leqq m$. Hence using the inequality $|\phi(z) - \phi(x)| \geqq d$ for $z \in \partial\mathcal{D}$ and $x \in \Gamma$, it follows by induction, that

$$\left|\left(\frac{d}{dx}\right)^n \left[\frac{g(x)\sin[\pi\phi(x)/h]}{\phi(z)-\phi(x)}\right]\right| \leqq C_5 h^{-n} \tag{4.4.10}$$

with C_5 a constant depending only on m, ϕ, g, and d. Using this estimate as well as the inequalities (4.2.7) on the nth derivative of (4.4.1), we get

$$\begin{aligned} &|F^{(n)}(x) - \left(\frac{d}{dx}\right)^n \sum_{k=-\infty}^{\infty} \frac{F(z_k)}{g(z_k)} g(x)S(k,h)\circ\phi(x) \\ &\leqq \frac{C_6 h^{-n} N(\phi' F/g)}{\sinh(\pi d/h)} \end{aligned} \tag{4.4.11}$$

with C_6 a constant depending only on m, ϕ, g, d, and F. By assumption, $F/g \in \mathbf{L}_\alpha(\mathcal{D})$, and combining with (4.4.7), we get

$$\begin{aligned} &\left|\sum_{|k|>N} \frac{F(z_k)}{g(z_k)} \left(\frac{d}{dx}\right)^n g(x)S(k,h)\circ\phi(x)\right| \\ &\leqq \frac{2C_2C_4}{\alpha} h^{-n-1} e^{-\alpha N h}. \end{aligned} \tag{4.4.12}$$

We then combine (4.4.11) and (4.4.12) and take h as in (4.4.4) to get (4.4.5). ■

The following result is useful for approximating F and some derivatives of F in $\mathbf{L}^2(\Gamma)$.

Theorem 4.4.3 *Let d, d', $\mathcal{D}$, and $\mathcal{D}'$ be defined as in Theorem 4.1.2, let $(1/\phi')'$ be uniformly bounded in $\mathcal{D}'$, and let $F/g \in \mathbf{L}_\alpha(\mathcal{D}')$. If h is selected as in (4.4.4), then for $n = 0, 1, 2, \cdots$, we have*

$$\left(\int_\Gamma \left| \left(\frac{d}{dx}\right)^n \left\{ \frac{F(x)}{g(x)} - \sum_{k=-N}^{N} \frac{F(z_k)}{g(z_k)} S(k,h) \circ \phi(x) \right\} \right|^2 |\phi'(x)|^{1-2n} |dx| \right)^{1/2}$$

$$\leqq C_n N^{(2n+1)/4} e^{-(\pi d \alpha N)^{1/2}} \tag{4.4.13}$$

where C_n is a constant depending only on n.

Proof. Replace x by $\psi(y)$, z by $\psi(\zeta)$, $F(x)$ by $f(y)$, and $g(x)$ by $\gamma(y)$ in (4.4.1), to get

$$\begin{aligned} &\frac{f(y)}{\gamma(y)} - \sum_{k=-\infty}^{\infty} \frac{f(kh)}{\gamma(kh)} S(k,h)(y) \\ &= \frac{\sin(\pi y/h)}{2\pi i} \int_{\partial \mathcal{D}_{d'}} \frac{[f(\zeta)/\gamma(\zeta)]\, d\zeta}{(\zeta - y)\sin(\pi\zeta/h)}. \end{aligned} \tag{4.4.14}$$

Now, since $F/g \in \mathbf{L}_\alpha(\mathcal{D}') \subset \mathbf{L}_\alpha(\mathcal{D})$, it follows that $f/\gamma \in \mathbf{L}_\alpha(\mathcal{D}_d)$, and therefore, by Corollary 3.5.2, we have

$$\begin{aligned} &\left(\int_\Gamma \left| \left(\frac{d}{dy}\right)^n \left\{ \frac{f(y)}{\gamma(y)} - \sum_{k=-N}^{N} \frac{f(kh)}{\gamma(kh)} S(k,h)(y) \right\} \right|^2 dy \right)^{1/2} \\ &\leqq C_n N^{(2n+1)/4} e^{-(\pi d \alpha N)^{1/2}}. \end{aligned} \tag{4.4.15}$$

Recall now, that if $(1/\phi')'$ is uniformly bounded in $\mathcal{D}'$, then it follows by Problem 4.1.8 that $(F/g)^{(n)}/[\phi']^n \in \mathbf{L}_\alpha(\mathcal{D})$. Hence $\{(F/g)^{(n)}/[\phi']^n\}^2 \in \mathbf{L}_{2\alpha}(\mathcal{D})$, and therefore, $\phi'\{(F/g)^{(n)}/[\phi']^n\}^2 \in \mathbf{H}^1(\mathcal{D})$. That is,

$$[\phi']^{1/2}\{(F/g)^{(n)}/[\phi']^n\} \in \mathbf{H}^2(\mathcal{D}).$$

The result (4.4.13) therefore follows by induction on n. ∎

Example 4.4.4 $\Gamma = (a,b)$. In this case ϕ, ϕ', ψ and z_k are given by Equations (4.2.32) and the domain $\mathcal{D}$ is the eye-shaped region defined in (1.7.11) and illustrated in Figure 1.7.4.c. Let $f \in \mathcal{H}(\mathcal{D})$ be such that $f, f', \cdots, f^{(m)}$ are continuous in $\mathcal{D}$, and $f^{(m)} \in \mathbf{L}_\alpha(\mathcal{D})$. We set

$$F = f - p_m \tag{4.4.16}$$

with p_m recursively defined by

$$p_0(x) = \left\{ \begin{array}{l} A_0 \left(\frac{b-x}{b-a}\right) + B_0 \left(\frac{x-a}{b-a}\right) \\ A_0 = f(a), B_0 = f(b) \end{array} \right\} \tag{4.4.17}$$

and

$$\begin{array}{l} p_{k+1}(x) = \\ \quad p_k(x) + A_{k+1} \left(\frac{b-x}{b-a}\right)^{k+2} \left(\frac{x-a}{b-a}\right)^{k+1} + B_{k+1} \left(\frac{b-x}{b-a}\right)^{k+1} \left(\frac{x-a}{b-a}\right)^{k+2} \end{array} \tag{4.4.18}$$

$$A_{k+1} = \frac{(b-a)^{k+1}}{(k+1)!} \left[f^{(k+1)}(a) - p_k^{(k+1)}(a)\right]$$

$$B_{k+1} = \frac{(b-a)^{k+1}}{(k+1)!} \left[f^{(k+1)}(b) - p_k^{(k+1)}(b)\right].$$

It follows that taking

$$g(x) = (x-a)^m (b-x)^m \tag{4.4.19}$$

that the functions F and g then satisfy the conditions of Theorems 4.4.1 and 4.4.2. Indeed, using the identities

$$\begin{aligned} \left(\frac{d}{dx}\right)^n g(x) &= \sum_{k=0}^{\infty} \binom{n}{k} \left[(x-a)^m\right]^{(k)} \left[(b-x)^m\right]^{(n-k)} \\ \phi'(x) &= \frac{1}{x-a} + \frac{1}{b-x} = \frac{b-a}{(b-x)(x-a)} \\ \phi^{(j)}(x) &= (j-1)! \left[\frac{(-1)^{j-1}}{(x-a)^j} + \frac{1}{(b-x)^j}\right], \quad j \geqq 1. \end{aligned} \tag{4.4.20}$$

Moreover, using the identity

$$\frac{\phi^{(n)}(x)}{\phi'(x)^n} = \frac{(n-1)!}{(b-a)^n} \left[(-1)^{n-1}(b-x)^n + (x-a)^n\right] \tag{4.4.21}$$

which is valid for $n = 1, 2, 3, \cdots$, we may readily show that for all such n,

$$\left|\phi^{(n)}(x)\right| \leqq (n-1)! \phi'(x)^n, a \leqq x \leqq b. \tag{4.4.22}$$

These results yield the inequalities

$$\left|\left(\frac{d}{dx}\right)^n [\phi(z)-\phi(x)]^{-1}\right| \leqq C\,[(b-x)(x-a)]^{-n} \;\; (n \geqq 1)$$

$$\left|\left(\frac{d}{dx}\right)^n \sin[\pi\phi(x)/h]\right| \leqq Ch^{-n}\,[(b-x)(x-a)]^{-n} \;\; (n \geqq 0)$$

$$\left|\left(\frac{d}{dx}\right)^n S(k,h)\circ\phi(x)\right| \leqq Ch^{-n}\,[(b-x)(x-a)]^{-n}. \tag{4.4.23}$$

Hence, if $f^{(m)} \in \mathbf{Lip}_\alpha(\mathcal{D})$, then F defined by (4.4.16) satisfies the inequality

$$|F(x)/\,[(b-x)(x-a)]^m| \leqq C\,[(b-x)(x-a)]^\alpha \tag{4.4.24}$$

on (a,b), which implies that $F \in \mathbf{L}_\alpha(\mathcal{D})$. By taking h as in (4.4.4), we thus get, for $n = 0, 1, \cdots, m$,

$$\left|F^{(n)}(x) - \left(\frac{d}{dx}\right)^n \sum_{k=-N}^{N} F\left(\frac{a+b\,e^{kh}}{1+e^{kh}}\right)\left\{\left(\frac{1+e^{kh}}{b-a}\right)^2 \cdot \frac{(b-x)(x-a)}{e^{kh}}\right\}^m S(k,h)\circ\phi(x)\right| \tag{4.4.25}$$

$$\leqq C_5 N^{n+1/2} e^{-(\pi d\alpha N)^{1/2}}$$

for $a \leqq x \leqq b$, with C_5 a constant which does not depend on N.

Example 4.4.5 $\Gamma = (0,\infty)$. In this case we take $\phi(x) = \log x$, so that $\phi'(x) = 1/x$, $z_k = e^{kh}$. The domain $\mathcal{D}$ is the sector defined by Equation (1.7.10), and illustrated in Figure 1.7.4.b. Let $f \in \mathbf{H}^1(\mathcal{D})$ be such that $f, f', \cdots, f^{(m)}$ are continuous and bounded in $\mathcal{D}$, such that $f^{(m)}$ is of class $\mathbf{Lip}_\alpha$ on the region $\{z \in \mathcal{D} : |z| < R\}$ and such that $f(z) - f(\infty) = \mathcal{O}(|z|^{-\alpha})$ in $\mathcal{D}$. We set

$$F(x) = f(x) - \sum_{k=0}^{m+1} a_k x^k/(1+x)^{m+1} \tag{4.4.26}$$

with

$$a_0 = f(0), \quad a_{m+1} = f(\infty)$$

$$a_n = \frac{f^{(n)}(0)}{n!} - \sum_{j=1}^{n} \frac{(-1)^j (m+1)_j a_{n-j}}{j!}, \quad n = 1, \cdots, m. \tag{4.4.27}$$

We may then take

$$g(x) = [x/(1+x)]^m \tag{4.4.28}$$

and for this function, the function F satisfies $\phi' F/g \in \mathbf{H}^1(\mathcal{D})$, and

$$\left|\frac{F(x)}{g(x)}\right| \leqq Cx^\alpha/(1+x)^{2\alpha}, 0 < x < \infty. \tag{4.4.29}$$

Furthermore, it follows readily, that for some constant C_m depending only on m,

$$\left|F^{(n)}(x)\right| \leqq \begin{cases} C_m x^{m+\alpha-n}, & 0 < x \leqq 1 \\ C_m x^{-n-\alpha}, & 1 \leqq x < \infty \end{cases} \tag{4.4.30}$$

for $n = 0, 1, \cdots, m$. Next, we also have the identity

$$\left(\frac{d}{dx}\right)^n \frac{1}{\log z - \log x} = \frac{1}{x^n} p_{n+1}\left(\frac{1}{\log z - \log x}\right) \tag{4.4.31}$$

where $p_{n+1}(u)$ is a polynomial of degree $n+1$ in $u = (\log z - \log x)^{-1}$ which is bounded since $|\log z - \log x| \geqq d$.

Next, we have

$$\left|\left(\frac{d}{dx}\right)^n e^{it\log x}\right| = x^{-n} \prod_{k=0}^{n-1} \left(t^2 + k^2\right)^{1/2} \tag{4.4.32}$$

which yields [1]

$$\begin{aligned} &\left|\left(\frac{d}{dx}\right)^n \sin\left[\pi(\log x)/h\right]\right| \\ &\quad = \left|\Im\left(\frac{d}{dx}\right)^n e^{i\pi(\log x)/h}\right| \leqq \left|\left(\frac{d}{dx}\right)^n e^{i\pi(\log x)/h}\right| \\ &\quad = x^{-n}\prod_{k=0}^{n-1}\left[\left(\frac{\pi}{h}\right)^2 + k^2\right]^{1/2} \leqq (n-1)!x^{-n}\left(\frac{\pi}{h}\right)^n C'_m, \end{aligned} \tag{4.4.33}$$

and

[1]Here we define $(n-1)! = 1$ if $n = 0$.

$$\begin{aligned}
&\left|\left(\frac{d}{dx}\right)^n S(k,h)\circ\log x\right| \\
&\quad = \left|\frac{h}{2\pi}\int_{-\pi/h}^{\pi/h} e^{-ikht}\left(\frac{d}{dx}\right)^n e^{it\log x}\,dt\right| \\
&\quad \leqq x^{-n}\prod_{k=0}^{n-1}\left[\left(\frac{\pi}{h}\right)^2+h^2\right] \leqq (n-1)!x^{-n}\left(\frac{\pi}{h}\right)^n C'_m,
\end{aligned} \tag{4.4.34}$$

with C'_m a positive constant, depending only on m. Since moreover, for $0 \leqq n \leqq m$,

$$\begin{aligned}
&\left|\left(\frac{d}{dx}\right)^n x^m(1+x)^{-m}\right| \\
&= \left|m\sum_{k=0}^{n}\binom{n}{k}\frac{(m+k)!}{(m-k)!}(-1)^{n-k}x^{m-k}(1+x)^{-(m+n-k)}\right| \\
&\leqq C''_m\left(\frac{x}{1+x}\right)^{m-n},
\end{aligned} \tag{4.4.35}$$

we have by (4.3.34) and (4.3.35), that

$$\begin{aligned}
&\left|\left(\frac{d}{dx}\right)^n g(x)S(k,h)\circ(\log x)\right| \\
&\quad \leqq C'_m C''_m\sum_{k=0}^{n}\binom{n}{k}(k-1)!x^{-k}\left(\frac{\pi}{h}\right)^k\left(\frac{x}{1+x}\right)^{m-(n-k)} \\
&\quad \leqq C'_m C''_m D_m\left(\frac{\pi}{h}\right)^n\left(\frac{x}{1+x}\right)^{m-n}
\end{aligned} \tag{4.4.36}$$

with D_m a constant depending only on m. Taking h as in (4.4.4) we thus get the formula

$$\begin{aligned}
&\sup_{0<x<\infty}\left|F^{(n)}(x)-\left(\frac{d}{dx}\right)^n\sum_{k=-N}^{N}F(e^{kh})\left[\frac{x(1+e^{kh})}{e^{kh}(1+x)}\right]^m S(k,h)\circ(\log x)\right| \\
&\qquad \leqq C'N^{(n+1)/2}e^{-(\pi d\alpha N)^{1/2}}, \quad n=0,1,\cdots,m,
\end{aligned} \tag{4.4.37}$$

with C' a constant depending only on m, d, α, and f.

Example 4.4.6 $\Gamma = (0, \infty)$. We consider here the case when f is no longer analytic and bounded in the sector (1.7.10) of the previous example, but rather only in the bullet-shaped region $\mathcal{D}$ defined in (1.7.12), and illustrated in Figure 1.7.4.d. Let $f \in \mathbf{H}^1(\mathcal{D})$ be such that each of $f, f', \cdots, f^{(m)}$ is of class $\mathbf{Lip}_\alpha$ in the region $\{z \in \mathcal{D} : |z| < R\}$, and such that $f(z) - f(\infty) = \mathcal{O}(e^{-\alpha|z|})$ in $\mathcal{D}$. We then define F by

$$F(x) = f(x) - \sum_{k=0}^{m+1} a_k (1 - e^{-x})^k \tag{4.4.38}$$

where

$$a_0 = f(0), \quad a_n = \frac{1}{n!}\left[f^{(n)}(0) - \sum_{k=1}^{n-1} a_k b_{n,k}\right], \quad n = 1, 2, \cdots, m$$
$$b_{n,k} = -\sum_{j=1}^{k} \binom{k}{j} j^n (-1)^{n-j}, \quad a_{m+1} = f(\infty) - \sum_{k=0}^{m} a_k. \tag{4.4.39}$$

Since $\phi(z) = \log(\sinh z)$, and

$$\phi'(x) = \frac{1 + e^{-2x}}{1 - e^{-2x}} \tag{4.4.40}$$

it follows that

$$|\phi'(x)| \leqq (1 - e^{-x})^{-1}, \quad 0 < x < \infty \tag{4.4.41}$$

and also, by taking

$$g(x) = (1 - e^{-x})^m \tag{4.4.42}$$

it follows that $\phi' F/g \in \mathbf{H}^1(\mathcal{D})$. Furthermore the selections (4.4.39) for a_n, $n = 0, 1, \cdots, m+1$ now show that (since $0 < \alpha \leqq 1$)

$$\left|\frac{F(x)}{g(x)}\right| \leqq C_m^{(0)} \begin{cases} x^\alpha & \text{if } 0 < x \leqq 1 \\ e^{-\alpha x} & \text{if } 1 \leqq x < \infty \end{cases} \tag{4.4.43}$$

and also, that

$$\left|F^{(n)}(x)\right| \leqq C_m^{(1)} \begin{cases} x^{m+\alpha-n} & \text{if } 0 < x \leqq 1 \\ e^{-\alpha x} & \text{if } 1 \leqq x < \infty \end{cases} \tag{4.4.44}$$

where $C_m^{(0)}$ and $C_m^{(1)}$ depend only on m. We also readily verify that for $n = 0, 1, \cdots, m$

$$\left|\left(\frac{d}{dx}\right)^n (1 - e^{-x})^m\right| \leqq C_m^{(2)} \left(1 - e^{-x}\right)^{m-n} \tag{4.4.45}$$

with $C_m^{(2)}$ depending only on m, and that

$$\left|\frac{\phi^{(n)}(z)}{[\phi'(z)]^n}\right| \leqq (n-1)!. \tag{4.4.46}$$

It then follows using this inequality after differentiating $e^{it\phi(x)}$ n times with respect to x, then taking the absolute value with respect to t over the ranges $-\pi/h \leqq t \leqq \pi/h$, that

$$\begin{aligned} &\left|\left(\frac{d}{dx}\right)^n \sin\left[\pi\phi(x)/h\right]\right| \leqq C_m^{(3)} \left(\frac{\pi}{h}\right)^n (1-e^{-x})^{-n} \\ &\left|\left(\frac{d}{dx}\right)^n S(k,h)\circ\phi(x)\right| \leqq C_m^{(3)} \left(\frac{\pi}{h}\right)^n (1-e^{-x})^{-n} \end{aligned} \tag{4.4.47}$$

with $C_m^{(3)}$ a constant depending only on m. Similarly, we can also use (4.4.46) to establish the inequality

$$\left|\left(\frac{d}{dx}\right)^n \frac{1}{\phi(z)-\phi(x)}\right| \leqq C_m^{(4)} (1-e^{-x})^{-n} \tag{4.4.48}$$

with $C_m^{(4)}$ a constant depending only on m.

The last four inequalities now yield

$$\begin{aligned} &\left|\left(\frac{d}{dx}\right)^n \frac{\sin[\pi\phi(x)/h]}{2\pi i[\phi(z)-\phi(x)]}\right| \leqq \frac{C_m^{(3)} C_m^{(4)} 2^m (1-e^{-x})^{-n}}{2\pi} \left(\frac{\pi}{h}\right)^n \\ &\left|\left(\frac{d}{dx}\right)^n g(x) S(k,h)\circ\phi(x)\right| \leqq C_m^{(2)} C_m^{(3)} 2^m (1-e^{-x})^{m-n} \left(\frac{\pi}{h}\right)^n. \end{aligned} \tag{4.4.49}$$

Theorem 4.4.2 then yields

$$\begin{aligned} &\sup_{0<x<\infty} \left| F^{(n)}(x) - \left(\frac{d}{dx}\right)^n \sum_{k=-N}^{N} F(z_k) \left[\frac{1-e^{-x}}{1-e^{-z_k}}\right]^m S(k,h)\circ\phi(x)\right| \\ &\leqq C_m^{(5)} N^{(n+1)/2} e^{-(\pi d\alpha N)^{1/2}}, \quad n = 0, 1, \cdots, m \end{aligned} \tag{4.4.50}$$

with $C_m^{(5)}$ a constant depending only on m, d, α, and f, and with $z_k = \log\left[e^{kh} + \sqrt{1+e^{2kh}}\right]$.

•

Problems for Section 4.4

4.4.1 Verify the inequality (4.4.22), using (4.4.21).

4.4.2 Prove the inequalities (4.4.23).

4.4.3 Why is (4.4.24) satisfied?

4.4.4 Show that if $a_0, a_1, \cdots, a_m$ are selected as in (4.4.27), then

$$F^{(n)}(\infty) = 0, F^{(n)}(0) = 0, n = 0, 1, \cdots, m.$$

4.4.5 Prove the inequality (4.4.29).

4.4.6 Prove the inequality (4.4.30).

4.4.7 Prove the identity (4.4.32).

4.4.8 Recalling (4.4.35), use the identity

$$\left| \left(\frac{d}{dx} \right)^n x^m (1+x)^{-m} \right| = S$$

with

$$S \equiv \left| m \sum_{k=0}^{n} \binom{n}{k} \frac{(m+k)!}{(m-k)!} (-1)^{n-k} x^{m-k} (1+x)^{-(m+n-k)} \right|$$

to prove that

$$S \leqq C_m'' \left(\frac{x}{1+x} \right)^{m-n}.$$

4.4.9 Prove that if $a_0, a_1, \cdots, a_{m+1}$ are selected as in Equation (4.4.39), then F defined as in (4.4.38) satisfies

$$F^{(n)}(0) = F^{(n)}(\infty) = 0, n = 0, 1, \cdots, m.$$

4.4.10 Prove that (4.4.43) holds.

4.4.11 Prove that (4.4.44) holds.

4.4.12 Prove the inequality (4.4.45).

4.4.13 Prove the inequality (4.4.48).

4.4.14 Let d, d', $\mathcal{D}$, and $\mathcal{D}'$ be defined as in Theorem 4.1.2, let $1/\phi'$ be uniformly bounded in $\mathcal{D}'$, and let $[\phi']^{1/2} F \in \mathbf{L}_\alpha(\mathcal{D}')$. Prove that both F and F' belong to $\mathbf{H}^2(\mathcal{D})$.

4.5 Indefinite Integral Over Γ

We shall give a general formula for approximating

$$(\mathcal{J}F)(v) \equiv \int_a^v F(u)\,du, \quad v \in \Gamma, \tag{4.5.1}$$

with Γ defined as in Definition 4.1.1, and with the indefinite integral taken over Γ. To this end we make the following assumption:

Assumption 4.5.1 *Let ϕ denote the conformal map of $\mathcal{D}'$ onto $\mathcal{D}_{d'}$, and also of $\mathcal{D}$ onto $\mathcal{D}_d$, with $0 < d \leqq d'$. Let $\alpha' > 0$, let $F/\phi' \in \mathbf{L}_{\alpha'}(\mathcal{D}')$, and corresponding to a positive number β, define α by $\alpha = \min(\alpha', \beta)$. Set $d = \min(d', \pi/(2\beta) - \delta)$ with δ an arbitrary positive number in the range $0 < \delta < \pi/(2\beta)$.*

Theorem 4.5.2 *Let Assumption 4.5.1 be satisfied. Let N be a positive integer, and let σ_k and h be defined by (1.10.30) and (3.1.27) respectively. Then there exists a positive number K, independent of N, such that*

$$\sup_{v\in\Gamma}\left|(\mathcal{J}F)(v) - \frac{e^{\beta\phi(v)}}{2\cosh[\beta\phi(v)]}h\sum_{|k|\leqq N}\frac{F(z_k)}{\phi'(z_k)} - h\sum_{|k|\leqq N}\cdot\right.$$
$$\left.\cdot\left\{\sum_{|\ell|\leqq N}\sigma_{k-\ell}\left[\frac{F(z_\ell)}{\phi'(z_\ell)} - \frac{\beta}{2\cosh^2(\beta\ell h)}h\sum_{|j|\leqq N}\frac{F(z_j)}{\phi'(z_j)}\right]\right\}S(k,h)\circ\phi(v)\right|$$
$$\leqq K N^{1/2} e^{-(\pi d\alpha N)^{1/2}}. \tag{4.5.2}$$

Proof. Let us consider for $v \in \Gamma$, the difference

$$\Delta(v) = \int_a^v F(u)du - \frac{e^{\beta\phi(v)}}{2\cosh[\beta\phi(v)]}\int_\Gamma F(u)du. \tag{4.5.3}$$

Setting $u = \psi(t)$, $v = \psi(x)$, $(\Delta\circ\psi)(x) = g(x)$, and $(F\circ\psi)(t)\psi'(t) = f(t)$ we get

$$g(x) = \int_{-\infty}^x f(t)dt - \frac{e^{\beta x}}{2\cosh(\beta x)}\int_{\mathbf{R}} f(t)dt. \tag{4.5.4}$$

Now, under the assumption that $F/\phi' \in \mathbf{L}_{\alpha'}(\mathcal{D}')$, it follows that, as a function of t, $f \in \mathbf{L}_{\alpha'}(\mathcal{D}_{d'})$ with $\mathbf{L}_{\alpha'}(\mathcal{D}_{d'})$ defined as in Definition 3.1.5. Hence f satisfies the conditions of Theorem 3.6.1. Moreover, after replacing u by $\psi(t)$ and v by $\psi(x)$, Equation (4.5.2) becomes identical to Equation (3.6.8). Hence the statement of Theorem 4.5.2 follows from Theorem 3.6.1, which we already proved. ■

By paraphrasing Corollaries 3.6.6, 3.6.7, and 3.6.8, we get the following results.

Corollary 4.5.3 *Let $F/\phi' \in \mathbf{L}_\alpha(\mathcal{D})$, with $0 < \alpha \leqq 1$, and $0 < d < \pi$, and let $\rho = e^\phi$. Let σ_k be defined as in Equation (1.10.30), and let $h = [\pi d/(\alpha N)]^{1/2}$, with N a positive integer. Then there exists a positive constant K, independent of N, such that*

$$\sup_{x\in\Gamma}\left| \int_a^x F(t)\,dt - \frac{\rho(x)}{1+\rho(x)} h \sum_{|k|\leqq N} \frac{F(z_k)}{\phi'(z_k)} - h \sum_{|k|\leqq N} \cdot \right.$$
$$\left. \cdot \left\{ \sum_{|\ell|\leqq N} \sigma_{k-\ell} \left[\frac{F(z_\ell)}{\phi'(z_\ell)} - \frac{e^{\ell h}}{(1+e^{\ell h})^2} h \sum_{|j|\leqq N} \frac{F(z_j)}{\phi'(z_j)} \right] \right\} S(k,h)\circ\phi(x) \right|$$
$$\leqq K\, N^{1/2}\, e^{-(\pi d \alpha N)^{1/2}}. \tag{4.5.5}$$

Corollary 4.5.4 *Let $F/\phi' \in \mathbf{L}_\alpha(\mathcal{D})$, with $\alpha > 0$, $d > 0$, let $\delta_k^{(-1)}$ be defined as in (1.10.30), and let $h = [\pi d/(\alpha N)]^{1/2}$. Then there exists a constant, K, which is independent of N, such that*

$$\left| \int_a^{z_k} F(t)\,dt - h \sum_{\ell=-N}^{N} \delta_{k-\ell}^{(-1)} \frac{F(z_\ell)}{\phi'(z_\ell)} \right| \leqq K e^{-(\pi d \alpha N)^{1/2}}. \tag{4.5.6}$$

It may at times be desirable to approximate in the spaces $\mathbf{H}^1(\mathcal{D})$, or in $\mathcal{M}_{\alpha,\beta}(\mathcal{D})$, where α and β are positive numbers. To this end, we state the following result, which may be readily established by proceeding as in the above proofs, and we therefore omit the proof, leaving it to Problem 4.5.5.

Corollary 4.5.5 *Let N be a positive integer, define M by $M = [\beta/\alpha N]$, let $\delta_k^{(-1)}$ be defined as in (1.10.30), and let $I^{(-1)}$ denote the matrix*

$$I^{(-1)} = \left[\delta_{k-\ell}^{(-1)}\right], \quad k,\ell = -M,\cdots,N, \tag{4.5.7}$$

define the diagonal matrix $D(1/\phi') = \mathrm{diag}[1/\phi'(z_{-M}),\cdots,1/\phi'(z_N)]$, set $h = [\pi d/(\beta N)]^{1/2}$, and $A = hI^{(-1)}D(1/\phi')$. Let $F \in \mathbf{H}^1(\mathcal{D})$, set $\hat{F} = (F(z_{-N}),\cdots,F(z_N))^T$, define $(\mathcal{J}F)(x) = \int_a^x F(t)\,dt$, set $\hat{G} = (G_{-N},\cdots,G_N)^T = A\hat{F}$, and then define $\mathcal{J}_N F$ by

$$(\mathcal{J}_N F)(x) = \frac{\rho(x)G_N}{1+\rho(x)} - \sum_{k=-M}^{N} \left(G_k - \frac{e^{kh}G_N}{1+e^{kh}} \right) S(k,h)\circ\phi(x) \tag{4.5.8}$$

(a) Then

$$\sup_{x \in \Gamma} |(\mathcal{J} F)(x) - (\mathcal{J}_N F)(x)| \to 0 \tag{4.5.9}$$

as $N \to \infty$.
(b) If $F/\phi' \in \mathbf{L}_{\alpha,\beta}(\mathcal{D})$, *then*

$$\sup_{x \in \Gamma} |(\mathcal{J} F)(x) - (\mathcal{J}_N F)(x)| \leqq K_1 N^{1/2} e^{-(\pi d \alpha N)^{1/2}}. \tag{4.5.10}$$

where K_1 *is a constant that is independent of* N.

•

Problems for Section 4.5

4.5.1 Suppose that d' and α' are exactly known. How should β and δ be selected so that the constant $(\pi d \alpha)^{1/2}$ in (4.5.2) is as large as possible?

4.5.2 Give an explicit statement of Theorem 4.5.2 for each of the following cases of Γ, ϕ, and z_k of the examples following Theorem 4.4.2 in Sec. 4.4.

4.5.3 Give a detailed proof of Corollary 4.5.3.

4.5.4 Give a detailed proof of Corollary 4.5.4.

4.5.5 Give a detailed proof of Corollary 4.5.5. [Hint: Use the results of Theorems 4.1.3 and 4.2.5.]

4.5.6 State and prove a theorem analogous to Theorem 3.6.1 for approximating $\int_x^b F(t)\,dt$.

4.5.7 In the notation of Corollary 4.5.4, prove that

$$\left| \int_{z_k}^b F(t)\,dt - h \sum_{\ell=-N}^{N} \delta_{\ell-k}^{(-1)} F(z_\ell)/\phi'(z_\ell) \right| \leqq K\, e^{-(\pi d \alpha N)^{1/2}}.$$

4.5.8 Let $F/\phi' \in \mathbf{L}_\alpha(\mathcal{D})$, with $\alpha > 0$ and $d > 0$. Let $(I^{(-1)})^T$ denote the transpose of $I^{(-1)}$, and let $G_k = \int_{z_k}^b F(t)\,dt$, $\hat{G} = (G_{-N}, \cdots, G_N)^T$, $F_k = F(z_k)$, and $\hat{F} = (F_{-N}, \cdots, F_N)^T$. Let $D(1/\phi')$ be defined as in Corollary 4.5.5. Set $B = h(I^{(-1)})^T D(1/\phi')$, and then set $\hat{E} = (E_{-N}, \cdots, E_N)^T = B\hat{F}$.

(a) Prove that

$$\hat{G} = \hat{E} + \hat{\varepsilon}$$

with $\hat{\varepsilon} = (\varepsilon_{-N}, \cdots, \varepsilon_N)^T$, and with $|\varepsilon_j| \leqq K \exp[-(\pi d\alpha N)^{1/2}]$.

(b) Prove that if $G(x) = \int_x^b F(t)\,dt$, $0 < \alpha \leqq 1$, and $0 < d < \pi$, then there exists a constant K_1 independent of N, such that for all $x \in \Gamma$,

$$\left| G(x) - \frac{E_{-N}}{1+\rho(x)} - \sum_{k=-N}^{N} \left(E_k - \frac{E_{-N}}{1+e^{kh}} \right) S(k,h) \circ \phi(x) \right|$$
$$\leqq K_1 N^{1/2} e^{(\pi d \alpha N)^{1/2}}.$$

4.5.9 (a) Use the method of this section to obtain an approximation of the incomplete Gamma function

$$\gamma(1/3, x) = \int_0^x e^{-t} t^{-2/3}\, dt$$

accurate to 6 decimals on $[0, \infty]$.

(b) Use the method of this section to obtain an approximation of the incomplete Gamma function

$$\Gamma(2/3, x) = \int_x^\infty e^{-t} t^{-1/3}\, dt$$

accurate to 6 decimals on $[0, \infty]$.

4.5.10 Prove that the constant K_1 in (4.5.10) may be assumed to be bounded by a constant C_2 times $\int_{\partial \mathcal{D}} |F(z)\,dz|$, with C_2 independent of F and $\mathcal{D}$.

4.6 Indefinite Convolution Over Γ

In this section we discuss the approximation of the convolutions p and q of functions f and g, defined by the integrals

$$\begin{aligned} p(x) &= \int_a^x f(x-t)\, g(t)\, dt, \quad x \in (a,b), \\ q(x) &= \int_x^b f(t-x)\, g(t)\, dt, \quad x \in (a,b), \end{aligned} \tag{4.6.1}$$

where $\Gamma = (a,b)$ is either all of $\mathbb{R}$, or else it is a subinterval of $\mathbb{R}$. Evidently, effective methods of approximating p and q in (1.2) can then be used to yield an accurate approximation to the definite convolution integral

$$p(x) + q(x) = \int_a^b f(|x - t|)\, g(t)\, dt. \tag{4.6.2}$$

Integrals of the type (4.6.1) are often evaluated via the use of an inaccurate integration procedure, such as the trapezoidal rule, or Euler's method of integrating a first-order differential equation. Another procedure is to use a definite integration formula after, e.g. for the case of $a = 0$, replacing x by xt in the integral expression for $p(x)$ in (4.6.1). This latter method then requires the evaluation of g at the points xt_j where t_j are the fixed points of the quadrature formula over (a, b). Thus different evaluations of g are required for each value of x. The method of the present section provides both formulas of arbitrary high accuracy, and it uses values of g at a fixed set of points on (a, b), to yield an approximation of either p or q over the entire interval (a, b). The present method also provides an improvement over the Fast Fourier Transform method of approximating convolutions of the form (4.6.2) in the case of finite or semi-infinite intervals, in the case when either f or g has singularities at one or both endpoints of its interval of definition, or in the case when the convolution (p or q) of f and g has singularities at one or both of the endpoints of (a, b).

Whereas the solution to an ordinary or partial differential equation can frequently be expressed as a convolution integral of a Green's function and a known function, due to the lack of existence of methods of obtaining accurate approximations of (indefinite and definite) convolution-type integrals, such transformations have to date been used more often to study properties of the solution than to obtain accurate approximations of the solution. After a differential equation is transformed into such a convolution form, the method of this section then provides an explicit procedure for obtaining an approximate solution to the differential equation. More generally, if the (linear or nonlinear) differential equation is transformed into a convolution-type integral equation via the use of Green's functions, the method of this section enables us to replace the resulting integral equation by a system of (linear or nonlinear) algebraic equations, whose solution yields an accurate approximate solution to the differential equation.

4.6.1 Approximation Procedure

In this subsection we describe the collocation procedure for approximating convolution integrals and for obtaining explicit approximations of the functions p and q defined in (4.6.1). While we describe the explicit procedure for approximating both p and q, we give a detailed proof only for the approximation of p, since the proof for the approximation of q is almost identically the same.

We assume that Definition 4.1.1 is in force, i.e., that $\phi : \mathcal{D} \to \mathcal{D}_d$, with $0 < d < \pi$, that $\Gamma = \phi^{-1}(\mathbb{R}) = (a, b) \subseteq \mathbb{R}$, that $z_k = \phi^{-1}(kh)$, that $\rho = e^{\phi}$

as well as the notations of $\mathbf{L}_{\alpha,\beta}(\mathcal{D})$ and $\mathbf{M}_{\alpha,\beta}(\mathcal{D})$. The space of functions $\mathbf{H}^1(\mathcal{D})$ will be denoted simply by $\mathbf{X} = \mathbf{X}(\mathcal{D})$, and the space $\mathbf{H}^\infty(\mathcal{D})$ by $\mathbf{Y} = \mathbf{Y}(\mathcal{D})$. If $g \in \mathbf{X}$, we define the norm of g by $\|g\|_{\mathbf{X}} = \sup_{x\in(a,b)} |g(x)/\phi'(x)|$, while if $g \in \mathbf{Y}$, we define $\|g\|_{\mathbf{Y}} = \sup_{x\in(a,b)} |g(x)|$. Given $w \in \mathbf{X}$, we use the notation $\mathcal{I}w$ to denote the indefinite integral of w. It may be shown, for example (see Section 4.1) that if $g/\phi' \in \mathbf{L}_{\alpha',\beta'}(\mathcal{D})$ (respectively, $g \in \mathbf{X}$), then $\mathcal{I}g \in \mathbf{M}_{\alpha,\beta}(\mathcal{D})$, $\alpha = \min(1,\alpha')$, and $\beta = \min(1,\beta')$ (respectively, $\mathcal{I}g \in \mathbf{Y}$). Conversely, if $u \in \mathbf{M}_{\alpha,\beta}(\mathcal{D}')$, (resp., $u \in \mathbf{Y}(\mathcal{D}')$) then $u'/\phi' \in \mathbf{L}_{\alpha,\beta}(\mathcal{D})$ (respectively, $u'/\phi' \in \mathbf{Y}(\mathcal{D})$), where $\mathcal{D}$ and $\mathcal{D}'$ are related as in Section 4.1. Since $0 < d < \pi$, and since $\alpha \in (0,1]$, and $\beta \in (0,1]$, it follows that $\mathbf{M}_{\alpha,\beta}(\mathcal{D}) \subset \mathbf{Y}$. We shall also use the numbers $\delta_k^{(-1)}$ defined as in (1.10.30).

While more precise assumptions on f, g, p, and q will be given below, we shall assume for now that $g \in \mathbf{X}$, and that f is analytic in a domain $\mathcal{D}_f$, with ϕ_f denoting a conformal mapping of $\mathcal{D}_f$ onto $\mathcal{D}_d$, and $\phi_f : (0,c) \to \mathbb{R}$, c being an arbitrary number on the interval $[b-a,\infty]$. We will assume that both p and q belong to $\mathbf{M}_{\alpha,\beta}(\mathcal{D}')$, with $\mathcal{D}' = \phi^{-1}(\mathcal{D}_{d'})$, and where $0 < d < d'$. Corresponding to a positive integer N, we take $M = [\beta N/\alpha]$, where $[x]$ denotes the greatest integer in x, we set $m = M + N + 1$, and we determine h via the formula $h = \{\pi d/(\beta N)\}^{1/2}$.

Let $I^{(-1)}$ be a square matrix of order m having $\delta_{i-j}^{(-1)}$ as its (i,j)th element. As in the previous section, define square matrices A_m and B_m by

$$A_m = hI^{(-1)}D(1/\phi'), \quad B_m = h\{I^{(-1)}\}^T D(1/\phi') \tag{4.6.3}$$

where the superscript "T" denotes the transpose, and where for any function u defined and continuous on Γ, we let $D(u)$ be the diagonal matrix defined by $D(u) = \mathrm{diag}[u(z_{-M}), \cdots, u(z_N)]$. Throughout this section, when referring to the Laplace transform of a function f we will mean the usual Laplace transform, $\hat{f}(s) = \int_0^c e^{-st} f(t)\,dt$, where c is defined as above, whereas when referring to the Laplace transform in quotations, i.e., the "Laplace transform," will mean the function F defined by $F(s) = \hat{f}(1/s)$, or

$$F(s) = \int_0^c f(t)\, e^{-t/s} dt. \tag{4.6.4}$$

We shall assume that the "Laplace transform" exists for some $c \in [b-a,\infty]$, for all s on the right half of the complex plane $\Omega^+ = \{z \in \mathbb{C} : \Re z > 0\}$. Although these conditions suffice for purposes of defining the convolution approximations, in Subsection 4.6.4 we shall require more stringent conditions on the "Laplace transform" for purposes of obtaining an accurate error bound on the approximation. For example, in Subsection 4.6.4 we shall require that the above defined "Laplace transform" should exist for some $c \in [2(b-a),\infty]$, and for all $s \in \Omega^+$. We shall also see below, that the eigenvalues of each of the matrices A_m and B_m defined in (2.2) belong to $\overline{\Omega^+}$ (actually it has been shown by direct computation that these eigenvalues belong to Ω^+ for $1 \leqq m \leqq 512$). Thus, if non-singular matrices X_m, Y_m, and complex numbers $s_{m,j}$ are determined such that

$$\begin{aligned} A_m &= X_m \mathrm{diag}[s_{m,-M}, \cdots, s_{m,N}] X_m^{-1} \\ B_m &= Y_m \mathrm{diag}[s_{m,-M}, \cdots, s_{m,N}] Y_m^{-1} \end{aligned} \tag{4.6.5}$$

then square matrices $F(A_m)$ and $F(B_m)$ may be defined via the equations

$$\begin{aligned} F(A_m) &= X_m \mathrm{diag}[F(s_{m,-M}), \cdots, F(s_{m,N})] X_m^{-1} \\ F(B_m) &= Y_m \mathrm{diag}[F(s_{m,-M}), \cdots, F(s_{m,N})] Y_m^{-1}. \end{aligned} \tag{4.6.6}$$

Now, define column vectors $\mathbf{g}_m$, $\mathbf{p}_m$, and $\mathbf{q}_m$ by

$$\begin{aligned} \mathbf{g}_m &= (g(z_{-M}), \cdots, g(z_N))^T \\ \mathbf{p}_m &= (p_{-M}, \cdots, p_N)^T = F(A_m)\mathbf{g}_m \\ \mathbf{q}_m &= (q_{-M}, \cdots, q_N)^T = F(B_m)\mathbf{g}_m. \end{aligned} \tag{4.6.7}$$

The component p_j of the vector $\mathbf{p}_m$ (respectively, q_j of the vector $\mathbf{q}_m$) approximates the value $p(x)$ (respectively, $q(x)$) at the Sinc point $x = z_j$.

The next step of the procedure is to describe a Sinc basis, $\{\omega_j\}$, such that the vectors $\mathbf{p}_m$ and $\mathbf{q}_m$ determined as in (4.6.7) may be used to approximate the functions $p(x)$ and $q(x)$ throughout (a, b), where $p(x)$ and $q(x)$ were defined as in (4.6.1). To this end, we set

$$\begin{aligned} \gamma_j(x) &= S(j,h) \circ \phi(x), \quad j = -M, \cdots, N \\ \omega_j(x) &= \gamma_j(x), \quad j = -M+1, \cdots, N-1 \\ \omega_{-M} &= [1 + e^{-Mh}] \left[\frac{1}{1+\rho} - \sum_{j=-(M-1)}^{N} \frac{\gamma_j}{1+e^{jh}} \right] \\ \omega_N &= [1 + e^{-Nh}] \left[\frac{\rho}{1+\rho} - \sum_{j=-M}^{N-1} \frac{e^{jh}\gamma_j}{1+e^{jh}} \right]. \end{aligned} \tag{4.6.8}$$

The functions ω_j defined in (4.6.8) satisfy the relations $\omega_j(z_k) = 1$ if $j = k$ while $\omega_j(z_k) = 0$ if $j \neq k$.

The approximations of p and q on (a, b) then take the form

$$\begin{aligned} p &\approx \sum_{j=-M}^{N} p_j \omega_j \\ q &\approx \sum_{j=-M}^{N} q_j \omega_j. \end{aligned} \tag{4.6.9}$$

The basis functions defined in (4.6.8) suffice for purposes of uniform norm approximation over (a, b). They are also convenient for purposes of uniform norm approximation in that the same basis functions can then be used to approximate *both* functions p and q, and hence also the sum, $p+q$. On the other hand, it may at times be convenient to replace one of the last two basis functions in (4.6.8) by an ordinary Sinc basis function given in the second line of (4.6.8), if it is desirable that the final approximation vanish at an endpoint of the interval, since these ordinary Sinc basis functions vanish at both endpoints of the interval of approximation. For example, for purposes of approximating p on (a,b) via the formula in (4.6.9), an equally accurate uniform norm approximation obtains, if instead of the basis in (4.6.8), we select the basis

$$\begin{aligned}\omega_j(x) &= S(j,h)\circ\phi(x), \quad j=-M,\cdots,N-1\\ \omega_N &= [1+e^{-Nh}]\left[\frac{\rho}{1+\rho}-\sum_{j=-M}^{N-1}\frac{e^{jh}\omega_j}{1+e^{jh}}\right].\end{aligned} \tag{4.6.10}$$

Furthermore, by using (4.6.10), the resulting approximation as defined in (4.6.9) then vanishes at the endpoint a. An obviously similar modification of (4.6.8) is possible for purposes of getting an approximation of q, which vanishes at b.

4.6.2 Formula Derivation

It suffices to consider only the case of $p(x)$ as defined in (4.6.1), since the case of $q(x)$ can be treated in exactly the same way. Let $\mathcal{J}w$ be defined for $w\in\mathbf{X}$ by

$$(\mathcal{J}w)(x)=\int_a^x w(t)\,dt. \tag{4.6.11}$$

It then follows that for $n=1,2,\cdots$

$$(\mathcal{J}^n w)(x)=\int_a^x \frac{(x-t)^{n-1}}{(n-1)!}w(t)\,dt. \tag{4.6.12}$$

The length of the interval (a,b) will at the outset be assumed to be finite. It is then convenient to take $\|w\|=(b-a)^{-1}\int_a^b|w(t)|\,dt$, since this choice of norm yields the simple inequality

$$\|\mathcal{J}w\|=\frac{1}{b-a}\int_a^b\left|\int_a^x w(t)\,dt\right|\,dx\leqq(b-a)\|w\| \tag{4.6.13}$$

which implies that

$$\begin{aligned}\|\mathcal{J}^n\| &\leqq \frac{(b-a)^n}{n!}\\ \|\mathcal{J}^n\| &\leqq \|\mathcal{J}\|^n\leqq(b-a)^n.\end{aligned} \tag{4.6.14}$$

By using the *Bromwich formula* for the inversion of the Laplace transform, and then converting to the "Laplace transform" via replacement of s by $1/s$ (a transformation which transforms Ω^+ to itself) the expression (4.6.1) for $p(x)$ may be written in the alternate forms

$$\begin{aligned} p(x) &= -\int_a^x \frac{1}{2\pi i}\int_{-i\infty}^{i\infty} e^{(x-t)/s} F(s) s^{-2} ds\, g(t)\, dt \\ &= -\frac{1}{2\pi i}\int_{-i\infty}^{i\infty} w(s,x)\, ds \end{aligned} \tag{4.6.15}$$

where

$$w(s,x) = \int_a^x s^{-2} e^{(x-t)/s} F(s)\, g(t)\, dt. \tag{4.6.16}$$

Now, if $|s| > b - a$, and we use the notations of (4.6.11) and (4.6.12), as well as the inequality (4.6.13), we get

$$\begin{aligned} w(s,x) &= \int_a^x \sum_{n=0}^{\infty} \frac{(x-t)^n}{n!\, s^{n+2}}\, g(t)\, dt \\ &= \left(\frac{\mathcal{J}}{s^2} \sum_{n=0}^{\infty} \frac{\mathcal{J}^n}{s^n} F(s)\, g \right)(x) \\ &= \left(\frac{\mathcal{J}}{s} (s - \mathcal{J})^{-1}\, F(s)\, g \right)(x). \end{aligned} \tag{4.6.17}$$

By analytic continuation as a function of s, it then follows that the identity

$$\int_a^x s^{-2} e^{(x-t)/s} F(s)\, g(t)\, dt = \left(\frac{\mathcal{J}}{s} (s - \mathcal{J})^{-1}\, F(s)\, g \right)(x) \tag{4.6.18}$$

holds not only for all $s \in \mathbb{C}$ such that $|s| > b - a$, but also in the larger, *resolvent set* of $\mathcal{J}$, (see Section 6.2.1) excluding the point $s = 0$. Here, the resolvent set of $\mathcal{J}$ is the set $\{s \in \mathbb{C} : (s - \mathcal{J})^{-1}$ exists and is bounded$\}$. The resolvent set of $\mathcal{J}$ can be more closely identified, upon setting

$$(u, v) \equiv \int_a^b u(x)\overline{v(x)}\, dx. \tag{4.6.19}$$

It follows, in this notation, that

$$\Re(\mathcal{J}u, u) = \frac{1}{2}\left| \int_a^b u(x)\, dx \right|^2 \geqq 0. \tag{4.6.20}$$

Hence, the resolvent set of $\mathcal{J}$ includes the set $\{s \in \mathbb{C} : \Re s < 0\}$, as well as the set $\{\Re s \in \mathbb{C} : |s| > b - a\}$.

Substitution of (4.6.18) into (4.6.15) yields an *operator valued integral* called the *Dunford integral,*

$$p(x) = -\lim_{\varepsilon \to 0+} \left(\frac{1}{2\pi i} \int_{-i\infty}^{i\infty} \frac{\mathcal{J}}{s-\varepsilon} (s - \varepsilon - \mathcal{J})^{-1} F(s)\, ds\, g \right)(x). \quad (4.6.21)$$

If (a, b) is a finite interval, then it follows that $F(s)$ is bounded for all sufficiently large $s \in \mathbb{C}$. It is also well-known that the classical Laplace transform $\hat{f}(s) \equiv \int_0^c e^{-st} f(t)\, dt \to 0$ as $s \to \infty$, so that $F(0) = \hat{f}(\infty) = 0$. The expression (4.6.21) thus enables us to deduce the formula

$$p(x) = (F(\mathcal{J})\, g)(x) \quad (4.6.22)$$

which is valid for bounded intervals (a, b).
Recall now the definition (4.6.4) of F, which implies that $F = F_1 + F_2$, where $F_2(s) = \int_{b-a}^c e^{-t/s} f(t)\, dt$. It follows from this, and the Bromwich inversion formula, that if $t < b - a$, then $(2\pi i)^{-1} \int_{-i\infty}^{i\infty} e^{t/s} F_2(s)\, s^{-2} ds = 0$. The formula (4.6.22) still holds in the case when $c > b - a$, since in that case, in view of the above remarks, and upon recalling that $g(t) = 0$ for $t > b - a$, it follows that $F_2(\mathcal{J})g = 0$, and therefore $F(\mathcal{J})g = F_1(\mathcal{J})g$.

Next, assume that (a, b) is an unbounded interval, and assume for this case that the function p defined in (4.6.1) exists for fixed f and g belonging to $\mathbf{X}$. The expression (4.6.22) can then again be established to be valid, by repeating the above argument for each of a sequence of nested finite intervals $(a_j, b_j) \subset (a_{j+1}, b_{j+1})$, where $a_j \to a$, and $b_j \to b$, as $j \to \infty$. We furthermore need to assume that $F(s)/s^2 = o(1/s)$ as $s \to \infty$ in Ω^+, in order to be able to apply the residue theorem to get (4.6.22) from (4.6.21). On the other hand, if $F(s) \sim \gamma s$ as $s \to \infty$, where γ is a constant, then $F(s) - \gamma s = o(s)$ as $s \to \infty$, i.e., we can assume that (4.6.22) holds, even if $F(s) = \mathcal{O}(s)$ as $s \to \infty$.

Now define linear operators V_m and Π_m as follows. For $g \in \mathbf{X}$, $V_m g = (g(z_{-M}), \cdots, g(z_N))^T$, and for a given vector $\mathbf{c} = (c_{-M}, \cdots, c_N)^T$, define $\Pi_m \mathbf{c}$ by $(\Pi_m \mathbf{c})(x) = \sum_{j=-M}^{N} c_j \omega_j(x)$, where the basis functions ω_j are defined in (4.6.8). Then, define $\mathcal{J}_m$ on $\mathbf{X}$ by $\mathcal{J}_m = \Pi_m A_m V_m$, where the matrix A_m is defined as in (4.6.5). Since $V_m \Pi_m$ is the unit matrix of order m, then recalling the above definition (2.5) of $F(A_m)$, it is consistent and also convenient to define $F(\mathcal{J}_m)$ by $F(\mathcal{J}_m) = \Pi_m F(A_m) V_m$. The operator $F(\mathcal{J}_m)$ can then be defined, of course, in the same way that the operator $F(\mathcal{J})$ was defined above. The expression (4.6.22) can thus be replaced by the approximating one

$$p(x) \approx (F(\mathcal{J}_m)\, g)(x) = (\Pi_m F(A_m) V_m g)(x). \quad (4.6.23)$$

The convergence and error of this approximation are discussed in the following subsection.

4.6.3 Convergence

We consider a convergence proof only for the case of $\int_a^x (\cdot)\,dt$, since the considerations for $\int_x^b (\cdot)\,dt$ are essentially the same.

It may be readily shown that the eigenvalues of the matrix A_m defined at the end of Subsection 4.6.3 lie in the closure of the right-half plane, Ω^+. To this end, let us set $I^{(-1)} = H + S$, where, in view of (1.10.30), each of the m^2 entries of H is the number $1/2$, and where S is a skew-symmetric Toeplitz matrix. Hence, if $\mathbf{c} = (c_{-M}, \cdots, c_N)^T$ is an arbitrary complex vector of order m,

$$\mathbf{c}^* I^{(-1)} \mathbf{c} = \frac{1}{2}\left| \sum_{j=-M}^{N} c_j \right|^2 + \mathbf{c}^* S \mathbf{c}. \tag{4.6.24}$$

Since the term $\mathbf{c}^* S\mathbf{c}$ is purely imaginary, it is clear that the eigenvalues of $I^{(-1)}$ all lie in the closure of the right half plane. Indeed, it has already been shown by numerical computation, that the real part of each eigenvalue of every matrix $I^{(-1)}$ of orders 1 to 512 is positive. This size of m usually suffices to guarantee 25 significant figures of accuracy in the approximation of p or q by the above outlined procedure in Section 4.6.1. *It will thus be assumed that m is such that every eigenvalue of $I^{(-1)}$ is positive.* Next, since $A_m = I^{(-1)} D$ where $D = D(1/\phi')$, where each entry d_j of the diagonal matrix D is positive, and since the eigenvalues of A_m are the same as the eigenvalues of $D^{1/2} A_m D^{-1/2} = D^{1/2} I^{(-1)} D^{1/2}$, it follows that

$$\mathbf{c}^* D^{1/2} I^{(-1)} D^{1/2} \mathbf{c} = \frac{1}{2}\left| \sum_{j=-M}^{N} c_j d_j \right|^2 + \mathbf{c}^* D^{1/2} S D^{1/2} \mathbf{c}, \tag{4.6.25}$$

where S is defined as in (4.6.24) above. The term involving the sum on the right-hand side of (4.6.25) is non-negative, whereas the inner product term involving the matrix S is purely imaginary, which again leads to the conclusion that the eigenvalues of A_m lie on the closure of the right-half plane. Indeed, the above discussion shows that the real parts of the eigenvalues of A_m are positive if and only if the real parts of the eigenvalues of $I^{(-1)}$ are positive. In view of the above assumptions about the real parts of the eigenvalues of $I^{(-1)}$, *it will be assumed that m is such that the real part of every eigenvalue of A_m is positive.* It therefore follows that the matrix $F(A_m)$ is well defined for such m.

In the next lemma we establish the convergence of $F(\mathcal{J}_m)g$ to $F(\mathcal{J})g$, as well as the continuity of the *Fréchet derivative* of F. To establish our notation, we set

$$\begin{aligned} \Delta &= \mathcal{J} - \mathcal{J}_m \\ \mathcal{J}(t) &= \mathcal{J}_m + t\Delta \end{aligned} \tag{4.6.26}$$

We may then deduce by proceeding as for (4.6.21), that

$$F[\mathcal{J}(t)]g = -\frac{1}{2\pi i}\int_{-i\infty}^{i\infty} \frac{\mathcal{J}(t)}{s}\,[s-\mathcal{J}(t)]^{-1}\,F(s)\,ds\,g. \tag{4.6.27}$$

Using (4.6.27) we readily derive the expression

$$\frac{d}{dt}F[\mathcal{J}(t)]g = -\frac{1}{2\pi i}\int_{-i\infty}^{i\infty} [s-\mathcal{J}(t)]^{-1}\,\Delta\,[s-\mathcal{J}(t)]^{-1}\,F(s)\,ds\,g. \tag{4.6.28}$$

We then find that

$$\begin{aligned}
&[F(\mathcal{J})-F(\mathcal{J}_m)]g \\
&= \int_0^1 \frac{d}{dt}F[\mathcal{J}(t)]\,dt\,g \qquad (4.6.29)\\
&= -\frac{1}{2\pi i}\int_{-i\infty}^{i\infty}\int_0^1 [s-\mathcal{J}(t)]^{-1}\,\Delta\,[s-\mathcal{J}(t)]^{-1}\,dtF(s)\,ds\,g \\
&= -\frac{1}{2\pi i}\int_{-i\infty}^{i\infty} [s-\mathcal{J}_m]^{-1}\,\Delta\,[s-\mathcal{J}]^{-1}\,F(s)\,ds\,g.
\end{aligned}$$

Moreover, since $\mathcal{J}$ and $\mathcal{J}_m$ are positive operators on (a,b) it follows, by Taylor's theorem, for F real valued on $(0,b-a)$, and g real valued on (a,b), there exists a $\tau \in [0,1]$, such that

$$[F(\mathcal{J})-F(\mathcal{J}_m)]g = \frac{d}{d\tau}F[\mathcal{J}(\tau)]g \tag{4.6.30}$$

(see Problem 4.6.2). This mean value result motivates the following lemma.

Lemma 4.6.1 Let $\Re s \leqq -c < 0$. Then, for $g \in \mathbf{X}$, we have
(a) As $m \to \infty$,

$$\varepsilon_m \equiv \left\|[s-\mathcal{J}]^{-1}\Delta[s-\mathcal{J}_m]^{-1}g\right\|_{\mathbf{Y}} \to 0; \tag{4.6.31}$$

(b) As $m \to \infty$, and for any fixed $t \in [0,1]$,

$$\left\|\left\{[s-\mathcal{J}]^{-1}\Delta[s-\mathcal{J}]^{-1} - [s-\mathcal{J}(t)]^{-1}\Delta[s-\mathcal{J}(t)]^{-1}\right\}g\right\|_{\mathbf{Y}} = o(\varepsilon_m), \tag{4.6.32}$$

where ε_m is defined as in (4.6.31).

Proof. Set

$$\begin{aligned}
w &= \frac{\mathcal{J}}{s}[s-\mathcal{J}]^{-1}g \\
w_t &= \frac{\mathcal{J}(t)}{s}[s-\mathcal{J}(t)]^{-1}g.
\end{aligned} \tag{4.6.33}$$

The expression for w can be written in explicit form, namely,

$$w(x) = \frac{1}{s^2} \int_a^x e^{(x-t)/s} g(t)\, dt. \tag{4.6.34}$$

Taking the difference between w and w_t as given in (4.6.33) yields

$$s[w - w_t] - \mathcal{J}(t)[w - w_t] = [\mathcal{J} - \mathcal{J}(t)][w + g/s]. \tag{4.6.35}$$

Since the *spectra*, (i.e., the set in $\mathbb{C}$ not belonging to the resolvent set) of both $\mathcal{J}$ and $\mathcal{J}_m$ lie in the right half plane, the same is true of $\mathcal{J}(t)$, and it thus follows that $\|[s - \mathcal{J}(t)]^{-1}\| \leqq 1/c$. A norm bound on $w - w_t$ is thus given by

$$\|w - w_t\|_{\mathbf{Y}} \leqq \frac{\|[\mathcal{J} - \mathcal{J}(t)](w + g/s)\|_{\mathbf{Y}}}{c} \tag{4.6.36}$$

where w is defined in (4.6.34). The first inequality (4.6.31) now follows, by Corollary 4.5.5.

To prove the inequality (4.6.32), we start with the following identity, which is valid for all $t \in [0, 1]$:

$$\begin{aligned} &\left\{[s - \mathcal{J}]^{-1}\Delta[s - \mathcal{J}]^{-1} - [s - \mathcal{J}(t)]^{-1}\Delta[s - \mathcal{J}(t)]^{-1}\right\} g \\ &= \left[[s - \mathcal{J}]^{-1} + [s - \mathcal{J}(t)]^{-1}\right] \Delta \left\{[s - \mathcal{J}]^{-1} - [s - \mathcal{J}(t)]^{-1}\right\} g. \end{aligned} \tag{4.6.37}$$

Now, if w and w_t are defined as in (4.6.33), we have

$$w - w_t = \left\{[s - \mathcal{J}]^{-1} - [s - \mathcal{J}(t)]^{-1}\right\} g \tag{4.6.38}$$

and moreover, since by assumption we have $\Re s \leqq -c < 0$, and since the spectrum of $\mathcal{J}(t)$ lies on Ω^+ for all $t \in [0, 1]$, we have

$$\|[s - \mathcal{J}]^{-1}\| \leqq \frac{1}{c}, \quad \|[s - \mathcal{J}(t)]^{-1}\| \leqq \frac{1}{c}. \tag{4.6.39}$$

The result (4.6.32) now follows from (4.6.31), (4.6.38) and (4.6.39). ∎

The following result follows directly, from Lemma 4.6.1, by, for example, the approximation of $F(s)$ by $F(s + 2c_m)$, with c_m playing the role of c in (4.6.36), and then selecting a sequence $\{c_m\}$ of such positive numbers, such that c_m converges monotonically to 0 as $m \to \infty$, and such that the right-hand side of (4.6.36) with c replaced by c_m also approaches 0 as $m \to \infty$.

Corollary 4.6.2 *Let the conditions of Lemma 4.6.1 be satisfied, and for a given* $g \in \mathbf{X}$, *let* $F(\mathcal{J})g \in \mathbf{Y}$. *Then,*

(a) As $m \to \infty$

$$\|\{F(\mathcal{J}) - F(\mathcal{J}_m)\}g\|_{\mathbf{Y}} \to 0. \tag{4.6.40}$$

(b) There exists a constant c which is independent of N, such that

$$\|\{F(\mathcal{J}) - F(\mathcal{J}_m)\}g\|_{\mathbf{Y}} \le c \left\| \left[\frac{d}{dt} F[\mathcal{J}(t)]g\right]_{t=1} \right\|_{\mathbf{Y}}. \tag{4.6.41}$$

Corollary 4.6.2 establishes the convergence of $F(\mathcal{J}_m)g$ to $F(\mathcal{J})g$, and it also provides a mechanism for obtaining a bound on the norm $\|\{F(\mathcal{J}) - F(\mathcal{J}_m)\}g\|_{\mathbf{Y}}$. Assumption 4.6.3 below enables us to prove the rapid convergence of $F(\mathcal{J}_m)g$ to $F(\mathcal{J})g$. These assumptions are what we require for our particular proof of convergence. They are somewhat more stringent than those required to define the approximation $F(\mathcal{J}_m)g$.

To aid in clarifying our notation, Definition 4.1 will continue to be in force, and in particular we let $\mathcal{D}$ and $\mathcal{D}'$ be related as in Definition 4.1. In order to introduce properties of the function f, we assume the existence of a function ϕ_f which maps the interval $[0, c]$ onto $\mathbb{R}$, where $c \ge 2(b-a)$, and which conformally maps the domain $\mathcal{D}_f$ onto $\mathcal{D}_d$. We will also use the notation $\rho_f = e^{\phi_f}$.

Assumption 4.6.3 *Let the "Laplace transform" of f defined as in (4.6.4) with $c \ge 2(b-a)$ exist for all s in the right-half plane Ω^+, and let $F(s) = \mathcal{O}(s)$ as $s \to \infty$ in Ω^+. Let $g \in \mathbf{X}$, and let α, β, α_f, and β_f be positive constants such that $0 < \alpha \le 1$, and $0 < \beta \le 1$.*
Set

$$P(r, \tau) = \int_a^\tau f(r + \tau - \eta) g(\eta)\, d\eta \tag{4.6.42}$$

and assume that $P(r, \cdot) \in \mathbf{M}_{\alpha,\beta}(\mathcal{D}')$, uniformly, for $r \in [0, b-a]$, and also, that

$$|P_r(r, \tau)| \le c_1 \frac{[\rho_f(r)]^{\alpha_f} \phi_f'(r)}{[1 + \rho_f(r)]^{\alpha_f + \beta_f}} \tag{4.6.43}$$

for all $r \in [0, b-a]$, and for all $\tau \in \mathcal{D}$, with c_1 a constant independent of r and τ.

The upper limit $c \ge 2(b-a)$, is required in Assumption 4.6.3, since the assumption requires that P be an analytic function of x, uniformly for $r \in [0, b-a]$; if c in (4.6.4) were taken to be $b - a$, then our function f as reproduced in our proof of Theorem 4.6.4 below would vanish when the argument of f would be greater than $b - a$, and it would then not be possible for P to be analytic as was assumed.

Theorem 4.6.4 *Let p and q be defined as in (4.6.1), let N be a positive integer, let $M = [\beta N/\alpha]$, set $m = M + N + 1$, let A_m and B_m be defined* *as in (4.6.3), and let F, V_m and Π_m be defined as in (4.6.23). Let $h = [\pi d/(\beta N)]^{1/2}$. If P satisfies Assumption 4.6.3, then there exists a constant c which is independent of N, such that*

$$\begin{aligned} \|p - \Pi_m F(A_m) V_m g\|_{\mathbf{Y}} &\leq cN^{1/2} e^{-(\pi d \beta N)^{1/2}} \\ \|q - \Pi_m F(B_m) V_m g\|_{\mathbf{Y}} &\leq cN^{1/2} e^{-(\pi d \beta N)^{1/2}}. \end{aligned} \tag{4.6.44}$$

Proof. We remark at the outset, that the function $P(r,\tau)$ satisfies Assumptions 4.6.3, then so does the function

$$Q(r,\tau) = \int_\tau^b f(r+\eta-\tau) g(\eta)\, d\eta. \tag{4.6.45}$$

We therefore only prove the first inequality in (4.6.44), since the proof of the second inequality is almost exactly the same.

We shall use (4.6.28) and (4.6.41) in Corollary 4.5.5.
Replacing s by $1/s$ in (4.6.28), and recalling that $F(s) = \hat{f}(1/s)$, we get

$$E_p \equiv \left[\frac{d}{dt} F[\mathcal{J}(t)] g\right]_{t=1} = \frac{1}{2\pi i} \int_{-i\infty}^{i\infty} (1 - s\mathcal{J})^{-1} \Delta (1 - s\mathcal{J})^{-1} \hat{f}(s)\, ds\, g \tag{4.6.46}$$

with Δ defined as in (4.6.26). Now it readily follows that if $g \in \mathbf{X}$, then for any $s \notin \Omega^+$, $s \neq 0$, we have

$$[(1 - s\mathcal{J})^{-1} \hat{f}(s) g](\tau) = \frac{\partial}{\partial \tau} \left\{ \int_a^\tau e^{s(\tau-\eta)} \hat{f}(s) g(\eta)\, d\eta \right\} \tag{4.6.47}$$

and applying this result once more, we get

$$\begin{aligned} &[(1 - s\mathcal{J})^{-1} \Delta (1 - s\mathcal{J})^{-1} \hat{f}(s) g](x) \\ &= \frac{\partial}{\partial x} \int_a^x e^{s(x-t)} \Delta_\tau \left\{ \frac{\partial}{\partial \tau} \int_a^\tau e^{s(\tau-\eta)} \hat{f}(s) g(\eta)\, d\eta \right\} (t)\, dt, \end{aligned} \tag{4.6.48}$$

where the subscript τ on Δ indicates that Δ operates with respect to the variable τ. We now apply the operation $(2\pi i)^{-1} \int_{-i\infty}^{i\infty} [\cdots]\, ds$ to each side of (4.6.48), to get

$$E_p(x) = \frac{\partial}{\partial x} \int_a^x \Delta_\tau \left\{ \frac{\partial}{\partial \tau} \int_a^\tau f(x - t + \tau - \eta)\, g(\eta)\, d\eta \right\} (t)\, dt. \tag{4.6.49}$$

(Towards proving the second inequality in (4.6.44), we remark at this point, that if $(\mathcal{J}g)(x) = \int_x^b g(t)\,dt$, and if $\mathcal{J}_m = \Pi_m B_m V_m$, with Π_m, B_m and V_m defined as in Assumption 2.12, then the result for q corresponding to the result (4.6.49) for p is

$$E_q(x) = \frac{\partial}{\partial x}\int_x^b \Delta_\tau \left\{ \frac{\partial}{\partial \tau}\int_\tau^b f(t-x+\eta-\tau)\,g(\eta)\,d\eta \right\}(t)\,dt \qquad (4.6.50)$$

where $\Delta_\tau = \mathcal{J} - \mathcal{J}_m$ operates with respect to the variable τ.)
Let us write

$$E_p = E_p^{(1)} + E_p^{(2)} \qquad (4.6.51)$$

where

$$\begin{aligned} E_p^{(1)}(x) &= \Delta_\tau \left\{ \frac{\partial}{\partial \tau}\int_a^\tau f(\tau - v) g(v)\,dv \right\}(x) = \Delta_\tau \left\{ P_\tau(0,\cdot) \right\}(x), \\ E_p^{(2)}(x) &= \int_a^x \Delta_\tau \left\{ P_{x\tau}(x-t,\cdot)\right\}(t)\,dt. \end{aligned} \qquad (4.6.52)$$

By assumption, $P(r,\cdot) \in \mathbf{M}_{\alpha,\beta}(\mathcal{D}')$ uniformly with respect to $r \in [0, b-a]$, and it therefore follows that $P_\tau(r,\cdot)/\phi' \in \mathbf{L}_{\alpha,\beta}(\mathcal{D})$, uniformly for $r \in [0, b-a]$, where $\mathcal{D}$ is defined in terms of $\mathcal{D}'$ as in Definition 4.1.1. Hence it follows immediately, by Corollary 4.5.5, that there exists a constant c_1, independent of N, such that

$$\left\| E_p^{(1)} \right\|_{\mathbf{Y}} \le c_1 N^{1/2} e^{-(\pi d \beta N)^{1/2}}. \qquad (4.6.53)$$

We next bound $E_p^{(2)}(x)$. In essence, it will suffice to show that for all $r \in [0, b-a]$ and $\tau \in \mathcal{D}$, we would have the inequality

$$|P_{r\tau}(r,\tau)| \le c_2 \frac{[\rho_f(r)]^{\alpha_f}\phi_f'(r)}{[1+\rho_f(r)]^{\alpha_f+\beta_f}} \frac{|\rho(\tau)|^\alpha \phi'(\tau)}{[1+|\rho(\tau)|]^{\alpha+\beta}} \qquad (4.6.54)$$

with c_2 a constant independent of τ and r. For, if (4.6.54) were to hold, then by Corollary 4.5.5 we would get

$$|\{\Delta_\tau P_{r\tau}(r,\tau)\}(t)| \le c_3 \frac{[\rho_f(r)]^{\alpha_f}\,\phi_f'(r)}{[1+\rho_f(r)]^{\alpha_f+\beta_f}} N^{1/2} e^{-(\pi d\beta N)^{1/2}}.$$

with c_3 a constant independent of r, τ, and N.
It would then follow that

$$
\begin{aligned}
\left|E_p^{(2)}(x)\right| &= \left|\int_a^x \{\Delta_\tau P_{x\tau}(x-t,\tau)\}(t)dt\right| \\
&\le c_3 \int_a^x \frac{[\rho_f(x-t)]^{\alpha_f}}{[1+\rho_f(x-t)]^{\alpha_f+\beta_f}} \phi_f'(x-t)\, dt\ N^{1/2} e^{-(\pi d\beta N)^{1/2}} \\
&\le c_3 \int_0^c \frac{[\rho_f(r)]^{\alpha_f}}{[1+\rho_f(r)]^{\alpha_f+\beta_f}} \phi_f'(r)\, dr\ N^{1/2} e^{-(\pi d\beta N)^{1/2}} \qquad (4.6.55) \\
&= c_3 \frac{\Gamma(\alpha_f)\Gamma(\beta_f)}{\Gamma(\alpha_f+\beta_f)} N^{1/2} e^{-(\pi d\beta N)^{1/2}},
\end{aligned}
$$

and we would then get (4.6.44) by combining (4.6.51), (4.6.53), and (4.6.55). It remains to derive (4.6.54). Since $P(r,\cdot) \in \mathbf{M}_{\alpha,\beta}(\mathcal{D}')$ uniformly with respect to $r \in [0, b-a]$, it follows that $P_\tau(r,\cdot)/\phi' \in \mathbf{L}_{\alpha,\beta}(\mathcal{D})$ uniformly with respect to $r \in [0, b-a]$. Since for $r \in [0, b-a]$ we also have $|P_r(\cdot,\tau)/\phi_f'| \le c_1[\rho_f]^{\alpha_f}/[1+\rho_f]^{\alpha_f+\beta_f}$, uniformly for $\tau \in \mathcal{D}$, with c_1 a constant independent of τ and r (see Problem 4.5.10), the result (4.6.54) follows. ■

4.6.4 Applications

The discretization procedure of this section has many applications.

Example 4.6.5 *Miscellaneous Examples.* For example, we can approximate the many special functions that can be expressed as convolution integrals, such as a hypergeometric function or a confluent hypergeometric function. Abel integral equations can be dealt with directly. Convolution Fredholm integral equations over a finite interval (a, b), over a semi-infinite interval (a, b), or over all of $\mathbb{R}$ can similarly be dealt with, by using two convolution approximations, one over (a, x) and the other over (x, b). We can also invert Laplace transforms. For example, if $f(0) = 0$, then it follows that

$$
f(x) = \int_0^x f'(t)\, dt = \int_0^x f'(x-t)\, dt. \qquad (4.6.56)
$$

Letting $F(s)$ denote the "Laplace transform" of f, then the Laplace transform of f' is $G(s) \equiv F(s)/s$ and taking $g(t) \equiv 1$, it therefore follows in the notation of (4.6.1), (4.6.22) and (4.6.23), that

$$
f(x) = (G(\mathcal{J})\, g)(x) \approx (\Pi_m G(A_m)\mathbf{1})(x), \qquad (4.6.57)
$$

where $\mathbf{1} = (1, 1, \cdots, 1)$.

We can also readily approximate the classical *feedback control* equation

$$
\hat{p}(s) = \frac{\hat{G}(s)}{1 + \hat{H}(s)\hat{G}(s)} \hat{g}(s) \qquad (4.6.58)
$$

where the "hat" functions denote standard Laplace transforms, and where a frequently occurring situation in practice is that g as well as the "hat" functions $\hat{G}$ and $\hat{H}$ are known, and p is to be determined as a function of time t. Upon setting

$$K(s) = \frac{\hat{G}(1/s)}{1 + \hat{H}(1/s)\hat{G}(1/s)} \tag{4.6.59}$$

the method of this section immediately yields the approximation

$$p(t) \approx (\Pi_m K(A_m) V_m g)\,(t). \tag{4.6.60}$$

Example 4.6.6 *Sinc Approximation of a Three Dimensional Convolution Integral.* The procedure outlined in the present example illustrates how well the one-dimensional formulas of this section adapt to the approximate solution of multidimensional convolution equations. The separation of variables technique illustrated here makes it possible to reduce the approximate solution of the multidimensional problem to a finite sequence of one-dimensional matrix multiplications. Although the resulting finite sequence of matrix multiplications can be efficiently carried out on a sequential computer, its final algorithmic layout is such that all of the computation can be carried out extremely easily on a parallel machine. Inasmuch as nearly all solutions of partial differential and integral equations can be explicitly expressed as a multidimensional convolution integral, or as a multidimensional convolution integral equation, due to its efficiency and ease of adaptability to multidimensional problems, the method of this section should provide a powerful technique for solving such problems. Although the illustration which follows is for a three-dimensional problem, the reader will readily deduce that the process extends equally easily to any multidimensional convolution integral.

The problem which we shall consider is

$$p(x_1, x_2, x_3) = \int_{a_3}^{x_3} \int_{x_2}^{b_2} \int_{a_1}^{x_1} f(x_1 - \xi_1, \xi_2 - x_2, x_3 - \xi_3)\, g(\xi_1, \xi_2, \xi_3)\, d\xi_1 d\xi_2 d\xi_3 \tag{4.6.61}$$

where the approximation is sought over the region $\mathbf{B} = \Pi_{i=1}^{3} \otimes (a_i, b_i)$, and with $(a_i, b_i) \subseteq \mathbb{R}$. The reader is instructed to pay particular attention to the method of separation of variables made possible by the one-dimensional convolution approximation, as well as to how easily the final algorithm can be adapted to parallel computation. In order to guarantee some accuracy in the final approximation, we shall simply assume, without going into detail, that the function p belongs to the class $\mathbf{M}_{\alpha_j, \beta_j}(\mathcal{D}'_j)$ with respect to each variable x_j, for all fixed values of the other variables, each in its respective interval of definition, $j = 1, 2, 3$. Examples for the determination of such multidimensional spaces of approximation may be found in Sections 6.5,

6.6, and 7.3 of this text. We shall assume that the mappings $\phi_j : \mathcal{D}'_j \to \mathcal{D}_{d'}$, with $0 < d < d' < \pi$ have been determined. We furthermore assume that positive integers N_j and M_j as well as positive numbers h_j $(j = 1, 2, 3)$ have been selected such that $[N_1 \beta_1] = [N_2 \beta_2] = [N_3 \beta_3]$, such that $M_j = [\beta_j N_j / \alpha_j]$, where $[\cdot]$ denotes the greatest integer function, and such that $h_j = \{\pi d / (\beta_j N_j)\}^{1/2}$. These definitions ensure that we get the same order of accuracy of approximation $\varepsilon = \mathcal{O}(N_1^{1/2} e^{-(\pi d \beta_1 N_1)^{1/2}})$ in each variable. We set $m_j = M_j + N_j + 1$, and we define the Sinc points by $z_\ell^{(j)} = \phi_j^{-1}(\ell h_j)$, for $\ell = -M_j, \cdots, N_j$; $j = 1, 2, 3$. Next, we determine matrices A_j, X_j, S_j, and X_j^{-1} such that

$$
\begin{aligned}
A_j &= h_j I_{m_j}^{(-1)} D(1/\phi'_j) &&= X_j S_j X_j^{-1}, \quad j = 1, 3 \\
A_2 &= h_2 \left(I_{m_2}^{(-1)} \right)^T D(1/\phi'_2) &&= X_2 S_2 X_2^{-1}.
\end{aligned}
\tag{4.6.62}
$$

In (6.6.62), $I_{m_j}^{(-1)}$ is defined as in (4.6.3), and the S_j are diagonal matrices,

$$
S_j = \operatorname{diag}[s_{-M_j}^{(j)}, \cdots, s_{N_j}^{(j)}]. \tag{4.6.63}
$$

The stable determination of the matrices X_j *as well as* their inverses X_j^{-1} is probably the most difficult part of the method; we hope this solution procedure will interest some experts in numerical linear algebra, once the ease of implementation and power of the remaining part of the procedure become better known.

Arbitrarily taking $c_j \in [2(b_j - a_j), \infty]$, we set

$$
\begin{aligned}
F(s^{(1)}, y, z) &= \int_0^{c_1} f(x, y, z)\, e^{-x/s^{(1)}}\, dx \\
G(s^{(1)}, s^{(2)}, z) &= \int_0^{c_2} F(s^{(1)}, y, z)\, e^{-y/s^{(2)}}\, dy \\
H(s^{(1)}, s^{(2)}, s^{(3)}) &= \int_0^{c_3} G(s^{(1)}, s^{(2)}, z)\, e^{-z/s^{(3)}}\, dz.
\end{aligned}
\tag{4.6.64}
$$

We mention at this point, that the functions F and G defined in (4.6.64) are introduced here solely for purposes of understanding the method of separation of variables described below. Only the function H defined in the last line of (4.6.64) is required in the final algorithm.

We now illustrate the method of separation of variables. To this end, we first rewrite (4.6.61) in the notationally more convenient form

$$
p(x, y, z) = \int_{a_3}^{z} \int_{y}^{b_2} \int_{a_1}^{x} f(x - \xi, \eta - y, z - \zeta)\, g(\xi, \eta, \zeta)\, d\xi\, d\eta\, d\zeta. \tag{4.6.65}
$$

Discretization with respect to x. We set

$$\mathbf{g}(\eta,\zeta) = \left(g(z_{-M_1}^{(1)},\eta,\zeta),\cdots,g(z_{N_1}^{(1)},\eta,\zeta)\right)^T \tag{4.6.66}$$

and we then *define* a vector $\mathbf{p}(y,z)$ by

$$\mathbf{p}(y,z) = \int_{a_3}^{z}\int_{y}^{b_2} F(A_1,\eta-y,z-\zeta)\mathbf{g}(\eta,\zeta)\,d\eta\,d\zeta \tag{4.6.67}$$

where A_1 and F are defined in (4.6.62) and (4.6.64) respectively.
Using the diagonalization identity $A_1 = X_1 S_1 X_1^{-1}$ given in (4.6.62), it now follows from (4.6.67) that

$$\mathbf{p}(y,z) = X_1 \int_{a_3}^{z}\int_{y}^{b_2} \begin{pmatrix} F(s_{-M_1}^{(1)},\eta-y,z-\zeta) & & \\ & \ddots & \\ & & F(s_{N_1}^{(1)},\eta-y,z-\zeta) \end{pmatrix} X_1^{-1}\mathbf{g}(\eta,\zeta)\,d\eta\,d\zeta. \tag{4.6.68}$$

The expression (4.6.68) motivates the transformations

$$\mathbf{h}(\eta,\zeta) = X_1^{-1}\mathbf{g}(\eta,\zeta), \quad \mathbf{q}(y,z) = X_1^{-1}\mathbf{p}(y,z). \tag{4.6.69}$$

Thus, denoting the components of $\mathbf{h}$ and $\mathbf{q}$ by h_i and q_i respectively, with $i = -M_1,\cdots,N_1$, Equation (4.6.68) reduces to the *decoupled* set of scalar equations

$$q_i(y,z) = \int_{a_3}^{z}\int_{y}^{b_2} F(s_i^{(1)},\eta-y,z-\zeta)h_i(\eta,\zeta)\,d\eta\,d\zeta. \tag{4.6.70}$$

Discretization with respect to y. We set

$$\mathbf{h}_i(\zeta) = \left(h_i(z_{-M_2}^{(2)},\zeta)\cdots h_i(z_{N_2}^{(2)},\zeta)\right)^T \tag{4.6.71}$$

and we then *define* a vector $\mathbf{q}_i(z)$ by

$$\mathbf{q}_i(z) = \int_{a_3}^{z} G(s_i^{(1)},A_2,z-\zeta)\mathbf{h}_i(\zeta)\,d\zeta, \tag{4.6.72}$$

where A_2 and G are defined in (4.6.62) and (4.6.64) respectively.
Using the diagonalization identity $A_2 = X_2 S_2 X_2^{-1}$ given in (4.6.62), it now follows from (4.6.72) that

$$\mathbf{q}_i(z) = X_2 \int_{a_3}^{z} \begin{pmatrix} G(s_i^{(1)},s_{-M_2}^{(2)},z-\zeta) & & \\ & \ddots & \\ & & G(s_i^{(1)},s_{N_2}^{(2)},z-\zeta) \end{pmatrix} X_2^{-1}\mathbf{h}_i(\zeta)\,d\zeta. \tag{4.6.73}$$

This last expression motivates the transformations

$$\mathbf{k}_i(\zeta) = X_2^{-1}\mathbf{h}_i(\zeta), \quad \mathbf{r}_i(z) = X_2^{-1}\mathbf{q}_i(z). \tag{4.6.74}$$

Denoting the components of $\mathbf{r}_i$ and $\mathbf{k}_i$ by $r_{i,j}$ and $k_{i,j}$ respectively, with $i = -M_1, \cdots, N_1$ and $j = -M_2, \cdots, N_2$, Equation (4.6.73) reduces to the decoupled set of scalar equations

$$r_{i,j}(z) = \int_{a_3}^{z} G(s_i^{(1)}, s_j^{(2)}, z - \zeta) k_{i,j}(\zeta)\, d\zeta. \tag{4.6.75}$$

Discretization with respect to z. We now set

$$\mathbf{k}_{i,j} = \left(k_{i,j}(z_{-M_3}^{(3)}) \cdots, k_{i,j}(z_{N_3}^{(3)})\right)^T \tag{4.6.76}$$

and we then *define* a vector $\mathbf{r}_{i,j}$ by

$$\mathbf{r}_{i,j} = H(s_i^{(1)}, s_j^{(2)}, A_3)\mathbf{k}_{i,j} \tag{4.6.77}$$

where A_3 and H are defined in (4.6.62) and (4.6.64) respectively. Using the diagonalization identity $A_3 = X_3 S_3 X_3^{-1}$ given in (4.6.62), it now follows from (4.6.77) that

$$\mathbf{r}_{i,j} = X_3 \begin{pmatrix} H(s_i^{(1)}, s_j^{(2)}, s_{-M_3}^{(3)}) & & \\ & \ddots & \\ & & H(s_i^{(1)}, s_j^{(2)}, s_{N_3}^{(3)}) \end{pmatrix} X_3^{-1}\mathbf{k}_{i,j}. \tag{4.6.78}$$

The expression (4.6.78) motivates the transformations

$$\mathbf{m}_{i,j} = X_3^{-1}\mathbf{k}_{i,j}, \quad \mathbf{t}_{i,j} = X_3^{-1}\mathbf{r}_{i,j}. \tag{4.6.79}$$

Denoting the components of $\mathbf{m}_{i,j}$ and $\mathbf{t}_{i,j}$ by $m_{i,j,k}$ and $t_{i,j,k}$ respectively, with $i = -M_1, \cdots, N_1$; $j = -M_2, \cdots, N_2$; and $k = -M_3, \cdots, N_3$, Equation (4.6.78) yields the decoupled set of scalar equations

$$t_{i,j,k} = H(s_i^{(1)}, s_j^{(2)}, s_k^{(3)}) m_{i,j,k}. \tag{4.6.80}$$

By assumption, the $m_{i,j,k}$ are known at this point, and (4.6.80) then determines the $t_{i,j,k}$. The second equation in (4.6.79) next determines the vectors $\mathbf{r}_{i,j}$. The second equation in (4.6.74) is then used to determine the vectors $\mathbf{q}_i(z)$ at the Sinc points $z_j^{(3)}$. The second transformation of (4.6.69) is next used to determine the vector $\mathbf{p}(y,z)$ at the set of Sinc points $(y,z) = (z_j^{(2)}, z_k^{(3)})$. We can thus recover the complete array of values $p(x,y,z)$ at the set of Sinc points $(z_i^{(1)}, z_j^{(2)}, z_k^{(3)})$. The whole procedure is illustrated succinctly via the following algorithm. In this algorithm we use the notation, $\mathbf{h}_{i,\cdot,k} = (h_{i,-M_2,k}, \cdots, h_{i,N_2,k})^T$. We again emphasize the obvious ease of adaptation of this algorithm to parallel computation.

Algorithm 4.6.7

1. Form the array $z_i^{(j)}$ and $\frac{d}{dx}\phi^{(j)}(x)$ at $x = z_i^j$ for $j = 1, 2, 3$, and $i = -M_j, \cdots, N_j$; and then form the array $[g_{i,j,k}] = [g(z_i^{(1)}, z_j^{(2)}, z_k^{(3)})]$.
2. Determine A_j, S_j, X_j, and X_j^{-1} for $j = 1, 2, 3$, as defined in (4.6.62).
3. Form (See (4.6.69)) [DONE] ↑

$$\mathbf{h}_{\cdot,j,k} = X_1^{-1}\mathbf{g}_{\cdot,j,k} \qquad \mathbf{p}_{\cdot,j,k} = X_1\mathbf{q}_{\cdot,j,k}$$

4. Form ↘ (See (4.6.74)) ↑

$$\mathbf{k}_{i,\cdot,k} = X_2^{-1}\mathbf{h}_{i,\cdot,k} \qquad \mathbf{q}_{i,\cdot,k} = X_2\mathbf{r}_{i,\cdot,k}$$

5. Form ↘ (See (4.6.79)) ↑

$$\mathbf{m}_{i,j,\cdot} = X_3^{-1}\mathbf{k}_{i,j,\cdot} \qquad \mathbf{r}_{i,j,\cdot} = X_3\mathbf{t}_{i,j,\cdot}$$

6. Form ↘ (See (4.6.80)) ↑

$$t_{i,j,k} = H(s_i^{(1)}, s_j^{(2)}, s_k^{(3)})\, m_{i,j,k}.$$

Once the numbers $p_{i,j,k}$ have been computed, we can then use these numbers to approximate p on the region B via the use of a Sinc basis; upon setting $\rho^{(\ell)} = e^{\phi^{(\ell)}}$, we can define the functions

$$\begin{aligned}
\gamma_i^{(\ell)} &= S(i, h_\ell) \circ \phi^{(\ell)}, \quad \ell = 1, 2, 3; \quad i = -M_\ell, \cdots, N_\ell \\
\omega_i^{(\ell)} &= \gamma_i^{(\ell)}, \quad \ell = 1, 2, 3; \quad i = -M_\ell + 1, \cdots, N_\ell - 1 \\
\omega_{-M_\ell}^{(\ell)} &= \frac{1}{1 + \rho^{(\ell)}} - \sum_{j=-M_\ell+1}^{N_\ell} \frac{1}{1 + e^{jh_\ell}} \gamma_j^{(\ell)} \\
\omega_{N_\ell}^{(\ell)} &= \frac{\rho^{(\ell)}}{1 + \rho^{(\ell)}} - \sum_{j=-M_\ell}^{N_\ell - 1} \frac{e^{jh_\ell}}{1 + e^{jh_\ell}} \gamma_j^{(\ell)}.
\end{aligned} \tag{4.6.81}$$

We then get the approximation

$$p(x, y, z) \approx \sum_{i=-M_1}^{N_1} \sum_{j=-M_2}^{N_2} \sum_{k=-M_3}^{N_3} p_{i,j,k}\omega_i^{(1)}(x)\omega_j^{(2)}(y)\omega_k^{(3)}(z). \tag{4.6.82}$$

Assuming a one-dimensional error of the order of $\varepsilon = N_1^{1/2} e^{-(\pi d \beta_1 N_1)^{1/2}}$, we may expect the approximation in (4.6.82) to have an error of the order of $\{\log(N_1)\}^2\varepsilon$ (see Problem 4.6.3).

It is also relatively simple to obtain an estimate of the complexity, i.e., the total amount of work required to achieve an error ε when carrying out the computations of the above algorithm on a sequential machine. By taking $\alpha_j = \beta_j = \alpha$, and then selecting $M_j = N_j = N$, it follows, that the error in the final approximation is roughly of the order of $\varepsilon = e^{-(\pi d \alpha N)^{1/2}}$. The amount of work required to factor the matrices A_j into the form $X_j S_j X_j^{-1}$ is of the order of $(2N+1)^3$, and the amount of work in the totality of the matrix-vector multiplications in steps 3 to 6 is of the order of $(2N+1)^4$. Hence the complexity is $\mathcal{O}([2N+1]^4) = \mathcal{O}([\log(\varepsilon)]^8)$. Similarly, the complexity of collocation of for solving a ν-dimensional problem by the above procedure is $\mathcal{O}([\log(\varepsilon)]^{2\nu+2})$.

•

Problems for Section 4.6

4.6.1 (a) Prove that if $I^{(-1)}$ is similar to a diagonal matrix, and $\phi' > 0$ on Γ, then there exists a non-singular matrix X such that $X^{-1} I^{(-1)} D(1/\phi') X$ is a diagonal matrix.

(b)* (Research Problem) Prove that if $\phi' > 0$ on Γ, and λ is an eigenvalue of $I^{(-1)} D(1/\phi')$ then $\Re \lambda > 0$.

(c) Prove that if $\phi' > 0$ on Γ, and λ is an eigenvalue of $I^{(-1)} D(1/\phi')$ then $\bar{\lambda}$ is also an eigenvalue.

4.6.2 Prove the mean value result of Equation (4.6.30).

4.6.3 Use the method of Section 4.6.2 to obtain an approximation of the Dawson-type integrals

$$(i) \quad D_i(x) = \int_0^x e^{-t} t^{-1/2} (x-t)^{-1/3} dt$$

and

$$(ii) \quad D_{ii}(x) = \int_x^\infty e^{-t} t^{-1/2} (t-x)^{-1/3} dt$$

accurate to 6 decimals on $[0, \infty]$.

4.6.4 Obtain a bound on the error of the approximation (4.6.82). [Hint: See Section 6.6.3.]

Historical Remarks on Chapter 4

Transformations have in the past been used by a number of authors, as an aid for surmounting the numerical integration difficulties due to singularities of the integrand (see *Goodrich and Stenger* [GiS1], *Moran* [Mn1,Mn2],

Sag and Szekeres [SgSz1], *Schwartz* [Stz1], *Stenger* [S14, S16], *Sugihara* [Su1,Su2,Su3]), *Takahasi and Mori* [TaMo1, TaMo2, Mo1, Mo2].

By combining the identities of the Cardinal function with Definition 4.1.1, we have obtained a unifying approach for simultaneously circumventing the difficulties that singularities cause in interpolation, quadrature, the evaluation of transforms, the solution of differential equations, etc., an approach which has given rise to many new formulas for approximation over infinite and semi-infinite intervals and even over contours [S19].

The identification of the class $\mathbf{L}_{\alpha,\beta}(\mathcal{D})$ is a direct extension of the growth relation defined in Definition 3.1.5; this identification enables us to obtain a uniform approach to bounding the error for all of the approximations. Moreover, this class enables us to conveniently approximate functions in frequently occurring practical problems, which are analytic in an open set, and of class $\mathbf{Lip}_\alpha$ on its closure. My personal conjecture is that in real life, physically occurring quantities such as potential, density, pressure, stress, conductivity, speed, etc., are always of class $\mathbf{Lip}_\alpha$, with $\alpha > 1/2$.

The class $\mathbf{L}_{\alpha,\beta}(\mathcal{D})$ is readily connected with a Hardy space of functions $\mathbf{H}^p(\mathcal{D})$ for which the best approximation rate of the form $\mathcal{O}(\exp[-cn^{1/2}])$ have been extensively studied (see *Andersson* [Ad1], *Andersson and Bojanov* [AdBoj1], *Bojanov* [Boj1, Boj2], *Burchard and Höllig* [BuH1], *Sikorski* [Si1], *Sikorski and Stenger* [SiS1], *Stenger* [S14, S16, S21], and Wilderotter [Wi1]). The constant c in our own estimates of the rate of convergence $\mathcal{O}(\exp[-cn^{1/2}])$ which we obtained in this chapter is not as large as that of the optimal rate possible, as predicted in [AdBoj1] and [Wi1]. It is not known at this time whether an appreciable improvement can be made on the constants obtained in this chapter. On the other hand, the interpolation and quadrature points which yield the optimal rates predicted in [AdBoj1] and [Wi1] are also not known explicitly. Moreover, it is doubtful that the optimal formulas (if sufficiently different from the Sinc formulas of this chapter) will be as replete with interconnecting and useful identities as the Sinc formulas.

The important log[sinh] transformation (4.2.55) is due to *Lund* [L1]. The double exponential transformation method exhibited in Example 4.2.13 has been extensively studied by Takahasi and Mori [TaMo2], by Mori [Mo1–Mo3], by Sugihara [Su2], by Toda and Ono [ToO1], and by Murota and Iri [MuIr1].

Proofs for the error of the indefinite integral formula (4.5.2) have previously been carried out by *Kearfott* [K1] and by *Haber* [H3]; in Section 4.5 we have given a lengthier proof than those in [K1] and [H3], in order to obtain sharper bounds.

Bialecki's procedure (see Theorem 3.1.9, Problems 3.1.9, 3.2.6, 3.4.11, 4.2.7, and 4.3.10) offers an alternative method of deriving formulas for principal-value-type integrals.

All of the theorems of this chapter are originally due to the author [S14,S16,S19,S24]. Some of the results—especially those involving the uniform convergence of the approximations on subsets of $\mathcal{D}$—are new. The new convolution procedure of Section 4.6 was discovered by the author during the final stages of writing this text, and while it has already been tested on some problems, it has not yet been tested on the solution of partial differential and integral equations to the extent that the other methods of the text have been tested. The technique appears to be easy to implement, and to have considerable breadth of applicability, even to parallel computation. The student and user may therefore find it interesting and profitable to become familiar with the technique.

5

Sinc-Related Methods

5.1 Introduction

The formulas of the previous chapter are all related to the Cardinal function representation

$$C(f,h)\circ\phi(x) = \sum_{k=-\infty}^{\infty} F(z_k)\, S(k,h)\circ\phi(x), \tag{5.1.1}$$

with $z_k = \phi^{-1}(kh)$. We recall here, that $C(F,h)\circ\phi$ both interpolates F on $\Gamma = \phi^{-1}(\mathbb{R})$, and it also accurately approximates it in a suitable space of functions. In the present chapter we examine some related methods of approximation. This chapter may be termed *the unfinished chapter*, in the sense that many other formulas are possible by each approach presented here, although we do not present or study the majority of these as we have done in the case of the Cardinal expansion; indeed, a thorough study, as done in the case of Sinc methods in this text would require the writing of at least an additional text. We have merely chosen to point out some connections of Sinc formulas with other formulas, thus opening doors to new avenues of research, and to present some esthetically pleasing formulas which may have practical value as well.

Two classes of approximations are given: interpolating and non-interpolating. The first involves approximations of the form

$$F(x) \approx F_h(x) \equiv \sum_{k=-\infty}^{\infty} F(z_k)\sigma_k(x) \tag{5.1.2}$$

with the z_k the same as those in Equation (5.1.1), but the new basis functions σ_k differing from the $S(k,h)\circ\phi$ basis used in (5.1.1). It then turns out, in the case of the basis σ_k being such that F_h in (5.1.2) interpolates F at the points z_k, the magnitude of the difference between F_h and F is roughly of the same order of magnitude as the magnitude of the difference between $C(F,h)\circ\phi$ and F, although this may not be the case when the σ_k are such that F_h does not interpolate F.

In Section 5.2, we study the approximations of the first kind, for which

$$\sigma_k(x) = \frac{h}{\pi}\frac{\sin\{\pi[\phi(x)-kh]/h\}}{\phi'(z_k)(x-z_k)}. \tag{5.1.3}$$

While the resulting approximation F_h has some disadvantages compared to $C(F,h)\circ\phi$, such as, for example, the loss of orthogonality of the basis, it does, nevertheless, enable us to derive the practically important and accurate approximation

$$\frac{P.V.}{\pi i} \int_\Gamma \frac{F(t)}{t-x} dt \approx \frac{h}{\pi i} \sum_{k=-N}^{N} F(z_k) \frac{\cos\{\pi[\phi(x) - kh]/h\} - 1}{\phi'(z_k)(x - z_k)} \tag{5.1.4}$$

which is better than (4.3.16), since (4.3.16) approximates $\mathcal{S}(F)$ as defined in (4.3.8), and not the Hilbert transform on the left-hand side of (5.1.4). In Section 5.2 we then also derive a Sinc-like method for $\mathbf{L}^1$ approximation on Γ.

In Section 5.3 we derive an interpolation formula of the form (5.1.2) over $(0, \infty)$ using the basis $\{\sigma_k\}$, with

$$\begin{aligned} \sigma_k(x) &= \frac{(-1)^k \Phi_h(x)}{(x - e^{kh})\Phi_h'(e^{kh})} \\ \Phi_h(x) &= \frac{x-1}{x+1} \prod_{j=1}^{\infty} \frac{(x - e^{jh})(1/x - e^{jh})}{(x + e^{jh})(1/x + e^{jh})}. \end{aligned} \tag{5.1.5}$$

The function Φ_h may be expressed as an elliptic function. The zeros of Φ_h are the same as the zeros of $\sin\{\pi[\log(x)]/h\}$, and, as mentioned above, the bounds $\sup_{x \in (0.\infty)} |F(x) - F_h(x)|$ and $\sup_{x \in (0,\infty)} |F(x) - C(F,h) \circ \phi(x)|$ are roughly of the same order of magnitude. Although the resulting approximation F_h has some disadvantages compared with the scheme $C(F,h) \circ \phi$, it also has some advantages—one being that the poles of Φ_h make it possible to obtain an explicit expression in the inversion of a Laplace transform, using F_h to approximate the Laplace transform and then applying the Bromwich integral formula (see Equation (5.4.3) below). This latter procedure is illustrated in Section 5.4. In Section 5.5 we illustrate yet another important byproduct of this same F_h which results upon truncation of the infinite product in (5.1.5); in this case we get rational function approximations of F with error that is roughly of the same order of magnitude as that of $C_N(F,h) \circ \phi$, with $C_N(F,h)$ defined as in (3.1.3).

In Section 5.6 we discuss various extrapolation algorithms, with particular emphasis on the role of the Thiele algorithm. Here we state explicit criteria which enable us to decide when to use the Aitken-Neville, when to use the Thiele algorithm, and of how to tailor these algorithms to enable us to evaluate the limit of a given sequence of numbers.

In Section 5.7 we derive a family of simple explicit rational functions to approximate the Heaviside, the filter, and the impulse function. Many of the Sinc formulas of the previous sections of this text could in fact be derived by means of these explicit rational functions.

In Section 5.8 we simultaneously consider a family of non-interpolating approximations of the form (5.1.2), over $\mathbb{R}$, i.e., with $\phi(x) = x$ and $z_k = kh$, where for an arbitrary fixed number β on the interval $(0, 1)$, σ_k takes one of the forms

$$\sigma_k(x) = \begin{cases} \dfrac{h^{1-\beta}}{\sqrt{2\pi}} e^{-h^{-2\beta}(x-kh)^2/2} \\ \dfrac{h^{1+\beta}}{\pi[(x-kh)^2 + h^{2\beta}]} \\ \dfrac{h^{1-\beta}}{2} e^{-h^{-\beta}|x-kh|} \\ \dfrac{h^{1-\beta}}{\pi \cosh[h^{-\beta}(x-kh)]} . \end{cases} \tag{5.1.6}$$

Although none of the resulting approximations F_h are interpolating, they have one advantage over Sinc approximations, namely, they all converge to F as $h \to 0$ whenever F is continuous and bounded on $\mathbb{R}$. Also, F_h reduces to interesting special forms in the cases when F is periodic, or has compact support on $\mathbb{R}$. The approximant $C(F, h) \circ \phi$ may be thought of as being *elastic*, in the sense that a small change in any of the values of $\{F(z_k)\}$ alters this approximant over all of Γ, whereas the approximant F_h which uses any one of the bases (5.1.6) is more *plastic* in behavior, similar to a spline interpolant, in that a small change in one of the values of $\{F(z_k)\}$ appreciably changes F_h only in a neighborhood of the point z_k where the corresponding value $F(z_k)$ is changed.

•

5.2 Variations of the Sinc Basis

Let us use the notation of Chapter 4, and let us define the functions

$$\begin{aligned} s_k(x) &= \frac{h}{\pi} \frac{\sin\{\pi[\phi(x) - kh]/h\}}{\phi'(z_k)(x - z_k)} \\ t_k(x) &= \frac{h}{\pi} \frac{\cos\{\pi[\phi(x) - kh]/h\} - 1}{\phi'(z_k)(x - z_k)} . \end{aligned} \tag{5.2.1}$$

If k and l are integers, we have

$$\begin{aligned}
s_k(z_l) &= \begin{cases} 0 \text{ if } k \neq l \\ 1 \text{ if } k = l \end{cases} \\
t_k(z_l) &= \begin{cases} \dfrac{h}{\pi}\dfrac{1-(-1)^{k-l}}{\phi'(z_k)(z_k - z_l)} & \text{if} \quad k \neq l \\ 0 & \text{if} \quad k = l. \end{cases}
\end{aligned} \tag{5.2.2}$$

We shall prove the following theorem.

Theorem 5.2.1 *Let $F \in \mathbf{H}^1(\mathcal{D})$. Then, for all $x \in \Gamma$, we have*

$$\begin{aligned}
\varepsilon_1(x) &\equiv F(x) - \sum_{k=-\infty}^{\infty} F(z_k) s_k(x) \\
&= \frac{\sin\{\pi\phi(x)/h\}}{2\pi i} \int_{\partial\mathcal{D}} \frac{F(z)\,dz}{(z-x)\sin\{\pi\phi(z)/h\}},
\end{aligned} \tag{5.2.3}$$

and

$$\begin{aligned}
\varepsilon_2(x) &\equiv \frac{P.V.}{\pi i} \int_\Gamma \frac{F(t)}{t-x}dt - \frac{1}{i}\sum_{k=-\infty}^{\infty} F(z_k)\, t_k(x) \\
&= \frac{1}{2\pi}\int_{\partial\mathcal{D}} F(z) \frac{[\exp\{i\pi\phi(z)\operatorname{sgn}\Im\phi(z)/h\} - \cos\{\pi\phi(x)/h\}]}{(z-x)\sin\{\pi\phi(z)/h\}}dz.
\end{aligned} \tag{5.2.4}$$

Moreover

$$\begin{aligned}
\varepsilon_{3,k} &\equiv \int_\Gamma F(t)\, s_k(t)\, dt - \frac{h\, F(z_k)}{\phi'(z_k)} \\
&= \frac{ih\,(-1)^k}{2\pi\phi'(z_k)} \int_{\partial\mathcal{D}} F(z) \frac{\exp\{i\pi\phi(z)\operatorname{sgn}\Im\phi(z)/h\}}{z - z_k}dz.
\end{aligned} \tag{5.2.5}$$

Furthermore, we have for $x \in \Gamma$

$$|\varepsilon_1(x)| \leq \frac{1}{2\pi\sinh(\pi d/h)} N(F, \mathcal{D}, x) \tag{5.2.6}$$

$$|\varepsilon_2(x)| \leq \frac{e^{-\pi d/h} + 1}{2\pi\sinh(\pi d/h)} N(F, \mathcal{D}, x) \tag{5.2.7}$$

and also

$$|\varepsilon_{3,k}| \leqq \frac{h\, e^{-\pi d/h}}{2\pi} N(F, \mathcal{D}, z_k) \tag{5.2.8}$$

where in each of the last three equations,

$$N(F, \mathcal{D}, x) = \int_{\partial\mathcal{D}} \left| \frac{F(z)}{z - x} dz \right|. \tag{5.2.9}$$

Proof. Let us first replace z by ξ in (4.2.1), and then let us replace $F(\xi)$ by

$$\mathcal{F}(\xi, x) = \frac{\phi(\xi) - \phi(x)}{\xi - x} F(\xi) \tag{5.2.10}$$

with x and ξ fixed on Γ, to get the identity

$$\frac{F(\xi)}{\phi'(\xi)} \frac{\phi(\xi) - \phi(x)}{\xi - x} - \sum_{k=-\infty}^{\infty} \frac{F(z_k)}{\phi'(z_k)} \frac{kh - \phi(x)}{z_k - x} S(k,h) \circ \phi(\xi)$$

$$= \frac{\sin\{\pi\phi(x)/h\}}{2\pi i} \int_{\partial\mathcal{D}} F(\zeta) \frac{\phi(\zeta) - \phi(x)}{[\phi(\zeta) - \phi(\xi)](\xi - x)\sin\{\pi\phi(\zeta)/h\}} d\zeta \tag{5.2.11}$$

which holds for all $F \in \mathbf{H}^1(\mathcal{D})$. Now setting $\xi = x$ in this identity, we get (5.2.3).

In order to prove (5.2.4) we multiply both sides of (5.2.11) by $\phi'(\xi)/[\phi(\xi) - \phi(x)]$, integrate over Γ, and use the identities

$$\int_{\mathbb{R}} \frac{\sin(\pi\xi/h)}{\xi - z} d\xi = \pi \exp\{\frac{i\pi z}{h} \operatorname{sgn}\Im z\},\ z \in \mathbb{C},\ z \notin \mathbb{R} \tag{5.2.12}$$

and

$$\frac{P.V.}{\pi} \int_{\mathbb{R}} \frac{\sin\{\pi\xi/h\}}{\xi - x} d\xi \quad = \quad \cos\{\pi x/h\}, x \in \mathbb{R}, \tag{5.2.13}$$

thus yielding the desired result.

The result (5.2.5) follows directly from (4.2.2), if in (4.2.2) we replace F by $s_k F$, then use the definition for s_k given in (5.2.1), and note that $\sin\{\pi[\phi(x) - kh]/h\} = (-1)^k \sin\{\pi\phi(x)/h\}$.

Finally, the error bounds (5.2.6) through (5.2.8) follow directly from estimating the contour integrals in (5.2.3) through (5.2.5). ∎

We remark that under our assumptions on F, the right-hand sides of (5.2.3) and (5.2.4) may become unbounded as x approaches an end-point of Γ. In such a case the formulas (5.2.3) through (5.2.5) must be interpreted in the sense of a relative error. Of course, an absolute bound may be possible, i.e., (5.2.3) and (5.2.4) will hold for all $x \in \Gamma$, if $M(F, \mathcal{D}) \equiv \sup_{(x \in \Gamma)} N(F, \mathcal{D}, x) < \infty$, with $N(F, \mathcal{D}, x)$ defined in (5.2.9).

In general, we must exercise caution in the evaluation of the approximating sums in (5.2.3) and (5.2.4). For example, if $x = z_\ell$, then $\varepsilon_1(z_\ell) = 0$, which poses no problem. On the other hand, $\varepsilon_2(z_\ell)$ is not necessarily zero, and we have the approximation

$$
\begin{aligned}
(\mathcal{S}F)(z_\ell) &\equiv \frac{P.V.}{\pi i} \int_\Gamma \frac{F(t)}{t - z_\ell} dt \\
&\approx \frac{h}{\pi} \sum_{k \neq \ell} \frac{F(z_k)}{\phi'(z_k)} \frac{1 - (-1)^{k-\ell}}{z_k - z_\ell}.
\end{aligned}
\tag{5.2.14}
$$

The approximation of the sum on the right-hand side of (5.2.14) can be carried out to within a *relative error* of δ by means of Algorithm 5.2.2. The approximation of $F(x)$, $x \neq z_\ell$, via (5.2.3), can be carried out by a similar algorithm.

Algorithm 5.2.2: Evaluation of (5.2.14) to Within δ Relative Error

$k = \ell - 1$
$S_1 = 2F(z_k)/[\phi'(z_k)(z_k - z_\ell)]$
$u = v = 1, w = |S_1|$
$(\clubsuit)$ $k = k - 2$
$T = 2F(z_k)/[\phi'(z_k)(z_k - z_\ell)]$
$S_1 = S_1 + T$
$u = v, v = w, w = |T|$
$(u + v + w)/|S_1| : \delta$
$(\geq) \longrightarrow (\clubsuit)$
$(<)$
$k = \ell + 1$
$S_2 = 2F(z_k)/[\phi'(z_k)(z_k - z_\ell)]$
$u = v = 1, w = |S_2|$
$(\spadesuit)$ $k = k + 2$
$T = 2F(z_k)/[\phi'(z_k)(z_k - z_\ell)]$
$S_2 = S_2 + T$
$u = v, v = w, w = |T|$
$(u + v + w)/|S_2| : \delta$
$(\geq) \longrightarrow (\spadesuit)$
$(<)$
$S = (h/\pi)(S_1 + S_2).$

The following lemma provides results which are convenient for the purpose of truncating the infinite series in (5.2.3) and (5.2.4).

Lemma 5.2.2 *Under any one of the transformations ϕ defined in Example 1.7.5, we have*

$$\sup_{x\in\Gamma} |s_k(x)| \leqq e^h \tag{5.2.15}$$

and

$$\sup_{x\in\Gamma} |t_k(x)| \leqq e^h. \tag{5.2.16}$$

Proof. (i)(Proof of (5.2.15)). We separate the proof into two parts: (a) on the interval $[z_{k-1}, z_{k+1}]$; and (b) on the remainder of the interval, $\Gamma \setminus [z_{k-1}, z_{k+1}]$.
(i)-(a): On the interval $[z_{k-1}, z_{k+1}]$, we use the inequality

$$|\sin\{\pi[\phi(x) - kh]/h\}| \leqq \pi|\phi(x) - kh|/h. \tag{5.2.17}$$

Hence, on this interval, we have

$$\begin{aligned} |s_k(x)| &\leqq \left| \frac{\phi(x) - kh}{\phi'(z_k)(x - z_k)} \right| \\ &\leqq \max_{\xi\in[z_{k-1},z_{k+1}]} \left| \frac{\phi'(\xi)}{\phi'(z_k)} \right| \leqq e^h \end{aligned} \tag{5.2.18}$$

as may be shown (see Problem 5.2.4) for each of the transformations in Example 1.7.5.
(i)-(b): On the interval $\Gamma\setminus[z_{k-1}, z_{k+1}]$, we use the inequality $|\sin\{\pi[\phi(x) - kh]/h\}| \leqq 1$. Then, setting $\psi(x) = \phi^{-1}(x)$, we have

$$\begin{aligned} |s_k(x)| &\leqq \max_{x=z_{k-1},z_{k+1}} \frac{h/\pi}{|\phi'(z_k)(x - z_k)|} \\ &= \max_{\xi_1\in[(k-1)h,(k+1)h]} \frac{1/\pi}{|\phi'(z_k)\psi'(\xi_1)|} \\ &= \max_{\xi\in[z_{k-1},z_{k+1}]} \left| \frac{(1/\pi)\phi'(\xi)}{\phi'(z_k)} \right| \leqq e^h \end{aligned} \tag{5.2.19}$$

with the last inequality following similarly to the last one in (5.2.18).
(ii)-(a): On the interval $[z_{k-1}, z_{k+1}]$, we have

$$\begin{aligned}
|t_k(x)| &= \left| \frac{h[1-\cos\{\pi[\phi(x)-kh]/h\}]}{\pi\phi'(z_k)(x-z_k)} \right| \\
&= \left| \frac{\int_{z_k}^{x} \phi'(t)\sin\{\pi[\phi(t)-kh]/h\}\,dt}{\phi'(z_k)(x-z_k)} \right| \qquad (5.2.20) \\
&\leqq \max_{\xi\in[z_{k-1},z_{k+1}]} \left| \frac{\phi'(\xi)}{\phi'(z_k)} \right| \leqq e^h
\end{aligned}$$

where in the second line of (5.2.20) we have again used the inequality $|\sin x| \leqq |x|$, and where the final inequality follows as the previous ones did.

(ii)-(b): On $\Gamma \setminus [z_{k-1} z_{k+1}]$, we have $|1-\cos\{\pi[\phi(x)-kh]/h\}| \leqq 2$. Hence, for all $x \in \Gamma \setminus [z_{k-1}, z_{k+1}]$, we have

$$\begin{aligned}
|t_k(x)| &\leqq \max_{x=z_{k-1},z_{k+1}} \frac{2h/\pi}{|\phi'(z_k)(x-z_k)|} \\
&\leqq \frac{2e^h}{\pi} \leqq e^h.
\end{aligned} \qquad (5.2.21)$$

This completes the proof of Lemma 5.2.2. ■

Lemma 5.2.3 *Let $\mathcal{D}$, ϕ, and Γ respectively denote any one of the regions, the corresponding mapping, and the corresponding contour defined as in Example 1.7.5, and for some number $\alpha \in (0,1)$, let $F \in \mathbf{L}_\alpha(\mathcal{D})$. If $N(F,\mathcal{D},x)$ is defined as in (5.2.9), then $N(F,\mathcal{D},x)$ is uniformly bounded for all $x \in \Gamma$.*

Proof. This is left to Problem 5.2.4. ■

Theorem 5.2.4 *Let $\mathcal{D}, \phi$ and Γ respectively denote any one of the regions, mappings and contours defined as in Example 1.7.5, and for some number $\alpha \in (0,1)$, let $F \in \mathbf{L}_\alpha(\mathcal{D})$. Let N denote a positive integer, and let h be selected by the formula (3.1.24). Then there exist constants C_1 and C_2, independent of N, such that for all $x \in \Gamma$,*

$$|F(x) - \sum_{k=-N}^{N} F(z_k) s_k(x)| \leqq C_1 N^{1/2}\, e^{-(\pi d \alpha N)^{1/2}} \qquad (5.2.22)$$

and

$$\left| \frac{P.V.}{\pi i} \int_\Gamma \frac{F(t)}{t-x} dt - \sum_{k=-N}^{N} F(z_k) t_k(x) \right| \leqq C_2 N^{1/2} e^{-(\pi d \alpha N)^{1/2}}. \qquad (5.2.23)$$

Proof. The proof is left to Problem 5.2.5. ∎

We remark that while the interpolation method (5.2.22) appears to be as good as the method (4.2.17) it does in fact have some drawbacks. In the first case, neither of the functions s_k nor t_k go to zero at the endpoints of Γ. Secondly, if we let $\mathcal{P}_N F$ denote either of the sums $\sum_{k=-N}^{N} F(z_k) s_k$ or $\sum_{k=-N}^{N} F(z_k) t_k$, then $\mathcal{P}_N$ is of the order of N, whereas the corresponding norms for (4.2.17) and (4.3.16) are $\mathcal{O}(\log N)$. The formula (5.2.23) is nevertheless very important, since it is very useful to approximate the Hilbert transform defined in (1.3.6).

Let us recall that the Sinc approximation defined, for example, as in Section 3.1 converges to a function $f \in \mathbf{H}^1(\mathcal{D}_d)$ with respect to the $\mathbf{L}^p(\mathbb{R})$ norm, for all $p \in [2, \infty]$. It does not converge in $\mathbf{L}^1(\mathbb{R})$. Similarly after transformation to Γ, in the notation of Definition 4.1.1, the Sinc approximations e.g. of $F \in \mathbf{L}_\alpha(\mathcal{D})$ by use of the basis functions $S(k,h) \circ \phi$ do not belong to $\mathbf{L}_\alpha(\mathcal{D})$. The Sinc-like method of approximation described in the following theorem may be used to remedy this situation.

Theorem 5.2.5 *Let $\phi' F \in \mathbf{H}^1(\mathcal{D})$, where in the notation of Definition 4.1.1, $\phi : \mathcal{D} \to \mathcal{D}_d$, with $0 < d < \pi$. Then for all $z \in \mathcal{D}$,*

$$\begin{aligned} &\frac{\sin\{\pi\phi(x)/h\}}{2\pi i} \int_{\partial\mathcal{D}} \frac{F(\zeta)\,\phi'(\zeta)\,d\zeta}{\sinh\{\phi(\zeta) - \phi(z)\}\,\sin\{\pi\phi(\zeta)/h\}} \\ &\qquad = F(z) - \frac{h}{\pi} \sum_{k \in \mathbf{Z}} \frac{F(z_k)\,\sin\{\pi[\phi(x) - kh]/h\}}{\sinh[\phi(x) - kh]}. \end{aligned} \tag{5.2.24}$$

Let $F \in \mathbf{L}_\alpha(\mathcal{D})$, N a positive integer $h = [\pi d/(\alpha N)]^{1/2}$, and define F_N by

$$F_N(x) = \frac{h}{\pi} \sum_{k=-N}^{N} \frac{F(z_k)\,\sin\{\pi[\phi(x) - kh]/h\}}{\sinh[\phi(x) - kh]}. \tag{5.2.25}$$

Then there exists a constant C, independent of N, such that

$$\left(\int_\Gamma |F(x) - F_N(x)|^p \,|dx| \right)^{1/p} \leqq C N^{1/2} e^{-(\pi d \alpha N)^{1/2}} \tag{5.2.26}$$

for all $p \in [2, \infty]$.

Proof. The proof uses techniques similar to those in Chapters 3 and 4, and we therefore leave it to Problem 5.2.7. ∎

•

Problems for Section 5.2

5.2.1 Verify (5.2.12) and (5.2.13).

5.2.2 Verify (5.2.14).

5.2.3 Write an explicit algorithm of the form of Algorithm 5.2.2, to evaluate the sum in (5.2.3) to within a relative error δ, given an arbitrary $x \in \Gamma$.

5.2.4 Prove Lemma 5.2.3.

5.2.5 Prove Theorem 5.2.4.

5.2.6 Let $0 < \alpha \leqq 1$, let $g \in \mathbf{L}_\alpha(\mathcal{D}_d^1)$, with $\mathcal{D}_d^1$ is defined as in (1.7.10) and with $g(z) = f(z) - f(0)/(1+z)$, take $\phi(z) = \log(z)$, $z_k = e^{kh}$, and let N denote a positive integer. Prove that there exists a constant, c, is independent of N, such that

$$\sup_{x \in (0,\infty)} \left| \mathcal{S}f(x) - f(0)\frac{\log(1/x)}{\pi i(1+x)} - \frac{1}{i}\sum_{j=-N}^{N} g(z_j)\, t_j(x) \right|$$

$$\leqq c\, N^{1/2} e^{-(\pi d \alpha N)^{1/2}}.$$

5.2.7 (a) Prove Theorem 5.2.5.

(b)* Prove that F_N converges to F in $\mathbf{L}^1(\Gamma)$.

5.3 Elliptic Function Interpolants

We shall here derive an interpolation formula which is accurate for functions in $\mathbf{H}^1(\mathcal{D})$ with $\mathcal{D}$ the sector $\mathcal{D}^1$ defined in (1.7.10). We recall here that the Sinc interpolant for this region was generated by the function $\sin\{(\pi/h)\log z\}$, which has zeros at the points $z = e^{jh}, j = 0, \pm1, \pm2, \ldots$ (see Example 4.2.10). Setting

$$\Phi_h(z) = \frac{z-1}{z+1} \prod_{j=1}^{\infty} \frac{(z - e^{jh})(\frac{1}{z} - e^{jh})}{(z + e^{jh})(\frac{1}{z} + e^{jh})} \tag{5.3.1}$$

we see that the infinite product converges, and that the resulting function also has zeros at the points e^{jh}, for all integer values of j.
Taking

$$q_1 = e^{-h/2} \tag{5.3.2}$$

with q_1 defined in (1.7.20), we see (compare Problem 1.7.6) that

$$\Phi_h(z) = k^{1/2}\text{sn}\left[\frac{K_1}{\pi}\log z; k\right]. \tag{5.3.3}$$

Some properties of Φ_h and some estimates on the parameters K_1 and k as functions of h are recorded in the following two lemmas.

Lemma 5.3.1 *Let Φ_h be given by (5.3.1). Then:*
(a) For all $x \in \mathbb{R}$,

$$|\Phi_h(ix)| = 1; \tag{5.3.4}$$

(b) For all $z \in \mathbb{C}$,

$$\Phi_h(-z) = \frac{1}{\Phi_h(z)}; \tag{5.3.5}$$

(c) For all $x \in (0, \infty)$,

$$|\Phi_h(x)| \leqq k^{1/2}; \tag{5.3.6}$$

(d) For $j = 0, \pm 1, \pm 2, \cdots$,

$$\Phi_h'(e^{jh}) = \frac{k^{1/2}K_1}{\pi}(-1)^j e^{-jh}; \tag{5.3.7}$$

and
(e) If $\theta \in \mathbb{R}$, and $j = 0, \pm 1, \pm 2, \cdots$, then

$$k^{1/2} \leqq |\Phi_h(e^{(j+1/2)h}e^{i\theta})| \leqq k^{-1/2}. \tag{5.3.8}$$

Proof. Part (a): The relation (5.3.4) follows at once from (5.3.1), since each of the factors $(it - e^{jh})/(it + e^{jh})$ has modulus 1 for $t \in \mathbb{R}$.
Part (b): This follows from (5.3.1), by inspection.
Part (c): The relation (5.3.6) is a consequence of the identity (5.3.3), and of the fact that $\text{sn}(u; k)$ maps $\mathbb{R}$ onto $[-1, 1]$ (see Problem 1.7.5 (b), (f)).
Part (d): This result is a consequence of the fact that $\Phi_h(-z) = 1/\Phi_h(z)$, the identity (5.3.3), the fact that $\text{sn}(u; k)$ has a pole at $u = 2mK + (2n + 1)iK_1$ with residue $(-1)^m k^{-1}$, and the chain rule of differentiation. We omit the straightforward details.
Part (e): This follows from the first identity in part (a) of Problem 1.7.5, combined with part (c) of Problem 1.7.5. ∎

Lemma 5.3.2 *Let $z = t\, e^{i\theta}$, with $t > 0$ and $|\theta| = d$, and with $0 < d < \pi$. Then*

$$\prod_{j=-\infty}^{\infty} \left|\frac{e^{jh} + z}{e^{jh} - z}\right| = \exp\left\{\frac{\pi}{h}(\frac{\pi}{2} - d) + \delta\right\} \tag{5.3.9}$$

with

$$\delta = \sum_{j=1}^{\infty} \frac{\sinh\{\pi^2 j(1-2d/\pi)/h\} \cos\{(2\pi j/h)\log t\}}{j \cosh(\pi^2 j/h)}. \tag{5.3.10}$$

In particular,

$$|\delta| \leqq \delta_1 \equiv \left| \frac{e^{-2\pi d/h}}{1-e^{-2\pi d/h}} - \frac{e^{-2\pi(\pi-d)/h}}{1-e^{-2\pi(\pi-d)/h}} \right|. \tag{5.3.11}$$

Proof. The identities (5.3.9) and (5.3.10) follow from Problem 1.7.6. Inequality (5.3.11) follows if we note that each coefficient of $\cos\{2\pi j \log(t)\}$ in (5.3.10) has the same sign. Taking absolute values, replacing each cosine term by 1, and then replacing $j\cosh(\pi^2 j/h)$ in the denominator by $(1/2)\exp(\pi^2 j/h)$, we can sum the resulting two series explicitly, to get (5.3.11). ■

We note, in passing, that if we take $d = \pi/2$ in (5.3.11), then $\delta = 0$.

Lemma 5.3.3 *Let k be be related to q_1 through (1.7.20) and (5.3.2). Then:*
(a) The estimate

$$k \leqq 4e^{-\pi^2/h} \tag{5.3.12}$$

is valid for all $h > 0$;
(b) As $h \to 0$,

$$k = e^{-\pi^2/h}\left[4 + \mathcal{O}(e^{-2\pi^2/h})\right]; \tag{5.3.13}$$

(c) As $h \to 0$,

$$\frac{K_1 k^{1/2}}{\pi} = \frac{2\pi}{h} e^{-\pi^2/(2h)} \left[1 + \mathcal{O}(e^{-2\pi^2/h})\right]. \tag{5.3.14}$$

Proof. Part (a): By Problem 1.7.7 (e), we have

$$\log(k^{1/2}) = -\frac{\pi^2}{2h} + \sum_{m=1}^{\infty} \frac{(-1)^{m-1}}{m} + \delta_2 \tag{5.3.15}$$

with

$$\delta_2 = 2\sum_{m=1}^{\infty} \frac{(-1)^m}{m} \frac{e^{-2m\pi^2/h}}{1+e^{-2m\pi^2/h}}. \tag{5.3.16}$$

Since δ_2 is clearly negative, and since $\sum_{m=1}^{\infty}[(-1)^m/m] = \log(2)$, (5.3.12) follows.

Part (b): We see from (5.3.16), that

$$|\delta_2| < 2e^{-2\pi^2/h} = \mathcal{O}(e^{-\pi^2/h}). \tag{5.3.17}$$

and hence (5.3.13) follows from (5.3.15).
Part (c): This part of the Lemma follows from (5.3.13) and parts (f) and (g) of Problem 1.7.7. ■

Theorem 5.3.4 *Let $0 < d \leqq \pi/2$, let $\mathcal{D} = \mathcal{D}_d^1$, with $\mathcal{D}_d^1$ defined by Equation (1.7.10), and let $g \in \mathbf{H}^1(\mathcal{D})$. Then, for all $s \in \mathcal{D}$,*

$$g(s) - \frac{\pi\Phi_h(s)}{K_1 k^{1/2}} \sum_{n=-\infty}^{\infty} \frac{(-1)^n e^{nh} g(e^{nh})}{s - e^{nh}} = \varepsilon_h(s) \tag{5.3.18}$$

with

$$\varepsilon_h(s) = \frac{\Phi_h(s)}{2\pi i} \int_{\partial\mathcal{D}} \frac{g(z)}{(z-s)\Phi_h(z)} dz. \tag{5.3.19}$$

Moreover, if $s = re^{i\theta}$, with $r > 0$ and $|\theta| < d$, then

$$|\varepsilon_h(s)| \leqq e^{-\pi(d-|\theta|)/h} \left[1 + \mathcal{O}(e^{-2\pi d/h})\right] N_1(g, \mathcal{D}, s) \tag{5.3.20}$$

with

$$N_1(g, \mathcal{D}, s) = \int_{\partial\mathcal{D}} \left| \frac{g(z)}{z-s} dz \right|. \tag{5.3.21}$$

Proof. Let W_N denote the region

$$W_N = \left\{ z \in \mathbb{C} : |\arg(z)| < d, e^{-(N+1/2)h} < |z| < e^{(N+1/2)h} \right\} \tag{5.3.22}$$

the positive integer N having been chosen sufficiently large so that $s \in W_N$. Let us set

$$\varepsilon_h^{(N)}(s) = \frac{\Phi(s)}{2\pi i} \int_{\partial W_N} \frac{g(z)}{(z-s)\Phi_h(z)} dz. \tag{5.3.23}$$

We may then note, by (5.3.8) through (5.3.11), that $1/\Phi_h(z)$ is uniformly bounded on ∂W_N, independently of N. Applying Lemma 5.3.1 (d), we can evaluate the contour integral via summation of the residues, to get

$$\varepsilon_h^{(N)}(s) = g(s) - \frac{\pi\Phi_h(s)}{K_1 k^{1/2}} \sum_{n=-N}^{N} \frac{(-1)^n e^{nh} g(e^{nh})}{s - e^{nh}}. \tag{5.3.24}$$

It now follows, due to the uniform boundedness of $1/\Phi_h(z)$ on W_N, and since $g \in \mathbf{H}^1(\mathcal{D})$, that we may let $N \to \infty$ in (5.3.24), to get (5.3.19), with $\varepsilon_h(s) = \lim_{(N\to\infty)} \varepsilon_h^{(N)}(s)$.
Let us now establish (5.3.20). Clearly, $\log|\Phi_h(s)|$ is harmonic in $\Omega^+ \cap \Omega_+$, (see (1.4 22) and (1.5.6)), and moreover, by (5.3.4) and (5.3.6), we have

$$\begin{aligned} \log|\Phi_h(s)| &= 0 \quad \text{if} \quad \arg(s) = \pi/2 \\ \log|\Phi_h(s)| &\leqq \log(k^{1/2}) \quad \text{if} \quad \arg(s) = 0. \end{aligned} \tag{5.3.25}$$

Setting $s = re^{i\theta}$, with $r > 0$, it follows that the function Ψ defined by

$$\Psi(s) = (1 - 2\theta/\pi)\log(k^{1/2}) \tag{5.3.26}$$

is also harmonic in $\Omega^+ \cap \Omega_+$, and by the maximum principle, we have $\log|\Phi_h(s)| \leqq \Psi(s)$, for all $s \in \Omega^+ \cap \Omega_+$. That is, we have

$$|\Phi_h(s)| \leqq k^{1/2-\theta/\pi} \tag{5.3.27}$$

for all $s = re^{i\theta} \in \Omega^+ \cap \Omega_+$. Using (5.3.12), we thus find that for all $s = re^{i\theta} \in \mathcal{D}$, we have

$$|\Phi_h(s)| \leqq 2e^{-\pi^2[1/2-|\theta|/\pi]/h}. \tag{5.3.28}$$

We now apply this result to bound the integral in (5.3.19). By Lemma 5.3.2, it follows, taking $\theta = \arg(s)$, that

$$\frac{\sup_{\arg(s)=\theta}|\Phi_h(s)|}{\inf_{\arg(z)=\pm d}|\Phi_h(z)|} = e^{-\pi(d-|\theta|)/h}\left[1 + \mathcal{O}(e^{-2\pi d/h})\right]. \tag{5.3.29}$$

Recalling that $N_1(g, \mathcal{D}, s)$ is bounded for $s \in \mathcal{D}$, we get (5.3.20). ■

The previous theorem may be extended to approximation over a contour Γ, by means of Definition 4.1.1, and then it may be used to yield approximation formulas over several special contours by means of the specific maps, which were introduced in Example 1.7.5. Truncation of the infinite sum in (5.3.18) is also possible, via the introduction of the class of functions $\mathbf{L}_\alpha(\mathcal{D})$ defined in Definition 4.1.1. Since the proofs of these results are similar to those of Sinc approximations carried out in Chapters 3 and 4, we defer the presentation of these results to the problems at the end of this section, as well as to the next two sections of this chapter. We shall, however, present the following result.

Theorem 5.3.5 *Let $\mathcal{D}$, Γ, z_j, ρ, and $\mathbf{L}_\alpha(\mathcal{D})$ be defined as in Definition 4.1.1. Let $g \in \mathbf{L}_\alpha(\mathcal{D})$, let N be a positive integer, and let $h = [\pi d/(\alpha N)]^{1/2}$. Then there exists a constant C, independent of N, such that*

$$\sup_{s\in\Gamma}\left|g(s)-\frac{h}{2\pi}e^{\pi^2/(2h)}\sum_{j=-N}^{N}\frac{(-1)^j e^{jh}g(z_j)\Phi_h\circ\phi(s)}{\rho(s)-e^{jh}}\right| \tag{5.3.30}$$
$$\leqq CN^{1/2}e^{-(\pi d\alpha N)^{1/2}}.$$

Proof. We leave the proof of this Theorem to Problem 5.3.5. ■

•

Problems for Section 5.3

5.3.1 What boundary value problem does the function Ψ of Equation (5.3.26) solve over the region $\Omega^+ \cap \Omega_+$?

5.3.2 In Theorem 5.3.4, let $d = \pi/2$, let $g \in \mathbf{H}^2(\Omega^+)$, and set

$$\|g\| = \lim_{c\to 0}\left(\int_{\mathbb{R}}|g(c+iy)|^2 dy\right)^{1/2}.$$

Let $c > 0$, $y \in \mathbb{R}$, and set

$$u(y) = \frac{g(c+iy)}{\Phi_h(iy)}$$
$$v(y) = \frac{\Phi_h(c+iy)}{2\pi i}\int_{\mathbb{R}}\frac{u(t)\,dt}{t-ic-y}.$$

(a) Prove the inequalities

$$\|u\|^2 \equiv \int_{\mathbb{R}}|u(y)|^2 dy = \int_{\mathbb{R}}|g(c+iy)|^2 dy \leqq \|g\|^2$$
$$\|v\|^2 \leqq \int_{\mathbb{R}}|\Phi_h(c+iy)\,u(y)|^2 dy \leqq \|u\|^2.$$

(b) Prove that given any $\varepsilon > 0$, $Y > 0$, there exists $h > 0$, such that if $-Y \leqq y \leqq Y$, then $|\Phi_h(c+iy)| < \varepsilon^2/2$.

(c) Prove that $\int_{\mathbb{R}}|v(y)|^2 dy \to 0$ as $h \to 0$. [Hint: (i) Choose $Y > 0$, such that $\int_{|y|>Y}|u(y)|^2 dy < (\varepsilon^2/2)\|u\|^2$. (ii) Then $\|v\|^2 = \int_{|y|\leqq Y}|v(y)|^2 dy + \int_{|y|>Y}|v(y)|^2 dy \leqq \varepsilon^2\|u\|^2$.]

5.3.3 Prove that if $\mathcal{D}$ is defined as in Theorem 5.3.4, if α is some positive number, if $g \in \mathbf{L}_\alpha(\mathcal{D})$, with $\mathbf{L}_\alpha(\mathcal{D})$ defined as in Definition 4.1.1, and if $N_1(g, \mathcal{D}, s)$ is defined as in (5.3.21), then $\sup_{(0<s<\infty)} N_1(g, \mathcal{D}, s) < \infty$.

(d) Conclude that if ε_h is defined as in (5.3.19), then for every $c > 0$, we have

$$\int_{\mathbb{R}} |\varepsilon_h(c + it)|^2 dt \to 0$$

as $h \to 0$.

5.3.4 Prove that the functions

$$\frac{\pi}{k^{1/2} K_1} \frac{(-1)^n e^{nh} \Phi_h(s)}{(s - e^{nh})}$$

are uniformly bounded by a constant which is independent of n and h on the interval $0 < s < \infty$.

5.3.5 Prove Theorem 5.3.5. [Hint: Use Lemma 4.2.4, Lemmas 5.3.1 to 5.3.3, the result of Problem 5.3.4, and then follow the steps of the proof of Theorem 4.2.5.]

5.3.6 Extend the statement of Theorem 5.3.4 to the general situation of Definition 4.1.1, and then prove the newly stated theorem.

5.3.7 Give the specific form of the statement of Theorem 5.3.5 for each of the transformations of Example 1.7.5.

5.4 Inversion of the Laplace Transform

In this section we derive an explicit formula to approximate the inverse of the Laplace transform, based on the explicit inversion of the infinite sum in (5.3.18), via the use of the Bromwich inversion integral. This procedure is possible since the only singularities of the interpolating approximant of g in (5.3.18) are simple poles at which the residues can be explicitly determined. This fortuitous property is not shared, for example, by the Sinc interpolant (4.2.52).

Let $f \in \mathbf{L}^2(0, \infty)$, and let g denote the Laplace transform of f, i.e., using the notation of Section 1.5,

$$g(s) = (\mathcal{L}f(s) = \int_0^\infty e^{-st} f(t)\, dt. \tag{5.4.1}$$

It then follows, as was shown in Section 1.5, that $g \in \mathbf{Hol}(\Omega^+)$, with Ω^+ the right-half plane, and that

$$\|g\|^2 = \lim_{\xi \to 0+} \int_{-\infty}^{\infty} |g(\xi + i\eta)|^2 d\eta = 2\pi \|f\|^2. \tag{5.4.2}$$

That is, $g \in \mathbf{H}^2(\Omega^+)$. We also record the Bromwich inversion formula given in Section 1.5, namely,

$$f(t) = (\mathcal{L}^{-1} g)(t) = \lim_{\xi \to 0+} \frac{1}{2\pi i} \int_{\xi - i\infty}^{\xi + i\infty} g(s)\, e^{st} ds. \tag{5.4.3}$$

In practice, the problem of inversion of the Laplace transform is to find a good approximation of f on $(0, \infty)$, given g. One popular procedure is to apply the trapezoidal rule to the integral in (5.4.3). However, this procedure requires complex values of g, and these are not always easy to obtain, particularly in applications such as, for example, the solution of a partial differential equation via separation of variables, in which the use of the Laplace transform can play a fundamental role, provided that one can reliably invert it. The procedure described in the theorem which follows has an advantage in this respect, in that the inversion formula given requires only the values of g on the interval $(0, \infty)$. We mention here that yet another formula requiring only the values of g on $(0, \infty)$ is described in Section 6.9. The formula given in the present section converges for a larger class of functions than the one in Section 6.9, and as a consequence, for most applications it converges more slowly than that in Section 6.9.

Theorem 5.4.1 *Given $g \in \mathbf{H}^2(\Omega^+)$, there exists a unique $f \in \mathbf{L}^2(0, \infty)$ such that $f = \mathcal{L}^{-1} g$. Let $\delta_h(g)$ be defined by the formula*

$$\delta_h(g)(t) = f(t) - \frac{h e^{\pi^2/(2h)}}{2\pi} \cdot$$
$$\cdot \sum_{m=-\infty}^{\infty} \left\{ (-e^h)^m g(e^{mh}) \left[\frac{h e^{\pi^2/(2h)}}{2\pi} \sum_{n=-\infty}^{\infty} \frac{(-e^h)^n \exp(-t e^{nh})}{e^{mh} + e^{nh}} \right] \right\}. \tag{5.4.4}$$

Then, for every positive number c, we have

$$\int_0^{\infty} e^{-2ct} |\delta_h(g)(t)|^2 dt \to 0 \tag{5.4.5}$$

as $h \to 0^+$.

Proof. Let us set

$$u(y) = \frac{g(0^+ + iy)}{\Phi_h(iy)}. \tag{5.4.6}$$

Then $u \in \mathbf{L}^2(\mathbb{R})$, and if we also set

$$u_+(s) = \frac{1}{2\pi i} \int_{-\infty}^{\infty} \frac{u(y)}{iy - s} i dy \tag{5.4.7}$$

then it follows by Theorem 1.4.6, Corollary 1.4.7, and Parseval's theorem, that $u_+ \in \mathbf{H}^2(\Omega^+)$, and moreover, that

$$\lim_{\xi \to 0} \int_{-\infty}^{\infty} |u_+(\xi + iy)|^2 dy \leqq \|u\|^2 = \|g\|^2. \tag{5.4.8}$$

Now, comparing (5.4.6) and (5.3.24), it follows that

$$u_+(s) = \frac{1}{\Phi_h(s)} \left[g(s) - \frac{\pi \Phi_h(s)}{K_1 k^{1/2}} \sum_{m=-\infty}^{\infty} \frac{(-e^h)^m g(e^{mh})}{s - e^{mh}} \right]. \tag{5.4.9}$$

Recall now by the representation (5.3.1) and the relation (5.3.4), that $\Phi_h \in \mathbf{Hol}(\Omega^+)$, and that $|\Phi_h(s)| \leqq 1$ for all $s \in \Omega^+$. Hence $u_+ \Phi_h \in \mathbf{H}^2(\Omega^+)$. Thus there exists a unique $\delta_h^*(g) \in \mathbf{L}^2(0,\infty)$, such that $u_+ \Phi_h = \mathcal{L}\{\delta_h^*(g)\}$. Using (5.3.5) and (5.3.7) and then proceeding as in the proof of Theorem 5.3.4, we get

$$\delta_h^*(g)(t) = f(t) - \frac{\pi^2}{K_1^2 k} \sum_{m,n=-\infty}^{\infty} \frac{(-e^h)^{m+n} \exp\left(-te^{nh}\right) g(e^{mh})}{e^{mh} + e^{nh}}. \tag{5.4.10}$$

Let us now recall, by Section 1.5, that for every $c > 0$ we have

$$u_+(c + iy)\Phi_h(c + iy) = \int_0^{\infty} e^{iyt} \left\{ e^{-ct} \delta_h^*(g)(t) \right\} dt. \tag{5.4.11}$$

Since $u_+ \in \mathbf{H}^2(\Omega^+)$, it follows by Problem 5.3.2, that

$$\int_{\mathbb{R}} |u_+(c + iy)\Phi_h(c + iy)|^2 \, dy \to 0 \tag{5.4.12}$$

as $h \to 0$. But by Parseval's theorem, we have

$$\int_{\mathbb{R}} |u_+(c + iy)\Phi_h(c + iy)|^2 \, dy = 2\pi \int_0^{\infty} e^{-2ct} \left|\delta_h^*(g)(t)\right|^2 dt \to 0 \tag{5.4.13}$$

as $h \to 0$, for every $c > 0$. Using (5.3.14), we arrive at (5.4.5). This completes the proof of Theorem 5.4.1. ∎

It is a simple fact that all methods of inversion of the Laplace transform are ill-posed; the ill-posedness of Laplace transform inversion via (5.4.4) is is revealed by the huge factor, $e^{\pi^2/h}$, in front of the double sum in (5.4.4). This factor shows that we must accurately evaluate the double sum in double precision, in order to get a single precision approximation of f. Also, since the statement of Theorem 5.4.1 holds for *every* f in the large class of functions $\mathbf{L}^2(0,\infty)$, the double infinite sum approximation to f given in (5.4.4) is likely to converge more slowly than that of (6.9.32), at least in the case of $f \in \mathbf{L}_\alpha(\mathcal{D})$, and with $\mathcal{D}$ the region $\mathcal{D}_d^1$ of Fig. 1.7.4 (b). Our numerical examples demonstrate that this is indeed the case.

•

PROBLEMS FOR SECTION 5.4

5.4.1 Let the conditions of Theorem 5.4.1 be satisfied. Use the results of Problem 3.2.2 to show that if $x > 0$, then as $h \to 0$,

$$\frac{he^{\pi^2/(2h)}}{2\pi} \sum_{n=-\infty}^{\infty} \frac{(-e^h)^n \exp(-te^{nh})}{x + e^{nh}}$$
$$= [1 + o(1)]\frac{1}{2\pi}\int_0^\infty \left\{\frac{e^{itu}u^{-i\pi/h}}{u+ix} + \frac{e^{-itu}u^{i\pi/h}}{u-ix}\right\} du$$

and

$$\frac{he^{\pi^2/(2h)}}{2\pi} \sum_{m=-\infty}^{\infty} \frac{(-e^h)^m g(e^m h)}{x + e^{mh}}$$
$$= [1 + o(1)]\frac{1}{2\pi}\int_0^\infty \left\{\frac{g(-iu)u^{1-i\pi/h}}{u+ix} + \frac{g(iu)u^{1+i\pi/h}}{u-ix}\right\} du.$$

5.4.2 Use formula 5.4.4 to approximate $f(t)$, given $g(s)$, with f and g related as in the following table:

$f(t)$	$g(s)$
e^{-t}	$(s+1)^{-1}$
$t\,e^{-t}$	$(s+1)^{-2}$
$e^{-t}\sin(t)$	$(s^2+2s+2)^{-1}$
$t^{4/3}e^{-4t/3}$	$\Gamma(4/3)(s+4/3)^{-7/3}$.

5.4.3 Under what conditions is the formula

$$g(s) \approx h \sum_{m=-\infty}^{\infty} e^{mh} f(e^{mh}) \exp(-s\,e^{mh})$$

accurate to approximate $g = \mathcal{L}f$, given f?

5.4.4 Prove that the Laplace transform of

$$f_h(t) = \frac{\pi^2}{K_1^2 k} \sum_{m,n=-\infty}^{\infty} \frac{(-e^h)^{m+n} \exp(-t\, e^{nh})\, g(e^{mh})}{e^{mh} + e^{nh}}$$

is

$$g_h(s) = \frac{\pi \Phi_h(s)}{K_1 k^{1/2}} \sum_{m=-\infty}^{\infty} \frac{(-e^h)^m g(e^{mh})}{s - e^{mh}}.$$

5.5 Sinc-Like Rational Approximation

In this section we use the elliptic function procedure from Section 5.3 to derive a family of interpolating rational approximations. The rationals which we shall construct have accuracy equivalent to that of Sinc approximation, in the very spaces in which we have shown Sinc methods to be effective. These rationals do not seem to have any particular direct advantages over Sinc methods—indeed, they seem to have some disadvantages, since they possess neither the orthogonality, nor the many explicit identities that are known for Sinc methods. On the other hand, there is a practically useful aspect to the presentation of this section. Namely, it enables us to tell *a priori* when we can expect rational approximation to work, which is, to a large extent, an art at which only a relatively small number of professional users excel. A specific property of rational functions which has not yet been developed for Sinc approximation is that one can perform extrapolation to the limit with rational functions. For practical purposes, a function which can be accurately approximated by a low degree polynomial can itself be identified with that polynomial, and hence the process of polynomial extrapolation will be accurate wherever there exists a polynomial which accurately approximates the function. The same is true of rational extrapolation, which can be expected to be accurate whenever and wherever it is possible to accurately approximate the function by a low-degree rational approximation. The results of this section will thus illustrate when rational extrapolation methods have an advantage over polynomial extrapolation methods, which is the case, for example, if we want to find the value of a function at ∞, or if we want to know the value of a function at a point where it has a singularity. For example, we will be able to tell *a priori* when we can expect the well known Thiele algorithm to work. A more detailed discussion of these topics is given in Subsection 5.6.5.

Let us begin by making some technical estimates enabling us to truncate the infinite product in (5.3.1), reducing it to a rational function.

Lemma 5.5.1 *Let M and N denote positive integers, and let us set*

$$P(z) = \prod_{j=-M}^{N} \frac{z + e^{jh}}{z - e^{jh}}. \tag{5.5.1}$$

Let us further set $z = t\,e^{i\theta}$, *with* $t = |z|$, *and let us denote* $|\theta| = d$. *Let* δ_1 *be defined as in (5.3.11).*
(a) If $0 < d \leqq \pi/2$, *then*

$$|P(z)| \leqq \exp\{\pi(\pi/2 - d)/h + \delta_1\}. \tag{5.5.2}$$

(b) If $\pi/2 \leqq d < \pi$, *if*

$$M^{1/2} e^{-Mh} \leqq t \leqq N^{-1/2} e^{Nh} \tag{5.5.3}$$

and if

$$Q = \exp\{\delta_1 + (M^{-1/2} + N^{-1/2})\pi^2/(4h)\} \tag{5.5.4}$$

then

$$Q^{-1} \exp\{\pi(\pi/2 - d)/h\} \leqq |P(z)| \leqq Q \exp\{\pi(\pi/2 - d)/h\}. \tag{5.5.5}$$

(c) If $d = \pi$, *and if* t *is as in (5.5.3), then*

$$|P(z)| \leqq 2 \exp\{-[1 - (M^{-1/2} + N^{-1/2})]\, \pi^2/(2h)\}. \tag{5.5.6}$$

Proof. Let m and n denote arbitrary integers such that $m \leqq n$, and set

$$W(m, n, z) = \prod_{j=m}^{n} \left| \frac{z - e^{jh}}{z + e^{jh}} \right|. \tag{5.5.7}$$

Then, with Φ_h defined as in (5.3.1), we have

$$|P(z)| = |\Phi_h(z)|^{-1} W(-\infty, -M-1, z)\, W(N+1, \infty, z). \tag{5.5.8}$$

Let us now consider, in turn, each part of Lemma 5.5.1.
Part (a): If $d \leqq \pi/2$, then for any $a > 0$, we have $|(z-a)/(z+a)| \leqq 1$ for all $z \in \Omega^+$, so that for $z \in \Omega^+$, each W-factor in (5.5.8) is bounded by 1. Hence, in this case, we have $|P(z)| \leqq |\Phi_h(z)|^{-1}$, and hence (5.5.2) follows from Lemma 5.3.2.
Part (b): If $\pi/2 \leqq d < \pi$, then with $\omega = \pi - d$, we have $|(z - e^{jh})/(z + e^{jh})| = |(1 + te^{i\omega}/e^{jh})/(1 - te^{i\omega}/e^{jh})| \leqq 1$; also, since $t \leqq N^{-1/2}e^{Nh}$, we have $|t\,e^{i\omega}/e^{jh}| < 1$ for $j > N$, and hence

$$
\begin{aligned}
\log W(N+1, \infty, z) & \\
&= \Re \sum_{m=0}^{\infty} \frac{2}{2m+1} \sum_{j=N+1}^{\infty} (t\, e^{i\omega}/e^{jh})^{2m+1} \\
&\leqq \sum_{m=0}^{\infty} \frac{2}{2m+1} \sum_{j=N+1}^{\infty} (t/e^{jh})^{2m+1} \\
&= \sum_{m=0}^{\infty} \frac{1}{2m+1} \left(\frac{t\, e^{-h/2}}{e^{Nh}} \right)^{2m+1} \frac{1}{\sinh[(m+1/2)h]} \\
&\leqq \frac{1}{N^{1/2} h} \sum_{m=0}^{\infty} \frac{2}{(2m+1)^2} \\
&= \frac{\pi^2}{4\, N^{1/2} h}
\end{aligned}
\tag{5.5.9}
$$

where, to get the last line from the second-to-last, we used the estimates $(t\, e^{-h/2}/e^{Nh})^{2m+1} \leqq (1/N^{1/2})^{2m+1} \leqq N^{1/2}$, and also $\sinh[(m+1/2)h] \geqq (m+1/2)h$.
Similarly, we also get

$$
\log W(-\infty, -M-1, z) \leqq \frac{\pi^2}{4M^{1/2}h}. \tag{5.5.10}
$$

Thus (5.5.5) follows from Lemma 5.3.2, and the inequalities (5.5.9) and (5.5.10).

Part (c): Finally, if $d = \pi$, then using (5.3.6) and (5.3.12), we get

$$
|P(z)| \leqq 2\, e^{-\pi^2/(2h)} W(-\infty, -M-1, z)\, W(N+1, \infty, z). \tag{5.5.11}
$$

Proceeding as for part (b) of the above proof, we get (5.5.6). ∎

Next, we shall require some technical estimates of some integrals of functions belonging to certain $\mathbf{H}^p$ spaces, in order to derive rational approximants for functions in such spaces, as described in Theorem 5.5.3 below.

Lemma 5.5.2 *Let $x \in (0, \infty)$, let $q \in (1, \infty)$, and let us set*

$$
G_q^q(x) = \frac{x^{1-q}}{\pi} \int_0^{\infty} (1 + x^2 t^2)^{(q-2)/2} (1 + t^2)^{-q/2} dt \tag{5.5.12}
$$

and also,

$$H_q^q(d,x) = \frac{1}{2\pi}\int_{\partial\mathcal{D}} \left|\frac{z+1}{z-x}\right|^q \frac{|dz|}{|1+z|^2} \tag{5.5.13}$$

with $\mathcal{D}$ defined as in Equation (1.7.10). Then

$$G_q^q(x) \leqq C_q \begin{cases} x^{1-q} & if \quad 0 < x \leqq 1 \\ x^{-1} & if \quad x \geqq 1 \end{cases} \tag{5.5.14}$$

where

$$C_q = \begin{cases} \pi^{-1/2}\Gamma((q-1)/2)/\Gamma(q/2)/2 & \text{if} \quad 1 < q \leqq 2 \\ 1 & \text{if} \quad q \geqq 2. \end{cases} \tag{5.5.15}$$

Moreover, there exists a constant $C(q,d)$, such that if $0 < d \leqq \pi/2$, then

$$H_q^q(d,x) \leqq C(q,d) G_q^q(x). \tag{5.5.16}$$

Proof. Let us note that if $0 < x \leqq 1$, then

$$\left[\frac{1+x^2t^2}{1+t^2}\right]^{(q-2)/2} \leqq \begin{cases} (1+t^2)^{1-q/2} & \text{if} \quad 1 < q \leqq 2 \\ 1 & \text{if} \quad q \geqq 2. \end{cases} \tag{5.5.17}$$

Hence, if $0 < x \leqq 1$, then

$$G_q^q(x) \leqq \begin{cases} \dfrac{x^{1-q}}{\pi}\displaystyle\int_0^\infty (1+t^2)^{-q/2}dt & \text{if} \quad 1 < q \leqq 2 \\ \dfrac{x^{1-q}}{\pi}\displaystyle\int_0^\infty (1+t^2)^{-1}dt & \text{if} \quad q \geqq 2. \end{cases} \tag{5.5.18}$$

After change of variable via the transformation $t = [\tau/(1-\tau)]^{1/2}$, which maps $(0,\infty)$ to $(0,1)$, and using the result of Problem 1.1.5 (g), we get (5.5.14) and (5.5.15), in the case $0 < x \leqq 1$. If $x > 1$, we replace t by $1/t$ in (5.5.12) to get

$$G_q^q(x) = \frac{1}{\pi x}\int_0^\infty (1+t^2/x^2)^{(q-2)/2}(1+t^2)^{-q/2}dt. \tag{5.5.19}$$

A comparison of this identity with (5.5.12) shows that the same argument which we used above in the case of (5.5.18) may also be applied to (5.5.19). Hence (5.5.14) and (5.5.15) hold for all $x \in (0,\infty)$.

Next, let us note, that

$$H_q^q(\pi/2,x) = \frac{1}{\pi}\int_0^\infty (1+t^2)^{(q-2)/2}(x^2+t^2)^{-q/2}dt. \tag{5.5.20}$$

Replacing t by xt in this equation, we see, upon comparison with (5.5.12), that $H_q^q(\pi/2, x) = G_q^q(x)$.
Now, recalling the definition (1.7.10) of $\mathcal{D}$, we see that if $z \in \mathcal{D}$, then $z = t\,e^{\pm id}$, with t and d defined as in Lemma 5.5.1, and hence

$$H_q^q(d, x) = \frac{1}{\pi} \int_0^\infty \left|t + e^{-id}\right|^{q-2} \left|t - xe^{-id}\right|^{-q} dt. \tag{5.5.21}$$

Since $x > 0$, $t > 0$, and $0 < d \leqq \pi/2$, we have

$$\begin{aligned} \left|t - x\,e^{-id}\right|^2 &= t^2 + x^2 - 2xt\,\cos(d) \\ &\geqq (t^2 + x^2)[1 - \cos(d)] \\ &= 2\,(t^2 + x^2)\,\sin^2(d/2) \end{aligned} \tag{5.5.22}$$

and also,

$$\begin{aligned} t^2 + 1 &\leqq t^2 + 2t\cos(d) + 1 \\ &= \left|t + e^{-id}\right|^2 \\ &\leqq (t^2 + 1)[1 + \cos(d)] \\ &= 2\,(t^2 + 1)\,\cos^2(d/2). \end{aligned} \tag{5.5.23}$$

Hence, we have that

$$\left|t - x\,e^{-id}\right|^{-q} \leqq 2^{-q/2}(t^2 + x^2)^{-q/2}\,\sin^q(d/2) \tag{5.5.24}$$

and also, that

$$\left|t + e^{-id}\right|^{q-2} \leqq \begin{cases} (1 + t^2)^{q/2-1} & \text{if} \quad 1 < q \leqq 2 \\ 2^{q/2-1}(1 + t^2)^{q/2-1}\cos^{q-2}(d/2) & \text{if} \quad q \geqq 2. \end{cases} \tag{5.5.25}$$

Replacing the left-hand sides of (5.5.24) and (5.5.25) by their right-hand sides in (5.5.21), and using the results of Problem 1.1.5 (h), we get (5.5.16), in which

$$\pi C(q, d) = \begin{cases} \{2^{1/2}\sin(d/2)\}^{-q} & \text{if} \quad 1 < q \leqq 2 \\ 2^{-1}\sin^{-q}(d/2)\cos^{q-2}(d/2) & \text{if} \quad q \geqq 2. \end{cases} \tag{5.5.26}$$

■

Theorem 5.5.3 *Let $0 < d \leqq \pi/2$, let $f \in \mathbf{Hol}(\mathcal{D})$, with $\mathcal{D}$ defined in (1.7.10), and for $p \in (1, \infty)$, let*

$$\|f\|_p \equiv \left(\int_{\partial\mathcal{D}} |f(z)|^p |1 + z|^{2p-2} |z|^{-p} |dz|\right)^{1/p} < \infty. \tag{5.5.27}$$

Let $q = p/(p-1)$, let N be a positive integer, and let us set

$$h = \pi \left(\frac{q}{2N}\right)^{1/2} \tag{5.5.28}$$

$$\begin{aligned} \beta(z) &= \prod_{j=-N}^{N} \frac{z - e^{jh}}{z + e^{jh}} \\ B(z) &= \frac{z}{1+z}\beta(z) \end{aligned} \tag{5.5.29}$$

$$\eta_N(x) = f(x) - \sum_{j=-N}^{N} \frac{f(e^{jh})\,B(x)}{(x - e^{jh})\,B'(e^{jh})}. \tag{5.5.30}$$

Then, there exists a constant C, independent of N, such that for all $x \in (0, \infty)$,

$$|\eta_N(x)| \leqq C\,N^{1/(2q)} \exp\{-d(2N/q)^{1/2}\}\,\|f\|_p. \tag{5.5.31}$$

Proof. It is evident, by inspection, that η_N defined by (5.5.30) also has the representation

$$\eta_N(x) = \frac{B(x)}{2\pi i} \int_{\partial\mathcal{D}} \frac{f(z)}{(z-x)\,B(z)} dz. \tag{5.5.32}$$

By Lemma 5.5.1 (a), we have

$$\sup_{z \in \partial\mathcal{D}} |\beta(z)|^{-1} \leqq \exp\{\pi(\pi/2 - d)/h + \delta_1\} \tag{5.5.33}$$

with δ_1 defined as in (5.3.11). Hence, setting $A = \exp(\delta_1)$, taking absolute values of each term on (5.5.32), and then applying Hölder's inequality, using the definition (5.5.27) for $\|f\|_p$, and also using (5.5.33), we get

$$|\eta_N(x)| \leqq A \frac{x}{1+x} |\beta(x)| \, \exp\{\pi(\pi/2 - d)/h\} H_q(d, x) \|f\|_p. \tag{5.5.34}$$

Now, using (5.5.5), and assuming that if $x > 0$, then $|x - e^{jh}|/|x + e^{jh}| < 1$, we get

$$|\beta(x)| \leqq \begin{cases} 2\,\exp\{-\pi^2/(2h)[1 - N^{-1/2}]\} \ \text{if} \ N^{1/2}e^{-Nh} \leqq x \leqq N^{1/2}e^{Nh} \\ 1 \ \text{if} \ 0 < x < N^{1/2}e^{-Nh}, \ \text{or if} \ N^{-1/2}e^{Nh} < x < \infty. \end{cases} \tag{5.5.35}$$

Substituting (5.5.35) into (5.5.34), and using (5.5.14) and (5.5.16), we get the estimate

$$\frac{x}{1+x}|\beta(x)|H_q(d,x) \leqq C_q C(q,d) \cdot$$

$$\cdot \begin{cases} 2\exp\{-\pi^2(1-N^{-1/2})/(2h)\} & \text{if } N^{1/2}e^{-Nh} \leqq x \leqq N^{-1/2}e^{Nh} \\ (N^{1/2}e^{-Nh})^{1/q} & \text{otherwise.} \end{cases} \tag{5.5.36}$$

The inequality (5.5.31) now follows from this result. This completes the proof of Theorem 5.5.3. ∎

Lemma 5.5.4 *(a) Let us define $H(\alpha,\beta,x)$ for fixed numbers $\alpha \in (0,1)$ and $\beta \in (0,1)$, and arbitrary $x \in (0,\infty)$ by*

$$G(\alpha,\beta,x) = \frac{1}{\pi}\int_0^\infty t^{\alpha-1}(1+t^2)^{(1-\alpha-\beta)/2}(x^2+t^2)^{-1/2}dt. \tag{5.5.37}$$

Then

$$G(\alpha,\beta,x) \leqq A_1 \begin{cases} x^{\alpha-1} & \text{if } 0 < x \leqq 1 \\ x^{-\beta} & \text{if } 1 \leqq x < \infty. \end{cases} \tag{5.5.38}$$

with $A_1 = \max\{A(\alpha,\beta), A(\beta,\alpha)\}$, $A(\alpha,\beta)$ given by

$$A(\alpha,\beta) = \frac{\Gamma(\alpha/2)}{2\pi}\begin{cases} \Gamma(\beta/2)\,\Gamma((\alpha+\beta)/2) & \text{if } \alpha+\beta \leqq 1 \\ \pi^{-1/2}\Gamma((1-\alpha)/2) & \text{if } \alpha+\beta > 1. \end{cases} \tag{5.5.39}$$

(b) Let $0 < d \leqq \pi/2$, let $\mathcal{D}$ be defined as in (1.7.10), and let us set

$$H(\alpha,\beta,d,x) = \frac{1}{\pi}\int_{\partial\mathcal{D}} |z|^{\alpha-1}(1+|z|)^{1-\alpha-\beta}|z-x|^{-1}|dz|. \tag{5.5.40}$$

Then the following inequality holds for all $x \in (0,\infty)$:

$$H(\alpha,\beta,d,x) \leqq \frac{1}{\sin(d/2)}G(\alpha,\beta,x). \tag{5.5.41}$$

Proof. Part (a): Replacing t by xt in (5.5.37), we get

$$G(\alpha,\beta,x) = \frac{x^{\alpha-1}}{\pi}\int_0^\infty t^{\alpha-1}(1+t^2)^{-1/2}(1+x^2t^2)^{(1-\alpha-\beta)/2}dt. \tag{5.5.42}$$

Let us now examine this equation in the case $0 < x \leqq 1$. Here we have

$$(1+x^2t^2)^{(1-\alpha-\beta)/2} \leqq \begin{cases} (1+t^2)^{(1-\alpha-\beta)/2} & \text{if} \quad \alpha+\beta \leqq 1 \\ 1 & \text{if} \quad \alpha+\beta > 1. \end{cases} \tag{5.5.43}$$

Using the estimates (5.5.43) in (5.5.42), and using the results of Problem 1.1.5 (h), we get

$$G(\alpha,\beta,x) \leqq x^{\alpha-1} A(\alpha,\beta) \tag{5.5.44}$$

with $A(\alpha,\beta)$ given above in (5.5.39).
Next, replacing t by x/t in (5.5.37), we get

$$G(\alpha,\beta,x) \leqq \frac{x^{-\beta}}{\pi} \int_0^\infty t^{\beta-1}(1+t^2)^{-1/2}(1+t^2/x^2)^{(1-\alpha-\beta)/2} dt. \tag{5.5.45}$$

A comparison of (5.5.45) with (5.5.42) shows that after replacing x by $1/x$ and interchanging the roles of α and β, in (5.5.45), we can make the exact same arguments for (5.5.45) that we made for (5.5.42). We thus find, that if $x \geqq 1$, then

$$G(\alpha,\beta,x) \leqq x^{-\beta} A(\beta,\alpha) \tag{5.5.46}$$

with $A(\alpha,\beta)$ given by (5.5.39). Hence, defining A_1 as in Lemma 5.5.4, the result (5.5.38) follows.
Part (b): Let us next consider the integral $H(\alpha,\beta,d,x)$ given in (5.5.40). Since $z \in \partial\mathcal{D}$, we have $z = t\,e^{\pm id}$, so that $1+|z| \leqq 2^{1/2}(1+t^2)^{1/2}$. We also use the estimate (5.2.22), to get

$$\begin{aligned} H(\alpha,\beta,d,x) &= \frac{1}{\pi}\int_0^\infty t^{\alpha-1}(1+t)^{1-\alpha-\beta}\left|t\,e^{\pm id}-x\right|^{-1} dt \\ &\leqq \frac{1}{\sin(d/2)} G(\alpha,\beta,x). \end{aligned} \tag{5.5.47}$$

This completes the Proof of Lemma 5.5.4. ∎

In analogy with Sinc approximation on Γ for $f \in \mathbf{L}_{\alpha,\beta}(\mathcal{D})$, we next consider rational approximation on this space.

Theorem 5.5.5 *Let $0 < d \leqq \pi/2$, let $\mathcal{D}$ be defined by Equation (1.7.10), let $\alpha \in (0,1)$ and $\beta \in (0,1)$ be fixed positive numbers, and let $f \in \mathbf{L}_{\alpha,\beta}(\mathcal{D})$. Set $\delta = \min(\alpha,\beta)$, $\gamma = \max(\alpha,\beta)$, let n be a positive integer, and let h be selected by the formula*

$$h = \frac{\pi}{(2\delta n)^{1/2}}. \tag{5.5.48}$$

Let integers M *and* N *be defined by*

$$\begin{array}{lllllll} M &=& n, & N &=& \lceil(\alpha/\beta)n\rceil & \text{if} \ \delta = \alpha \\ M &=& \lceil(\beta/\alpha)n\rceil, & N &=& n & \text{if} \ \gamma = \alpha. \end{array} \tag{5.5.49}$$

Set

$$\begin{aligned} \beta(z) &= \prod_{j=-M}^{N} \frac{z-e^{jh}}{z+e^{jh}} \\ B(z) &= \frac{z}{1+z}\beta(z) \end{aligned} \tag{5.5.50}$$

$$\eta_n(x) = f(x) - \sum_{j=-M}^{N} \frac{f(e^{jh})\, B(x)}{(x-e^{jh})\, B'(e^{jh})}. \tag{5.5.51}$$

Then there exists a constant C*, independent of* n*, such that for all* $x \in (0,\infty)$*,*

$$|\eta_n(x)| \leqq C n^{\gamma/2} \exp\{-d(2\delta n)^{1/2}\}. \tag{5.5.52}$$

Proof. Clearly, $\eta_n(x)$ is also defined by the right-hand side of (5.5.32). Taking absolute values of each term in this contour integral, and as a consequence of $f \in \mathbf{L}_{\alpha,\beta}(\mathcal{D})$, replacing $|f(z)|$ by $C_1|z|^{\alpha}/(1+|z|)^{\alpha+\beta}$, we get, for $0 < x < \infty$,

$$|\eta_n(x)| \leqq A_1 C_1 \frac{x}{1+x} |\beta(x)| H(\alpha,\beta,d,x) \tag{5.5.53}$$

with the constant A_1 defined in Lemma 5.5.4, and with $H(\alpha,\beta,d,x)$ defined by (5.5.40). Using Lemma 5.5.4, we get

$$\frac{x}{1+x} H(\alpha,\beta,d,x) \leqq A_1 A_2 \left\{ \begin{array}{lll} x^{\alpha} & \text{if} & 0 < x \leqq 1 \\ x^{-\beta} & \text{if} & x \geqq 1 \end{array} \right. \tag{5.5.54}$$

where A_2 is defined as (5.5.39). Now, if $M^{1/2}e^{-Mh} \leqq x \leqq N^{-1/2}e^{Nh}$, then $|\beta(x)|$ is bounded by proceeding as on the right-hand side of (5.5.35), whereas if $0 < x \leqq M^{1/2}e^{-Mh}$, or if $x \geqq N^{-1/2}e^{Nh}$, then $|\beta(x)| \leqq 1$. Hence we have, for all $x \in (0,\infty)$, that

$$|\beta(x)| \frac{x}{1+x} H(\alpha,\beta,d,x) \leqq C_2 n^{\gamma/2} e^{-\delta n h} \tag{5.5.55}$$

with C_2 a constant. The result (5.5.52) then follows, selecting h by (5.5.48). ∎

We can now extend the above results to the case of a contour Γ, in the general setting of Definition 4.1.1. To this end, we let $\mathcal{D}$, z_j, ϕ, ρ, and Γ be defined as in Definition 4.1.1, and for some positive integers M and N, we set

$$\begin{aligned} \beta(z) &= \prod_{j=-M}^{N} \frac{\rho(z) - e^{jh}}{\rho(z) + e^{jh}} \\ B(z) &= \frac{\rho(z)}{1+\rho(z)}\beta(z), \end{aligned} \tag{5.5.56}$$

and

$$\eta_n(x) = f(x) - \sum_{j=-M}^{N} \frac{f(z_j)\, e^{jh} \phi'(z_j)\, B(x)}{[\rho(x) - e^{jh}]\, B'(z_j)}. \tag{5.5.57}$$

The proofs of Theorems 5.5.3 and 5.5.5 may then be repeated, almost verbatim, to enable us to establish Theorems 5.5.6 and 5.5.7, which follow. We omit the proofs of these theorems, deferring them to the problems at the end of this section.

Theorem 5.5.6 *Let* $0 < d \leqq \pi/2$, *let* $f \in \mathbf{Hol}(\mathcal{D})$, *let* $p \in (1, \infty)$, *and let*

$$\|f\|_p = \left(\frac{1}{2\pi}\int_{\partial\mathcal{D}} |f(z)|^p |1+\rho(z)|^{2p-2} |\rho(z)|^{1-p} |\phi'(z)\, dz|\right)^{1/p} < \infty. \tag{5.5.58}$$

Let $q = p/(p-1)$, *and for* N *a positive integer, let us set*

$$h = \pi\left(\frac{q}{2N}\right)^{1/2} \tag{5.5.59}$$

and let $\eta_n(x)$ *be defined as in (5.5.57), with* $M = N$. *Then there exists a positive constant* C, *independent of* N, *such that for all* $x \in (0, \infty)$,

$$|\eta_n(x)| \leqq C N^{1/(2q)} \exp\{-d(2N/q)^{1/2}\} \|f\|_p. \tag{5.5.60}$$

Theorem 5.5.7 *Let* $0 < d \leqq \pi/2$, *let* $\alpha \in (0,1)$ *and* $\beta \in (0,1)$ *denote fixed numbers, and let* $f \in \mathbf{L}_{\alpha,\beta}(\mathcal{D})$. *Set* $\delta = \min(\alpha, \beta)$, $\gamma = \max(\alpha, \beta)$, *let* n *be a positive integer, and let* h *be selected by the formula*

$$h = \frac{\pi}{(2\delta n)^{1/2}}. \tag{5.5.61}$$

Let integers M *and* N *be defined by*

$$\begin{array}{llllll} M = n, & N = \lceil(\alpha/\beta)n\rceil & \text{if} & \delta & = & \alpha \\ M = \lceil(\beta/\alpha)n\rceil, & N = n & \text{if} & \gamma & = & \alpha. \end{array} \tag{5.5.62}$$

Let $\eta(x)$ be defined by (5.5.57). Then there exists a constant C, independent of n, such that for $x \in (0, \infty)$,

$$|\eta_n(x)| \leqq C n^{\gamma/2} \exp\{-d(2\delta n)^{1/2}\}. \tag{5.5.63}$$

The above theorems enable us to get explicit rational approximations in the $\rho(x)$ variable, for each of the transformations of Example 1.7.5.

Example 5.5.8 *Rational Approximation on* $(-1,1)$. Let $\mathcal{D} = \mathcal{D}_d^2$, as defined in (1.7.11). In this case we have

$$\phi(z) = \log\left(\frac{1+z}{1-z}\right), \quad \rho(z) = \frac{1+z}{1-z}, \quad z_j = \frac{e^{jh}-1}{e^{jh}+1}. \tag{5.5.64}$$

(i) Let us first consider the case of an arbitrary $d \in (0, \pi/2)$. The class $\mathbf{L}_{\alpha,\beta}(\mathcal{D}_d^2)$ is given by

$$\mathbf{L}_{\alpha,\beta}(\mathcal{D}_d^2) = \{f \in \mathbf{Hol}(\mathcal{D}_d^2) : |f(z)| \leqq C|1+z|^{\alpha}|1-z|^{\beta}\} \tag{5.5.65}$$

with C some constant. The terms $B(z)$ and $\eta(x)$ take the form

$$\begin{aligned} B(z) &= (1-z^2) \prod_{j=-M}^{N} \frac{z-z_j}{1-z_j z} \\ \eta(x) &= f(x) - \sum_{j=-M}^{N} \frac{f(z_j)\,B(x)}{(x-z_j)B'(z_j)}. \end{aligned} \tag{5.5.66}$$

Upon selecting δ, γ, h, M, and N as in (5.5.61) and (5.5.62), we may bound η as in (5.5.63).

(ii) The special case of $d = \pi/2$ is of theoretical interest. Let $\mathcal{U} = \mathcal{D}_{\pi/2}^2$ denote the unit disc in $\mathbb{C}$, let $p \in (1, \infty)$, let f be analytic in $\mathcal{U}$, so that F defined by $F(z) = f(z)/(1-z^2)$ is in the Hardy space $H^p(\mathcal{U})$, i.e., such that

$$\|f\| = \lim_{r \to 1^-} \left(\frac{1}{2\pi}\int_0^{2\pi} |F(re^{i\theta})|^p d\theta\right)^{1/p} < \infty \tag{5.5.67}$$

and let $q = p/(p-1)$, $h = \pi[q/(2N)]^{1/2}$, let z_j be defined as in (5.5.64), $B(z)$ and $\eta(x)$ (with $M = N$) as in (5.5.66). Then there exists a constant C depending only on p, such that

$$\sup_{-1<x<1} |\eta(x)| \leqq C N^{1/(2q)} \exp\{-\pi[N/(2q)]^{1/2}\}\|f\|. \tag{5.5.68}$$

We may observe, by inspection of (5.5.66) that these rational approximations are rational functions in the x variable.

Example 5.5.9 *Approximation of Decaying Oscillatory Functions on* $(0, \infty)$. Whereas the approximation schemes of Theorems 5.5.3 and 5.5.5 are well-suited for functions that are analytic in a sector and have algebraic decay to zero at 0 and at ∞, the scheme of the present example is suited for functions that are analytic and bounded only in the "bullet-shaped" strip, $\mathcal{D}_d^3$, defined in (1.7.12). This situation occurs frequently in the case of Fourier transforms, which may decay to zero at an algebraic rate at 0 and (ideally) at an exponential rate at ∞. In this case, let $0 < \alpha < 1$, $0 < \beta < 1$, and set

$$\phi(z) = \log[\sinh(z)], \quad \rho(z) = \sinh(z), \quad z_j = \log\left[e^{jh} + \sqrt{1 + e^{2jh}}\right]. \tag{5.5.69}$$

The class $\mathbf{L}_{\alpha,\beta}(\mathcal{D}_d^3)$ is given by

$$\mathbf{L}_{\alpha,\beta}(\mathcal{D}_d^3) = \left\{ f \in \mathbf{Hol}(\mathcal{D}_d^3) : |f(z)| \leqq C \left(\frac{|z|}{1+|z|}\right)^{\alpha} e^{-\beta \Re z} \right\} \tag{5.5.70}$$

with C some constant. The terms $B(z)$ and $\eta(x)$ take the form

$$\begin{aligned} B(z) &= \frac{\sinh(z)}{1+\sinh(z)} \prod_{j=-M}^{N} \frac{\sinh(z) - e^{jh}}{\sinh(z) + e^{jh}} \\ \eta(x) &= f(x) - \sum_{j=-M}^{N} \frac{f(z_j) e^{jh} \sqrt{1 + e^{-2jh}}\, B(x)}{[\sinh(x) - e^{jh}] B'(z_j)}. \end{aligned} \tag{5.5.71}$$

Upon selecting δ, γ, h, M, and N as in (5.5.61) and (5.5.62), we may bound η as in (5.5.63).

We note, here, that the rational approximants in (5.5.71) are in the variable $\sinh(x)$.

Example 5.5.10 *Rational Approximation on* $\mathbb{R}$, *the Exponentially Decaying Case.* Let $\mathcal{D} = \mathcal{D}_d$, as defined in (1.7.9), and set

$$\phi(z) = z, \quad \rho(z) = e^z, \quad z_j = jh. \tag{5.5.72}$$

The class $\mathbf{L}_{\alpha,\beta}(\mathcal{D}_d)$ is given by

$$\mathbf{L}_{\alpha,\beta}(\mathcal{D}_d) = \left\{ f \in \mathbf{Hol}(\mathcal{D}_d) : |f(z)| \leqq C \frac{e^{\alpha \Re z}}{1 + e^{(\alpha+\beta)\Re z}} \right\}. \tag{5.5.73}$$

where C is some constant. The terms $B(z)$ and $\eta(x)$ take the form

$$
\begin{aligned}
B(z) &= \frac{e^z}{1+e^z} \prod_{j=-M}^{N} \frac{e^z - e^{jh}}{e^z + e^{jh}} \\
\eta(x) &= f(x) - \sum_{j=-M}^{N} \frac{f(z_j) e^{jh} B(x)}{(e^x - e^{jh}) B'(jh)}.
\end{aligned}
\tag{5.5.74}
$$

Upon selecting δ, γ, h, M, and N as in (5.5.61) and (5.5.62), we may bound η as in (5.5.63).

Note, we now have rational approximation in the variable e^x.

Example 5.5.11 *Approximation of Algebraically Decaying Functions on* $\mathbb{R}$. As illustrated in the previous example, the identity map is suitable for constructing rationals in e^x on $\mathbb{R}$, which are suitable for approximating functions that may be oscillatory, but decay exponentially on $\mathbb{R}$. The following method of rational approximation may be suitable if the function f is analytic in the "hourglass" shaped region $\mathcal{D}_d^4$ of (1.7.13), which decays algebraically, rather than exponentially on $\mathbb{R}$. In this case, let $0 < \alpha < 1$, $0 < \beta < 1$, and set

$$
\phi(z) = \log[z + \sqrt{1+z^2}], \quad \rho(z) = z + \sqrt{1+z^2}, z_j = \sinh(jh). \tag{5.5.75}
$$

The class $\mathbf{L}_{\alpha,\beta}(\mathcal{D}_d^3)$ may be described by

$$
\mathbf{L}_{\alpha,\beta}(\mathcal{D}_d^4) = \left\{ f \in \mathbf{Hol}(\mathcal{D}_d^4) : |f(z)| \leqq \left\{ \begin{array}{ll} C(1+|z|)^{-\alpha} & \text{if} \quad \Re z \leqq 0 \\ C(1+|z|)^{-\beta} & \text{if} \quad \Re z \geqq 0 \end{array} \right. \right\}, \tag{5.5.76}
$$

with C some constant. The terms $B(z)$ and $\eta(x)$ take the form

$$
\begin{aligned}
B(z) &= \frac{z+\sqrt{1+z^2}}{1+z+\sqrt{1+z^2}} \prod_{j=-M}^{N} \frac{z+\sqrt{1+z^2} - e^{jh}}{z+\sqrt{1+z^2}+e^{jh}} \\
\eta(x) &= f(x) - \sum_{j=-M}^{N} \frac{f(z_j) e^{jh} B(x)}{(x+\sqrt{1+x^2} - e^{jh}) \cosh(jh) B'(z_j)}.
\end{aligned}
\tag{5.5.77}
$$

Upon selecting δ, γ, h, M, and N as in (5.5.61) and (5.5.62), we may bound η as in (5.5.63).

We note, here, that the rational approximants in (5.5.71) are in the variable $x + \sqrt{1+x^2}$.

Let us add a brief summary comment regarding the above derived rational approximants. Although these approximants appear to be as accurate as the Sinc methods constructed in Chapter 4, the Sinc formulas are easier to evaluate, and they have many other advantages over these rational function interpolants. The chief practical role of the constructed rationals appear to be their existence, since they guarantee us to be able to apply rational extrapolation to the limit.

Let us conclude with some remarks on the accuracy of rational versus polynomial approximation. Let $R > 1$ and let f be analytic and uniformly bounded by M in $\mathcal{E}_R$, with $\mathcal{E}_R$ defined in (1.7.6). Let $\mathbf{P}_n$ denote the family of all polynomials of degree n. One might expect that the sequence of numbers γ_n defined by

$$\gamma_n = \inf_{p_n \in \mathbf{P}_n, q_n \in \mathbf{P}_n} \sup_{-1<x<1} \left| f(x) - \frac{p_n(x)}{q_n(x)} \right| \tag{5.5.78}$$

to approach zero considerably more rapidly than the numbers α_n defined in (2.3.11), as $n \to \infty$. Although this is usually the case in applications, the following theorem, whose proof has been omitted, demonstrates that this is not always so.

Theorem 5.5.12 (Szabados:) *There exist functions $f \in \mathbf{H}^\infty(\mathcal{E}_R)$ such that*

$$\lim_{n\to\infty} \sup_{k>n} \{\alpha_k\}^{1/k} = \lim_{n\to\infty} \sup_{k>n} \{\gamma_k\}^{1/k} . \tag{5.5.79}$$

■

It is evident, in view of the results of Problem 2.4.4, that if z is an interior point of a simply connected domain of analyticity of a function g, for which we are given data $\{(w_j, g(w_j))\}$, and if we want to approximate $g(z)$ using n values of this data, then we can use the Aitken–Neville algorithm to get an approximation within an error bounded by $C\, e^{-cn}$, with C and c positive constants independent of n. A knowledge of this rapid rate of convergence enables us, for practical purposes, to identify g with a low-degree polynomial, and it guarantees in practice that the process of approximating $g(z)$ to the limit via polynomial extrapolation will be efficient. Theorem 5.5.11 above states that rational extrapolation to the limit may not be more efficient than polynomial extrapolation to the limit. On the other hand, if z is a boundary point of the region of analyticity of g, then we may not be able to accurately approximate g via a polynomial which uses values of g in the interior of the region of analyticity of g, (see Problem 2.1.4) and hence we cannot expect to accurately approximate $g(z)$ in this case, via the use of polynomial extrapolation. It may, nevertheless, still be possible to achieve accurate rational extrapolation in this case, and this is precisely when rational extrapolation methods have an advantage over polynomial extrapolation methods.

These points are further elaborated upon in Section 5.6.5.

•

Problems for Section 5.5

5.5.1 Give a detailed proof of Theorem 5.5.6.

5.5.2 Give a detailed proof of Theorem 5.5.7.

5.5.3 Find the explicit rational approximations of this section, along with bounds for the approximation error for the following functions over the given intervals:

$$(1-x^2)^{2/3}(1+x^2)^{1/2}, \quad -1 < x < 1$$
$$(1+x)^{\alpha}(1-x)^{\beta}\log(x), \quad \alpha > 0,\ \beta > 0,\ -1 < x < 1$$
$$\frac{x^{1/3}}{1+x}, \quad 0 < x < \infty$$
$$e^{-x^2}, \quad -\infty < x < \infty$$
$$\frac{x^{1/3}e^{ia\sqrt{1+x^2}}}{\sqrt{1+x^2}}, \quad \Im a > 0,\ \ 0 < x < \infty$$
$$\frac{\cos(x)}{\cosh(x)}, \quad -\infty < x < \infty and$$

and

$$\frac{2x+1}{1+x^2}\log(1+2x^2), \quad -\infty < x < \infty.$$

5.5.4 Proceed along the lines of Problem 4.2.6 (b) and (c) to construct yet another (rational) approximation to $|x|^{\alpha}$ on $[-1, 1]$, accurate to within the bound given in Problem 4.2.6 (b).

5.6 Rationals and Extrapolation

Although many methods to extrapolate the limit exist in applications, we shall confine our discussion here mainly to polynomial and rational function methods. We have already presented polynomial algorithms, including the Aitken–Neville algorithm, and the approximation by Chebyshev polynomials, in Chapter 2. In the present chapter we briefly describe the Padé method, the Aitken Δ^2 process, the epsilon algorithm; and, most importantly, the Thiele algorithm. The Padé method is frequently used, in the form of a ratio of two polynomials of equal degree, (which is denoted, in this section, by $[n/n]$) to extrapolate a function f at ∞. We therefore present this algorithm here, and we also present Aitken's Δ^2 process, along with the epsilon algorithm since these algorithms can be used to generate the fraction $[n/n]$. On the other hand, relatively little is known regarding when $[n/n]$ does in fact yield an accurate approximation of the desired limit. After studying this section, the reader will be able to tell *a priori* when the Thiele algorithm is expected to work, and moreover, he/she will be able to prescribe a convergent variant of a Thiele algorithm for a large class of convergent sequences.

The Aitken–Neville algorithm given in Example 2.3.5 (which, except for computational layout, is equivalent to the algorithmically more suitable and popular Richardson algorithm) and its extension given in Problem 2.4.4 provide a powerful method to extrapolate to the limit. As part (b) of Problem 2.4.4 shows, when used properly, the convergence of these algorithms to the true value of the function they approximate can be very rapid. Such rapid convergence occurs precisely when the extrapolation point (e.g., z in the Aitken–Neville algorithm, of Example 2.3.5) is an interior point of the region of analyticity of the function we want to approximate. In this case we can generally expect an $\mathcal{O}(e^{-cn})$ error, as $n \to \infty$, c being a positive constant. On the other hand, the Aitken–Neville algorithm will fail totally if the extrapolation point is not in the region of analyticity of the function we want to approximate. An interesting and frequently occurring situation in application transpires if the extrapolation point is on the boundary of the region of analyticity of the function we want to approximate. In this case, polynomial approximation of the function on an interval or contour containing the extrapolation point will converge very slowly, such as, for example, at the rate $\mathcal{O}(n^{-c'})$, c' a positive constant, and as a consequence, the Aitken–Neville extrapolation procedure will also converge very slowly, (particularly if c' is small) so that the computed results may be practically useless. On the other hand, the convergence of approximation by the ratio of two polynomials of degree n can still occur at the rate $\mathcal{O}(e^{-c'' n^{1/2}})$, which is considerably faster than polynomial approximation, and we may thus expect a procedure based on rational extrapolation to the limit to converge quite rapidly. In the problems at the end of this section we shall present some well-chosen examples to illustrate the use of extrapolation methods based on analyticity properties of the function we wish to approximate.

5.6.1 Padé Approximation

One of the best known methods to construct a rational function approximation is via the *Padé method.* We start with a given function f analytic in a neighborhood of $z = 0$, so that we may write

$$f(z) = \sum_{k=0}^{\infty} a_k z^k. \tag{5.6.1}$$

Letting $\mathbf{P}_n$ denote the set of all polynomials of degree n, we attempt to determine a rational function $r(x) = p(x)/q(x)$ with $p \in \mathbf{P}_m$, $q \in \mathbf{P}_n$, such that

$$f(z) - r(z) = \mathcal{O}(z^\ell) \tag{5.6.2}$$

with ℓ as large as possible. We may note here, that p has $m+1$ free parameters, and q has $n+1$, so that r has $m+n+1$ free ones (why not $m+n+2$?). Hence we may expect to have $\ell \geqq m+n+1$. Unfortunately this is not always possible, as seen by taking $f(z) = z$, $m = 0$, and $n = 1$. Thus, instead of posing the problem as in (5.6.2), we adopt a linearized version of this equation. Given f, m, and n, we select $p \in \mathbf{P}_m$, and $q \in \mathbf{P}_n$, such that

$$q(z)f(z) - p(z) = \mathcal{O}(z^{m+n+1}) \tag{5.6.3}$$

as $z \to 0$. The approximation p/q determined to satisfy Equation (5.6.3) is called the *Padé approximant* to f of type (m,n), and it is denoted by $[m/n](z)$. In the special case of $n = 0$, we get the Taylor polynomial:

$$[m/0](z) = \sum_{k=0}^{m} a_k z^k. \tag{5.6.4}$$

It is easy to show that the Padé approximant $[m/n]$ is unique. For if

$$\begin{aligned} (qf - p)(z) &= \mathcal{O}(z^{m+n+1}), \\ (Qf - P)(z) &= \mathcal{O}(z^{m+n+1}), \end{aligned} \tag{5.6.5}$$

as $z \to 0$, with p and P in $\mathbf{P}_m$ and q and Q in $\mathbf{P}_n$, then multiplying the first equation in (5.6.5) by Q and the second by q and subtracting, we get

$$(pQ - Pq)(z) = \mathcal{O}(z^{m+n+1}) \tag{5.6.6}$$

as $z \to 0$. However, both pQ and Pq belong to $\mathbf{P}_{m+n}$, and therefore (5.6.6) implies that the left-hand side of (5.6.6) is identically zero, i.e., $p/q = P/Q$.

Let us now return to (5.6.3), in which we take

$$p(z) = \sum_{k=0}^{m} p_k z^k, \quad q(z) = \sum_{k=0}^{n} q_k z^k. \tag{5.6.7}$$

Equating equal powers of z, we thus arrive at the system of equations

$$\begin{bmatrix} a_m & a_{m-1} & \cdots & a_{m-n+1} \\ a_{m+1} & a_m & \cdots & a_{m-n+2} \\ \vdots & \vdots & \ddots & \vdots \\ a_{m+n-1} & a_{m+n-2} & \cdots & a_m \end{bmatrix} \begin{bmatrix} q_1 \\ q_2 \\ \vdots \\ q_n \end{bmatrix} = \begin{bmatrix} c_{m+1} \\ c_{m+2} \\ \vdots \\ c_{m+n} \end{bmatrix}. \tag{5.6.8}$$

Notice that the coefficient matrix of this system is a Toeplitz matrix. If the determinant of this matrix is non-zero, then we can readily solve for the q_j. Once the q_j have been obtained, we can immediately get the p_j solving the equations

$$\begin{aligned} p_0 &= c_0 \\ p_1 &= c_1 + q_1 c_0 \\ p_2 &= c_2 + q_1 c_1 + q_2 c_0 \\ \vdots & \quad \ddots \\ p_m &= c_m + \sum_{k=1}^{\min(m,n)} q_k c_{m-k}. \end{aligned} \tag{5.6.9}$$

We have thus proved the following theorem:

Theorem 5.6.1 *Let $p \in \mathbf{P}_m$ and $q \in \mathbf{P}_n$ be determined by the requirements (5.6.3) and (5.6.7). If the determinant of the matrix in (5.6.8) is non-zero, then p and q exist, and as $z \to 0$, $[m/n] = p/q$ satisfies the relation*

$$f(z) - [m/n](z) = \mathcal{O}(z^{m+n+1}). \tag{5.6.10}$$

We have thus described an explicit procedure for determining the entries of the *Padé* table. This table is a doubly infinite array of rational functions which is displayed in the form

$$\begin{array}{cccc} [0/0] & [1/0] & [2/0] & \cdots \\ [0/1] & [1/1] & [2/1] & \cdots \\ [0/2] & [1/2] & [2/2] & \cdots \\ \ddots & \ddots & \ddots & \ddots \end{array} \tag{5.6.11}$$

While Padé approximations are frequently used in applications, it is not always possible to tell *a priori* when a Padé approximation converges to a function. For example, Wallin's theorem, which we do not prove here, states:

Theorem 5.6.2 (Wallin): *There exists an entire function f for which the sequence of Padé approximations $\{[n/n]\}_{n=0}^{\infty}$ is unbounded at every point of $\mathbb{C} \setminus 0$.*

■

In the notation of (5.6.7), the successive ratios of the numbers p_n/q_n are frequently used to approximate the value $f(\infty)$.

5.6.2 Continued Fractions

A continued fraction provides a very convenient tool to approximate numbers, or to evaluate a rational function. The theory of continued fractions is a very rich and beautiful theory, toward which almost every famous mathematician of the 18th and 19th century made contributions.

The *nth convergent* of a continued fraction is defined by

$$c_n = b_0 + \cfrac{a_1}{b_1 + \cfrac{a_2}{b_2 + \cfrac{a_3}{b_3 + \ddots + \cfrac{a_n}{b_n}}}} \tag{5.6.12}$$

The tabular form of (5.6.12) is awkward to write, and it is more customary to abbreviate this to the form

$$c_n = b_0 + \frac{a_1}{b_1+}\ \frac{a_2}{b_2+}\ \cdots\ \frac{a_n}{b_n}. \tag{5.6.13}$$

Euler's procedure giving an algorithm to evaluate a continued fraction is to define A_k and B_k by

$$\begin{aligned} A_{-1} &= 1 & B_{-1} &= 0 \\ A_0 &= b_0 & B_0 &= 1 \\ A_k &= b_k A_{k-1} + a_k A_{k-2} & B_k &= b_k B_{k-1} + a_k B_{k-2} \quad k \geqq 1. \end{aligned} \tag{5.6.14}$$

The A_k and B_k defined in (5.6.14) are related to the c_n in (5.6.13) as described in the following theorem.

Theorem 5.6.3 *If A_k and B_k are defined for $k \geqq 1$ by (5.6.14), and if c_n is defined in (5.6.13), then*

$$c_n = \frac{A_n}{B_n}. \tag{5.6.15}$$

Proof. We prove by induction on n the identity $c_n = A_n/B_n$ is valid for all n. The cases $n = 0$ and $n = 1$ are evident, by inspection. Thus, assuming that (5.6.15) holds for some $n \geqq 1$, we get the $n+1$ result replacing b_n in (5.6.13) by $b_n + a_{n+1}/b_{n+1}$. Since b_n does not occur in A_j or in B_j, $j \leqq n-1$, we find, replacing b_n by $b_n + a_{n+1}/b_{n+1}$ in the assumed valid identity

$$\begin{aligned}\frac{A_{n+1}}{B_{n+1}} &= \frac{(b_n + a_{n+1}/b_{n+1})A_{n-1} + a_n A_{n-2}}{(b_n + a_{n+1}/b_{n+1})B_{n-1} + a_n B_{n-2}} \\ &= \frac{b_{n+1}A_n + a_{n+1}A_{n-1}}{b_{n+1}B_n + a_{n+1}B_{n-1}}.\end{aligned} \tag{5.6.16}$$

This completes the proof. ■

Let us next summarize some of the above results relating continued fractions and series.

Theorem 5.6.4 *Let A_k and B_k be defined as in (5.6.14), so that*

$$c_n = \frac{A_n}{B_n} = b_0 + \frac{a_1}{b_1+}\,\frac{a_2}{b_2+}\,\cdots\,\frac{a_n}{b_n}. \tag{5.6.17}$$

Then

$$\begin{aligned}\frac{A_0}{B_0} &= b_0, \\ \frac{A_n}{B_n} - \frac{A_{n-1}}{B_{n-1}} &= \frac{(-1)^{n-1}}{B_n B_{n-1}} a_1 a_2 \cdots a_n, \quad n \geqq 1.\end{aligned} \tag{5.6.18}$$

Moreover,

$$\begin{aligned}\frac{A_n}{B_n} &= \frac{A_0}{B_0} + \sum_{k=1}^{n}\left(\frac{A_k}{B_k} - \frac{A_{k-1}}{B_{k-1}}\right) \\ &= b_0 + \sum_{k=1}^{n} \frac{(-1)^{k-1}}{B_{k-1}B_k} a_1 a_2 \cdots a_k.\end{aligned} \tag{5.6.19}$$

Given the sequences of numbers $\{A_k, B_k\}$, we can also readily recover the numbers a_k and b_k, via the simultaneous solution of each of the equations in the last line of (5.6.14).

Several procedures are known to convert an infinite series into a continued fraction. The simplest of these, readily verified via the use of the previous theorem, is the following. Let us set

$$c_n = \sum_{k=0}^{n} u_k, \quad n = 0, 1, 2, \cdots. \tag{5.6.20}$$

We wish to determine a_k and b_k, $k = 0, 1, \cdots, n$, so that if A_k and B_k are defined as in (5.6.14), then $c_n = A_n/B_n$, $n = 0, 1, 2, \cdots$. Since $A_0 = b_0$, and $B_0 = 1$, we have

$$\frac{A_0}{B_0} = b_0 = u_0. \tag{5.6.21}$$

Next, we have by (5.6.18), that

$$\frac{A_1}{B_1} - \frac{A_0}{B_0} = \frac{a_1}{B_0 B_1} = u_1. \tag{5.6.22}$$

If we also take $B_1 = 1$, then we clearly get $a_1 = u_1$. For convenience, let us attempt to take $B_j = 1$ for all j. Then, we have, by (5.6.18), that

$$(-1)^{k-1} a_1 a_2 \cdots a_k = u_k, \tag{5.6.23}$$

and solving these in turn yields the solution

$$a_k = -\frac{u_k}{u_{k-1}}. \tag{5.6.24}$$

Now combining this result with our above assumption of $B_k = 1$ for all k in the second expansion of (5.6.14), we get $b_k = 1 + a_k$ for all $k \geqq 1$, so that

$$b_k = 1 + \frac{u_k}{u_{k-1}}. \tag{5.6.25}$$

Summing up, we find, that if we define a_k and b_k by the equations

$$\begin{aligned} b_0 &= u_0 \\ a_1 &= u_1 \\ b_1 &= 1 \\ a_k &= -u_k/u_{k-1} \quad k = 2, 3, \cdots \\ b_k &= 1 + u_k/u_{k-1}, \quad k = 2, 3, \cdots \end{aligned} \tag{5.6.26}$$

then

$$b_0 + \frac{a_1}{b_1 +} \frac{a_2}{b_2 +} \cdots \frac{a_n}{b_n} = \sum_{k=0}^{n} u_k. \tag{5.6.27}$$

As is the case with Padé approximation, it is usually not possible to tell *a priori* when a continued fraction converges. The explicit expression (5.6.19) is often helpful toward this end. Results, such as that of the following theorem are known in some special cases.

Theorem 5.6.5 *The infinite continued fraction*

$$b_0 + \frac{1}{b_1 +} \frac{1}{b_2 +} \cdots \tag{5.6.28}$$

with the $b_i > 0$ for all i, converges if and only if the sum $\sum_{i=1}^{\infty} b_i$ diverges.

Proof. We leave the proof of this theorem to Problem 5.6.13 at the end of this Section. ∎

5.6.3 The Epsilon Algorithm and Aitken's Δ^2 Process

In this section we briefly present two other well-known algorithms to accelerate the convergence of a sequence $\{S_n\}_{n=0}^{\infty}$, namely, *Aitken's* Δ^2 *process*, and *Wynn's epsilon algorithm.*

The very popular Aitken Δ^2 process is usually taught to students in undergraduate numerical analysis. This algorithm enables us to generate a new sequence, $\{S'_n\}_{n=0}^{\infty}$ from the sequence $\{S_n\}_{n=0}^{\infty}$ by means of the formula

$$S'_n = S_{n+1} + \Delta\left(\frac{1}{\Delta S_n}\right) \tag{5.6.29}$$

with $\Delta u_n = u_{n+1} - u_n$. This algorithm is usually used to accelerate sequences $\{S_n\}_{n=0}^{\infty}$ converging to a limit S, and having the following property: *there exists a number* $c \neq 1$ *such that*

$$\lim_{n\to\infty} \frac{S_{n+1} - S}{S_n - S} = c. \tag{5.6.30}$$

This algorithm can also be used to generate a sequence of rational functions approximating a function f, as seen in the problems at the end of this section.

The epsilon algorithm was developed by Wynn as a means to easily compute the determinants in Shank's transformation, which transforms the sequence $\{S_n\}_{n=0}^{\infty}$ into a set of sequences $\{e_k(S_n)\}_{k,n=0}^{\infty}$ by means of the formula

$$e_k(S_n) = \frac{H_{k+1}(S_n)}{H_k(\Delta^2 S_n)} \tag{5.6.31}$$

with

$$H_k(u_n) = \begin{vmatrix} u_n & u_{n+1} & \cdots & u_{n+k-1} \\ u_{n+1} & u_{n+2} & \cdots & u_{n+k} \\ \vdots & \vdots & & \vdots \\ u_{n+k-1} & u_{n+k} & \cdots & u_{n+2k-1} \end{vmatrix}. \tag{5.6.32}$$

Using the epsilon algorithm, we develop a table of numbers $\{\varepsilon_k^{(n)}\}$ by means of the formulas

$$\begin{aligned} &\varepsilon_{-1}^{(n)} = 0, \quad \varepsilon_0^{(n)} = S_n, && n = 0, 1, \cdots, \\ &\varepsilon_{k+1}^{(n)} = \varepsilon_{k-1}^{(n+1)} + (\varepsilon_k^{(n+1)} - \varepsilon_k^{(n)})^{-1}, && k, n = 0, 1, \cdots. \end{aligned} \tag{5.6.33}$$

These numbers $\varepsilon_k^{(n)}$ are simply related to the numbers $e_k(S_n)$ defined in (5.6.31) by the expression

$$\varepsilon_{2k}^{(n)} = e_k(S_N), \quad k, n = 0, 1, \cdots. \tag{5.6.34}$$

Inspection of (5.6.29), (5.6.31) and (5.6.32) shows that when $k = 1$, then Shank's transformation reduces to the Aitken Δ^2 process, i.e., that the numbers generated by Aitken's Δ^2 process are just the numbers $\varepsilon_2^{(n)}$. The ε algorithm can also be used to generate some of the entries in the Padé table. For example, it may be shown (see Problem 5.6.11) that if we set $S_n = \sum_{j=0}^{n} c_j x^j$, then

$$[n/n + k] = \varepsilon_{2k}^{(n)}. \tag{5.6.35}$$

Experience has shown that Aitken's Δ^2 process and the ε algorithm are powerful procedures to compute the limits of sequences that alternate in sign. The Chebyshev acceleration algorithm (see Section 5.6.4) also seems to be aptly suited for this purpose. The Thiele algorithm of Section 5.6.5 is more suited to the computation of the limits of slowly convergent monotonic sequences.

5.6.4 Chebyshev Acceleration

Gustafson's Chebyshev Acceleration procedure is an effective way to compute slowly convergent oscillatory series. Let us assume that we want to compute the sum of a slowly convergent series, such as

$$S = \sum_{m=0}^{\infty} a_m z^m. \tag{5.6.36}$$

If $Q_n(t)$ is an arbitrary known polynomial of degree n in t, we may define a polynomial $P_{n-1}(t)$ of degree $n - 1$ in t, by the identity

$$\frac{1}{1 - tz} = P_{n-1}(t) + \frac{Q_n(t)}{(1 - tz)Q_n(1/z)}. \tag{5.6.37}$$

Evidently, for fixed t, $P_{n-1}(t)$ is a rational function in the z variable.

Hence, if the a_m in (5.6.36) are defined by

$$a_m = \int_0^1 t^m w(t)\, dt, \tag{5.6.38}$$

then we have

$$S = \int_0^1 \frac{w(t)}{1 - tz} dt = \int_0^1 P_{n-1}(t)\, w(t)\, dt + \varepsilon_n, \tag{5.6.39}$$

with the error ε_n given by

$$\varepsilon_n = \frac{1}{Q_n(1/z)} \int_0^1 \frac{Q_n(t)\, w(t)}{1 - tz} dt. \tag{5.6.40}$$

In practice, one attempts to select Q_n so that the error ε_n is small. An excellent choice of Q_n working especially well for slowly convergent and oscillating series is $Q_n(t) = T_n(2t-1)$, with T_n the Chebyshev polynomial of degree n. The method has been found to be particularly useful to sum slowly convergent Cardinal series expansions.

5.6.5 Thiele's Algorithm

The Thiele algorithm may be viewed with respect to Padé's approximation the same way Newton's, or Lagrange's polynomial is viewed with respect to Taylor's polynomial, in the sense that both Taylor's polynomial and Padé's approximation procedure get all of their data at a single point, whereas both Lagrange's polynomial and the rational function produced via Thiele's algorithm interpolate a function at all the distinct points given.

Let us now describe the *Thiele's algorithm* for constructing a rational function that interpolates at $m+1$ distinct points in the plane. Assuming that the given data is $\{(x_j, y_j)\}_{j=0}^{m}$, we define the entries ρ_i^j in the ρ-*Table* by means of the following equations:

$$\begin{aligned}
\rho_0^j &= y_j, & j &= 0, 1, \cdots, m; \\
\rho_1^j &= \frac{x_{j+1} - x_j}{\rho_0^{j+1} - \rho_0^j}, & j &= 0, 1, \cdots, m-1; \\
\rho_i^j &= \rho_{i-2}^{j+1} + \frac{x_{j+i} - x_j}{\rho_{i-1}^{j+1} - \rho_{i-1}^j}, & &\left\{ \begin{array}{lcl} j &=& 0, 1, \cdots, m-i; \\ i &=& 2, 3, \cdots, m. \end{array} \right.
\end{aligned} \tag{5.6.41}$$

Assuming that all of the numbers ρ_i^j in (5.6.41) exist, we may define a rational function $r(x)$ by means of the continued fraction representation:

$$r(x) = \rho_0^0 + \frac{x - x_0}{\rho_1^0 +} \; \frac{x - x_1}{\rho_2^0 - \rho_0^0 +} \; \cdots \; \frac{x - x_{m-1}}{\rho_m^0 - \rho_{m-2}^0}. \tag{5.6.42}$$

It is easily shown that $r(x_j) = y_j$. If $m = 2n$, with n an integer, then the function $r(x)$ has the form

$$r(x) = \frac{p_n(x)}{q_n(x)} \tag{5.6.43}$$

with both p_n and q_n polynomials of degree at most n in x, while if $m = 2n+1$, it has the form

$$r(x) = \frac{p_{n+1}^*(x)}{q_n^*(x)} \tag{5.6.44}$$

with p_{n+1}^* a polynomial of degree at most $n+1$ in x and q_n^* a polynomial of degree at most n in x.

The form of $r(x)$ is very important for extrapolating to the limit in the case $m = 2n$. In this case, we have

$$r(x) = \frac{\rho_{2n}^0 x^n + c_1 x^{n-1} + \cdots + c_n}{x^n + d_1 x^{n-1} + \cdots + d_n}, \tag{5.6.45}$$

with the c_j and d_j some constants, and with ρ_{2n}^0 the single entry in the last row of (5.6.41). We note, in this case, that

$$\rho_{2n}^0 = \lim_{x \to \infty} r(x), \tag{5.6.46}$$

i.e., if the y_j in (5.6.41) are values of a function $f(x)$; that is, if $y_j = f(x_j)$. Then, under suitable conditions, we may expect the approximation

$$\rho_{2n}^0 = f(\infty) \tag{5.6.47}$$

to be accurate. But what are these conditions?

Conditions of Applicability of Thiele Algorithm

In view of the results of Section 5.5 we may now summarize the conditions under which we may expect Thiele's algorithm to yield an accurate approximation of a function g at a boundary point b of its region of analyticity. By "accurate," we mean that using $\mathcal{O}(N)$ values of g, we can approximate $g(b)$ to within an $\mathcal{O}(e^{-cN^{1/2}})$ error, with c a positive constant, depending for practical purposes neither on N, nor on the particular set of given data $\{w_j, g(w_j)\}$.

- We can expect Thiele's algorithm to be accurate if f defined by $f(z) = g(z) - [g(0) + z\, g(\infty)]/(1+z)$ satisfies the conditions of Theorem 5.5.5, and if the points x_j belong to the sector $\mathcal{D} = \mathcal{D}_d^1$, as defined in Equation (1.7.10).

- The first step in the generalization of Thiele's algorithm is to observe that the algorithm can be applied after making a simple linear transformation. Let $0 < \omega_1 < \omega_2$, and let g be a function analytic in a sector $S = S(w_0, \omega_1, \omega_2) = \{w \in \mathbb{C} : \omega_1 < \arg(w - w_0) < \omega_2\}$. Let the given data $\{(w_j, g(w_j))\}$ be such that the w_j all lie in S, and assume that after making the transformation $w = w_0 + z\, e^{i\omega}$, with $\omega = (\omega_1 + \omega_2)/2$, the function $f(z) = g(z) - [g(0) + z\, g(\infty)]/(1+z)$ satisfies the conditions of Theorem 5.5.5, with $f(z) = g(w_0 + z\, e^{i\omega})$. Setting $x_j = (w_j - w_0)\, e^{-i\omega}$, we can expect to accurately approximate $g(\infty)$ by applying Thiele's algorithm to the data $\{(x_j, y_j)\} = \{(x_j, g(w_j))\}$.

- We next consider the use of Thiele's algorithm after applying a bilinear transformation, i.e., a transformation of the form $w(z) = (az + b)/(cz+d)$ (its inverse transformation being $z(w) = (Aw+B)/(Cw+D)$). Let the given data $\{(w_j, g(w_j))\}$ be such that the points w_j all lie in an "eye-shaped" region $\mathcal{D}$, whose boundary consists of two intersecting circular arcs, and the extrapolation problem is to use this data to approximate the value $g(b)$ at a corner point b of $\mathcal{D}$. Such an eye-shaped region $\mathcal{D}$ can always be transformed into the sector $\mathcal{D}_d^1$ by means of a transformation of the form $w(z)$, and moreover, $w(z)$ can be determined such that $z(b) = \infty$. Assume it is known that the transformed function $F(z) = g(w(z))$ is such that $f(z) = F(z) - [F(0) + z\,F(\infty)]/(1 + z)$ satisfies the conditions of Theorem 5.5.5. We may then expect to obtain an accurate approximation of $g(b)$ by applying Thiele's algorithm to the data $\{(x_j, y_j)\} = \{z(w_j), g(w_j)\}$.

- Finally, let us consider the use of Thiele's algorithm in the general setting of Definition 4.1.1. To this end, we assume, in the notation of Definition 4.1.1, that the given data $\{(w_j, g(w_j))\}$ is such that the points w_j all lie in a simply connected domain, $\mathcal{D}$, which is conformally mapped by ϕ onto $\mathcal{D}_d$, that $g \in \mathbf{M}_\alpha(\mathcal{D})$, and that the problem is to approximate $g(b)$, with $b = \psi(\infty) = \phi^{-1}(\infty)$. In view of Theorem 5.5.7, we may then apply Thiele's algorithm (5.6.41) to the data $\{(x_j, y_j)\} = \{(\rho(w_j), g(w_j))\}$, with $\rho = e^\phi$, to get an accurate approximation of $g(b)$.

Aitken–Neville vs. Thiele

Let us again make some remarks regarding the use of Aitken–Neville's algorithm versus the use of Thiele's algorithm. We propose, for purposes of optimal efficiency, that the choice should depend on our knowledge of the approximation of the function generating the data, if such knowledge is available. Let us assume that g is analytic in $\mathcal{D}$, and that the given data is of the form $\{(w_j, g(w_j))\}$, with each w_j in $\mathcal{D}$, and that we want to know the value of g at an interior point b of $\mathcal{D}$. Then, from Problem 2.4.3, we know that for any compact subregion $\mathcal{D}' \subset \mathcal{D}$, such that $b \in \mathcal{D}'$, there exists a sequence of polynomials $\{p_n(w)\}$, with p_n of degree n, such that for all $w \in \mathcal{D}'$, we have

$$\sup_{w \in \mathcal{D}'} |g(w) - p_n(w)| \leqq C e^{-cn}, \tag{5.6.48}$$

with C and c positive constants, independent of n. Hence, by applying an n-point Aitken–Neville algorithm, we can expect to approximate $g(b)$ to within a maximum error $C' e^{-cn}$, with c as in (5.6.48), and with C' independent of n. On the other hand, whereas there may in fact exist a sequence of rational functions which converges to g in $\mathcal{D}'$ substantially more rapidly than a sequence of polynomials, we know by Theorem 5.5.11 that this need not necessarily be the case. Indeed, our experiments have shown that Thiele's algorithm nearly always requires more work to approximate $g(b)$ to the same accuracy as that achievable by Aitken–Neville's algorithm, whenever b is a point within the region of analyticity of g. However, if b is on the boundary of the region of analyticity of g, then, as the results of Section 5.5 show, rational approximation can have a marked advantage over polynomial approximation (see for example, Problem 2.1.4 and and the problems at the end of this section). In these situations it is much more advantageous to use Thiele's algorithm than Aitken–Neville's algorithm.

Tailoring Thiele to a Sequence

Let us now discuss the choice of an extrapolation method to evaluate the limit of a given sequence $\{S_j\}_{j=0}^{\infty}$. We assume, of course, that the limit

$$S = \lim_{j \to \infty} S_j \tag{5.6.49}$$

exists and is finite. Now, let α be a positive number, and let

$$S_j - S \equiv c_j = \mathcal{O}(e^{-j\alpha}), \quad j \to \infty. \tag{5.6.50}$$

We know, by Theorem 3.1.8, that there exists a function $g \in \mathbf{L}_\alpha(\mathcal{D}_d)$, such that $g(j) = c_{|j|}$, $j = 0, 1, 2, \cdots$. By Example 5.5.10, we may accurately approximate the function $f \equiv g + S$ on $\mathbb{R}$ by rationals in e^x, and hence we may effectively approximate $S = f(\infty)$ by Thiele's algorithm (5.6.41), in which we take $\{(x_j, y_j)\} = \{(e^j, S_j)\}$. Incidentally, Aitken–Neville's algorithm (in which we evaluate $p(0)$ using the sequence $\{(x_j, p(x_j)\} = \{(1/(j+1), S_j)\}$, the Aitken Δ^2 process, and the epsilon algorithm (which are both applied directly to the sequence $\{S_j\}$), as well as the Chebyshev algorithm (in which we can take $z = e^{-\alpha}$) are all accurate to evaluate the limit of this sequence.

Let us also discuss explicitly the important case of $\rho(x) = x$. We suppose, again, that the sequence $\{S_j\}$ has a limit S, and that it satisfies (5.6.50). Then, of course, we know, from Theorem 4.2.14 that there exists a function $g \in \mathbf{L}_\alpha(\mathcal{D})$, with $\mathcal{D} = \mathcal{D}_d^1$ defined in (1.7.10), such that $g(e^j) = c_{|j|} = S - S_j$. By Theorem 5.5.7, the function $f(x) = g(x) + S$ which belongs to $\mathbf{M}_\alpha(\mathcal{D})$ may thus be accurately approximated by a rational function of x on $[0, \infty]$, and hence Thiele's algorithm (5.6.41) above, in which we take $\{(x_j, y_j)\} = \{(e^j, S_j)\}$, may be used to accurately approximate $S = f(\infty)$. Now, as a consequence of the fact that functions in $\mathbf{M}_\alpha(\mathcal{D}_d^1)$ can be accurately approximated by rational functions in x, Thiele's algorithm is accurate for such functions, even though the points x_j in Thiele's algorithm are not Sinc points. That is, Thiele's algorithm, in which we take $\{(x_j, y_j)\} = \{(j, S_j)\}$, will accurately compute the limit S even for sequences $\{S_j\}$ that are *monotonic*, and satisfy, instead of (5.6.50), a relation such as

$$\frac{1}{S_j - S} \sim c j^\alpha, \quad j \to \infty, \tag{5.6.51}$$

with c some constant which is not necessarily known.

Let us next discuss the evaluation of limits of sequences $\{S_j\}$ which satisfy neither of the order relations (5.6.50) or (5.6.51). Let us assume for sake of simplicity, that the given sequence $\{S_j'\}$ is positive, and monotonically decreasing, and that it is sufficiently *well-behaved*, such that we have an idea of the rate at which it approaches its limiting value S'. Specifically, let us assume that we know explicitly a strictly increasing function u analytic in the interior of some region $\mathcal{D}_d^1$ and which provides a strictly monotonic map from $[a, \infty]$ onto $[b, \infty]$, with $a \in \mathbb{R}$ and $b > 0$, such that

$$u\left(\frac{1}{S_j' - S'}\right) \sim c\, j^{\alpha}, \quad j \to \infty \tag{5.6.52}$$

with c some (not necessarily known) constant. Evidently then, the sequence $\{S_j\}$, with $S_j = S' + [u(1/(S_j' - S'))]^{-1}$ satisfies the relation (5.6.51) above; i.e., we can use Thiele's algorithm (5.6.41), taking $\{x_j, y_j)\} = \{(j, S_j)\}$, to accurately compute the limit of the sequence $\{S_j\}$. This means there is a rational function of x, which on an interval $[a, \infty]$ accurately approximates the values S_j at $x = j$. Assuming that we also know explicitly the inverse function $v = u^{-1}$, however, implies that this rational is a rational function of the variable $v(x)$ which accurately approximates S_j' at $v(j)$. By the argument of the previous paragraph, we may therefore identify the data $\{(v(j), S_j)\}$ with coordinate values of a function belonging to $\mathbf{M}_\alpha(\mathcal{D}_d^1)$, and we may therefore apply the Thiele algorithm (5.6.41) to the data $\{(x_j, y_j)\} = \{v(j), S_j')\}$ to approximate S'. For example, if $S_j - S = \log(j+1)$, then we could take $u(x) = e^x - 2$, and $v(x) = \log(x+2)$.

Numerical Stability

Next, we will discuss the influence of the given data $\{(x_j, y_j)\}$ on the numerical outcome of the polynomial or rational function extrapolation methods discussed above. It is a simple fact that there exists *bad data* which, due to ill conditioning, can destroy the accuracy of the results of polynomial and rational extrapolation to the limit. Based on our experience, if the data is not contaminated with noise, then this type of failure almost never occurs for problems stemming from applications, provided these algorithms are used properly, as described above. That is, for nearly all practical problems involving the use of Thiele's algorithm to extrapolate the limit, the number of data points required to achieve a desired accuracy is sufficiently small so that the computations have been completed before ill conditioning takes over and destroys the accuracy. Hence, although the algorithms may be ill conditioned, the effect of this ill conditioning is almost always harmless.

There is, nevertheless, a problem regarding the termination of Thiele's algorithm. Typically, when Thiele's algorithm is applied, and convergence occurs in one of the odd rows of the Thiele table, several, but not all, of these entries will be excellent approximations to the desired limit. This row of numbers in the Thiele table will typically have a form such as

$$u_1,\ u_2,\ u_3,\ v_4,\ u_5,\ v_6,\ u_7,\ u_8,\ u_9,\ u_{10},\ v_{11} \tag{5.6.53}$$

with the numbers u_i all close to one another, whereas the numbers v_j differ by orders of magnitude from the u_i. A good automated algorithm would eliminate the v_j and pick the best of the u_i as the desired limit. This can, in fact, be achieved via a simple process of ℓ^1-averaging, which follows:

Given a set of m real numbers $a_1,\ a_2,\ \cdots,\ a_m$, the ℓ^1-average of these numbers is the number $\overline{a}$ defined by

$$\overline{a} = \min_{a \in \mathbf{R}} \sum_{i=1}^{m} |a - a_i|. \tag{5.6.54}$$

Now assume that the a_j are ordered, such that $a_1 \leqq a_2 \leqq \cdots \leqq a_m$. Then $\overline{a}$, the ℓ^1-average of these numbers is given by

$$\overline{a} = \begin{cases} a_n & \text{if} \quad m = 2n-1 \\ t & \text{if} \quad m = 2n, \quad \text{for arbitrary } t \in [a_n, a_{n+1}]. \end{cases} \tag{5.6.55}$$

If the a_i are complex, then the determination of $\overline{a}$ in (5.6.54) is a nonlinear problem which is not easily tractable, and for practical purposes, to get an accurate approximation of the expected limit in the Thiele algorithm, it suffices to take separate ℓ^1-averages of both the real and the imaginary parts of the a_i, and to let these results be the respective real and imaginary parts of the extrapolated limit.

The process of ℓ^1-averaging described above can then be used effectively to determine the extrapolated limit, via Thiele's algorithm. We first form the ℓ^1-average of a particular odd row of the Thiele table which contains at least p entries (say $p \geqq 5$). Letting ε be some positive number, we terminate the extrapolation process, taking the ℓ^1-average as the extrapolated limit, whenever more than half (say) of the entries in a row are within ε of the ℓ^1-average of the numbers in that row. We have tested this procedure in applications, and found it to be effective.

Example 5.6.6 The integral

$$f(\infty) \equiv \int_0^\infty t^{-1/\pi} \cos\left[\left(t^6 + t\right)^{1/6}\right] dt \tag{5.6.56}$$

is not easily approximated by any known quadrature method. On the other hand, for any $z \in \Omega^+$, with Ω^+ the right-half plane, the integral

$$f(z) \equiv \int_0^\infty e^{-t/z} t^{-1/\pi} \cos\left[\left(t^6 + t\right)^{1/6}\right] dt \tag{5.6.57}$$

may be accurately approximated by use of the "log sinh" quadrature formula (4.2.61) (Why?). Clearly, f defined by (5.6.57) belongs to $\mathbf{Hol}(\Omega^+)$, and moreover, it may be easily shown that for all z sufficiently large, and $z \in \Omega^+$, we have

$$|f(z) - f(\infty)| \leqq C|z|^{-1/4}. \tag{5.6.58}$$

Hence we may use Thiele's algorithm on a few accurately evaluated values of $f(z)$ to accurately approximate $f(\infty)$. On the other hand polynomial extrapolation (at points $1/z$) fails, since $f(1/z)$ has a singularity at $1/z = 0$.

Other examples which illustrate the power of this procedure may be found in the problems at the end of this section.

•

PROBLEMS FOR SECTION 5.6

5.6.1 Prove that if $r(x)$ is defined as in (5.6.42), then $r(x_j) = y_j$.

5.6.2 Verify the representations (5.6.42) through (5.6.44).

5.6.3 Let f be defined by

$$f(x) = (x+1)^{1/2}(1-x)^{-1/3}\log[(1+x)/2].$$

Let z_j be defined as in (5.5.64), and let $4N+1$ points (x_j, ρ_0^j) be defined by the formulas

$$\begin{aligned}
(x_{2j}, \rho_0^{2j}) &= (z_{-N+j}, \frac{1 - z_{-N+j}^2}{f(z_{-N+j})}) && j = 0, 1, \cdots, 2N \\
(x_{2j-1}, \rho_0^{2j-1}) &= (1/z_{-N+j-1}, 0) && j = 0, 1, \cdots, N \\
(x_{2j-1}, \rho_0^{2j-1}) &= (1/z_{-N+j}, 0) && j = 0, 1, \cdots, N
\end{aligned}$$

(a) Use these $4N+3$ values to determine a rational function $r(x)$, via the Thiele algorithm.

(b) Show that if $r(x)$ is determined as in (a), then

$$R(x) = \frac{1-x^2}{r(x)} = \sum_{j=-N}^{N} \frac{f(z_j)\, B(x)}{(x - z_j)\, B'(z_j)},$$

with $B(x)$ is defined as in (5.5.29), with $M = N$.

(c) Use this procedure to accurately approximate $f(x)$, for $x = -0.9,\ -0.8,\ -0.7, \cdots, 0,\ 0.1,\ \cdots, 0.9$.

5.6.4 Let f be defined by (5.6.57). Prove the relation (5.6.58).

5.6.5 Let g be given by

$$g(x) = x^{-1/e}(1-x)^{-\pi/2}\log x,$$

and set

$$f(N) = \frac{1}{N} \sum_{j=1}^{N} g\left(\frac{2j-1}{2N}\right).$$

(a) How large does N have to be, so that $|f(\infty) - f(N)| < 10^{-9}$?

(b) Evaluate $f(10)$, $f(12)$, $f(14)$, $f(16)$, $f(18)$, $f(20)$, $f(22)$, and use these values in the Thiele algorithm to approximate $f(\infty)$.

(c) Evaluate $f(2^k)$, $k = 3,\ 4,\ 5,\ 6,\ 7,\ 8,\ 9$, and use these values in the Thiele algorithm to approximate $f(\infty)$.

(d) Compare the procedures used in (b) and (c), and explain why both procedures work so well, whereas polynomial extrapolation (at either the points $1/10,\ 1/12,\ \cdots,\ 1/22$, or else the points 2^{-k}, $k = 3,\ 4,\ \cdots,\ 9$) fails.

(e) Evaluate $\int_0^1 g(x)\,dx$ via Equation (4.2.37), taking $a = 0$, $b = 1$, $\alpha = 1 - 1/e$, $d = \pi$, and for $M = N = 20,\ 40,\ 60$.

(f) How would you evaluate $f(\infty)$?

5.6.6 Define $f(N)$ by

$$f(N) = \sum_{j=1}^{N} \frac{1}{1 + j^{4/3}}.$$

(a) How large does N have to be, so that $|f(N) - f(\infty)| < 10^{-9}$?

(b) Evaluate $f(N)$ for $N = 10,\ 12,\ 14,\ 16,\ 18,\ 20,\ 22$, and use these values in the Thiele algorithm to approximate $f(\infty)$.

(c) How accurate is the the result obtained in (b) ?

(d) Explain why the numerical procedure used in (b) works so well, whereas polynomial extrapolation (at the points $1/N$) fails.

5.6.7 Let $y(x)$ denote the solution of the differential equation problem

$$\begin{aligned} y' = u(x)\,y + v(x), \quad 0 < x < 1, \\ y(0) = 0, \end{aligned}$$

with $u(x) = (1 - x)^{-1/3}/(1 + x^2)$, and $v(x) = (1 - x)^{-1/4}$.

(a) Let $f(N) = y_N$, with y_j determined for $j = 1,\ 2,\ \cdots,\ N$ by the scheme

$$y_j = [1 + u((j-1)/N)]y_{j-1} + v((j-1)/N).$$

and prove that $y(1) = \lim_{(N \to \infty)} f(N)$.

(b) Use the scheme in part (a) of this problem to evaluate $f(N)$ for $N = 4,\ 8,\ 16,\ 32,\ 64$, and then use these values in the Thiele algorithm to get an accurate approximation of $f(\infty) = y(1)$.

(c) Explain why the method used in part (b) of this problem works so well, whereas polynomial extrapolation at the points $1/N$, and the classical Romberg scheme fail.

5.6.8 Given the sequence of numbers $A_0,\ B_0,\ A_1,\ B_1,\ \cdots$, with $B_0 = 1$, show that, by (5.6.16), we can determine the a_i (for $i \geqq 1$) and the b_i for $i \geqq 0$ by means of the equations

$$b_0 = A_0, \quad a_1 = A_1 - B_1 A_0, \quad b_1 = B_1$$

$$a_i = \frac{A_i B_{i-1} - A_{i-1} B_i}{A_{i-2} B_{i-1} - A_{i-1} B_{i-2}}, \quad b_i = \frac{A_{i-2} B_i - A_i B_{i-2}}{A_{i-2} B_{i-1} - A_{i-1} B_{i-2}}, \quad i \geqq 2.$$

5.6.9 Define a sequence of numbers p_i by the formulas

$$p_1 = \frac{a_1 z}{b_1}, \quad 1 + p_2 = \left[1 + \frac{a_2 z}{b_1 b_2}\right]^{-1},$$

$$1 + p_i = \left[1 + \frac{a_i z (1 + p_{i-1})}{b_{i-1} b_i}\right]^{-1}, \quad i = 3, 4, \cdots.$$

(a) Establish the identity

$$\frac{a_1 z}{b_1 +}\ \frac{a_2 z}{b_2 +} \cdots \frac{a_n z}{b_n} = \sum_{i=1}^{n} \prod_{j=1}^{i} p_j .$$

(b) Define $A_1,\ B_1,\ A_2,\ B_2$, so that the recurrence relation $B_i = b_i B_{i-1} + a_i z\, B_{i-2}$ holds, and such that

$$\frac{A_n}{B_n} = \sum_{i=1}^{n} \prod_{j=1}^{i} p_j .$$

(c) Use this result to verify the following identities:

$$\frac{B_{i-1}}{B_{i-2}} = -\frac{a_i z}{b_i} \left\{ \frac{1}{p_i} + 1 \right\}$$

$$\frac{B_{i-2}}{B_{i-1}} = \frac{1}{b_i} \left\{ 1 + p_{i-1} \right\}.$$

5.6.10 (a) Verify Equation (5.6.45).

(b) How can Aitken's Δ^2 process be used to generate a rational function ?

5.6.11 Let a sequence $\{S_n\}$ be defined by

$$S_n = S + \sum_{j=1}^{M} c_j w_j^n,$$

with c_j and w_j as arbitrary complex numbers, and with w_j distinct.

(a) Prove that $\varepsilon_{2M}^{(n)} = S$.

(b) Prove that if $|w_k| > |w_{k+1}|$ for all non-negative integers k, then $\varepsilon_{2k}^{(n)} - S = \mathcal{O}(w_{k+1}^n)$, as $n \to \infty$.

5.6.12 Verify that $P_{n-1}(t)$ defined as in (5.6.37) is indeed a polynomial of degree $n-1$ in t, and is a rational function of z.

5.6.13 Prove Theorem 5.6.5.

5.6.14 Prove the result of Equation (5.6.35).

5.7 Heaviside and Filter Rationals

Modern design of electronic filters, such as low-pass, high-pass, and band-pass filters includes the construction of a rational function which satisfies the desired specifications for cutoff frequencies, pass-band gain, transition bandwidth, and stop-band attenuation. Thus elementary and explicit methods to construct rational function approximations to the χ^*-function ($\chi^*(w) = 1$ if $-1 < w < 1$, 0 if $|w| > 1$) and the sgn function ($\operatorname{sgn} w = 1$ if $w > 1, 0$ if $w = 0, -1$ if $w < 0$) are important in applications. Known elementary analytical methods lead to rational functions, such as

$$\frac{1}{1 + \varepsilon^2 T_n^2(w)} \quad \text{or} \quad \frac{1}{1 + w^{2n}}$$

with $T_n(w)$ the nth degree Chebyshev polynomial. More advanced methods use elliptic functions to derive a rational function approximation which has equi-ripple characteristic in both the pass and stop bands, but these Solotareff-type methods require the evaluations of elliptic functions.

In this section we shall derive rational function approximations, for the Heaviside, the filter, and the impulse functions. Our main motivation for these derivations are the simplicity of the results; the formulas depend on a single parameter, N, which determines the degree and accuracy. Another advantage of the formulas are their simple relationships with one another; letting $H_N(w)$ denote the approximation of the Heaviside function, we have $\chi_N(w) = H_N[(1+w)/(1-w)]$ the approximation to the χ^* function, and $H_N'(w)$, the derivative of $H_N(w)$, the approximation to the impulse (delta) function.

5.7.1 Approximation of the Heaviside Function

Let h be a positive number, and for $a \in \mathbb{R}$, set

$$\operatorname{sgn} a = \begin{cases} 1 & \text{if} \quad a > 0 \\ 0 & \text{if} \quad a = 0 \\ -1 & \text{if} \quad a < 0 \end{cases} \tag{5.7.1}$$

Let us define $\tilde{H}(w)$, $H(w)$, and $H_N(w)$, for $w \in \mathbb{R}$, by

$$\tilde{H}(w) = \frac{1}{2} + \frac{1}{2}\operatorname{sgn} w \tag{5.7.2}$$

$$H(w) = \frac{1}{2} + \frac{hw}{\pi} \sum_{k=-\infty}^{\infty} \frac{e^{(k+1/2)h}}{w^2 + e^{(2k+1)h}} \tag{5.7.3}$$

and

$$H_N(w) = \frac{1}{2} + \frac{hw}{\pi} \sum_{k=-N}^{N-1} \frac{e^{(k+1/2)h}}{w^2 + e^{(2k+1)h}} \tag{5.7.4}$$

where in (5.7.4), N is a positive integer.

Let us note at the outset, that if G denotes any one of the three functions $\tilde{H}, H$, or H_N then

$$G\left(\frac{1}{w}\right) = G(w) \tag{5.7.5}$$

and

$$G(-w) + G(w) = 1. \tag{5.7.6}$$

The first result which we shall prove is

Theorem 5.7.1 *If $h > 0$ and $w \in \mathbb{R}$, then*

$$|\tilde{H}(w) - H(w)| \leq \sum_{n=1}^{\infty} \frac{1}{\cosh\left(\frac{n\pi^2}{h}\right)} < \frac{2e^{-\pi^2/h}}{1 - e^{-\pi^2/h}} \tag{5.7.7}$$

Proof. We apply Poisson's summation formula (1.6.25), taking

$$f(t) = \frac{we^{t+h/2}}{\pi[w^2 + e^{2t+h}]} \tag{5.7.8}$$

then we have, as shown in Problem 1.4.5,

$$F(x) = \int_{\mathbb{R}} e^{ixt} f(t)\, dt = \frac{e^{ix \log|we^{-h/2}|}}{2\cosh(\pi x/2)} \operatorname{sgn}(w) \tag{5.7.9}$$

and so

$$\frac{hw}{\pi} \sum_{m=-\infty}^{\infty} \frac{e^{(m+1/2)h}}{w^2 + e^{(2m+1)h}} = \frac{1}{2} \operatorname{sgn} w \sum_{n=-\infty}^{\infty} \frac{e^{2i(n\pi/h)\log|we^{-h/2}|}}{\cosh(\pi^2 n/h)}. \tag{5.7.10}$$

Hence, in view of the notation of (5.7.2) and (5.7.3), we have, using (5.7.10), that

$$|H(w) - \tilde{H}(w)| \leqq \sum_{n=1}^{\infty} \frac{1}{\cosh\left(\frac{\pi^2 n}{h}\right)} \leqq \frac{2e^{-\pi^2/h}}{1 - e^{-\pi^2/h}} \cong 2e^{-\pi^2/h}. \tag{5.7.11}$$

■

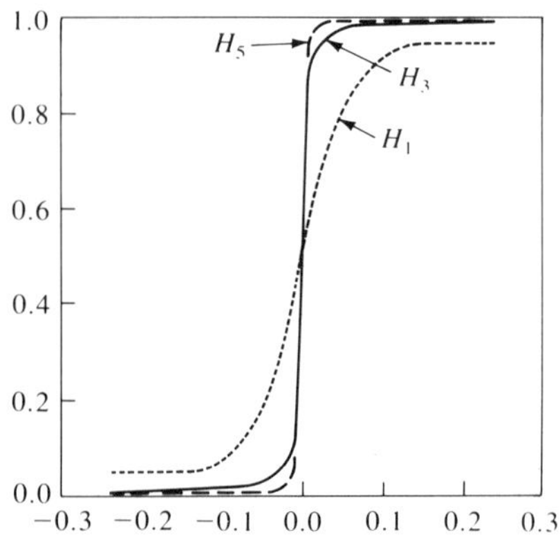

Figure 5.7.1: The Rational Heaviside Function

We now estimate the convergence rate of the series defining $H(w)$. To this end, we set $\delta_N(w) = H(w) - H_N(w)$, with $H_N(w)$ defined by (5.7.4), so that

$$\delta_N(w) = \frac{hw}{\pi} \sum_{|n| \geqq N} \left\{ \frac{e^{(n+1/2)h}}{w^2 + e^{(2n+1)h}} + \frac{e^{(n+1/2)h}}{1 + w^2 e^{(2n+1)h}} \right\}. \tag{5.7.12}$$

It then follows that

$$\delta_N(1/w) = \delta_N(w) \tag{5.7.13}$$

and

$$\delta_N(-w) = -\delta_N(w). \tag{5.7.14}$$

Theorem 5.7.2 *Let $\tilde{H}(w)$, $H(w)$ and $H_N(w)$ be defined by (5.7.3) through (5.7.5). Then, for $w \in \mathbb{R}$, we have*

$$|H_N(w) - H(w)| < \frac{1}{\pi}(|w| + \frac{1}{|w|})e^{-Nh} \tag{5.7.15}$$

$$|\tilde{H}(w) - H_N(w)| < \frac{1}{\pi}\left(|w| + \frac{1}{|w|}\right)e^{-Nh} + 2e^{-\pi^2/h}. \tag{5.7.16}$$

Moreover, taking

$$h = \frac{\pi}{N^{1/2}} \tag{5.7.17}$$

we have

$$|H_N(w) - \tilde{H}(w)| < \frac{1}{\pi}\left[|w| + 2\pi + \frac{1}{|w|}\right]e^{-\pi N^{1/2}}. \tag{5.7.18}$$

Proof. From the relations (5.7.13) and (5.7.14), we see that it suffices to consider the case $w \geqq 1$. In this case we have

$$\begin{aligned} H(w) &= H_N(w) + \frac{hw}{\pi}\sum_{m=N}^{\infty}\left\{\frac{e^{(m+1/2)h}}{w^2 + e^{(2m+1)h}} + \frac{e^{(m+1/2)h}}{1 + w^2 e^{(2m+1)h}}\right\} \\ &< H_N(w) + \frac{h}{\pi}\sum_{m=N}^{\infty}\left(w + \frac{1}{w}\right)e^{-(m+1/2)h} \\ &= H_N(w) + \frac{1}{\pi}\left(w + \frac{1}{w}\right)\frac{h/2}{\sinh(h/2)}e^{-Nh} \\ &< H_N(w) + \frac{1}{\pi}\left(w + \frac{1}{w}\right)e^{-Nh}. \end{aligned} \tag{5.7.19}$$

This proves (5.7.15). The result (5.7.16) comes from (5.7.11) and (5.7.15). Finally (5.7.18) results from substituting (5.7.17) into (5.7.16). ■

Let a be > 1, let us set

$$\varepsilon = \frac{1}{a + \sqrt{a^2 - 1}}. \tag{5.7.20}$$

Hence if $-1/\varepsilon < w < -\varepsilon$ or else if $\varepsilon < w < 1/\varepsilon$, then $|w| + 1/|w| < 2a$. This establishes

Corollary 5.7.3 *If $\tilde{H}$ and H_N are defined by (5.7.2) and (5.7.4) and if h is given by (5.7.17), then for $-1/\varepsilon < w < -\varepsilon$ or for $\varepsilon < w < 1/\varepsilon$, with ε defined by (5.7.20), with $a > 1$, we have*

$$|H_N(w) - \tilde{H}(w)| < \left(\frac{2a + 2\pi}{\pi}\right)e^{-\pi N^{1/2}}. \tag{5.7.21}$$

5.7.2 Approximation of the Filter Function

Let us replace w in (5.7.2), (5.7.3), and (5.7.4) by $(1+w)/(1-w)$, and let us denote the new functions by $\tilde{\chi}(w)$, $\chi(w)$ and $\chi_N(w)$, i.e. we set

$$\tilde{\chi}(w) = \begin{cases} 0 & \text{if} \quad w < -1 \\ \frac{1}{2} & \text{if} \quad w = -1 \\ 1 & \text{if} \quad -1 < w < 1 \\ \frac{1}{2} & \text{if} \quad w = 1 \\ 0 & \text{if} \quad w > 1, \end{cases} \tag{5.7.22}$$

$$\chi(w) = \frac{1}{2} + \frac{h}{\pi}\left(1-w^2\right) \sum_{k=-\infty}^{\infty} \frac{e^{(k+1/2)h}}{(1+w)^2 + (1-w)^2 e^{(2k+1)h}} \tag{5.7.23}$$

and

$$\chi_N(w) = \frac{1}{2} + \frac{h}{\pi}\left(1-w^2\right) \sum_{k=-N}^{N-1} \frac{e^{(k+1/2)h}}{(1+w)^2 + (1-w)^2 e^{(2k+1)h}}. \tag{5.7.24}$$

Then, paraphrasing Theorems 5.7.1 and 5.7.2, we have

Theorem 5.7.4 *Let $\tilde{\chi}$, χ, and χ_N be defined by (5.7.22), (5.7.23), and (5.7.24) as above. Then for $h > 0$, and N a positive integer, we have for all $w \in \mathbb{R}$,*

$$|\tilde{\chi}(w) - \chi(w)| \leqq \sum_{n=1}^{\infty} \frac{1}{\cosh(\frac{n\pi^2}{h})} < \frac{2e^{-\pi^2/h}}{1-e^{-\pi^2/h}} \cong 2e^{-\pi^2/h}, \tag{5.7.25}$$

$$|\chi(w) - \chi_N(w)| \leqq \frac{2}{\pi}\left|\frac{1+w^2}{1-w^2}\right| e^{-Nh}, \tag{5.7.26}$$

and

$$|\tilde{\chi}(w) - \chi_N(w)| \leqq \frac{2}{\pi}\left|\frac{1+w^2}{1-w^2}\right| e^{-Nh} + 2e^{-\pi^2/h}. \tag{5.7.27}$$

Moreover, taking h as in (5.7.17), we have

$$|\tilde{\chi}(w) - \chi_N(w)| \leqq \frac{2}{\pi}\left[\frac{(1+w^2)}{|1-w^2|} + \pi\right] e^{-\pi N^{1/2}}. \tag{5.7.28}$$

Note, that whereas the bound on the differences (5.7.15), (5.7.16), and (5.7.18) is unbounded for $w = 0$ and $w = \infty$, the corresponding differences (5.7.26), (5.5.27), and (5.7.28) are unbounded for $w = \pm 1$. Paraphrasing Corollary 5.7.3 we also have

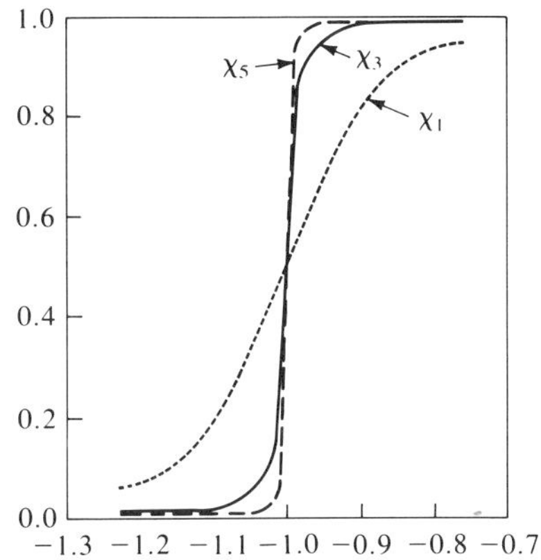

Figure 5.7.2: The Rational Filter Function

Corollary 5.7.5 *Let* $\tilde{\chi}$ *and* χ_N *be defined by (5.7.22) and (5.7.24) respectively, and let* h *be given by (5.7.17). Let* $a > 1$, *let* δ *be determined by the expression*

$$\delta = \left(\frac{a-1}{a+1}\right)^{1/2}. \tag{5.7.29}$$

If either $-\delta \leqq w < \delta$, *or if* $w \in \mathbb{R}$ *and* $1/\delta < |w| < \infty$, *then*

$$|\tilde{\chi}(w) - \chi_N(w)| \leqq \frac{2}{\pi}[a+\pi]e^{-\pi N^{1/2}}. \tag{5.7.30}$$

5.7.3 Approximation of the Delta Function

The derivative of $\tilde{H}$ defined in (5.7.2) is the Dirac delta function, i.e., $\tilde{H}'(w) = \delta(w)$, which satisfies the relation

$$\int_{\mathbb{R}} f(w)\tilde{H}'(w)dw = f(0) \tag{5.7.31}$$

for all functions f continuous on $\mathbb{R}$. We may expect, then, that $H_N'(w)$ is also a close approximation to $\tilde{H}'(w)$, in the sense of (5.7.31). We shall see below that this is in fact the case, provided f is not only continuous, but also bounded and absolutely integrable over $\mathbb{R}$. Indeed, if f is not absolutely integrable over $\mathbb{R}$ we can get some rather disturbing results, such as, for example, by taking $f(w) \equiv 1$, for all $w \in \mathbb{R}$, we get

$$\int_{\mathbb{R}} f(w)H_N'(w)dw = \int_{\mathbb{R}} H_N'(w)dw = 0. \tag{5.7.32}$$

Although here we shall only study the approximation of the distribution $\tilde{H}'$ via H_N', we mention that for $k > 1$, $H_N^{(k)}$ also provides an accurate approximation of the distribution $\tilde{H}^{(k)}$.

We shall see below that $H'_N(0)$ is very large, and $H_N(w)$ is very small for $w \in D_N = \{w \in \mathbb{R} : R_N^{-1} < |w| < R_N\}$, with $R_N > 1$ and moreover $H'_N(0)$ and R_N increase very rapidly as N increases. On the other hand, $H'_N(w)$ becomes negative in the region $|w| > R_N$ and as $|w|$ increases in this region, $H'_N(w)$ first increases in magnitude, and then it approaches zero very slowly. One may thus ask about the behavior of $H'_N(w)$, e.g., whether there exists a sequence of numbers $\{\alpha_n\}$ such that $\alpha_n H'_N(\alpha_n w)$ approaches a fixed function as $N \to \infty$.

Previous statements of the above observations, and answers to the above questions are given in the following

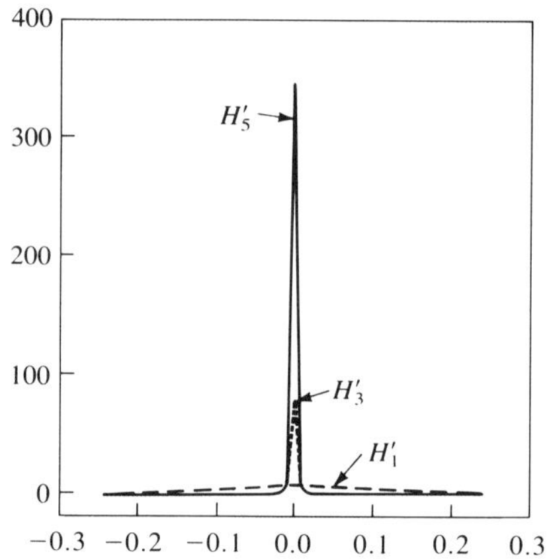

Figure 5.7.3: The Rational Delta Function

Theorem 5.7.6 *Let H_N be defined as in (5.7.4). Then H'_N, the derivative of H_N, is given by*

$$H'_N(w) = \frac{h}{\pi} \sum_{m=-N}^{N-1} e^{(m+1/2)h} \frac{e^{(2m+1)h} - w^2}{(e^{(2m+1)h} + w^2)^2}. \tag{5.7.33}$$

Setting

$$\alpha_N = e^{-\pi N^{1/2}}, \quad \beta_N = e^{\pi N^{1/2}} \tag{5.7.34}$$

and letting h be given by (5.7.17), we have, as $N \to \infty$

$$\alpha_N H'_N(\alpha_N w) = \frac{1}{\pi(1+w^2)} + \mathcal{O}(1/N^{1/2}) \tag{5.7.35}$$

uniformly, for $w \in \mathbb{R}$, and also,

$$\beta_N H'_N(\beta_N w) = -\frac{1}{\pi(1+w^2)} + \mathcal{O}(1/N^{1/2}) \tag{5.7.36}$$

uniformly for $w \in \mathbb{R} \setminus [-\varepsilon, \varepsilon]$, with ε an arbitrary positive number. Furthermore, if f is continuous and bounded on $\mathbb{R}$, and if $\int_{\mathbb{R}} |f(w)| dw < \infty$, then

$$\int_{\mathbb{R}} H_N(w) f(w) dw$$

$$= f(0) + \mathcal{O}\left(\omega(f; e^{-\frac{\pi}{2}N^{1/2}})\right) + \mathcal{O}(N^{-1/2}), \quad N \to \infty \tag{5.7.37}$$

with $\omega(f;\delta)$ *the modulus of continuity of* f*:*

$$\omega(f;\delta) = \sup_{x\in\mathbb{R}, h\in\mathbb{R}, |h|<\delta} |f(x+h) - f(x)|. \tag{5.7.38}$$

Proof. If α_N is given by (5.7.34), then we have

$$\alpha_N H_N'(\alpha_N w) = \frac{1}{\sqrt{N}} \sum_{k=1}^{2N} f\left\{\frac{2k-1}{2\sqrt{N}}; w\right\}, \tag{5.7.39}$$

with

$$f(x;w) = e^{\pi x} \frac{e^{2\pi x} - w^2}{(e^{2\pi x} + w^2)^2}. \tag{5.7.40}$$

Thus (5.7.39) is a Riemann approximation of the integral

$$\int_0^{2\sqrt{N}} f(x;w)\, dx = \frac{1}{\pi}\left(\frac{1}{1+w^2} - \frac{e^{2\sqrt{N}\pi}}{e^{4\sqrt{N}\pi} + w^2}\right). \tag{5.7.41}$$

After some computation, we get

$$\left|\frac{d}{dx} f(x;w)\right| \leqq 7\pi e^{-\pi x}. \tag{5.7.42}$$

Hence

$$\begin{aligned}
&\left|\frac{1}{\sqrt{N}} \sum_{k=1}^{2N} f\left\{\frac{2k-1}{2\sqrt{N}}; w\right\} - \int_0^{2\sqrt{N}} f(x;w)\, dx\right| \\
&= \frac{1}{\sqrt{N}} \left|\sum_{k=1}^{2N} \left[f\left\{\frac{2k-1}{\sqrt{N}}; w\right\} - \sqrt{N} \int_{(k-1)/\sqrt{N}}^{k/\sqrt{N}} f(x;w) dx\right]\right| \\
&\leqq \frac{1}{\sqrt{N}} \sum_{k=1}^{2N} \frac{1}{2\sqrt{N}} \max_{(k-1)/\sqrt{N} \leqq x \leqq k/\sqrt{N}} \left|\frac{d}{dx} f(x;w)\right| \\
&\leqq \frac{1}{2N} \sum_{k=1}^{2N} 7\pi e^{-\pi(k-1)/\sqrt{N}} \leqq \frac{7\pi}{N\left[1 - e^{-\pi/\sqrt{N}}\right]} \leqq \frac{7e^{\pi/\sqrt{N}}}{\sqrt{N}}.
\end{aligned} \tag{5.7.43}$$

By (5.7.39), (5.7.41), (5.7.42), and (5.7.43),

$$
\begin{aligned}
&\left|\alpha_N H_N'(\alpha_N w) - \frac{1}{\pi(1+w^2)}\right| \\
&\leq \frac{e^{2\pi\sqrt{N}}}{\pi[e^{4\pi\sqrt{N}} + w^2]} + \frac{7e^{\pi/\sqrt{N}}}{\sqrt{N}} \\
&= \mathcal{O}\left(\frac{1}{N^{1/2}}\right)
\end{aligned}
\tag{5.7.44}
$$

as $N \to \infty$, and with the bound on the extreme right-hand side of (5.7.44) clearly uniformly valid for $w \in \mathbb{R}$.

The considerations in the case of $\beta_N H'(\beta_N w)$ are similar to those of $\alpha_N H'(\alpha_N w)$, and we leave this part of the proof of Theorem 5.7.6 to Problem 5.7.1.

It only remains to prove (5.7.37). Now taking α_N as in (5.7.34), we have

$$
\begin{aligned}
&\int_{\mathbb{R}} H_N'(w)\, f(w)\, dw = \int_{\mathbb{R}} \alpha_N H_N'(\alpha_N w)\, f(\alpha_N w)\, dw \\
&= \int_{\mathbb{R}} \left[\alpha_N H_N'(\alpha_n w) - \frac{1}{\pi(1+w^2)}\right] f(\alpha_N w)\, dw + \int_{\mathbb{R}} \frac{\alpha_N}{\pi} \frac{f(v)}{\alpha_n^2 + v^2} dv \\
&= \mathcal{O}\left(\frac{1}{\sqrt{N}}\right) + \int_{\mathbb{R}} \frac{\alpha_N}{\pi} \frac{f(v)}{(\alpha_N^2 + v^2)} dv
\end{aligned}
\tag{5.7.45}
$$

On the other hand, it is readily shown that the integral on the extreme right-hand side of (5.7.45) is $f(0) + \delta_N$, with δ_N bounded by $\omega(f, \sqrt{\alpha_N}) + (2M/\pi)\sqrt{\alpha_N}$, and with M bounding f on $\mathbb{R}$. The details of the proof are left to Problem 5.7.2. ■

•

Problems for Section 5.7

5.7.1 Give a detailed proof of (5.7.36).

5.7.2 Give a detailed proof of (5.7.37).

5.7.3 * (a) Show that if $H_N(w)$ is defined by

$$
H_N(w) = \frac{hw}{\pi} \sum_{k=-N}^{N} \frac{e^{kh}}{w^2 + e^{2kh}},
$$

instead of by (5.7.4), if $\tilde{H}(w)$ is defined in (5.7.2), and if we set $H(w) = \lim_{(N\to\infty)} H_N(w)$, then all of the theorems of Section 5.7 remain valid (see Ikebe, Kowalski, and Stenger, [IKS1]).

(b) Is the form of H_N given in (5.7.3) better, for practical purposes, than the form given in part (a) of this problem ?

5.7.4 (a) Replace w by $(1+w)/(1-w)$ in the function H defined in part (b) of Problem 5.7.3, to get a new function:

$$\chi(w) = \frac{1}{2} + \frac{h(1-w^2)}{\pi} \sum_{k \in \mathbf{Z}} \frac{e^{kh}}{(1+w)^2 + (1-w)^2 e^{2kh}}.$$

(see Ikebe, Kowalski, and Stenger, [IKS1]).

(b) Now replace w by $(w+i)/(1+iw)$ in part (a) of this problem, to get a new function:

$$\hat{\chi}(w) = \frac{1}{2} + \frac{2h(1-w^2)}{2\pi i} \sum_{k \in \mathbf{Z}} \frac{e^{kh}}{(1+w)^2 - (1-w)^2 e^{2kh}}.$$

(c) Define $\hat{\chi}^*$ on the unit circle, by

$$\hat{\chi}^* = \begin{cases} 1 & \text{if} \quad 0 < \arg(w) < \pi \\ \frac{1}{2} & \text{if} \quad \arg(w) = 0, \pi \\ 0 & \text{if} \quad \pi < \arg(w) < 2\pi. \end{cases}$$

and show that, for all $w \in \mathbb{C}$, $|w| = 1$, we have

$$|\hat{\chi}(w) - \hat{\chi}^*(w)| \leq \frac{2\, e^{-\pi^2/h}}{1 - e^{-\pi^2/h}}.$$

(d) Find the poles and residues at the poles of $\hat{\chi}$, with $\hat{\chi}$ defined in part (b) of this problem.

(e) Deduce that if $f \in \mathbf{H}^1(\mathcal{U})$, with $\mathcal{U}$ denoting the unit disk in $\mathbb{C}$, then

$$\left| \int_{-1}^{1} f(x)\, dx - 2h \sum_{k \in \mathbf{Z}} \frac{e^{kh}}{(1+e^{2kh})^2} f\left(\frac{e^{kh}-1}{e^{kh}+1}\right) \right|$$
$$\leq \frac{2\, e^{-\pi^2/h}}{1 - e^{-\pi^2/h}} \|f\|_1.$$

(f)* Can the approach used in the above parts of this problem be used to derive all of the Sinc approximation formulas of Chapter 4?

5.8 Positive Base Approximation

In this section we derive a family of formulas using positive exponentials and rationals as bases. Various operations such as integration and Fourier transforms may be carried out explicitly on the new approximations. Moreover, these approximations converge to continuous functions; that is not the case for any of the previously constructed Sinc-like approximations. The approximations which we shall construct to approximate a function $f(t)$ use the values $f(0)$, $f(\pm h)$, $f(\pm 2h)$, $\cdots$, $h > 0$. We give bounds on the approximation error as simple functions of h, and these bounds illustrate the rate at which the approximants converge to f as $h \to 0$.

A well known polynomial expression to approximate a continuous function is the *Bernstein polynomial*,

$$B_n(f)(t) = \sum_{k=0}^{n} f\left(\frac{k}{n}\right)\binom{n}{k} t^k (1-t)^{n-k}. \tag{5.8.1}$$

Although $B_n(f)$ does not interpolate f at the points k/n, it nevertheless converges to f, uniformly on $[0,1]$, as $n \to \infty$, for $f \in C[0,1]$. It has thus become a frequently used tool for proving Weierstrass' approximation theorem.

Another formula, not as well known as those mentioned above, is the *Szasz formula*,

$$P(f,h)(t) = \sum_{k=0}^{\infty} f(kh) \frac{(t/h)^k}{k!} e^{-t/h}. \tag{5.8.2}$$

It may be shown that if f is uniformly bounded and continuous on $\mathbb{R}$, then for all finite M,

$$\sup_{0<t<M} |f(t) - P(f,h)(t)| \to 0 \tag{5.8.3}$$

as $h \to 0$. The function $P(f,h)$ does not interpolate f at the points kh, as was the case for $B_n(f)$. Both $B_n(f)$ and $P(f,h)$ play a role in probability theory.

Let us next discuss some Sinc-like results. Let $h > 0$, as above, let $\beta \in (0,1)$ be a positive number, and let us set

$$\begin{aligned} F_1(\beta,h)(t) &= h^{1-\beta} \sum_{k\in\mathbf{Z}} f(kh) \exp\{-h^{-\beta}|t-kh|\}, && (5.8.4)\\ F_2(\beta,h)(t) &= \frac{h^{1-\beta}}{\sqrt{\pi}} \sum_{k\in\mathbf{Z}} \frac{f(kh)}{\cosh\{h^{-\beta}(t-kh)\}}, && (5.8.5)\\ F_3(\beta,h)(t) &= \frac{h^{1-\beta}}{\sqrt{\pi}} \sum_{k\in\mathbf{Z}} f(kh) \exp\{-h^{-2\beta}(t-kh)^2\}, && (5.8.6) \end{aligned}$$

$$F_4(\beta, h)(t) \quad = \quad \frac{h^{1+\beta}}{\pi} \sum_{k \in \mathbf{Z}} \frac{f(kh)}{(t-kh)^2 + h^{2\beta}}. \tag{5.8.7}$$

In order to prove convergence of F_j to f, let us define the usual *modulus of continuity* of f by

$$w(f, \tau) = \sup_{x \in \mathbf{R}, y \in \mathbf{R}, |x-y| < \tau} |f(x) - f(y)|. \tag{5.8.8}$$

Theorem 5.8.1 *Let β be a constant such that $0 < \beta < 1$, and let α be a constant such that $0 < \alpha < \beta$. Let $F_j(\beta, h)$ be defined by (5.8.4) through (5.8.7). Then there exist constants A_j and B_j, $j = 1, 2, 3, 4$, that are independent of h, such that*

$$\begin{aligned} &\sup_{t \in \mathbf{R}} |f(t) - F_j(\beta, h)(t)| \\ &< A_j \phi_j(\beta, h) + B_j \psi_j(\alpha, \beta, h) + w(f, h^\alpha) \end{aligned} \tag{5.8.9}$$

with

$$\left\{ \begin{array}{l} \phi_1(\beta, h) = h^{2-2\beta} \\ \phi_2(\beta, h) = \exp\left(-\pi^2 h^{\beta-1}\right) \\ \phi_3(\beta, h) = \exp\left(-\pi^2 h^{2\beta-2}\right) \\ \phi_4(\beta, h) = \exp\left(-2\pi h^{\beta-1}\right) \\ \psi_1(\alpha, \beta, h) = \psi_2(\alpha, \beta, h) = \exp\left(-h^{\alpha-\beta}\right) \\ \psi_3(\alpha, \beta, h) = \exp\left(-h^{2\alpha-2\beta}\right) \\ \psi_4(\alpha, \beta, h) = h^{\beta-\alpha}. \end{array} \right. \tag{5.8.10}$$

Proof. Let g and G be integrable functions defined on $\mathbb{R}$, such that

$$G(x) = \int_{\mathbf{R}} e^{ixt} g(x) dt \tag{5.8.11}$$

and such that for all $x \in \mathbb{R}$,

$$\begin{aligned} &\lim_{h \to 0+} \{g(x+h) - 2g(x) + g(x-h)\} = 0 \\ &\lim_{h \to 0+} \{G(x+h) - 2G(x) + G(x-h)\} = 0. \end{aligned} \tag{5.8.12}$$

Then Poisson's summation formula states that for arbitrary $h > 0$,

$$h \sum_{k \in \mathbf{Z}} g(t - kh) = \sum_{k \in \mathbf{Z}} e^{2\pi i k t/h} G\left(\frac{2\pi k}{h}\right). \tag{5.8.13}$$

We present four functions, g_j, and their Fourier transforms, G_j, in Table 5.8.1, with $a > 0$ for all of these. We also include in this table the required special values of a, $0 < \beta < 1$.

j	$g_j(t)$	$G_j(x)$	a
1	$(a/2)\exp\{-a\|t\|\}$	$a^2/(a^2+x^2)$	$h^{-\beta}$
2	$a\{\pi\cosh at\}^{-1}$	$1/\cosh\{\pi x/(2a)\}$	$h^{-\beta}$
3	$(a/\pi)^{1/2}\exp(-at^2)$	$\exp\{-x^2/(4a)\}$	$h^{-2\beta}$
4	$a\{\pi(t^2+a^2)\}^{-1}$	$\exp\{-a\|x\|\}$	h^{β}

Table 5.8.1: Transform pairs and value of a

We shall then study the approximation of f via the formula

$$T_j(f,h)(t) = h\sum_{k\in\mathbf{Z}} f(kh)g_j(t-kh). \tag{5.8.14}$$

To this end we write

$$\begin{aligned} f(t) \quad &- \quad T_j(f,h)(t) \\ &= \quad f(t)\,[1-T_j(1,h)(t)] \\ &\quad +h\sum_{|kh-t|\leqq h^\alpha} \{f(t)-f(kh)\}\,g_j(t-kh) \\ &\quad +h\sum_{|kh-t|>h^\alpha} \{f(t)-f(kh)\}\,g_j(t-kh) \end{aligned} \tag{5.8.15}$$

with α a positive number $0<\alpha<\beta$. Let us denote the three terms on the right-hand side of (5.8.15) by $\varepsilon_j^{(1)}$, $\varepsilon_j^{(2)}$, and $\varepsilon_j^{(3)}$ in the respective order in which they appear, and let us assume that f is bounded by M on $\mathbb{R}$.

Noting that for each j $(j=1,2,3,4)$, the term corresponding to $k=0$ on the right-hand side of (5.8.13) having the value 1, we have:
For $j=1$:

$$\begin{aligned} |1-T_1(1,h)(t)| \quad &\leqq \sum_{|k|>0} \frac{a^2}{a^2+4k^2\pi^2/h^2} \\ &< 2a^2\sum_{k=1}^{\infty} \frac{1}{4k^2\pi^2/h^2} \\ &= \frac{a^2h^2}{12} \end{aligned} \tag{5.8.16}$$

since

$$\sum_1^\infty k^{-2} = \pi^2/6$$

for $j=2$:

$$|1 - T_2(1,h)(t)| \leqq \sum_{|k|>0} \left\{ \cosh\left(\frac{\pi^2 k}{ah} \right) \right\}^{-1}$$
$$< 4 \sum_{k=1}^{\infty} e^{-\pi^2 k/(ah)} = \frac{4e^{-\pi^2/(ah)}}{1 - e^{-\pi^2/(ah)}} \tag{5.8.17}$$

for $j = 3$:

$$|1 - T_3(1,h)(t)| \leqq \sum_{|k|>0} e^{-k^2\pi^2/(ah^2)}$$
$$< \frac{2e^{-\pi^2/(ah^2)}}{1 - e^{-\pi^2/(ah^2)}} \quad \text{and} \tag{5.8.18}$$

for $j = 4$:

$$|1 - T_4(1,h)(t)| < \sum_{|k|>0} e^{-2\pi a|k|/h}$$
$$= \frac{2e^{-2\pi a/h}}{1 - e^{-2\pi a/h}}. \tag{5.8.19}$$

Hence, for $j = 1, 2, 3$, and 4, $\varepsilon_j^{(1)}$ is bounded by M times the corresponding bound on the extreme right-hand sides of (5.8.16) to (5.8.19), with a selected as in Table 5.8.1.

Next, for $\varepsilon_j^{(2)}$, we have

$$\begin{aligned} \left|\varepsilon_j^{(2)}\right| &= \left| h \sum_{|kh-t| \leqq h^\alpha} \{f(t) - f(kh)\} g_j(t - kh) \right| \\ &\leqq w(f, h^\alpha) h \sum_{(kh-t) \leqq h^\alpha} g_j(t - kh) \\ &\leqq w(f, h^\alpha) \left[1 + |1 - T_j(1,h)(t)|\right] \end{aligned} \tag{5.8.20}$$

with $w(f, \tau)$ defined in (5.8.8).

Finally, for the case of $\varepsilon_j^{(3)}$, we have, since $g_j(t) > 0$, $g_j(-t) = g_j(t)$, that

$$\left|\varepsilon_j^{(3)}\right| \leqq 4M \int_{h^\alpha - h}^{\infty} g_j(t)\,dt. \tag{5.8.21}$$

Thus

$$\left|\varepsilon_1^{(3)}\right| \leqq 4M \exp\left(-h^{\alpha-\beta} + h^{1-\beta}\right) \quad \text{if } j = 1; \tag{5.8.22}$$

also

$$\begin{aligned} \left|\varepsilon_2^{(3)}\right| &\leqq 8M \tan^{-1}\left(e^{-h^{-\beta}(h^\alpha - h)}\right) \\ &\leqq 8M \exp\left(-h^{\alpha-\beta} + h^{1-\beta}\right) \quad \text{if } j = 2. \end{aligned} \tag{5.8.23}$$

Similarly,

$$\begin{aligned} \left|\varepsilon_3^{(3)}\right| &\leqq 4M \frac{h^{-\beta}}{\sqrt{\pi}} \int_{h^\alpha - h}^\infty \exp\left(-h^{-2\beta} u^2\right) du \\ &= \frac{2M}{\sqrt{\pi}} \int_{h^{-2\beta}(h^\alpha - h)^2}^\infty e^{-v} v^{-1/2} dv \\ &= O\left(\exp\left(-h^{2\alpha - 2\beta}\right)\right) \quad \text{as } h \to 0 \end{aligned} \tag{5.8.24}$$

and

$$\begin{aligned} \left|\varepsilon_4^{(3)}\right| &< \frac{4Mh^\beta}{\pi} \int_{h^\alpha - h}^\infty \frac{du}{u^2 + h^{2\beta}} \\ &= \frac{4M}{\pi} \tan^{-1} \frac{h^\beta}{h^\alpha - h} \\ &= O\left(h^{\beta - \alpha}\right). \end{aligned} \tag{5.8.25}$$

as $h \to 0$. ■

Let us mention, at this point, the identity

$$\frac{hy}{2} \sum_{k \in \mathbb{Z}} e^{-y|x - kh|} = \frac{\cosh\left\{hy\left(\frac{x}{h} - \left[\frac{x}{h}\right] - \frac{1}{2}\right)\right\}}{\sinh(hy/2)/(hy/2)} \tag{5.8.26}$$

(see Problem 1.6.5) which enables us to obtain yet another explicit expression for $F_1(\beta, h)$, in the case of f *periodic*. More specifically, if f is continuous and periodic on $\mathbb{R}$, with period T, then, taking $h = T/N$, N being a positive integer, we get

$$F_1(\beta, T/N)(t) = \tag{5.8.27}$$

$$\frac{1}{2}(T/N)^{1-\beta} \sum_{k=0}^{N-1} f\left(\frac{kT}{N}\right) \frac{\cosh\left\{T(N/T)^\beta\left(\frac{t}{T} - \frac{k}{N} - \left[\frac{t}{T} - \frac{k}{N}\right] - \frac{1}{2}\right)\right\}}{\sinh\left\{\frac{1}{2}T(N/T)^\beta\right\}}.$$

The formulas (5.8.4) through (5.8.7) will of course be more useful in practice in the cases where f has compact support, or $f(t)$ decreases to zero rapidly as $t \to \pm\infty$. We note, furthermore, that in such cases, we can:

(a) Explicitly express the Fourier transform of each F_j; and

(b) Explicitly express the indefinite integrals of F_1, F_2, and F_4.

Moreover, if f has compact support, then $F_3(\beta, h)$ is an entire function.

In Theorem 5.8.1, $w(f, \tau)$ is defined by (5.8.8). Just as we did in the case of $F_1(\beta, h)$ in (5.8.27) we can obtain another explicit (and useful!) expression for $F_4(\beta, h)$ in the case f periodic on $\mathbb{R}$. In this case we may use the identity

$$\frac{y}{\pi} \sum_{k \in \mathbb{Z}} \frac{1}{(x - kh)^2 + y^2} = \frac{\sinh(2\pi y/h)}{\cosh(2\pi y/h) - \cos(2\pi x/h)} \tag{5.8.28}$$

(see Problem 1.6.13) to deduce that if f has period T on $\mathbb{R}$, then taking $h = T/N$, N being a positive integer, we get

$$\begin{aligned} &F_4(\beta, T/N)(t) \\ &= \frac{T}{N} \sum_{k=0}^{N-1} f(kh) \frac{\sinh\left\{\frac{2\pi}{T}(T/N)^\beta\right\}}{\cosh\left\{\frac{2\pi}{T}(T/N)^\beta\right\} - \cos\left\{\frac{2\pi}{T}\left(t - \frac{kT}{N}\right)\right\}}. \end{aligned} \tag{5.8.29}$$

We remark also, that if $f(t) \to 0$ as $t \to \pm\infty$, the Fourier and Hilbert transforms of $F_4(\beta, h)$ may be explicitly expressed,

$$\begin{aligned} &\int_{\mathbb{R}} F_4(\beta, h)(t) e^{ixt} dt = h e^{-h^\beta |x|} \sum_{k \in \mathbb{Z}} f(kh) e^{ikhx} \\ &P.V. \int_{\mathbb{R}} \frac{F_4(\beta, h)(t)}{t - x} dt = h \sum_{k \in \mathbb{Z}} f(kh) \frac{(kh - x)}{(x - kh)^2 + h^{2\beta}}. \end{aligned} \tag{5.8.30}$$

Finally, using the identity

$$\sum_{k \in \mathbb{Z}} \frac{x - k}{(x - k)^2 + y^2} = \frac{\pi \sin(2\pi x)}{\cosh(2\pi y) - \cos(2\pi x)} \tag{5.8.31}$$

(see Problem 1.6.1) we find that if f is periodic, with period T on $\mathbb{R}$, then taking $h = T/N$, N being a positive integer, we get

$$\lim_{\varepsilon\to 0+} \int_{\varepsilon<|t-x|<1/\varepsilon} \frac{F_4(\beta,h)(t)}{t-x} dt \tag{5.8.32}$$

$$= -\frac{\pi T}{N} \sum_{k=0}^{N-1} f\left(\frac{kT}{N}\right) \frac{\sin\left\{\frac{2\pi}{T}\left(x-\frac{kT}{N}\right)\right\}}{\cosh\left\{\frac{2\pi}{T}(T/N)^\beta\right\} - \cos\left\{\frac{2\pi}{T}\left(x-\frac{kT}{N}\right)\right\}}.$$

•

Problems for Section 5.8

5.8.1 Let $f \in \mathbf{H}^\infty(\mathcal{D}_d)$ with $\mathcal{D}_d$ defined in (1.7.9). Which of the approximations, (5.8.4) through (5.8.7) can be expected to be accurate for such a function, f ?

5.8.2 Alter Weierstrass' original proof of the polynomial approximation theorem, making it completely constructive, via the use of $F_3(\beta,h)(t)$ and Theorem 5.8.1.

5.8.3 State and give a constructive proof of a theorem on the existence of a sequence of rational functions which converges to a given continuous function on a compact interval on $\mathbb{R}$, using $F_4(\beta,h)(t)$ and Theorem 5.8.1.

Historical Remarks on Chapter 5

The results of Section 5.2 were first obtained by *Elliott and Stenger*, [ES1]. The elliptic function approximant in Section 5.3 was first studied in *Stenger*, [S28]. The Laplace transform inversion procedure of Section 5.4 was first obtained by *Ang, Lund and Stenger*, [AnLS1]. The results of Section 5.5 were obtained by *Stenger* in [S28]. The deductions in Section 5.6 regarding when rational extrapolation via the Thiele algorithm is effective are mostly new, although some of these were already made in [S28]. The Heaviside and filter rational results of Section 5.7 are new, but they are closely related to those obtained by *Kowalski, Ikebe and Stenger*, [IKS1]. A different derivation of χ_N than the one obtained in Section 5.7 (see Problem 5.7.4) was previously obtained in *Stenger*, [S16], and summarized in Problem 1.7.8, using elliptic functions. Although we could have easily started with the results in [S16], we have chosen, instead, to bypass the use of elliptic functions and directly use Poisson's summation formula. While the approximations of Section 5.7 do not yield the optimal ones which solve Solotareff's problem (see, e.g., *Todd*, [Td1]), they are close to optimal, and lead to simple and explicit functional expressions. The results of Section 5.8 were first obtained in [S29].

6

Integral Equations

6.1 Introduction

In this chapter we discuss some basic properties of *Volterra and Fredholm integral equations*, and illustrate the use of Sinc methods to obtain an approximate solution of these equations.

Let us briefly illustrate some examples of integral equations which we shall consider.

The equations

$$\int_a^x K(x,t)f(t)dt = g(x), \quad a < x < b \tag{6.1.1}$$

and

$$f(x) - \lambda \int_a^x K(x-t)f(t)dt = g(x), \quad a < x < b \tag{6.1.2}$$

are *linear Volterra integral equations of the first and second kind* respectively. In these equations, the *kernel* $K(x,t)$ and the function $g(x)$ are given, and the problem is to determine or approximate the solution f. For example, the *Abel equation*

$$\int_a^x (x-t)^{-\alpha} f(t)dt = g(x), \quad a < x < b \tag{6.1.3}$$

is a Volterra integral equation of the first kind.

We may also consider a *nonlinear Volterra integral equation* of the form

$$f(x) - \lambda \int_a^x K(x,t,f(t))dt = g(x), \quad a < x < b. \tag{6.1.4}$$

Mainly, the *Fredholm integral equations* which we shall consider in this chapter are linear, and have the form

$$\int_{\mathcal{B}} K(x,t)f(t)dt = g(x), \quad x \in \mathcal{B} \tag{6.1.5}$$

or

$$f(x) - \lambda \int_{\mathcal{B}} K(x,t)f(t)dt = g(x), \quad x \in \mathcal{B} \tag{6.1.6}$$

with $\mathcal{B}$ a region in $\mathbb{R}^n$. *Equations (6.1.5) and(6.1.6) are Fredholm integral equations of the first and second kind*, respectively.

Examples of Fredholm integral equations of the first kind are the Laplace, Fourier, and Hilbert transforms. Examples of Fredholm integral equations of the second kind are *equations of convolution type*,

$$f(x) - \lambda \int_{\mathbb{R}} k(x-t) f(t) dt = g(x), \quad x \in \mathbb{R} \tag{6.1.7}$$

or

$$f(x) - \lambda \int_0^\infty k(x-t) f(t) dt = g(x), \quad 0 < x < \infty. \tag{6.1.8}$$

Here, Equation (6.1.8) is called a *Wiener–Hopf integral equation.* Obtaining its solution is considerably more difficult than obtaining the solution of (6.1.7).

An example of an important class of linear Fredholm integral equations which we shall explore in this chapter, is an integral equation from *potential theory*, over a surface S, having the form

$$f(\overline{r}) - \frac{1}{2\pi} \int\int_S \left\{ \frac{\partial}{\partial n(\overline{r})} \frac{1}{|\overline{r} - \overline{r}'|} \right\} f(\overline{r}') dA(\overline{r}') = g(\overline{r}), \overline{r} \in S. \tag{6.1.9}$$

Another example of an important Fredholm integral equation is the *Lippmann–Schwinger integral equation from scattering theory,*

$$f(\overline{r}) - k^2 \int\int\int_{\mathcal{B}} \frac{e^{ik|\overline{r} - \overline{r}'|}}{4\pi |\overline{r} - \overline{r}'|} \gamma(\overline{r}') f(\overline{r}') d\overline{r}' = g(\overline{r}), \overline{r} \in \mathcal{B}. \tag{6.1.10}$$

We shall also consider the solution of a *Cauchy singular integral equation* (CSIE) given on an arc Γ, and having the form

$$\begin{aligned} u(z)w(z) + v(z) \frac{P.V.}{\pi i} \int_\Gamma \frac{w(\xi)}{\xi - z} d\xi - \int_\Gamma k(z,\xi) w(\xi) d\xi \\ = g(z), \quad z \in \Gamma. \end{aligned} \tag{6.1.11}$$

Here we assume that u, v, k and g are given, and w is the unknown.

As our final example, we shall consider the solution of a special Fredholm integral equation of the first kind, namely, the Laplace transform equation, via the use of Fourier transforms, and DFT.

•

6.2 Mathematical Preliminaries

In this section we discuss various mathematical properties of integral equations. We may consider these equations as equations on a *Banach space*, after defining them in suitable *operator* form. In this form we are able to discuss the *existence of solutions*, the role of *compact operators*, *convergence of approximate solutions*, and we are also able to obtain a bound on the norm of the difference between the solutions of two different integral equations. In addition, we discuss *degenerate kernels*, and *eigensolutions*. Although our discussions are confined mainly to linear Fredholm equations of the second kind, we briefly discuss Volterra integral equations, and integral equations of the first kind.

6.2.1 Use of Functional Analysis

The operator notation is a very convenient tool for discussing an integral equation. We write

$$(I - \lambda\kappa)f = g \tag{6.2.1}$$

where the operator equation is defined on a Banach space $\mathbf{X}$, I is the identity operator on $\mathbf{X}$, and κf is defined by one of the expressions

$$(\kappa f)(x) = \int_a^x K(x,t)f(t)dt; \tag{6.2.2}$$

$$(\kappa f)(x) = \int_a^x K(x,t,f(t))dt; \tag{6.2.3}$$

or

$$(\kappa f)(x) = \int_{\mathcal{B}} K(x,t)f(t)dt \tag{6.2.4}$$

in the cases of (6.1.2), (6.1.4), and (6.1.6) respectively, and λ is a given complex number.

Every Banach space is a linear space $\mathbf{X}$ on which a norm $\|\cdot\|$ is defined, which has the properties:

(i) $\|f\| \geqq 0$ for all $f \in \mathbf{X}$, and $\|f\| = 0 \Leftrightarrow f = 0$;

(ii) $\|\alpha f\| = |\alpha|\|f\|$ for all $f \in \mathbf{X}$, $\alpha \in \mathbb{C}$;

and

(iii) $\|f+g\| \leqq \|f\| + \|g\|$.

A Banach space $\mathbf{X}$ is also a *complete* normed space. We shall describe the property of completeness in the sequel.

A sequence $\{f_n\}$, $f_n \in \mathbf{X}$, *converges* to a point $f \in \mathbf{X}$ if $\|f_n - f\| \to 0$ as $n \to \infty$, and we simply write $f_n \to f$ as $n \to \infty$. A sequence $\{f_n\}$ is called a *Cauchy sequence* if

$$\lim_{N\to\infty} \left\{ \sup_{m>N,n>N} \|f_m - f_n\| \right\} = 0.$$

Clearly, if $f_n \to f$ as $n \to \infty$, then $\{f_n\}$ is a Cauchy sequence, but a Cauchy sequence $\{f_n\}$ need not converge to a point $f \in \mathbf{X}$. The normed space $\mathbf{X}$ is a *complete normed space*, i.e. a *Banach space*, if every Cauchy sequence $\{f_n\}$ converges to a point f of $\mathbf{X}$.

In a Banach space $\mathbf{X}$, let $r > 0$, and let $\mathcal{B}(f_0; r) = \{f \in \mathbf{X} : \|f - f_0\| < r\}$ denote the (open) *ball* with center f_0 and radius r. A set $\mathcal{D} \subset \mathbf{X}$ is said to be open if every $f \in \mathcal{D}$ is the center of some (open) ball contained in $\mathcal{D}$.

Let $\mathbf{X}$ and $\mathbf{Y}$ be two Banach spaces, with norms $\|\cdot\|_{\mathbf{X}}$ and $\|\cdot\|_{\mathbf{Y}}$. A single valued function $g = Tf$ with *domain* $\mathcal{D} \subset \mathbf{X}$ and *range* $\mathcal{E} \subset \mathbf{Y}$ is called an *operator*, or *mapping* from $\mathcal{D}$ onto $\mathcal{E}$, and we write $T : \mathcal{D} \to \mathcal{E}$, if to every $f \in \mathcal{D}$ there corresponds a unique $g \in \mathcal{E}$, and every $g \in \mathcal{E}$ is the *image* of at least one $f \in \mathcal{D}$. The operator $T : \mathcal{D} \to \mathcal{E}$ is said to be *one-to-one* if to every $g \in \mathcal{E}$ there is a unique f in $\mathcal{D}$ such that $g = Tf$. In this case we write $f = T^{-1}g$, i.e., there exists a unique single-valued mapping $T^{-1} : \mathcal{E} \to \mathcal{D}$; moreover, $T(T^{-1}g) = g$ for every $g \in \mathcal{E}$, and $T^{-1}(Tf) = f$ for every $f \in \mathcal{D}$.

The mapping $T : \mathcal{D} \to \mathcal{E}$ from the domain $\mathcal{D}$ in a Banach space $\mathbf{X}$ to the range $\mathcal{E}$ in a Banach space $\mathbf{Y}$ is said to be *continuous* at $f_0 \in \mathcal{D}$ if given any $\varepsilon > 0$ there exists a $\delta > 0$, such that whenever $\|f - f_0\|_{\mathbf{X}} < \delta$ then $\|Tf - Tf_0\|_{\mathbf{Y}} < \varepsilon$. The mapping $T : \mathcal{D} \to \mathcal{E}$ is said to be *continuous on* $\mathcal{D}$ *if it is continuous at every* $f_0 \in \mathcal{D}$.

Theorem 6.2.1 (Contraction Mapping Principle) *Let* $\mathbf{X}$ *be a Banach space, and let* $T : \mathcal{D} \to \mathbf{X}$, *with* $\mathcal{D} \subset \mathbf{X}$, *such that for all* f *and* g *in* $\mathcal{D}$,

$$\|Tf - Tg\| \leqq \alpha \|f - g\| \tag{6.2.5}$$

with $0 < \alpha < 1$. *Let* $f_0 \in \mathcal{D}$ *be selected such that the closed ball* $\overline{\mathcal{B}}(f_0; r) \subset \mathcal{D}$, *with* $r \geqq \|f_0 - Tf_0\|/(1-\alpha)$, *and define* $\{f_n\}$ *by*

$$f_{n+1} = Tf_n, \quad n = 0, 1, 2, \cdots. \tag{6.2.6}$$

Then $\{f_n\}$ *converges to* $f^* \in \overline{\mathcal{B}}(f_0; r)$, *with* f^* *the unique solution in* $\overline{\mathcal{B}}(f_0; r)$ *of the equation* $f = Tf$. *Moreover, we have*

$$\|f^* - f_n\| \leqq \frac{\alpha}{1-\alpha}\|f_{n-1} - f_n\| \leqq \frac{\alpha^n}{1-\alpha}\|f_0 - Tf_0\|. \tag{6.2.7}$$

Proof. This is left to Problem 6.2.4 which is at the end of this section. ■

Let $\mathbf{X}$ and $\mathbf{Y}$ be Banach spaces, let $\mathcal{D} \subset \mathbf{X}$, $\mathcal{E} \subset \mathbf{Y}$. The mapping $T : \mathcal{D} \to \mathcal{E}$ is said to be *linear* if $\mathcal{D}$ is a linear subspace of $\mathbf{X}$, $\mathcal{E}$ is a linear subspace of $\mathbf{Y}$, and if $T(af + bg) = aTf + bTg$ for every f and g in $\mathcal{D}$ and for every a and b in $\mathbb{C}$ with, as before, $\mathbb{C}$ the complex plane. If $\mathcal{E} = \mathbb{C}$, then T is called a *linear functional.*

The linear mapping $T : \mathcal{D} \to \mathcal{E}$ is called *bounded* if there exists a positive number M, such that $\|Tf\|_{\mathbf{Y}} \leqq M\|f\|_{\mathbf{X}}$ for all $f \in \mathcal{D}$. We can then define $\|T\|$ by $\|T\| = \sup\{\|Tf\|_{\mathbf{Y}} : \|f\|_{\mathbf{X}} \leqq 1\}$. Moreover, the *linear mapping $T : \mathcal{D} \to \mathcal{E}$ is continuous if and only if it is bounded.* If T is a linear mapping and $\lambda \in \mathbb{C}$, the operator $(\lambda I - T)^{-1}$ is called the *resolvent* of T. The *resolvent set* of the linear mapping T is the set $\Lambda = \{\lambda \in \mathbb{C}$: the resolvent exists and is bounded$\}$. The complementary set $\mathbb{C} \setminus \Lambda$ is called the spectrum of the linear operator T.

Some examples of Banach spaces are $\mathbf{C}[\mathcal{B}]$, $\mathbf{L}^p(B)$, and $\mathbf{H}^p(\mathcal{B})$, where $\mathbf{C}[\mathcal{B}]$ denotes the family of all functions f that are continuous on a closed and bounded region $\mathcal{B}$ of $\mathbb{R}^n$, such that

$$\|f\| \equiv \sup_{x\in\mathcal{B}} |f(x)| < \infty.$$

$\mathbf{L}^p(B)$ $(1 \leqq p \leqq \infty)$ denotes the family of all measurable functions f defined on a region B of $\mathbb{R}^n$ such that $\|f\| \equiv (\int_{\mathcal{B}} |f(x)|^p dx)^{1/p} < \infty$, and $\mathbf{H}^p(\mathcal{D})$ is defined for some specific regions $\mathcal{D}$ in Section 1.8, for $1 \leqq p \leqq \infty$.

Example 6.2.2 Let us take $\mathbf{X} = \mathbf{C}[a, b]$, normed by

$$\|f\| = \sup_{(x\in(a,b))} |f(x)|.$$

Let T be defined by $Tf = \lambda\kappa f$, with κ defined as in (6.2.3), and with the function $K(x, t, f(t))$ in (6.2.3) assumed to have the following properties:

(a) The function

$$w(x) \equiv \int_a^x K(x, t, u(t))dt \tag{6.2.8}$$

is in $\mathbf{C}[a, b]$ for all u in the open ball $\mathcal{B}(g, M)$, with $g \in \mathbf{X}$ as in (6.1.4), and with M a positive number;

(b) For some number $\alpha \in (0, 1)$, we have

$$\sup_{x\in[a,b],u\in\mathcal{B}(g,M),v\in\mathcal{B}(g,M)} |\lambda| \left| \int_a^x [K(x,t,u(t)) - K(x,t,v(t))]\, dt \right| \leqq \alpha\|u-v\|; \tag{6.2.9}$$

and

(c) Let

$$\sup_{x\in(a,b)} |\lambda| \left| \int_a^x K(x,t,g(t))dt \right| < (1-\alpha)M. \tag{6.2.10}$$

Under these conditions, let $\{f_n\}$ be defined by

$$\begin{aligned} f_0(x) &= g(x) \\ f_{n+1}(x) &= \lambda \int_a^x K(x,t,f_n(t))dt + g(x), \quad n = 0,1,2,\cdots. \end{aligned} \tag{6.2.11}$$

Then it follows by Theorem 6.2.1 that $\{f_n\}$ converges to the unique solution in $\mathcal{B}(g,M)$ of Equation (6.1.4).

Example 6.2.3 Let us illustrate the application of Theorem 6.2.1 to the solution of linear Fredholm equations. Let $T = \lambda\kappa$, with κ a linear operator, defined, for example, by (6.2.4), mapping a subspace **Y** of a Banach space **X** into **X**, and such that $|\lambda|\ \|\kappa\| < 1$. Define $\{f_n\}$ by

$$\begin{aligned} f_0 &= g \\ f_{n+1} &= \lambda\kappa f_n + g, \quad n = 0,1,2,\cdots. \end{aligned} \tag{6.2.12}$$

If $g \in \mathbf{Y}$, then $\{f_n\}$ converges to the unique solution f^* of the equation $f - \lambda\kappa f = g$. Moreover,

$$\begin{aligned} \|f_n - f^*\| &\leqq \frac{|\lambda|\ \|\kappa\|}{1-|\lambda|\ \|\kappa\|}\|f_n - f_{n-1}\| \\ &\leqq \frac{(|\lambda|\ \|\kappa\|)^n}{1-|\lambda|\ \|\kappa\|}\|f_1 - f_0\|. \end{aligned} \tag{6.2.13}$$

We note in particular that

$$f_n = (I + \lambda\kappa + \lambda^2\kappa^2 + \cdots + \lambda^n\kappa^n)g \tag{6.2.14}$$

and this equation shows that the solution of (6.1.6) is an analytic function of λ in the region $\{\lambda \in \mathbb{C} : |\lambda| < 1/\|\kappa\|\}$.

Theorem 6.2.1 shows that if $\mathbf{X}$ is a Banach space and if $T : \mathbf{X} \to \mathbf{X}$ is a linear mapping such that $\|T\| < 1$, then $(I - T)^{-1}$ exists, and moreover, $\|(I - T)^{-1}\| \leqq (1 - \|T\|)^{-1}$. More generally, if $\mathbf{X}$ and $\mathbf{Y}$ are Banach spaces and $T : \mathbf{X} \to \mathbf{Y}$ is a linear mapping, then T^{-1} exists, provided that $Tf = 0$ if, and only if, $f = 0$. Clearly T^{-1} is a *linear operator.* Moreover, T^{-1} is bounded if and only if there exists a constant $c > 0$, such that $\|Tf\|_{\mathbf{Y}} \geqq c\|f\|_{\mathbf{X}}$ for every $f \in \mathbf{X}$, and the largest possible value of c such that this inequality holds is $c = 1/\|T^{-1}\|$.

6.2.2 Approximation, Convergence, and Error

We attempt to obtain an approximate solution of (6.2.1) by replacing (6.2.1) with the equation

$$(I - \lambda\kappa_n) f_n = g_n \tag{6.2.15}$$

in the hope of getting an f_n which is close to f when both $\|\kappa - \kappa_n\|$ and $\|g - g_n\|$ are "small". At the outset, we shall require:

Lemma 6.2.4 *Let S and T be linear operators defined on a Banach space $\mathbf{X}$ such that $S : \mathbf{X} \to \mathbf{X}$, $T : \mathbf{X} \to \mathbf{X}$. If S^{-1} exists on $\mathbf{X}$, and if*

$$\|S^{-1}(T - S)\| < 1 \tag{6.2.16}$$

then T^{-1} exists on $\mathbf{X}$; moreover,

$$\|T^{-1}\| \leqq \frac{\|S^{-1}\|}{1 - \|S^{-1}(T - S)\|} \tag{6.2.17}$$

and

$$\|S^{-1} - T^{-1}\| \leqq \|S^{-1}\| \ \|T^{-1}\| \ \|S - T\|. \tag{6.2.18}$$

Proof. The proof is deferred to Problem 6.2.7. ∎

Theorem 6.2.5 *Let κ and κ_n be bounded linear operators mapping a Banach space $\mathbf{X}$ into itself, let g and g_n belong to $\mathbf{X}$, and let λ be a complex scalar.*

(a) *If $\|(I - \lambda\kappa)^{-1}\| < \infty$, and if*

$$|\lambda| \ \|(I - \lambda\kappa)^{-1}(\kappa - \kappa_n)\| < 1, \tag{6.2.19}$$

then $(I - \lambda\kappa_n)^{-1}$ exists, and moreover,

$$\|(I - \lambda\kappa_n)^{-1}\| \leqq \frac{\|(I - \lambda\kappa)^{-1}\|}{1 - |\lambda|\|(I - \lambda\kappa)^{-1}(\kappa - \kappa_n)\|}. \tag{6.2.20}$$

(b) *If* $\|(I - \lambda\kappa_n)^{-1}\| < \infty$, *and if*

$$|\lambda| \; \|(I - \lambda\kappa_n)^{-1}(\kappa - \kappa_n)\| < 1, \tag{6.2.21}$$

then $(I - \lambda\kappa)^{-1}$ *exists, and moreover,*

$$\|(I - \lambda\kappa)^{-1}\| \leqq \frac{\|(I - \lambda\kappa_n)^{-1}\|}{1 - |\lambda| \; \|(I - \lambda\kappa_n)^{-1}(\kappa - \kappa_n)\|}. \tag{6.2.22}$$

If either of the conditions (a) or (b) are satisfied, then Equation (6.2.1) has a solution f in **X** *and Equation (6.2.15) has a solution f_n in* **X**, *with*

$$\begin{aligned} \|f - f_n\| \leqq |\lambda| \; \|(I - \lambda\kappa_n)^{-1}\| \; \|(I - \lambda\kappa)^{-1}\| \; \|\kappa - \kappa_n\| \; \|g\| \\ + \|(I - \lambda\kappa_n)^{-1}\| \;\; \|g - g_n\|. \end{aligned} \tag{6.2.23}$$

Proof. The inequalities (6.2.20) and (6.2.22) follow directly from Lemma 6.2.4. First, taking $S = (I - \lambda\kappa)$, and $T = (I - \lambda\kappa_n)$, we get (6.2.20) from (6.2.17), and similarly by reversing the roles of S and T we get (6.2.22). The inequality (6.2.23) follows directly from the identity

$$\begin{aligned} &(I - \lambda\kappa)^{-1}g - (I - \lambda\kappa_n)^{-1}g_n \\ &= \left[(I - \lambda\kappa)^{-1} - (I - \lambda\kappa_n)^{-1}\right] g + (I - \lambda\kappa_n)^{-1}(g - g_n). \end{aligned} \tag{6.2.24}$$

This completes the proof of Theorem 6.2.5. ■

Let **X** be a Banach space. A subset $\mathbf{Y} \subset \mathbf{X}$ is said to be *dense in* **X** if given any $f \in \mathbf{X}$ and any positive number ε, there exists a $g \in \mathbf{Y}$, such that $\|f - g\| < \varepsilon$.

Let **X** be a Banach space. A subset G of **X** is said to be *compact* if every infinite subset of G has an *accumulation point* in G, i.e., if G contains a Cauchy sequence. A *linear operator* $\kappa : \mathbf{X} \to \mathbf{X}$ is said to be *compact*, if for every sequence $\{f_n\}$, such that $f_n \in \mathbf{X}$ and $\|f_n\| \leqq M < \infty$ for all n, the sequence $\{\kappa f_n\}$ contains convergent subsequence $\{\kappa f_{n_k}\}$. It thus follows that a compact operator maps a bounded subset of a Banach space **X** onto a compact set.

Equation (6.2.1) given on a Banach space **X** is frequently reduced to (6.2.15) to attempt to find an approximation $f_n \in \mathbf{Y}$, with **Y** a dense subset of **X**. The same technique is used to replace κ by a compact operator κ_n, which maps **X** into **Y**. Regarding the convergence of κ_n to κ, we have:

Theorem 6.2.6 *Let $\kappa : \mathbf{X} \to \mathbf{X}$ be a bounded linear operator defined on a Banach space $\mathbf{X}$, and let $\{\kappa_n\}$ be a sequence of compact linear operators such that $\kappa_n : \mathbf{X} \to \mathbf{Y}$, with $\mathbf{Y}$ a dense subset of $\mathbf{X}$. If*
(i)

$$\sup_n \|\kappa_n\| = c < \infty, \tag{6.2.25}$$

and if
(ii)

$$\kappa_n f \to \kappa f \tag{6.2.26}$$

as $n \to \infty$, for all $f \in Y$, then $\kappa_n f \to \kappa f$ for all $f \in \mathbf{X}$. Moreover, κ is a compact linear operator on $\mathbf{X}$.

Proof. It is readily seen (Problem 6.2.6) that $\|\kappa\| = c$. Hence, if $f \in \mathbf{X}$ and $g \in \mathbf{Y}$, we have

$$\begin{aligned} \|\kappa f - \kappa_n f\| &\leqq \|\kappa f - \kappa g\| + \|\kappa g - \kappa_n g\| + \|\kappa_n g - \kappa_n f\| \\ &\leqq 2c\|f - g\| + \|\kappa g - \kappa_n g\|. \end{aligned} \tag{6.2.27}$$

Now, since $\kappa_n g \to \kappa g$, it follows that given any $\varepsilon > 0$, the last term on the extreme right-hand side of (6.2.27) is less than $\varepsilon/2$ for all n sufficiently large. Next, taking $g \in \mathbf{Y}$ such that $\|f - g\| < \varepsilon/(4c)$, it follows that the first term on the extreme right-hand side of (6.2.27) is also less than $\varepsilon/2$. Hence, $\|\kappa f - \kappa_n f\| < \varepsilon$, which proves that $\kappa_n f \to \kappa f$ as $n \to \infty$, for all $f \in \mathbf{X}$.

Next, we show that κ is compact. To this end, we start with an arbitrary bounded sequence $\{f_n\}$, $f_n \in \mathbf{X}$, $\|f_n\| \leqq C_1$, and proceed in a manner similar to the proof of Montel's Theorem. First, we pick a subsequence $\{f_n^{(1)}\}$ of $\{f_n\}$, which is such that $\{\kappa_1 f_n^{(1)}\}$ converges. Next, we pick a subsequence $\{f_n^{(2)}\}$ of $\{f_n^{(1)}\}$, such that $\{\kappa_2 f_n^{(2)}\}$ converges, and so on. Set $g_n = f_n^{(n)}$, and for arbitrary $m > n$, consider

$$\begin{aligned} \|\kappa g_n - \kappa g_m\| &\leqq \|\kappa g_n - \kappa_n g_n\| + \|\kappa_n g_n - \kappa_n g_m\| + \|\kappa_n g_m - \kappa g_m\| \\ &\leqq \|\kappa - \kappa_n\| \, [\|g_m\| \; + \; \|g_n\|] + \|\kappa_n\| \, |g_m - g_n\|. \end{aligned} \tag{6.2.28}$$

That is, since $\|g_m\| \leqq C_1$, $\|g_n\| \leqq C_1$, $\|\kappa_n\| \leqq C$, and since both $\|g_m - g_n\|$ and $\|\kappa - \kappa_n\|$ can be made as small as we please by taking m and n sufficiently large, it follows that $\{\kappa g_n\}$ is a Cauchy sequence. Thus, κ is a compact operator.

This completes the proof of Theorem 6.2.6. ■

6.2.3 Fredholm Alternative

The concept of compactness is important to obtain an approximate solution of equation (6.1.6) by replacing the kernel $K(x,t)$ by a *degenerate kernel*, i.e., by an equation of the form

$$f(x) - \lambda \int_{\mathcal{B}} \sum_{j=1}^{n} u_j(x) v_j(t) f(t) dt = g(x), \quad x \in \mathcal{B}, \tag{6.2.29}$$

with the u_j linearly independent functions on B. Indeed, we see from (6.2.29), that

$$f(x) = \lambda \sum_{j=1}^{n} u_j(x) c_j + g(x), \tag{6.2.30}$$

with

$$c_j = \int_{\mathcal{B}} v_j(t) f(t) dt. \tag{6.2.31}$$

Multiplying (6.2.30) by $v_i(x)$ and integrating over B we get

$$c_i - \lambda \sum_{j=1}^{n} a_{ij} c_j = b_i, \tag{6.2.32}$$

with

$$\begin{aligned} a_{ij} &= \int_{\mathcal{B}} v_i(x) u_j(x) dx, \\ b_i &= \int_{\mathcal{B}} v_i(x) g(x) dx. \end{aligned} \tag{6.2.33}$$

The system (6.2.32) is a system of linear algebraic equations for the unknowns $c_1, \cdots, c_n$; if this system has a solution, we immediately obtain the solution f of (6.2.29) given by (6.2.30). There is a unique solution if $\det(I - \lambda A) \neq 0$, with I the unit matrix, and $A = [a_{ij}]$. Since $\det(I - \lambda A)$ is a polynomial of degree $\leqq n$ in λ, there are at most n values of λ for which $\det(I - \lambda A) = 0$. These n special values are called the *characteristic values* of Equation (6.2.29).

Let us consider the special case $\det(I - \lambda A) = 0$. We gain much insight into this case by considering the homogeneous form of the adjoint equation corresponding to (6.2.29), i.e.,

$$\phi(x) - \overline{\lambda} \int_{\mathcal{B}} \sum_{j=1}^{n} \overline{v_j(x) u_j(t)} \phi(t) dt = 0. \tag{6.2.34}$$

Setting

$$\gamma_j = \int_B \overline{u_j(t)}\phi(t)dt \tag{6.2.35}$$

in (6.2.34), then multiplying the resulting equation by $\overline{u_i(x)}$ and integrating over B, we get the system of linear algebraic equations

$$(I - \overline{\lambda}A^*)\,\gamma = 0. \tag{6.2.36}$$

Here $A^* = [\overline{a_{ji}}]$ is the *adjoint* of the square matrix A, and γ is the vector, $\gamma = (\gamma_1, \cdots, \gamma_n)^T$. Clearly, the system (6.2.36) can have a nontrivial solution, if and only if, $\det(I - \overline{\lambda}A^*) = 0$. If $\det(I - \overline{\lambda}A^*) = 0$, then there exists a nontrivial solution of (6.2.34), given by

$$\phi(x) = \overline{\lambda}\sum_{i=1}^{n} \gamma_i \overline{v_i(x)}. \tag{6.2.37}$$

Now note that the equation $\det(I - \overline{\lambda}A^*) = 0$ has exactly the same solutions λ as the equation $\det(I - \lambda A) = 0$. Hence, setting $b = (b_1, \cdots, b_n)^T$, $c = (c_1, \cdots, c_n)^T$, it follows that if (6.2.29) has a solution f given by (6.2.30), then

$$0 = \gamma^*\,(I - \lambda A)\,c = \gamma^*\,b. \tag{6.2.38}$$

Equivalently, we must have

$$\int_B g(x)\overline{\phi(x)}dx = 0 \tag{6.2.39}$$

for every solution ϕ of (6.2.34). Conversely, if (6.2.39), and hence (6.2.38) holds for every solution ϕ of (6.2.34), then b is in the column space spanned by $I - \lambda A$, which implies that the equation $(I - \lambda A)\,c = b$ has a solution.

Summing up, we have:

Theorem 6.2.7 (Fredholm Alternative): *The given integral equation (6.2.29) has a unique solution for all g, if and only if, the homogeneous equation (6.2.34) has only $\phi(x) \equiv 0$ as a solution. The equation*

$$w(x) - \lambda\int_B \sum_{j=1}^{n} u_j(x)v_j(t)w(t)dt = 0 \tag{6.2.40}$$

has the same number of linearly independent solutions as (6.2.34). If (6.2.34) has a nontrivial solution ϕ, then (6.2.29) has a non-unique solution, if and only if

$$\int_B g(x)\overline{\phi(x)}dx = 0 \tag{6.2.41}$$

for all solutions ϕ of (6.2.34).

Let us now discuss the existence of eigensolutions and regular values of (6.1.6) with the notation of (6.2.1), κ given by (6.2.4), a compact operator. To this end, let us assume that given any $\varepsilon > 0$, we can find a degenerate (and compact) kernel $K^{(2)}(x,t)$ of the form

$$K^{(2)}(x,t) = \sum_{j=1}^{n} u_j(x) v_j(t), \tag{6.2.42}$$

such that if the finite dimensional operator $\kappa^{(2)}$ is defined by

$$\kappa^{(2)} f(x) = \int_{\mathcal{B}} K^{(2)}(x,t) f(t) dt, \tag{6.2.43}$$

and $\kappa^{(1)} = \kappa - \kappa^{(2)}$, then $\|\kappa^{(1)}\| < \varepsilon$. We can then rewrite (6.2.1) in the form

$$(I - \lambda \kappa^{(1)}) f = g + \lambda \kappa^{(2)} f. \tag{6.2.44}$$

Let us denote the right-hand side of (6.2.44) by $g^{(1)}$, and let us temporarily assume $g^{(1)}$ to be known. In this case, we know from Example 6.2.3 that if $\lambda \in \mathcal{D}(0; 1/\varepsilon) = \{\lambda \in \mathbb{C} : |\lambda| < 1/\varepsilon\}$, then $(I - \lambda \kappa^{(1)})^{-1}$ exists, and we can set

$$(I - \lambda \kappa^{(1)})^{-1} g^{(1)} = g^{(1)} + \lambda \mathcal{R}^{(1)} g^{(1)} \tag{6.2.45}$$

with the *resolvent* $\mathcal{R}^{(1)}$ given by

$$\mathcal{R}^{(1)} = \kappa^{(1)} + \lambda (\kappa^{(1)})^2 + \lambda^2 (\kappa^{(1)})^3 + \cdots. \tag{6.2.46}$$

Furthermore, it is readily seen that if $\lambda \in \mathcal{D}(0; 1/\varepsilon)$, then

$$(\mathcal{R}^{(1)} g^{(1)})(x) = \int_{\mathcal{B}} R^{(1)}(x,t,\lambda) g^{(1)}(t) dt \tag{6.2.47}$$

with

$$R^{(1)}(x,t,\lambda) = \sum_{j=1}^{\infty} \lambda^{j-1} K_j^{(1)}(x,t), \tag{6.2.48}$$

and (see Problem 6.2.8) with $K^{(1)}(x,t) = K(x,t) - K^{(2)}(x,t)$, the $K_j^{(1)}, j = 1, 2, \cdots$, defined by

$$\begin{aligned} K_1^{(1)}(x,t) &= K^{(1)}(x,t) \\ K_j^{(1)}(x,t) &= \int_{\mathcal{B}} K^{(1)}(x,\tau) K_{j-1}^{(1)}(\tau,t) d\tau, j \geqq 2. \end{aligned} \tag{6.2.49}$$

Hence, it follows that for a.e. x on B, $(\mathcal{R}^{(1)} g)(x)$ is an analytic function of $\lambda \in \mathcal{D}(0; 1/\varepsilon)$.

Using $\mathcal{R}^{(1)}$, we may therefore replace Equation (6.2.1) with κ given by

$$f - \lambda(I + \lambda\mathcal{R}^{(1)})\kappa^{(2)}f = (I + \lambda\mathcal{R}^{(1)})g. \tag{6.2.50}$$

We have thus been able to replace Equation (6.1.6) with a new integral equation having a degenerate kernel. Written in full, this equation takes the form

$$f(x) - \lambda \int_{\mathcal{B}} K^{(3)}(x,t,\lambda)f(t)dt = g^{(3)}(x,\lambda) \tag{6.2.51}$$

with

$$\begin{cases} K^{(3)}(x,t,\lambda) = \sum_{j=1}^{n} u_j(x,\lambda)v_j(t), \\ u_j(x,\lambda) = u_j(x) + \lambda \int_{\mathcal{B}} R^{(1)}(x,t,\lambda)u_j(t)dt, \\ g^{(3)}(x,\lambda) = g(x) + \lambda \int_{\mathcal{B}} R^{(1)}(x,t,\lambda)g(t)dt. \end{cases} \tag{6.2.52}$$

The solution of Equation (6.2.51) can be obtained in the same way as the solution of Equation (6.2.29). As in the case of (6.2.29), the solution to (6.2.51) is unique, except when λ is a characteristic value of the Equation (6.1.6). Now from our analyticity observations made about $\mathcal{R}^{(1)}$ it follows that for a.e. x on B the functions $u_j(x,\lambda)$ and $g^{(3)}(x,\lambda)$ are analytic in λ in $\mathcal{D}(0;1/\varepsilon)$. The characteristic values of Equation (6.1.6) are given by the solution of the equation $\det[I - \lambda A(\lambda)] = 0$ with $A(\lambda) = [a_{ij}(\lambda)]$, and with $a_{ij}(\lambda) = \int_{\mathcal{B}} v_i(x)u_j(x,\lambda)dx$. Thus $\det[I - \lambda A(\lambda)]$ is an analytic function of λ in $\mathcal{D}(0;1/\varepsilon)$, and hence the equation $\det[I - \lambda A(\lambda)] = 0$ can have at most a finite number of solutions (counting multiplicity) in any closed subset of $\mathcal{D}(0;1/\varepsilon)$. However, since ε was chosen arbitrarily, it follows that Equation (6.1.6) can have at most a finite number of characteristic values in the interior of any bounded disc in the complex plane.

Summing up, we have proved

Theorem 6.2.8 (Fredholm Alternative–General): *Let the kernel $K(x,t)$ in (6.1.6) be such that $(\kappa f)(x) = \int_{\mathcal{B}} K(x,t)f(t)dt$ maps the Banach space $\mathbf{X}$ onto $\mathbf{X}$ and such that $K(x,t)$ can be approximated by sequence of degenerate kernels*

$$\left\{K^{(n)}(x,t)\right\} = \left\{\sum_{j=1}^{n} u_j^{(n)}(x)v_j^{(n)}(t)\right\}.$$

Let $\kappa^{(n)}$ defined by

$$(\kappa^{(n)}f)(x) = \int_{\mathcal{B}} K^{(n)}(x,t)f(t)dt$$

be a compact operator from **X** *to* **X**, *such that* $\|\kappa - \kappa^{(n)}\| \to 0$ *as* $n \to \infty$. *Let* κ^* *be defined on* **X** *by* $(\kappa^* f)(x) = \int_{\mathcal{B}} \overline{K(t,x)} f(t)dt$. *Then Equation (6.1.6) has a solution, if and only if,*

$$\int_{\mathcal{B}} g(x)\overline{\phi(x)}dx = 0 \tag{6.2.53}$$

for every solution ϕ *of the equation*

$$(I - \lambda\kappa^*)\phi = 0. \tag{6.2.54}$$

Now, if $A(\lambda)$ is defined as in the proof of Theorem 6.2.7, then due to the fact that the function $\det[I - \lambda A(\lambda)]$ can have, at most, a finite number of zeros (counting multiplicity) in any bounded disc in the complex λ plane, it follows that the dimension of the invariant subspace corresponding to a bounded characteristic value λ_0 must also be finite, and indeed, the dimension of this invariant subspace must be exactly equal to the multiplicity of λ_0. Moreover, if λ_0 denotes a characteristic value, then there exists a positive number δ, depending on λ_0, such that no other characteristic value of (6.1.6) exists in the closed disc $\overline{\mathcal{D}(\lambda_0;\delta)}$.

If λ is not a characteristic value, then the solution of (6.2.1) with κ defined by (6.2.4) may be expressed in the form

$$f = g + \lambda\mathcal{R}g \tag{6.2.55}$$

with

$$(\mathcal{R}g)(x) = \int_{\mathcal{B}} R(x,t,\lambda)g(t)dt \tag{6.2.56}$$

and, where for almost every x and t on B, the resolvent $R(x,t,\lambda)$ is an analytic function of λ, except at the characteristic values of (6.1.6), where $\mathcal{R}(x,t,\lambda)$ has a pole. We may deduce by Kramer's rule applied to $K^{(3)}$ of (6.2.52) (Problem 6.2.9), that in a neighborhood of a characteristic value λ_0 of multiplicity m $R(x,t,\lambda)$ has the form

$$R(x,t,\lambda) = \sum_{j=1}^{m} \frac{r_j(x,t)}{(\lambda-\lambda_0)^j} + S(x,t,\lambda). \tag{6.2.57}$$

Furthermore, we may readily deduce that for $j = 1, 2, \cdots, m$,

$$\begin{aligned}&\int_{\mathcal{B}} r_j(x,t)S(x,t,\lambda)dt = 0,\\&\int_{\mathcal{B}} r_j(x,t)S(x,t,\lambda)dx = 0,\end{aligned} \tag{6.2.58}$$

i.e., that the functions r_j and S are orthogonal. This may also be established if, after reduction of the degenerate system (6.5.21) to a system of algebraic equations, we then transform the resulting matrix to Jordan-box form.

6.2.4 Perturbed Equations

Now let $\kappa(\zeta) : \mathbf{X} \to \mathbf{X}$ and $\kappa_0 = \kappa(0) : \mathbf{X} \to \mathbf{X}$, be two finite rank compact operators of the form

$$(\kappa(\zeta)f)(x) = \int_{\mathcal{B}} \sum_{j=1}^{m} u_j(x,\zeta)v_j(t,\zeta)f(t)dt \tag{6.2.59}$$

$$(\kappa_0 f)(x) = \int_{\mathcal{B}} \sum_{j=1}^{m} u_j(x,0)v_j(t,0)f(t)dt \tag{6.2.60}$$

where for almost every fixed $(t,x) \in \mathcal{B} \times \mathcal{B}$, the functions $u_j(x,\zeta)v_j(x,\zeta)$ are analytic in ζ in a disc $\mathcal{D}(0;\delta_1)$, with δ_1 a positive constant. Let λ_0 be a characteristic value of κ_0, and let ξ_0 be an eigenvector associated with λ_0. Let δ be a positive number that is sufficiently small, so that the closed disc $\overline{\mathcal{D}(\lambda_0;\delta)}$ contains no other characteristic values of κ_0. Then we can prove

Theorem 6.2.9 (Kato, Lancaster): (a) *If λ_0 has multiplicity 1, then for $|\zeta|$ sufficiently small, $\kappa(\zeta)$ has a characteristic value $\lambda_0(\zeta)$ in $\mathcal{D}(\lambda_0;\delta)$, which is the only characteristic value of $\kappa(\zeta)$ in the closed disc $\overline{\mathcal{D}(\lambda_0,\delta)}$. Moreover, $\lambda_0(\zeta)$ its associated eigenvector $\xi_0(\zeta) = \xi_0(x,\zeta)$ are analytic functions of ζ in some neighborhood of $\zeta = 0$ for almost every $x \in \mathcal{B}$.*
(b) *If λ_0 has multiplicity $p > 1$, then for $|\zeta|$ sufficiently small, $\kappa(\zeta)$ has exactly p characteristic values, counting multiplicity, in the disc $\mathcal{D}(\lambda_0,\delta)$, and no characteristic value on the boundary of this disc. Each characteristic value $\lambda_{0j}(\zeta)$, and, for almost every $x \in \mathcal{B}$, its associated eigenvector $\xi_{0j}(\zeta) = \xi_{0j}(x,\zeta)$ is an analytic function of $\zeta^{1/q}$ in a neighborhood of $\zeta = 0$, for some integer q such that $1 \leqq q \leqq p$. Furthermore*

$$\lambda_0^*(\zeta) = \frac{1}{p}\sum_{j=1}^{p} \lambda_{0j}(\zeta) \tag{6.2.61}$$

is an analytic function of ζ in a neighborhood of $\zeta = 0$, and there exist constants c_j, $j = 1,\cdots,p$, such that for almost every $x \in \mathcal{B}$,

$$\xi^*(\zeta) = \sum_{j=1}^{p} c_j \xi_{0j}(\zeta) \tag{6.2.62}$$

is an analytic function of ζ in a neighborhood of $\zeta = 0$.

Proof. This is a consequence of the explicit solution procedure (6.2.29) through (6.2.33), Kramer's rule, and Hurwitz's theorem. The details are left to Problem 6.2.11. ■

Corollary 6.2.10 *Let $\kappa : \mathbf{X} \to \mathbf{X}$ be a compact linear operator of the form (6.2.4), and let $\{\kappa_n\}$ be a sequence of compact linear operators of the form (6.2.59), such that $\delta_n \equiv \|\kappa - \kappa_n\| \to 0$ as $n \to \infty$. Let λ_0 be a characteristic value of K, such that λ_0 has multiplicity $p \geqq 1$, and let ξ_0 denote an eigenvector (or eigenfunction), associated to λ_0. Then there exists a sequence of positive numbers $\{\varepsilon_n\}$, such that $\varepsilon_n \to 0$ as $n \to \infty$, and such that for all n sufficiently large, κ_n has exactly p characteristic values $\lambda_{0j}^{(n)}(\zeta)$, $j = 1, \cdots, p$ in the disc $\mathcal{D}(\lambda_0; \varepsilon_n)$, and no characteristic value on the boundary of this disc. Let $\xi_{0j}^{(n)}(\zeta)$ denote a non-trivial eigenfunction associated to $\lambda_{0j}^{(n)}(\zeta)$. Then there exists a constant C, independent of n, such that for all n sufficiently large,*

$$
\begin{aligned}
&|\lambda_0 - \lambda_{0j}^{(n)}| \leqq C\delta_n^{1/p} \\
&|\xi_0 - \xi_{0j}^{(n)}| \leqq C\delta_n^{1/p} \\
&|\lambda_0 - \frac{1}{p}\sum_{j=1}^{p} \lambda_{0j}^{(n)}| \leqq C\delta_n.
\end{aligned}
\tag{6.2.63}
$$

Moreover, there exist constants $c_1, \cdots, c_p$, independent of n, such that for all n sufficiently large

$$
\left| \xi_o - \sum_{j=1}^{p} c_j \xi_{oj}^{(n)} \right| \leqq C\delta_n. \tag{6.2.64}
$$

Proof. The proof may be carried out by combining Theorem 6.2.5 and the results of Problem 6.2.12. The details are left to problem 6.2.13. ■

6.2.5 Tikhonov Regularization

Tikhonov regularization has proved to be a powerful practical tool to deal with Fredholm integral equations of the first kind, and with other *ill-posed*, linear or nonlinear integral equations. In the present section we illustrate the role of Tikhonov regularization, to deal with Fredholm integral equations of the first kind.

Let $\mathbf{X}$ be a *Hilbert space*, i.e., let $\mathbf{X}$ be a Banach space on which an *inner product* $(f, g) \in \mathbb{C}$ is defined for all f and g in $\mathbf{X}$, and for which $\|f\| = (f, f)^{1/2}$. The inner product (f, g) has the properties:

(a) For all f and g in $\mathbf{X}$,

$$(f,g) = \overline{(g,f)} \tag{6.2.65}$$

and
(b) For all f, g and h in $\mathbf{X}$, and for all α and β in $\mathbb{C}$,

$$(\alpha f + \beta g, h) = \alpha(f,h) + \beta(g,h). \tag{6.2.66}$$

The integral equations we shall consider take the form (6.1.5), with

$$\int_{\mathcal{B}} \int_{\mathcal{B}} |K(x,t)|^2 dx dt < \infty, \tag{6.2.67}$$

and with $g \in \mathbf{L}^2(B)$. In operator notation we may write (6.1.5) in the form

$$\kappa f = g, \tag{6.2.68}$$

with $\kappa : \mathbf{X} \to \mathbf{X}$ a compact operator.

Let the *adjoint-operator* κ^* be defined by the requirement that

$$(\kappa f, g) = (f, \kappa^* g) \tag{6.2.69}$$

for all f and g in $\mathbf{X}$. Tikhonov regularization of (6.2.68) involves replacing Equation (6.2.68) by the equation

$$(\alpha I + \kappa^* \kappa) f_\alpha = \kappa^* g \tag{6.2.70}$$

with α being a (usually small) positive number. Equation (6.2.70) is a Fredholm integral equation of the second kind, which has a solution f_α for all $g \in \mathbf{X}$.

More specifically, Tikhonov regularization enables us to replace (6.1.5) by the Fredholm integral equation of the second kind,

$$\alpha f_\alpha(x) + \int_{\mathcal{B}} G(x,t) f_\alpha(t) dt = g^*(x), \quad x \in \mathcal{B}, \tag{6.2.71}$$

with

$$\left\{ \begin{array}{rcl} G(x,t) & = & \displaystyle\int_{\mathcal{B}} \overline{K(\tau,x)} K(\tau,t) d\tau \\ g^*(x) & = & \displaystyle\int_{\mathcal{B}} \overline{K(\tau,x)} g(\tau) d\tau. \end{array} \right. \tag{6.2.72}$$

We next use convenient Hilbert space notation to illustrate the practical usefulness of the above Tikhonov regularization procedure. To this end, let $\{\omega_k\}$ be a complete orthonormal sequence in $\mathbf{X}$, i.e., $(\omega_k, \omega_\ell) = \delta_{k-\ell}$, with $\delta_{k-\ell}$ the Kronecker delta, and such that every $g \in \mathbf{X}$ has the expansion and norm,

$$g = \sum (g, \omega_k)\omega_k$$
$$\|g\| = (\sum |(g, \omega_k)|^2)^{1/2}. \tag{6.2.73}$$

Let $\kappa : \mathbf{X} \to \mathbf{X}$, and furthermore, for a given $f \in \mathbf{X}$, $f = \sum f_k \omega_k$, let κf be defined by

$$\kappa f = \sum \mu_k f_k \omega_k, \tag{6.2.74}$$

with the μ_k complex scalars satisfying the relation

$$\|\kappa\| = \left(\sum |\mu_k|^2\right)^{1/2} < \infty, \tag{6.2.75}$$

a necessary requirement for compactness. Thus μ_k is an eigenvalue, and ω_k is an eigenfunction of κ.

Using (6.2.73) and (6.2.74) we deduce the "formal solution" to (6.2.68), namely

$$f = \sum (g_k/\mu_k)\omega_k; \; g_k = (g, \omega_k). \tag{6.2.76}$$

This solution is in $\mathbf{X}$, provided that

$$\|f\|^2 = \sum |g_k/\mu_k|^2 < \infty, \tag{6.2.77}$$

a relation usually *not* satisfied. For example, a solution does not exist if some of the μ_k are zero, with the corresponding g_k non-zero. Even if f given by (6.2.76) is in $\mathbf{X}$, the equation (6.2.68) is *ill-posed.* In view of (6.2.75), we must have $\mu_k \to 0$ as $\kappa \to \infty$; on the other hand, adding a perturbation $\Delta g = \varepsilon\omega_m$ to the right-hand side of (6.2.68), we increase the square of the norm of the solution by $(\varepsilon^2 + 2\varepsilon g_m)/|\mu_m|^2$, and this increase may be made arbitrarily large by choosing m sufficiently large, even though ε may be small.

Let us now assume for simplicity, that in addition to κf being defined by (6.2.74), we also have

$$\kappa^* f = \sum \overline{\mu_k} f_k \omega_k. \tag{6.2.78}$$

Then (6.2.69) is satisfied. Using (6.2.74) and (6.2.78) in (6.2.70), we get the *regularized solution*

$$f_\alpha = \sum \frac{\overline{\mu_k}}{\alpha^2 + |\mu_k|^2} g_\kappa \omega_\kappa, \tag{6.2.79}$$

which is clearly in $\mathbf{X}$ for all $\alpha > 0$.

If none of the μ_k are zero, then as $\alpha \to 0$, $f_\alpha \to f$, with f given by (6.2.76), provided (6.2.77) is satisfied. In the case where some of the μ_k are zero, but with (6.2.77) still satisfied, where the sum in (6.2.77) extends only over the non-zero μ_k, then as $\alpha \to 0$, f_α converges to the f given in (6.2.76), the sum extending only over the non-zero μ_k. This special solution is the unique f of smallest norm among the set of all $f \in \mathbf{X}$ which minimize $\|\kappa f - g\|^2$.

Even if Equation (6.2.68) does have a unique solution $f \in \mathbf{X}$, due to the ill-posedness of this equation, it is usually convenient, in practice, to obtain a regularized approximate solutions f_α, by solving (6.2.70) for a small value of α. One thus has the problem of determining a suitable value of α. While the choice of α is important, it depends on the specific problem, the specific method of solution, and the error in g. We shall illustrate the selection of regularization parameters α in Section 6.9.

•

Problems for Section 6.2

6.2.1 Let $\mathbf{X}$ and $\mathbf{Y}$ be Banach spaces, and let $T : \mathbf{X} \to \mathbf{Y}$ be a linear operator. Prove that T is continuous, if and only if T is bounded.

6.2.2 let $T : \mathbf{X} \to \mathbf{Y}$ be a linear operator with $\mathbf{X}$ and $\mathbf{Y}$ Banach spaces. Prove that if there exists a positive constant c such that $\|Tf\|_{\mathbf{Y}} \geqq c\|f\|_{\mathbf{X}}$, for all $f \in \mathbf{X}$, then T^{-1} exists, and $\|T^{-1}\| \leqq c^{-1}$.

6.2.3 Prove Banach's theorem: If $\mathbf{X}$ is a Banach space and if $T : \mathbf{X} \to \mathbf{X}$, with $\|T\| < 1$, then $(I-T)^{-1}$ exists, and $\|(I-T)^{-1}\| \leqq (I - \|T\|)^{-1}$.

6.2.4 Prove Theorem 6.2.1. [Hint: Show that (a) $f_n \in \overline{\mathcal{B}(f_0; r)}$ for all n; (b) $\{f_n\}$ is a Cauchy sequence $\Rightarrow f_n \to f^*$; (c) $f^* = Tf^*$; and (d) Use (6.2.5) to establish the uniqueness of f^*.]

6.2.5 Prove that if T is defined as in (6.2.4), and if we set

$$\kappa_n = I + \kappa + \kappa^2 + \cdots + \kappa^{n-1}$$

then

$$\|(I - \kappa)^{-1} - \kappa_n\| \leqq \frac{\|\kappa\|^n}{1 - \|\kappa\|}.$$

6.2.6 Under the conditions of Theorem 6.2.6, prove that $\|\kappa\| = \sup\{\|\kappa x\| : \ \|x\| \leqq 1\} = c$.

6.2.7 Prove Lemma 6.2.4, starting with the identity

$$S^{-1}(T - S)T^{-1} = S^{-1} - T^{-1}.$$

6.2.8 Let κ be defined as in (6.2.4). Prove that for $j = 1, 2, \cdots$

$$(\kappa^j f)(x) = \int_{\mathcal{B}} K_j(x,t) f(t) dt$$

with $K_1(x,t) = K(x,t)$, and with

$$K_j(x,t) = \int_{\mathcal{B}} K(x,\tau) K_{j-1}(\tau,t) d\tau, \quad j \geqq 2.$$

6.2.9 Let $R(x,t,\lambda)$ be defined via (6.2.55) and (6.2.56). Prove that (6.2.57) and (6.2.58) hold.

6.2.10 Let **X** be defined as in Theorem 6.2.8, and let λ_0 be a characteristic value of equation (6.2.1). Prove that there exists a positive number δ such that the closed disc $\overline{\mathcal{D}(\lambda_0;\delta)}$ contains no other characteristic values of Equation (6.2.1).

6.2.11 Give a detailed proof of Theorem 6.2.9. [Hint: Prove that for $p > 1$, the p roots λ of the polynomial equation

$$(\lambda - \lambda_0)^p \left[a_{m-p}(\zeta)\lambda^{m-p} + \cdots + a_0(\zeta)\right] + \zeta \left[b_m(\zeta)\lambda^m + \cdots b_0(\zeta)\right] = 0$$

with the a_j and b_j analytic functions of ζ, are analytic functions of $\zeta^{1/p}$ in a neighborhood of $\zeta = 0$.]

6.2.12 Use (6.2.57) to deduce the identity

$$\frac{1}{2\pi i} \int_{\partial\mathcal{D}(\lambda_0,\delta)} \left[(I - \lambda\kappa)^{-1} f\right] (\lambda - \lambda_0)^{j-1} d\lambda$$

$$= \int_{\mathcal{B}} r_j(x,t) f(t) dt, \quad j = 1, \cdots, m.$$

6.2.13 Use the results of the previous problem and Theorem 6.2.5 to give a detailed proof of Corollary 6.2.10. [Hint: Use the bound given in Theorem 6.2.5 on the boundary of the contour integral expression of the previous problem.]

6.2.14 Prove that f_α given by (6.2.79) is in **X** for all $\alpha > 0$.

6.3 Reduction to Algebraic Equations

We shall briefly discuss the *Galerkin method*, the *Nyström method*, the *generalized inverse* of a matrix, and *numerical stability* of the solution of a system of linear algebraic equations.

6.3.1 Galerkin Method

The Galerkin method, which approximates a solution of an equation of the form (6.2.1) defined on a Banach space $\mathbf{X}$, is simple to describe. We start with a set of basis functions $\{\psi_k\}_{k=1}^n$, and we assume that for each $f \in \mathbf{X}$, we can determine a unique set of numbers $c_1, \cdots, c_n$ in $\mathbb{C}$, such that the *projection* $\mathcal{P}_n : \mathbf{X} \to \mathbf{X}$ is defined, with

$$\mathcal{P}_n f = \sum_{k=1}^{n} c_k \psi_k. \tag{6.3.1}$$

The Galerkin method then enables us to replace Equation (6.2.1) by the approximating one, namely

$$(I - \lambda \mathcal{P}_n \kappa) f_n = \mathcal{P}_n g, \tag{6.3.2}$$

with f_n given by the right-hand side of (6.3.1), and with the coefficients $c_1, \cdots, c_n$ now determined by solving the system of algebraic equations resulting from equating the coefficients of ψ_k on each side of (6.3.2). For example, if κ is linear, we arrive at a system of linear equations, and the procedure is similar to that used in (6.2.29) through (6.2.33).

In practice one attempts to select projections $\mathcal{P}_n$ which are easily dealt with, and for which $\|\kappa - \mathcal{P}_n \kappa\|$ and $\|g - \mathcal{P}_n g\|$ are small. We can then obtain a bound on the error of approximation to the solution f of (6.2.1) by f_n. In the case of κ linear, we can use Theorem 6.2.5 to obtain such a bound.

6.3.2 Nyström's Method

If it can be applied, *Nyström's method* is a convenient procedure to reduce a Fredholm integral equation of the form (6.1.6) to a system of algebraic equations. This method can be effectively applied in the case when the kernel $K(x,t)$ in (6.1.6) is continuous on $B \times B$. Applying a suitable quadrature formula

$$\int_B g(x)dx \approx \sum_{j=1}^{n} w_j g(x_j) \tag{6.3.3}$$

to the integral in Equation (6.1.6), we can reduce (6.1.6) to the equation

$$f(x) - \lambda \sum_{j=1}^{n} w_j K(x, x_j) f_j = g(x). \tag{6.3.4}$$

Taking $x = x_i, i = 1, 2, \cdots, n$, we may replace (6.3.4) by the system of algebraic equations

$$f_i - \lambda \sum_{j=1}^{n} w_j K(x_i, x_j) f_j = g(x_i). \tag{6.3.5}$$

After solving for $f_1, \cdots, f_n$, we thus use (6.3.4) to get the approximation

$$f(x) = g(x) + \lambda \sum_{j=1}^{n} w_j K(x, x_j) f_j. \tag{6.3.6}$$

Similarly, if $K(x, y, z)$ is continuous for all $(x, y) \in \mathcal{B} \times \mathcal{B}$ and all z is a suitable range of values containing the range of the solution, we may apply the formula (6.3.3) to reduce the nonlinear integral equation

$$f(x) - \lambda \int_{\mathcal{B}} K(x, t, f(t)) dt = g(x) \tag{6.3.7}$$

to the system of nonlinear algebraic equations

$$f_i - \lambda \sum_{j=1}^{n} w_j K(x_i, x_j, f_j) = g(x_i), \quad (i = 1, 2, \cdots, n) \tag{6.3.8}$$

and upon solving this for the f_i, we get the Nyström approximate solution to (6.3.7), namely

$$f(x) = g(x) + \lambda \sum_{j=1}^{n} w_j K(x, x_j, f_j). \tag{6.3.9}$$

6.3.3 The Generalized Inverse Procedure

Methods to solve linear and nonlinear algebraic equations are very important to the solution of integral equations. While this topic is too vast to be covered here, inasmuch as lengthy textbooks have been written on these subjects (see the Historical Remarks at the end of this chapter), we shall nevertheless summarize some properties of the *generalized inverse* of a matrix, due to its frequent applicability, as well as its power as a tool to study *Tikhonov regularization.* In addition, we shall briefly examine the effect of error in the elements of a linear system of algebraic equations on the solution of the corresponding integral equations, which gave rise to the linear system.

Definition 6.3.1 Let A be a given rectangular $m \times n$ matrix of elements $a_{ij} \in \mathbb{C}$, $1 \leqq i \leqq m$, $1 \leqq j \leqq n$. A matrix X is the generalized inverse of a matrix A, if and only if:

(i) $AXA = A$

(ii) $XAX = X$

(iii) $(AX)^* = AX$

(iv) $(XA)^* = XA$.

We shall denote the generalized inverse of A by $A^\dagger$.

Theorem 6.3.2 *Some properties of $A^\dagger$ are:*

(i) *If A is non-singular, then $A^\dagger = A^{-1}$;*

(ii) *If A is $m \times n$, then $A^\dagger$ is $n \times m$;*

(iii) $(A^*)^\dagger = (A^\dagger)^*$;

(iv) *If $A^* = A$ then $(A^\dagger)^* = A^\dagger$;*

(v) $(A^\dagger)^\dagger = A$

(vi) $0^\dagger = 0^T$, *with 0^T is the $n \times m$ matrix of zeros;*

(vii) *If $c \neq 0$, then $(cA)^\dagger = (1/c)A^\dagger$;*

(viii) *If A has full column rank, then $A^\dagger = (A^*A)^{-1}A^*$;*

(ix) *If A has full row rank, then $A^\dagger = A^*(AA^*)^{-1}$; and*

(x) *If U and V are unitary, then $(UAV)^\dagger = V^*A^\dagger U^*$.*

Proof. See Problem 6.3.2. ■

Theorem 6.3.3 *Every $m \times n$ matrix A has a unique generalized inverse.*

Proof. See Problem 6.3.3. ■

The generalized inverse of a matrix A arises naturally, in the process of finding a least squares approximation x to the problem

$$Ax = b, \tag{6.3.10}$$

with $A = [a_{ij}]_{(m\times n)}$ and $b = [b_i]_{(m\times 1)}$ given.

Theorem 6.3.4 *If the equation (6.3.10) has solutions, then they are given by*

$$x = \left\{A^{\dagger}b + \left(I - A^{\dagger}A\right)y : y \in \mathbb{C}^n\right\}. \tag{6.3.11}$$

Otherwise

$$x = A^{\dagger}b \tag{6.3.12}$$

is the unique vector of smallest ℓ^2 norm among all $n \times 1$ vectors z which minimize the ℓ^2 norm of $Ax - b$.

Proof. See Problem 6.3.4. ■

The *singular value decomposition* of an $m \times n$ matrix A, with $m \geqq n$ is a procedure which has been used successfully in practice to compute the generalized inverse of A. The procedure produces matrices U, S and V, such that

$$A = USV^* \tag{6.3.13}$$

with

$$S = \text{diag}\ \left(s_1, s_2, \cdots, s_n\right) \tag{6.3.14}$$

$U^*U = I_m, V^*V = I_n$, with I_m (resp. I_n) the unit matrix in $\mathbb{C}_m$ (resp. $\mathbb{C}_n$). The σ_i are the *singular values*; they are the non-negative square roots of the eigenvalues of A^*A (and also of AA^*). The columns of U are the eigenvectors of AA^* corresponding to its n eigenvalues, and the columns of V are the eigenvectors of A^*A. It is assumed, furthermore, that the σ_i are ordered so that $\sigma_1 \geqq \sigma_2 \geqq \cdots \geqq \sigma_n \geqq 0$. If the rank of A is $r < n$, then $\sigma_{r+1} = \cdots = \sigma_n = 0$. If $m < n$, then the property (iii) of Theorem 6.3.2 gives us a method to obtain $A^{\dagger}$ by considering A^*.

Once the singular value decomposition (6.3.13) has been obtained, $A^{\dagger}$ can be represented using the properties (vi) and (vii) of Theorem 6.3.2. We thus get

$$A^{\dagger} = VS^{\dagger}U^*, \tag{6.3.15}$$

with

$$S^{\dagger} = \text{diag}\ \left(s_1^{\dagger}, s_2^{\dagger}, \cdots, s_n^{\dagger}\right), \tag{6.3.16}$$

and with

$$s_i^{\dagger} = \begin{cases} 1/s_i & \text{if} \quad s_i \neq 0 \\ 0 & \text{if} \quad s_i = 0. \end{cases} \tag{6.3.17}$$

Due to the limited precision of computers it is frequently impossible to tell in practice whether a particular σ_i is, or is not, zero. We can gain some insight into this situation by considering, in place of (6.3.10), the problem,

$$(\alpha I + A^*A)x_\alpha = A^*b \tag{6.3.18}$$

which has a unique solution x_α for all $\alpha > 0$. By finding the singular value decomposition of both $\alpha I + A^*A$ and of A, we deduce that

$$x_\alpha = V(\alpha I + S^2)^{-1}SU^*b. \tag{6.3.19}$$

Comparing this with the solution $x = A^\dagger b$ of (6.3.10), which may also be written in the form

$$x = VS^\dagger U^*b, \tag{6.3.20}$$

we see that since $(\alpha I + S^2)^{-1}S \to S^\dagger$ as $\alpha \to 0$, we have $x_\alpha \to x$ as $\alpha \to 0$. Moreover, minimizing $\|x_\alpha - x\|^2$, we may further deduce that as $\alpha \to 0$,

$$\|x_\alpha - x\| = \mathcal{O}(\alpha) \quad \text{if} \quad x \in \text{Range}(A^*A) \tag{6.3.21}$$

$$\|x_\alpha - x\| = \mathcal{O}(\alpha^{1/2}) \quad \text{if} \quad x \in \text{Range}(A^*). \tag{6.3.22}$$

6.3.4 Errors in the Numerical Solution

Let us consider the norm of the error in the numerical solution of a system of linear equations. We assume that the exact linear system of equations which we want to solve takes the form

$$Ax = b \tag{6.3.23}$$

with A an $n \times n$ matrix and x and b being $n \times 1$ vectors. Let us assume that the elements of A and b are the exact matrix and vector obtained via a Galerkin or a Nyström procedure, but that we must approximate these via a quadrature formula. Thus, instead of solving (6.3.10), we end up solving a *perturbed linear system*

$$(A + \Delta A)(x + \Delta x) = b + \Delta b, \tag{6.3.24}$$

where ΔA and Δb respectively denote a matrix and vector of quadrature errors. In order to obtain a bound on $\|\Delta x\|/\|x\|$, let us furthermore assume that A^{-1} exists, and that $\|A^{-1}\| \; \|\Delta A\| < 1$.

Theorem 6.3.5 *If A^{-1} exists, and if $\|\Delta A\| \; \|A^{-1}\| < 1$, then*

$$\frac{\|\Delta x\|}{\|x\|} \leq \frac{c}{1 - c\frac{\|\Delta A\|}{\|A\|}} \left[\frac{\|\Delta b\|}{\|b\|} + \frac{\|\Delta A\|}{\|A\|} \right] \tag{6.3.25}$$

and

$$c = \|A\| \, \|A^{-1}\| \tag{6.3.26}$$

the condition number of A.

Proof. Let us subtract (6.3.23) from (6.3.24). Then, solving for Δx, we get

$$\Delta x = \left(I + A^{-1}\Delta A\right)^{-1} A^{-1} \left[\Delta b - (\Delta A) x\right]. \tag{6.3.27}$$

Next, taking norms of each side of (6.3.27) and then using the definition of c, we get

$$\frac{\|\Delta x\|}{\|x\|} \leqq \frac{c}{1 - c\frac{\|\Delta A\|}{\|A\|}} \left[\frac{\|\Delta b\|}{\|x\| \; \|A\|} + \frac{\|\Delta A\|}{\|A\|}\right]. \tag{6.3.28}$$

But from (6.3.23), we have $\|b\| \leqq \|A\| \, \|x\|$, and hence, we get (6.3.25). ■

Inequality (6.3.28) shows that for small $\|\Delta A\|$, we have, approximately,

$$\|\Delta x\| \leqq c \left[\frac{\|\Delta b\|}{\|b\|} + \frac{\|\Delta A\|}{\|A\|}\right]. \tag{6.3.29}$$

In general, if $10^{k-1} < c \leqq 10^k$, with k an integer, we may lose up to k significant figures of accuracy in the numerical solution of (6.3.23) for x. Hence, if k is close to, or exceeds the number of significant figures of the floating point computations of the computer, then the system (6.3.23) is said to be unstable. Otherwise (6.3.23) is said to be stable.

Using the Sinc methods of the previous chapter, we shall see in the examples of the later sections of this chapter that the Sinc quadrature errors $\|\Delta A\|$ and $\|\Delta b\|$ are, at most, as large as the Sinc–Galerkin approximation errors $\|\kappa - \mathcal{P}_n \kappa\|$ and $\|g - \mathcal{P}_n g\|$, which occur upon replacing (6.2.1) with (6.3.2).

•

Problems for Section 6.3

6.3.1 Discuss any method to bound the difference between the approximate solution (6.3.6) obtained by solving the system (6.3.5) and the exact solution to (6.1.6).

6.3.2 Prove Theorem 6.3.2.

6.3.3 (a) Let A be an $m \times n$ matrix of rank r. Prove that there exist non-singular matrices P and Q, of order m and n respectively, such that

$$PAQ = \begin{bmatrix} \alpha & 0 \\ 0 & 0 \end{bmatrix}$$

with α an $r \times r$ non-singular matrix.

(b) Let part (a) of Problem 6.3.3 be satisfied, let X be an $n \times m$ matrix, and define

$$Q^{-1}XP^{-1} = \begin{bmatrix} \beta & \gamma \\ \delta & \varepsilon \end{bmatrix}$$

with β, γ, δ and ε are matrices of dimension $r \times r$, $r \times (m - r)$, $(n - r) \times r$ and $(n - r) \times (m - r)$ respectively.

(b-i) Prove that $AXA = A$ if and only if $\beta = \alpha^{-1}$;

(b-ii) Prove that $XAX = X$ if and only if $\beta = \alpha^{-1}$ and $\varepsilon = \delta\alpha\gamma$;

(b-iii) Prove that $(AX)^* = AX$ if and only if $\beta = \alpha^{-1}$, $\gamma = -\alpha^{-1}P_1P_2^\dagger$, and $\varepsilon = -\delta P_1 P_2^\dagger$. Here $P = \begin{bmatrix} P_1 \\ P_2 \end{bmatrix}$, with P_1 is $r \times m$ matrix, P_2 an $(m - r) \times m$ matrix, and $P_2^\dagger$ the generalized inverse of P_2 as defined by Theorem 6.3.2-(ix)).

(b-iv) Prove that $(XA)^* = XA$ if and only if $\beta = \alpha^{-1}$, $\delta = -Q_2^\dagger Q_1 \alpha^{-1}$, and $\varepsilon = -Q_2^\dagger Q_1 \gamma$. Here $Q = [Q_1, Q_2]$, with Q_1 an $n \times r$ matrix, and $Q_2^\dagger$ the generalized inverse of Q_2 as defined by Theorem 6.3.2 (viii).

(b-v) Prove that $AXA = A$, $XAX = X$, $(AX)^* = AX$, and $(XA)^* = XA$ if and only if $\beta = \alpha^{-1}$, $\gamma = -\alpha^{-1}P_1P_2^\dagger$, $\delta = -Q_2^\dagger Q_1 \alpha^{-1}$, and $\varepsilon = Q_2^\dagger Q_1 \alpha^{-1} P_1 P_2^\dagger$.

6.3.4 Prove Theorem 6.3.4.

6.3.5 Prove (6.3.21).

6.3.6 Prove (6.3.22).

6.4 Volterra Integral Equations

The notation of Definition 4.1.1 will again be assumed to be in force. We shall consider the approximate solution, via Sinc methods, of the equation

$$T(f) \equiv f - \kappa f = g \tag{6.4.1}$$

for f given κ and g. In (6.4.1) κf is defined by

$$(\kappa f)(x) = \int_a^x K(x, t, f(t))dt, \quad x \in \Gamma, \tag{6.4.2}$$

with the integral taken along Γ.

6.4.1 Linear Volterra Equations

At the outset, let us consider the simple linear equation

$$Tf(x) = f(x) - \int_a^x K(t)\, f(t)\, dt = g(x), \quad x \in \Gamma. \tag{6.4.3}$$

The solution of Equation (6.4.3) can be explicitly expressed in terms of integrals, in the case of K and g scalar functions. It is

$$f(x) = T^{-1}g(x) = g(x) + \int_a^x \exp\left\{\int_t^x K(\tau)\, d\tau\right\} K(t)\, g(t)\, dt. \tag{6.4.4}$$

It follows at once, from Theorem 4.1.3, that if k and g are scalar functions, such that $K/\phi' \in \mathbf{L}_\alpha(\mathcal{D})$, and $g \in \mathbf{M}_\alpha(\mathcal{D})$, then the solution f given by (6.4.4) belongs to $\mathbf{M}_\alpha(\mathcal{D})$.

Let us now construct an approximate solution via Sinc methods. To this end we assume that $K/\phi' \in \mathbf{L}_\alpha(\mathcal{D})$, in the case of a scalar equation (respectively, each element k_{ij} of the matrix $K = [k_{ij}]$ is such, that $k_{ij}/\phi' \in \mathbf{L}_\alpha(\mathcal{D})$, if K is a matrix), and that $g \in \mathbf{M}_\alpha(\mathcal{D})$ (respectively, each element g_i of the vector $g = [g_i]$ is such, that $g_i \in \mathbf{M}_\alpha(\mathcal{D})$ if g is a vector). It is then readily seen that the solution f (resp., each element of the solution vector f) belongs to $\mathbf{M}_\alpha(\mathcal{D})$.

We first consider the approximate solution of (6.4.3). To this end, we present two alternate approaches, the first one being the direct discretization of (6.4.3) via the use of the indefinite integration formula (4.5.8), and the second one, the application of the indefinite integration formula (4.5.8) to (6.4.4).

To obtain a direct discretization of (6.4.3), we take $h = [\pi d/(\alpha N)]^{1/2}$. Let us assume, at the outset, that (6.4.3) is a scalar equation. Let u be a given function defined on Γ, then define $D(u)$ to be the diagonal matrix, $D(u) = \mathrm{diag}(u(z_{-N}), \cdots, u(z_N))$, set $\mathbf{g} = (g(z_{-N}), \cdots, g(z_N))^T$, $\mathbf{f} = (f_{-N}, \cdots, f_N)^T$, with f_j an approximation to the solution f of (6.4.3) at the Sinc point z_j, and let $I^{(}-1)$ be defined as in (4.5.7). Then, we arrive at the system of linear equations

$$\mathbf{T}\mathbf{f} \equiv \left(I - hI^{(-1)}D(K/\phi')\right)\mathbf{f} = \mathbf{g} \tag{6.4.5}$$

for the vector $\mathbf{f}$. We remark, that a similar system of $m(2N+1)$ equations results in the case when (6.4.3) is a system of equations, and the direct (or iterative, when possible) solution of this system will yield an approximation to the vector $\mathbf{f}$. Furthermore, it may be shown (see Problem 6.4.1) that the coefficient matrix $\mathbf{T}$ of $\mathbf{f}$ in (6.4.5) has a condition number which is independent of N for all N sufficiently large, and hence, this vector $\mathbf{f}$ will serve as an approximate solution that satisfies a theorem similar to Theorem 6.4.1 given below.

We derive yet another matrix expression for the solution of (6.4.3) when (6.4.3) is a scalar equation that uses the explicit integral expression (6.4.4). To this end, we first write (6.4.4) in the form

$$\begin{aligned} &[f(x) - g(x)] \exp\left\{-\int_a^x K(t)\,dt\right\} \\ &= \int_a^x \exp\left\{-\int_a^t K(\tau)\,d\tau\right\} K(t)\,g(t)\,dt. \end{aligned} \tag{6.4.6}$$

Let us first determine approximations κ_j of $\int_a^{z_j} K(\tau)\,d\tau$. We set $\mathbf{k} = (K(z_{-N}), \cdots, K(z_N))^T$, and we use the indefinite integration formula (4.5.8) to determine a vector $\tilde{\kappa} = (\kappa_{-N}, \cdots, \kappa_N)^T$ by means of the expression

$$\tilde{\kappa} = -h\,I^{(-1)}D(1/\phi')\,\mathbf{k}. \tag{6.4.7}$$

We then define a diagonal matrix E by

$$E = \text{diag}(\exp(\kappa_{-N}), \cdots, \exp(\kappa_N)). \tag{6.4.8}$$

Then, once again using (4.5.8) to approximate the integral with respect to t in (6.4.6), and using the notation of (6.4.8), we get the expression

$$\mathbf{f} = \mathbf{g} + h\,E^{-1}I^{(-1)}ED(K/\phi')\,\mathbf{g}, \tag{6.4.9}$$

with $\mathbf{f} = (f_{-N}, \cdots, f_N)^T$, and $\mathbf{g} = (g(z_{-N}), \cdots, g(z_N))^T$. We have thus proved the following result.

Theorem 6.4.1 *Let $K/\phi' \in \mathbf{L}_\alpha(\mathcal{D})$, let $g \in \mathbf{M}_\alpha(\mathcal{D})$, let the function f be defined as in (6.4.4), and let the vector $\mathbf{f} = (f_{-N}, \cdots, f_N)^T$ be defined as in (6.4.9). Then $f \in \mathbf{M}_\alpha(\mathcal{D})$, and there exists a constant C, independent of N, such that*

$$\sup_{x\in\Gamma}\left| f(x) - \frac{f_{-N} + \rho(x)\, f_N}{1+\rho(x)} + \sum_{k=-N}^{N}\left(f_j - \frac{f_{-N} + e^{jh} f_N}{1+e^{jh}}\right) S(j,h)\circ\phi(x)\right|$$
$$\leqq C\, N^{1/2} e^{-(\pi d \alpha N)^{1/2}}. \tag{6.4.10}$$

6.4.2 Non-Linear Equations via Neumann Series

Let us now examine the existence of an approximate solution of the nonlinear Volterra integral equation, Equation (6.4.1), in the space $\mathbf{X} = \mathbf{Hol}(\mathcal{D})\cap \mathbf{C}(\overline{\mathcal{D}})$, using a Neumann series. To this end, we shall assume the notation of Section 4.1, and we will also assume that κ and g have the following properties:

(a) $g \in \mathbf{M}_\alpha(\mathcal{D})$;

(b) Let r be a positive number, let $f^{(0)} \in \mathbf{X}$, let $\mathcal{B}(f^{(0)};r)$ be defined by

$$\mathcal{B}(f^{(0)};r) = \{u \in \mathbf{X} : \|u - f_0\|_\Gamma < r\} \tag{6.4.11}$$

with $\|u\|_\Gamma = \sup_{x\in\Gamma}|u(x)|$. Let $w = \kappa u + g \in \mathbf{M}_\alpha(\mathcal{D})$ for all $u \in \mathcal{B}(f^{(0)};r)$;

(c) There exists a number $\beta \in (0,1)$, such that for all $u \in \mathcal{B}(f^{(0)};r)$, $v \in \mathcal{B}(f^{(0)},r)$, we have

$$\|\kappa u - \kappa v\|_\Gamma \leqq \beta\|u - v\|_\Gamma; \tag{6.4.12}$$

(d) Let $\|\kappa f^{(0)} + g - f^{(0)}\|_\Gamma < (1-\beta) r$.

We remark that one can usually prove existence and uniqueness of the solution of (6.4.1) even if $\beta > 1$ (see Theorem 7.2.3). The restriction to the case of $\beta < 1$ is solely for the sake of simplicity in this discussion.

Theorem 6.4.2 *If assumptions (a) through (d) are satisfied, then starting with $f^{(0)} \in \mathbf{X}$, the sequence $\{f^{(n)}\}$ defined by*

$$f^{(n+1)} = \kappa f^{(n)} + g, \quad n = 0,\ 1,\ 2, \cdots \tag{6.4.13}$$

converges to the unique solution $f \in \mathcal{B}(f^{(0)};r)\cap\mathbf{M}_\alpha(\mathcal{D})$ of Equation (6.4.1).

Proof. The proof follows directly, from Theorem 6.2.1 (see also Example 6.2.2). ■

Let us next describe a Sinc approximation f_N to the solution $f \in \mathbf{M}_\alpha$ of Equation(6.4.1). To this end, given $u \in \mathbf{M}_\alpha$, we define $\mathcal{P}_N u$ by

$$(\mathcal{P}_N u)(x) = Lu(x) + \sum_{k=-N}^{N} [u(z_k) - (Lu)(z_k)] \cdot S(k,h) \circ \phi(x), \qquad (6.4.14)$$

with

$$Lu(x) = \frac{u(a) + \rho(x)\, u(b)}{1 + \rho(x)}. \qquad (6.4.15)$$

We then note that $\mathcal{P}_N u \in \mathbf{X}$. Instead of (6.4.1), we consider the equation

$$f_N = \mathcal{P}_N \kappa f_N + \mathcal{P}_N g, \qquad (6.4.16)$$

with f_N of the form

$$f_N(x) = \frac{c_{-N-1}}{1+\rho(x)} + \sum_{j=-N}^{N} c_j S(k,h) \circ \phi(x) + \frac{c_{N+1}\rho(x)}{1+\rho(x)}. \qquad (6.4.17)$$

Theorem 6.4.3 *Let $\mathcal{P}_N u$ and f_N be defined as in (6.4.14) through (6.4.16), let $h = [\pi d/(\alpha N)]^{1/2}$ and let assumptions (a) through (d) be satisfied. Starting with a particular $f_N^{(0)}$ of the form (6.4.17), define $f_N^{(m)}$, for $m = 1,\ 2,\ \cdots$, by*

$$f_N^{(m+1)} = \mathcal{P}_N \kappa f_N^{(m)} + \mathcal{P}_N g, \quad m = 0,\ 1,\ 2, \cdots. \qquad (6.4.18)$$

If N is sufficiently large, then $f_N^{(m)} \to f_N$ as $m \to \infty$, with f_N the unique solution of (6.4.16) and of the form (6.4.17). Moreover, there exists a constant C independent of N, such that if f denotes the exact solution of (6.4.1), then

$$\sup_{x \in \Gamma} |f(x) - f_N(x)| < C N^{1/2} e^{-(\pi d \alpha N)^{1/2}}. \qquad (6.4.19)$$

Proof. We defer the proof of this theorem to Problem 6.4.4. ■

At this point we identify two practically important special forms of κf and we discuss an appropriate method to evaluate $\mathcal{P}_N \kappa f_N^{(m)}$ in (6.4.19) and to solve (6.4.1) for each of them. Specifically, we shall assume that κf takes one of the forms (6.4.20) or (6.4.21), i.e., either

$$(\kappa f)(x) = \int_a^x (x-t)^{\gamma-1} w(x,t,f(t))dt, \qquad (6.4.20)$$

or else

$$(\kappa f)(x) = \int_a^x w(x, t, f(t))dt, \tag{6.4.21}$$

with γ a positive constant and with the function w in (6.4.20) and (6.4.21) satisfying either one of the assumptions (e-20) or (e-21), which we now state:

(e-20): *Let ϕ_x denote a conformal map of a portion $\mathcal{D}_x$ of $\mathcal{D}$ onto $\mathcal{D}_d$, such that if Γ_x denotes the portion of Γ having endpoints a and x, then $\phi_x^{-1}(\Gamma_x) = \mathbb{R}$. Furthermore, we shall assume that uniformly for all $x \in \Gamma$, and for all $f \in \mathcal{B}$, we have $w(x, \cdot, f(\cdot))/\phi_x' \in \mathbf{L}_{\alpha'}(\mathcal{D}_x)$ where $\alpha' = \alpha - \gamma > 0$.*

(e-21): *For the case of (6.4.13) we assume that $w(x, \cdot, f(\cdot))/\phi' \in \mathbf{L}_\alpha(\mathcal{D})$, uniformly for $x \in \Gamma$ and $f \in \mathcal{B}$.*

The quadrature formula (see Equation (4.2.29))

$$\int_a^x v(t)dt \approx h \sum_{\ell=-N}^{N} \frac{v(z_\ell(x))}{\phi_x'(z_\ell(x))} \tag{6.4.22}$$

may be used to approximate either (6.4.20) or (6.4.21), with $z_\ell(x) = \phi_x^{-1}(\ell h)$. The equations (6.4.18) are then reduced to algebraic equations, in which the right-hand sides can be explicitly evaluated, yielding a sequence of vectors of order $2N+1$, $\{\hat{f}_N^{(m)}\}_{m=1}^{\infty}$, with $f_N^{(m)} \to f_N$ as $m \to \infty$.

In the case of (6.4.21), we may apply either (6.4.22), or else we can use Theorem 4.5.2. In the latter case we get the approximation

$$\begin{aligned} \int_a^{z_j} w(z_j, t, f(t))dt &\approx \frac{\rho(z_j)T_j}{1+\rho(z_j)} \\ &+h \sum_{|\ell| \leq N} \sigma_{j-\ell} \left[\frac{w(z_j, z_\ell, f(z_\ell))}{\phi'(z_\ell)} - \frac{\rho(z_\ell)T_j}{1+\rho(z_\ell)} \right] \end{aligned} \tag{6.4.23}$$

with

$$T_j = h \sum_{\ell=-N}^{N} \frac{w(z_j, z_\ell, f(z_\ell))}{\phi'(z_\ell)}, \tag{6.4.24}$$

and with m sufficiently large so that the numbers $f^{(m)}(z_\ell)$ are sufficiently close to the numbers $f(z_\ell)$. The advantage of the integration scheme (6.4.23) to (6.4.24) over (6.4.22) is that the points z_ℓ are independent of x, and consequently, to compute the c_j in (6.4.17), f_N need only be known at the points z_ℓ. Once the c_j are computed, we can of course approximate $f_N(x)$ for other values of x using (6.4.17).

In this latter case we may also use (4.5.14), which enables us to reduce (6.4.18) to the sequence of algebraic equations

$$\begin{aligned} w_j^{(m+1)} &= h \sum_{\ell=-N}^{N} \delta_{j-\ell}^{(-1)} \frac{w(z_j, z_\ell, w_\ell^{(m)})}{\phi'(z_\ell)} + g(z_j), \\ j &= -N, \cdots, N, \\ m &= 0, 1, 2, \cdots, \end{aligned} \tag{6.4.25}$$

which begins with $w_j^{(0)} = g(z_j)$. After achieving convergence, we have $\lim_{k\to\infty} w_j^{(m)} = w_j$, and our approximation f_N takes the form

$$f_N(x) = \frac{w_{-N} + \rho(x)\, w_N}{1 + \rho(x)} + \sum_{j=-N}^{N} \left(w_j - \frac{w_{-N} + e^{jh} w_N}{1 + e^{jh}} \right) S(j,h) \circ \phi(x), \tag{6.4.26}$$

which is also uniformly bounded on Γ by the right-hand side of (6.4.17).

6.4.3 Non-Linear Equations by Newton's Method

Newton's method provides a particularly powerful tool to solve nonlinear problems, especially *stiff ordinary differential equations*, which we will encounter in Chapter 7. Let us consider, at the outset, the solution of the scalar problem

$$Tf(x) \equiv f(x) - \int_a^x k(t, f(t))\, dt = g(x) \tag{6.4.27}$$

for f, given k and g, by Newton's method. To this end, we again take our solution space to be $\mathbf{X} = \mathbf{Hol}(\mathcal{D}) \cap \mathbf{C}(\overline{\mathcal{D}})$, and we again define $\mathcal{B}(f^{(0)}; r)$ for $f^{(0)} \in \mathbf{X}$ and $r > 0$, by (6.4.11), with $\|f\| = \sup_{x\in\Gamma} |f(x)|$. We assume that $k_f(\cdot, f)$ and $k_{ff}(\cdot, f)$ exist, and for fixed $f^{(0)} \in \mathbf{X}$, we assume the following:

(i) $g \in \mathbf{M}_\alpha(\mathcal{D})$;

(ii) $k(\cdot, f)/\phi'$, $k_f(\cdot, f)/\phi'$, and $k_{ff}(\cdot, f)/\phi'$ all belong to $\mathbf{L}_\alpha(\mathcal{D})$ for all $f \in \mathcal{B}(f^{(0)}, r)$;

Recalling the definition of T given by (6.4.27), we may define the linear operator $T_f' : \mathbf{X} \to \mathbf{M}_\alpha(\mathcal{D})$, with $f \in \mathbf{X}$ by

$$(T_f')w(x) = w(x) - \int_a^x k_f(t, f(t))\, w(t)\, dt, \tag{6.4.28}$$

with $k_f(t, f(t))$ the partial derivative of $k(\cdot, f)$ with respect to f evaluated at $(t, f(t))$.

Comparing this equation with (6.4.3), we see from (6.4.4), that the inverse $(T_f')^{-1} : \mathbf{M}_\alpha(\mathcal{D}) \to \mathbf{X}$ of T_f' is given by

$$(T_f')^{-1} v(x) = v(x) + \int_a^x k_f(t, f(t)) \exp\left[\int_t^x k_f(\tau, f(\tau))\, d\tau \right] v(t)\, dt. \tag{6.4.29}$$

Let us now introduce constants β, η, and K, as follows:

$$\begin{aligned} \beta &= \exp\left[\left|\int_\Gamma \left|k_f(t, f^{(0)}(t))\, dt\right|\right|\right]; \\ \eta &= \sup_{x\in\Gamma} \left|(\tau'_{f^{(0)}})^{-1}\tau(f^{(0)})\right|; \quad \text{and} \\ K &= \sup_{f\in\mathcal{B}(f^{(0)};r)} \int_\Gamma |k_{ff}(t, f(t))\, dt|. \end{aligned} \tag{6.4.30}$$

(iii) We shall assume, in addition, that

$$\begin{aligned} \gamma &\equiv \beta\eta K < 1/2, \quad \text{and} \\ \frac{1-\sqrt{1-2\gamma}}{\gamma} &< r. \end{aligned} \tag{6.4.31}$$

Then we have

Theorem 6.4.4 (Kantorovich) *Starting with $f^{(0)} \in \mathbf{X}$, the sequence $\{f^{(m)}\}$, where*

$$f^{(m+1)} = f^{(m)} - (T'_{f^{(m)}})^{-1}(T(f^{(m)}) - g), \quad m = 0, 1, \cdots \tag{6.4.32}$$

is well-defined, $f^{(m)} \in \mathbf{M}_\alpha(\mathcal{D}) \cap \mathcal{B}(f^{(0)}; r)$ for every positive integer m, and $f^{(m)} \to f^*$, as $m \to \infty$, where $\tau(f^*) = 0$. Moreover,

$$\left\| f^{(m)} - f^* \right\| \leqq (2\gamma)^{2^m} \frac{\eta}{\gamma}. \tag{6.4.33}$$

Proof. Let us note at the outset, that

$$\begin{aligned} \left\|(T'_{f^{(0)}})^{-1}\right\| &= \sup_{w\in\mathbf{M}_\alpha(\mathcal{D}), \|w\|\leqq 1} \left\|(T'_{f^{(0)}})^{-1} w\right\| \\ &\leqq 1 + \sup_{x\in\Gamma} \int_a^x \left|k_f(t, f^{(0)}(t))\right| \exp\left[\left|\int_t^x \left|k_f(\tau, f^{(0)}(\tau))\, d\tau\right|\right|\right] |dt| \\ &= \exp\left[\left|\int_\Gamma \left|k_f(t, f^{(0)}(t))\, dt\right|\right|\right]. \end{aligned} \tag{6.4.34}$$

The remainder of the proof is the classical proof of this theorem, as originally given by Kantorovich. This part of the proof is left to Problem 6.4.5. ∎

Let us next turn to the discrete form of Newton's method, after application of Sinc collocation to (6.4.27) and either (6.4.28) or (6.4.29). To this end it is convenient to use as model the results obtained in Equations

(6.4.3) through (6.4.9) for linear equations. Thus, we first carry out Sinc collocation of the expression $T(f)$ given in Equation (6.4.27), yielding

$$\mathbf{T}(\mathbf{f}) \equiv \mathbf{f} - h\, I^{(-1)} D(1/\phi')\mathbf{k} = \mathbf{g}. \qquad (6.4.35)$$

In (6.4.35), $\mathbf{f} = (f_{-N}, \cdots, f_N)^T$, h, $I^{(-1)}$, and $D(\cdot)$ are defined as for (6.4.5), and $\mathbf{k} = (k(z_{-N}, f_{-N}), \cdots, k(z_N, f_N))^T$.

Sinc discretization provides two explicit methods to solve the equation $\mathbf{T}(\mathbf{f}) - \mathbf{g} = \mathbf{0}$, via Newton's method, starting with an initial vector $\mathbf{f}^{(0)}$. In the case of the first method, which is applicable to scalar equations of the form $\mathbf{T}(\mathbf{f}) - \mathbf{g} = \mathbf{0}$, we define a matrix $\mathbf{T}_f^{-1}$, via collocation of (6.4.29), based on the use of the indefinite integration formula (4.5.6) through (4.5.8) into (6.4.29), i.e.,

$$(\mathbf{T}'_f)^{-1} = I + h\, E^{-1} I^{(-1)} E D(k_f/\phi'), \qquad (6.4.36)$$

with $D(k_f/\phi') = \mathrm{diag}(k(z_{-N}, k_{-N})/\phi'(z_{-N}), \cdots, k(z_N, k_N)/\phi'(z_N))$, and with the matrix E defined by

$$\begin{aligned} (\kappa_{-N}, \cdots, \kappa_N)^T &= -h\, I^{(-1)} D(1/\phi')\, \mathbf{k}_f \\ E &= \mathrm{diag}(\kappa_{-N}, \cdots, \kappa_N). \end{aligned} \qquad (6.4.37)$$

Newton's method then takes the form

$$\mathbf{f}^{(m+1)} - \mathbf{f}^{(m)} = -(\mathbf{T}'_{\mathbf{f}^{(m)}})^{-1}(\mathbf{T}(\mathbf{f}^{(m)}) - \mathbf{g}). \qquad (6.4.38)$$

Next, in the case of a system of n equations, we first write Equation (6.4.27) in the form

$$TF(x) \equiv F(x) - \int_a^x K(t, F(t))\, dt = G(x), \qquad (6.4.39)$$

with $F(x) = (f_1(x), \cdots, f_n(x))^T$, $G(x) = (g_1(x), \cdots, g_n(x))^T$, and $K(t, F(t)) = (k_1(t, F(t)), \cdots, k_n(t, F(t)))^T$. Application of collocation via the use of Equations (4.5.6) through (4.5.8) now enables us to write (6.4.39) in the discrete, *Kronecker product* form

$$\mathbf{T}(\mathbf{F}) \equiv \mathbf{F} - I_1 \otimes \left\{ h\, I^{(-1)} D(1/\phi') \right\} \mathbf{K} = \mathbf{G}. \qquad (6.4.40)$$

Here, I_1 is the unit matrix of order n,

$$\begin{aligned} \mathbf{F} &= (f_{1,-N}, \cdots, f_{1,N}, f_{2,-N}, \cdots, f_{2,N}, \cdots, f_{n,N})^T; \\ \mathbf{G} &= (g_1(z_{-N}), \cdots, g_1(z_N), g_2(z_{-N}), \cdots, g_2(z_N), \cdots, g_n(z_N))^T; \\ \mathbf{K} &= \equiv (\mathbf{k}_1, \cdots, \mathbf{k}_n)^T; \\ \mathbf{k}_j &= (k_j(z_{-N}, F_{-N}), \cdots, k_j(z_N, F_N))^T; \\ F_\ell &= (f_{1,\ell}, \cdots, f_{n,\ell})^T. \end{aligned}$$

(6.4.41)

In addition, for a given rectangular $p \times q$ matrix $A = [a_{ij}]$ and an $r \times s$ matrix $B = [b_{ij}]$, we denote $A \otimes B$ to be the Kronecker product matrix $[a_{ij} B]$.

The *Fréchet derivative* equation corresponding to (6.4.39) is

$$(\mathbf{T}'_F) W(x) \equiv W(x) - \int_a^x J(t, F(t))\, W(t)\, dt \tag{6.4.42}$$

where, upon denoting w_i to be the Fréchet derivative of f_i, we set $W = (w_1, \cdots, w_n)^T$, and we define J to be the $n \times n$ matrix J_{ij}, with

$$J_{ij}(x, F(x)) = \frac{\partial k_i(x, F(x))}{\partial f_j}. \tag{6.4.43}$$

We now apply collocation to (6.4.42), using (6.4.6) through (6.4.8). Thus, we may define the jacobian matrix for a discrete Newton method by

$$\mathbf{T}'_{\mathbf{F}} = I_1 \otimes I_2 - I_1 \otimes \{h I^{(-1)} D(1/\phi')\}\, [B_{ij}], \tag{6.4.44}$$

with I_1 the unit matrix of order n, I_2 the unit matrix of order $2N+1$, $hI^{(-1)} D(1/\phi')$ defined as in (6.4.40) above, and with $i, j = 1, \cdots, n$), B_{ij} the diagonal matrix of order $2N+1$, defined by

$$B_{ij} = \operatorname{diag}(J_{ij}(z_{-N}, F_{-N}), \cdots, J_{ij}(z_N, F_N)). \tag{6.4.45}$$

It may be shown (see Problem 6.4.1) that for N sufficiently large, the matrix $\mathbf{T}'_{\mathbf{F}}$ in (6.4.44) is nonsingular, and has a uniformly bounded condition number. Newton's method for the vector

$$\mathbf{F}^{(m+1)} = (f_{1,-N}^{(m+1)}, \cdots, f_{1,N}^{(m+1)}, f_{2,-N}^{(m+1)}, \cdots, f_{2,N}^{(m+1)}, \cdots, f_{n,N}^{(m+1)})^T$$

then takes the form

$$\mathbf{T}'_{\mathbf{F}^{(m)}} \left(\mathbf{F}^{(m+1)} - \mathbf{F}^{(m)}\right) = -\mathbf{T}(\mathbf{F}^{(m)}) - \mathbf{G}. \tag{6.4.46}$$

•

Problems for Section 6.4

6.4.1 Let Af be defined by the right-hand side of (6.4.3), where $K/\phi' \in \mathbf{L}_\alpha(\mathcal{D})$, where $g \in \mathbf{L}_\alpha(\mathcal{D})$. Let $A^{-1} f$ be defined, in the case of a scalar equation (6.4.3), by (6.4.4).

(a) Set $\|K\| = \int_\Gamma |K(t)\, dt|$, $\|f\| = \sup_{x \in \Gamma} |f(x)|$. Prove that $\|A\| \leqq 1 + \|K\|$, $\|A^{-1}\| \leqq 1 + \|K\| \exp(\|K\|)$.

(b) Let A_h be defined by the coefficient of the vector f in (6.4.5). Prove, with the use of *Sinc* interpolation, that

$$\|A - A_h\| \leqq C\, N^{1/2} e^{-(\pi d \alpha N)^{1/2}},$$

where C is a constant that is independent of N.

(c) Let A_h^{-1} be defined by the coefficient of the vector g in (6.4.6). Prove, using *Sinc* interpolation, that for N sufficiently large, we have $\|A^{-1} - A_h^{-1}\| \leqq C_1\, N^{1/2} e^{-(\pi d \alpha N)^{1/2}}$, with C_1 a constant independent of N.

(d) Use the above results of (a) through (c) to deduce that if A_h is defined by the coefficient of the vector f in (6.4.5), then for all N sufficiently large, we have $\|A_h\|\, \|A_h^{-1}\| \leqq C_2$, with C_2 is a finite constant independent of N.

(e) Deduce that conclusion (d) holds true even if (6.4.3) is a system of several integral equations.

6.4.2 Obtain an *a-priori* bound as a function of k, on

$$\sup_{x \in \Gamma} |f_N(x) - f_N^{(k)}(x)|.$$

with $f_N^{(k)}$ defined in Theorem 6.4.2.

6.4.3 Let ϕ_x be defined by Assumption (e-20). Show that the choice $\phi_x(z) = \log \sinh[\{\phi(z) - \phi(x)\}^{-1}]$ is a suitable contender in this assumption.

6.4.4 Give a detailed proof of Theorem 6.4.3.

6.4.5 Let η, β, and K be positive numbers, such that $\gamma = \eta \beta K \leqq 1/2$.

(a) Define $t(v)$ by $t(v) = (K/2)v^2 - (1/\beta)v + \eta/\beta$. Define the recurrence relation $v_{m+1} = v_m - t(v_m)/t'(v_m)$, with $v_0 = 0$. Set $v^* = (1 - \sqrt{1 - 2\gamma})$, then use an induction process to prove each of the following relations, for $m \geqq 0$:

$$\begin{aligned}
|t'(v_{m+1})| &\geqq (1/2)|t'(v_m)|; \\
|v_{m+1} - v^*| &\leqq (2^{m-1}/\gamma)|v_m - v^*|^2; \\
|v_m - v^*| &\leqq 2^{-m}\left(1 - \sqrt{1 - 2\gamma}\right)^{2^m} \eta/\gamma; \text{ and} \\
|t(v_{m+1})| &= (K/2)|v_{m+1} - v_m|^2.
\end{aligned}$$

(b) Now, let β, η, and K be defined as in Theorem 6.4.4. Using the notation of Theorem 6.4.4, we have

$$T(f^{(m)}) + T'_{f^{(m)}}(f^{(m+1)} - f^{(m)}) = 0$$

for $m \geqq 0$, then conclude from the identity

$$T(f^{(m+1)}) = T(f^{(m)}) + T'_{f^{(m)}}(f^{(m+1)} - f^{(m)}) = 0$$

that

$$\|T(f^{(m+1)})\| \leqq (K/2)\|f^{(m+1)} - f^{(m)}\|^2.$$

(c) Conclude from parts (a) and (b) of this problem that

$$\|f^{(m+1)} - f^{(m)}\| \leqq |v_{m+1} - v_m|^2,$$

and use this fact, as well as the inequalities in parts (a) and (b) of this problem to prove Theorem 6.4.4.

6.4.6 Derive yet another Newton's method to solve a system of equations of the form (6.4.22), by first expressing the vector of the Fréchet derivative in the form (6.2.28), and then explicitly solving each i th linear differential equation for the component w_i using the scalar inversion formula (6.4.29), assuming that the components w_j $(j \neq i)$ in the i th equation are known.

6.4.7 (a) Prove that if conditions of Theorem 6.4.4 are satisfied, and if N is taken sufficiently large, then with $h = [\pi d/(\alpha N)]^{1/2}$, the method (6.4.38) converges to the unique solution of (6.4.35).

(b) Obtain a bound of the difference between the exact solution of (6.4.37) and that obtained via the use of (6.4.38), when this solution vector is converted to a Sinc approximation using (4.5.9).

6.4.8 (a) State and prove a theorem analogous to Theorem 6.4.4, involving the solution of Equation (6.4.39) by Newton's method.

(b) State and prove a theorem involving the convergence of the scheme (6.4.45).

6.4.9 Use the Section 4.6 to derive an alternate Sinc method to solve (6.4.1) in the case of κf defined by (6.4.20).

6.5 Potential Theory Problems

In this section we illustrate a Sinc method to obtain an approximate solution of a potential theory problem via the approximate solution of a related integral equation over the surface of a body. At the outset we discuss the simplest case of a single equation, developing and analyzing in detail the analyticity of the solution which will enable us to achieve the $\mathcal{O}(\exp(-c\,n^{1/2}))$ rate of convergence for its Sinc approximation. After completion of this endeavor, we briefly discuss a (more ordinary) case which requires several such equations to express the solution, one over every patch on the surface, with each patch selected such that the solution will have the requisite analyticity property over that patch, in order to yield the same approximation order over that patch, as that for the previously discussed case of a single equation. This criteria thus defines our method of *domain decomposition.* We end this section with an illustrative example.

6.5.1 Problem Description

We shall consider here the approximate solution of a class of Fredholm integral equations of the second kind having the form

$$f(\bar{r}) - \lambda \int\!\!\int_S K(\bar{r},\bar{r}')f(\bar{r}')dS(\bar{r}') = g(\bar{r}), \quad \bar{r} \in S. \tag{6.5.1}$$

Here $\bar{r}$ and $\bar{r}'$ are points of a two-dimensional surface S embedded in $\mathbb{R}^3$, $dS(\bar{r}')$ is the element of surface area at $\bar{r}' \in S$, λ is a complex number which may or may not be a characteristic value, and g is a given function defined on S. The kernel $K(\bar{r},\bar{r}')$ will usually have a *weak singularity* at $\bar{r}' = \bar{r}$, i.e., $K(\bar{r},\bar{r})$ may be infinite at all $\bar{r} \in S$, although $\int\int_S |K(\bar{r},\bar{r}')|dS(\bar{r}') < \infty$.

Equations of the type (6.5.1) frequently arise in potential theory. For example, the solution u of the *Neumann problem* in a region V in $\mathbb{R}^3$, i.e.,

$$\begin{aligned} \nabla^2 u = u_{\mu\mu} + u_{\sigma\sigma} + u_{\tau\tau} &= 0 \ \text{ in } \ V \\ \frac{\partial u}{\partial n} &= g \ \text{ on } \ S = \partial V \end{aligned} \tag{6.5.2}$$

with n the unit normal to S pointing to the exterior of V, and g a given function, can be represented in V by a *single-layer potential*

$$u(\bar{r}) = \int\!\!\int_S \frac{f(\bar{r}')}{4\pi|\bar{r} - \bar{r}'|} dS(\bar{r}'), \quad \bar{r} \in V \tag{6.5.3}$$

with $|\bar{r} - \bar{r}'|$ the Euclidean distance from $\bar{r}$ to $\bar{r}'$. Using Green's formula, it may now be shown that if S is sufficiently smooth, then f satisfies the integral equation

$$f(\overline{r}) + \frac{1}{2\pi} \int\int_S \left\{ \frac{\partial}{\partial n(\overline{r})} \frac{1}{|\overline{r} - \overline{r}'|} \right\} f(\overline{r}') dS(\overline{r}') = g(\overline{r}), \quad \overline{r} \in S. \tag{6.5.4}$$

Solving (6.5.4) for f on S, we may substitute this solution into (6.5.3) and evaluate $u(\overline{r})$ for any $\overline{r} \in V$.

Problems from applications typically have the property that the surface S consists of a relatively small number of "smooth" patches on each of which the function g is also smooth. For suitable K, such as that in (6.5.4), it may then be shown that the solution f is also smooth on the interior of every such patch. We shall give a precise definition of "smoothness" in terms of analyticity, in the section which follows. Our definition will enable us to identify these smoothness conditions which are always present in applications, and which are easily checked. These conditions will then enable us to easily determine requisite patches, and thus, to reduce the original integral equation to a system of integral equations, one over every such patch. Our analyticity assumptions will thus pave the way for proper *domain decomposition* of the problem, so as to retain accuracy in the Sinc function approximate solutions. At the outset, we shall consider the simplest problem, the solution over a single patch, for the sake of simplicity, and for deducing the correct convergence rate of Sinc approximation. It will then be shown that this simplified procedure carries over directly to the case of several patches. We then conclude with an explicit example, solving a two-patch problem.

6.5.2 Spaces for Sinc Approximation

Let us introduce some notations and some function spaces.

Let $Q = [-1,1] \times [-1,1]$, let $\mathcal{D}$ be defined in Equation (1.7.11), with $a = -1$, $b = 1$, and $0 < d \leqq \pi/2$, and let $\overline{\mathcal{D}}$ denote the closure of $\mathcal{D}$. Let Ω be defined by

$$\Omega = \left\{[-1,1] \times \overline{\mathcal{D}}\right\} \cup \left\{\overline{\mathcal{D}} \times [-1,1]\right\}. \tag{6.5.5}$$

Let $f : \Omega \to \mathbb{C}$, $f \in \mathbf{C}(\Omega)$. For fixed $x \in [-1,1]$ define $f^{(x)} : \overline{\mathcal{D}} \to \mathbb{C}$ by

$$f^{(x)}(y) = f(x,y). \tag{6.5.6}$$

For fixed $y \in [-1,1]$ define $f^{(y)} : \mathcal{D} \to \mathbb{C}$ by

$$f^{(y)}(x) = f(x,y). \tag{6.5.7}$$

Definition 6.5.1 *Let $f \in \mathbf{C}(\Omega)$, and let us consider the following assumptions:*

Assumption A_1: *f will be said to satisfy A_1 if*

$$f^{(x)} \in \ \mathbf{Hol}(\mathcal{D}) \ \ \textit{for all} \ \ x \in [-1,1]; \ \ \textit{and}$$

$$f^{(y)} \in \ \mathbf{Hol}(\mathcal{D}) \ \ \textit{for all} \ \ y \in [-1,1].$$

Assumption $A_{2\alpha}$**:** *f will be said to satisfy* $A_{2\alpha}$ *if f satisfies* A_1*, and if for some constant* $\alpha \in (0,1)$ *and for each fixed* $x \in [-1,1]$, $f^{(x)} \in \mathbf{Lip}_\alpha(\overline{\mathcal{D}})$, *and if for each fixed* $y \in [-1,1]$, $f^{(y)} \in \mathbf{Lip}_\alpha(\overline{\mathcal{D}})$.
Assumption $A_{3\alpha}$**:** *f will be said to satisfy* $A3\alpha$ *if g satisfies* $A_{2\alpha}$*, and if* $g \equiv 0$ *on* ∂Q*, with*

$$g(x,y) = (1-x^2)(1-y^2)f(x,y). \tag{6.5.8}$$

Note: *It is easily shown (Problem 6.5.1) that if f satisfies* $A3\alpha$*, then there exists a positive constant C such that*

$$|f(x,y)| \leqq C|(1-x^2)(1-y^2)|^{\alpha-1}, \quad (x,y) \in \Omega. \tag{6.5.9}$$

Let us set

$$\|f\|_\Omega = \sup_{(x,y)\in\Omega} |f(x,y)|, \tag{6.5.10}$$

and let us denote by

$$X_1 = \{f \ \text{ defined on } \ \Omega : f \ \text{ satisfies } \ A_1 \quad \text{and} \ \ \|f\|_\Omega < \infty\}$$

$$X_{2\alpha} = \{f \ \text{ defined on } \ \Omega : f \ \text{ satisfies } A_{2\alpha} \ \text{ and } \ \|f\|_\Omega < \infty\}$$

$$X3\alpha = \{f \ \text{ defined on } \ \Omega : f \ \text{ satisfies } \ A_{3\alpha} \ \text{ and } \ \|g\|_\Omega < \infty, \\ \text{with } \ g \ \text{ defined as in (6.5.8)}\}.$$

Note: Each of the spaces X_1, $X_{2\alpha}$ and $X_{3\alpha}$ are Banach spaces.

6.5.3 Sinc Approximation

We will now define a projection operator in $X_1 \cap X_{2\alpha}$, in order to get an accurate approximation. To this end, let N be a positive integer, let $h > 0$, and let (compare Example 5.2.8)

$$\begin{cases} x_{-N-1} = -1, \quad x_{N+1} = 1 \\ x_k = \dfrac{e^{kh}-1}{e^{kh}+1}, \quad k = -N, \cdots, N \\ \phi(x) = \log \dfrac{1+x}{1-x} \\ \psi_{-N-1}(x) = \dfrac{1-x}{2} - \displaystyle\sum_{j=-N}^{N} \dfrac{1-x_j}{2} S(j,h) \circ \phi(x) \\ \psi_{N+1}(x) = \dfrac{1+x}{2} - \displaystyle\sum_{j=-N}^{N} \dfrac{1+x_j}{2} S(j,h) \circ \phi(x) \\ \psi_k(x) = S(k,h) \circ \phi(x), \quad k = -N, \cdots, N. \end{cases} \tag{6.5.11}$$

Given $f \in X_1$, let us define T_1, T_2, Π_1 and Π_2 by

$$\begin{aligned} (T_1 f)(x) &= \frac{1-x}{2} f^{(y)}(-1) + \frac{1+x}{2} f^{(y)}(1) \\ (T_2 f)(y) &= \frac{1-y}{2} f^{(x)}(-1) + \frac{1+y}{2} f^{(x)}(1) \\ (\Pi_1 f)(x) &= \sum_{k=-N-1}^{N+1} f^{(y)}(x_k)\psi_k(x) \\ (\Pi_2 f)(y) &= \sum_{k=-N-1}^{N+1} f^{(x)}(x_k)\psi_k(y), \end{aligned} \tag{6.5.12}$$

and let us define $\mathcal{P}_N$ by

$$\begin{aligned} (\mathcal{P}_N f)(x,y) &= \Pi_1\{\Pi_2 f\}(x,y) \\ &= \sum_{m=-N-1}^{N+1} \sum_{n=-N-1}^{N+1} f(x_m, x_n)\psi_m(x)\psi_n(y). \end{aligned} \tag{6.5.13}$$

Theorem 6.5.2 *If f belongs to both X_1 and $X_{2\alpha}$, if N is a positive integer, and if $h = [\pi d/(\alpha N)]^{1/2}$, then*

$$\begin{aligned} &\sup_{(x,y)\in Q} |f(x,y) - (\mathcal{P}_N f)(x,y)| \\ &\leq [C_1 + C_2 \log N]\, N^{1/2} e^{-(\pi d \alpha N)^{1/2}}, \end{aligned} \tag{6.5.14}$$

with C_1 and C_2 constants independent of N.

Proof. Let us recall from Example 5.2.8, that if g is an arbitrary function in $\mathbf{L}_\alpha(\mathcal{D})$, with $\mathcal{D}$ given by Equation (1.7.11), $a = -1$ and $b = 1$, then with $h = [\pi d/(\alpha N)]^{1/2}$, we have

$$\begin{aligned}
&\sup_{x\in[-1,1]} |g(x) - (\Pi_1 g)(x)| \\
&= \sup_{x\in[-1,1]} \left|g(x) - (T_1 g)(x) - \sum_{k=-N}^{N} [g(z_k) - (T_1 g)(z_k)]\, S(k,h)\circ\phi(x)\right| \\
&\leqq C_3 N^{1/2} e^{-(\pi d\alpha N)^{1/2}}.
\end{aligned} \tag{6.5.15}$$

Hence, since both $f^{(x)} - T_1 f$ and $f^{(y)} - T_2 f$ are in $\mathbf{L}_\alpha(\mathcal{D})$, we have

$$\begin{aligned}
\sup_{(x,y)\in Q} |f^{(y)}(x) - (\Pi_1 f^{(y)})(x)| &\leqq C_4 N^{1/2} e^{-(\pi d\alpha N)^{1/2}}, \\
\sup_{(x,y)\in Q} |f^{(x)}(y) - (\Pi_2 f^{(x)})(y)| &\leqq C_5 N^{1/2} e^{-(\pi d\alpha N)^{1/2}}.
\end{aligned} \tag{6.5.16}$$

Now, from the identity

$$f - \mathcal{P}_N f = f - \Pi_1 f + \Pi_1(f - \Pi_2 f), \tag{6.5.17}$$

we get

$$\begin{aligned}
&\sup_{(x,y)\in Q} |f(x,y) - (\mathcal{P}_N f)(x,y)| \\
&\leqq \sup_{(x,y)\in Q} |f(x,y) - (\Pi_1 f^{(y)})(x)| + \sup_{(x,y)\in Q} \left|\Pi_1 \left[f(x,y) - (\Pi_2 f^{(x)})(y)\right]\right| \\
&\leqq \sup_{(x,y)\in Q} |f(x,y) - (\Pi_1 f^{(y)})(x)| + \|\Pi_1\| \sup_{(x,y)\in Q} |f(x,y) - (\Pi_2 f)(x,y)|
\end{aligned} \tag{6.5.18}$$

with

$$\begin{aligned}
\|\Pi_1\| &= \sup_{x\in[-1,1]} \left\{\frac{1+x}{2} + \frac{1-x}{2} + \sum_{k=-N}^{n} |S(k,h)\circ\phi(x)|\right\} \\
&= 1 + \sup_{x\in\mathbf{R}} \sum_{k=-N}^{N} |S(k,h)(x)| \\
&\leqq 1 + \frac{2}{\pi}(3 + \log N)
\end{aligned} \tag{6.5.19}$$

and with the last inequality obtained in Problem 3.1.5. Hence, using (6.5.16) and (6.5.19) in (6.5.18), we get (6.5.14). ■

6.5.4 Properties of the Integral Equation

We now consider a specific integral equation over Q, namely

$$(I - \lambda\kappa)f = g \tag{6.5.20}$$

with κ defined by

$$(\kappa f)(\overline{\rho}) = \int\int_Q K(\overline{\rho};\overline{\sigma})f(\overline{\sigma})d\overline{\sigma}, \tag{6.5.21}$$

with $\overline{\rho} = (x,y)$, $\overline{\sigma} = (\xi,\eta)$, $d\overline{\sigma} = d\xi d\eta$, and

$$K(\overline{\rho};\overline{\sigma}) = \frac{F(\overline{\rho};\overline{\sigma})}{|\overline{\rho}-\overline{\sigma}|}. \tag{6.5.22}$$

Our purpose of studying the mapping of κf is to show that this mapping is compact in a space of analytic functions. We will establish this in the case $f \in X_1$. This case is particularly interesting to us, since, as we shall prove, if $f \in X_1$, then $\kappa f \in X_{2\alpha}$, and κf can then be accurately approximated by the Sinc approximation procedure of the previous subsection.

We thus make the following assumptions on F:

Assumption A_F**:** Let α be as in Assumption A2α. F will be said to satisfy A_F if:

- $F(\cdot,y;\xi,\eta) \in \mathbf{Hol}(\mathcal{D}) \cap \mathbf{Lip}_\alpha(\mathcal{D})$ for all fixed $(y,\xi,\eta) \in [-1,1] \times \overline{\mathcal{D}} \times [-1,1]$;
- $F(x,\cdot;\xi,\eta) \in \mathbf{Hol}(\mathcal{D}) \cap \mathbf{Lip}_\alpha(\mathcal{D})$ for all fixed $(x,\xi,\eta) \in [-1,1] \times [-1,1] \times \overline{\mathcal{D}}$;
- $F(x,y;\cdot,\eta) \in \mathbf{Hol}(\mathcal{D}) \cap \mathbf{C}(\overline{\mathcal{D}})$ for all fixed $(x,y,\eta) \in \overline{\mathcal{D}} \times [-1,1] \times [-1,1]$; and
- $F(x,y;\xi,\cdot) \in \mathbf{Hol}(\mathcal{D}) \cap \mathbf{C}(\overline{\mathcal{D}})$ for all fixed $(x,y,\xi) \in \overline{\mathcal{D}} \times [-1,1] \times [-1,1]$.

Let F be normed by

$$\|F\|_{\Omega\times\Omega} = \max\left\{ \sup_{(x,\xi)\in\mathcal{D}\times\mathcal{D},(y,\eta)\in Q\times Q} |F(x,y;\xi,\eta)|, \right. \tag{6.5.23}$$

$$\left. \sup_{(y,\eta)\in\mathcal{D}\times\mathcal{D},(x,\xi)\in Q\times Q} |F(x,y;\xi,\eta)| \right\}.$$

Let us now give the proper definition of $(\kappa f)(x, y)$, which, when considered as a function of $(x, y) \in \mathcal{D} \times [-1, 1]$, does, in fact, provide the analytic continuation of κf defined on Q in (6.5.21) through (6.5.22). To this end, we take $x \in \mathcal{D}$, $f \in X_1$, and we set

$$
\begin{aligned}
L' &= \int_{-1}^{x} \frac{F(x, y; \xi, \eta)\, f(\xi, \eta)\, d\xi}{\{(x-\xi)^2 + a^2\}^{1/2}} \\
M' &= \int_{x}^{1} \frac{F(x, y; \xi, \eta)\, f(\xi, \eta)\, d\xi}{\{(\xi - x)^2 + a^2\}^{1/2}}
\end{aligned}
\tag{6.5.24}
$$

with $a = |y - \eta|$. In these definitions, we choose the branch of the square root of $\{(x-\xi)^2 + a^2\}$ approaching a as $\xi \to x$ from the part of $\mathcal{D}$ where $\Re\xi < \Re x$, and asymptotically equal to $x - \xi$ as $\Re\xi \to -\infty$, while we take the branch of the square root of $\{(\xi - x)^2 + a^2\}$ approaching a as $\xi \to x$ in the part of $\mathcal{D}$ where $\Re\xi > \Re x$, and asymptotically equal to $\xi - x$ as $\xi \to \infty$. Under our assumptions on F and f, it is clear that the value of L' is independent of the path of integration joining -1 and x, so long as each point ξ on the path satisfies $\Re\xi < \Re x$, and similarly in the case of M'. We may thus assume that we have a straight line path of integration with respect to ξ for L' as well as for M'. Having defined L' and M' in this manner, we define L, M and then κf by

$$
\begin{aligned}
L &= \int_{-1}^{1} L'\, d\eta, \quad M = \int_{-1}^{1} M'\, d\eta \\
(\kappa f)(x, y) &= L + M
\end{aligned}
\tag{6.5.25}
$$

Let us now deduce some properties of the operator κf defined in (6.5.24) through (6.5.25).

Lemma 6.5.3 *Let F satisfy A_F, and let $f \in X_1$. Let both x and x_1 belong to $\mathcal{D}$, let $\Re x \leqq \Re x_1$, and let $\mathcal{P}$ denote the line segment in the direction from x_1 to x. Then*

$$
I \equiv \left| \int_{-1}^{1} \int_{\mathcal{P}} \frac{F(x, y; \xi, \eta)\, f(\xi, \eta)}{\{(x-\xi)^2 + (y-\eta)^2\}^{1/2}} d\xi\, d\eta \right| \tag{6.5.26}
$$

$$
\leqq |x - x_1| \left[\pi + 2 + (2+\pi)\log(2) - \pi \log|x - x_1|\right] \|F\|_{\Omega\times\Omega} \|f\|_{\Omega} \tag{6.5.27}
$$

$$
\leqq 14\, \|F\|_{\Omega\times\Omega} \|f\|_{\Omega}. \tag{6.5.28}
$$

Proof. The definition (6.5.26) of I immediately yields

$$I \leqq \|F\|_{\Omega\times\Omega}\|f\|_{\Omega} \int_{-1}^{1} J\, d\eta, \tag{6.5.29}$$

with the norms of F and f defined in (6.5.23) and (6.5.10), respectively. Using the the notation

$$\begin{aligned} p &= |x - x_1| \\ \gamma &= \arg(x - x_1) \\ \xi &= x - (x - x_1)t, \quad 0 \leqq t \leqq 1, \\ a &= |y - \eta|, \end{aligned} \tag{6.5.30}$$

we have

$$\begin{aligned} J = &\int_{\mathcal{P}} \frac{|d\xi|}{|(x-\xi)^2 + a^2|^{1/2}} \\ &= p\int_0^1 |t^2p^2e^{2i\gamma} + a^2|^{-1/2} dt \\ &\leqq K, \end{aligned} \tag{6.5.31}$$

with

$$\begin{aligned} K &= \begin{cases} \sin^{-1}(p/a) & \text{if} \quad 0 < p \leqq a; \\ \pi/2 + \log\{p/a + [(p/a)^2 - 1]^{1/2}\} & \text{if} \quad 0 \leqq a < p. \end{cases} \\ &\leqq \begin{cases} \pi p/(2a) & \text{if} \quad 0 < p \leqq a; \\ \pi/2 + \log 2 + \log p - \log a & \text{if} \quad 0 \leqq a < p. \end{cases} \end{aligned} \tag{6.5.32}$$

Let us now recall that $y \in [-1, 1]$ is fixed and that $a = |y - \eta|$; then, integrating K with respect to η over $[-1, 1]$, we get

$$\begin{aligned} &\int_{-1}^{1} K\, d\eta \\ &\quad \leqq 2\int_0^p \{\pi/2 + \log 2 + \log p - \log a\}\, da + 2\int_p^{1+|y|} \{\pi p/(2a)\}\, da \\ &\quad \leqq \{\pi + 2 + (2+\pi)\log 2 - \pi \log p\}p, \end{aligned} \tag{6.5.33}$$

where the last result in (6.5.33) was obtained upon explicit integration of the extreme right-hand side of (6.5.32), and where we have used the results that if $y \in [-1, 1]$, then $1 + |y| \leqq 2$. This establishes the result (6.5.27).

The result (6.5.28) is obtained by finding the maximum of the right-hand side (6.5.27) with respect to $p \in [0,2]$; this maximum occurs for $p = 2$, and equals $13.055\cdots < 14$.
This completes the proof of Lemma 6.5.3. ■

Corollary 6.5.4 *Let $x \in \mathcal{D}$, let $\mathcal{P}_1$ denote the line segment joining -1 and x, and let $\mathcal{P}_2$ denote the line segment joining x and 1. Then*

$$\sup_{(x,y)\in\mathcal{D}\times[-1,1]} \int_{-1}^{1} \left\{ \int_{\mathcal{P}_1} \frac{|d\xi|}{|(x-\xi)^2 + (y-\eta)^2|^{1/2}} + \int_{\mathcal{P}_2} \frac{|d\xi|}{|(\xi - x)^2 + (y-\eta)^2|^{1/2}} \right\} d\eta \leqq 28. \tag{6.5.34}$$

Moreover,

$$\sup_{(x,y)\in\Omega} |(\kappa f)(x,y)| \leqq 28\|F\|_{\Omega\times\Omega}\|f\|_{\Omega}. \tag{6.5.35}$$

Proof. If we set $p = |1+x|$, $q = |1-x|$, then by Lemma 6.5.3 (see Equation (6.5.28)), we have

$$\sup_{(x,y)\in\mathcal{D}\times[-1,1]} \int_{-1}^{1} \left\{ \int_{\mathcal{P}_1} \frac{|F(x,y;\xi,\eta)\, f(\xi,\eta)|\,|d\xi|}{|(x-\xi)^2 + (y-\eta)^2|^{1/2}} \right\} d\eta \leqq 14\|F\|_{\Omega\times\Omega}\|f\|_{\Omega}. \tag{6.5.36}$$

Similarly, the second integral term in (6.5.34) is also bounded by the right-hand side of (6.5.36). Hence (6.5.34) follows.
Hence, it follows that

$$\sup_{(x,y)\in\mathcal{D}\times[-1,1]} |(\kappa f)(x,y)| \leqq 28\|F\|_{\Omega\times\Omega}\|f\|_{\Omega}. \tag{6.5.37}$$

Similarly, repeating the above proof with the roles of x and y interchanged, we get

$$\sup_{(x,y)\in[-1,1]\times\mathcal{D}} |(\kappa f)(x,y)| \leqq 28\|F\|_{\Omega\times\Omega}\|f\|_{\Omega}. \tag{6.5.38}$$

Hence (6.5.35) follows by combining (6.5.37) and (6.5.38). ■

Lemma 6.5.5 *If F satisfies A_F, and if $f \in X_1$, then $\kappa f \in X_1$.*

Proof. Let $y \in [-1,1]$, let $\varepsilon_1 > \varepsilon_2 > \varepsilon_3 \cdots$ denote a sequence of positive numbers, such that $\varepsilon_j \in (0,1)$ for every positive integer j, and such that $\varepsilon_j \to 0$ as $j \to \infty$, and define P and P_j by

$$\begin{aligned} P &= \{(t,\eta) : 0 \leqq t \leqq 1;\ -1 \leqq \eta \leqq 1\}, \\ P_j &= \{(t,\eta) \in P : t \geqq \varepsilon_j;\ |y-\eta| \geqq \varepsilon_j\}. \end{aligned} \tag{6.5.39}$$

We may then define K_j by

$$K_j(x,y) = \int\int_{P_j} \frac{(1+x)F(x,y;x-(1+x)t,\eta)\, f(x-(1+x)t,\eta)}{\{(1+x)^2t^2+(y-\eta)^2\}^{1/2}} dt\, d\eta. \tag{6.5.40}$$

Factoring the denominator function in the form

$$\{(1+x)^2t^2+(y-\eta)^2\}^{1/2} = \{[(1+x)t - i(y-\eta)][(1+x)t + i(y-\eta)]\}^{1/2}, \tag{6.5.41}$$

it is evident, since $\Re(1+x) > 0$ for all $x \in \mathcal{D}$, since y and η are real, and since $t \geqq \varepsilon_j > 0$, that this denominator does not vanish for any $x \in \mathcal{D}$. It thus follows that the integrand in (6.5.40) is an analytic function of x in $\mathcal{D}$, and hence, so is $K_j(x,y)$. Moreover, it follows from Lemma 6.5.3, that the sequence $\{K_j(\cdot,y)\}$ is a sequence of functions, each of which is continuous and uniformly bounded in $\mathcal{D}$, for every fixed $y \in [-1,1]$. This family is thus compact, by Montel's Theorem. Moreover, it is easily verified that the sequence $\{K_j(x,y)\}$ converges for all $x \in [-1/2, 1/2]$, and for every fixed $y \in [-1,1]$. Hence, by Vitali's Theorem, the sequence $\{K_j(\cdot,y)\}$ converges to the function $L(\cdot,y)$ with L defined in (6.5.25), and with $L(\cdot,y)$ analytic in $\mathcal{D}$. Finally, since each $K_j(\cdot,y)$ is continuous in $\overline{\mathcal{D}}$ the limit function $L(\cdot,y)$ is also continuous in $\overline{\mathcal{D}}$.
Proceeding similarly for the term M defined in (6.5.25), we deduce that $M(\cdot,y)$ is also analytic in $\mathcal{D}$ and continuous in $\overline{\mathcal{D}}$, for each fixed $y \in [-1,1]$. Next, interchanging the roles of x and y, and proceeding similarly, we are able to deduce the same analyticity and continuity properties of κ as a function of y, for each fixed $x \in [-1,1]$. Hence it follows that $\kappa f \in X_1$. ■

Theorem 6.5.6 *If F satisfies A_F, and if $f \in X_1$, then $\kappa f \in X_{2\alpha}$.*

Proof. Let F satisfy A_F, let $x_j \in \overline{\mathcal{D}}$, for $j = 1, 2$, and let us set

$$F_j = F(x_j, y; \xi, \eta), \quad r_j = \{(x_j - \xi)^2 + (y-\eta)^2\}^{1/2}. \tag{6.5.42}$$

Then, taking $f \in X_1$, denoting ∇ and J_j by

$$\begin{aligned} \nabla &= J_1 - J_2 \\ J_j &= \int_{-1}^{x_j} \frac{F_j f}{r_j} d\xi \end{aligned} \tag{6.5.43}$$

and assuming that $\Re x_1 \leqq \Re x_2$, we have

$$\nabla = I_1 + I_2 + I_3, \tag{6.5.44}$$

with

$$\begin{aligned} I_1 &= \int_{-1}^{x_1} \frac{(F_1 - F_2)f}{r_1} d\xi, \\ I_2 &= \int_{-1}^{x_1} F_2 f \left\{ \frac{1}{r_1} - \frac{1}{r_2} \right\} d\xi, \\ I_3 &= -\int_{x_1}^{x_2} \frac{F_2 f}{r_2} d\xi. \end{aligned} \tag{6.5.45}$$

We thus immediately have, by Lemma 6.5.3 and by our assumptions A_F on F, that there exist constants C_1 and C_3 such that the inequalities

$$\begin{aligned} \left| \int_{-1}^{1} I_1 d\eta \right| &\leqq C_1 |x_1 - x_2|^\alpha, \\ \left| \int_{-1}^{1} I_3 d\eta \right| &\leqq C_3 |x_1 - x_2| [\pi + 1 + (2 + \pi) \log 2 - \pi \log |x_1 - x_2|], \end{aligned} \tag{6.5.46}$$

hold for all $(x_1, x_2, y, \eta) \in \overline{\mathcal{D}} \times \overline{\mathcal{D}} \times [-1, 1] \times [-1, 1]$, and such that $\Re x_1 \leqq \Re x_2$. Also, by Lemma 6.5.5, the term $\int_{-1}^{1} J_j d\eta$ belongs to X_1 as a function of x_j, for $j = 1, 2$, so that, setting $I_2 = \nabla - I_1 - I_3$ and using (6.5.46) it follows that there exists a constant C_2, such that

$$\left| \int_{-1}^{1} I_2 d\eta \right| \leqq C_2 |x_1 - x_2|^\alpha \tag{6.5.47}$$

for all x_1 and x_2 on compact subsets of $\mathcal{D}$. Evidently, starting with a domain $\mathcal{D}' = \{z \in \mathbb{C} : |\arg\{(1 + z)/(b - z)\}| < d'\}$, $b > 1$, and $d' > d$, we would come to the same conclusion, that I_2 satisfies (6.5.47) for all x_1 and x_2 on compact subsets of $\mathcal{D}'$. However, since $\mathcal{D}'$ contains all of $\overline{\mathcal{D}}$ except the point -1, it follows that I_2 satisfies (6.5.47) for all x_1 and x_2 on $\overline{\mathcal{D}} \backslash \mathcal{D}(-1; r)$, for arbitrary $r \in (0, 1)$, with $\mathcal{D}(-1; r)$ the disc centered at -1 and of radius r. That is, we only need to check the smoothness of J_j as a function of x_j, when $x_j \in \overline{\mathcal{D}} \cap \mathcal{D}(-1; r)$. However, this behavior may be deduced from Lemma 6.5.3, which tells us that

$$\left| \int_{-1}^{1} J_j d\eta \right| \leqq |1 + x_j| [\pi + 2 + (2 + \pi) \log 2 - \pi \log |1 + x_j|] \|F\|_{\Omega \times \Omega} \|f\|_\Omega \tag{6.5.48}$$

for *all* $x_j \in \overline{\mathcal{D}}$. It thus follows from this, that there exists a constant $C > 0$, such that for all x_1 and x_2 in $\overline{\mathcal{D}}$ we have

$$|\nabla| \leqq C|x_1 - x_2|^\alpha. \tag{6.5.49}$$

This completes the proof of Theorem 6.5.6. ∎

6.5.5 Galerkin Approximation

Let us now return to the equation (6.5.20), and replace it by the finite-dimensional Galerkin system

$$f_N - \lambda \mathcal{P}_N \kappa f_N = \mathcal{P}_N g \tag{6.5.50}$$

with $\mathcal{P}_N$ defined in (6.5.13), and with

$$f_N(x,y) = \sum_{k=-N-1}^{N+1} \sum_{\ell=-N-1}^{N+1} c_{k\ell} \psi_k(x) \psi_\ell(y). \tag{6.5.51}$$

In view of Theorem 6.5.6 and our assumption of $g \in X_{2\alpha}$, it follows that if a solution f of Equation (6.5.20) exists, then $f \in X_{2\alpha}$.

Let us assume, henceforth, that Equation (6.5.20) has a solution f, even for the case when λ is a characteristic value of this equation.

We can deduce accurate rates on the order of convergence of the approximate eigenvalues and the approximate eigensolutions of (6.5.50) to those of (6.5.20). Thus, even if Equation (6.5.50) does not have a solution, we can determine a function f_N of the form (6.5.51) using the generalized inverse procedure (see Section 6.3.3). We can thus deduce the existence of a solution f of Equation (6.5.20), such that the solutions f_N determined in this manner satisfy $\|f - f_N\|_Q \to 0$ as $N \to \infty$.

By Theorem 6.2.5, we may deduce that if $[I - \lambda\kappa]^{-1}$ exists and is bounded on $\mathbf{C}[Q]$, then as a consequence of Theorem 6.5.2, Equation (6.5.50) has a solution f_N for all N sufficiently large, and moreover, we may deduce an accurate rate of convergence of f_N to f as $N \to \infty$. In the case of λ a characteristic value of Equation (6.5.20), Equation (6.5.50) may not have a solution. On the other hand, we can invoke Corollary 6.2.10 to deduce that the characteristic values and the corresponding eigensolutions of (6.5.50) in the homogeneous case $g = 0$ converge to those of (6.5.20) as $N \to \infty$, and once again, combining Corollary 6.2.10 and Theorem 6.5.2, we get

Theorem 6.5.7 (a) *If* $(I - \lambda\kappa)^{-1}$ *exists and is bounded on* X_1, *if* $g \in X_{2\alpha}$, *and if* h *is selected by the formula* $h = [\pi d/(\alpha N)]^{1/2}$, *then as* $N \to \infty$,

$$\|f - f_N\|_Q = \mathcal{O}[N^{1/2} \log(N) e^{-(\pi d \alpha N)^{1/2}}]. \tag{6.5.52}$$

(b) *Let λ be a characteristic value of multiplicity $r \geqq 1$ for the homogeneous equation corresponding to (6.5.20), and let f be an associated eigensolution. Then for all N sufficiently large, and $h = [\pi d/(\alpha N)]^{1/2}$, the homogeneous equation corresponding to (6.5.50) has characteristic values $\lambda_j^{(N)}$ and associated eigensolutions $f_N^{(j)}$ of the form (6.5.51), $j = 1, 2, \cdots, r$, such that as $N \to \infty$,*

$$|\lambda - \lambda_N^{(j)}| = \mathcal{O}[N^{1/(2r)} \log N e^{-(\pi d\alpha N)^{1/2}/r}],$$

$$\|f - \sum_{j=1}^{N} b_{Nj} f_N^{(j)}\|_Q = \mathcal{O}[N^{1/2} \log N e^{-(\pi d\alpha N)^{1/2}}], \tag{6.5.53}$$

with the $b_{Nj} (j = 1, 2, \cdots, r)$ suitably chosen constants.

(c) *Let λ be a characteristic value of the homogeneous equation corresponding to Equation (6.5.20), let $g \in X_{2\alpha}$, and let f be a solution of Equation (6.5.20), for that λ. Then, for all N sufficiently large, and with $h = [\pi d/(\alpha N)]^{1/2}$, there exist characteristic values $\lambda_N^{(j)}$ and associated eigensolutions $f_N^{(j)}$ of the homogeneous equation corresponding to (6.5.50), having the form (6.5.51) for $j = 1, 2, \cdots, r$, such that as $N \to \infty$,*

$$\|f - f_N^{(0)} - \sum_{j=1}^{r} b_{Nj} f_N^{(j)}\|_Q$$
$$= \mathcal{O}[N^{1/2} \log(N) e^{-(\pi d\alpha N)^{1/2}}], \tag{6.5.54}$$

with $f_N^{(0)}$ of the form (6.5.51), the generalized inverse solution to (6.5.50) and with $b_{Nj} (j = 1, 2, \cdots, r)$ suitably chosen constants.

6.5.6 Several Surface Patches—Domain Decomposition

We now consider for Equation (6.5.1) the extension of the Subsections 6.5.2 through 6.5.5 to the case of S, the union of more than one smooth surface patches. With reference to Equation (6.5.20), the boundaries of the patches are chosen to be either fixed singular points, or fixed curves on S along which either g or K is non-analytic, or else points or curves on S along which S is not smooth, e.g., where S comes to a point or has an edge. Assuming that the same conditions of smoothness are satisfied on each of the resulting patches as they were for the previously considered case of a single patch (namely Q), we shall be able to obtain the convergence rate for the approximate solutions that as are obtained for the case of a single patch.

Let us assume that the surface S defined in (6.5.1) may be represented as a union of "smooth" patches $S_1, \cdots, S_p$, whose interiors are disjoint. More specifically, let us assume that for each $j = 1, 2, \cdots, p$, S_j has the representation

$$S_j = \{(u^j(\overline{\sigma}), v^j(\overline{\sigma}), w^j(\overline{\sigma})) : \overline{\sigma} \in Q\} \tag{6.5.55}$$

with as before, $\overline{\sigma} = (\xi, \eta)$, and $Q = [-1,1] \times [-1,1]$. Each of the functions u^j, v^j and w^j is assumed to belong to X_1 defined in Definition 6.5.1. We also insist on an additional requirement of the mappings u^j, v^j and w^j, namely, that if S_j and S_ℓ $(j \neq \ell)$ have a common edge $\mathcal{L}$ given, for example, by

$$\begin{aligned}\mathcal{L} &= \left\{(u^j(1,\eta), v^j(1,\eta), w^j(1,\eta)) : -1 \leqq \eta \leqq 1\right\} \\ &= \left\{(u^\ell(-1,\eta), v^\ell(-1,\eta), w^\ell(-1,\eta)) : -1 \leqq \eta \leqq 1\right\},\end{aligned} \tag{6.5.56}$$

then for all $\eta \in (-1, 1)$

$$\begin{aligned} u^j(1,\eta) &= u^\ell(-1,\eta), \\ v^j(1,\eta) &= v^\ell(-1,\eta), \\ w^j(1,\eta) &= w^\ell(-1,\eta). \end{aligned} \tag{6.5.57}$$

This requirement combined with our interpolation scheme yields common points on the boundaries of the patches, thus enabling us to assign the same values in our Galerkin scheme.

Letting f and g be defined in (6.5.1), we may thus represent f and g on S_j by the formulas

$$\begin{aligned} f^j(\overline{\sigma}) &= f(u^j(\overline{\sigma}), v^j(\overline{\sigma}), w^j(\overline{\sigma})) \\ g^j(\overline{\sigma}) &= g(u^j(\overline{\sigma}), v^j(\overline{\sigma}), w^j(\overline{\sigma})). \end{aligned} \tag{6.5.58}$$

That is, if $f^j(\overline{r})$, (with $\overline{r} = (u^j, v^j, w^j)$) denotes f defined on S_j, we may write (here $f^j(\overline{r}) \equiv f^j(\overline{\sigma})$)

$$f(\overline{r}) = \sum_{j=1}^{p} \xi_j(\overline{r}) f^j(\overline{r}) \tag{6.5.59}$$

with $\xi_j(\overline{r})$, the characteristic function taking the value 1 on the interior of S_j, $1/2$ on the interior of any curve $\mathcal{L}$ which is the boundary of both S_j and S_ℓ, and $1/q$ at a corner where q patches share a common boundary point.

Using (6.5.55) through (6.5.59) in (6.5.1), we arrive at a system of p integral equations over Q, namely

$$f^i(\overline{\rho}) - \lambda \sum_{j=1}^{p} \int\int_Q K_j^i(\overline{\rho};\overline{\sigma}) f^j(\overline{\sigma}) d\overline{\sigma} = g^i(\overline{\rho}), \overline{\rho} \in Q \tag{6.5.60}$$

$$i = 1, 2, \cdots, p$$

with $\overline{\rho} = (x, y)$, $\overline{\sigma} = (\xi, \eta)$, $d\overline{\sigma} = d\xi d\eta$, and

$$K_i^i(\overline{\rho};\overline{\sigma}) = \frac{F_i^i(\overline{\rho};\overline{\sigma})}{|\overline{\rho} - \overline{\sigma}|}, \quad i = 1, 2, \cdots, p$$
$$K_j^i(\overline{\rho};\overline{\sigma}) = F_j^i(\overline{\rho};\overline{\sigma}), \quad j \neq i, \quad i \cdot j = 1, 2, \cdots, p. \tag{6.5.61}$$

While there are many possible mappings (6.5.55) of Q onto S_j, we shall assume that these mappings were selected such that the functions $F_i^i(\overline{\rho};\overline{\sigma})$ in (6.5.61) satisfy the same conditions as F did in Section 6.5.4, namely $F_i^i(\overline{\rho};\overline{\sigma})$ satisfies A_F. While this assumption implies the uniform boundedness of the $F_i^i(\overline{\rho};\overline{\sigma})$ as a function of $\overline{\sigma}$ on Q, for all $\overline{\rho} \in \Omega$, we must allow the $F_j^i(\overline{\rho},\overline{\sigma})(i \neq j)$ to become unbounded when $\overline{\rho} \in \partial Q$. This requirement stems from our assumption that the kernel $K(\overline{r},\overline{r}')$ in (6.5.1) has a weak singularity at $\overline{r} = \overline{r}'$ in the case S_i and $S_j(i \neq j)$ share a common point or boundary arc, $\overline{r} \in S_i$ may be identical to $\overline{r}' \in S_j$. Thus, $F_j^i(\overline{\rho};\overline{\sigma})$ will be unbounded for such regions. We then assume that if $i \neq j$, then F_j^i belongs to $X_{2\alpha}$ with respect to $\overline{\rho}$ for fixed $\overline{\sigma}$ in the interior of Q. Furthermore, we shall assume that as a function of $\overline{\sigma}$, $F_i^i(\overline{\rho};\overline{\sigma}) \in X_1$ uniformly, $\overline{\rho} \in Q$, while for $i \neq j$, $F_j^i(\overline{\rho};\overline{\sigma})$ satisfies $A3\alpha$ with respect to $\overline{\sigma}$, uniformly, $\overline{\rho} \in Q$.

It then follows that each of the operators κ_j^i defined by

$$(\kappa_j^i f)(\overline{\rho}) = \int\int_Q K_j^i(\overline{\rho};\overline{\sigma}) f(\overline{\sigma}) d\overline{\sigma} \tag{6.5.62}$$

satisfies the same conditions as κf did in subsections 6.5.2 through 6.5.5. Furthermore, K_j^i satisfies the same conditions as K did in Corollary 6.5.4. Hence, it follows that if the system (6.5.60) has a solution vector $(f^1, \cdots, f^p)$ then $f^i \in X_{2\alpha}$, $(i = 1, 2, \cdots, p)$.

Defining $\mathcal{P}_N$ as in (6.5.13) and setting

$$f_N^i(\overline{\rho}) = \sum_{m=-N-1}^{N+1} \sum_{n=-N-1}^{N+1} c_{mn}^i \psi_m(x) \psi_n(y), \tag{6.5.63}$$

we replace (6.5.60) by the Galerkin system

$$f_N^i(\overline{\rho}) - \lambda \sum_{j=1}^{p} \mathcal{P}_N \int\int_Q K_j^i(\overline{\rho};\overline{\sigma}) f_n^j(\overline{\sigma}) d\overline{\sigma} = \mathcal{P}_N g(\overline{\rho}), \tag{6.5.64}$$

$$\overline{\rho} \in Q, \quad i = 1, 2, \cdots, p.$$

In view of Theorem 6.5.2, it thus follows that the order of approximation of f_N^i to f^i is the same as that of f_N to f. For definiteness, let us write

$$
\begin{aligned}
f(\overline{\rho}) &= \sum_{j=1}^{p} \xi_j(\overline{r}(\overline{\rho}))f^j(\overline{\rho}) \\
f_N(\overline{\rho}) &= \sum_{j=1}^{p} \xi_j(\overline{r}(\overline{\rho}))f_N^j(\overline{\rho})
\end{aligned}
\tag{6.5.65}
$$

with f^j and f_N^j defined in (6.5.58) and (6.5.63) respectively, and with $\xi_j(\overline{r})$ defined in (6.5.59). Then we have

Corollary 6.5.8 *Theorem 6.5.7 holds for the functions $f(\overline{\rho})$ and $f_N(\overline{\rho})$ defined in (6.5.65).*

6.5.7 An Explicit Example

Let us present an explicit example to illustrate the theory of this section. More specifically, we shall solve the problem (6.5.2), with V the unit sphere, and u a single-layer potential of the form (6.5.3). We thus obtain equation (6.5.4). Equation (6.5.4) is indeed of the form (6.5.1), with the kernel $K(\overline{r},\overline{r}') = (\partial/\partial n(\overline{r}))|\overline{r}-\overline{r}'|^{-1}$ having a weak singularity at $\overline{r}=\overline{r}'$, and moreover, $\int\int_S |K(\overline{r},\overline{r}')|dS(\overline{r}') < \infty$.

Using our standard operator notation, Equation (6.5.4) may be written in the form

$$
(I - \lambda\kappa)f = g, \tag{6.5.66}
$$

with

$$
(\kappa f)(\overline{r}) = \int\int_S \left\{ \frac{\partial}{\partial n(\overline{r})} \frac{1}{|\overline{r}-\overline{r}'|} \right\} f(\overline{r}')dS(\overline{r}'), \quad \overline{r} \in S. \tag{6.5.67}
$$

The adjoint operator κ^* is then given by

$$
(\kappa^* f)(\overline{r}) = \int\int_S \left\{ \frac{\partial}{\partial n(\overline{r}')} \frac{1}{|\overline{r}-\overline{r}'|} \right\} f(\overline{r}')dS(\overline{r}'), \overline{r} \in S. \tag{6.5.68}
$$

Since

$$
\int\int_S \left\{ \frac{1}{\partial n(\overline{r}')} \frac{1}{|\overline{r}-\overline{r}'|} \right\} dS(\overline{r}') = -2\pi, \tag{6.5.69}
$$

it follows that $\lambda = -1/(2\pi)$ is a characteristic value, and $f(\overline{r}) \equiv 1$ is the corresponding eigenfunction of the homogeneous problem

$$(I - \lambda\kappa^*)f = 0. \tag{6.5.70}$$

According to the Fredholm alternative, in order for (6.5.66) to have a solution, the function g must be orthogonal (in the $\mathbf{L}^2(S)$ sense) to $f = 1$, i.e., we must have

$$\int\int_S g(\overline{r}')dS(\overline{r}') = 0. \tag{6.5.71}$$

Now in the case of our problem, S is the unit sphere, $|\overline{r}| = |\overline{r}'| = 1$, and therefore

$$\begin{aligned} \frac{\partial}{\partial n(\overline{r})}\frac{1}{|\overline{r}-\overline{r}'|} &= \nabla_{\overline{r}}\frac{1}{|\overline{r}-\overline{r}'|}\cdot\hat{r} \\ &= -\frac{1}{2|\overline{r}-\overline{r}'|}, \\ \frac{\partial}{\partial n(\overline{r}')}\frac{1}{|\overline{r}-\overline{r},|} &= -\frac{1}{2|\overline{r}-\overline{r}'|}. \end{aligned} \tag{6.5.72}$$

Hence $\kappa = \kappa^*$, i.e., κ is self-adjoint. Hence $\lambda = -1/(2\pi)$ is a characteristic value, and $f(\overline{r}) \equiv 1$ is the corresponding eigenfunction to that eigenvalue of κ.

Next, setting

$$\overline{r} = (\mu, \sigma, \tau), \quad \overline{r}' = (\mu', \sigma', \tau'). \tag{6.5.73}$$

we get

$$dS(\overline{r}) = \frac{d\mu\, d\sigma}{\sqrt{1-\mu^2-\sigma^2}}. \tag{6.5.74}$$

Let us now take

$$g(\overline{r}) = 2(1 - 3r^2), \tag{6.5.75}$$

with $r = |\overline{r}|$. Splitting S into the upper hemisphere S_+ and the lower hemisphere S_-, we get

$$\begin{aligned}
&\int\int_S g(\overline{r})dS(\overline{r}) \\
&= \int\int_{S_+} g(\overline{r})dS(\overline{r}) + \int\int_{S_-} g(\overline{r})dS(\overline{r}) \\
&= 2\int\int_{S_+} g(\overline{r})dS(\overline{r}) \qquad (6.5.76) \\
&= 4\int_{\sigma=-1}^{1}\int_{\mu=-\sqrt{1-\sigma^2}}^{\sqrt{1-\sigma^2}} \frac{[1-3(1-\mu^2-\sigma^2)]}{\sqrt{1-\mu^2-\sigma^2}} d\mu d\sigma \\
&= 0.
\end{aligned}$$

The condition (6.5.71) is therefore satisfied for our function g.

The above calculations enable us to easily check the solution f of (6.5.66). Substituting (6.5.72) into (6.5.67), and using (6.5.4), we get

$$\begin{aligned}
f(\overline{r}) + \frac{1}{2\pi}\int\int_S \left\{\frac{\partial}{\partial_n(\overline{r})}\frac{1}{|\overline{r}-\overline{r}'|}\right\} f(\overline{r}')dS(\overline{r}') &= \frac{g(\overline{r})}{2} \\
f(\overline{r}) - \frac{1}{4\pi}\int\int_S \frac{1}{|\overline{r}-\overline{r}'|} f(\overline{r}')dS(\overline{r}') &= \frac{g(\overline{r})}{2} \qquad (6.5.77) \\
f(\overline{r}) &= \frac{u(\overline{r})+g(\overline{r})}{2}.
\end{aligned}$$

The symmetry of the function $g(\overline{r})$ enables us to take advantage of some simplifications. To this end, let $\overline{r}_+ \in S_+$, $\overline{r}_- \in S_-$, with S_+ (rsp. S_-) the upper and lower hemispheres of S. Thus, we have, for $\tau > 0$, $\tau' > 0$,

$$\begin{aligned}
&\overline{r}_+ = (\mu,\sigma,\tau), \quad \overline{r}_- = (\mu,\sigma,-\tau), \\
&\overline{r}'_+ = (\mu',\sigma',\tau'), \quad \overline{r}'_- = (\mu',\sigma',-\tau'), \\
&f_+(\overline{r}) = f(\overline{r})/_{\overline{r}\in S_+}, \qquad (6.5.78) \\
&f_-(\overline{r}) = f(\overline{r})/_{\overline{r}\in S_-}.
\end{aligned}$$

We may thus rewrite (6.5.4) in the form

$$f_+(\overline{r}_+) + \int\int_{S_+} \left\{ \frac{\partial}{\partial n(\overline{r}_+)} \frac{1}{2\pi|\overline{r}_+ - \overline{r}'_+|} \right\} f_+(\overline{r}'_+) dS(\overline{r}'_+)$$

$$+ \int\int_{S_-} \left\{ \frac{\partial}{\partial n(\overline{r}_+)} \frac{1}{2\pi|\overline{r}_+ - \overline{r}'_-|} \right\} f_-(\overline{r}'_-) dS(\overline{r}'_-) = g(\overline{r}_+)$$

$$f_-(\overline{r}_-) + \int\int_{S_+} \left\{ \frac{\partial}{\partial n(\overline{r}_-)} \frac{1}{2\pi|\overline{r}_- - \overline{r}'_+|} \right\} f_+(\overline{r}'_+) dS(\overline{r}'_+)$$

$$+ \int\int_{S_-} \left\{ \frac{\partial}{\partial n(\overline{r}_-)} \frac{1}{|\overline{r}_- - \overline{r}'_-|} \right\} f_-(\overline{r}'_-) dS(\overline{r}'_-) = g(\overline{r}_-). \tag{6.5.79}$$

But since

$$g(\overline{r}_+) = g(\overline{r}_-)$$

$$\frac{\partial}{\partial n(\overline{r}_+)} \frac{1}{|\overline{r}_+ - \overline{r}'_+|} = \frac{\partial}{\partial n(\overline{r}_-)} \frac{1}{|\overline{r}_- - \overline{r}'_-|} \tag{6.5.80}$$

$$\frac{\partial}{\partial n(\overline{r}_+)} \frac{1}{|\overline{r}_+ - \overline{r}'_-|} = \frac{\partial}{\partial n(\overline{r}_-)} \frac{1}{|\overline{r}_- - \overline{r}'_+|}$$

it follows that

$$f_+(\overline{r}_+) = f_-(\overline{r}_-). \tag{6.5.81}$$

In view of (6.5.78), (6.5.80) and (6.5.81), the pair of equations (6.5.79) reduces to the single equation

$$f_+(\overline{r}_+) - \frac{1}{4\pi} \int\int_{S_+} \left\{ \frac{1}{|\overline{r}_+ - \overline{r}'_+|} + \frac{1}{|\overline{r}_- - \overline{r}'_+|} \right\} f_+(\overline{r}'_+) dS(\overline{r}'_+)$$
$$= g(\overline{r}_+). \tag{6.5.82}$$

We now transform S_+ to Q by means of the transformation

$$\mu = \sin(\pi x/2)$$

$$\sigma = \cos(\pi x/2) \sin(\pi y/2) \tag{6.5.83}$$

$$\tau = \sqrt{1 - \mu^2 - \sigma^2} = \cos(\pi x/2) \cos(\pi y/2).$$

We then get

$$dS(\overline{r}'_+) = (1 - \mu^2 - \nu^2)^{-1/2} d\mu d\sigma = (\pi^2/4) \cos(\pi x/2) dx\, dy$$

and

$$f(\overline{\rho}) - \frac{1}{4\pi}\int\int_Q \frac{K(\overline{\rho};\overline{\sigma})}{|\overline{\rho}-\overline{\sigma}|} f(\overline{\sigma})d\overline{\sigma} = g(\overline{\rho}), \quad \overline{\rho} \in Q, \tag{6.5.84}$$

with

$$\left\{\begin{aligned} &\overline{\rho} = (x,y), \quad \overline{\sigma} = (\xi,\eta), \quad d\overline{\sigma} = d\xi d\eta \\ &K(\overline{\rho};\overline{\sigma}) = |\overline{\rho}-\overline{\sigma}|\cos(\pi x/2)\left\{\frac{1}{|\overline{r}_+ - \overline{r}'_+|} + \frac{1}{|\overline{r}_- - \overline{r}'_+|}\right\} \\ &|\overline{r}_+ - \overline{r}'_+|^2 = 2\{1 - \sin(\pi x/2)\sin(\pi\xi/2) \\ &\qquad\qquad - \cos(\pi x/2)\cos(\pi\xi/2)\cos[\pi(y-\eta)/2]\} \\ &|\overline{r}_- - \overline{r}'_+|^2 = 2\{1 - \sin(\pi x/2)\sin(\pi\xi/2) \\ &\qquad\qquad + \cos(\pi x/2)\cos(\pi\xi/2)\cos[\pi(y-\eta)/2]\}. \end{aligned}\right. \tag{6.5.85}$$

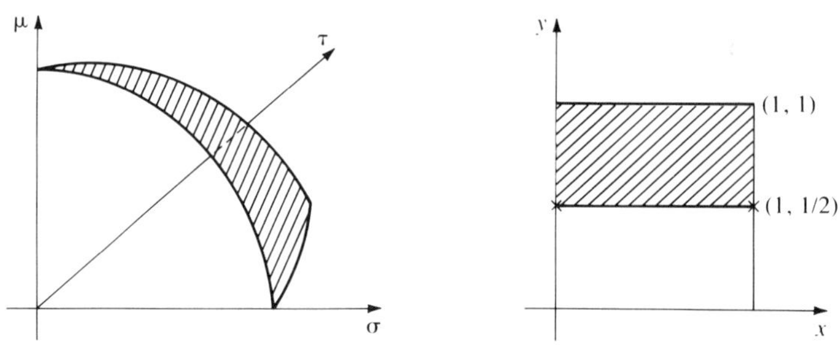

Figure 6.5.1: Transformation under (6.5.83).

Notice the transformation of the positive quadrant $\{(\mu,\sigma) : 0 \leqq \mu \leqq 1,\ 0 \leqq \sigma \leqq \sqrt{1-\mu^2}\}$ onto the square by means of (6.5.83), as illustrated in Fig. 6.5.1. It is particularly important to note that the single point $(\mu,\gamma) = (1,0)$ is transformed onto the segment $\{(x,y) : x = 1, -1 \leqq y \leqq 1\}$.

At this point we can use a Galerkin approximation of the form (6.5.63), i.e., we approximate f by

$$f_N(\overline{\rho}) = \sum_{j=-N-1}^{N+1}\sum_{k=-N-1}^{N+1} c_{jk}\psi_j(x)\psi_k(y), \tag{6.5.86}$$

with the functions ψ_j defined in (6.5.11). We can then replace (6.5.84) by its Galerkin approximation

$$f_N(\overline{\rho}) - \mathcal{P}_N \int\int_Q \frac{K(\overline{\rho};\overline{\sigma})}{|\overline{\rho}-\overline{\sigma}|} f_N(\overline{\sigma})d\overline{\sigma} = (\mathcal{P}_N g)(\overline{\rho}), \quad \overline{\rho} \in Q, \tag{6.5.87}$$

with $\mathcal{P}_N$ defined in (6.5.13).

It is worthwhile to take advantage of additional symmetry properties of g, namely (see (6.5.75),

$$g(\pm\mu, \pm\sigma, \tau) = g(\mu, \sigma, \tau). \tag{6.5.88}$$

In view of this and our observations on the transformation (6.5.83), it follows that the solution f of (6.5.84) also satisfies the symmetry relations (6.5.88), and furthermore, $f(\pm 1, y) =$ constant. Hence, the coefficients c_{jk} in (6.5.86) satisfy the identities

$$\begin{aligned} c_{\pm j,\pm k} &= c_{jk}, \quad j,k = -(N+1),\cdots,N+1 \\ c_{N+1,k} &= c_{N+1,0}, \quad k = -N-1,\cdots,N+1. \end{aligned} \tag{6.5.89}$$

Because of these relations, f_N reduces to the form

$$f_N(x,y) = \sum_{(j,k)\in J} c_{jk} b_{jk}(x,y) \tag{6.5.90}$$

with

$$J = \{(x_j, y_k) : 0 \leqq j \leqq N, 0 \leqq k \leqq N+1\} \cup (1,0) \tag{6.5.91}$$

and with

$$\begin{aligned} b_{00}(x,y) &= \psi_0(x)\psi_0(y) \\ b_{N+1,0}(x,y) &= [\psi_{N+1}(x) + \psi_{-N-1}(x)] \textstyle\sum_{k=1}^{N+1} [\psi_k(y) + \psi_{-k}(y)] \\ b_{j,N+1}(x,y) &= [\psi_{N+1}(y) + \psi_{-N-1}(y)] [\psi_j(x) + \psi_{-j}(x)] \\ & \qquad j = 1,2,\cdots,N \\ b_{jk}(x,y) &= [\psi_j(x) + \psi_{-j}(x)] [\psi_k(y) + \psi_{-k}(y)] \\ & \qquad j,k = 1,2,\cdots,N. \end{aligned} \tag{6.5.92}$$

Symmetry has thus made it possible to reduce the number of unknown coefficients in (6.5.86) from $(2N+3)^2$ to $N^2 + 3N + 3$ in (6.5.90). For example, if $N = 1$, then $(2N+3)^2 = 25$, while $N^2 + 3N + 3 = 7$; if $N = 2$, then $(2N+3)^2 = 49$, while $N^2+3N+3 = 13$. The basis functions $b_{jk}(x, y)$ of the Galerkin scheme are the coefficients of the unknowns c_{jk} in (6.5.90). The projection operator $\mathcal{P}_N$ may then be defined by

$$(\mathcal{P}_N g)(x, y) = \sum_{j,k \in J} g(x_j, x_k) b_{jk}(x, y) \tag{6.5.93}$$

with x_j defined in (6.5.11) and J in (6.5.91). It then follows that (6.5.87) is still valid for this definition of $\mathcal{P}_N$.

Let us now solve (6.5.87), with $\mathcal{P}_N$ defined in (6.5.93), and with $h = 1/\sqrt{N}$. We first consider the case $N = 1$. For this case, we have

$$x_j = \frac{e^j - 1}{e^j + 1}, \quad j = 0, 1; \quad x_2 = 1. \tag{6.5.94}$$

The points $\overline{\rho} = \{(x_j, x_k) : (j, k) \in J\}$ and the corresponding points $\{(\mu, \sigma)\}$ defined by (6.5.83) are then given in Table 6.5.1, in which we have used the notation $s = \sin(\pi x_1/2)$, $c = \cos(\pi x_1/2)$.

j	$\overline{\rho}_j = (x, y)_j$	$(\mu, \sigma, \tau)_j$
1	$(0, 0)$	$(0, 0, 1)$
2	$(0, x_1)$	$(0, s, c)$
3	$(0, 1)$	$(0, 1, 0)$
4	$(x_1, 0)$	$(s, 0, c)$
5	(x_1, x_1)	(s, cs, c^2)
6	$(x_1, 1)$	$(s, c, 0)$
7	$(1, 0)$	$(1, 0, 0)$

Table 6.5.1: Points $(x, y) \in Q$ and $(\mu, \sigma, \tau) \in S_+$ for $N = 1$.

We momentarily delay our description to evaluate the integrals in (6.5.87) to the following Section, 6.5.8.

The eigenvalues of the linear system resulting from (6.5.87) in the case $N = 1$ are

$$\begin{aligned} &\tilde{\lambda}_1^{-1} = .037, \quad \tilde{\lambda}_2^{-1} = .291, \quad \tilde{\lambda}_3^{-1} = .293, \\ &\tilde{\lambda}_4^{-1} = .777, \quad \tilde{\lambda}_5^{-1} = .801, \quad \tilde{\lambda}_6^{-1} = .886, \\ &\tilde{\lambda}_7^{-1} = .934. \end{aligned} \tag{6.5.95}$$

The eigenvector corresponding to $\tilde{\lambda}_1^{-1}$ is

$$\tilde{v}_1 = (.387, .391, .386, .388, .393, .394, .393)^T.$$

For the exact solution (with $\tilde{\lambda}_1^{-1} = 0$) we would have

$$v_1 = \gamma_1(1, 1, 1, 1, 1, 1, 1)^T,$$

where γ_1 is an arbitrary constant. The calculated values of f_1 and the exact values $f = (u + 2g)/2$ are given in Table 6.5.2, with $u = \mu^2 + \sigma^2 - 2\tau$.

$\overline{\rho}_i$	f_1	f	$f_1 - f$
$\overline{\rho}_1$	1.057	.024	1.033
$\overline{\rho}_1$	1.388	.398	.990
$\overline{\rho}_3$	.722	−.270	.992
$\overline{\rho}_4$	.721	−.270	.991
$\overline{\rho}_5$	.196	−.795	.991
$\overline{\rho}_6$	1.333	.398	.935
$\overline{\rho}_7$	1.394	.398	.996

Table 6.5.2: Solution and Error for $N = 1$.

Please note, that if f_1 were exact, then each entry in the third column of Table 6.5.1 would be a "1".

In the case $N = 2$, $h = 1/\sqrt{N} = 1/\sqrt{2}$, we set

$$x_j = \frac{e^{j/\sqrt{2}} - 1}{e^{j/\sqrt{2}} + 1}, j = 0, 1, 2; \quad x_3 = 1. \tag{6.5.96}$$

Then, setting $s_j = \sin(\pi x_j/2)$, $c_j = \cos(\pi x_j/2)$, $j = 1, 2$, the points $\overline{\rho}_j$ and the corresponding points (μ, σ, τ) defined in (6.5.83) are given in Table 6.5.3.

The corresponding values of f_2, the exact values of f and the differences $f_2 - f$, whose exact values are "1" are given in Table 6.5.4.

j	$\overline{\rho}_j = (x, y)_j$	$(\mu, \sigma, \tau)_j$
1	$(0, 0)$	$(0, 0, 1)$
2	$(0, x_1)$	$(0, s_1, c_1)$
3	$(0, x_2)$	$(0, s_2, c_2)$
4	$(0, 1)$	$(0, 1, 0)$
5	$(x_1, 0)$	$(s_1, 0, c_1)$
6	(x_1, x_1)	$(s_1, c_1 s_1, c_1^2)$
7	(x_1, x_2)	$(s_1, c_1 s_2, c_1 c_2)$
8	$(x_1, 1)$	$(s_1, c_1, 0)$
9	$(x_2, 0)$	$(s_2, 0, c_2)$
10	(x_2, x_1)	$(s_2, c_2 s_1, c_1 c_2)$
11	(x_2, x_2)	$(s_2, c_2 s_2, c_2^2)$
12	$(x_2, 1)$	$(s_2, c_2, 0)$
13	$(1, 0)$	$(1, 0, 0)$

Table 6.5.3: Points $(x, y) \in Q$ and $(\mu, \gamma, \varepsilon) \in S_+$ for $N = 2$.

$\overline{\rho}_i$	f_2	f	$f_2 - f$
$\overline{\rho}_1$	.02027	−.7958	.9985
$\overline{\rho}_2$	.5117	−.4873	.9990
$\overline{\rho}_3$	.9995	.0012	.9983
$\overline{\rho}_4$	1.3972	.3979	.9993
$\overline{\rho}_5$	.5114	−.4873	.9987
$\overline{\rho}_6$	.7436	−.2585	1.0021
$\overline{\rho}_7$	1.1050	.1037	1.0013
$\overline{\rho}_8$	1.3856	.3979	.9977
$\overline{\rho}_9$	.9994	.0012	.9982
$\overline{\rho}_{10}$	1.1049	.1037	1.0012
$\overline{\rho}_{11}$	1.3856	.3979	.9977
$\overline{\rho}_{12}$	1.3983	.3979	1.0004
$\overline{\rho}_{13}$	1.3976	.3979	.9997

Table 6.5.4: Solutions and Error for $N = 2$.

Note that the maximum error in the computed solutions is about 2 in the third decimal.

Let us now assume that we have obtained f_N, with f_N defined in (6.5.90) through (6.5.92), and let us consider approximating f by f_N to obtain an approximate solution u_N of u to the problem (6.5.2), using the formula (6.5.3).

Using the transformation (6.5.83), we thus have, in view of the symmetry of f_N, that

$$u_N(\overline{r}) = \frac{\pi}{16}\int\int_Q f_N(\xi,\eta)\cos(\pi\xi/2)\left\{\frac{1}{|\overline{r}-\overline{r}'|}+\frac{1}{|\overline{r}^*-\overline{r}'|}\right\}\cdot d\xi d\eta$$

$$\overline{r} = (\mu,\sigma,\tau), \quad \overline{r}^* = (\mu,\sigma,-\tau)$$

$$\overline{r}' = \overline{r}'(\xi,\eta) = (\sin(\pi\xi/2), \sin(\pi\eta/2)\cos/\pi\xi/2), \cos(\pi\xi/2)\cos(\pi\eta/2)) \tag{6.5.97}$$

and under the assumption that $\overline{r} \in V$, with V the unit ball, $V = \{\overline{r} = (\mu,\sigma,\tau) : \mu^2+\sigma^2+\tau^2 < 1\}$. By direct application of repeated Sinc quadrature over Q, i.e., using the formula (see (4.2.32) and (4.2.37))

$$\int\int_Q g(\xi,\eta)d\xi d\eta = h^2 \sum_{m=-N-1}^{N+1}\sum_{n=-N-1}^{N+1} w_m w_n g(\zeta_m,\zeta_n)$$
$$\zeta_n = \frac{e^{nh}-1}{e^{nh}+1}, \quad w_m = \frac{2e^{mh}}{(1+e^{mh})^2}, \tag{6.5.98}$$

and using the notation $\overline{r}'_{jk} = \overline{r}'(\zeta_j, \zeta_k)$, we get the approximation

$$u_N(\overline{r}) \approx \frac{\pi h^2}{4} \left\{ \frac{c_{0,0}}{4} \left(\frac{1}{|\overline{r} - \overline{r}'_{0,0}|} + \frac{1}{|\overline{r}^* - \overline{r}'_{0,0}|} \right) \right.$$

$$+ w_{N+1} \sum_{j=1}^{N} w_j \cos(\pi\zeta_j/2) c_{j,N+1} \cdot \left(\frac{1}{|\overline{r} - \overline{r}_{j,N+1}|} + \frac{1}{|\overline{r}^* - \overline{r}'_{j,N+1}|} \right)$$

$$\left. + \sum_{j=1}^{N} \sum_{k=1}^{N} w_j w_k \cos(\pi\zeta_j/2) c_{jk} \cdot \left(\frac{1}{|\overline{r} - \overline{r}'_{jk}|} + \frac{1}{|\overline{r}^* - \overline{r}'_{jk}|} \right) \right\}. \tag{6.5.99}$$

Since

$$\frac{\pi}{16} \int\!\!\int_Q \cos(\pi\xi/2) \left\{ \frac{1}{|\overline{r} - \overline{r}'|} + \frac{1}{|\overline{r}^* - \overline{r}'|} \right\} d\xi d\zeta = 1 \tag{6.5.100}$$

it follows that if V denotes the unit ball, then

$$\|u - u_N\|_V \leqq \|f - f_N\|_Q \leqq C N^{1/2} \log(N) e^{-(\pi d \alpha N)^{1/2}}. \tag{6.5.101}$$

Let us now consider the difference between u_N given by (6.5.97) and the sum on the right-hand side of (6.5.99). To this end, we note that

$$\int\!\!\int_Q F(\overline{\sigma}) d\overline{\sigma} - h^2 \sum_{m=-N-1}^{N+1} \sum_{n=-N-1}^{N+1} w_m w_n F(\sigma_{mn}) = A + B, \tag{6.5.102}$$

with

$$A = \int\!\!\int_Q F(\overline{\sigma}) d\overline{\sigma} - h^2 \sum_{m=-\infty}^{\infty} \sum_{n=-\infty}^{\infty} w_m w_n F(\overline{\sigma}_{mn})$$

$$B = h \sum_{m=-\infty}^{\infty} w_m \left\{ h \sum_{n=-\infty}^{\infty} w_n F(\overline{\sigma}_{mn}) - h \sum_{|n| \leqq N+1} w_n F(\overline{\sigma}_{mn}) \right\}$$

$$+ h \sum_{|n| \leqq N+1} w_n \left\{ h \sum_{m=-\infty}^{\infty} w_m F(\overline{\sigma}_{mn}) - h \sum_{|m| \leqq N+1} w_m F(\overline{\sigma}_{mn}) \right\} \tag{6.5.103}$$

and with F the integrand in (6.5.97), and $\sigma_{mn} = (\zeta_m, \zeta_n)$. The term B in (6.5.102) is $\mathcal{O}(e^{-Nh}) = \mathcal{O}(e^{-(\pi d \alpha N)^{1/2}})$ (see Section 5.2). However, $A = \mathcal{O}(e^{-2\pi\delta/h})$ (see Equation (5.2.6)), with

$$\delta = \min\left(d, \inf_{\overline{r}' \in \partial V} |\overline{r} - \overline{r}'|\right),$$

and with d the same as in Definition 6.5.1, so that $e^{-\pi d/h} = e^{-(\pi d\alpha N)^{1/2}}$. Thus, if $\delta \geqq d/2$, the absolute value of the difference between the exact solution u and the approximation given in (6.5.99) is of the same order of magnitude as the right-hand side of (6.5.101). On the other hand, if $\delta < d/2$, then the difference between u_N and its approximation on the right-hand side of (6.5.99) is of order $e^{-\pi\delta/h} = \left\{e^{-(\pi d\alpha N)^{1/2}}\right\}^{2\delta/d}$, which may not be an acceptable error. We can, nevertheless, accurately evaluate the integral in (6.5.97), as follows. Let $\overline{r}'' = (\mu'', \sigma'', \tau'')$ denote the point of ∂V on the ray from the origin through $\overline{r}$ (i.e., $\overline{r}'' = \overline{r}/|\overline{r}|$) and let $\overline{\rho}$ denote the point of Q which is mapped onto $\overline{r}''$ via the transformation (6.5.83). We can then get an accurate approximation of u_N by approximating the integral in (6.5.97) via the procedure described in Section 6.5.8.

6.5.8 Kernel Singularities and Integration

Let us now consider the evaluation of the integrals in (6.5.64), required in the Galerkin approximations. At the outset, we consider the case when the number of surface patches is $p = 1$, in which case the integrals in (6.5.64) take the form

$$I = \int\int_Q \frac{K(\overline{\rho}; \overline{\sigma})}{|\overline{\rho} - \overline{\sigma}|} w(\overline{\sigma}) d\overline{\sigma} \tag{6.5.104}$$

with $w(\xi, \eta) = \psi_m(\xi)\psi_n(\eta)$. We thus split Q, depending on the location of $\overline{\rho}$, as follows:

(i) If $\overline{\rho}$ is a corner point of Q, then we do not need to split Q. We simply use the formula (See (4.2.32) and (4.2.37))

$$\begin{aligned}
&\int_a^b \int_c^d \frac{K(\overline{\rho}; \overline{\sigma})}{|\overline{\rho} - \overline{\sigma}|} w(\xi, \eta) d\overline{\sigma} \quad (\overline{\sigma} = (\xi, \eta)) \\
&\approx h^2 (b-a)(d-c) \sum_{j=-N-1}^{N+1} \sum_{k=-N-1}^{N+1} w_j w_k \frac{K(\overline{\rho}; (\xi_j; \eta_k))}{|\overline{\rho} - (\xi_j, \eta_k)|} w(\xi_j, \eta_k) \\
&\xi_j = \frac{a + be^{jh}}{1 + e^{jh}}, \quad \eta_k = \frac{c + de^{kh}}{1 + e^{kh}}, \quad w_j = \frac{e^{jh}}{(1 + e^{jh})^2}
\end{aligned} \tag{6.5.105}$$

in which we take $a = c = -1$, $b = d = 1$.

(ii) If $\overline{\rho}$ is an edge of Q, but $\overline{\rho}$ is not a corner point, we split Q into *two* regions, Q_1 and Q_2 by drawing a line $\mathcal{L}$ through Q, $\mathcal{L}$ passing through $\overline{\rho}$ and perpendicular to the edge containing $\overline{\rho}$. Each region Q_1 and Q_2 is thus a rectangular region. Denoting the corners of either one of the regions by (a,c), (b,c), (b,d) or (a,d), we approximate the integral over each of these regions by the formula given in (6.5.101).

(iii) If $\overline{\rho}$ is an interior point of Q, we split Q into four rectangular regions by drawing two perpendicular lines through Q, one parallel to the ξ-axis and the other parallel to the η-axis. The integrals over each of the rectangular regions are then approximated by the formula in (6.5.105).

Under our assumptions, each integrand in (6.5.104) is in $X_{3\alpha}$, with $\alpha = 1/2$, and thus the approximations (6.5.105), may be expected to be very accurate.

Next, let us consider the case of $p > 1$, with p given in (6.5.59). Let K_j^i be defined as in (6.5.61) and (6.5.64), and consider the integral

$$J_{ij} = \int\int_Q K_j^i(\overline{\rho};\overline{\sigma})w(\overline{\sigma})d\overline{\sigma}, \tag{6.5.106}$$

with $w(\overline{\sigma}) = \psi_m(\xi)\psi_n(\eta)$. Corresponding to the cases (i), (ii) and (iii) above, we now have the cases $(i)_p$, $(ii)_p$ and $(iii)_p$ where:

$(i)_p$ $\overline{\rho}$ is a corner point of Q. In this case we proceed exactly as in (i) above.

$(ii)_p$ $\overline{\rho}$ is an edge of Q, but not a corner point. In this case we split Q as in (ii) above, if $K_j^i(\overline{\rho},\overline{\sigma})$ is an unbounded function of $\overline{\sigma}$ on Q. If K_j^i is bounded on Q, it is unnecessary to split Q.

$(iii)_p$ $\overline{\rho}$ is an interior point of Q. We assume, now, that Q corresponds to the surface patch S_j. If $K_j^i(\overline{\rho};\overline{\sigma})$ is unbounded on Q (i.e., we must have $i = j$, then we split Q as in (iii) above). In this case we also recommend splitting those integrals of the form

$$I_{j\ell} = \int\int_Q K_\ell^j(\overline{\rho};\overline{\sigma})w(\overline{\sigma})d\overline{\sigma}, \quad \ell \neq j, \tag{6.5.107}$$

for which the surface S_i shares a common edge with S_j. For every such integral, Q is split into two regions. This split is carried out as follows. If we momentarily denote by Q_j (respectively Q_i) the region Q which is mapped by (6.5.55) onto S_j (rsp. S_i) then the boundary $\mathcal{L}$ between S_i and S_j corresponds to the edge $\mathcal{L}_i$ of Q_i and $\mathcal{L}_j$ of Q_j. One of the perpendicular lines through $\overline{\rho} \in Q_j$ hits $\mathcal{L}_j$ at $\overline{\rho}_j$ (say) which corresponds to $\overline{\rho}_i$ on $\mathcal{L}_i$. By drawing a line through Q_i, and perpendicular to $\mathcal{L}_i$, we thus split Q_i into two regions. We then use the formula (6.5.99) to approximate the integral in (6.5.106) over each of these rectangular regions.

•

PROBLEMS FOR SECTION 6.5

6.5.1 Prove that if f satisfies $A_{3\alpha}$, then inequality (6.5.9) holds in Q. [Hint: Use the result of Problem 4.2.6.]

6.5.2 Derive Equation (6.5.4) from the representation (6.5.3) in the case of V the unit sphere.

6.5.3 (a) Let $0 < \tau < 1$. Expand each term on the right-hand side of the identity

$$|1 - \tau e^{i\theta}| = (1 - \tau e^{i\theta})^{1/2}(1 - \tau e^{-i\theta})^{1/2}$$

in powers of τ and integrate termwise, to get the identity

$$(i) \qquad \int_0^{2\pi} \frac{d\theta}{|1 - \tau e^{-i\theta}|} = 2\pi \sum_{n=0}^{\infty} a_n \tau^{2n} = 4\,K(\tau),$$

with $a_n = \{\Gamma(n + 1/2)/n!\}^2 \pi$, (see Problem 1.1.5) and with $K(\tau)$ defined in (1.7.20).

(b) Use Stirling's formula (see Problem 1.1.5 (f)) to conclude that $n a_n \to 1/\pi$ as $n \to \infty$, with a_n defined in (i) above.

(c) Prove that if a_n is defined as in (i) above, then $n a_n$ is an increasing function of n.

(d) Conclude from parts (b) and (c) of this problem that if $0 < \tau < 1$, then

$$K(\tau) \leqq \frac{\pi}{2} + \sum_{n=1}^{\infty} \frac{\tau^{2n}}{2n}.$$

(e) Use the result in (d) to deduce that

$$\int_0^1 K(\tau)\,d\tau \leqq \frac{\pi}{2} + \sum_{n=1}^{\infty} \frac{1}{2n(2n+1)} = \frac{3\pi + 1}{6} < 2$$

$$\int_0^1 \tau\,K(\tau)\,d\tau \leqq \frac{\pi}{4} + \sum_{n=1}^{\infty} \frac{1}{2n(2n+2)} = \frac{\pi + 1}{4} < \frac{5}{4}$$

and that if $0 < \delta < 1$,

$$\int_{\delta}^{1} \frac{K(\tau)}{\tau}\, d\tau \leqq \frac{\pi}{2} \log(1/\delta) + \sum_{n=1}^{\infty} \frac{1}{(2n)^2} = \frac{\pi}{2} \log(1/\delta) + \frac{\pi^2}{24}$$
$$< 2\log(1/\delta) + 5/8.$$

6.5.4 Use the results of Problem 6.5.3 to prove that if f satisfies A_F, and if $f \in C(\overline{Q})$, then $\kappa f \in \mathbf{Lip}_\alpha(\overline{Q})$.

6.5.5 Give a detailed proof of Lemma 6.5.5, via differentiation of κf.

6.5.6 Prove in detail, each of the identities in (6.5.76).

6.5.7 **(a)** What is the value of α for the example considered in Section 6.5.7?

(b) Hence, what is d, so that $h = 1/\sqrt{N}$ yields the optimal error?

(c) Obtain an estimate on the error in the Galerkin solution as $N \to \infty$.

6.5.8 Repeat the solution of the example in Section 6.5.6, for $N = 1$ and 2, using the transformation

$$\mu = x$$
$$\sigma = y\sqrt{1 - x^2}$$
$$\tau = \sqrt{1 - x^2}\sqrt{1 - y^2}$$

and taking $h = 1.6/\sqrt{N}$.

6.5.9 **(a)** What is the value of α if the transformation in Problem 6.5.8 is used?

(b) What is d?

(c) Obtain an estimate on the convergence rate of the Galerkin scheme as $N \to \infty$.

6.5.10 Prove identity (6.5.69).

6.5.11 Prove that if either $F(x, y; \xi, \eta) = |\xi|$, and $f \equiv 1$, or else if $F \equiv 1$ and $f = |\xi|$, and if κf is defined in (6.5.21), then $(\kappa f)(\cdot, y) \notin \mathbf{Hol}(\mathcal{D})$, for any $y \in [-1, 1]$. Conclude that there exist combinations of functions F and f, such that if either the assumptions A_F are not satisfied, but $f \in X_1$, or else if these assumptions are satisfied, but $f \notin X_1$, then $\kappa f \notin X_1$.

6.5.12 Let S be defined as in Subsection 6.5.7, and for $\overline{r} = (x, y, z) \in S$, let g be defined by

$$g(x,y,z) = \begin{cases} z^{1/2} & \text{if} \quad z > 0, \\ (-z)^{1/2} & \text{if} \quad z < 0. \end{cases}$$

Find an approximate solution f of Equation (6.5.66) re-working the equations of Subsection 6.5.7 the case of this new function g. In particular, note the following:

(a) g does in fact satisfy (6.5.76) and (6.5.77);

(b) Conclude, however, that instead of (6.5.81), one obtains

$$f_+(\overline{r}_+) = -f_-(\overline{r}_-).$$

(c) This result enables us to replace (6.5.79) by an equation of the form (6.5.82), but with a different kernel. What is the new form of (6.5.82)?

(d) How does the fact that the present g satisfies $g \in X_{2\alpha}$ affect the choice of h as a function of N?

(e) Solve the new integral equation, using an approximation of the form (6.5.86). However, one can replace the upper limits '$N + 1$' in each sum in (6.5.86) by N. Why?

(f) In consideration of (d), obtain a 5 point numerical solution with $N = 1$, and a 9 point numerical solution with $N = 2$.

(g) Conclude that whereas it was unnecessary to split S into two separate patches in the example solved in the text, it is, nevertheless necessary to split S into two separate patches in the example of this problem.

6.5.13 Use the method of Section 4.6 and the identity

$$\int_0^\infty \int_0^\infty \exp\{-\frac{t}{r} - \frac{\tau}{s}\} \frac{dt\, d\tau}{\sqrt{t^2 + \tau^2}} = \frac{rs}{2\mu} \log \left\{ \frac{\mu + r}{\mu - r} \frac{mu + s}{\mu - s} \right\},$$

where $\mu = \sqrt{r^2 + s^2}$, in order to develop an efficient scheme for evaluating integrals of the form

$$\int_{-1}^1 \int_{-1}^1 \frac{F(\xi, \eta)\, d\xi\, d\eta}{\sqrt{(x - \xi)^2 + (y - \eta)^2}}.$$

6.6 Reduced Wave Equation on a Half-Space

In this section we discuss a Sinc-integral equation method of solution of the reduced wave equation

$$\nabla^2 u(\bar{r}) + k^2 u(\bar{r}) = -k^2 f(\bar{r}) u(\bar{r}), \quad \bar{r} \in V, \tag{6.6.1}$$

V the half-space

$$V = \left\{ \bar{r} = (x, y, z) : \bar{\rho} \equiv (x, y) \in \mathbb{R}^2, \quad z \in \mathbb{R}_+ \right\}, \tag{6.6.2}$$

$\mathbb{R}_+ = (0, \infty)$. The equation (6.6.1) stems from the wave equation,

$$\nabla^2 p(\bar{r}, t) = \frac{1}{c^2(\bar{r})} \frac{\partial^2 p(\bar{r}, t)}{\partial t^2} \tag{6.6.3}$$

with $p(\bar{r}, t)$ the sound pressure at $(\bar{r}, t)$, for $\bar{r} \in V$, and (time) $t > 0$, and with $c(\bar{r})$ the sound speed at the point $\bar{r}$. Denoting $w = 2\pi \times$ frequency and setting

$$\begin{aligned} k &= w/c_0 \\ p(\bar{r}, t) &= u(\bar{r}) e^{-iwt} \\ f(\bar{r}) &= \frac{c_0^2}{c^2(\bar{r})} - 1 \end{aligned} \tag{6.6.4}$$

we get (6.6.1) from (6.6.3). We shall assume that c_0 and f are such that $f(\bar{r}) \to 0$ as $|\bar{r}| \to \infty$, and $f(\bar{r}) = 0$ for $\bar{r} \notin V$.

Inasmuch as the equations (6.6.1) and (6.6.3) are only approximations to actual physical phenomena, and since one always has damping in real life, we shall assume that

$$\begin{aligned} k &= k_0 (c + is) \\ f &= f_R + i\, f_I \end{aligned} \tag{6.6.5}$$

with $k_0 > 0$, and $s > 0$, and $c^2 + s^2 = 1$. The integral equation approach of this section thus enables us to deal with a complex $c(\bar{r})$ in (6.6.3), which would not be so easy if we had to approximate the time derivative in (6.6.3) via a central difference approximation, since the resulting equation would then be unstable. That is, when $c(\bar{r})$ is complex, the integral equation approach of this section eliminates the solution of (6.6.3), which increases in magnitude a function of time, but which is not present in a physical problem. We also give a uniqueness proof, if $c > 0$, and if the assumed real-valued functions f_R and f_I defined on V satisfy the hypothesis

$$\left(f_R^2 + f_I^2 \right)^{1/2} < 2cs \tag{6.6.6}$$

in V.

6.6.1 Problem Description

Given $\overline{r}$ and $\overline{r}'$ in $\mathbb{R}^3$, the function

$$G_k(\overline{r}-\overline{r}') = \frac{e^{ik|\overline{r}-\overline{r}'|}}{4\pi|\overline{r}-\overline{r}'|} \tag{6.6.7}$$

satisfies the equation

$$\nabla^2_{\overline{r}'} G_k(\overline{r}-\overline{r}') + k^2 G_k(\overline{r}-\overline{r}') = -\delta(\overline{r}-\overline{r}'), \tag{6.6.8}$$

in the distribution sense, that is, if g is a continuous function defined on $\mathbb{R}^3$, then

$$\int\int\int_{\mathbb{R}^3} \left[\nabla^2_{\overline{r}'} G_k(\overline{r}-\overline{r}') + k^2 G_k(\overline{r}-\overline{r}')\right] g(\overline{r}')d\overline{r}' = -g(\overline{r}). \tag{6.6.9}$$

Next, let u be a function defined on $\mathbb{R}^3$, such that u satisfies Equation (6.6.1). Under appropriate integrability conditions on u, we may formally apply Green's theorem in the form

$$\begin{aligned} &\int\int_{\mathbb{R}^2} \{G_k(\overline{r}-\overline{r}')\nabla_{\overline{r}'} u(\overline{\rho}',0) - u(\overline{\rho}',0)\nabla_{\overline{r}'} G_k(\overline{r}-\overline{r})\} \cdot \hat{n} d\overline{\rho}' \\ &= \int\int\int_V \{G_k(\overline{r}-\overline{r}')\nabla^2_{\overline{r}'} u(\overline{r}') - u(\overline{r}')\nabla^2_{\overline{r}'} G_k(\overline{r}-\overline{r}')\} \, d\overline{r}' \end{aligned} \tag{6.6.10}$$

with $\hat{n} = -\hat{z}$, the unit vector pointing in the direction of the negative z-axis. In view of (6.6.1) and (6.6.8), we thus get

$$\begin{aligned} u^{in}(\overline{r}) &\equiv \int\int_{\mathbb{R}^2} \left\{ u(\overline{\rho}',0)\frac{\partial}{\partial z} G_k(\overline{r}-\overline{r}') - G_k(\overline{r}-\overline{r})\frac{\partial u}{\partial z'}(\overline{\rho}',0)\right\} d\overline{\rho} \\ &= u(\overline{r}) - k^2 \int\int\int_V G_k(\overline{r}-\overline{r}')f(\overline{r}')u(\overline{r}')d\overline{r}'. \end{aligned} \tag{6.6.11}$$

Here u^{in} is the source, which in applications is most frequently taken to be

$$u^{in}(\overline{r}) = \frac{\exp\{ik|\overline{r}-\overline{r}_0|\}}{4\pi|\overline{r}-\overline{r}_0|}, \quad \overline{r}_0 \notin \overline{V}, \tag{6.6.12}$$

with $\overline{V}$ the closure of V. We mention here that the source u^{in} defined by (6.6.11) (as well as the source u^{in} defined by (6.6.12)), always satisfies in V the equation $\nabla^2 u^{in} + k^2 u^{in} = 0$.

Equation (6.6.11) may be written in operator form as

$$u - \kappa u = u^{in} \tag{6.6.13}$$

with u^{in} defined in (6.6.12), and with κu given by

$$(\kappa u)(\bar{r}) = k^2 \int\int\int_V G_k(\bar{r} - \bar{r}')f(\bar{r}')u(\bar{r}')d\bar{r}'. \tag{6.6.14}$$

For suitable functions v, we shall also define the operators κ_0 and κ_1 by means of the equations

$$(\kappa_0 v)(\bar{r}) = k^2 \int\int\int_V G_k(\bar{r} - \bar{r}')v(\bar{r}')d\bar{r}' \tag{6.6.15}$$

and

$$\begin{aligned}(\kappa_1 v)(\bar{r}) &= k^2 f(\bar{r}) \int\int\int_V G_k(\bar{r} - \bar{r}')v(\bar{r}')d\bar{r}' \\ &= f(\bar{r})(\kappa_0 v)(\bar{r}).\end{aligned} \tag{6.6.16}$$

The forward problem corresponding to (6.6.13) is to find u in V, given u^{in} and f in V. The inverse problem is to find f in V measuring the solution u of (6.6.13) on ∂V for each source u^{in} in a given family of such sources. We remark that once u is known in V, Equation (6.6.13) may be used to determine $u(\bar{r})$ for any $\bar{r} \in \mathbb{R}^3$.

If one multiplies both sides of Equation (6.6.13) by f, one gets the equation

$$v - \kappa_1 v = v^{in}, \tag{6.6.17}$$

with $v = fu$, $v^{in} = fu^{in}$, and with κ_1 is defined in (6.6.16). Rather than solving (6.6.13) for u, it is more convenient to solve (6.6.17) for v if we want to solve the inverse problem (i.e., to determine f by measuring u on ∂V for many different sources u^{in}) via an iteration scheme. The main reason for this is if v is approximated via a Sinc basis approximant (see (6.6.29)), such as $v = \sum c_j \psi_j$, the moments $\kappa_0 \psi_j$ may be evaluated independently of f, once and for all. Moreover, if Sinc basis functions are used, the function u can then be approximated on $\mathbb{R}^3$ by means of the identity

$$u = \kappa_0 v + u^{in} \tag{6.6.18}$$

via simple collocation in $\mathbb{R}^3 \backslash V$ or using the previously determined moments $\kappa_0 \psi_j$ in V.

Let us next determine an accurate approximate solution of (6.6.17), and following this, an accurate approximation of u via (6.6.18).

6.6.2 Spaces for Sinc Approximation

At the outset, it is convenient to define suitable spaces of functions in which we can effectively approximate f, fu^{in}, $\kappa_1 v$, and the solutions u and v of

(6.6.13) and (6.6.17) respectively. To this end, let $\mathcal{D}_d$ be defined in Equation (1.7.9), let $\mathcal{D}$ be defined in Equation (1.7.12), and let us define Ω_d by

$$\Omega_d = (\mathbb{R} \times \mathbb{R} \times \mathcal{D}) \cup (\mathcal{D}_d \times \mathbb{R} \times \mathbb{R}_+) \cup (\mathbb{R} \times \mathcal{D}_d \times \mathbb{R}_+), \tag{6.6.19}$$

and let $\overline{\Omega}_d$ denote the closure of Ω_d. Let $u : \overline{\Omega}_d \to \mathbb{C}$, and for fixed:

$$\begin{aligned}
&(x,y) \in \overline{\mathbb{R}}^2, \quad \text{define } u^{(xy)} : \overline{\mathcal{D}} \to \mathbb{C} \text{ by} \\
&u^{(xy)}(z) = u(x,y,z); \\
&(y,z) \in \overline{\mathbb{R}} \times \overline{\mathbb{R}}_+, \quad \text{define } u^{(yz)} : \overline{\mathcal{D}}_d \to \mathbb{C} \text{ by} \\
&u^{(yz)}(x) = u(x,y,z); \\
&(z,x) \in \overline{\mathbb{R}}_+ \times \overline{\mathbb{R}}, \quad \text{define } u^{(zx)} : \overline{\mathcal{D}}_d \to \mathbb{C} \text{ by} \\
&u^{(zx)}(y) = u(x,y,z).
\end{aligned} \tag{6.6.20}$$

We also set

$$v(x,y,z) = u(x,y,z) - u(x,y,0)e^{-z}, \tag{6.6.21}$$

and we define $v^{(xy)}$, $v^{(yz)}$ and $v^{(zx)}$ the way $u^{(xy)}$, $u^{(yz)}$ and $u^{(zx)}$ were defined above.

Definition 6.6.1 *Let u be continuous and bounded on $\overline{\Omega}_d$.*

(i) *Let X_1 denote the set of all functions u, such that each of the following conditions is satisfied:*

$$\begin{aligned}
&u^{(xy)} \in \mathbf{Hol}(\mathcal{D}) \text{ for all } (x,y) \in \overline{\mathbb{R}}^2; \\
&u^{(yz)} \in \mathbf{Hol}(\mathcal{D}_d) \text{ for all } (y,z) \in \overline{\mathbb{R}} \times \overline{\mathbb{R}}_+; \text{ and} \\
&u^{(zx)} \in \mathbf{Hol}(\mathcal{D}_d) \text{ for all } (z,x) \in \overline{\mathbb{R}}_+ \times \overline{\mathbb{R}}.
\end{aligned} \tag{6.6.22}$$

(ii) *Let $\alpha \in (0,1)$ denote a fixed constant, and let v be defined in terms of u by (6.6.21). Let $X_{2\alpha}$ denote the family of all functions $u \in X_1$ such that each of the following conditions is satisfied:*

$$\begin{aligned}
&v^{(xy)} \in \mathbf{L}_\alpha(\mathcal{D}); \\
&v^{(yz)} \in \mathbf{L}_\alpha(\mathcal{D}_d); \\
&v^{(zx)} \in \mathbf{L}_\alpha(\mathcal{D}_d).
\end{aligned} \tag{6.6.23}$$

(iii) *Let $w : \Omega_d \to \mathbb{C}$, and let η be defined in terms of w, by*

$$\eta(x,y,z) = w(x,y,z)\frac{\sinh(z)}{1+\sinh(z)} \tag{6.6.24}$$

Furthermore, let

$$\begin{aligned} w^{(xy)} &\in \mathbf{Hol}(\mathcal{D}) \\ w^{(yz)} &\in \mathbf{Hol}(\mathcal{D}_d) \\ w^{(zx)} &\in \mathbf{Hol}(\mathcal{D}_d) \end{aligned} \tag{6.6.25}$$

and for all $(x,y,z) \in \Omega_d$, let

$$|\eta(x,y,z)| \leqq \frac{C|z|^\alpha}{1+|z|^\alpha} e^{-\alpha(|x|+|y|+|z|)}. \tag{6.6.26}$$

with C a constant. The set of all such w, such that w satisfies each of these conditions, will be denoted by $X_{3\alpha}$.

Note: Each of the spaces X_1, $X_{2\alpha}$ and $X_{3\alpha}$ are Banach spaces, normed respectively by (see (6.6.24) for the case of $X_{3\alpha}$)

$$\begin{aligned} \|u\| &= \sup_{\overline{r}\in\Omega_d} |u(\overline{r})| \\ \|v\| &= \sup_{\overline{r}\in\Omega_d} |v(\overline{r})| \\ \|w\| &= \sup_{\overline{\rho}\in\Omega_d} |\eta(\overline{\rho})|. \end{aligned} \tag{6.6.27}$$

6.6.3 Sinc Approximation

We now consider *Sinc approximation in $X_{2\alpha}$*. To this end, we select a positive integer N, determine $h > 0$ and set

$$\begin{aligned} &\phi(z) = \log[\sinh(z)] \\ &z_\ell^* = \log\left[e^{\ell h} + (1+e^{2\ell h})^{1/2}\right], \quad \ell = -N, \cdots, N \\ &z_{-N-1}^* = 0 \\ &\psi_\ell(z) = S(\ell,h)\cdot\phi(z), \quad \ell = -N, \cdots, N \\ &\psi_{-N-1}(z) = e^{-z} - \sum_{\ell=-N}^{N} \exp(-z_\ell^*)\psi_\ell(z). \end{aligned} \tag{6.6.28}$$

We then define $\mathcal{P}_N u$ for given $u \in X_1$ by

$$(\mathcal{P}_N u)(x,y,z) =$$

$$\sum_{j=-N}^{N} \sum_{k=-N}^{N} \sum_{\ell=-N-1}^{N} u(jh, kh, z_\ell^*)\, S(j,h)(x)\, S(\ell,h)(y)\, \psi_\ell(z) \tag{6.6.29}$$

Note: $\mathcal{P}_N u \in X_1$ for all $u \in X_1$.

Theorem 6.6.2 *Let $u \in X_{2\alpha}$, let N denote a positive integer, let $h = [\pi d/(\alpha N)]^{1/2}$, and let $\mathcal{P}_N u$ be defined in (6.6.29). Then, there exist constants C_1, C_2, and C_3, independent of N, such that*

$$\sup_{(x,y,z)\in V} |u(x,y,z) - \mathcal{P}_N u(x,y,z)|$$

$$\leqq \left[C_1 + C_2 \log N + C_3 (\log N)^2\right] N^{1/2} e^{-(\pi d \alpha N)^{1/2}}. \tag{6.6.30}$$

Proof. We may write

$$u - \mathcal{P}_N u = A + B + C \tag{6.6.31}$$

with

$$A = \left\{ u^{(yz)}(x) - \sum_{j=-N}^{N} u^{(yz)}(jh) S(j,h)(x) \right\}$$

$$B = \sum_{j=-N}^{N} S(j,h)(x) \left\{ u^{(z,jh)}(y) - \sum_{k=-N}^{N} u^{(z,jh)}(kh) S(k,h)(y) \right\}$$

$$C = \sum_{j=-N}^{N} \sum_{k=-N}^{N} S(j,h)(x) S(k,h)(y) \cdot$$
$$\cdot \left\{ u^{(jh,kh)}(z) - \textstyle\sum_{\ell=-N-1}^{N} u^{(jh,kh)}(z_\ell^*) \psi_\ell(z) \right\}. \tag{6.6.32}$$

Now, by assumption, $u^{(yz)} \in \mathbf{L}_\alpha(\mathcal{D}_d)$, $u^{(zx)} \in \mathbf{L}_\alpha(\mathcal{D}_d)$ and $v^{(xy)} \in \mathbf{L}_\alpha(\mathcal{D})$, with v defined in terms of u, by (6.6.21). Hence, each of the terms in braces in (6.6.32) is bounded by $CN^{1/2} e^{-(\pi d \alpha N)^{1/2}}$ with C a constant independent of N (see Theorem 3.1.7 and Example 4.2.11). By Problem 3.1.5 we also have

$$\sum_{j=-N}^{N} |S(j,h)(x)| \leqq \frac{2}{\pi}(3 + \log N)$$

$$\sum_{j=-N}^{N} \sum_{k=-N}^{N} |S(j,h)(x) S(k,h)(y)| \leqq \left\{\frac{2}{\pi}(3 + \log N)\right\}^2 . \tag{6.6.33}$$

Inequality (6.6.30) thus follows. ∎

6.6.4 Properties of the Integral Equation

Let us now study the action of the operators κ, κ_0 and κ_1 defined in (6.6.14), (6.6.15) and (6.6.16) on the spaces just defined in the previous subsection. In the following theorem, Ω denotes the region

$$\Omega = (\mathbb{R} \times \mathbb{R} \times \mathcal{D}_d) \cup (\mathcal{D}_d \times \mathbb{R} \times \mathbb{R}) \cup (\mathbb{R} \times \mathcal{D}_d \times \mathbb{R}).$$

Clearly, then $\Omega \supset \Omega_d$, with Ω_d defined in (6.6.19).

Theorem 6.6.3 *Let $k = k_0 e^{i\theta}$, with $k_0 > 0$, and $0 < \theta < \pi$. Let κ, κ_0 and κ_1 be defined in (6.6.14), (6.6.15) and (6.6.16) respectively. Let $u \in X_1$, and let $f \in X_{2\alpha}$. Then:*
(i) $\kappa_0 u \in \mathbf{L}^\infty(\overline{\Omega})$;
(ii) $\kappa_0 u \in X_1$;
(iii) $\kappa_0 u \in \mathbf{Lip}_\alpha(\Omega_d)$; *and*
(iv) $\kappa_1 u \in X_{2\alpha}$.
Moreover, if $u \in X_{2\alpha}$, and if $k_0 \sin(\theta) > 3^{1/2}\alpha$, then $\kappa_0 u \in X_{2\alpha}$.

Proof of Theorem 6.6.3: Let us set $a^2 = (y-y')^2 + (z-z')^2$, with y, y', z and z' in $\mathbb{R}$, and with $a > 0$. If u is as stated above, then the integral

$$I \equiv \int_{\mathbb{R}} \frac{\exp[ik\{(x-x')^2 + a^2\}^{1/2}]}{\{(x-x')^2 + a^2\}^{1/2}} u(x', y', z')\, dx' \tag{6.6.34}$$

is clearly well-defined for $x \in \mathbb{R}$, provided we take the positive square root of $\{(x-x')^2 + a^2\}$ for all $x' \in \mathbb{R}$. On the other hand, if we take $x = \xi + i\eta$, with $\xi \in \mathbb{R}$, $\eta \in \mathbb{R}$, and $\eta \neq 0$, and we use a single branch of one of the square roots of $\{(x-x')^2 + a^2\}$, then the integral in (6.6.34) no longer exists for all $u \in X_1$, regardless of which branch of this square root we pick!

In order to circumvent this difficulty, let $u \in X_1$, let $x = \xi + i\eta \in \mathcal{D}_d$, with $\xi \in \mathbb{R}$, $-d < \eta < d$, and for $M > |x|$, let us consider the integral

$$I(M,x) \equiv \int_{\Gamma(M,x)} \frac{\exp[ik\{(x-x')^2 + a^2\}^{1/2}]}{\{(x-x')^2 + a^2\}^{1/2}} u(x', y', z')\, dx' \tag{6.6.35}$$

with

$$\begin{aligned}\Gamma(M,x) = \{x' \in \mathbb{C} : x' &= s,\ -\infty \leqq s \leqq -M; \\ x' &= -M + it\eta,\ 0 \leqq \eta \leqq 1; \\ x' &= i\eta + \tau,\ -M \leqq \tau \leqq \xi\}.\end{aligned} \tag{6.6.36}$$

In view of the analyticity of the integrand as a function of x', it clearly follows that $I(M,x)$ is independent of M, and we can thus simplify the contour integral by letting $M \to \infty$. Hence, we may carry out the analytic continuation of I defined in (6.6.34) as a function of x via the expressions

$$\begin{aligned} I^- &\equiv \int_{-\infty}^{\xi} \frac{\exp[ik\{(\xi-\tau)^2 + a^2\}^{1/2}]}{\{(\xi-\tau)^2 + a^2\}^{1/2}} u(\tau + i\eta, y', z')\, d\tau \\ I^+ &\equiv \int_{\xi}^{\infty} \frac{\exp[ik\{(\tau-\xi)^2 + a^2\}^{1/2}]}{\{(\tau-\xi)^2 + a^2\}^{1/2}} u(\tau + i\eta, y', z')\, d\tau \\ I &= I^- + I^+. \end{aligned} \tag{6.6.37}$$

In both integrals I^+ and I^-, τ varies over $\mathbb{R}$, while (y', z') is a fixed point in $\mathbb{R}^2$. We take in the first integral the branch of the square root which is asymptotically equal to $-\tau$ as $\tau \to -\infty$, and in the second integral the branch which is asymptotically equal to τ as $\tau \to \infty$. If $0 < \theta < \pi$ in the definition $k = k_0 e^{i\theta}$, it follows that for $a > 0$, both integrals I^+ and I^- converge absolutely, and moreover, as $\eta = \Im x \to 0$, the term I defined in (6.6.37) becomes identical to the one defined in (6.6.34). We may thus extend the definition of $\kappa_0 u$ to complex values of x by means of the formula

$$(\kappa_0 u)(x,y,z) = k^2 \int_{\mathbb{R}_+} \int_{\mathbb{R}} I dy'\, dz'. \tag{6.6.38}$$

We remark that a procedure similar to the one above will enable us to appropriately define the operator κ_0, if we consider $\kappa_0 u$ as a function of the complex variable y for $(x,z) \in \mathbb{R}^2$, or as a function of the complex variable z for $(x,y) \in \mathbb{R}^2$. Let us now state and prove a sequence of Lemmas which will enable us to verify each of the statements of Theorem 6.6.3.

Lemma 6.6.4 *Let $u \in X_1$. Then*

$$\sup_{\overline{r} \in \Omega} |(\kappa_0 u)(\overline{r})| \leqq \frac{1}{\sin^2(\theta)} \|u\|_{\Omega}. \tag{6.6.39}$$

Proof. We have

$$\begin{aligned}\sup_{\bar{r}\in\Omega} |(\kappa_0 u)(\bar{r})| \\ &\leqq \frac{k_0^2\|u\|_\Omega}{4\pi}\int\int\int_{\mathbf{R}^3} \frac{e^{-k_0 r \sin(\theta)}}{r} d\bar{r} \\ &\leqq \frac{1}{\sin^2(\theta)}\|u\|_\Omega.\end{aligned} \tag{6.6.40}$$

■

Let us now formally differentiate each of the integrals in (6.6.37) with respect to ξ, to get

$$\frac{\partial I}{\partial \xi} = \int_{\mathbf{R}} \frac{e^{ikr}}{R^2}\left\{ik - \frac{1}{R}\right\}(\xi - \tau)\, u(\tau + i\eta, y', z')\, d\tau \tag{6.6.41}$$

with $R = \{(\xi-\tau)^2+(y-y')^2+(z-z')^2\}^{1/2}$. Using the inequality $|\xi-\tau| \leqq R$ in (6.6.41), and then integrating with respect to y' and z', thus representing $(\partial/\partial x)(\kappa u)$, we get

$$\left|\frac{\partial}{\partial x}(\kappa_0 u)(x,y,z)\right| \leqq k_0^2\|u\|_\Omega \int\int\int_V \frac{e^{-k_0\sin(\theta) r}}{4\pi r}\left(k_0 + \frac{1}{r}\right) d\bar{r}. \tag{6.6.42}$$

We may now explicitly evaluate the integral in (6.6.42). Moreover, proceeding similarly with respect to the variables y and z, we arrive at

Lemma 6.6.5 *If $u \in X_1$, then $\kappa u \in X_1$. Moreover,*

$$\left\{\begin{array}{ll} \sup\limits_{(x,y,z)\in\mathcal{D}_d\times\mathbf{R}\times\mathbf{R}_+} & \left|\frac{\partial(\kappa_0 u)(x,y,z)}{\partial x}\right| \\ \sup\limits_{(x,y,z)\in\mathbf{R}\times\mathcal{D}_d\times\mathbf{R}_+} & \left|\frac{\partial(\kappa_0 u)(x,y,z)}{\partial y}\right| \\ \sup\limits_{(x,y,z)\in\mathbf{R}\times\mathbf{R}\times\mathcal{D}} & \left|\frac{\partial(\kappa_0 u)(x,y,z)}{\partial z}\right| \end{array}\right\} \leqq \frac{k_0}{\sin(\theta)}\left\{1 + \frac{1}{s}\right\}\|u\|_\Omega. \tag{6.6.43}$$

with s is defined in (6.6.5).

Next, we prove

Corollary 6.6.6 *If $u \in X_1$, then $\kappa_0 u \in X_1 \cap \mathbf{Lip}_\alpha(\Omega_d)$.*

Proof. It follows directly from Lemma 6.6.5, that $\kappa_0 u \in \mathbf{Lip}_1(\Omega_d) \subset \mathbf{Lip}_\alpha(\Omega_d)$. ■

Lemma 6.6.7 *If $u \in X_1$, then $\kappa_1 u \in X_{2\alpha}$.*

Proof. We have seen from Corollary 6.6.6, that $\kappa_0 u \in \mathbf{Lip}_\alpha(\Omega_d)$. Since $\kappa_1 u = f\kappa_0 u$, and since $f \in X_{2\alpha}$, it follows, via a simple computation, that $\kappa_1 u \in X_{2\alpha}$. ∎

Lemma 6.6.8 *If $u \in X_{2\alpha}$, and if $k_0 \sin(\theta) > 3^{1/2}\alpha$, then $\kappa_0 u \in X_{2\alpha}$.*

Proof. In view of Corollary 6.6.6, we only need to show that uniformly, for all $\overline{r} = (x, y, z) \in \Omega$, we have

$$|(\kappa_0 u)(x,y,z)| \leqq Ce^{-\alpha(|x|+|y|+|z|)}. \tag{6.6.44}$$

with C a constant. To this end, we have for $x = \xi + i\eta \in \mathcal{D}_d$, $\xi \in \mathbb{R}$, $y \in (-d, d)$, $(x, y, z) \in \Omega$, I defined in (6.6.37), and $u \in X_{2\alpha}$, that

$$\begin{aligned} |I| &\leqq \int_{\mathbb{R}} \frac{e^{-k_0 \sin(\theta) R}}{R} |u(x' + i\eta, y', z')|\, dx' \\ &\leqq \int_{\mathbb{R}} \frac{e^{-k_0 \sin(\theta) R}}{R} e^{-\alpha(|x'|+|y'|+|z'|)}\, dx', \end{aligned} \tag{6.6.45}$$

with $R = \{(\xi - x')^2 + (y - y')^2 + (z - z')^2\}^{1/2}$. Thus

$$\begin{aligned} &|(\kappa_0 u)(x,y,z)| \leqq \\ &\qquad k_0^2 \int\!\!\int\!\!\int_{\mathbb{R}^3} \frac{e^{-k_0 \sin(\theta) R}}{4\pi R} e^{-\alpha(|x'|+|y'|+|z'|)}\, dx'\, dy'\, dz'. \end{aligned} \tag{6.6.46}$$

Now split $\mathbb{R}^3$, the region of integration in the integral (6.6.46), into two regions, B and $B' = \mathbb{R}^3 \setminus B$, where $B = \{(x', y', z') : |x' - \xi| \leqq 1, |y' - y| \leqq 1, |z' - z| \leqq 1\}$. Then $|\kappa_0 u(x, y, z)| \leqq J_1 + J_2$, with J_1 the part of the integral in (6.6.46) over B', and J_2 the part over B. On B', we have $R \geqq 1$. By Schwarz's inequality, we have $R = \{(\xi - x')^2 + (y - y')^2 + (z - z')^2\}^{1/2} \geqq (|\xi - x'| + |y - y'| + |z - z'|)/\sqrt{3}$. Hence,

$$\begin{aligned} &J_1 \\ &\leqq k_0^2 \int\!\!\int\!\!\int_{B'} e^{-(k_0 \sin(\theta)/\sqrt{3})(|\xi - x'| + |y - y'| + |z - z'|)} e^{-\alpha\,(|x'|+|y'|+|z'|)}\, dx'\, dy'\, dz' \\ &\leqq k_0^2 \int\!\!\int\!\!\int_{\mathbb{R}^3} e^{-(k_0 \sin(\theta)/\sqrt{3})(|\xi - x'| + |y - y'| + |z - z'|)} e^{-\alpha\,(|x'|+|y'|+|z'|)}\, dx'\, dy'\, dz' \\ &\leqq C' e^{-\alpha(|x|+|y|+|z|)} \end{aligned} \tag{6.6.47}$$

where, by assumption, we have $k_0 \sin(\theta)/\sqrt{3} > \alpha$, and the last inequality in (6.6.47) follows from the previous one by explicit integration. Thus C' is a constant depending only on $k_0 \sin(\theta)$, and α. Next, on B we have, for example, that $|y'| \geqq |y| - |y - y'| \geqq |y| - 1$. Hence, we have

$$\begin{aligned} J_2 & \\ &\leqq k_0^2 \int\int\int_B \frac{e^{-(k_0 \sin(\theta)R}}{4\pi R} e^{-\alpha(|x'|+|y'|+|z'|)} dx'dy'dz' \qquad (6.6.48) \\ &\leqq k_0^2 e^{3\alpha - \alpha(|x|+|y|+|z|)} \int\int\int_{\mathbb{R}^3} \frac{e^{-(k_0 \sin(\theta)R}}{4\pi R} dx'dy'dz' \\ &\leqq C' e^{-\alpha(|x|+|y|+|z|)} \end{aligned}$$

where C' is a constant depending only on $k_0, \sin(\theta)$, and α. The inequalities (6.6.47) and (6.6.48) show that $\kappa_0 u \in X_{2\alpha}$. ■

Note: The above proof shows that if $k_0 \sin(\theta)/\sqrt{3} > \alpha$, then for all $\overline{r} \in \Omega_d$, we have

$$|(\kappa u)(\overline{r})| \leqq C_2 \exp\{-\alpha(|x| + |y| + |z|)\} \, \|u\|_{\Omega_d}, \qquad (6.6.49)$$

with C_2 a constant independent of $u \in X2$.

At this point we conclude, that in proving the above lemmas, we also have completed the proof of Theorem 6.6.3. ■

We now take stock of the fact that we have established that κ is a compact operator from X_1 to $X_{2\alpha}$. It thus follows that Equation (6.6.13) is a linear Fredholm integral equation which has a unique solution for all but a possibly countably infinite set of values of k. One may at times wish to establish uniqueness as *a priori*. To this end, we prove the following theorem.

Theorem 6.6.9 *Let $s > 0$ and let $k = k_0(c+is)$ with $k_0 > 0$, $k_0 s > 3^{1/2}\alpha$, and $c^2 + s^2 = 1$. Let $f = f_R + i f_I \in X2\alpha$, with f_R and f_I real-valued functions such that $\left(f_R^2 + f_I^2\right)^{1/2} < 2|cs|$ in V. Then the problem*

$$w(\overline{r}) - k^2 \int\int\int_V G_k(\overline{r} - \overline{r}') f(\overline{r}') w(\overline{r}') d\overline{r}' = 0 \qquad (6.6.50)$$

has only the trivial solution $w = 0$ in X_1.

Proof. We recall that $f \equiv 0$ on $\mathbb{R}^3 \backslash V$, so that if a non-trivial function w satisfies (6.6.50), then

$$\nabla^2 w + k^2(1 + f)w = 0 \;\text{ in }\; \mathbb{R}^3. \qquad (6.6.51)$$

Note that the definition of w may be extended to $\mathbb{R}^3 \backslash V$ by means of Equation (6.6.50). Setting $\mathcal{B}_R \equiv \{\bar{r} \in \mathbb{R}^3 : |\bar{r}| < R\}$, we know by Gauss' Theorem, that

$$\begin{aligned}\iiint_{\mathcal{B}_R} \overline{w}(\bar{r}')\nabla^2 w(\bar{r})d\bar{r}' + \iiint_{\mathcal{B}_R} |\nabla w(\bar{r}')|^2 d\bar{r}' \\ = \iint_{\partial\mathcal{B}_R} w(\bar{r}')\frac{\partial w(\bar{r}')}{\partial|\bar{r}'|}dS(\bar{r}')\end{aligned} \tag{6.6.52}$$

Using (6.6.51) in (6.6.52), and taking imaginary parts of each side of the resulting equation, we get

$$\begin{aligned}M_L &\equiv \iiint_{\mathcal{B}_R} |w(\bar{r}')|^2 \left\{2cs + (c^2 - s^2)f_I(\bar{r}') + 2csf_R(\bar{r}')\right\} d\bar{r}' \\ &= -\frac{c}{k_o}\iint_{\partial\mathcal{B}_R} \left\{|w(\bar{r}')|^2 - \Im\left[w(\bar{r}')\left(\frac{\partial w}{\partial r}(\bar{r}') - ik\, w(\bar{r}')\right)\right]\right\} dS(\bar{r}').\end{aligned} \tag{6.6.53}$$

Now, as it was shown in the proof of Lemma 6.6.8, if $w \in X_2$, then $\kappa w \in X_{2\alpha}$. Hence, (Problem 6.6.2) w satisfies the radiation condition

$$\iint_{\partial\mathcal{B}_R} \left|\frac{\partial w}{\partial r}(\bar{r}') - ikw(\bar{r}')\right|^2 dS(\bar{r}') \to 0 \tag{6.6.54}$$

as $R \to \infty$, as well as the relation

$$\iint_{\partial\mathcal{B}_R} |v(\bar{r}')|^2 dS(\bar{r}') \to 0 \tag{6.6.55}$$

as $R \to \infty$.

Hence, we have

$$\lim_{R\to\infty} T_R = 0 \tag{6.6.56}$$

with T_R the right-hand side of equation (6.6.53). Consequently, the left-hand side, M_L, must also approach *zero* as $R \to \infty$. Assume now that $2|cs| - |f| \geqq \varepsilon > 0$ in V. Then, we have

$$|2cs + (c^2 - s^2)f_I + 2csf_R| \geqq 2|cs| - |f| \geqq \varepsilon, \tag{6.6.57}$$

and hence we have

$$M_L \geqq \varepsilon \lim_{R\to\infty} \iiint_{\mathcal{B}_R} |w(\bar{r}')|^2 d\bar{r}' > 0. \tag{6.6.58}$$

This contradiction proves that equation (6.6.50) cannot have a non-trivial solution $w \in X_1$.

This completes the proof of Theorem 6.6.9. ■

Note: If the conditions of Theorem 6.6.9 are satisfied, then Equation (6.6.17) has a unique solution $v \in X_{2\alpha}$, and a fortiori, Equation (6.6.13) has a unique solution $u \in X_{2\alpha}$.

6.6.5 Galerkin Approximation

Since the solution v of (6.6.17) is in $X_{2\alpha}$, we shall, for the sake of variety, use the approximation described in Theorem 6.6.2, i.e., we take

$$v_N = \sum_{j=-N}^{N} \sum_{k=-N}^{N} \sum_{\ell=-N}^{N} c_{jk\ell} S(j,h)(x) S(k,h)(y) \psi_\ell(z), \tag{6.6.59}$$

with the ψ_ℓ defined in (6.6.28). We then replace (6.6.17) with the approximating equation,

$$v_N - (\mathcal{P}_N \kappa_1) v_N = \mathcal{P}_N v_N^{in} \tag{6.6.60}$$

with v_N given in (6.6.59) and with $\mathcal{P}_N$ defined in (6.6.29). We recall also that $v^{in} = f u^{in}$, with u^{in} defined in (6.6.12).

Theorem 6.6.10 *Let the statement of Theorem 6.6.3, yielding the conclusions (i) and (ii) be satisfied, and let the conditions of Theorem 6.6.9 hold. If h is selected by the formula $h = \left[\pi d/(\alpha N)\right]^{1/2}$, then Equation (6.6.60) has a unique solution v_N for all N sufficiently large, and there exists a constant C, independent of N, such that for all such N we have*

$$\sup_{\bar{r} \in V} |v(\bar{r}) - v_N(\bar{r})| \leqq C\, N^{1/2} (\log N)^2 e^{-(\pi d \alpha N)^{1/2}}, \tag{6.6.61}$$

with v the unique solution of Equation (6.6.17). If in addition we have $k_o s > 3^{1/2}\alpha$, let u_N be defined by

$$u_N = \kappa_0 v_N + u^{in} \tag{6.6.62}$$

with v_N defined in (6.6.60). Then there exists a constant C_1 which is independent of N, such that if u denotes the unique solution of Equation (6.6.13), then

$$\sup_{\bar{r} \in V} |u(\bar{r}) - u_N(\bar{r})| \leqq C_1 N^{1/2} (\log N)^2 e^{-(\pi d \alpha N)^{1/2}}. \tag{6.6.63}$$

Proof. Let us first note that if $u \in X_1$, then $\mathcal{P}_N u \in X_1$. Since $v \in X_1$ implies that $v \in \mathbf{C}(V)$, it follows by Theorem 6.6.3 that $\kappa_0 v_n \in X_1$, and that $\kappa_1 v_N \in X_{2\alpha}$. Hence, by Corollary 6.6.4 and Theorem 6.2.9, it follows that since Equation (6.6.17) has a unique solution v, Equation (6.6.60) has a unique solution v_N of the form (6.6.59) for all sufficiently large N. The inequality (6.2.23) now yields (6.6.61). Equation (6.6.62) follows from (6.6.18), while (6.6.63) is then a consequence of the result of Theorem 6.6.3, namely, if $k_0 s > 3^{1/2}\alpha$, then $u \in X_{2\alpha}$. ■

Note: Although it would be possible we shall not consider the case of eigensolutions for Equation (6.6.17). In applications such as seismic inversion, or ultrasonic tomography one always has damping of solutions, i.e., $\Im k > 0$; in this situation it can be shown that the phenomenon of eigensolutions does not occur.

6.6.6 Evaluation of Moment Integrals

Let us now consider in detail, *the evaluation of the integrals* in the term $\kappa_1 v_N$. That is, setting

$$b_{\ell,m,n}(\bar{r}) = S(\ell,h)(x)S(m,h)(y)\psi_n(z) \tag{6.6.64}$$

with ψ_n defined in (6.6.28), we shall require accurate approximations of the integrals

$$I_{\ell,m,n}(\bar{r}) = \int\int\int_V G_k(\bar{r}-\bar{r}')b_{\ell,m,n}(\bar{r}')d\bar{r}'. \tag{6.6.65}$$

In order to reduce (6.6.60) to a system of linear algebraic equations, we need to determine $\mathcal{P}_N \kappa_1 v_N$, and to this end we only require the values of $I_{\ell mn}(\bar{r})$ for $\bar{r} = (\ell' h, m' h, z^*_{n'})$, with $-N \leqq \ell' \leqq N$, $-N \leqq m' \leqq N$, and $-N-1 \leqq n' \leqq N$.

As a consequence of the identity

$$S(m+n,h)(x) = S(m,h)(x-nh) \tag{6.6.66}$$

we immediately find that (Problem 6.6.3(a))

$$I_{\ell,m,n}(\ell' h, m' h, z) = I_{\ell-\ell',m-m',n}(0,0,z). \tag{6.6.67}$$

Trivially, we also have (Problem 6.6.3(b))

$$\begin{aligned} I_{\ell m,-n}(0,0,z) &= I_{\ell,m,n}(0,0,z) \\ I_{-\ell,m,n}(0,0.z) &= I_{\ell,m,n}(0,0,z) \\ I_{\ell,-m,n}(0,0,z) &= I_{\ell,m,n}(0,0,z). \end{aligned} \tag{6.6.68}$$

Hence, we only need to calculate $I_{\ell mn}(0,0,z^*_n)$ for $0 \leqq \ell \leqq m \leqq 2N$, $n = -N-1, -N, \cdots, N$.

By using Parseval's Theorem over $\mathbb{R}^2$ in (6.6.65), (1.10.10), as well as the identity

$$\int\int_{\mathbf{R}^2} e^{i\overline{\lambda}\cdot\overline{\rho}'} \frac{\exp\left\{ik\left[|\overline{\rho}-\overline{\rho}'|^2+(z-z')^2\right]^{1/2}\right\}}{4\pi\left[|\overline{\rho}-\overline{\rho}'|^2+(z-z')^2\right]^{1/2}} d\overline{\rho}'$$

$$= \frac{e^{i\mu|z-z'|}}{2\mu} i\, e^{i\overline{\lambda}\cdot\overline{\rho}} \tag{6.6.69}$$

with

$$\mu^2 = k^2 - |\overline{\lambda}|^2, \quad \Im\mu > 0, \quad \overline{\lambda} = (\xi, \eta), \tag{6.6.70}$$

we get

$$\begin{aligned} I_{\ell mn}(0,0,z) &= \frac{h^2}{4\pi^2}\int_{-\pi/h}^{\pi/h}\int_{-\pi/h}^{\pi/h} e^{ih\overline{\lambda}\cdot(\ell,m)}. \\ &\quad \cdot\left\{\int_0^\infty \frac{i\, e^{i\mu|z-z'|}}{2\mu}\psi_n(z')dz'\right\} d\overline{\lambda}. \end{aligned} \tag{6.6.71}$$

Setting $\lambda = |\overline{\lambda}|$, it is clear from (6.6.70) that the integral in braces in (6.6.71) depends only on λ, and we denote it by $g(z, \lambda)$. That is, setting

$$\begin{aligned} \rho &= |\overline{\rho}| = h\{\ell^2 + m^2\}^{1/2} \\ g(z,\lambda) &= \frac{i}{2\mu}\int_0^\infty e^{i\mu|z-z'|}\psi_n(z')dz', \end{aligned} \tag{6.6.72}$$

we get the simplified expression

$$I_{\ell,m,n}(0,0,z) = \frac{h^2}{4\pi^2}\int_{-\pi/h}^{\pi/h}\int_{-\pi/h}^{\pi/h} g(z,\overline{\lambda})e^{i\overline{\lambda}\cdot\overline{\rho}}d\overline{\lambda}. \tag{6.6.73}$$

We now split $I_{\ell,m,n}$ into two parts

$$I_{\ell,m,n} = I_{\ell,m,n}^{(A)} - I_{\ell,m,n}^{(B)} \tag{6.6.74}$$

with

$$\begin{aligned} I_{\ell,m,n}^{(A)} &= \int\int_A g(z,\lambda)e^{i\overline{\lambda}\cdot\overline{\rho}}d\overline{\lambda} \\ I_{\ell,m,n}^{(B)} &= \int\int_B g(z,\lambda)e^{i\overline{\lambda}\cdot\overline{\rho}}d\overline{\lambda}, \end{aligned} \tag{6.6.75}$$

and with

$$A = \{\overline{\lambda} = (\xi, \eta) : |\overline{\lambda}| \leqq \pi/h\}$$
$$B = \{\overline{\lambda} = (\xi, \eta) : -\pi/h \leqq \xi \leqq \pi/h, -\pi/h \leqq \eta \leqq \pi/h\} \setminus A. \tag{6.6.76}$$

Then, we have

$$\begin{aligned} I^{(A)}_{\ell,m,n} &= \int_0^{\pi/h} \lambda g(z,\lambda) \int_0^{2\pi} e^{i\lambda\rho\cos\theta} d\theta \; d\lambda, \\ &= 2\pi \int_0^{\pi/h} \lambda g(z,\lambda) J_0(\lambda\rho) d\lambda, \end{aligned} \tag{6.6.77}$$

with $J_0(x)$ the zero order Bessel function (see Problem 1.1.4).

After setting

$$\begin{aligned} \overline{\lambda} &= (\cos\omega, \sin\omega)\lambda \quad (\lambda = |\overline{\lambda}|), \\ e^{i\overline{\lambda}\cdot\overline{\rho}} &= e^{i\lambda\rho\cos(\theta-\omega)}, \\ &= \sum_{n=-\infty}^{\infty} i^n e^{in(\theta-\omega)} J_n(\lambda\rho), \end{aligned} \tag{6.6.78}$$

we get

$$\begin{aligned} I^{(B)}_{\ell,m,n} &= \sum_{n=-\infty}^{\infty} i^n e^{in\theta} \int\int_{\mathcal{B}} \lambda g(z,\lambda) J_n(\lambda\rho) e^{-in\,\omega} d\omega \, d\lambda \\ &= \sum_{n=-\infty}^{\infty} i^n e^{in\theta} \int_{\pi/h}^{\sqrt{2}\pi/h} \lambda g(x,\lambda) J_n(\lambda\rho)\, W_n(\lambda) d\lambda \end{aligned} \tag{6.6.79}$$

with

$$W_n(\lambda) = \begin{cases} 0 \text{ if } n \neq 4s, \quad s \text{ an integer}, \\ 8\cos^{-1}\left(\dfrac{\pi}{\lambda h}\right) \quad \text{if } n = 0 \\ \dfrac{2}{s}\sin\left[4s\cos^{-1}\left(\dfrac{\pi}{\lambda h}\right)\right] \quad \text{if } n = 4s, \quad s = \pm 1, \pm 2, \cdots. \end{cases} \tag{6.6.80}$$

Hence, collecting terms, we get

$$I_{j,k,\ell}(0,0,z) = I^{(A)}_{j,k,\ell} - I^{(B)}_{j,k,\ell} \tag{6.6.81}$$

with

$$
\begin{aligned}
I^{(A)}_{j,k,\ell} &= \frac{h^2}{2\pi}\int_0^{2^{1/2}\pi/h} \lambda J_0(\lambda\rho) g(z,\lambda)\, d\lambda \\
I^{(B)}_{j,k,\ell} &= \frac{h^2}{2\pi}\int_{\pi/h}^{2^{1/2}\pi/h} \lambda g(z,\lambda) \cdot \\
&\quad \cdot \left\{ 8 J_0(\lambda\rho)\cos^{-1}\left(\frac{\pi}{\lambda h}\right) \right. \\
&\quad \left. + 4\sum_{s=1}^{\infty} s^{-1} J_{4s}(\lambda\rho)\cos(4s\theta)\sin\left[4s\cos^{-1}\frac{\pi}{\lambda h}\right]\right\} d\lambda.
\end{aligned} \tag{6.6.82}
$$

6.6.7 Numerical Evaluation of Solution

Let us briefly remark on the evaluation of u via (6.6.18). If z is near 0, we recommend approximating u via the procedure of Section 6.6.6, i.e., we use the approximation

$$
u(\bar{r}) \approx u_N(\bar{r}) \equiv k^2 \sum_{\ell=-N}^{N}\sum_{m=-N}^{N}\sum_{n=-N-1}^{N} c_{\ell,m,n} I_{\ell,m,n}(\bar{r}) + u^{in}(\bar{r}) \tag{6.6.83}
$$

with $I_{\ell,m,n}(\bar{r})$ evaluated as in Section 6.6.6, and the $c_{\ell mn}$ obtained by solving the linear system of equations corresponding to (6.6.60). On the other hand, if $z \notin V$, $|z| \geqq d$, we may use Sinc quadrature after substitution into (6.6.18) of v_N given by (6.6.59). In this case, we get the approximation

$$
u(\bar{r}) \approx u_N(\bar{r}) \equiv k^2 h^3 \sum_{\ell=-N}^{N}\sum_{m=-N}^{N}\sum_{n=-N-1}^{N} c_{\ell,m,n} K_{\ell,m,n}(\bar{r}) + u^{in}(\bar{r}) \tag{6.6.84}
$$

with

$$
K_{\ell,m,n}(\bar{r}) = \frac{\exp\{ik|\bar{r} - (\ell h, mh, z_n^*)|\}}{4\pi|\bar{r} - (\ell h, mh, z_n^*)|}(1 + e^{-2\ell h})^{-1/2}. \tag{6.6.85}
$$

The result (6.6.84) is accurate to within an error of order $N^{1/2}(\log N)^2 e^{-(\pi d\alpha N)^{1/2}}$, since the integrand in (6.6.18) is then clearly in $X_{3\alpha}$.

6.6.8 An Explicit Example

A simulated example will serve to illustrate the procedure of this section. We consider the problem

$$\nabla^2 u + k^2 u = -k^2 f u \tag{6.6.86}$$

in $V = \mathbb{R}^2 \times \mathbb{R}_+$, in which we take

$$\begin{aligned} k &= k_0(3^{1/2}/2 + i/2), \\ f(\overline{r}) &= 2\left[\frac{i}{kR} + \left(\frac{i}{kR}\right)^2\right], \qquad (6.6.87) \\ R &= |\overline{r} - (0,0,-z_0)|, \quad z_0 > 0. \end{aligned}$$

Then, in the case of $k_0 > 2$ and $z_0 \geqq 1$, the condition $|f(\overline{r})| \leqq 2|cs| = 3^{1/2}/2$ is satisfied, and so, given a suitable u^{in}, the problem (6.6.86) will then have a unique solution in V. Let us take $z_0 = \pi/2$, $k_0 = 3$.

One solution to (6.6.86) is

$$u(\overline{r}) = \frac{e^{ikR}}{4\pi R^2}, \tag{6.6.88}$$

with R defined in (6.6.87). According to (6.6.11), the u^{in} which yields (6.6.88) as the solution to (6.6.86) is given by

$$u^{in}(\overline{r}) = \int\int_{\mathbb{R}^2} \left\{ u(\overline{\rho}', 0)\frac{\partial G_k(\overline{r} - \overline{r}')}{\partial z} - \frac{\partial u}{\partial z'}(\overline{\rho}', 0) G_k(\overline{r} - \overline{r}')\right\} d\overline{\rho}'. \tag{6.6.89}$$

Using Fourier transforms over $\mathbb{R}^2$ and the result (6.6.71), we also have

$$u^{in}(\overline{r}) = \frac{1}{4\pi}\int\int_{\mathbb{R}^2} e^{-i\overline{\lambda}\cdot\overline{\rho} + i\mu z}\left\{\tilde{u}(\overline{\lambda}, 0) + \frac{i}{\mu}\tilde{u}_z(\overline{\lambda}, 0)\right\} d\overline{\lambda} \tag{6.6.90}$$

with $\overline{\rho} = (x, y)$, $\overline{\rho}' = (x', y')$, and with

$$\tilde{u}(\overline{\lambda}, z) = \int\int_{\mathbb{R}^2} e^{i\overline{\lambda}\cdot\overline{\rho}} u(\overline{\rho}, z) d\overline{\rho}. \tag{6.6.91}$$

Clearly, the solution u given by (6.6.88) is in $X_{2\alpha}$. Consequently, by proceeding as in the proof of Theorem 6.6.3, it follows that u^{in} defined by (6.6.90) is also in $X_{2\alpha}$. Hence, we may expect the approximation

$$u^{in}(x,y,z) \approx \mathcal{P}_N u^{in}(x,y,z)$$
$$= \sum_{\ell=-N}^{N} \sum_{m=-N}^{N} \sum_{n=-N-1}^{N} u^{in}(\ell h, mh, z_n^*) \cdot \qquad (6.6.92)$$
$$\cdot S(\ell,h)(x) S(m,h)(y) \psi_n(z)$$

to be accurate.

The numbers $u^{in}(\ell h, mh, z_n^*)$ may be obtained by proceeding as in the previous subsection.

Replacing $u(\overline{\rho},0)$ and $u_z(\overline{\rho},0)$ in (6.6.14) by their $(2N+1)^2$-point Sinc approximation over $\mathbb{R}^2$, and then taking the Fourier transforms of these approximations via (6.6.92), we get the expansions

$$\tilde{u}(\overline{\lambda},0) \approx (\widetilde{\mathcal{P}_N u})(\overline{\lambda},0) = h^2 \sum_{m,n=-N}^{N} u(mh,nh,0) e^{ih(m,n)\cdot\overline{\lambda}}$$
$$(6.6.93)$$
$$\tilde{u}_z(\overline{\lambda},0) \approx (\widetilde{\mathcal{P}_N u})(\overline{\lambda},0) = h^2 \sum_{m,n=-N}^{N} u_z(mh,nh,0) e^{ih(m,n)\cdot\overline{\lambda}},$$

which are zero outside the square having its center at $(0,0)$ and length of side equal to $2\pi/h$. Substituting these terms into (6.6.90) and replacing $\overline{\rho}$ by (jh,kh), it is clear that we only need to evaluate

$$J_{\ell,m}^{(A)}(s,z) = \frac{h^2}{4\pi} \int\int_A e^{i\mu z} \mu^{-s} e^{ih(\ell,m)\cdot\overline{\lambda}} d\overline{\lambda}$$
$$(6.6.94)$$
$$J_{\ell,m}^{B}(s,z) = \frac{h^2}{4\pi} \int\int_B e^{i\mu z} \mu^{-s} e^{ih(\ell,m)\cdot\overline{\lambda}} d\overline{\lambda}$$

where A and B are defined, as in (6.6.76), to get

$$u^{in}(jh,kh,z) \approx \sum_{m,n=-N}^{N} \Big[u(mh,nh,0) \Big\{ J_{j-m,k-n}^{(A)}(0,z)$$
$$+ J_{j-m,k-n}^{(B)}(0,z) \Big\} \qquad (6.6.95)$$
$$+ i u_z(mh,nh) \Big\{ J_{j-m,k-n}^{(A)}(1,z)$$
$$+ J_{j-m,k-n}^{(B)}(1,z) \Big\}.$$

The expansions for $J_{\ell,m}^{(A)}$ and $J_{\ell,m}^{(B)}$

$$J_{\ell,m}^{(A)}(s,z) = \frac{h^2}{2\pi}\int_0^{2^{1/2}\pi/h} \lambda J_0(\lambda\rho) e^{i\mu z}\mu^{-s}\,d\lambda$$

$$J_{\ell m}^{(B)}(s,z) = \frac{h^2}{2\pi}\int_{\pi/h}^{2^{1/2}\pi/h} \lambda e^{i\mu z}\mu^{-s}.$$

$$\cdot\left\{8J_0(\lambda\rho)\cos^{-1}\left(\frac{\pi}{\lambda h}\right) + 4\sum_{s=1}^{\infty} J_{4s}(\lambda\rho)\frac{\cos(4s\theta)\sin\left[4s\cos^{-1}\frac{\pi}{\lambda\rho}\right]}{s}\right\}d\lambda. \tag{6.6.96}$$

For this problem, we may take $d = \pi/2$, $\alpha = 1$, so that $h = \pi/(2N)^{1/2}$. Taking $N = 2$, we get nearly 3 accurate digits in the approximate solution. •

PROBLEMS FOR SECTION 6.6

6.6.1 Prove (6.6.33).

6.6.2 Use Theorem 4.1.2 and Lemma 6.6.8 to prove that if $f \in X_{2\alpha}$ and if $k_0 s > 3^{1/2}\alpha$ then the integrals in (6.6.54) and (6.6.55) approach zero as $R \to \infty$.

6.6.3 **(a)** Use the identity (6.6.92) to prove (6.6.93).

(b) Verify each of the identities in (6.6.94).

6.6.4 Write a computer program to solve the equation (6.6.77) in $V = \mathbb{R}^2 \times \mathbb{R}^+$, for the case when k and f are defined as in (6.6.4).

6.6.5 Show that the number of elements $I_{\ell,m,n}(0,0,z_{n'})$ such that $0 \leqq \ell \leqq m \leqq 2N$, and $-N-1 \leqq n, n' \leqq N$ is $4(N+1)^2[2N^2+5N+4] \sim 8N^4$.

6.6.6 Set

$$\begin{array}{rclrclrcl}
\overline{\rho} &=& (x,y), & \overline{\rho}' &=& (x',y'), & \overline{\lambda} &=& (\mu,\nu) \\
\rho &=& |\overline{\rho}|, & \rho' &=& |\overline{\rho}'|, & \lambda &=& |\overline{\lambda}| \\
\overline{r} &=& (\overline{\rho},z), & \overline{r}' &=& (\overline{\rho}',z'), & G(\overline{r},\overline{r}') &=& \frac{\exp\{ik|\overline{r}-\overline{r}'|\}}{4\pi|\overline{r}-\overline{r}'|},
\end{array}$$

with $\Im k > 0$. Prove that

$$\int\int_{\mathbb{R}^2} e^{i\overline{\lambda}\cdot\overline{\rho}} G(\overline{r},\overline{r}')\,dx\,dy = ie^{i\overline{\lambda}\cdot\overline{\rho}'}\frac{e^{i|z-z'|\sqrt{k^2-\lambda^2}}}{2\sqrt{k^2-\lambda^2}},$$

with the branch of the square root chosen such that $\Im\sqrt{k^2 - \lambda^2} > 0$.

6.6.7 Let $\overline{r} \in \mathbb{R}^3$, $\overline{R} \in \mathbb{R}^3$, $k = k_0 e^{i\theta}$, with $k_0 = |k|$, and where $\theta = \arg(k)$. Set $r = |\overline{r}|$, $R = |\overline{R}|$, and

$$p = \{(\overline{r} + i\overline{R}) \cdot ((\overline{r} + i\overline{R})\}^{1/2}, \quad p_0 = |p|,$$

and where we select the branch of the square root such that $p \to r$ as $r \to \infty$. Prove that

$$\Re\{ikp\} \leqq k_0 \{R - r\sin(\theta)\}.$$

6.6.8 Let Equation (6.6.11) be given, with u^{in} defined in (6.6.12), and with $f \in \mathbf{Lip}_{2\alpha}(\mathcal{D})$, $(0 < 2\alpha \leqq 1)$ in a neighborhood of each point $r \in V$.

(a) Prove that there exist numbers θ_1 and θ_2, with $0 \leqq \theta_1 < \theta_2 < \pi$, such that for each fixed $\overline{r} \in V$, $\overline{r} \neq \overline{r_0}$, $\overline{r} \neq \overline{r_1}$, and $|\overline{r}| < \infty$, the solution $u = u(k, \overline{r_0}, \overline{r})$ of (6.6.11) has the property that $u(k, \overline{r_0}, \overline{r})$ is an analytic, uniformly bounded, and non-zero function of k in the sector

$$\mathcal{S} = \{k \in \mathbb{C} : 0 < |k| < \infty,\ \theta_1 < \arg k < \theta_2\}.$$

(b) Assume that $u(k, \overline{r_0}, \overline{r})$ is measured at $\overline{r} = \overline{r_1}$. Prove that

$$w(k) \equiv \frac{1}{ik} \log \left\{ \frac{u(k, \overline{r_0}, \overline{r_1})}{u(k, \overline{r_0}, \overline{r_0})} \right\}$$

is analytic and uniformly bounded in $\mathcal{S}$. (One may assume here that the location point of the source, $\overline{r_0}$, is slightly outside of $\overline{V}$, i.e., that the two values of $\overline{r_0}$ in the denominator of this last expression differ slightly.)

(c)* Prove that

$$w(k) = \int_{\mathcal{P}} \sqrt{1 + f^2(\overline{r})}\, ds + \mathcal{O}(|k|^{-\alpha}), \quad k \to \infty \text{ in } \mathcal{S},$$

with $\mathcal{P}$ the *ray path* connecting $\overline{r_0}$ and $\overline{r}_1$. Conclude that $w \in \mathbf{M}_\alpha(\mathcal{S})$.

(d) Recall the results of Section 5.6.5 to conclude that if $w(k)$ is known for a sufficient number of values of $k \in \mathcal{S}$, then we may compute $w(\infty) = \int_{\mathcal{P}} \sqrt{1 + f^2}\, ds$ via the Thiele algorithm.

(e)* Can this procedure be used to reconstruct $f \in V$, based on measuring $u(k, \overline{r_0}, \overline{r_1})$ at several values of $\overline{r_1} \in \partial V$ for each fixed

$\overline{r_0} \in \partial V$, and repeating the experiment, for several values of $\overline{r_0}$ on a line in ∂V?

(f)* In practice one encounters noisy data, which destroys the accuracy of the Thiele algorithm. How would you circumvent the effects of noise?

6.6.9 (a) Let $\overline{x} = (x^1, x^2, x^3)$, $\overline{y} = (y^1, y^2, y^3)$, with $x^i \in \mathbb{R}$, $y^i \in \mathbb{R}$, and let $a \in \mathbb{C}$, and set $\hat{y} = \overline{y}/y$. Prove that

$$\int\int_{\mathbb{R}} e^{a\overline{x}\cdot\overline{y}} d\Omega(\hat{y}) = 4\pi \frac{\sinh(a|\overline{x}||\overline{y}|)}{a|\overline{x}||\overline{y}|}.$$

(b) Use the result in part (a) of this problem to obtain an explicit series solution to the expression

$$u(\overline{x}) = \int\int\int_{\mathbb{R}^3} \frac{e^{-|\overline{y}|^2}}{4\pi|\overline{x}-\overline{y}|} d\overline{y}.$$

6.6.10 Set $S_m(x) = S(m,h)(x)$, and prove that

$$\int\int\int_{\mathbb{R}^3} \frac{e^{-|\overline{y}|^2} S_\ell(y^1) S_m(y^2) S_n(y^3)}{|\overline{x}-\overline{y}|} d\overline{y}$$
$$= e^{-h^2(\ell^2+m^2+n^2)} \int\int\int_{\mathbb{R}^3} \frac{S_\ell(y^1) S_m(y^2) S_n(y^3)}{|\overline{x}-\overline{y}|} d\overline{y} + \mathcal{O}(e^{-\pi d/(2h)}),$$

where $d > 0$ but it is otherwise arbitrary.

6.6.11 * (*Research Problem*) Use the identity (6.6.69) to reduce the integral equation (6.6.11) into an integral equation in one variable only, for $\overline{\lambda} \in \mathbb{R}^2$ and then use the method of Section 4.6 to derive a parallel algorithm to solve it.

6.7 Cauchy Singular Integral Equations

In this section we shall demonstrate the application of Sinc methods to the solution of Cauchy singular integral equations (CSIE).

With the exception of special cases, classical quadrature formulas based on polynomial precision do not fare well when used to approximate Hilbert transforms over an interval. The main difficulty stems from the fact that it is not easy to deduce the exact nature of the singularity of a Hilbert transform at the end-point of an interval. Although CSIE methods are quite powerful, this difficulty, and a general lack of understanding of complex variables, are largely to blame for how relatively little CSIE methods are being used in applications. On the other hand, the approximate solution of CSIE is appropriate via Sinc methods, since it is nearly always possible to reduce a CSIE problem stemming from applications, to one involving the approximation of a function belonging to the class "$\mathbf{Lip}_\alpha \cap \{\text{analytic}\}$", for which the previously derived Sinc methods fare so well.

6.7.1 The Problem

In this section we adopt the notation of Sections 1.3 and 4.1. We consider the approximate solution of the CSIE

$$u\omega + v\mathcal{S}\omega + \mathbf{k}_1\omega = g_1, \tag{6.7.1}$$

for ω, with $\Gamma = \Gamma(a,b)$ an open arc of finite length, having endpoints at a and b, with $t \in \Gamma$, $\mathcal{S}(t)$ defined in (1.3.5), and

$$(\mathbf{k}_1\omega)(t) = \int_\Gamma \mathbf{k}_1(t,\tau)\,\omega(\tau)\;d\tau. \tag{6.7.2}$$

Furthermore, we assume that if $\mathcal{D}$ is defined as in Definition 4.1.1, then each of the functions $u, v, k_1(t,\cdot)$ $(t \in \Gamma)$ and $k_1(\cdot,\tau)$ $(\tau \in \Gamma)$ belong to $\mathbf{Hol}(\mathcal{D}) \cap \mathbf{Lip}_{\alpha'}(\overline{\mathcal{D}})$, with $\alpha' \in (0,1]$ a positive constant. Moreover, we assume that

$$r(z) \equiv u^2(z) + v^2(z) \neq 0, z \in \mathcal{D}. \tag{6.7.3}$$

We seek a solution ω to Problem (6.7.1), $\omega \in \mathbf{H}^1(\mathcal{D})$.

The various numerical methods that are used to obtain an approximate solution to Problem (6.7.1) may be split into two families: (i) those that involve direct collocation of Equation (6.7.1), and (ii) those that first transform Equation (6.7.1) into a Fredholm integral equation, and then numerically solve this Fredholm integral equation. Whereas the methods (i) seem to be simpler to implement on a computer, the theory for the convergence of these methods was not fully developed at the time this text was written, and we shall thus present an approximate solution procedure along the lines of (ii), using Nyström's method.

6.7.2 The Method of Regularization

Textbooks on CSIE commonly reduce a CSIE to a Fredholm integral equation by a procedure which they call the *method of regularization.* We now describe this method of regularization.

Let u and v be defined as in (6.7.1), let $F_1 \in \mathbf{Hol}(\mathcal{D}) \cap \mathbf{Lip}_{\alpha'}(\overline{\mathcal{D}})$, and let us consider the problem of determining a function $w \in \mathbf{H}^1(\mathcal{D})$ such that

$$u(t)\;w(t) + v(t)\;(\mathcal{S}w)(t) = F_1(t), \quad t \in \Gamma. \tag{6.7.4}$$

Let us assume that (6.7.4) has a solution $w \in \mathbf{H}^1(\mathcal{D})$, and that W is defined in terms of w by (1.3.2). We then replace w and $\mathcal{S}w$ in (6.7.4), and using the Plemelj formulas, get

$$u(W_+ - W_-) + v(W_+ + W_-) = F_1. \tag{6.7.5}$$

After rearrangement of terms, Equation (6.7.5) reduces to the non-homogeneous Riemann–Hilbert problem (1.3.33), with

$$G = \frac{u - v}{u + v}, \quad g = \frac{F_1}{u + v}. \tag{6.7.6}$$

We may thus define the function Y, the numbers γ_j, the integers κ_j, $j = 1, 2$, the index κ, the polynomial Π, and the canonical function Z as in (1.3.29) through (1.3.32). We shall assume, moreover, for the sake of simplicity, that

$$-1 < \alpha_j - \kappa_j < 0. \tag{6.7.7}$$

That is, we shall omit the more complicated case $\alpha_j - \kappa_j = 0$, deferring it to Problem 6.7.3 which is at the end of this section.

We also define the *fundamental solution* X on Γ in terms of the canonical function Z by the equation

$$X(t) = [u(t) + v(t)]Z_+(t) = [u(t) - v(t)]Z_-(t). \tag{6.7.8}$$

We can readily show, then (see Problem 6.7.1) that

$$X(t) = r(t)^{1/2}\, \Pi(t)\, e^{Y(t)}, \tag{6.7.9}$$

with $r(t)$ defined as in (6.7.3), and with

$$\begin{aligned} Y &= \mathcal{S}(\log G) \\ \Pi(t) &= (t - a)^{-\kappa_1}\, (t - b)^{-\kappa_2}. \end{aligned} \tag{6.7.10}$$

It follows that since both u and v belong to the class $\mathbf{Hol}(\mathcal{D}) \cap \mathbf{Lip}_{\alpha'}(\overline{\mathcal{D}})$, and since $r(z) \neq 0$ in $\overline{\mathcal{D}}$, that $G \in \mathbf{Hol}(\mathcal{D}) \cap \mathbf{Lip}_{\alpha'}(\overline{\mathcal{D}})$. Hence Y, too, has an analytic continuation into $\mathcal{D}$ (recall Sections 1.3 and 4.3), and moreover, $Y \in \mathbf{Lip}_{\alpha'}$ everywhere in $\overline{\mathcal{D}}$ except at $z = a$ or at $z = b$. Hence, X may be continued analytically into $\mathcal{D}$, and by choosing the integers κ_j as in (6.7.7), it is clear that $X(z) \neq 0$ at a or b. It readily follows that $X \in \mathbf{Hol}(\mathcal{D})$, that $X \neq 0$ in $\overline{\mathcal{D}}$, that $X^{-1} \in \mathbf{Hol}(\mathcal{D})$, and that X^{-1} is uniformly bounded in $\mathcal{D}$.

Recalling the results of Sections 1.3.3 (iii) and 1.3.4, we find that the solution w of (6.7.4) is thus given by

$$w = \frac{u\,F_1}{X} - v\,\mathcal{S}\left(\frac{F_1}{X}\right) + v\,P_{\kappa-1}, \tag{6.7.11}$$

with $P_{\kappa-1}$ an arbitrary polynomial of degree $\kappa - 1$ ($P_{\kappa-1} = 0$ if $\kappa < 1$).

The process of regularization in CSIE is to apply the above procedure (6.7.4) through (6.7.11) to the CSIE (6.7.1), after setting $F_1 = g_1 - \mathbf{k}_1\omega$. Applying this procedure we are thus led to the Fredholm integral equation

$$\omega - \mathbf{k}_2\,\omega = g_2, \tag{6.7.12}$$

with the function g_2 and the operator $\mathbf{k}_2$ defined by

$$\begin{aligned} g_2 &= \frac{u\,g_1}{r} - \frac{v\,X}{r}\mathcal{S}\left(\frac{g_1}{X}\right) + \frac{v\,X}{r}\,P_{\kappa-1}, \\ (\mathbf{k}_2\omega)(t) &= \int_\Gamma k_2(t,\tau)\,\omega(\tau)\,d\tau, \\ k_2(t,\tau) &= \frac{v\,X}{r}(t)\mathcal{S}\left(\frac{k_1(\cdot,\tau)}{X}\right)(t) - \frac{u(t)}{r(t)}\,k_1(t,\tau), \end{aligned} \tag{6.7.13}$$

with $P_{\kappa-1}$ an arbitrary polynomial of degree $\kappa - 1$, and with $P_{\kappa-1} = 0$ if $\kappa < 1$. Furthermore, if $\kappa < 0$, then in order for Equation (6.7.1) to have a solution, we shall require the additional conditions (recall (1.3.37))

$$\int_\Gamma \frac{t^{j-1}[\mathbf{k}_1\,\omega - g_1](t)}{X(t)}\,dt = 0, \quad j = 1, 2, \cdots, -\kappa. \tag{6.7.14}$$

Later in this section, we shall see that the solution ω of (6.7.1) belongs to $\mathbf{H}^1(\mathcal{D})$. Indeed, ω will be bounded in $\mathcal{D}$, except possibly at a and b. Inasmuch as the unboundedness of ω presents a difficulty from the viewpoint of its uniform approximation on Γ, we make the *Dow-Elliott transformation*

$$\omega = \frac{X\,f}{r} \tag{6.7.15}$$

in (6.7.12), to get the new Fredholm integral equation

$$(\mathbf{I} - \mathbf{k})\,f = g, \tag{6.7.16}$$

with

$$\begin{aligned} (\mathbf{k}\,f)(t) &= \int_\Gamma k(t,\tau)f(\tau)\,d\tau \\ k(t,\tau) &= \frac{X(\tau)}{r(\tau)}\left[v(t)\,\mathcal{S}\left(\frac{k_1(\cdot,\tau)}{X}\right)(t) - \frac{u(t)}{X(t)}k_1(t,\tau)\right] \\ g &= \frac{u\,g_1}{r} - v\,\mathcal{S}\left(\frac{g_1}{r}\right) + v\,P_{\kappa-1}. \end{aligned} \tag{6.7.17}$$

In the next subsection we shall study the properties of Equation (6.7.16), thus showing that Sinc methods may indeed be used to obtain accurate approximate solutions for problem (6.7.1). We shall deduce analyticity properties of $k(t,\tau)$ and g at the endpoints of Γ, which will make it possible to deduce analyticity and boundedness of the solution f in $\mathcal{D}$ and *a fortiori*, to enable an accurate Sinc approximation to f, as well as an accurate bound on the approximation error.

6.7.3 Properties of the Fredholm Equation

Let us now deduce some properties of the function $g(t)$ and the kernel $k(t,\tau)$, defined in (6.7.17).

Lemma 6.7.1 *Let X be defined as in (6.7.9), and with reference to the definitions of α_j and κ_j, set*

$$\sigma_j = \kappa_j - \alpha_j, \quad j = 1, 2. \tag{6.7.18}$$

Then $0 < \sigma_j < 1$, $X \in \mathbf{Hol}(\mathcal{D})$, and there exist positive constants C_1 and C_2, such that for all $z \in \mathcal{D}$,

$$C_1 |z-a|^{-\sigma_1} \leqq |X(z)| \leqq C_2 |z-a|^{-\sigma_2}. \tag{6.7.19}$$

Proof. The inequalities $0 < \sigma_j < 1$ follow from our assumptions (6.7.7). The property, $X \in \mathbf{Hol}(\mathcal{D})$, was established earlier in this section. The estimates (6.7.19) follow from (1.3.26) to (1.3.27) (see also Problem 1.3.6), since if (1.3.27) holds in a neighborhood of a or b, then

$$Y(z) = (\alpha_1 + i\beta_1)\ \log\ (z-a) + (\alpha_2 + i\beta_2) \log\ (z-b) + V(z), \tag{6.7.20}$$

with V holomorphic and bounded in $\mathcal{D}$. Under our assumptions on u and v, the ratio X/Z is uniformly bounded and non-vanishing in $\overline{\mathcal{D}}$. Thus, both Z and X may be bounded from above and below, as in (6.7.19). This completes the proof of Lemma 6.7.1. ■

Let us now recall that the functions u, v, g_1, and $k_1(\cdot,\cdot)$ (k_1 with respect to each variable in turn, with the other variable fixed on $\overline{\Gamma}$) belong to $\mathbf{Hol}(\mathcal{D}) \cap \mathbf{Lip}_{\alpha'}(\overline{\mathcal{D}})$, and let us set

$$\alpha = \min(\alpha', \sigma_1, \sigma_2). \tag{6.7.21}$$

Then $g_1/X \in \mathbf{Hol}(\mathcal{D}) \cap \mathbf{Lip}_\alpha(\overline{\mathcal{D}})$, and moreover, since by Lemma 6.7.1 we have $g_1/X = 0$ at a and b, it follows from Problem 1.3.5, that $\mathcal{S}(g_1/X) \in \mathbf{Hol}(\mathcal{D}) \cap \mathbf{Lip}_\alpha(\overline{\mathcal{D}})$. Thus, g defined by (6.7.17) belongs to $\mathbf{Hol}(\mathcal{D}) \cap \mathbf{Lip}_\alpha(\overline{\mathcal{D}})$. Similarly, it follows by inspection of the expression defining $k(t,\tau)$ in

(6.7.17) that the function in square brackets belongs to $\mathbf{Hol}(\mathcal{D}) \cap \mathbf{Lip}_\alpha(\overline{\mathcal{D}})$, both as a function of t for each fixed τ on $\overline{\Gamma}$, and as a function of τ for each fixed t on $\overline{\Gamma}$. Also, by Lemma 6.7.1, $X \in \mathbf{H}^1(\mathcal{D})$. These properties, together with the results of Problem 6.7.3 readily establish

Theorem 6.7.2 *Let* $\mathbf{k}$ *and* g *be defined as in (6.7.17). Then* $\mathbf{k}\, f + g \in \mathbf{Hol}(\mathcal{D}) \cap \mathbf{Lip}_\alpha(\overline{\mathcal{D}})$ *for all* $f \in \mathbf{C}(\overline{\Gamma})$.

Since $\mathbf{k}$ is a compact operator from $\mathbf{C}(\overline{\Gamma})$ to $\mathbf{Hol}(\mathcal{D}) \cap \mathbf{Lip}_\alpha(\overline{\mathcal{D}})$, we may invoke the results of Section 6.2, to conclude that if $(\mathbf{I} - \mathbf{k})^{-1}$ exists, then Equation (6.7.16) has a solution f, $f \in \mathbf{Hol}(\mathcal{D}) \cap \mathbf{Lip}_\alpha(\overline{\mathcal{D}})$. If $\kappa = 0$, then the solution f is unique. If $\kappa > 0$, then f has κ degrees of freedom, since the function g in (6.7.17) then depends on the κ coefficients of the polynomial $P_{\kappa-1}$. If $\kappa < 0$, then a unique solution exists, provided the $-\kappa$ equations (6.7.14) are satisfied, with $\omega = X\, f/r$; for if these equations are not satisfied, then our method of solution fails, and the resulting function ω does not satisfy the CSIE (7.7.1).

Other interesting possibilities may occur, for example, in the case $\mathbf{I} - \mathbf{k}$ not invertible. In this case, it is convenient to rewrite (6.7.16) in the form

$$(\mathbf{I} - \mathbf{k})\, f = g_0 + v\, P_{\kappa-1}, \tag{6.7.22}$$

since g defined by (6.7.16) may be written as the sum of two terms, a multiple of $P_{\kappa-1}$, and a term g_0 independent of $P_{\kappa-1}$. Since $(\mathbf{I} - \mathbf{k})$ is not invertible, both $\mathbf{k}$ and $\mathbf{k}^*$ have eigensolutions. Let us assume, for definiteness, that each has a total number of p independent eigensolutions, i.e.,

$$\left\{ \begin{array}{lcl} (\mathbf{I} - \mathbf{k})\, f_{0j} & = & 0 \\ (\mathbf{I} - \mathbf{k}^*)\, F_{0j} & = & 0 \end{array} \right. \quad j = 1, 2, \cdots, p. \tag{6.7.23}$$

According to the Fredholm alternative, it then follows that if $\kappa \geqq p$, Equation (6.7.22) will still have a solution, provided that $P_{\kappa-1}$ is selected so that

$$\int_\Gamma [g_0(t) + v(t)\, P_{\kappa-1}(t)]\overline{F_{0j}(t)}\, dt = 0, \quad j = 1, 2, \cdots, p. \tag{6.7.24}$$

Choosing $P_{\kappa-1}$ so that (6.7.24) is satisfied, the general solution of (6.7.22) is given by

$$f = \sum_{j=1}^{p} c_j\, f_{0j} + f_1, \tag{6.7.25}$$

with the coefficients c_j arbitrary, and with f_1 depending on $\kappa - p$ arbitrary (still undetermined) coefficients of $P_{\kappa-1}$.

In the case $\kappa < p$, Equation (6.7.22) may still have a p- parameter solution, provided the conditions (6.7.24) can still be satisfied. If the conditions

(6.7.24) cannot be satisfied, then Equation (6.7.22) does not have a solution. However, it may still be desirable in applications to determine a *generalized inverse* solution in this case, i.e.,

$$f_3 = (\mathbf{I} - \mathbf{k}^\dagger)\, g, \tag{6.7.26}$$

where this expression provides a concise definition for the function f_3, the function of smallest $\mathbf{L}^2(\Gamma)$-norm belonging to the set of all solutions F_3 which minimize the expression

$$\int_\Gamma |(\mathbf{I} - \mathbf{k})F_3(t) - g(t)|^2 |dt|. \tag{6.7.27}$$

6.7.4 Approximation via Nyström's Method

Upon taking $h = \{2\pi d/(\alpha N)\}^{1/2}$, the Nyström solution to (6.7.16) is given by

$$f(t) \approx f_N(t) \equiv g(t) + h \sum_{m=-N}^{N} \frac{k(t, z_m)}{\phi'(z_m)}\, f_j \tag{6.7.28}$$

with the z_m denoting the Sinc points, $z_m = \phi^{-1}(mh)$, with ϕ the conformal map of $\mathcal{D}$ onto $\mathcal{D}_d$ and also $\phi : \Gamma \to \mathbb{R}$, and with the numbers f_k determined by the solution of the linear system of equations

$$f_n - h \sum_{m=-N}^{N} \frac{k(z_n, z_m)}{\phi'(z_m)}\, f_m = g(z_m), \quad m = -N, -N+1, \cdots, N. \tag{6.7.29}$$

Theorem 6.7.3 *With reference to Equation (6.7.1), let both $k_1(\cdot, t)$ and $k_1(t, \cdot)$ for all $t \in \overline{\Gamma}$, as well as the functions u, v and g_1 belong to* $\mathbf{Hol}(\mathcal{D}) \cap \mathbf{Lip}_{\alpha'}(\overline{\mathcal{D}})$, *with $0 < \alpha' \leqq 1$. Let $u^2 - v^2 \neq 0$ in $\overline{\mathcal{D}}$, and let integers κ_j be determined such that (1.3.29) is satisfied. Let σ_j be determined by (6.7.18), and let $\alpha = \min(\alpha', \sigma_1, \sigma_2) > 0$. Let $(\mathbf{I} - \mathbf{k})^{-1}$ exist, let h be given by $h = \{2\pi d/(\alpha N)\}^{1/2}$, with N a positive integer, and let $f_{-N}, f_{-N+1}, \cdots, f_N$ be the solution of the linear system (6.7.29). If $f_N(z)$ is defined as in (6.7.28), and if $f = (\mathbf{I} - \mathbf{k})^{-1} g$, then there exists a constant C, independent of N, such that*

$$\sup_{t\in\Gamma} |f(t) - f_N(t)| < C \exp\{-(2\pi d \alpha N)^{1/2}\}. \tag{6.7.30}$$

The chief difficulty of the above approximation is the evaluation of $X(t)$ and the kernel $k(t, \tau)$ as a function of t. To this end, we recommend using Equation (5.2.22), Algorithm 5.2.2, and Equation (5.2.23).

•

Problems for Section 6.7

6.7.1 Describe an explicit algorithm based on the above theory and the formulas of Section 5.2 to solve a CSIE over the interval (-1,1).

6.7.2 Use the approximation procedure of Problem 5.2.6 to approximate the Hilbert transform over $(0, \infty)$, to describe an explicit algorithm to solve a CSIE over $(0, \infty)$.

6.7.3 * Discuss the solution of the problem (6.7.1) in the case of at least one of the α_j an integer.

6.8 Convolution-Type Equations

In this section we shall present the theory of the method of Wiener and Hopf, as well as a method for solving the simpler convolution equation over the real line $\mathbb{R}$. We shall develop procedures for the approximate solution of such equations, and prove the convergence of our approximate solutions to the actual solution of the corresponding problems, in the case when an actual solution exists. While the material of this section can, in fact, be related to the procedure developed in Sections 1.3 and 6.7, we have chosen to give a self-contained presentation here, which is independent of that theory.

6.8.1 The Problems and Theoretical Solutions

Two convolution-type equations frequently arise in applications, namely

$$f(t) - \int_{\mathbb{R}} k(t-\tau)\, f(t)\, d\tau = g(t), \quad t \in \mathbb{R}, \tag{6.8.1}$$

and

$$f(t) - \int_0^\infty k(t-\tau)\, f(\tau)\, d\tau = g(t), \quad t \in \mathbb{R}^+. \tag{6.8.2}$$

These are linear integral equations for f, given g and k. The first equation over the real line $\mathbb{R}$ is considerably simpler than the second, a Wiener–Hopf equation over $\mathbb{R}^+ = (0, \infty)$. In practical problems the given functions usually belong to some $\mathbf{L}^p$ class of functions at intervals on which they are defined, with $1 \leqq p < \infty$, and while the final algorithm which we shall derive in this section will also be applicable in those cases, the theory which we shall present will be that of the theoretically simplest $\mathbf{L}^1$ case.

Let us now consider Equation (6.8.1), the simpler of the above two. Assuming that both k and g belong to $\mathbf{L}^1(\mathbb{R})$, we take the Fourier transform of each side of (6.8.1) to formally get the equation

$$F\,(1-K)=G, \tag{6.8.3}$$

with F, K and G the Fourier transforms of f, k, and g respectively, defined as in Section 1.4. For example, the Fourier transform F of f is given by

$$F(x)=\int_{\mathbb{R}} f(t)\,e^{ixt}dt. \tag{6.8.4}$$

It is notationally convenient to let $\mathbf{L}$ denote $\mathbf{L}^1(\mathbb{R})$, and to let $\mathbf{W}$ denote the set of all Fourier transforms F of functions $f \in \mathbf{L}$ as defined in (6.8.4).

Theorem 6.8.1 *Let both k and g belong to $\mathbf{L}$, and let K and G denote their respective Fourier transforms. If $K(x) \neq 1$ on $\mathbb{R}$, then there exists a unique function $f \in \mathbf{L}$ satisfying Equation (6.8.1), and given on $\mathbb{R}$ by*

$$f(t)=\frac{1}{2\,\pi}\int_{\mathbb{R}}\frac{e^{-ixt}G(x)}{1-K(x)}\,dx. \tag{6.8.5}$$

Proof. Let us assume, for the moment, that there exists a function $f \in \mathbf{L}$ satisfying (6.8.1). Then, taking the Fourier transform of each side of (6.8.1), we get (6.8.3). This result gives us the Fourier transform of f, i.e., f is then given by (6.8.5) via the use of the inverse transform formula, (see (1.4.8)). Conversely, if $K(x) \neq 1$ on $\mathbb{R}$, then, setting $w(z) = 1/(1-z)$, it is clear that the function w is analytic everywhere except at $z = 1$. By assumption, $K(x) \neq 1$; it therefore follows from the Wiener–Levy Theorem (Theorem 1.4.10) that $w(K) - w(0) = w(K) - 1 \in \mathbf{W}$. Hence, by Theorem 1.4.9, $[w(K) - 1]\,G \in \mathbf{W}$, and therefore $[w(K) - 1]\,G + G = G/(1-K) \in \mathbf{W}$. That is, the function f defined by (6.8.5) is in $\mathbf{L}$, and moreover, this function f solves (6.8.1). ■

Let us now consider problem (6.8.2). It is convenient, in this case, to let $\mathbf{L}$ and $\mathbf{W}$ be defined as above. Furthermore we also let $\mathbf{L}_+$ (resp. $\mathbf{L}_-$) denote all those functions f belonging to $\mathbf{L}$ such that $f(t) = 0$ if $t < 0$ (resp. $f(t) = 0$ if $t > 0$), and we let $\mathbf{W}_+$ (resp. $\mathbf{W}_-$) denote the set of all Fourier transforms of functions in $\mathbf{L}_+$ (resp. $\mathbf{L}_-$).

Theorem 6.8.2 *In Equation (6.8.2), let $k \in \mathbf{L}$, and let $g \in \mathbf{L}_+$. If $K(x) \neq 1$ on $\mathbb{R}$, with K the Fourier transform of k, and if*

$$\operatorname{ind}(1-K) \equiv \int_{\mathbb{R}} d\,\log[1-K(x)] = 0 \tag{6.8.6}$$

then Equation (6.8.2) has a unique solution $f \in \mathbf{L}_+$.

Proof. The simplicity of the solution of Equation (6.8.1) suggests that we attempt the same approach for Equation (6.8.2). Since Equation (6.8.2) is only defined on $\mathbb{R}_+$, we may extend its definition to $\mathbb{R}$ by adding to the right-hand side a function $s \in \mathbf{L}_-$, which enables us to extend the definition of (6.8.2) to all of $\mathbb{R}$. That is, the new equation, defined on all of $\mathbb{R}$ takes the form

$$f(t) - \int_{\mathbb{R}} k(t-\tau)\, f(\tau)\, d\tau = g(t) + s(t). \tag{6.8.7}$$

Since $s \in \mathbf{L}_-$, Equation (6.8.7) is clearly equivalent to (6.8.2) for $t > 0$. At this point the function $s \in \mathbf{L}_-$ is unknown; however, once the function $f \in \mathbf{L}_+$ is known, then $s \in \mathbf{L}_-$ becomes defined by taking $t < 0$ in (6.8.7). Let us assume, at the outset, that there exists a function $f \in \mathbf{L}_+$ which solves Equation (6.8.7). Then, taking the Fourier transform of each side of (6.8.7), we arrive at the equation

$$F_+(1-K) = G_+ + S_-, \tag{6.8.8}$$

with $F_+ \in \mathbf{W}_+$, $K \in \mathbf{W}$, $G_+ \in \mathbf{W}_+$, and $S_- \in \mathbf{W}_-$. Consider now, the function $w(z) = -\log[1-z]$, which has its domain of analyticity a Riemann surface S. An arc Γ starting and ending at $z = 0$ will remain on the same sheet S, provided Γ does not pass through the point $z = 1$, and provided $\operatorname{ind}(e^w) = \int_\Gamma d\log(1-z) = 0$. Identifying Γ to be the curve $\{z \in \mathbb{C} : z = K(x), x \in \mathbb{R}\}$, it follows according to the Wiener–Levy Theorem that since $K(x) \neq 1$ for all $x \in \mathbb{R}$, and since (6.8.6) is satisfied, that

$$\Phi \equiv -\log[1-K] \in \mathbf{W}. \tag{6.8.9}$$

From the definition of the classes of functions $\mathbf{W}$, $\mathbf{W}_+$ and $\mathbf{W}_-$, there exist unique functions $\Phi_+ \in \mathbf{W}_+$, and $\Phi_- \in \mathbf{W}_-$, such that

$$\Phi_+ + \Phi_- = \Phi. \tag{6.8.10}$$

Hence, according to the Wiener–Levy Theorem, since the exponential function is analytic in $\mathbb{C}$, it follows that

$$\begin{aligned} K_+ &\equiv e^{-\Phi_+} - 1 \in \mathbf{W}_+, \\ K_- &\equiv e^{\Phi_-} - 1 \in \mathbf{W}_-. \end{aligned} \tag{6.8.11}$$

Substituting these results into (6.8.8), we get

$$F_+\, e^{-\Phi_+} = e^{\Phi_-}[G_+ + S_-]. \tag{6.8.12}$$

Now, letting $\mathcal{P}$ be defined as in (1.4.18), we use the unique decomposition of a function in $\mathbf{W}$ of Theorem 1.4.6, i.e.,

$$\mathcal{P}\{e^{\Phi_-} G_+\} + (I - \mathcal{P})\{e^{\Phi_-} G_+\} = e^{\Phi_-} G_+ \tag{6.8.13}$$

to get the expression

$$F_+ e^{-\Phi_+} - \mathcal{P}\{e^{\Phi_-} G_+\} = (I - \mathcal{P})\{e^{\Phi_-} G_+\} + e^{\Phi_-} S_-. \tag{6.8.14}$$

At this point we may observe that the left-hand side of Equation (6.8.14) is in $\mathbf{W}_+$, while the right-hand side is in $\mathbf{W}_-$. By the unique decomposition of a function in $\mathbf{W}$ into a sum of a function in W_+ plus a function in W_-, it follows that both sides of Equation (6.8.14) must vanish identically, i.e., that

$$F_+ = e^{\Phi_+} \mathcal{P}\{e^{\Phi_-} G_+\}. \tag{6.8.15}$$

Conversely, if F_+ is defined by the right-hand side of (6.8.15), it follows readily via the Wiener–Levy Theorem, that $F_+ \in \mathbf{W}_+$. Backtracking from (6.8.15) to (6.8.14), and so on, back to (6.8.8), it follows that F_+ satisfies (6.8.8), i.e., that the corresponding inverse transform $f \in \mathbf{L}_+$ satisfies (6.8.2). ∎

Note: If $\text{ind}(1 - K) \neq 0$, then we recommend transforming the upper-half plane onto the unit disc, and obtaining a factorization as described in Section 1.3. In this case there are several possibilities which are analogous to the non-zero index case discussed in Sections 1.3 and 6.7.

Note: The expression (6.8.15) may be reduced to (see Problem 6.8.1) the form

$$F_+ = \mathcal{P}\left\{\frac{G_+}{1 - K} - K_+(1 - \mathcal{P})\, K_- G_+\right\}, \tag{6.8.16}$$

which will be seen below to be computationally more convenient.

6.8.2 Approximate Solution

The method of Wiener and Hopf described in the above constructive proof of Theorem 6.8.2 thus provides an elegant procedure to solve Equation (6.8.2). However, its success in applications is based on explicitly determining functions $K_+ \in \mathbf{W}_+$ and $K_- \in \mathbf{W}_-$, such that $(1 - K)^{-1} = (1 + K_+)(1 + K_-)$. Inasmuch as this usually cannot be done explicitly in applications, we extend the applicability in this subsection we extend the applicability range of the Wiener and Hopf method by providing a simple explicit procedure based on Theorem 3.4.1, to determine accurate approximations of $K_\pm$. We are thus able to construct an explicit approximation to the solution of Equation (6.8.2), and we state and give (nontrivial) proofs on the convergence of these approximations.

We shall assume that the Fourier transforms K, G (or G_+, in the case of (6.8.2)) are given. Otherwise these transforms can be approximated via the use of appropriate integration formulas of Sections 3.3 or 4.2 of this text.

At the outset, we examine the numerical approximation of the solution of (6.8.1), as given in (6.8.5).

Theorem 6.8.3 *Let the assumptions of Theorem 6.8.1 be satisfied, and in addition, let G, K, and $K/(1-K)$ belong to $\mathbf{H}^2(\mathcal{D}_d)$, for some $d > 0$, with $\mathcal{D}_d$ defined in (1.7.9). Let f denote the solution to (6.8.1), let N denote a positive integer, let $f_N^{(h)}(t)$ and $f^{(h)}(t)$ be defined for $h > 0$, N and $t \in \mathbb{R}$ by the expressions*

$$\begin{aligned} f_N^{(h)}(t) &= g(t) + \frac{h}{2\pi} \sum_{j=-N}^{N} \frac{K(jh)\, G(jh)\, e^{-ijht}}{1 - K(jh)}, \quad \text{if } |t| < \pi/h, \\ f_N^{(h)}(t) &= g(t) \ \text{ if } |t| > \pi/h, \\ f^{(h)}(t) &= \lim_{N \to \infty} f_N^{(h)}(t). \end{aligned} \tag{6.8.17}$$

Then there exists a constant C, independent of h such that

$$\|f - f^{(h)}\|_s \leqq \frac{C}{\sinh(\pi d/h)}. \tag{6.8.18}$$

for all $s \in [2, \infty]$. If, in addition, either k or g belong to $\mathbf{H}^2(\mathcal{D}_\alpha)$, for some $\alpha > 0$, then taking $h = [\pi d/(\alpha N)]^{1/2}$, there exists a constant C_1, independent of N, such that

$$\|f - f_N^{(h)}\|_2 \leqq C_1 N^{1/2} e^{-(\pi d \alpha N)^{1/2}}, \tag{6.8.19}$$

for all $s \in [2, \infty]$.

Proof. Since G, K, and $K/(1-K)$ belong to $\mathbf{H}^2(\mathcal{D}_d)$, the conditions of Theorem 3.3.1 are satisfied, and we may therefore use the bounds (3.3.4) and (3.1.13), to get (6.8.18). If either g or k belongs to $\mathbf{H}^2(\mathcal{D}_\alpha)$, then, by Theorem 1.8.4, the corresponding Fourier transform G or K is bounded on $\mathbb{R}$ by $C_2/\cosh(\alpha x)$, with C_2 a constant. Also, since both g and k belong to $\mathbf{L}^1(\mathbb{R})$, the corresponding functions G and K are uniformly bounded on $\mathbb{R}$. The result (6.8.19) now follows by Theorem 3.3.1. ∎

Let us next turn to the approximate solution of Equation (6.8.2). This is a non-trivial problem, since we are making our approximations in the space of Fourier transforms, and since there is no readily available practical test to determine whether or not a given function does in fact belong to $\mathbf{W}$.

For example, if $\Phi \in \mathbf{W}$, and we form a sequence of piecewise constant approximations $\{\Phi_n\}$, which converges to Φ in some norm, none of the functions Φ_n belongs to $\mathbf{W}$, and furthermore there may not even exist a function w belonging to $\mathbf{L}^p(\mathbb{R})$, for any $p \in [1, \infty)$, whose Fourier transform is, e.g., $\exp\{\mathcal{P}\Phi_n\} - 1$. Since functions belonging to $\mathbf{L}^2(\mathbb{R})$ are always Fourier transforms, we shall assume that the function $\Phi = -\log(1 - K)$ in (6.8.2) belongs to both $\mathbf{L}^1(\mathbb{R})$ and $\mathbf{L}^2(\mathbb{R})$, to make possible the convergence proof which follows. This is, in fact, the case, if the kernel function k belongs to both $\mathbf{L}^1(\mathbb{R})$ and to $\mathbf{L}^2(\mathbb{R})$. However, even with this added simplicity, the convergence proof of the approximations is non-trivial.

Lemma 6.8.4 *Let $k \in \mathbf{L}^1(\mathbb{R})$, let Φ be defined as in (6.8.9), and set $K_+ = \exp\{\mathcal{P}\,\Phi\} - 1$, and $K_- = \exp\{(1 - \mathcal{P})\Phi\} - 1$, with $\mathcal{P}$ defined as in Theorem 1.4.6. Let $\Phi^{(h)}$ be a family of functions, depending on a parameter $h > 0$, such that*

$$\|\Phi - \Phi^{(h)}\|_s \to 0 \tag{6.8.20}$$

as $h \to 0$, for $s = 2$ and $s = \infty$, and set $K_+^{(h)} = \exp\{\mathcal{P}\,\Phi^{(h)}\} - 1$, and $K_- = \exp\{(1 - \mathcal{P})\Phi^{(h)}\} - 1$. Then

$$\|K_+ - K_+^{(h)}\|_q \to 0, \quad \textit{and} \quad \|K_- - K_-^{(h)}\|_q \to 0 \tag{6.8.21}$$

as $h \to 0$, for all $q \in [2, \infty)$, and in particular, we have $K_\pm^{(h)} \in \mathbf{L}^2(\mathbb{R}) \cap \mathbf{L}^q(\mathbb{R})$, for all $q \in [2, \infty)$.

Proof. Let us set $\eta = \Phi - \Phi^{(h)}$, and $\delta = K_+ - K_+^{(h)}$. We then obtain the identity

$$\delta = \left[e^{(1/2)(\Phi + \mathcal{S}\Phi)} - 1\right]\left[1 - e^{-(1/2)(\eta + \mathcal{S}\eta)}\right] + 1 - e^{-(1/2)(\eta + \mathcal{S}\eta)}, \tag{6.8.22}$$

with the operator $\mathcal{S} = \mathcal{P} - (I - \mathcal{P})$ denoting the Hilbert transform over $\mathbb{R}$. Letting $q \in [2, \infty)$, taking the q-norm of each side of (6.8.22), and using Hölder's inequality on the first term on the right-hand side of (6.8.22), we get

$$\begin{aligned} \|\delta\|_q &= \left\|e^{(1/2)(\Phi + \mathcal{S}\Phi)} - 1\right\|_{\alpha q} \left\|1 - e^{-(1/2)(\eta + \mathcal{S}\eta)}\right\|_{\beta q} \\ &+ \left\|1 - e^{-(1/2)(\eta + \mathcal{S}\eta)}\right\|_q, \end{aligned} \tag{6.8.23}$$

with $1 < \alpha < \infty$, and $1/\alpha + 1/\beta = 1$. The first term, $e^{(1/2)(\Phi + \mathcal{S}\Phi)} - 1 = e^{\mathcal{P}\Phi} - 1$ is clearly in $\mathbf{W}_+$, and hence is uniformly bounded on $\mathbb{R}$. We use the inequality $|e^x - 1| \leqq |x|\, e^{|x|}$, to find that for any $s \geqq 2$,

$$\left\| 1 - e^{-(1/2)(\eta + \mathcal{S}\eta)} \right\|_s \leqq \| e^{|(1/2)\eta|} \|_\infty \left(\| \frac{1}{2}\eta \|_s + \| \frac{1}{2} \mathcal{S}\eta \, e^{|(1/2)\mathcal{S}\eta|} \|_s \right). \tag{6.8.24}$$

Since η converges to zero in $\mathbf{L}^2(\mathbb{R}) \cap \mathbf{L}^\infty(\mathbb{R})$ as $h \to 0$, the term $e^{|(1/2)\eta|}$ is clearly uniformly bounded. Let us now recall Pichorides' Theorem (Theorem 1.2.1), which in our case states that if $2 \leqq s < \infty$, and if $\eta \in \mathbf{L}^s(\mathbb{R})$, then

$$\| \mathcal{S}\eta \|_s \leqq \cot(\frac{\pi}{2s}) \| \eta \|_s \leqq \frac{2s}{\pi} \| \eta \|_s. \tag{6.8.25}$$

Applying this result to the term $\| \frac{1}{2} \mathcal{S}\eta \, e^{|(1/2)\mathcal{S}\eta|} \|_s$, and by proceeding as in Problem 6.8.2 we find that if $\|\eta\|_\infty$ is small enough so that $s\, e \, \|\eta\|_\infty < \pi$, then

$$\| \frac{1}{2} \mathcal{S}\eta \, e^{|(1/2)\mathcal{S}\eta|} \|_s) \leqq [\pi \, \Gamma(s+1)]^{1/s} \, e \, (\pi - s \, e \, \|\eta\|_\infty)^{-1-1/s} \|\eta\|_s. \tag{6.8.26}$$

Using this bound in (6.8.24), we thus readily obtain a bound on $\|\delta\|_q = \|K_+ - K_+^{(h)}\|_q$, and similarly one for $\|K_- - K_-^{(h)}\|_q$, for any $q \in [2, \infty)$. These bounds take the form

$$\left\{ \begin{array}{c} \|K_+ - K_+^{(h)}\|_q \\ \|K_- - K_-^{(h)}\|_q \end{array} \right\} \leqq C(q) \, \|\Phi - \Phi^{(h)}\|_q \tag{6.8.27}$$

with $C(q)$ a constant depending only on q. ■

Let us now set

$$F_+^{(h)} = \mathcal{P} \left\{ \frac{G_+}{1 - K} - K_+^{(h)} (1 - \mathcal{P}) \, K_-^{(h)} G_+ \right\}. \tag{6.8.28}$$

Proceeding as in the proof of the above lemma, it is readily shown that if q is an arbitrary number such that $2 \leqq q < \infty$, if $\alpha > 1$, and if $\beta = \alpha/(\alpha - 1)$, then letting $F_+^{(h)}$ be defined by (6.8.28), we have

$$\|F_+^{(h)}\|_q \leqq \frac{1}{2}(1 + \frac{\beta q}{\pi}) \, \|K_+\|_{\alpha q} \|K_-\|_{\beta q} \|G_+\|_\infty. \tag{6.8.29}$$

In view of the previous lemma, and of the fact that $G_+ \in \mathbf{W}$, it now follows that $F_+^{(h)} \in \mathbf{L}^q(\mathbb{R})$, for all $q \in [2, \infty)$. In particular, $F_+^{(h)}$ is the Fourier transform of a function $f^{(h)} \in \mathbf{L}^2(\mathbb{R})$. Moreover, using (6.8.28), we find that $f^{(h)}(t)$ is given for $t > 0$ by the expression

$$\begin{aligned} f^{(h)}(t) &= g(t) + \frac{1}{2\pi} \int_{\mathbb{R}} e^{-ixt} \frac{G_+(x)\, K(x)}{1 - K(x)}\, dx \\ &- \frac{1}{4\pi^2} \int_{-\infty}^{0} \left(\int_{\mathbb{R}} e^{-iy(t-u)} K_+^{(h)}(y)\, dy \right) \cdot \\ &\cdot \left(\int_{\mathbb{R}} e^{-ixu} K_-^{(h)}(x)\, G_+(x)\, dx \right) du. \end{aligned} \tag{6.8.30}$$

The expression (6.8.30) is, of course, incorrect for $t < 0$, since we want $f^{(h)}(t) = 0$ if $t < 0$.

Theorem 6.8.5 *Let the conditions of Theorem 6.8.2 and Lemma 6.8.4 be satisfied, let $f \in \mathbf{W}_+ \cap \mathbf{L}^2(\mathbb{R})$ denote the unique solution to Equation (6.8.2), and let $f^{(h)}$ be defined as in (6.8.30). Then for any constant $\alpha > 1$, we have*

$$\|f - f^{(h)}\|_2 \leqq C\, \|\Phi - \Phi^{(h)}\|_2^{1/\alpha} \|\Phi - \Phi^{(h)}\|_\infty^{1-1/\alpha}, \tag{6.8.31}$$

with C a constant depending only on α.

Proof. Combining (6.8.16) and (6.8.28), we get

$$\|F_+ - F_+^{(h)}\|_2 = \|\mathcal{P}\{K_+(I - \mathcal{P})(K_- G_+) - K_+^{(h)}(I - \mathcal{P})(K_-^{(h)} G_+)\}\|_2. \tag{6.8.32}$$

From Parseval's Theorem and (1.4.17), we have for any $H \in \mathbf{L}^2(\mathbb{R})$, that both $\|\mathcal{P}H\|_2$ and $\|(1 - \mathcal{P})H\|_2$ are bounded by $\|H\|_2$, and so

$$\|F_+ - F_+^{(h)}\|_2 \leqq \left\{ \begin{array}{l} \|K_+(I - \mathcal{P})(K_- G_+) - K_+^{(h)}(I - \mathcal{P})(K_-^{(h)} G_+)\|_2 \\ \\ \|(K_+ - K_+^{(h)})(I - \mathcal{P})(K_-^{(h)} G_+)\|_2 + \|K_+(I - \mathcal{P})((K_- - K_-^{(h)}) G_+)\|_2. \end{array} \right. \tag{6.8.33}$$

At this point we recall that $G_+ \in \mathbf{W}$, so that $G_+ \in \mathbf{L}^\infty(\mathbb{R})$. Also, by (6.8.25), we have for any $q \geqq 2$ and $H \in \mathbf{H}^q(\mathbb{R})$, that $\|(I - \mathcal{P})H\|_q = \|(1/2)(I - \mathcal{S})H\|_q \leqq (1/2)(1 + 2q/\pi)\|H\|_q$. Using these results, substituting (6.8.29) for (6.8.23), and then using Parseval's Theorem to deduce that $\|f - f^{(h)}\|_2 = \|F_+ - F_+^{(h)}\|_2/(2\pi)^{1/2}$, we get (6.8.31). ∎

We next describe an explicit numerical scheme to approximate $\mathcal{P}\Phi$ and $(I - \mathcal{P})\Phi$, and also the solution of (6.8.2). Let K and G satisfy the conditions of Theorem 6.8.3, and let us make the approximation

$$\Phi \approx \Phi^{(h)} \equiv \sum_{k=-\infty}^{\infty} \Phi(kh)\, S(k,h). \tag{6.8.34}$$

Then, setting

$$\begin{aligned} K_+^{(h)} &= \exp\{\mathcal{P}\Phi^{(h)}\} - 1 \\ K_-^{(h)} &= \exp\{(I-\mathcal{P})\Phi^{(h)}\} - 1 \end{aligned} \tag{6.8.35}$$

with, from Section 3.4,

$$\begin{aligned} \mathcal{P}\Phi^{(h)} &= \sum_{j=-\infty}^{\infty} \Phi(jh)\, T_+(j,h) \\ (I-\mathcal{P})\Phi^{(h)} &= \sum_{j=-\infty}^{\infty} \Phi(jh)\, T_-(j,h) \end{aligned} \tag{6.8.36}$$

and with

$$\begin{aligned} T_+(j,h)(x) &= \frac{e^{i\pi(x-jh)/h} - 1}{2\pi i(x-jh)/h} \\ T_-(j,h)(x) &= \frac{1 - e^{-i\pi(x-jh)/h}}{2\pi i(x-jh)/h}. \end{aligned} \tag{6.8.37}$$

We also define U and V by the expressions

$$\begin{aligned} U &= \sum_{j=-\infty}^{\infty} K_+^{(h)}(jh)\, T_+(j,h) \\ V &= \sum_{j=-\infty}^{\infty} K_-^{(h)}(jh)\, T_-(j,h). \end{aligned} \tag{6.8.38}$$

In view of these definitions, in the case $t > 0$, our approximate solution $\tilde{f}^{(h)}(t)$ now takes the form

$$\begin{aligned} \tilde{f}^{(h)}(t) &= g(t) + \frac{1}{2\pi}\int_{\mathbf{R}} \frac{G_+(x)\, K(x)\, e^{-ixt}}{1-K(x)}\, dx \\ &- \frac{1}{4\pi^2}\int_{-\infty}^{0} \left(\int_{\mathbf{R}} e^{-i\xi(t-\eta)} U(\xi)\, d\xi\right) \left(\int_{\mathbf{R}} e^{-ix\eta} V(x)\, G_+(x)\, dx\right) d\eta, \end{aligned} \tag{6.8.39}$$

while for $t < 0$ we set $\tilde{f}^{(h)}(t) = 0$. We remark here, that Equation (6.8.39) is obtained from (6.8.16), upon replacing $G_+/(1-K)$ by $G_+ + G_+K/(1-K)$; in this case, the inverse Fourier transform of G_+ can be explicitly expressed, and the numerical evaluation of the inverse Fourier transform of $G_+K/(1-K)$ can now be carried out more efficiently via the trapezoidal rule, as described in Section 3.3. That is, we shall use the approximation

$$\begin{aligned} &\frac{1}{2\pi}\int_{\mathbb{R}} \frac{G_+(x)\,K(x)\,e^{-ixt}}{1-K(x)}\,dx \\ &\approx \begin{cases} \dfrac{h}{2\pi}\displaystyle\sum_{j=-\infty}^{\infty} \frac{G_+(jh)\,K(jh)}{1-K(jh)} e^{-ijht} & \text{if} \quad t \in (0,\pi/h) \\ 0 & \text{if} \quad t \notin (0,\pi/h). \end{cases} \end{aligned} \tag{6.8.40}$$

Similarly, U and V defined by (6.8.38) are the Fourier transforms over $\mathbb{R}$ of the functions

$$\begin{aligned} u(t) &= \begin{cases} \dfrac{h}{2\pi}\displaystyle\sum_{j=-\infty}^{\infty} K_+(jh)\,e^{-ijht} & \text{if} \quad t \in (0,\pi/h) \\ 0 & \text{if} \quad t \notin (0,\pi/h) \end{cases} \\ v(t) &= \begin{cases} \dfrac{h}{2\pi}\displaystyle\sum_{j=-\infty}^{\infty} K_-(jh)\,e^{-ijht} & \text{if} \quad t \in (-\pi/h,0) \\ 0 & \text{if} \quad t \notin (-\pi/h,0). \end{cases} \end{aligned} \tag{6.8.41}$$

Using (6.8.40) and (6.8.41) in (6.8.39), we thus arrive at our sought-after, explicit expression to approximate the solution f of Equation (6.8.2), namely,

$$\begin{aligned} \tilde{f}^{(h)}(t) &= g(t) + \frac{h}{2\pi}\sum_{j=-\infty}^{\infty} \frac{G_+(jh)\,K(jh)}{1-K(jh)} e^{-ijht} \\ &\quad - \frac{h^2}{4\pi^2}\left(\frac{\pi}{h} - t\right) \sum_{j=-\infty}^{\infty} K_+^{(h)}(jh)\,K_-^{(h)}(jh)\,G_+(jh)\,e^{-ijht} \\ &\quad - \frac{ih}{4\pi^2}\sum_{j=-\infty}^{\infty} K_+^{(h)}(jh)\,e^{-ijht} \sum_{\ell \neq j} \frac{[1-(-1)^{\ell-j}e^{-i(\ell-j)ht}]}{\ell - j} \cdot \\ &\quad \cdot K_-^{(h)}(\ell h)\,G_+(\ell h) \quad \text{if} \quad t \in (0,\pi/h) \\ \tilde{f}^{(h)}(t) &= g(t) \quad \text{if} \quad t > \pi/h \\ \tilde{f}^{(h)}(t) &= 0 \quad \text{if} \quad t < 0. \end{aligned} \tag{6.8.42}$$

Lemma 6.8.6 *Let let* $\kappa \in \mathbf{L}^1(\mathbb{R})$, *with* $\kappa(t) = k(t)\, e^{d|t|}$. *Let* $\operatorname{ind}(1-K) = \int_{\mathbb{R}} d\log[1-K(x)] = 0$, *and let* $1 - K(x) \neq 0$ *for all* x *on an arbitrary compact subset of* $\mathcal{D}_d$. *If* K_+ *and* K_- *are defined as in (6.8.11), then all three functions,* $\log(1-K), K_+$, *and* K_- *belong to* $\mathbf{H}^2(\mathcal{D}_d) \cap \mathbf{H}^\infty(\mathcal{D}_d)$.

Proof. Since $\kappa \in \mathbf{L}^1(\mathbb{R})$, it follows from Theorem 1.8.4, that the Fourier transform K of k is holomorphic and uniformly bounded in $\mathcal{D}_d$. Furthermore, it follows from the Wiener–Levy Theorem, that $\Psi_y \in \mathbf{W}$, with $\Psi_y(x) \equiv -\log[1 - K(x+iy)]$. In particular, it therefore follows that $\Phi = -\log(1-K)$ is uniformly bounded and analytic in the region $\{z \in \mathcal{D}_d : |\Re z| < M\}$, for all $M > 0$. Therefore, it follows that $\int_{-M}^{M} |\Psi_y(x)|^2\, dx < \infty$, for all $y \in [-d, d]$. By the Riemann–Lebesgue lemma, we may take M sufficiently large, so that $|K(z)| < 1/2$ in the region $\{z \in \mathcal{D}_d : |\Re z| > M\}$. We then have

$$
\begin{aligned}
&\int_{|x|>M} |\Psi_y(x)|^2 dx \leqq \\
&\int_{|x|>M} |K(x+iy)|^2 [1 + |K(x+iy)|/2 + |K(x+iy)|^2/3 + \cdots]^2 dx \\
&\quad \leqq (\log 2)^2 \int_{|x|>M} |K(x+iy)|^2 dx < \infty.
\end{aligned}
\tag{6.8.43}
$$

Hence $\Psi_y \in \mathbf{W} \cap \mathbf{L}^2(\mathbb{R}$ for all $y \in [-d, d]$, and therefore, $\log(1-K) \in \mathbf{H}^2(\mathcal{D}_d) \cap \mathbf{L}^\infty(\mathcal{D}_d)$. By Theorem 1.8.4 and the Wiener–Levy Theorem, there exists a function $\omega \in \mathbf{L}^2(\mathbb{R})$ such that $\int_{\mathbb{R}} |\omega(t)\, e^{d|t|} dt < \infty$, and such that the Fourier transform of ω is $-\log(1-K)$. Since, moreover, $\mathcal{P}\Psi_y \in \mathbf{W}$ for all $y \in [-d, d]$, it follows that $\mathcal{P}\Psi_y \in \mathbf{H}^2(\mathcal{D}_d) \cap \mathbf{H}^\infty(\mathcal{D}_d)$ for all $y \in [-d.d]$. Consequently, proceeding as in (6.8.43), it follows that both K_+ and K_- belong to $\mathbf{H}^2(\mathcal{D}_d) \cap \mathbf{H}^\infty(\mathcal{D}_d)$. ∎

Corollary 6.8.7 *Let the conditions of Lemma 6.8.6 be satisfied, let* f *denote the solution of (6.8.2), and let* $f^{(h)}$ *be defined as in (6.8.30). Then there exists a constant* C *independent of* h, *such that*

$$\|f - f^{(h)}\|_2 \leqq C\, e^{-\pi d/h}. \tag{6.8.44}$$

Proof. This is a consequence of Lemma 6.8.6, Theorem 6.8.5, and Theorem 3.1.3. ∎

Theorem 6.8.8 *Let the conditions of Lemma 6.8.6 be satisfied, let* f *denote the solution of (6.8.2), and let* $\tilde{f}^{(h)}$ *be defined as in (6.8.42). Then given any* $\varepsilon \in (0, d)$, *there exists a constant* C_ε *depending only on* ε, *such that*

$$\|f - \tilde{f}^{(h)}\|_2 \leqq C_\varepsilon e^{-\pi(d-\varepsilon)/h}. \tag{6.8.45}$$

Proof. Since by Lemma 6.8.6 each of the functions $\log(1-K), K_+$, and K_- belong to $\mathbf{H}^2(\mathcal{D}_d) \cap \mathbf{H}^\infty(\mathcal{D}_d)$, it follows by Theorem 3.4.1 that if the functions $K_\pm^{(h)}$ are defined as in (6.8.35), then both of these functions belong to $\mathbf{H}^2(\mathcal{D}_{d-\varepsilon}) \cap \mathbf{H}^\infty(\mathcal{D}_{d-\varepsilon})$, independently of h. Hence, if U and V are defined as in (6.8.38), it follows from Theorem 3.4.1 that there exists a constant C'_ε such that

$$\begin{aligned} \|K_+^{(h)} - U\|_p &\leqq C'_\varepsilon \exp\{-\pi(d-\varepsilon)/h\} \\ \|K_-^{(h)} - V\|_p &\leqq C'_\varepsilon \exp\{-\pi(d-\varepsilon)/h\} \end{aligned} \tag{6.8.46}$$

for $p = 2$ and also for $p = \infty$. The remainder of the proof now follows proceeding as for (6.8.33). We skip the details, leaving these to Problem 6.8.4. ■

In practical applications it will almost always occur that $f(0^+) \neq 0$, f the solution to Problem (6.8.2). Since $f(t) = 0$ for $t < 0$, the Fourier transform $F_+(x)$ of f will decrease to zero slowly, as $x \to \pm\infty$. As a consequence, the coefficients in the Fourier series representation of $\tilde{f}^{(h)}$ in (6.8.42) will usually decrease to zero relatively slowly, and a large number of terms will be required to get an accurate approximation of f. For example, we expect $K_\pm^{(h)}(jh)$ to decrease roughly like $|j|^{-1}$ as $j \to \pm\infty$. Thus, in order to alleviate this difficulty, we point the interested reader to Problem 6.8.5 for further research work on the Wiener–Hopf method.

6.8.3 Explicit Examples

Let us solve these two problems

$$f(t) - \lambda \int_{\mathbb{R}} e^{-a|t-\tau|} f(\tau)\, d\tau = e^{-b|t|}, \quad t \in \mathbb{R}, \tag{6.8.47}$$

and also

$$f(t) - \lambda \int_0^\infty e^{-a|t-\tau|} f(\tau)\, d\tau = e^{-bt}, \quad t > 0, \tag{6.8.48}$$

both exactly, and via the approximate methods of this section. In these equations we assume that $a > 0$, $b > 0$, and $\lambda < a/2$.

Let us first consider (6.8.47). Setting $c = |a^2 - 2\lambda a|^{1/2}$, and then taking Fourier transforms of each side of (6.8.47), we get

$$F(x) = \frac{2\,b\,(x^2+a^2)}{(x^2+b^2)(x^2+c^2)} \tag{6.8.49}$$

$$= \frac{2b}{x^2+b^2} + \frac{2b\,(a^2-c^2)}{(x^2+b^2)(x^2+c^2)} \tag{6.8.50}$$

$$= \frac{2b}{x^2+b^2} - \frac{a^2-c^2}{c(b^2-c^2)}\left\{c\frac{2b}{x^2+b^2} - b\frac{2c}{x^2+c^2}\right\}. \tag{6.8.51}$$

We now easily find the inverse Fourier transform of the expression (6.8.51), to find that for $t \in \mathbb{R}$,

$$f(t) = e^{-b|t|} - \frac{a^2-c^2}{c(b^2-c^2)}\{c\,e^{-b|t|} - b\,e^{-c|t|}\}. \tag{6.8.52}$$

Next, applying the approximate inversion formula (6.8.17) to (6.8.48), we find that for $t \in \mathbb{R}$,

$$f^{(h)}(t) = \tag{6.8.53}$$

$$\begin{cases} e^{-b|t|} - \dfrac{2b(a^2-c^2)\,h}{2\pi} \displaystyle\sum_{j=-\infty}^{\infty} \frac{e^{-ijht}}{(j^2h^2+b^2)(j^2h^2+c^2)}, & \text{if} \quad |t| < \pi/h, \\ e^{-b|t|} & \text{if} \quad |t| > \pi/h. \end{cases}$$

Using the result of Problem 1.6.5 (b), we can now deduce the error bound (Problem 6.8.6)

$$|f(t) - f^{(h)}(t)| \leqq C\,e^{-\pi d/h}, \tag{6.8.54}$$

with C a constant independent of h. The Fourier series in (6.8.53) converges slowly, since neither k nor g belong to $\mathbf{H}^2(\mathcal{D}_\alpha)$, for any $\alpha > 0$, and hence we cannot apply the second part of Theorem 6.8.3, to obtain a sharp bound on the error. However, we can choose N *a priori*, so that the finite sum in (6.8.17) is accurate to within an error of order $\exp(-\pi d/h)$ (see Problem 6.8.8)). For example, taking $a = b = 1$, $\lambda = -1/2$, $h = -0.5$ and $N = 20$, the computed $f^{(h)}$ is accurate to within an error bounded by .004 on $\mathbb{R}$. Also, convergence acceleration (for example, the ε algorithm, or Chebyshev acceleration) may be applied to (6.8.53).

Let us now turn to the solution of Equation (6.8.48). Adding $s \in \mathbf{L}_-$ to the right-hand side, thus extending the equation to $\mathbb{R}$, and then taking Fourier transforms, we get

$$F_+(x)\left[\frac{x^2+c^2}{x^2+a^2}\right] = \frac{i}{x+ib} + S_-, \tag{6.8.55}$$

with $F_+ \in \mathbf{W}_+$, $S_- \in \mathbf{W}_-$, and with c defined as above. In this special case, $1 - K(x) = (x^2 + c^2)/(x^2 + a^2)$ is a rational function for which we can factor both numerator and denominator polynomials explicitly. We may note furthermore, in anticipation to applying the Wiener–Levy Theorem, that $(x+ic)/(x+ia) = 1 + i(c-a)/(x+ia)$ with $i(c-a)/(x+ia) \in \mathbf{W}_+$, and also, that $(x-ia)/(x-ic) = 1 + i(c-a)/(x-ic)$ with $i(c-a)/(x-ic) \in \mathbf{W}_-$. We thus get

$$F_+(x) \left[\frac{x+ic}{x+ia} \right] = \frac{x-ia}{x-ic} \left[\frac{i}{x+ib} + S_- \right]. \tag{6.8.56}$$

That is,

$$F_+(x) \left[\frac{x+ic}{x+ia} \right] - \frac{a+b}{b+c} \frac{i}{x+ib} = T_-, \tag{6.8.57}$$

with $T_- \in \mathbf{W}_-$. Furthermore, since the left-hand side of (6.8.57) belongs to $\mathbf{W}_+$, both sides of (6.8.57) must vanish, and therefore, we find that

$$F_+(x) = \frac{i}{x+ib} + \frac{a-c}{c^2-b^2} \left\{ (a+c) \frac{i}{x+ib} - (a+b) \frac{i}{x+ic} \right\}. \tag{6.8.58}$$

Taking the inverse Fourier transform, we now get

$$f(t) = e^{-bt} + \frac{a-c}{c^2-b^2} \left\{ (a+c) e^{-bt} - (a+b) e^{-ct} \right\}. \tag{6.8.59}$$

Let us now find the approximate solution as defined in (6.8.42). To begin with, we make the approximation

$$\begin{aligned} \Phi(x) &\equiv \log[1 - K(x)] = \log\left[\tfrac{x^2+a^2}{x^2+c^2} \right] \approx \Phi^{(h)}(x) \\ \Phi^{(h)}(x) &\equiv \sum_{j=-\infty}^{\infty} \log\left[\frac{j^2h^2 + a^2}{j^2h^2 + c^2} \right] S(j,h)(x). \end{aligned} \tag{6.8.60}$$

We shall require each of the sequences $\{K_{\pm}^{(h)}(jh)\}$, with $K_+^{(h)} = \exp\{\mathcal{P}\Phi^{(h)}\} - 1$, and $K_-^{(h)} = \exp\{(I - \mathcal{P})\Phi^{(h)}\} - 1$. Using (6.8.36) and the definitions of $T_\pm$ in (6.8.37), we get

$$\begin{aligned} \mathcal{P}\Phi^{(h)}(\ell h) &= \frac{1}{2}\Phi(\ell h) + \frac{h}{2\pi i} \sum_{j \neq \ell} \Phi(jh) \frac{[(-1)^{\ell-j} - 1]}{\ell - j} \\ (I - \mathcal{P})\Phi^{(h)}(\ell h) &= \frac{1}{2}\Phi(\ell h) - \frac{h}{2\pi i} \sum_{j \neq \ell} \Phi(jh) \frac{[(-1)^{\ell-j} - 1]}{\ell - j}. \end{aligned} \tag{6.8.61}$$

Having found these numbers, we can readily evaluate the functions U and V defined in (6.8.38), and substitute their values into (6.8.42) to get an approximation $f^{(h)}$. For example, taking $a = b = 1$, $\lambda = -1/2$, $h = 0.5$, and $N = 20$, and replacing all lower and upper limits in the in the sums in (6.8.42) by $-N$ and N respectively, the computed $f^{(h)}$ obtained is accurate to within an error of .005 on $(0, \infty)$.

•

Problems for Section 6.8

6.8.1 Show that expression (6.8.15) may be reduced to the one given in (6.8.16).

6.8.2 Let $s \geqq 2$, and let $\eta \in \mathbf{L}^q(\mathbb{R})$ for all $q \in [2, \infty]$.

(a) Set $H = \|(1/2)\mathcal{S}\eta\, e^{|(1/2)\mathcal{S}\eta|}\|_s$, and prove that

$$H^s \leqq \int_{\mathbb{R}} \sum_{n=0}^{\infty} \frac{|(1/2)\mathcal{S}\eta(t)|^{s+n}\, s^n}{n!}\, dt.$$

(b) Conclude using inequality (6.8.25), that

$$H^s \leqq \int_{\mathbb{R}} \sum_{n=0}^{\infty} \frac{|\eta(t)|^{s+n}\,(s+n)^{s+n}\, s^n}{\pi^{s+n} n!}\, dt.$$

(c) Use the result $(s+n)^{s+n} \leqq \Gamma(s+n+1)\, e^{s+n}$, which can be obtained from the relations

$$(s+n)^{s+n} e^{-(s+n)} = (s+n)^{s+n+1} \int_1^{\infty} e^{-(s+n)t}\, dt)$$

$$\leqq (s+n)^{s+n+1} \int_0^{\infty} (t\, e^{-t})^{s+n}\, dt = \Gamma(s+n+1),$$

to deduce that

$$H^s \leqq \Gamma(s+1) \int_{\mathbb{R}} \sum_{n=0}^{\infty} \frac{(s+1)_n\, s^n\, e^{s+n}\, |\eta(t)|^{s+n}}{\pi^{s+n} n!}\, dt.$$

(d) Now use the identities $\|\eta\|_{s+n}^{s+n} \leqq \|\eta\|_s^s\, \|\eta\|_\infty^n$ and

$$(1-x)^{-s-1} = \sum_{n=0}^{\infty} \frac{(s+1)_n}{n!}\, x^n, \ |x| < 1,$$

to conclude that if $se\|\eta\|_\infty < \pi$, then

$$H \leqq [\pi\, \Gamma(s+1)]^{1/s}\, e\, (\pi - s\, e\, \|\eta\|_\infty)^{-1-1/s} \|\eta\|_s.$$

6.8.3 Use the last inequality in the previous problem to obtain an explicit estimate for the constant $C(q)$ in (6.8.27).

6.8.4 Complete the details of the proof of Theorem 6.8.8.

6.8.5 * (Research Problem) (a) Determine the function $H \in \mathbf{W}_-$ such that if we add H to the solution F_+ of (6.8.16), then the resulting function decreases rapidly to zero as $x \to \pm\infty$.

(b) Is it possible to derive a constructive transformation of Problem (6.8.2) into Problem (6.8.1), such that the solution to the resulting Problem (6.8.1) is the same on $(0, \infty)$ is the same as the solution to Problem (6.8.2), and such that all functions in the resulting Problem (6.8.1) are smooth on $\mathbb{R}$?

6.8.6 Use the result of Problem 1.6.5 (b) to derive the bound (6.8.53).

6.8.7 (a) Obtain a numerical solution to the problem

$$f(t) - \frac{1}{2\pi} \int_{\mathbb{R}} e^{-(t-\tau)^2} f(\tau)\, d\tau = e^{-2t^2}, \quad t \in \mathbb{R}.$$

(b) Obtain a numerical solution to the problem

$$f(t) - \frac{1}{2\pi} \int_0^\infty e^{-(t-\tau)^2} f(\tau)\, d\tau = e^{-2t^2}, \quad t \in (0, \infty).$$

6.8.8 Use the result of Problem 3.1.6 to determine an N such that when the infinite sum $\sum_{-\infty}^{\infty}$ in (6.8.53) is replaced by $\sum_{-N}^{N}$ then the resulting computed $f^{(h)}$ will differ from f by the bound given in (6.8.53).

6.8.9 * (a) Discuss the solution of Equations (6.8.1) and (6.8.2) using the method of Section 4.6, combined with one of the approximation methods of Theorem 6.9.5 to approximate the Laplace transform.

(b) Solve the equations of Problem 6.8.7 via the method of Section 4.6.

6.9 The Laplace Transform and Its Inversion

Let f, g, and $\mathcal{L}(f)$ be related as in the equation

$$g(s) = (\mathcal{L}f)(s) \equiv \int_0^\infty e^{-st} f(t) dt. \tag{6.9.1}$$

In this section we discuss some procedures to approximate g given f, and also to approximate f, given g. The approximation of g is carried out via direct application of Sinc quadrature, using values of f on the interval $(0,\infty)$. The inverse problem, i.e., the approximation of f, given g, is ill-posed (see Problem 1.5.2), and in this section we use special regularization to circumvent the difficulties due to this ill-posedness. The *data* g in (6.9.1) may in fact be contaminated with error, i.e., there may exist an exact equation

$$g^e = \mathcal{L} f^e, \tag{6.9.2}$$

and while the functions g^e and f^e are not known, we shall assume that we know g, as well as a positive number ε, with

$$\mathcal{N}_1(g - g^e) \leqq \varepsilon, \tag{6.9.3}$$

and with in (6.9.3) and throughout the remainder of this section, the *norm* $\mathcal{N}_q(w)$ defined for $q \in \mathbb{R}$, and $w : (0,\infty) \to \mathbb{C}$ by

$$\mathcal{N}_q(w) = \left(\int_0^\infty s^q |w(s)|^2 ds \right)^{1/2}. \tag{6.9.4}$$

We denote by $\mathbf{N}_q(0,\infty)$ the space of all functions $w : (0,\infty) \to \mathbb{C}$ such that $\mathcal{N}_q(w) < \infty$.

It is easily verified that the operator $\mathcal{L}$ defined above is self-adjoint, and instead of applying a numerical procedure directly to the integral equation (6.9.1) of the first kind, one makes the inversion problem tractable via the standard method of regularization. This standard method of regularization involves replacing (6.9.1) by the regularized equation (recall Subsection 6.2.5)

$$\beta f_\beta + \mathcal{L}^* \mathcal{L} f_\beta = \mathcal{L}^* g, \tag{6.9.5}$$

obtained by applying the operator $\mathcal{L}^*$ to both sides of (6.9.1), and then adding βf to the left-hand side, $\beta > 0$ the regularization parameter.

Let us now derive an equivalent Fourier transform version of (6.9.5). To this end, let us first transform Equation (6.9.1) into a convolution equation via the transformations

$$\begin{cases} t = e^u, \quad s = e^{-v} \\ F(u) = f(e^u), \quad G(v) = e^{-v} g(e^{-v}) \\ H(u) = e^{-u} \exp(-e^{-u}). \end{cases} \tag{6.9.6}$$

The norm $\mathcal{N}_q$ defined in (6.9.4) and the corresponding space of functions $\mathbf{N}_q(0,\infty)$ are motivated by the transformation (6.9.6), which enables us to invert equation (6.9.1) in the Fourier transform domain. Thus, by means of equation (6.9.1), the usual $\mathbf{L}^2$ norm in the Fourier transform domain becomes the $\mathcal{N}_1$ norm in the Laplace transform space, $\mathbf{N}_1(0,\infty)$, containing the function g of (6.9.1), and it becomes the $\mathcal{N}_{-1}$ norm in the space $\mathbf{N}_{-1}(0,\infty)$ containing the function "f" of (6.9.1).

After substitution of (6.9.6) into (6.9.1) we obtain the convolution equation

$$G(v) = \int_{\mathbb{R}} H(v-u)\, F(u)\, du. \tag{6.9.7}$$

Taking the Fourier transform of both sides, i.e., using the formula

$$\hat{G}(x) = \int_{\mathbb{R}} G(v) e^{ixv}\, dv, \tag{6.9.8}$$

we get the simple "solution"

$$\hat{F}(x) = \frac{\hat{G}(x)}{\hat{H}(x)}. \tag{6.9.9}$$

Unfortunately, this procedure has a pitfall if we want to use this expression to approximate $\hat{F}$. The difficulty is revealed by examining $\hat{H}$, the Fourier transform of H, i.e.,

$$\begin{aligned} \hat{H}(x) &= \int_{\mathbb{R}} H(u) e^{ixu}\, du \\ &= \int_{\mathbb{R}} e^{-u} \exp(-e^{-u}) e^{ixu}\, du \\ &= \Gamma(1-ix), \end{aligned} \tag{6.9.10}$$

with $\Gamma(\cdot)$ Euler's gamma function. It follows from (6.9.10), that $\Gamma(1+ix) = \overline{\Gamma(1-ix)}$ for $x \in \mathbb{R}$, and moreover, for $x \in \mathbb{R}$,

$$|\hat{H}(x)|^2 = |\Gamma(1+ix)|^2 = \Gamma(1+ix)\Gamma(1-ix) = \frac{\pi x}{\sinh(\pi x)}. \tag{6.9.11}$$

Thus $|\Gamma(1-ix)| = \{\pi x/\sinh(\pi x)\}^{1/2}$, and $\hat{H}(x) = \Gamma(1-ix)$ approaches zero very rapidly as $x \to \pm\infty$. Since $\hat{H}$ appears in the denominator of (6.9.9), it is apparent that small errors in $\hat{G}(x)$ or in $\hat{H}(x)$ for large x become very large errors in $\hat{F}(x)$, i.e., whereas (6.9.9) is a theoretically exact expression, it is ill-posed as a formula to compute $\hat{F}$.

We thus replace (6.9.9) by a formula more stable using *regularization*. At the outset, we multiply both the numerator and denominator of (6.9.9) by $\overline{\hat{H}(x)}$. We then add a small positive number β to the denominator. Thus, instead of solving (6.9.9) for $\hat{F}$, we solve the equation

$$\hat{F}_\beta(x) = \frac{\overline{\hat{H}(x)}\hat{G}(x)}{\beta + |\hat{H}(x)|^2} \tag{6.9.12}$$

for $\hat{F}_\beta$. We mention here (Problem 6.9.4) that we could also have obtained (6.9.12) directly, substituting (6.9.6) in (6.9.5).

Let us state and prove a lemma giving a bound on the operator $(\beta + \mathcal{L}^*\mathcal{L})^{-1}\mathcal{L}^*$.

Lemma 6.9.1 *Let $\beta > 0$. The operator*

$$\mathcal{T}_\beta \equiv (\beta + \mathcal{L}^*\mathcal{L})^{-1}\mathcal{L}^*, \tag{6.9.13}$$

with $\mathcal{T}_\beta : \mathbf{N}_1(0,\infty) \to \mathbf{N}_{-1}(0,\infty)$, and $\mathbf{N}_q(0,\infty)$ defined in the paragraph following Equation (6.9.4), is a bounded linear operator, with norm

$$\|\mathcal{T}_\beta\| \leqq \frac{1}{2\,\beta^{1/2}}. \tag{6.9.14}$$

Proof. Let us make the substitutions (6.9.6) in Equation (6.9.5), where in (6.9.5) the function g is now assumed to belong to $\mathbf{N}_1(0,\infty)$. Then, taking the Fourier transform of both sides of the resulting equation, and solving for $\hat{F}_\beta$, we get Equation (6.9.12), $\hat{G} \in \mathbf{L}^2(\mathbb{R})$. Now since for $x \in \mathbb{R}$, $|\hat{H}(x)|^2 = \pi x/\sinh(\pi x)$, it follows that $|\hat{H}(x)|$ decreases monotonically from 1 to zero as $|x|$ increases from 0 to ∞. Moreover, the maximum value of the function $t/(\beta + t^2)$ for $0 \leqq t \leqq 1$ is $1/(2\beta)^{1/2}$ if $0 < \beta \leqq 1$ and $1/(\beta+1)$ if $\beta > 1$, and furthermore, $1/(2\beta)^{1/2} \geqq 1/(\beta+1)$ for all $\beta > 0$. Hence it follows that

$$\sup_{x\in\mathbb{R}} \frac{|\hat{H}(x)|}{\beta + |\hat{H}(x)|^2} \leqq \frac{1}{2\,\beta^{1/2}}. \tag{6.9.15}$$

However, taking $\mathbf{L}^2$-norms of each side of (6.9.12), and using Parseval's Theorem and the transformation (6.9.6), it follows that the bound on the right-hand side of (6.9.15) also bounds $\|\mathcal{T}_\beta\|$. ∎

Lemma 6.9.2 *Let f denote the solution of (6.9.1), and let β and f_β be defined as in (6.9.5). Let F be defined as in (6.9.6), and let $\hat{F}$ denote the Fourier transform of F. Then*

$$\mathcal{N}_{-1}(f - f_\beta) = (2\pi)^{-1/2} \left\| \frac{\beta\,\hat{F}}{\beta + |\hat{H}|^2} \right\|_2 . \tag{6.9.16}$$

Proof. The result (6.9.16) follows directly by subtracting (6.9.12) from (6.9.9), then using Parseval's Theorem and the transformations (6.9.6). ∎

Note: We remark at this point, that given $g \in \mathbf{N}_1(0,\infty)$, it is easily shown that the right-hand side of (6.9.16) approaches zero as $\beta \to 0$. Unfortunately, this convergence to 0 can be very slow, as a function of β, as

$\beta \to 0$, for many functions f that possess a Laplace transform. We shall see below that this rate of convergence depends on the smoothness of f on $(0, \infty)$.

Lemma 6.9.3 *Let f^e and g^e be related as in (6.9.2), let (6.9.3) be satisfied for some $\varepsilon \geqq 0$, let $\hat{F}^e$ denote the Fourier transforms of f^e, and let f_β be defined as in (6.9.5). Then*

$$\mathcal{N}_{-1}(f^e - f_\beta) \leqq \frac{\varepsilon}{2\beta^{1/2}} + (2\pi)^{-1/2} \left\| \frac{\beta\,\hat{F}^e}{\beta + |\hat{H}|^2} \right\|_2 . \tag{6.9.17}$$

Proof. The proof is similar to the proof of Lemma 6.9.2, and we omit it.

■

We shall present only one method to approximate the solution f_β defined by Equation (6.9.5). Our method involves application of the trapezoidal rule and the discrete Fourier transform (DFT) to (6.9.12), thus deriving a $2N + 1$-term Sinc series approximation $f_{\beta,N}$ of f_β. One may of course also solve for $f_{\beta,N}$ applying, for example, a Sinc–Galerkin scheme to (6.9.5), thus reducing (6.9.5) to a system of algebraic equations via Sinc quadrature. This alternate direct method is described in the problems (Problem 6.9.11). It requires more computer time than the Fourier transform–DFT method. Each of the given inversion methods only use values of $g(s)$ for $s \in (0, \infty)$.

We thus seek the following estimates:

$$\begin{aligned} \mathcal{N}_{-1}(f - f_\beta) &\leqq \; ? \\ \mathcal{N}_{-1}(f_\beta - f_{\beta,N}) &\leqq \; ? \\ \\ \mathcal{N}_{-1}(f^e - f_\beta) &\leqq \; ? \\ \mathcal{N}_{-1}(f^e - f_{\beta,N}) &\leqq \; ? \end{aligned} \tag{6.9.18}$$

with $\mathcal{N}_q$ is defined as in (6.9.4). These bounds then enable us to determine an appropriate value for the regularization parameter β.

The space of functions that have Laplace transforms is too big for purposes of obtaining accurate approximations for all functions in the space, and in order to guarantee some accuracy via our proposed Sinc approximation methods of the form (4.2.52) or (4.2.58) , we first restrict this space somewhat, by demanding that f have some analyticity, as described below, as well as some restricted growth behavior at 0 and at ∞. For the case of the approximation of f we cannot achieve an $\mathcal{O}(\exp\{-cN^{1/2})$ accuracy via $(2N + 1)$-point Sinc approximation if $f(t)$ does not have the requisite asymptotic behavior as $t \to 0$, and also as $t \to \infty$, as determined by (4.2.53). As a remedy, we propose subtracting from g a function g_2 whose inverse transform is explicitly known, such that the function $g_1 = g - g_2$ has the correct asymptotic form as required by (4.2.53). To this end, the following result is useful.

Lemma 6.9.4 *Let w be a function such that for each fixed $\lambda > 0$, $w(\lambda x)/w(x) \to 1$ as $x \to \infty$, let $\rho \in \mathbb{R}$, and let $g = \mathcal{L}f$. Then each of the relations*

$$g(s) \sim s^{-\rho} w(1/s), \quad s \to 0 \quad (\text{resp.} g(s) \sim s^{-\rho} w(s) \quad s \to \infty), \tag{6.9.19}$$

and

$$\begin{gathered} \int_0^t f(\tau)\, d\tau \sim \frac{t^\rho w(t)}{\Gamma(\rho+1)}, \quad t \to \infty \\ \left(\text{resp.} \int_0^t f(\tau)\, d\tau \sim \frac{t^\rho w(1/t)}{\Gamma(\rho+1)}, \quad t \to 0\right) \end{gathered} \tag{6.9.20}$$

implies the other.

Proof. The proof is left to Problem 6.9.1. ∎

It is practically convenient to concentrate first on achieving the correct asymptotic behavior of f at $t = 0$. If we can deduce from Lemma 6.9.4 that the appropriate asymptotic conditions governed by (4.2.53) are not satisfied, then, assuming the inverse transform of $s^{-\rho} w(s)$ to be known, w defined as in Lemma 6.9.4, we recommend to apply to g_1 the inversion procedures described in this section instead of to g, $g_1(s) = g(s) - A\, s^{-\rho} w(s)$, and A, ρ, and w selected so that $\mathcal{L}^{-1} g_1(t)$ satisfies the appropriate inequality for $t \to 0$, as required by one of (4.2.53) or (4.2.59). Once this requirement is achieved, we may find that even though g or g_1 is known explicitly, Lemma 6.9.4 may not give any information on the asymptotic behavior of $f(t)$, or of $f_1(t) = \mathcal{L}^{-1} g_1(t)$ as $t \to \infty$. To this end, if e.g. $g_1 \in \mathbf{H}^2(\Omega^+)$, Ω^+ the right-half plane, then by Section 1.5, $f_1 \in \mathbf{L}^2(0, \infty)$, and moreover, for $a > 0$, the inverse transform of $g(s + a)$ is $e^{-at} f_1(t)$, which *does* have the requisite asymptotic behavior as $t \to \infty$. These asymptotic conditions may, in fact, fall out naturally, if instead of checking whether $g \in \mathbf{H}^2(\Omega^+)$, we check whether $s\, g \in \mathbf{H}^2(\Omega^+)$, or whether $t^{-1} f \in \mathbf{H}^2(S_d)$, S_d a sector defined by (1.7.10). As we shall see, these conditions guarantee, by Theorem 4.2.2 (a) that the Sinc series do converge to f on $(0, \infty)$. If $g \in \mathbf{H}^2(\Omega^+)$, the function g may already be in the correct form, such that $t\, f$ already satisfies the appropriate condition (4.2.53) (recall the approximation of f/ϕ' via (4.2.1)). Finally, we mention that if the inversion methods of this section fail, the slower convergent inversion method of Equation (6.9.1), presented in Section 5.4 does converge for all $g \in \mathbf{H}^2(\Omega^+)$, and does not require any *a priori* asymptotic behavior on $f(t)$ as $t \to 0$ or as $t \to \infty$.

Let us first describe procedures to approximate $\mathcal{L}f$. To this end, we use the Sinc quadrature formula (4.2.54) or (4.2.60), depending on our knowledge of analyticity of f.

Theorem 6.9.5 *(a) Let* $t\, f \in \mathbf{L}_\alpha(S_d)$, *with* $0 < d \leqq \pi/2$, $\alpha > 0$, *and with* S_d *the sector* $\mathcal{D}_d^1$ *defined in (1.7.10). Let* N *be a positive integer and define* h *by*

$$h = \left(\frac{2\pi d}{\alpha N}\right)^{1/2}. \tag{6.9.21}$$

Then there exists a positive number C_1, *independent of* N, *such that for all* $s \in [0, \infty]$,

$$\begin{aligned} &\left| g(s) - h \sum_{k=-N}^{N} \exp\{kh - se^{kh}\} f(e^{kh}) \right| \\ &\qquad < C_1 e^{-(2\pi d\alpha N)^{1/2}}. \end{aligned} \tag{6.9.22}$$

(b) Let $\tanh(t)\, f \in \mathbf{L}_\alpha(B_d)$, *with* $0 < d \leqq \pi/2$, $\alpha > 0$, *with* B_d *the bullet-shaped region* $\mathcal{D}_d^3$ *defined by (1.7.12). Let* N *be a positive integer and define* h *and* z_k *by*

$$\begin{aligned} h &= \left(\frac{2\pi d}{\alpha N}\right)^{1/2} \\ z_k &= \log\{e^{kh} + (1 + e^{2kh})^{1/2}\}. \end{aligned} \tag{6.9.23}$$

Then there exists a positive number C_1, *independent of* N, *such that for all* $s \in [0, \infty]$,

$$\begin{aligned} &\left| g(s) - h \sum_{k=-N}^{N} (1 + e^{-2kh})^{-1/2} \exp\{-sz_k\} f(z_k) \right| \\ &\qquad < C_1 e^{-(2\pi d\alpha N)^{1/2}}. \end{aligned} \tag{6.9.24}$$

Note: If we select $h = \{\pi d/(\alpha N)\}^{1/2}$ instead of (6.9.21), then the resulting error on the right-hand side of (6.9.22) is replaced by $C_1 \exp\{-(\pi d\alpha N)^{1/2}\}$. This latter choice of h corresponds to our selection of h to approximate $\mathcal{L}^{-1}$. A similar remark applies to part (b) of Theorem 6.9.5.

Proof. **Part (a):** Since f is an analytic function in S_d, and since for $t\,f \in \mathbf{L}_\alpha(S_d)$, f satisfies an inequality of the form (4.2.53) in S_d (where we take $\beta = \alpha$ in (4.2.53)), and since $0 < d \leqq \pi/2$, the function f_1 defined for each fixed $s \in [0, \infty]$ by $f_1(t) = e^{-st} f(t)$ satisfies the same conditions as f. Hence the result (6.9.22) follows from (4.2.54).
Part (b): This is similar to part (a), and we omit it. ■

Considering (6.9.12), we need to find accurate approximations of the quantities $\hat{G}$ and $\overline{\hat{H}}$. To this end, we recommend the application of a Sinc quadrature procedure, i.e., the trapezoidal rule. The function $\hat{G}(x)$ may be approximated with the trapezoidal formula applied to its Fourier transform expression (6.9.8). We may also approximate $\overline{\hat{H}}(x)$ this way, applying the trapezoidal formula to the expression

$$\overline{\hat{H}}(x) = \int_{\mathbb{R}} e^u \exp(-e^u)\, e^{ixu} du. \tag{6.9.25}$$

obtained from the Fourier transform of $\hat{H}$, replacing x by $-x$. Now for $h > 0$, h to be determined later, and $x \in \mathbb{R}$, let $\hat{G}_N(x)$ and $\hat{H}_N(x)$ be defined for $-\pi/h \leqq x \leqq \pi/h$ by

$$\begin{aligned} \hat{G}_N(x) &= h\left\{\frac{1}{2}G(-Nh)e^{-iNhx} + \frac{1}{2}G(Nh)e^{iNhx} + \sum_{|k| \leqq N-1}^{N-1} G(kh)e^{ikhx}\right\} \\ &\quad \text{if } |x| \leqq \pi/h \\ \hat{G}_N(x) &= 0 \quad \text{if } |x| > \pi/h, \end{aligned} \tag{6.9.26}$$

and, with $Q(x)$ denoting $H(-x)$,

$$\begin{aligned} \overline{\hat{H}}_N(x) &= h\left\{\frac{1}{2}Q(-Nh)e^{-iNhx} + \frac{1}{2}Q(Nh)e^{iNhx} + \sum_{|k| \leqq N-1}^{N-1} Q(kh)e^{ikhx}\right\} \\ &\quad \text{if } |x| \leqq \pi/h \\ \overline{\hat{H}}_N(x) &= 0 \quad \text{if } |x| > \pi/h. \end{aligned} \tag{6.9.27}$$

Next, setting

$$\hat{F}^*_{\beta,N} = \frac{\hat{G}_N \overline{\hat{H}}_N}{\beta + |\hat{H}|^2}, \tag{6.9.28}$$

we shall construct a discrete Fourier transform approximation $\hat{F}_{\beta,N}$ of $\hat{F}^*_{\beta,N}$ of the form (recall Sections 2.2 and 3.3)

$$\hat{F}_{\beta,N}(x) = \begin{cases} h \sum_{k=-N}^{N} c_{\beta,k} e^{ikhx}, & \text{if} \quad -\pi/h < x < \pi/h \\ 0, & \text{if} \quad x \in \mathbb{R}, \ |x| > \pi/h. \end{cases} \tag{6.9.29}$$

To this end, let r and w be defined on the interval $[-\pi/h, \pi/h]$, let us define the inner product (see Problems 2.2.8 and 2.2.9) by

$$< r, w >_N = \frac{1}{2N} \sum_{m=-N}^{N} a_m r(x_m) \overline{w(x_m)} \tag{6.9.30}$$

with $x_m = m\pi/(Nh)$, $a_{\pm N} = 1/2$, and $a_j = 1$ if $|j| < N$. In terms of this notation, we have

$$h\, c_{\beta,k} = < \hat{F}^*_{\beta,N}, \psi_k >_N, \tag{6.9.31}$$

where $\psi_k(x) = e^{ikhx}$. Finally, the identity (1.10.10), taken together with (6.9.28) and (6.9.6) yields the approximation

$$f_{\beta,N}(t) = \sum_{k=-N}^{N} c_{\beta,k} S(k,h)(\log t). \tag{6.9.32}$$

Let us now obtain a bound on the approximation error.

Lemma 6.9.6 *Let β, $\hat{G}$ and $\overline{\hat{H}}$ be determined as in (6.9.8), (6.9.10) and (6.9.15), and let $\hat{G}_N$ and $\hat{H}_N$ be determined as in (6.9.26) to (6.9.27). Let $\alpha \in (0,1]$, let $sg \in \mathbf{L}_\alpha(\Omega^+)$, Ω^+ the right-half plane, and let $\delta \in (0,\alpha)$. Then there exist constants C_1 and C_2 independent of β, N, and h, such that*

$$\left\| \frac{\hat{G}\overline{\hat{H}} - \hat{G}_N \overline{\hat{H}_N}}{\beta + |\hat{H}|^2} \right\|_2 \leqq \frac{C_1 e^{-\pi(\pi-\delta)/(2h)} + C_2 e^{-(\alpha-\delta)Nh}}{2\beta^{1/2}}. \tag{6.9.33}$$

Proof. Since $sg \in \mathbf{L}_\alpha(\Omega^+)$, it follows readily, that G defined in (6.9.6) belongs to $\mathbf{L}_\alpha(\mathcal{D}_{\pi/2})$, $\mathcal{D}_{\pi/2}$ defined as in (1.7.9). Thus $G \in \mathbf{H}^p(\mathcal{D}_{\pi/2})$ for all $p \in [1,\infty]$, and it therefore follows by Theorem 1.8.4, that $\hat{G}$, the Fourier transform of G, can be expressed as

$$\hat{G}(x) = \frac{\hat{W}(x)}{\cosh(\pi x/2)}, \tag{6.9.34}$$

$\hat{W}$ the Fourier transform of a function belonging to $\mathbf{L}^p(\mathbb{R})$, and also, that $\hat{G} \in \mathbf{H}^p(\mathcal{D}_\alpha)$, for all $p \in (1,\infty]$. In particular, it follows from (6.9.34), since $\sinh(\pi x)/(\pi x) \leqq \cosh(\pi x)$ for all $x \in \mathbb{R}$, that

$$|\hat{G}(x)| \leqq C_1 \left(\frac{\pi x}{\sinh(\pi x)} \right)^{1/2} = C_1 |\overline{\hat{H}(x)}|, \tag{6.9.35}$$

C_1 a constant. Hence $\hat{G}_N$ defined in (6.9.26) also satisfies an inequality of the form (6.9.35) (with possibly a slight increase in the constant C_1). Thus, we find that the left-hand side of (9.6.33) may be written

$$\begin{aligned} &\left\| \frac{\overline{\hat{H}}[\hat{G} - \hat{G}_N] + \hat{G}_N[\overline{\hat{H}} - \overline{\hat{H}_N}]}{\beta + |\hat{H}|^2} \right\|_2 \\ &\leqq \sup_{x \in \mathbb{R}} \frac{|\overline{\hat{H}(x)}|}{\beta + |\hat{H}(x)|^2} [\|\hat{G} - \hat{G}_N\|_2 + C_1 \|\overline{\hat{H}} - \overline{\hat{H}}_N\|_2] \\ &\leqq \frac{1}{2\beta^{1/2}} [\|\hat{G} - \hat{G}_N\|_2 + C_1 \|\overline{\hat{H}} - \overline{\hat{H}}_N\|_2] \end{aligned} \tag{6.9.36}$$

where in the last line we have invoked the result of Lemma 6.9.1. Since both G and $\overline{H}$ belong to $\mathbf{H}^p(\mathcal{D}_{\pi/2-\delta/2})$ for all $p \in [1, \infty]$, and since both of these functions are bounded on $\mathbb{R}$ by $C \exp[-(\alpha - \delta)|x|]$, the statement of Lemma 6.9.6 follows by application of Theorem 3.3.1 (see (3.3.11) and (3.2.8)). ■

Lemma 6.9.7 *Let $d \in (0,1)$, and let $\beta > 0$. Then there exists a constant C, independent of d and β, such that*

$$\sup_{|\Im z| \leqq d} \left| \beta + \frac{\pi z}{\sinh(\pi z)} \right|^{-1} \leqq \frac{C}{(1-d)\beta}. \tag{6.9.37}$$

Proof. We omit the proof, which is outlined in Problem 6.9.8. ■

Let us now return to bounding the error in $f_{\beta,N}$. Application of the results of Sections 2.2 and 3.3 now yields

Theorem 6.9.8 *Let $\alpha \in (0,1]$, $\beta > 0$, and let $s\, g \in \mathbf{L}_\alpha(\Omega^+)$, Ω^+ the right half plane. Let f_β be defined in (6.9.5), let $f_{\beta,N}$ be defined as in (6.9.32), and let $\delta \in (0, \alpha)$. Then there exist constants C_1, C_2, and C_3, independent of N, β, and h, such that*

$$\begin{aligned} &\mathcal{N}_{-1}(f_\beta - f_{\beta,N}) \leqq \\ &\frac{C_1}{\beta} e^{-(\alpha-\delta)Nh} + \frac{C_2}{\beta^{1/2}} e^{-\pi(\pi-\delta)/(2h)}. \end{aligned} \tag{6.9.38}$$

Note: This result shows that $\mathcal{N}_{-1}(f_\beta - f_{\beta,N}) \to 0$ as $N \to \infty$, for each fixed β. Judicious selections of h, N and β are made below, in other theorems and lemmas of this section.

The proof which follows provides an illustration of the relationship previously encountered in Section 3.3, between the DFT of Fourier series methods of Section 2.2, and the Fourier series approximation of the Fourier transform via the trapezoidal rule.

Proof. Throughout this proof, we use the notation C and C_j to denote generic constants, independent of β, N, and h.
Let us recall some of our previous deductions made in this section. Since $sg \in \mathbf{L}_{\alpha'}(\Omega^+)$, Ω^+ the upper half plane, G defined in (6.9.6) belongs to $\mathbf{L}_{\alpha'}(\mathcal{D}_{\pi/2})$. Thus $G \in \mathbf{H}^p(\mathcal{D}_{\pi/2})$, for all $p \in [1, \infty]$, and by Theorem 1.8.4, the Fourier transform $\hat{G} \in \mathbf{H}^p(\mathcal{D}_\alpha)$, for all $p \in (1, \infty]$, and also, $|\hat{G}(x)| \leqq C\exp(-\pi|x|/2)$ for all $x \in \mathbb{R}$. Similarly, H defined in (6.9.6) satisfies $|H(u)| \leqq C\exp(-|u|)$ for all $u \in \mathbb{R}$, $H \in \mathbf{H}^p(\mathcal{D}_{\pi/2-\delta})$, for all $p \in [1, \infty]$, $\overline{\hat{H}} \in \mathbf{H}^p(\mathcal{D}_\alpha)$, for all $x \in \mathbb{R}$, and

$$|\overline{\hat{H}(x)}| \leqq Ce^{-(\pi/2-\delta)|x|}. \tag{6.9.39}$$

Combining these results with the result of Lemma 6.9.7, it follows that $\hat{F}_\beta = \hat{G}\overline{\hat{H}}/[\beta + |\hat{H}|^2] \in \mathbf{H}^p(\mathcal{D}_d)$ for all $p \in [1, \infty]$, and also, that

$$|\hat{F}_\beta(x)| \leqq \frac{C}{\beta} e^{-(\pi-\delta)|x|} \tag{6.9.40}$$

for all $x \in \mathbb{R}$. Hence, once again appealing to Theorem 1.8.4, it follows that $F_\beta \in \mathbf{H}^p(\mathcal{D}_{\pi-\delta})$ for all $p \in (1, \infty]$, and that

$$|F_\beta(u)| \leqq \frac{C}{\beta} e^{-(\alpha-\delta)|u|}. \tag{6.9.41}$$

for all $u \in \mathbb{R}$.
We have already defined $\hat{F}_\beta$, $\hat{F}^*_{\beta,N}$, and $\hat{F}_{\beta,N}$ respectively in (6.9.12), (6.9.28), and (6.9.29) above. We shall also require the following definition of $\hat{F}_{\beta,h}$, namely,

$$\hat{F}_{\beta,h}(x) = \begin{cases} h \displaystyle\sum_{k=-\infty}^{\infty} F_\beta(mh) e^{ikhx}, & \text{if} \quad |x| < \pi/h \\ 0, & \text{if} \quad x \in \mathbb{R}, \ |x| > \pi/h. \end{cases} \tag{6.9.42}$$

We remark that while the function F_β is not explicitly known at this point, our use of (6.9.42) is solely to obtain a bound for the norm of $f_\beta - f_{\beta,N}$. The existence of F_β satisfying the above properties thus suffices. The difference between F_β and $F_{\beta,N}$ may thus be expressed by the equation

$$\hat{F}_\beta - \hat{F}_{\beta,N} = \eta_1 + \eta_2 + \eta_3 + \eta_4. \tag{6.9.43}$$

In defining these quantities, it is convenient to use the notation

$$K_N(x) = \sum_{m=-N}^{N} e^{imhx} \tag{6.9.44}$$

in terms of which the Fourier polynomial expression (6.9.29) may be combined via the inner product definition (6.9.30) to yield the more concise expression

$$F_{\beta,N}(x) = \begin{cases} < \hat{F}^*_{\beta,N}, K_N(x-\cdot) >_N, & \text{if } \ |x| \leqq \pi/h \\ 0, & \text{if } \ |x| > \pi/h. \end{cases} \tag{6.9.45}$$

We may thus define the quantities η_j in (6.4.43) by the equations

$$\begin{aligned} \eta_1(x) &= \hat{F}_\beta(x) - \hat{F}_{\beta,h}(x) \ \text{ if } \ x \in \mathbb{R} \\ \eta_2(x) &= \begin{cases} \hat{F}_{\beta,h}(x) - < \hat{F}_{\beta,h}, K_N(x-\cdot) >_N & \text{if } \ |x| \leqq \pi/h \\ 0 & \text{if } \ |x| > \pi/h \end{cases} \\ \eta_3(x) &= \begin{cases} < \hat{F}_{\beta,h} - \hat{F}_\beta, K_N(x-\cdot) >_N & \text{if } \ |x| \leqq \pi/h \\ 0 & \text{if } \ |x| > \pi/h \end{cases} \\ \eta_4(x) &= \begin{cases} < \hat{F}_\beta - \hat{F}^*_{\beta,N}, K_N(x-\cdot) >_N & \text{if } \ |x| \leqq \pi/h \\ 0 & \text{if } \ |x| > \pi/h. \end{cases} \end{aligned} \tag{6.9.46}$$

By applying Theorem 3.3.1, we immediately have

$$\|\eta_1\|_2 \leqq \frac{C}{\beta} e^{-\pi(\pi-\delta)/h}. \tag{6.9.47}$$

Next, using the result of Problem 2.2.9 (b), we get

$$\|\eta_2\|_2 \leqq \frac{C}{\beta} e^{-(\alpha-\delta)Nh}. \tag{6.9.48}$$

Combining the result of Theorem 3.3.1 and Problem 2.2.8 (iv), we get

$$\|\eta_3\|_2 \leqq \frac{C}{\beta} e^{-\pi(\pi-\delta)/h}. \tag{6.9.49}$$

Finally, using Lemma 6.9.6 and Problem 2.2.8 (iv), we have

$$\|\eta_4\|_2 \leqq \frac{C_1}{\beta^{1/2}} e^{-\pi(\pi-\delta)/(2h)} + \frac{C_2}{\beta^{1/2}} e^{-(\alpha-\delta)Nh}. \tag{6.9.50}$$

Adding the above bounds, we observe that the bounds for $\|\eta_1\|_2$ and $\|\eta_3\|_2$ may be united into a single term, which can be discarded, since

$\exp\{-\pi(\pi-\delta)/h\}/\beta = [\exp\{-\pi(\pi-\delta)/(2h)\}/\beta^{1/2}]^2$, i.e., it it dominated by one of the error terms bounding $\|\eta_4\|$. The term $(C_2/\beta^{1/2})e^{-(\alpha-\delta)Nh}$ may be also be discarded, since it is dominated by the bound for η_2. Using Parseval's Theorem and the transformations (6.9.6) we thus arrive at inequality (6.9.38). ∎

The next lemma gives us an idea about the dependence of the difference between f and f_β on the analyticity properties of f.

Lemma 6.9.9 *(a) Let $d > 0$, and let either $t^{-1/2}f \in \mathbf{H}^2(S_d)$, or else, let $f \in \mathbf{L}_\alpha(\mathcal{D}_d)$, α a positive number. Then there exists a constant C, independent of β, such that*

$$\mathcal{N}_{-1}(f - f_\beta) \leqq C\beta^{d/\pi}. \tag{6.9.51}$$

(b) Let each of the following inequalities be satisfied:

$$\begin{aligned} \mathcal{N}_{-1}(f) \quad & < \quad \infty \\ \mathcal{N}_{-1}\left(\left\{\frac{t\,d}{dt}\right\}^m f\right) \quad & < \quad \infty, \end{aligned} \tag{6.9.52}$$

m a positive integer. Let $0 < \beta \leqq 1/2$. Then there exists a positive constant C independent of β, such that

$$\mathcal{N}_{-1}(f - f_\beta) \leqq \frac{C}{\{|\log(\beta)|\}^m}. \tag{6.9.53}$$

Proof. Part (a): The proof is a consequence of Problems 6.9.6 and 6.9.7, and we omit it.

Part (b): Using (6.9.6) and Parseval's Theorem, it is readily seen that if both of the inequalities in (6.9.52) are satisfied, then $\|\hat{F}\|_2 < \infty$, and also $\|x^{2m}\hat{F}\|_2 < \infty$. Let $x_m > 0$ be determined by the solution of the equation $(*)$ $K_\beta(x) = x^{-m}$, with $K_\beta(x) \equiv \beta/[\beta + |\hat{H}(x)|^2]$. Since $|\hat{H}(x)|^2$ is an even function of x, so is $K_\beta(x)$; moreover, since $|\hat{H}(x)|$ decreases (strictly) monotonically from 1 to 0 as $|x|$ increases (strictly) monotonically from 0 to ∞, $K_\beta(x)$ increases (strictly) monotonically from $\beta/(\beta+1)$ to 1, as $|x|$ increases (strictly) from 0 to ∞. Since x^{-m} is (strictly) monotonically decreasing as $|x|$ as (strictly) monotonically increases from 0 to ∞, there is a unique positive number x_m which satisfies the above equation $(*)$. Moreover, it is easily shown that for $\beta \leqq 1/2$ we have $x_m > C_1/|\log(\beta)|$, with C_1 a positive constant independent of β. Then, for $|x| < x_m$, we have $K_\beta(x) \leqq x_m^{-m}$, while if $|x| > x_m$, then $K_\beta(x) < (|x|/x_m)^m$. Hence, recalling that $\hat{F} - \hat{F}_\beta = \beta\,\hat{F}/(\beta + |\hat{H}|^2)$, squaring each side of this equation, using the inequalities just established, integrating each term over the subset of $\mathbb{R}$ over which the inequalities are satisfied, replacing each of the integrals over subintervals of $\mathbb{R}$ by an integral of the same function over $\mathbb{R}$, and finally using Parseval's Theorem and the transformations (6.9.6), we get (6.9.53). ■

Corollary 6.9.10 *Let $d > 0$, let f^e and g^e be defined in (6.9.2), and let the inequality (6.9.3) be satisfied. Let either $t^{-1/2} f^e \in \mathbf{H}^2(S_d)$, or else, let $f^e \in \mathbf{L}_\alpha(\mathcal{D}_d)$, α a positive number. Let f satisfy the same conditions as f^e. Then there exists a constant C, independent of ε and β, such that*

$$\mathcal{N}_{-1}(f^e - f_\beta) \leqq \frac{\varepsilon}{2\,\beta^{1/2}} + C\beta^{d/\pi}. \tag{6.9.54}$$

Proof. The proof follows from Lemmas 6.9.3 and 6.9.9. ■

We have omitted the obvious analogous statement of Corollary 6.9.10, resulting from part (b) of Lemma 6.9.9 (see Problem 6.9.13).

In the above equations we did not specify any possible connections of N, h, and β. In the following Theorem we specify a connection, based on the bounds obtained.

Theorem 6.9.11 *Let the conditions of Theorem 6.9.8 and Lemma 6.9.9 be satisfied. If we select h and β according to the formulas*

$$\begin{aligned} h &= \pi \left\{ \frac{\pi + d}{(\pi + 2d)(\alpha - \delta)N} \right\}^{1/2} \\ \beta &= \exp\left\{ -\pi^2 \left[\frac{(\alpha - \delta)N}{(\pi + d)(\pi + 2d)} \right]^{1/2} \right\}, \end{aligned} \tag{6.9.55}$$

then there exists a constant C independent of N, such that

$$\mathcal{N}_{-1}(f - f_{\beta,N}) \leqq C \exp\left\{ -\pi d \left[\frac{(\alpha - \delta)N}{(\pi + d)(\pi + 2d)} \right]^{1/2} \right\}. \tag{6.9.56}$$

Proof. The straightforward proof follows upon eliminating β and h from the equations

$$\frac{e^{-(\alpha-\delta)Nh}}{\beta} = \frac{e^{-\pi^2/(2h)}}{\beta^{1/2}} = \beta^{d/\pi} \tag{6.9.57}$$

which we obtain from Equations (6.9.16), (6.9.38) and (6.9.50), after replacing the positive number δ in (6.9.38) and (6.9.50) by zero. This replacement can be justified, since $\delta \in (0, \alpha)$ was otherwise arbitrary, and since the final answer involves $\alpha - \delta$, in which we can absorb the error made by our replacing $\alpha - \delta$ by α. The details are left to Problem 6.9.9, and we omit them. ■

Theorem 6.9.11 suggests that if the assumptions of the theorem are satisfied, then we can compute $f_{\beta,N}$ to arbitrary accuracy. This is indeed the case, provided we can achieve arbitrary high computer precision. On the other hand, only limited accuracy of approximation of f^e by $f_{\beta,N}$ is possible in the case when $\varepsilon > 0$, ε defined in (6.9.3). We give a precise statement of this result in the next theorem. Note that there is no point in taking N larger than the value recommended in the statement of the theorem. We mention that even if ε defined in (6.9.3) is zero, we are generally limited to a certain attainable precision on a computer, and taking ε as the attainable precision, the next theorem then tells us the accuracy to which we can then expect to compute $f_{\beta,N}$ on that computer.

Theorem 6.9.12 *Let f^e and ε be defined in Lemma 6.9.3, and let the conditions of Theorem 6.9.8 and Corollary 6.9.10 be satisfied. If we first select β by the formula*

$$\beta = \varepsilon^{\frac{2\pi}{\pi+2d}}, \tag{6.9.58}$$

next, select N to be the smallest integer such that

$$\exp\left\{-d\left[\frac{(\pi+d)(\alpha-\delta)N}{\pi+2d}\right]^{1/2}\right\} \leqq \varepsilon^{\frac{2d}{\pi+2d}}, \tag{6.9.59}$$

and finally, select h as in (6.9.55), then there exists a constant C independent of ε, such that

$$\mathcal{N}_{-1}(f^e - f_{\beta,N}) \leqq C\,\varepsilon^{\frac{2d}{\pi+2d}}. \tag{6.9.60}$$

Proof. Using (6.9.38) and (6.9.54), we get

$$\begin{aligned} \mathcal{N}_{-1}(f^e - f_{\beta,N}) &\leqq \mathcal{N}_{-1}(f^e - f_\beta) + \mathcal{N}_{-1}(f_\beta - f_{\beta,N}) \\ &\leqq \frac{\varepsilon}{2\beta^{1/2}} + C_0\beta^{d/\pi} + \frac{C_1}{\beta}e^{-(\alpha-\delta)Nh} + \frac{C_2}{\beta^{1/2}}e^{-\pi(\pi-\delta)/(2h)}. \end{aligned} \tag{6.9.61}$$

Now, each of the four terms on the right-hand side of this inequality will be of the same order of magnitude, if we set

$$\frac{\varepsilon}{\beta^{1/2}} = \beta^{d/\pi} = \frac{1}{\beta}e^{-(\alpha-\delta)Nh} = \frac{e^{-\pi(\pi-\delta)/(2h)}}{\beta^{1/2}}. \tag{6.9.62}$$

Solving these equations, we get

$$\beta = \varepsilon^{2\pi/(\pi+2d)} = e^{-\pi(\alpha-\delta)Nh/(\pi+d)} = \exp\{-\pi^2(\pi-\delta)/[(\pi+2d)h]\}$$

$$h = \{\pi(\pi-\delta)(\pi+d)/[(\alpha-\delta)(\pi+2d)N]\}^{1/2}. \tag{6.9.63}$$

We then select N such that (6.9.59) holds. The remaining details of the proof are left to Problem 6.9.10. ■

We may note the interesting special case of (6.9.59), when $d \geqq \pi/2$, i.e., when f and f^e are both Laplace transforms. In this case the error bound on the right-hand side of (6.9.59) is $C\,\varepsilon^{1/2}$.

Example 6.9.13 Let us attempt the inversion of the function $g_1(s) = [\Gamma'(2.5) - \Gamma(2.5)\log(s)]/s^{2.5}$. In this case we see from Lemma 6.9.4 that $f(t) \sim t^{3/2}\log(t)$ as $t \to 0$, and also as $t \to \infty$. The behavior at $t = 0$ is acceptable, whereas the behavior at $t = \infty$ is not. Replacing s by $s+1$ amounts to multiplying $f(t)$ by e^{-t}. Hence, let us consider approximating the inverse transform of $g(s) = [\Gamma'(2.5) - \Gamma(2.5)\log(s+1)]/(s+1)^{2.5}$, with Γ' the derivative of the gamma function. The exact inverse of $g(s)$ is $f(t) = t^{3/2}e^{-t}\log(t)$. In this case, $\alpha = 3/2$ (approximately), and $d = \pi/2$. Thus, substituting these values into the equations of Theorem 6.9.11, we get $h = \pi/(2N)^{1/2}$, $\beta = \exp\{-\pi(N/2)^{1/2}\}$, and a bound on $\mathcal{N}_{-1}(f - f_{\beta,N})$ is $\mathcal{O}(\exp\{-(\pi/2)(N/2)^{1/2}\})$. These results correspond, approximately to our solution, which we computed for $N = 10$ and $N = 20$ via the above described algorithm.

•

PROBLEMS FOR SECTION 6.9

6.9.1 * Prove Lemma 6.9.4. [Hint: See Feller, [Fe1, p.44], and Widder, [Wi1, Ch. 5]].

6.9.2 Prove Post's formula, i.e., if $g = \mathcal{L}f$, $g \in \mathbf{H}^2(\Omega^+)$, then for all $t > 0$,

$$f(t) = \lim_{n\to\infty} \frac{(-1)^n}{n!}\left(\frac{n}{t}\right)^{n+1} g^{(n)}\left(\frac{n}{t}\right).$$

[Hint: See [Wi1, Ch. 7]].

6.9.3 (a) Given that $g(s) = -(2/\pi)\,(s^2+1)^{-1/2}\log[s+(s^2+1)^{1/2}]$ (so that g is the Laplace transform of the Bessel function Y_0), use Theorem 6.9.1 to show that $g = \mathcal{L}f$, with f satisfying the relation $f(t) \sim (2/\pi)[\log t + const]$, as $t \to 0$.

(b) Show that, if $g_1(s) = g(s) + (2/\pi)\{(s+1)^{-1}[\log(s+1)+\log 2]\}$, g defined as in (a), then $g_1 = \mathcal{L}f_1$, with $f_1(t) = \mathcal{O}(t\log t)$, as $t \to 0$.

6.9.4 Make the substitutions (6.9.6) in (6.9.13) to get (6.9.15).

6.9.5 Use the identity $\hat{G} = \hat{H}\hat{F}$ (See Equation 6.9.9) and the definitions for $\mathcal{N}_q$ and $|\hat{H}|^2$ to prove that

$$\mathcal{N}_1(g) \leqq \mathcal{N}_{-1}(f)$$
$$\|\hat{G}\|_\infty \leqq \left(\frac{\pi}{2}\right)^{1/2} \mathcal{N}_{-1}(f).$$

6.9.6 Prove that if $x \in \mathbb{R}$, then $e^{-|x|} \leqq x/\sinh(x)$.

6.9.7 (a) Let d and β be positive constants, and set

$$I = \int_0^\infty \frac{e^{-2dx}}{(\beta + e^{-\pi x})^2}\, dx.$$

(b) Prove that

$$I = \frac{\beta^{2d/\pi - 2}}{2d(1+\beta)^{2d/\pi}} F(2d/\pi - 1, 2d/\pi; 2d/\pi + 1; \frac{1}{1+\beta}),$$

F the hypergeometric function. Now prove that

$$I \leqq \begin{cases} \dfrac{\beta^{2d/\pi-2}}{2d(1+\beta)^{2d/\pi}} & \text{if} \quad 0 < 2d \leqq \pi \\ \dfrac{\beta^{2d/\pi-2}}{\pi(1+\beta)^{2d/\pi}} \dfrac{\pi - 2d}{\sin(\pi - 2d)} & \text{if} \quad \pi < 2d < 2\pi. \end{cases}$$

(c) Complete the details of the proof of Lemma 6.9.9.

6.9.8 (a) Prove that the equation $\beta + \pi x/\sinh(\pi x) = 0$ where $\beta > 0$, has a countable set of roots z and $\bar{z} \in \mathbb{C}$, and that the root in the upper half plane having smallest imaginary part is

$$z = \frac{1}{\pi}\left[\log\left(\frac{2}{\beta}\log\frac{2}{\beta}\right) + \pi i\right]\left[1 + \frac{1}{\log\frac{2}{\beta}} + o\left(\frac{1}{\log\frac{2}{\beta}}\right)\right], \quad \beta \to 0.$$

(b) Prove that if $\beta + \pi x/\sinh(\pi x) = 0$, then $x \notin \{z \in \mathbb{C} : |\Im z| \leqq 1\}$ for β sufficiently small.

(c) Prove that if for $\varepsilon \in (0,1)$ we have $x \in \{z \in \mathbb{C} : |\Im x| \leqq 1 - \varepsilon\}$, then there exists a positive constant C_ε independent of β, such that $|\beta + \pi x/\sinh(\pi x)|^{-1} \leqq C_\varepsilon/\beta$.

6.9.9 Give a detailed proof of Theorem 6.9.11.

6.9.10 Give a detailed proof of Theorem 6.9.12.

6.9.11 Let $h = \{\pi d/(\alpha N)\}^{1/2}$, let $f \in \mathbf{L}_\alpha(\mathcal{D}_d)$, and let the sequence of numbers $\{f_m\}_{-N}^{N}$ be determined by the solution of the linear system of equations

$$\beta f_m + h \sum_{k=-N}^{N} \frac{e^{kh} f_k}{e^{kh} + e^{mh}} = h \sum_{k=-N}^{N} e^{kh} g(e^{kh}) e^{-kh-mh}.$$

Let $f_{\beta,N}$ be defined on $(0, \infty)$ by

$$f_{\beta,N}(t) = \sum_{k=-N}^{N} f_m S(k,h)(\log t).$$

Then, there exists a constant C, independent of N and β, such that

$$\mathcal{N}_{-1}(f - f_{\beta,N}) \leqq \frac{C}{\beta} N^{1/2} e^{-(\pi d \alpha N)^{1/2}}.$$

[Hint: Bound the approximation

$$\mathcal{N}_{-1}(\{\mathcal{L}^*\mathcal{L} - \mathcal{P}_N \mathcal{L}^* \mathcal{L}\} f_\beta)$$

by $\delta_N \mathcal{N}_{-1}(f_\beta)$, with $\mathcal{P}_N$ the projection onto the space spanned by $\{S(k,h)(\log t)\}_{-N}^{N}$, and then combine Lemma 6.9.1 and Theorem 6.2.5.]

6.9.12 Supply the omitted details of the proof of part (b) of Lemma 6.9.9. [Hint: See Ang, Lund and Stenger, AnLS1.]

6.9.13 Corollary 6.9.10 is related to part (a) of Lemma 6.9.9. State and prove a result analogous to Corollary 6.9.10, but related to part (b) of Lemma 6.9.9.

6.9.14 Take $g \equiv 1$ in the definition of p in (4.6.1), and then study the use of the method of Section 4.6 to approximate $\int_a^x f(t)\,dt$, and also $f(t)$, given the Laplace transform of f.

Historical Remarks on Chapter 6

Sections 6.2 & 6.3 contain excerpts from *Atkinson* [At1], *Anselone* [Al1], *Ben–Israel* [B–I1], *Elliott* [E1,E2], *Kato* [Ka1], *Lancaster* [La1], *Linz* [Lz1], *Noble* [No1], and *Tikhonov & Arsenin* [TiAr1].

The results of Section 6.4 are, in part, due to *Riley* [Ri1], and, partly, are new.

The potential problem of Section 6.5 was studied first in *Schwing's* Ph.D. thesis [Sw1], although the convergence proofs given here are different, and some of the estimates sharper.

A Sinc approximate solution procedure for the reduced wave equation problem of Section 6.6 was previously studied by *Johnson, Zhou, Tracy, Berggren, and Stenger* [JZTBS1]. The present development contains a number of new results, and is considerably different from previous presentations.

The results of Section 6.7 were previously obtained by *Bialecki and Stenger* [BiS1].

Section 6.8 consists of excerpts from *Stenger* [S11, S12, S13], as well as some new results, such as the approximate solution of Equation (6.8.1), Corollary 6.8.7, and Theorem 6.8.8.

The Laplace transform inversion Section 6.9 is based on one of the methods presented in the paper of *Ang, Lund and Stenger* [AnLS1], although the present development goes somewhat further than that in [AnLS1]. These authors were unaware, at the time of publication of the paper [AnLS1], that the transformation (6.9.6) and the result (6.9.9) were published previously by Gardner, Gardner, Laush and Meinke [GdGdLuMi1], who then attempted to solve for $\hat{F}$ via FFT. The authors Gardner, Gardner, Laush and Meinke did not use regularization, nor did they express their final solution as a Sinc approximation, in the form (6.9.32). Rather, they attempted to circumvent the numerical difficulties caused by the the ill-posed expression (6.9.9) by experiments involving replacement of the right-hand side of (6.9.9) by zero outside of various compact intervals.

7

Differential Equations

Our goal in this chapter is to familiarize the reader with the use of Sinc methods for solving ordinary and partial differential equations. As for the case of integral equations considered in the previous chapter, the reader will find that replacing differential equations by their discrete Sinc approximations is straightforward and simple. Probably the most important advantage of Sinc methods for solving initial value problems over classical methods, such as Adams–type methods, or Runge–Kutta methods, is that Sinc methods are easily implemented and give good accuracy for problems with singularities and boundary layers, which are stiff. Also, they can produce accurate approximate solutions to problems over infinite or semi-infinite intervals. As in the previous chapter, the implementation of Sinc procedures is carried out along with adequate descriptions of the spaces of analytic functions, which are nearly always present in applications, and in which the Sinc methods perform close to optimally. Our presentation for solving differential equations by Sinc methods is cast in the terminology of the properties of solutions which we require, namely, that a solution should be both analytic and also belong to the class $\mathbf{Lip}_\alpha$ in the domain of analyticity in order to get accurate error bounds. Solutions of differential equations arising in applications nearly always possess these properties. Our aim is to present the Sinc methods in their simplest form, and thus we have presented all of the results of this chapter in the space $\mathbf{L}_\alpha(\mathcal{D})$, rather than the more complicated (although no more difficult) space $\mathbf{L}_{\alpha,\beta}(\mathcal{D})$. We mention, at the outset, that considerable saving of computer time and effort may, at times, be achieved in the Sinc approximation of particular problems by dealing with the space $\mathbf{L}_{\alpha,\beta}(\mathcal{D})$, rather than with the space $\mathbf{L}_\alpha(\mathcal{D})$.

We also briefly discuss singularities in the solutions of partial differential equations resulting from irregular shapes of boundaries of regions. A knowledge of the precise nature of these singularities is unnecessary for purposes of obtaining sharp bounds on the error of Sinc approximation, yet our description given here of the nature of these singularities is sufficiently precise to enable us to choose the best Sinc approximation for the close to optimal solution of such problems within the described spaces of analytic functions. In most instances, we shall be able to use the tools already established in previous chapters for setting up the Sinc approximations, and for identifying the correct function spaces.

In Section 7.1, we discuss the solution of initial value problems, via the use of the indefinite integration formula which was derived in Section 4.5, give proofs of convergence, discuss non-stiff and stiff problems, and illustrate the application of the method on some examples.

In Section 7.2, we derive discrete Sinc approximations for solving ordinary differential equation boundary value problems, i.e., the *Sinc–Galerkin method.* We restrict our derivations solely to the case of a general, linear second order equation, and we prove the convergence of these methods when applied to the solution of such equations. We also discuss the application of the methods to non-linear equations, the handling of more complicated boundary conditions, and we illustrate the application of the methods on specific examples. Thus, if a solution is both analytic and of class $\mathbf{Lip}_\alpha$, then a $2N+1$-point Sinc approximation produces an error which is roughly of the order of $e^{-(\pi d \alpha N)^{1/2}}$).

In Section 7.3, we briefly discuss the type of analyticity which we require for solution of partial differential equations by Sinc methods, thus enabling the straightforward application of previously derived Sinc methods for ordinary differential equations to partial differential equations, and yet retain the $\mathcal{O}(\exp[-c\,N^{1/2}])$ convergence rate. We then describe an elementary transformation which provides the key to domain decomposition and parallel computation, for using Sinc methods to solve partial differential equations over non-product regions which may have curvilinear boundaries, and which enable us to reduce the original partial differential equation over the original given domain to a small number of partial differential equations over product regions, and yet retain this $\mathcal{O}(\exp[-c\,N^{1/2}])$ rate of convergence. We also discuss singularities of solution of partial differential equations due to corners in domains. We conclude this section, and chapter, with the presentation of several examples illustrating the use of Sinc methods for solving partial differential equations.

Let us briefly discuss the amount of work required to solve a differential equation whose solution is analytic in a domain, and belongs to the class $\mathbf{Lip}_\alpha$ in that domain. For example, to solve a second order, linear boundary value problem to within an error ε via a Sinc method, we need to solve a system of linear equations of order $2N+1$, having a coefficient matrix which is Toeplitz, for practical purposes, and so the amount of work required to solve such a system is $\mathcal{O}(N^{2+\gamma})$, where $\gamma \in (0,1)$ is otherwise arbitrary. Thus, the amount of work required to get a solution which is accurate to within ε is $\mathcal{O}(\{\log(1/\varepsilon)\}^{4+c})$, where c is an arbitrary positive number. Using the classical finite difference approximation $\delta^2 f(x) = (f(x+h)-2f(x)+f(x-h))/h^2$ to approximate $f''(x)$, and $f(x+h)-f(x-h))/(2h)$ to approximate $f'(x)$, we need to solve a tridiagonal system of order N, which we may consider to be Toeplitz, for practical purposes, for which the error of approximation of a solution which is analytic and of class $\mathbf{Lip}_\alpha$ is $\mathcal{O}(N^{-2\alpha})$, and the amount of work required to solve such a system is $\mathcal{O}(N) = \mathcal{O}((1/\varepsilon)^{1/(2\alpha)})$. Similar simple analysis enables us to establish the *complexity*, that is, the amount of work required to achieve a certain accuracy, in other situations.

Generally, it is preferable to use that particular method which requires the least amount of work to achieve a given accuracy.

After studying this chapter, the reader who has had some experience with solving partial differential equations by other methods, will thus find that Sinc methods for solving partial differential equations are unique, in that they possess *all* of the following properties:

- All three approaches, Galerkin, spectral, or collocation methods, may be used to replace the (linear or non-linear) partial differential equation by an explicit system of algebraic equations. All three approaches yield a similar systems of algebraic equations, and each system yields a solution to the same order of accuracy;

- Domain decomposition is based on identifying regions of analyticity of the solution, and these can nearly always be determined *a priori*. One can thus set up a global Sinc approximation of the solution over every domain in the decomposition. Also, infinitely large ranges of domains or curvilinear boundaries do not appreciably increase the total amount of work required for achieving a solution to a desired accuracy;

- If we adopt the Sinc approach, then writing down the explicit system of algebraic equations corresponding to a partial differential equation is always straightforward and simple, regardless of whether the equations are linear or non-linear, and moreover, due to the accuracy of the Sinc method, the size of this system of equations is relatively small;

- The Sinc method automatically carries out close to optimal point selection in every domain in the decomposition, and hence the effects of corners in domains, or singularities in the solution or in the given data are taken care of automatically. Thus, one does not need to bother with constructing special approximating elements to take care of special singularities in the solution. As a consequence, the computer code for the whole problem, which is relatively short, can be written based on a single parameter N, where N corresponds to the order of the solution in each variable, and the computer can then determine the N required to meet a pre-specified accuracy of the solution. The error in the solution, as a function of the parameter N is $\mathcal{O}(\exp\{-cN^{1/2}\})$, this rate being a close to optimal rate when the exact nature of a singularity in the solution is not explicitly known;

- The dependence of the solution of non-linear partial differential equations (such as, e.g., Navier–Stokes equations) on *time*, is also expressed globally, via the use of Sinc collocation of derivatives and Sinc indefinite integration, or via the use of the Sinc convolution procedure of Section 4.6. If the differential equations are non-stiff, the resulting system of non-linear algebraic equations may be solved iteratively, essentially whenever the theoretical method of successive approximation converges, and this procedure yields both an effective and rapidly convergent method for solving the non-linear system, and also circumvents the instability problems that one typically encounters in finite-difference methods. The same approsimating procedures may also be used to solve highly stiff problems (using Newton, instead of Neumann iteration);

- Upon reaching this part of the text, the reader will have found that the family of Sinc methods is replete with accurate approximation formulas, which may be used to provide quick remedies for a broad range of difficulties that one encounters in solving partial differential equations (such as the evaluation of integrals, the approximation of functions or derivatives, the approximation and inversion of Fourier, Laplace, Hilbert and Mellin transforms). Moreover, each of the $2N+1$-point formulas uses the same Sinc points, it uses the same function values, and it has the same order of accuracy. In the case of linear boundary value problems, the matrices are always Toeplitz, and FFT methods are ideally suited to solve the problems, since FFT belongs to the family of Sinc methods. The Sinc interpolation formula then enables the explicit evaluation of the solution at any point in the region, with the same order of accuracy to which it was computed at the Sinc points;

- Although the relatively small system of approximating algebraic equations resulting by application of the Sinc–Galerkin method is full (as opposed to sparse) for the case of an ordinary differential equation, it is, in fact, sparse for the case of a partial differential equation.

The notation of Definition 4.1.1 will be used throughout this chapter.

7.1 ODE–IVP

In this section we shall consider the approximate solution, by Sinc methods, of ordinary differential equation (ODE) initial value problems (IVP). Our aim will be to approximate the solutions of such problems in the space $\mathbf{M}_\alpha(\mathcal{D})$ as defined in Section 4.1, in order to be able to establish the $\mathcal{O}(e^{-cn^{1/2}})$ rate of convergence of the error via an n-point Sinc approximation. We shall also interlace some classical some new results, in order to help us achieve our aim. As already seen in the previous chapter, the study of an integral equation corresponding to a differential equation provides an effective approach for determining the analyticity of solutions. Indeed, in this chapter we shall rely heavily on results about analyticity established in Chapter 6.

We also recall some known results about analyticity of solutions of differential equations in the problems at the end of this section.

7.1.1 Linear Initial Value Problems

At the outset, we examine a first-order linear initial value problem of the form

$$\begin{aligned} y' &= K(t)\,y + g(t), \quad t \in \Gamma, \\ y(a) &= y_a, \end{aligned} \tag{7.1.1}$$

which may represent either a single, scalar equation or a system of order n. In the latter case y and g are column vectors of order n, and K is a square matrix of order n. In referring to K (or g) as being analytic in a domain, or belonging to the space $\mathbf{L}_\alpha(\mathcal{D})$, we shall imply that all of the components of K (or g) are analytic in that domain, or that they all belong to the space $\mathbf{L}_\alpha(\mathcal{D})$. We begin by citing a well-established result from the classical theory of linear ordinary differential equations, which we state in the terminology of Definition 4.1.1.

Theorem 7.1.1 *Let both K/ϕ' and g/ϕ' belong to $\mathbf{L}_\alpha(\mathcal{D})$. Then the initial value problem (7.1.1) has a unique solution y on Γ which may be continued analytically into $\mathcal{D}$.*

Proof. The proof follows upon application of analytic continuation to the solution constructed as outlined in Problem 7.1.1. We omit the details. ∎

The initial value problem (7.1.1), can be readily transformed into the integral equation problem

$$y(x) = \int_a^x [K(t)\,y(t) + g(t)]dt + y_a, \tag{7.1.2}$$

and from Theorem 6.4.1 we know that if (7.1.2) is a scalar equation, then it has a solution $y \in \mathbf{M}_\alpha(\mathcal{D})$ whenever both K/ϕ' and g/ϕ' belong to $\mathbf{L}_\alpha(\mathcal{D})$. This observation applies equally to the case when K is a matrix, and g and y are vectors, provided that we interpret the statement $K/\phi' \in \mathbf{L}_\alpha(\mathcal{D})$ to mean that each element of the matrix K/ϕ' belongs to $\mathbf{L}_\alpha(\mathcal{D})$, and similarly for the vector g. The scalar equation (7.1.1) (or (7.1.2)) can be explicitly solved, to yield the solution

$$y(x) = \exp\left\{\int_a^x K(\tau)\,d\tau\right\}\left[\int_a^x \exp\left\{-\int_a^t K(\tau)\,d\tau\right\} g(t)\,dt + y_a\right]. \tag{7.1.3}$$

As in the case of linear Volterra integral equations discussed in Section 6.4, we now discuss two procedures based on collocation, for solving Equation (7.1.1), the first, by applying the integration formula (4.5.9) to approximate the integral in (7.1.2), while the second is based on approximating each of the indefinite integrals in (7.1.3). In both cases we select $h = [\pi d/(\alpha N)]^{1/2}$ and we form approximations y_j to the solution $y(x)$ evaluated at the Sinc points z_j. Furthermore, given an arbitrary function u defined on Γ, we shall use the notation $D(u) = \mathrm{diag}(u(z_{-N}), \cdots, u(z_N))$, and we shall use the definition of $I^{(-1)}$ as given in (4.5.7).

By the first procedure we reduce Equation (7.1.2) to a system of $2N+1$ linear equations

$$[I - hI^{(-1)}D(K/\phi')]\mathbf{y} = hI^{(-1)}D(1/\phi')\mathbf{g} + \mathbf{y}_a, \tag{7.1.4}$$

where $\mathbf{y} = (y_{-N}, \cdots, y_N)^T$, where $\mathbf{g} = (g(z_{-N}), \cdots, g(z_N))^T$, and where $\mathbf{y}_a$ denotes the vector of $2N+1$ constant values, $\mathbf{y}_a = (y_a, \cdots, y_a)^T$.

By the second method, we proceed as in Equations (6.4.6) through (6.4.9). We thus set $\mathbf{k} = (K(z_{-N}), \cdots, K(z_N))^T$, and we then use the formula (4.5.9) to approximate the values of the integral $\int_a^x K(t)\,dt$ at the Sinc points $x = z_j$, thus defining a vector $\tilde{\kappa} = (\kappa_{-N}, \cdots, \kappa_N)^T$ by means of the formula

$$\tilde{\kappa} = -hI^{(-1)}D(1/\phi')\mathbf{k}. \tag{7.1.5}$$

We then define a diagonal matrix E by means of the expression

$$E = \mathrm{diag}(\exp(\kappa_{-N}), \cdots, \exp(\kappa_N)). \tag{7.1.6}$$

Then, once again using (4.5.9) to approximate the integral with respect to t in (7.1.3), we get the explicit expression

$$\mathbf{y} = E^{-1}[hI^{(-1)}ED(1/\phi')\mathbf{g} + \mathbf{y}_a]. \tag{7.1.7}$$

As already already in Section 6.4, both of the above approaches are important, since taken together in view of the accuracy of Sinc indefinite integration of functions w such that $w/\phi' \in \mathbf{L}_\alpha(\mathcal{D})$, they enable us to deduce that the matrix multiplying the vector $\mathbf{y}$ in (7.1.4) has a condition number that is uniformly bounded independently of N for all N sufficiently large (see Problem 6.4.1). Hence, in each case, we obtain an approximate solution, $y_N(x)$ of the differential equation problem (7.1.1), of the form

$$y_N(x) = \frac{y_{-N} + \rho(x)y_N}{1+\rho(x)} + \sum_{j=-N}^{N}\left(y_j - \frac{y_{-N} + e^{jh}y_N}{1+e^{jh}}\right) S(j,h)\circ\phi(x), \tag{7.1.8}$$

and the error satisfies the inequality

$$\sup_{x\in\Gamma} |y(x) - y_N(x)| \leqq C N^{1/2} e^{-(\pi d \alpha N)^{1/2}}, \tag{7.1.9}$$

where, if the y_j are obtained via the solution of (7.1.4), then C is bounded, independently of N for all N sufficiently large, whereas, if the y_j are obtained via (7.1.7), then C may be selected independently of N for all $N > 0$.

However, as was previously noted in Section 6.4, we can further deduce that if (7.1.1) is a system of equations of order n, such that $K/\phi' \in \mathbf{L}_\alpha(\mathcal{D})$ and $g/\phi' \in \mathbf{L}_\alpha(\mathcal{D})$. In this case the integral equation representation (7.1.2) is still valid, although we cannot in general find an explicit solution y of the form (7.1.3). Applying the Sinc indefinite integration formula (4.5.9) to (7.1.2), we get

$$\left(I_1 \otimes I_2 - I_1 \otimes \{hI^{(-1)} D(1/\phi')\}[K_{ij}]\right) \mathbf{y} = I_1 \otimes \{hI^{(-1)} D(1/\phi')\}\mathbf{g} + \mathbf{y}_a, \tag{7.1.10}$$

where the notation "$\otimes$" denotes the Kronecker product, defined in (6.4.40). The matrices I_1 and I_2 are unit matrices of order n and $2N+1$ respectively, $[K_{ij}]$ denotes a $n \times n$ block of diagonal matrices K_{ij} of order $2N+1$, defined by

$$K_{ij} = \operatorname{diag}(k_{ij}(z_{-N}), \cdots, k_{ij}(z_N)),$$

with k_{ij} the (i,j)th element of the matrix K, and

$$\mathbf{y} = (\eta_1, \cdots, \eta_n)^T, \quad \eta_i = (\eta_{i,-N}, \ddots, \eta_{i,N})^T,$$

with η_{ij} approximating the component $y_i(x)$ of the vector y at the Sinc point $x = z_j$. The vector $\mathbf{g}$ is defined similarly, while $\mathbf{y}_a$ denotes the vector of order $(2N+1)n$ given by

$$\mathbf{y}_a = (y_{1a}, \cdots, y_{1a}, y_{2a}, \cdots, y_{2a}, \cdots, y_{na})^T.$$

It is interesting to note the sparsity of the coefficient matrix of the vector $\mathbf{y}$ in (7.1.10). The above two methods of solution of the scalar equation (7.1.1) enable us to deduce

Theorem 7.1.2 *Let* $\mathbf{T}$ *denote the coefficient matrix of the vector* $\mathbf{y}$ *of the system of equations (7.1.10). If* h *is selected by the formula* $h = [\pi d/(\alpha N)]^{1/2}$, *then the condition number of* $\mathbf{T}$ *is uniformly bounded, independently of* N, *for all* N *sufficiently large.*

Proof. The proof of this theorem is left to Problem 7.1.4. ∎

Example 7.1.3 Let us consider the Sinc solution of

$$y' = \frac{iy}{\sqrt{1-x^2}}, \quad , -1 < x < 1, \quad y(-1) = -i \tag{7.1.11}$$

where $i = \sqrt{-1}$. The exact solution of this problem is

$$y(x) = \sqrt{1-x^2} + ix. \tag{7.1.12}$$

In this case, we take as our mapping ϕ the function $\phi(x) = \log\{(1+x)/(1-x)\}$, so that $1/\phi'(x) = (1/2)(1-x^2)$. Since the function $K(x) = i/\sqrt{1-x^2}$ satisfies $K(x)/\phi'(x) = \sqrt{1-x^2}$, we have $K/\phi' \in \mathbf{L}_\alpha(\mathcal{D})$, with $\alpha = 1/2$ and $\mathcal{D} = \{z \in \mathbb{C} : |\arg\{(1+z)/(1-z)\}| < \pi - \varepsilon\}$, $\varepsilon \in (0, \pi)$, arbitrary. Hence, taking $h = \pi(2/N)^{1/2}$, and solving the system (7.1.4), we get a solution

$$\begin{aligned} y_N(x) &= \frac{(1-x)y_{-N} + (1+x)y_N}{2} \\ &+ \sum_{j=-N}^{N} \left(y_j - \frac{y_{-N} + e^{jh} y_N}{1 + e^{jh}} \right) S(j,h) \circ \log\left(\frac{1+x}{1-x}\right), \end{aligned} \tag{7.1.13}$$

which is accurate to four decimals on $[-1, 1]$ with $N = 20$, and to six decimals with $N = 40$. Similar accuracy is obtained using the method (7.1.7).

i. Non-Stiff Problems

We may also attempt to solve the linear systems of equations (7.1.4) by successive approximations, that is, by means of the iterative scheme

$$\mathbf{y}^{(m+1)} = hI^{(-1)} D(K/\phi')\mathbf{y}^{(m)} + hI^{(-1)} D(1/\phi')\mathbf{g} + \mathbf{y}_a, \tag{7.1.14}$$

$\mathbf{y}^{(0)}$ being the null vector. Similarly, we may attempt to solve the system (7.1.10) by means of the scheme

$$\begin{aligned} &(I_1 \otimes I_2)\, \mathbf{y}^{(m+1)} \\ &= \left(I_1 \otimes \{hI^{(-1)} D(1/\phi')\}[K_{ij}]\right) \mathbf{y}^{(m)} + I_1 \otimes \{hI^{(-1)} D(1/\phi')\}\mathbf{g} + \mathbf{y}_a, \end{aligned} \tag{7.1.15}$$

$\mathbf{y}^{(0)}$ being the null vector.

Theorem 7.1.4 *If the length of the contour Γ is finite, if $\mathbf{y}^{(0)}$ is the null vector, then the sequence of vectors $\mathbf{y}^{(m)}$ defined by (7.1.14) (respectively by (7.1.15) for a system of equations) converges to the solution of (7.1.4) (respectively, to the exact solution of the system of equations (7.1.10)) for all N sufficiently large.*

Proof. The details of the proof are left to Problem 7.1.5. ■

In practice it is nearly always more efficient to use the *Seidel iteration*, instead of the Neumann iteration described in (7.1.14) or (7.1.15). In using the Seidel version, we let A denote the matrix multiplying the vector $\mathbf{y}^{(m)}$ in (7.1.15), and we set $A = L + U$, where the matrix L has only zero entries above the diagonal, and where U has only zero entries on and below the diagonal. Then, letting $\mathbf{b}$ be the vector on the right-hand side of (7.1.10), the Seidel iteration scheme then takes the form

$$(I - L)\mathbf{y}^{(m+1)} = U\mathbf{y}^{(m)} + \mathbf{b} \tag{7.1.16}$$

The procedures (7.1.14), (7.1.15) and (7.1.16) are practical, provided that the corresponding problem (7.1.1) is *non-stiff*. Let us give a practical notion of *stiffness*.

Definition 7.1.5 *Let $N = N_0$ in (7.1.14) be the smallest value of N, with which we can achieve a desired accuracy in the computed solution of Equation (7.1.1). We shall then say that the corresponding equation (7.1.10) is non-stiff with respect to the iteration scheme (7.1.15) (respectively, (7.1.16)), provided that the sequence of vectors $\mathbf{y}^{(m)}$ defined by (7.1.15) (respectively, by (7.1.16)) with $N = N_0$ converges to the solution of Equation (7.1.10). Otherwise, we shall say that Equation (7.1.10) is stiff.*

It is easily seen that this notion of stiffness also depends on the length of Γ. For example, if Γ is the interval $(0, T)$, and we take as our mapping the function $\phi(z) = \log[z/(T - z)]$, then, taking the ℓ^∞ norm, and recalling that $\delta_j^{(-1)}$ defined by (1.10.30) satisfies the inequality $|\delta_j^{(-1)}| \leqq 1.1$ (see Lemma 3.6.5), we have

$$\begin{aligned}
\left\|hI^{(-1)}D(K/\phi')\right\|_\infty &= \max_i h \sum_{j=-N}^{N} |\delta_{i-j}^{(-1)} K(z_j)/\phi'(z_j)| \\
&\leqq 1.1\, h \sum_{j=-N}^{N} |K(z_j)/\phi'(z_j)| \\
&\approx 1.1 \int_0^T |K(t)|\, dt \\
&\leqq 1.1\, T \sup_{t\in\Gamma} |K(t)| \\
&= 1.1\, T\, \|K\|_\infty .
\end{aligned} \tag{7.1.17}$$

It follows, therefore, that we can always achieve convergence of the scheme (7.1.14) (or (7.1.15)) by starting with a Γ of sufficiently small length.

ii. Stiff Problems

In applications, especially those involving the solution of partial differential equations, one frequently faces problems that are so stiff it is necessary to make the length of Γ unreasonably small to achieve convergence of the scheme (7.1.15) (or (7.1.16)). In such instances, it is worthwhile to keep in mind that the solution of the system (7.1.10) is accurate, regardless of whether or not the Equation (7.1.10) is stiff. We should then solve the system (7.1.10) directly, by use of a trusted linear equation solver. The case of $\Gamma = (0, \infty)$ is a frequently occurring, important, special case of this type. The following lemma gives assurance that this procedure will actually work.

Lemma 7.1.6 *Let the matrix K in the system of n equations (7.1.1) be such that all eigenvalues of K/ϕ' have negative real parts on Γ. Then the matrix multiplying the vector* $\mathbf{y}$ *in (7.1.10) is non-singular.*

Proof. The proof is outlined in Problem 7.1.9. ∎

We remark that if Γ is an interval, then $\phi' > 0$ on Γ. Therefore the conditions of Lemma 7.1.6 are satisfied, if all of the eigenvalues of K have negative real parts on Γ.

Example 7.1.7 Let us consider the solution of the equation

$$\frac{dy}{dt} = \begin{pmatrix} -11 & 9 \\ 9 & -11 \end{pmatrix} y \tag{7.1.18}$$

which is to be solved as a function of $t \in (0, \infty)$ subject to the initial condition $y(0) = (2, 0)^T$. The exact solution of this problem is

$$y(t) = \begin{pmatrix} 1 \\ 1 \end{pmatrix} e^{-2t} + \begin{pmatrix} 1 \\ -1 \end{pmatrix} e^{-20t}. \tag{7.1.19}$$

The coefficients of equation (7.1.18) are, in fact, analytic everywhere. On the other hand, since the point ∞ is a boundary point of the region $\mathcal{D}$, we take $\phi(t) = \log[\sinh(t)]$, and we let $\mathcal{D}$ be defined as in (1.7.12), with $d = \pi/2$, this being the largest d for which this mapping function is one-to-one. Since the solution is analytic in every compact subset of $\mathcal{D}$ (so that it belongs to $\mathbf{Lip}_1$ there), and since in this case we can deduce that the solution decreases at the rate $\mathcal{O}(e^{-2t})$ as $t \to \infty$ in $\mathcal{D}$, we may take $\alpha = 1$. Hence, we may take $h = \pi/(2N)^{1/2}$, in the system of equations

$$\begin{pmatrix} \mathbf{y}_1 \\ \mathbf{y}_2 \end{pmatrix} = \begin{pmatrix} hI^{(-1)}D(1/\phi') & \mathbf{0} \\ \mathbf{0} & hI^{(-1)}D(1/\phi') \end{pmatrix} \cdot \\ \cdot \begin{pmatrix} D(-11) & D(9) \\ D(9) & D(-11) \end{pmatrix} \begin{pmatrix} \mathbf{y}_1 \\ \mathbf{y}_2 \end{pmatrix} + \begin{pmatrix} \mathbf{c}_1 \\ \mathbf{c}_2 \end{pmatrix} \tag{7.1.20}$$

where $\mathbf{y}_i = (y_{i,-N}, \cdots, y_{i,N})^T$, $i = 1, 2$, $D(9)$ is a diagonal matrix of order $2N + 1$, all of whose elements are 9 and similarly for $D(-11)$, and $\mathbf{c}_1$ (respectively, $\mathbf{c}_2$) is a vector of order $2N + 1$ all of whose entries are 2 (respectively, 0). The iterative solution of (7.1.20) is futile for values of $N < 100$, but its direct solution is straightforward, and is guaranteed to work, according to Lemma 7.1.6. Solving for the numbers $y_{i,j}$, $i = 1, 2$, $j = -N, \cdots, N$, and setting $\eta_j = (y_{1,j}, y_{2,j})^T$, we get the approximation

$$y_N^T(t) = \frac{\eta_{-N} + \rho(t)\eta_N}{1 + \rho(t)} + \sum_{j=-N}^{N} \left(\eta_j - \frac{\eta_{-N} + e^{jh}\eta_N}{1 + e^{jh}} \right) S(j,h) \circ \log[\sinh(t)] \tag{7.1.21}$$

which has the error

$$\|y - y_N\|_\Gamma \leqq C N^{1/2} e^{-\pi(N/2)^{1/2}} \tag{7.1.22}$$

with C a constant independent of N. The norm in (7.1.22) is defined by taking the sup-norm of each component of the difference $y - y_N$ over Γ, and then taking the maximum value of these numbers. For example, with $N = 20$, we get a solution which is accurate to five decimals on $(0, \infty)$.
The mapping function $\log[\sinh(t)]$ is more convenient here than the mapping function $\log(t)$, since the Neumann series for the former converges for a larger range of initial values than the latter, as a result of the fact that the eigenvalue of largest modulus of the matrix $hI^{(-1)}D(1/\phi')$ is considerably smaller in the case $\phi(t) = \log[\sinh(t)]$ than in the case $\phi(t) = \log(t)$. If we had used the latter mapping, $\log(t)$, then we could still have taken $\alpha = 1$, and $d = \pi/2$, since the solution is unbounded outside the closure of the sector $\mathcal{D} = \{t \in \mathbb{C} : |\arg(t)| < \pi/2\}$.
We remark here that if we had solved problem (7.1.18) over an arbitrary finite interval rather than over $(0, \infty)$, we could also have taken $\alpha = 1$, but

we then could have taken d arbitrarily large, since the solution is an entire function. We wish to emphasize that in solving (7.1.18) over $(0, \infty)$ via the use of the mapping function $\phi(t) = \log[\sinh(t)]$, we *must* take $0 < d \leqq \pi/2$ and the choice $\alpha = 1$ is satisfactory, since the solution decays sufficiently rapidly, at the rate $\mathcal{O}(e^{-2t})$ as $t \to \infty$. On the other hand, the solutions of the problem

$$\frac{dy}{dt} = \begin{pmatrix} -41/4 & 39/4 \\ 39/4 & -41/4 \end{pmatrix} y \tag{7.1.23}$$

which is to be solved subject to the initial conditions $y(0) = (2, 0)^T$, are also entire functions, and to solve this problem over a finite interval, it suffices to take $\alpha = 1$ and d arbitrarily large. However, inspection of the exact solution, which is

$$y(t) = \begin{pmatrix} 1 \\ 1 \end{pmatrix} e^{-(1/2)t} + \begin{pmatrix} -1 \\ 1 \end{pmatrix} e^{-20t} \tag{7.1.24}$$

shows that, upon solving (7.1.23), via the use of the $\log[\sinh(t)]$ map, the value $\alpha = 1$ is not the optimal choice of α for determining $h = \pi/(2\alpha N)^{1/2}$ for this problem, but $\alpha = 1/2$ is. By choosing $\alpha = 1$ we achieve the $\mathcal{O}(N^{1/2} e^{-\pi (N/8)^{1/2}})$ rate of convergence (see the proof of Theorem 3.1.7). This rate is not unduly pessimistic, although it is not as good as the optimal one gotten by determining h using $\alpha = 1/2$, namely, $\mathcal{O}(N^{1/2} e^{-(\pi/2) N^{1/2}})$. A method for the algorithmic determination of a close to optimal h based on computed solution values for small N is an unsolved problem, the solution of which would be an extremely important contribution towards the implementation of Sinc initial value methods (as well as the Sinc–Galerkin methods found in the later sections of this chapter) in automated algorithms.

iii. Unbounded Solutions

An important advantage of Sinc methods is that they accurately compute solutions of differential equations in the neighborhood of singularities. The *variations of constants* method is a well-known device of differential equations which makes it possible to study and compute unbounded solutions. To illustrate, let us consider, once again, the scalar case of equation (7.1.1), where now $K/\phi' \notin \mathbf{L}_\alpha(\mathcal{D})$, but is such that we can form the decomposition $K = K_1 + K_2$, where a solution $u(x)$ to the problem $u' = K_1 u$ can be explicitly expressed. We can set $y(x) = u(x)\, \eta(x)$ to transform (7.1.1) into the integral equation problem

$$\eta(x) = (u(x))^{-1} \left[\int_a^x [K_2(t)\, u(t)\, \eta(t) + g(t)] dt + y_a \right] \tag{7.1.25}$$

This form is a special case of the equation

$$\eta(x) = \int_a^x [K(x,t)\,u(t)\,\eta(t) + g(t)]dt + w(x) \tag{7.1.26}$$

which has a solution $\eta \in \mathbf{M}_\alpha(\mathcal{D})$, provided that K is such that $\int_a^x k(x,t)\,\eta(t)\,dt \in \mathbf{M}_\alpha(\mathcal{D})$ whenever $\eta \in \mathbf{M}_\alpha(\mathcal{D})$, provided that $g/\phi' \in \mathbf{M}_\alpha(\mathcal{D})$, and provided that $w \in \mathbf{M}_\alpha(\mathcal{D})$. Thus, although we cannot approximate the original solution of the differential equation, if we write the solution y in the form $y = u\,\eta$, where u is an unbounded factor, we may be able to accurately approximate the factor η. A similar procedure was used in Section 6.7 to approximate the solution of a Cauchy singular integral equation.

The remarks of the previous paragraph once again apply to the case of a system of equations, but in order to be able to apply the variation of constants method as described above, it is necessary that K_1 be a diagonal matrix, a situation which is not always easy to achieve. A great deal of effort has been spent on the part of mathematicians to derive factorizations of this type, in order to find the solution of a differential equation in the neighborhood of a singular point. Two of these procedures, to obtain a solution in the neighborhood of two important types of singularities, namely, poles of order 1 and higher order, respectively, are summarized in Problems 7.1.2 and 7.1.3 at the end of this section. Another example of this type is given in Problem 7.1.10. We also give an example of the approximation of the solution of an equation of the form (7.1.25) via Sinc methods, in Problem 7.1.10.

7.1.2 Non-Linear Initial Value Problems

The model differential equation of this section is the problem

$$\begin{aligned} \frac{dy}{dx} &= F(x,y), \quad x \in \Gamma \\ y(a) &= y_a. \end{aligned} \tag{7.1.27}$$

Analyticity properties of solutions of non-linear initial value problems of type (7.1.27) are not as easy to predict as those for linear equations. For example, the solution $y(x)$ of the initial value problem $y' = 1+y^2$, $y(0) = 0$, is $y(x) = \tan(x)$, which exists on the interval $0 \leqq x < \pi/2$, but which becomes unbounded at $x = \pi/2$. We can nevertheless guarantee analyticity in a small neighborhood of the point a, as is exhibited by the following well-known result.

Theorem 7.1.8 *Let $r > 0$, $R > 0$, let $B = \mathcal{D}(a;r) \times \mathcal{D}(y_a;R)$, let $F(x,y)$ be an analytic function of both variables $(x,y) \in B$, and let $|F(x,y) - F(a,y_a)| \leqq M$ in B, M a finite, positive number. Then the initial value problem (7.1.27) has a unique solution $y(x)$ which is analytic in the disc $\mathcal{D}(a;\rho)$, with*

$$\rho = r\left(1 - e^{-\frac{R}{2Mr}}\right). \tag{7.1.28}$$

Proof. We leave the details of the proof of this theorem to Problem 7.1.6. ■

Just as in the previous section, we proceed at once to transform the problem (7.1.27) into the Volterra integral equation problem

$$y(x) = (Sy)(x) \equiv \int_a^x F(t, y(t))\, dt + y_a. \tag{7.1.29}$$

We know by Theorem 7.1.8, that if $F(x,y)$ is a holomorphic function of (x,y) which is bounded by M in a region B as defined in the theorem, then the problem (7.1.29) has a unique analytic solution $y(x)$ in a neighborhood of a. Let us next state conditions which will guarantee that the solution y of (7.1.29) exists and belongs to $\mathbf{M}_\alpha(\mathcal{D})$.

Let $\mathbf{X}$ denote the Banach space $\mathbf{Hol}(\mathcal{D}) \cap \mathbf{C}(\mathcal{D})$, with which we associate the norm for $y \in \mathbf{X}$ by $\|y\| = \sup_{z\in\Gamma} |y(z)|$. Corresponding to a function $g \in \mathbf{X}$, let $r > 0$, and let $\mathcal{B}(g;r)$ denote the family of all functions $y \in \mathbf{X}$ such that $\|y - g\| < r$.

Assumption 7.1.9 *We now make the following assumptions about Sy:*

(a) *$Sy \in \mathcal{B}(g;r) \cap \mathbf{M}_\alpha(\mathcal{D})$ for all $y \in \mathcal{B}(g;r)$ (We remark that this condition may be easily tested by checking to see if $F(\cdot, y(\cdot))/\phi'(\cdot) \in \mathbf{L}_\alpha(\mathcal{D})$, see Theorem 4.1.3.);*

(b) *There exists a positive number $\beta \in (0,1)$ such that for all u and v in $\mathcal{B}(g;r)$ we have $\|Su - Sv\| \leqq \beta\,\|u - v\|$; and*

(c) *$\|Sg - g\| < (1 - \beta) r$.*

Let a sequence of functions $\{y^{(m)}\}$ be defined for $x \in \Gamma$ by

$$\begin{aligned} y^{(0)}(x) &= g(x), \\ y^{(m+1)}(x) &= Sy^{(m)}(x), \quad m = 0,\ 1,\ \cdots. \end{aligned} \tag{7.1.30}$$

Theorem 7.1.10 *Let Assumptions 7.1.9 be satisfied, and let the sequence of functions $\{y^{(m)}\}$ be defined for $x \in \Gamma$ by Equation (7.1.30). Then $y^{(m)} \to Y$, where Y is the unique solution in the ball $\mathcal{B}(g;r)$ of the initial value problem (7.1.27). Moreover, for any fixed positive integer m, we have $\|y^{(m)} - Y\| \leqq \beta^m r$.*

Proof. The proof is a straightforward application of the Contraction Mapping Principle, and is omitted (see Theorem 6.2.1, and Example 6.2.2). ■

Let us now state another version of Theorem 7.1.10, which exhibits the more rapid rate of convergence that is typical of initial value problems. We start by reformulating Assumption 7.1.9.

Assumption 7.1.11 *Let the following conditions be satisfied:*

(a) $Sy \in \mathcal{B}(g;r) \cap \mathbf{M}_\alpha(\mathcal{D})$ *for all* $y \in \mathcal{B}(g;r)$

(b) *There exists a function* ω *defined on* Γ, *such that for all* $t \in \Gamma$ *we have* $|\omega(t)/\phi'(t)| \leqq C\,\rho(t)^\alpha/[1+\rho(t)]^{2\alpha}$, *with* C *a constant, and* ρ *defined as in Definition 4.1.1, such that for all* u *and* v *in* $\mathcal{B}(g;r)$ *and all* $t \in \Gamma$ *we have*

$$|F(t,u(t)) - F(t,v(t))| \leqq |\omega(t)|\,|u(t)-v(t)| \tag{7.1.31}$$

(c) $\|Sg - g\| < r\,e^{-\Omega(b)}$ *for all* $x \in \Gamma$, *with* $\Omega(x)$ *defined by* $\Omega(x) = \int_a^x |\omega(t)\,dt|$, *and with the integration taken along the contour* Γ.

Theorem 7.1.12 *If Assumptions 7.1.11 are satisfied, and if* $\{y^{(m)}\}$ *is defined as in (7.1.30), then* $y^{(m)} \to Y$ *as* $m \to \infty$, *with* Y *the unique solution in* $\mathcal{B}(g;r)$ *of the initial value problem (7.1.27). Moreover, for any fixed positive integer* m, *we have*

$$\|y^{(m)} - Y\| \leqq \frac{[\Omega(b)]^m}{m!} r. \tag{7.1.32}$$

Proof. : We leave the straightforward proof of this theorem to Problem 7.1.7. ■

Let us now turn to the approximate solution of (7.1.27) by application of the Sinc indefinite integration formula (4.5.9) to approximate the indefinite integral in (7.1.29). To this end, we select $h = [\pi d/(\alpha N)]^{1/2}$, and we form approximations y_j to the solution $y(x)$ evaluated at the Sinc points z_j as solutions of the system of $2N+1$ non-linear equations

$$\mathbf{y} = \mathbf{S}(\mathbf{y}) \equiv h I^{(-1)} D(1/\phi')\mathbf{f} + \mathbf{y}_a, \tag{7.1.33}$$

where $\mathbf{y} = (y_{-N}, \cdots, y_N)^T$, $\mathbf{f} = (F(z_{-N}, y_{-N}), \cdots, F(z_N, y_N))^T$ and where $\mathbf{y}_a$ denotes the vector of $2N+1$ constant values, $\mathbf{y}_a = (y_a, \cdots, y_a)^T$.

At the outset, we consider a solution of the system (7.1.33) by successive approximations, i.e.,

$$\mathbf{y}^{(m+1)} = \mathbf{S}(\mathbf{y}^{(m)}), \quad m = 0, 1, 2, \cdots, \tag{7.1.34}$$

where, with the notation of either Theorem 7.1.9 or Theorem 7.1.11, $\mathbf{y}^{(0)} = (g(z_{-N}), \cdots, g(z_N))^T$.

Theorem 7.1.13 *Let either Assumptions 7.1.9 or 7.1.11 be satisfied. Then the sequence* $\{\mathbf{y}^{(m)}\}$ *defined (7.1.34) converges for all* N *sufficiently large, to a unique vector* $\mathbf{y} = (y_{-N}, \cdots, y_N)^T$. *Moreover, if we define* $y_N(x)$ *by*

$$y_N(x) = \frac{\rho(x) y_N}{1+\rho(x)} + \sum_{j=-N}^{N} \left(y_j - \frac{e^{jh} y_N}{1+e^{jh}} \right) S(j,h) \circ \phi(x), \qquad (7.1.35)$$

and if Y *denotes the unique solution of Equation (7.1.27) as defined in either Theorem 7.1.10 or Theorem 7.1.12, then there exists a constant* C *independent of* N, *such that*

$$\sup_{x \in \Gamma} |Y(x) - y_N(x)| \leqq C N^{1/2} e^{-(\pi d \alpha N)^{1/2}}. \qquad (7.1.36)$$

It is worthwhile to mention that as in the case of linear equations, in practice it is nearly always more efficient to use the *Seidel* iteration, instead of the Neumann iteration described in (7.1.34). In using the Seidel version, we evaluate the equations of the system $\mathbf{y} = \mathbf{S}\mathbf{y}$ one at a time, starting at the top and working down. We then use the newly computed component of $\mathbf{y}$ on the right-hand side of the next equation immediately after it has been computed.

Finally, we mention that the above iteration scheme is practical only for non-stiff problems. We recommend Newton's method (described below) for stiff problems.

Example 7.1.14 Let us apply the method (7.1.34) to the problem

$$\begin{aligned} \frac{dy}{dx} &= \frac{y(1-y)}{2\,y-1}, \quad 0 < x < b, \\ y(0) &= \tfrac{5}{6} \end{aligned} \qquad (7.1.37)$$

where b will be specified below, whose exact solution is

$$y(x) = 1/2 + \sqrt{1/4 - 5e^{-x}/36}. \qquad (7.1.38)$$

We first solve the problem (7.1.37) on the interval $(0,1)$, using $\phi(x) = \log[x/(1-x)]$. There is no suggestion of any singularity in the solution from the form of the differential equation, and so we take $\alpha = 1$. We also take $d = \pi/2$, assuming, at the outset, that the solution is analytic in the unit disc. This, then, determines $h = [\pi d/(\alpha N)]^{1/2} = \pi/(2N)^{1/2}$. It then turns out that the iteration scheme (7.1.34) converges for all positive integers N, For example, with $N = 20$, we are able to compute the exact solution of (7.1.37) correct to five decimals on $[0,1]$.
Next, we solve (7.1.37) on $(0,\infty)$, using the map $\phi(x) = \log[\sinh(x)]$, but also using the same values for α and d. In this case the iteration scheme (7.1.34) diverges, for example, for $N \leqq 10$ but converges for $N \geqq 20$. For

$N = 20$, we get nearly five decimals accuracy of the solution on the interval $[0, \infty]$. We remark that in this latter case, about 160 iterations of the scheme (6.1.34) are required to achieve convergence. By using the Seidel version, only about 20 iterations are required. Using Newton's method (see below), we achieve convergence in 8 iterations.

The choices of α and d which we used in the above example determine the optimal value of h, and we arrived at this optimal value by inspection of the exact solution. If the optimal value of h is not known, then by taking $h = \gamma/N^{1/2}$, γ a positive constant, the resulting error of the approximation converges at the rate $\mathcal{O}(\exp[-\delta N^{1/2}])$, with δ a related positive constant. If γ is smaller than the optimal value, then the computed solution is more accurate (than that of the approximation with optimal γ) around the mid-range of the interval, but it is less accurate near the ends of the interval, while if γ is larger than the optimal value, then we get worse accuracy around the mid-range of the interval, but better accuracy near the ends of the interval. This property may provide a practical method for selecting a close to optimal h depending on N in an automatic program of this method of approximate solution.

We now turn to the case when (7.1.27) denotes a system of n non-linear differential equations, in which $y = (y_1(x), \cdots, y_n(x))^T$, and $F(x, y) = (F_1(x, y), \cdots, F_n(x, y))^T$. In this case we can convert to a Volterra integral equation, in the very notation of (7.1.29). We could then again state and prove theorems analogous to Theorem 7.1.10 and 7.1.12, with only slight modifications of the assumptions preceding these theorems, such as changing the norm from $\|y\| = \sup_{(x \in \Gamma)} |y(x)|$, which was used in the scalar case, to

$$\|y\| = \max_{i=1,\cdots,n} \left(\sup_{x \in \Gamma} |y_i(x)|, \right) \tag{7.1.39}$$

with $y(x) = (y_1(x), \cdots, y_n(x))^T$.
We discretize Equation (7.1.29) by the application of the indefinite integration formula (4.5.9) to (7.1.29), to arrive at the system of non-linear algebraic equations

$$\mathbf{y} - \mathbf{S}\mathbf{y} = \mathbf{y}_a, \tag{7.1.40}$$

where

$$\mathbf{S}\mathbf{y} = I_1 \otimes \left\{ h\, I^{(-1)} D(1/\phi') \right\} \mathbf{f}. \tag{7.1.41}$$

Here, I_1 is the unit matrix of order n,

$$\begin{array}{rcl} \mathbf{y} & = & (\eta_1, \cdots, \eta_n)^T \\ \eta_j & \equiv & (y_{j,-N}, \cdots, y_{j,N}) \\ \mathbf{f} & = & (\kappa_1, \cdots, \kappa_n)^T \\ \kappa_j & \equiv & (F_j(z_{-N}, Y_{-N}), \cdots, F_j(z_N, Y_N)) \\ Y_\ell & = & (y_{1,\ell}, \cdots, y_{n,\ell})^T \\ \mathbf{y}_a & = & (y_{1,a}, \cdots, y_{n,a})^T, \end{array} \tag{7.1.42}$$

where $y_{j,\ell}$ is used to approximate $y_j(x)$ at the Sinc point $x = z_\ell$, and where $y_{i,a}$ denotes the row vector of $(2N+1)$ constant values $y_i(a)$. We can then consider solving the system (7.1.40) via the successive approximation scheme (7.1.34).

Towards this end, we first consider a definition of stiffness, as in the case of linear initial value problems. Namely, *we shall say that the system (7.1.40) is not stiff, if the sequence of vectors $\mathbf{y}^{(m)}$ defined by (7.1.34) converges, starting with some vector $\mathbf{y}^{(0)}$. Otherwise, we shall say that the system (7.1.40) is stiff.*

Thus, if the system (7.1.40) is not stiff, then the sequence of vectors $\mathbf{y}^{(m)}$ defined as in (7.1.34) does converge to the exact solution of (7.1.40). Moreover, the resulting constant vector $\mathbf{y}$ of order $n(2N+1)$ can then be used to define an approximate solution vector $y_N(x)$, by setting

$$y_N(x) = \frac{\rho(x)}{1+\rho(x)} Y_N + \sum_{j=-N}^{N} \left[Y_j - \frac{e^{jh}}{1+e^{jh}} Y_N \right] S(j,h) \circ \phi(x), \tag{7.1.43}$$

with the Y_j defined as in (7.1.42), and upon defining $\|y\|$ as in (7.1.39) it follows that

$$\|y - y_N\| \leqq C N^{1/2} e^{-(\pi d \alpha N)^{1/2}}, \tag{7.1.44}$$

with C a constant independent of N.

In the case when the system (7.1.40) is stiff, it is worthwhile to keep in mind that this system of equations does, in fact, have a solution with the property where the resulting approximation defined in (7.1.42) is accurate to within the bound given in (7.1.44). It may be worthwhile, in such instances, to solve the system (7.1.40) by Newton's method, which we now briefly describe.

Set $T = I - \mathbf{S}$, where I denotes the unit matrix of order $n(2N+1)$, and where $\mathbf{S}$ is defined as in (7.1.40). The *Fréchet derivative* equation corresponding to (7.1.40) is defined by the expression

$$(\mathbf{T}'_{\mathbf{y}})\eta \equiv \left[I_1 \otimes I_2 - I_1 \otimes \{h I^{(-1)} D(1/\phi')\}[B_{ij}] \right] \eta \tag{7.1.45}$$

where, upon denoting $\eta_{i,\ell}$ to be the Fréchet derivative of $y_{i,\ell}$, we set $\eta = (\eta_{1,-N}, \cdots, \eta_{1,N}, \eta_{2,-N}, \cdots, \eta_{n,N})^T$. Also, letting $F_{i,j}(x, y_1, \cdots, y_n)$ denote the derivative of $F_i(x, y_1, \cdots, y_n)$ with respect to y_j, and letting Y_j be defined as in (7.1.42), we define the diagonal matrix B_{ij} in (7.1.45) by $B_{ij} = \mathrm{diag}(F_{i,j}(z_{-N}, Y_{-N}), \cdots, F_{i,j}(z_N, Y_N))$.

Newton's method, starting with an initial vector $\mathbf{y}^{(0)}$, then takes the form

$$\mathbf{T}'_{\mathbf{y}^{(m)}} \left(\mathbf{y}^{(m+1)} - \mathbf{y}^{(m)}\right) = -\mathbf{T}(\mathbf{y}^{(m)}) + \mathbf{y}_a. \tag{7.1.46}$$

We end this subsection with an example, illustrating that Newton's method may also be applied to the solution of linear equations, and also, illustrating that, at times, it may be advantageous to select h by a formula other than $(\pi d/[\alpha N])^{1/2}$.

Example 7.1.15 Consider the solution of the initial value problem

$$\begin{aligned} y'' - y + 2/\cosh^3(x) = 0, \quad x \in \mathbb{R}, \\ y(-\infty) = y'(-\infty) = 0, \end{aligned} \tag{7.1.47}$$

which has the exact solution, $y(x) = 1/\cosh(x)$. Let us first transform this second-order equation into a system of two first-order equations,

$$\begin{aligned} y' &= z \\ z' &= y - 2/\cosh^3(x), \end{aligned} \tag{7.1.48}$$

which is to be solved, subject to the initial conditions $(y(-\infty), z(-\infty))^T = (0,0)^T$. One may note, by inspection of either (7.1.47) or (7.1.48), that the solutions of these equations are necessarily analytic in the strip $\mathcal{D}_{\pi/2}$, and moreover, that we must have $y(x) = \mathcal{O}(e^{-|x|})$, as $x \to \pm\infty$. Thus, by taking ϕ to be the identity map, $d = \pi/2$, and $\alpha = 1$, we would then be able to construct a solution to within an error of the order of $\mathcal{O}(N^{1/2} e^{-(\pi d \alpha N)^{1/2}})$ as $N \to \infty$. On the other hand, we note that the coefficients of either (7.1.47) or (7.1.48) are analytic in the region $\mathcal{D}^4_{\pi/2}$ as defined in (1.7.13), and hence we may take $\phi = \log\{x + \sqrt{1 + x^2}\}$. Since the Sinc points are now given by $z_k = \sinh(kh)$, it follows, in view of the rate of decrease of the solution y as $x \to \pm\infty$, that the truncation error, using an approximation of the form (7.1.43) is $\mathcal{O}(\exp(-\sinh(Nh))) = \mathcal{O}(\exp(-(1/2)e^{Nh}))$. Upon taking $N = 20$, and solving the equation $\exp\{-\pi^2/(2h)\} = \exp\{-(1/2)e^{Nh}\}$, we get $h \approx 0.19$. Then, upon solving (7.1.48) with this choice of h and N, Newton's method converges in two iterations to a solution with maximum error of about 7×10^{-11} on $\mathbb{R}$.

•

Problems for Section 7.1

7.1.1 Given the system of differential equations

$$(1) \qquad \frac{d\mathbf{w}}{dz} = A(z)\,\mathbf{w}, \quad \mathbf{w}(a) = \mathbf{w}_a,$$

where for all z in the disc $\mathcal{D}(a;R)$ of center a and radius R we have

$$A(z) = \sum_{k=0}^{\infty} A_k (z-a)^k$$

and where $A_0, A_1, A_2, \cdots$ are constant $n \times n$ matrices.

(a) Show that if a sequence of vectors $\mathbf{w}_k$ are determined sequentially by the equations

$$\mathbf{w}_{k+1} = \frac{1}{k+1} \sum_{j=0}^{k} A_{k-j}\mathbf{w}_j, \quad k = 0, 1, 2, \cdots$$

then

$$\mathbf{w}(z) = \sum_{j=0}^{\infty} \mathbf{w}_j (z-a)^j$$

is, in fact, the unique solution of equation (1) above.

(b) Show that this series representation of $\mathbf{w}(z)$ converges for all $z \in \mathcal{D}(a;R)$.

(c) Show that the system (1) has the equivalent integral equation representation

$$\mathbf{w}(z) = \mathbf{w}_a + \int_a^z A(t)\,\mathbf{w}(t)\,dt.$$

(d) Prove that we can also find a power series solution in a neighborhood of ∞, provided that the system of equations takes the form

$$\frac{d\mathbf{w}}{dz} = z^{-2} A(z)\mathbf{w}, \quad \mathbf{w}(\infty) = \mathbf{w}_a,$$

and provided that the domain of analyticity of A contains the region $\mathcal{D}'_R = \{z \in \mathbb{C} : |z| > R\}$, i.e., for all $z \in \mathcal{D}'_R$ we have

$$A(z) = \sum_{k=0}^{\infty} A_k z^{-k}.$$

(e) Show that if K and g satisfy the conditions of Theorem 7.1.1, then the solution y to the problem (7.1.1) belongs to $\mathbf{M}_\alpha(\mathcal{D})$.

7.1.2 * Let $R > 0$, let $\mathcal{D}'_R$ denote the region $\{z \in \mathbb{C} : |z| > R\}$, and let the system of differential equations

$$(2a) \qquad \frac{d\mathbf{w}}{dz} = z^{-1} A(z) \mathbf{w}$$

be given, where each element of the $n \times n$ matrix A is analytic in $\mathcal{D}'_R$, where it has a convergent series expansion of the form

$$A(z) = \sum_{k=0}^{\infty} A_k z^{-k}.$$

The *Canonical form* of the system (2a) has the form

$$(2b) \qquad \frac{d\mathbf{v}}{dz} = z^{-1} B(z) \mathbf{v},$$

where

$$B(z) = \sum_{k=0}^{\infty} B_k z^{-k}$$

for all $|z| > R' \geqq R$, and where B_0 is the Jordan canonical form of A_0, with every Jordan block being identified by its distinct eigenvalue, λ_j, and where $0 \leqq \Re \lambda_j < 1$.

(a) Prove that the transformation from (2a) to (2b) can always be carried out by means of a transformation of the form

$$\mathbf{w} = \sum_{j=-M}^{N} T_j z^j$$

with the T_j matrices, and with M and N some non negative integers.

(b) Prove that the system (2b) has a series solution matrix of the form

$$(2c) \qquad V(z) = \left(\sum_{k=0}^{\infty} V_k z^{-k} \right) \exp[B_0 \log(z)]$$

with $V_0 = I$ (the unit matrix) and with V_k given by the unique solution of

$$k\, V_k + B_0 V_k - V_k B_0 = - \sum_{\ell=1}^{k} B_\ell V_{k-\ell}.$$

Moreover, the series in (2c) converges for all $z \in \mathcal{D}'_{R'}$.

(c) Show that if $\mathbf{v}_m$ denotes a partial sum of the block of vectors corresponding to the eigenvalue λ_j in the series solution (2c), that is,

$$\mathbf{v}_m(z) = \left(\sum_{k=0}^{m-1} \mathbf{v}_k z^{-k} \right) \exp\left[(\lambda_j I_{\mu_j} + J_{\mu_j}) \log(z) \right],$$

and we define η by

$$\eta(z) = \mathbf{v}(z) - \mathbf{v}_m(z),$$

then η satisfies the integral equation

$$\begin{aligned} \eta(z) \quad = \quad & - \int_z^\infty \exp\left[B_0 \log(z/t)\right] \left[t^{-2} B_1(t)\, \eta(t) \right. \\ & \left. - R_m(t) \exp\left[B_0 \log(t)\right]\right] dt, \end{aligned}$$

with

$$R_m(z) \exp\left[(\lambda_j I_{\mu_j} + J_{\mu_j}) \log(z) \right] = -\frac{d}{dz}\eta(z) - z^{-1} B(z) \eta(z)$$

7.1.3 * Let ν denote a positive integer, and consider the system of differential equations

$$\text{(3a)} \qquad \frac{d\mathbf{w}}{dz} = z^{\nu-1} A(z)\, \mathbf{w},$$

with $A(z)$ defined as in in Problem 7.1.2 above, and with ν a positive integer. If the eigenvalues of the matrix A_0 are distinct, then the system (3a) has the canonical form

$$\text{(3b)} \qquad \frac{d\mathbf{v}}{dz} = \left[z^{\nu-1} C'(z) + z^{-2} B(z) \right] \mathbf{v},$$

with

$$C'(z) = \sum_{k=0}^{\nu} C_k z^{-k}$$

and where, for arbitrary positive integer s,

$$B(z) = \sum_{k=0}^{s-1} B_k z^{-k} + B_s(z)\, z^{-s},$$

the elements of $B_s(z)$ being holomorphic and bounded for all $|z| > R'$. Moreover, the matrix $C'(z)$ is a diagonal matrix.

(a) Prove that there exist $n \times n$ matrices $T_0,\ T_1, \cdots, T_\nu$, such that when the transformation

$$\mathbf{w} = T(z)\,\mathbf{v} = \left(\sum_{k=0}^{\nu} T_k z^{-k}\right) \mathbf{v}$$

is applied to (3a), we obtain the system in canonical form (3b).

(b) Prove that the system of equations (3b) has a formal solution of the form

$$(3c) \qquad V(z) = U(z)\, \exp[C(z)],$$

with

$$C(z) = \sum_{k=0}^{\nu-1} \frac{C_k}{\nu - k} z^{\nu-k} + C_\nu \log(z),$$

where $U(z)$ has the formal (and generally non-convergent) series expansion

$$(3d) \quad U(z) = I + \sum_{k=1}^{\infty} U_k z^{-k}$$

which is obtained by substituting (3c) into (3b) and then equating equal powers of z.

(c) Derive a Volterra integral equation, involving integration from z to ∞, for the difference between the $j\,th$ vector of the matrix U in (3d) and a partial sum of an m-term asymptotic approximation of this vector, in the case when the $j\,th$ element of C_0 is an extreme point of the smallest closed and convex polygon containing all of the elements of C_0.

(d) Derive a system of two Volterra integral equations, each involving integration from z to ∞, but each integration extending to infinity in a different direction, for the difference between the $j\,th$ vector of the matrix U in (3d) and a partial sum of an m-term asymptotic approximation of this vector, in the case when the $j\,th$ element of C_0 is an interior point of the smallest closed and convex polygon containing all of the elements of C_0. [Hint: See Stenger, S3.]

7.1.4 Give a detailed proof of Theorem 7.1.2.

7.1.5 Prove that if the length of the contour Γ is finite, then taking y_0 to be the null vector, the sequence of vectors $\mathbf{y}^{(m)}$ thus produced by the iteration scheme (7.1.15) converges to the exact solution of the system (7.1.10) for all N sufficiently large.

7.1.6 Give a detailed proof of Theorem 7.1.8. [Hint: By repeated differentiation of both sides of the differential equation $y' = F(x, y)$, prove using the coefficient bounds given by Taylor's Theorem that for $k = 1, 2, 3, \cdots$, we have $|y^{(k)}(a)| \leqq |Y^{(k)}(0)|$, with $Y(x)$ the solution of the differential equation

$$(6) \qquad \frac{dY}{dx} = G(x, Y)$$

and with

$$G(x, Y) = \frac{M}{\left(1 - \frac{x}{r}\right)\left(1 - \frac{Y}{R}\right)}.$$

Now find the explicit solution of Equation (6), subject to the initial condition $Y(0) = 0$, and deduce that this solution is analytic in $\mathcal{D}(0; \rho)$, with ρ given in (7.1.28).]

7.1.7 Prove Theorem 7.1.12.

7.1.8 Prove that only a negligible $\mathcal{O}(e^{-(\pi d \alpha N)^{1/2}})$ error is introduced if we define $y_N(x)$ by

$$y_N(x) = \frac{y_a + \rho(x) y_N}{1 + \rho(x)} + \sum_{j=-N}^{N} \left[y_j - \frac{y_a + e^{jh} y_N}{1 + e^{jh}} \right] S(j, h) \circ \phi(x)$$

instead of by (7.1.8).

7.1.9 Give a detailed proof of Lemma 7.1.6. [Hint: Conclude the result from the fact that if λ is an eigenvalue of $I^{(-1)}$, then $\Re \lambda > 0$ (see Problem 3.6.8).]

7.1.10 *Complementing the Limitation of Asymptotic Expansions.* Let $\mathcal{D}$ be a simply connected domain, with which we associate the notation of Section 4.1, such that Γ has the properties that its endpoints are a and b, where $b = \infty$, and also, that $\Re x \leqq \Re t$ for all points x and t on Γ in the order a, x, t, b . Given the differential equation

$$(1) \quad y''(x) + [1 + Q(x)]y(x) = 0, \quad x \in \Gamma,$$

with $Q/\phi' \in \mathbf{L}_\alpha(\mathcal{D})$, let U be a given function, such that if R is defined by

$$(2) \qquad R = -[U'' - 2U' - QU],$$

then $R/[\phi']^2 \in \mathbf{L}_\alpha(\mathcal{D})$.

(i) Prove (see Olver [O1]) that the Problem (1) has a solution $y = (U + V)e^{-x}$, with V a solution to the integral equation

$$(3) \quad V(x) = \int_x^\infty \frac{1 - e^{2x-2t}}{2} [Q(t)V(t) + R(t)]\, dt.$$

(ii) Use the method of Section 4.6 to describe an explicit procedure for approximating the solution to (3) on Γ to any preassigned accuracy.

Remark: In this problem, the term $U(x)e^{-x}$ is an approximate solution, which may be obtained by asymptotic methods, such as the *WKBJ* method (see Olver [O1]), or by the procedure of Problem 7.1.3. It is usually not possible to produce a solution to arbitrary high accuracy on Γ via an asymptotic method. The procedure of the present problem enables us to circumvent this limitation of asymptotics.

7.1.11 Solve the problem (7.1.47) in "scalar form", via the use of the method of Section 4.6, that is, by using the expression

$$\mathbf{y} = \left[hI^{(-1)}D(1/\phi')\right]^2 \mathbf{g},$$

with $\mathbf{g} = (g(z_{-N}), \cdots, g(z_N))^T$, and $g(z) = -2/\cosh^3(z)$. How does this method compare with the one described in Example 7.1.15? Discuss this result in light of the method of Section 4.6.

7.2 ODE–BVP

In this section we derive basic formulas for solving ordinary differential equation (ODE) boundary value problems (BVP). For the sake of simplicity of presentation, we shall almost exclusively consider only the cases of second order equations. At the outset, below, we present a sketch of the general Galerkin method for obtaining an approximate solution to an ODE–BVP, and we state the general linear ODE–BVP which is the model differential equation that is used in the sections which follow, for deriving the *Sinc–Galerkin methods.* In Subsection 7.2.1 we derive the Sinc–Galerkin method which illustrates best that Galerkin and collocation procedures are the same when using a Sinc basis to approximate the solution of a differential equation. This collocation method is probably the simplest to derive, and to implement. In Subsection 7.2.2 we use an integration by parts procedure to derive the Sinc–Galerkin method which was obtained first, historically, and which has an accuracy equal to that of the Sinc–Galerkin method of Subsection 7.2.1. This method complements the one of Subsection 7.2.1, in that, when we take the average of this method and the one in Subsection 7.2.1, we obtain yet a third method, which we present in Subsection 7.2.3. This third method has the same accuracy as the other two, but it has the desirable feature that its convergence can be easily established. In Subsection 7.2.4 we prove the convergence of this third method, and we also demonstrate that the first two methods then also converge, under the same conditions. In Subsection 7.2.5 we illustrate how the Sinc–Galerkin method may be used to replace a non-linear ODE–BVP by an explicit system of non-inear algebraic equations. In Subsection 7.2.6 we discuss the treatment of non-homogeneous boundary conditions when using the Sinc–Galerkin method. In Subsection 7.2.7 we derive the *Symmetric Sinc–Galerkin method*, which has the advantage of transforming a self-adjoint differential equation into a self-adjoint algebra problem. Finally, in Subsection 7.2.8 we present several illustrative examples of the application of Sinc–Galerkin methods to solve ODE–BVP.

We will now briefly describe the general *Galerkin method* that is typically used to solve a linear or non-linear differential equation of the form

$$Ly(x) = 0, \quad x \in \Gamma, \quad y(a) = y(b) = 0. \tag{7.2.1}$$

It is customary to assume for Problem (7.2.1) that there exists a suitable *Hilbert space* H, on which the *inner product* (f, g) is defined. It is then assumed that (Lf, g) is defined and finite for all functions f and g on a suitable subset B of H containing the solution to (7.2.1). A set of basis functions $\{\psi_j\}$, where $\psi_j \in B$, is then used to approximate the solution y to (7.2.1), by means of a linear combination of the form

$$y_n = \sum_{j=1}^{n} w_j \psi_j, \tag{7.2.2}$$

for which the coefficients w_j are determined by solving the system of equations

$$(Ly_n, \psi_j) = 0, \quad j = 1, \cdots, n. \tag{7.2.3}$$

The system of equations (7.2.3) will be linear or non-linear depending on whether the equation $Ly = 0$ is linear or non-linear in y.

In applications, we would like the above procedure to have the following properties:

(i) It is not unduly difficult to deduce the explicit system of equations defined by (7.2.3);
(ii) The resulting system of equations (7.2.3) is neither unduly difficult to solve, nor extremely ill-conditioned; and
(iii) It is not unduly difficult to estimate the accuracy of the resulting approximation y_n defined by (7.2.2) and (7.2.3). Moreover, it is desirable that this approximation be reasonably accurate.

We shall see in what follows, that the Sinc–Galerkin methods which we derive in this section *do* satisfy all of the above criteria. In addition, we shall see that when using a Sinc basis, Galerkin and collocation become synonymous.

It is convenient, for purposes of deriving Sinc–Galerkin methods, to start with the model problem

$$\begin{aligned} (Ly)(x) &= y''(x) + \mu(x)y'(x) + \nu(x)y(x) - \sigma(x) = 0, \quad x \in \Gamma, \\ y(a) &= y(b) = 0, \end{aligned} \tag{7.2.4}$$

and to use the notation of Section 4.1. We shall approximate the solution of the problem (7.2.4) with a linear combination of Sinc functions, that is,

$$y(x) \approx y_N(x) \equiv \sum_{j=-N}^{N} w_j S(j,h) \circ \phi(x). \tag{7.2.5}$$

7.2.1 Sinc–Galerkin and Collocation

While it is possible to proceed under rather weak assumptions, it is more straightforward and convenient to discuss (7.2.4) under the assumption that it has a solution $y \in \mathbf{L}_\alpha(\mathcal{D})$. It will furthermore be seen below, that a convenient inner product is

$$(f,g) \equiv \int_\Gamma \frac{f(x)\overline{g(x)}}{\phi'(x)} dx. \tag{7.2.6}$$

Assumption 7.2.1 *We assume for the differential equation (7.2.4) that μ/ϕ', $(1/\phi')'$, and $\nu/[\phi']^2$ belong to $\mathbf{H}^\infty(\mathcal{D})$, that $\sigma/[\phi']^2 \in \mathbf{L}_\alpha(\mathcal{D})$, and that the Problem (7.2.4) has a unique solution $y \in \mathbf{L}_\alpha(\mathcal{D})$.*

Throughout this section we shall take

$$h = \left(\frac{\pi d}{\alpha N}\right)^{1/2} \tag{7.2.7}$$

with N as in (7.2.5). Upon replacing y in the differential equation (7.2.4) by y_N as in (7.2.5), multiplying the resulting equation by $\{h/\phi'(z)\}^2$, and then setting $z = z_k$, we get the *collocation result*

$$\frac{h^2\, Ly(z_k)}{[\phi'(z_k)]^2}$$

$$= h^2 \frac{y_N''(z_k)}{[\phi'(z_k)]^2} + h^2 \frac{\mu(z_k) y_N'(z_k)}{[\phi'(z_k)]^2} + h^2 \frac{\nu(z_k) y_N(z_k)}{[\phi'(z_k)]^2} - h^2 \frac{\sigma(z_k)}{[\phi'(z_k)]^2} = 0. \tag{7.2.8}$$

We see now, that from the viewpoint of making a connection with the integration by parts procedure presented in the next subsection, the result (7.2.8) can also be obtained, if after replacing y in (7.2.4) by y_N of (7.2.5), we multiply the resulting equation by $h[\phi'(x)]^{-1} S(k,h) \circ \phi(x)$, integrate over Γ, and then use the approximation given in Corollary 4.2.15.
Using the explicitly form of $y_N(x)$ of (7.2.5), we recall the notation $\delta_k^{(m)}$ as defined in (1.10.19), and then rewrite (7.2.8) as

$$\sum_{j=-N}^{N} \left\{ \delta_{j-k}^{(2)} + h \left[\frac{\mu(z_k)}{\phi'(z_k)} + \frac{\phi''(z_k)}{[\phi'(z_k)]^2} \right] \delta_{j-k}^{(1)} + h^2 \frac{\nu(z_k)}{[\phi'(z_k)]^2} \delta_{j-k}^{(0)} \right\} w_j$$
$$= h^2 \frac{\sigma(z_k)}{[\phi'(z_k)]^2}. \tag{7.2.9}$$

Now, since $\delta_{j-k}^{(2)} = \delta_{k-j}^{(2)}$, $\delta_{j-k}^{(1)} = -\delta_{k-j}^{(1)}$, and $\delta_{j-k}^{(0)} = \delta_{k-j}^{(0)}$, and since $\phi''/[\phi']^2 = -(1/\phi')'$, we may rewrite the above equation in the form

$$\sum_{j=-N}^{N} \left\{ \delta_{k-j}^{(2)} + h \left[\left(\frac{1}{\phi'} \right)' - \frac{\mu}{\phi'} \right] (z_k) \delta_{k-j}^{(1)} + h^2 \frac{\nu(z_k)}{[\phi'(z_k)]^2} \delta_{k-j}^{(0)} \right\} w_j$$
$$= h^2 \frac{\sigma(z_k)}{[\phi'(z_k)]^2}. \tag{7.2.10}$$

We now rewrite these equations in matrix form. Corresponding to a given function u defined on Γ, we use the notation
$D(u) = \mathrm{diag}(u(z_{-N}), \cdots, u(z_N))$, we set $I^{(m)} = [\delta_{i-j}^{(m)}]$, where $\delta_{i-j}^{(m)}$ denotes the (i,j)th element of the matrix $I^{(m)}$, and we define A and $\mathbf{p}$ by

$$\begin{aligned}
A &= I^{(2)} + hD\left(\left(\frac{1}{\phi'} \right)' - \frac{\mu}{\phi'} \right) I^{(1)} + h^2 D\left(\frac{\nu}{[\phi']^2} \right), \\
\mathbf{p} &= h^2 D\left(\frac{1}{[\phi']^2} \right) \mathbf{s}, \\
\mathbf{s} &= (\sigma(z_{-N}), \cdots, \sigma(z_N))^T.
\end{aligned} \tag{7.2.11}$$

We then get the system of equations

$$A\mathbf{w} = \mathbf{p} \tag{7.2.12}$$

for the vector $\mathbf{w}$, the solution of which yields the coefficients w_j in the approximation $y_N(x)$ of $y(x)$.

7.2.2 Integration by Parts

The integration by parts procedure which we shall describe in this subsection was, in fact, first obtained historically. To this end, it is convenient to make Assumption 7.2.2, which is stronger than Assumption 7.2.1. We mention, however, that this stronger assumption is made solely to simplify the proof of convergence of the integration by parts procedure. We could also have proved convergence assuming only Assumption 7.2.1, although the proof would then be more complicated in detail.

Assumption 7.2.2 *In addition to Assumption 7.2.1 being satisfied, assume that both of the functions $(\mu/\phi')'/\phi'$ and $(1/\phi')''/\phi'$ belong to $\mathbf{H}^\infty(\mathcal{D})$.*

We may observe, by Theorem 4.1.2 and Problem 4.1.7, that if Assumption 7.2.1 is satisfied in a domain $\mathcal{D}'$, then Assumption 7.2.2 is satisfied in a domain $\mathcal{D}$ which is related to $\mathcal{D}'$, as in Theorem 4.1.2. Assumption 7.2.2 is made in order to simplify the proof of convergence of Sinc–Galerkin methods. The same bounds on the errors which we shall obtain under Assumption 7.2.2 could be obtained via a more complicated proof, assuming only Assumption 7.2.1.

As in the previous subsection, we again replace y by y_N in the differential equation (7.2.4), we multiply the differential equation (7.2.4) by $h[\phi'(x)]^{-1}S(k,h)\circ\phi(x)$ and integrate over Γ. However, instead of using Corollary 4.2.15 directly, we first integrate by parts in order to replace derivatives of y_N by y_N. We thus get

$$\begin{aligned}
&h\int_\Gamma \frac{S(k,h)\circ\phi(x)}{\phi'(x)} y_N''(x)dx \\
&\quad = h\left(\frac{S(k,h)\circ\phi(x)}{\phi'(x)}\right) y_N'(x)\Big|_a^b - h\int_\Gamma \left(\frac{S(k,h)\circ\phi(x)}{\phi'(x)}\right)' y_N'(x)dx \\
&\quad = -h\left(\frac{S(k,h)\circ\phi(x)}{\phi'(x)}\right)' y_N(x)\Big|_a^b + h\int_\Gamma \left(\frac{S(k,h)\circ\phi(x)}{\phi'(x)}\right)'' y_N(x)dx \\
&\quad = h\int_\Gamma \left(\frac{S(k,h)\circ\phi(x)}{\phi'(x)}\right)'' y_N(x)dx,
\end{aligned} \tag{7.2.13}$$

since it is readily seen from Problem 4.1.7 that the boundary terms vanish. By applying the approximation of Corollary 4.2.15 to the last integral expression in (7.2.13), we now get the approximation

$$h \int_\Gamma \frac{y_N''(x) S(k,h) \circ \phi(x)}{\phi'(x)}\, dx$$
$$= \sum_{j=-N}^{N} \left\{ \delta_{k-j}^{(2)} + h \delta_{k-j}^{(1)} (1/\phi')'(z_j) + h^2 \delta_{k-j}^{(0)} (1/\phi')''(z_j)/\phi'(z_j) \right\} w_j. \tag{7.2.14}$$

Similarly, we find under Assumption 7.2.2, and the identity in Corollary 4.2.15, that

$$h \int_\Gamma \frac{\mu(x) y_N'(x) S(k,h) \circ \phi(x)}{\phi'(x)}\, dx$$
$$= -h \sum_{j=-N}^{N} \left\{ \delta_{k-j}^{(1)} \frac{\mu(z_j)}{\phi'(z_j)} + \delta_{k-j}^{(0)} \left(\frac{\mu'(z_j)}{[\phi'(z_j)]^2} + \frac{\mu(z_j)}{\phi'(z_j)} \left(\frac{1}{\phi'}\right)' (z_j) \right) \right\} w_j. \tag{7.2.15}$$

No integration by parts is required for the remaining terms of the equation $h \int_\Gamma L y_N(x) [\phi'(x)]^{-1} S(k,h) \circ \phi(x) dx = 0$. Upon collecting terms, we arrive at the system of equations

$$B\mathbf{w} = \mathbf{p} \tag{7.2.16}$$

where $\mathbf{w}$ and $\mathbf{p}$ are defined as in (7.2.10), and where

$$B = I^{(2)} + hI^{(1)} D\left(\left(\frac{1}{\phi'}\right)' - \frac{\mu}{\phi'} \right) + h^2 D\left(\left(\frac{1}{\phi'}\right)'' \frac{1}{\phi'} - \left(\frac{\mu}{\phi'}\right)' \frac{1}{\phi'} + \frac{\nu}{[\phi']^2} \right). \tag{7.2.17}$$

7.2.3 Collocation and Integration by Parts

It requires relatively little effort at this point to derive one additional system of equations for the vector $\mathbf{w}$, namely

$$C\mathbf{w} = \mathbf{p}, \tag{7.2.18}$$

where $\mathbf{w}$ and $\mathbf{p}$ are defined as in (7.2.10), and where the matrix C is the average of the matrices A defined in (7.2.11) and B defined as in (7.2.17), that is,

$$C = \frac{1}{2}(A+B) =$$
$$I^{(2)} + \frac{h}{2}\left(DI^{(1)} + I^{(1)}D\right) + \frac{h^2}{2}D\left(\left(\left(\frac{1}{\phi'}\right)'' - \left(\frac{\mu}{\phi'}\right)'\right)\frac{1}{\phi'} + \frac{2\nu}{[\phi']^2}\right);$$
$$D = D\left(\left(\frac{1}{\phi'}\right)' - \frac{\mu}{\phi'}\right).$$
(7.2.19)

The derivation of the matrix C is valid under Assumption 7.2.2. It will be seen in Subsection 7.2.4 that the matrix C plays a particularly important role for purposes of establishing the convergence of y_N to y, where y_N, defined as in (7.2.5), is determined by any one of the three equations, $A\mathbf{w} = \mathbf{p}$, $B\mathbf{w} = \mathbf{p}$, or $C\mathbf{w} = \mathbf{p}$.

7.2.4 Convergence of Sinc–Galerkin Methods

Given a vector $\mathbf{w} = (w_{-N}, \cdots, w_N)^T$, we shall define $\|w\|$ throughout this section by

$$\|\mathbf{w}\| = \left(\sum_{k=-N}^{N} |w_k|^2\right)^{1/2}. \tag{7.2.20}$$

Also, throughout this section, c, c_1, $c_2, \cdots$, will denote generic constants independent of N.

Let us first sketch the layout of our proof.

At the outset, we shall require accurate estimates of the quantities

$$\|A\mathbf{y} - \mathbf{p}\|, \quad \|B\mathbf{y} - \mathbf{p}\|, \quad \|C\mathbf{y} - \mathbf{p}\|, \tag{7.2.21}$$

where, with $y(x)$ denoting the exact solution of (7.2.4), we set $\mathbf{y} = (y(z_{-N}), \cdots, y(z_N))^T$. However, we shall first examine in detail only the third equation in (7.2.21), the equation $C\mathbf{y} = \mathbf{p}$. Once bounds on the quantities in (7.2.21) are known, and we identify a vector $\mathbf{w}_C = (w_{C,-N}, \cdots, w_{C,N})^T$, as the solution to the problem $C\mathbf{w}_C = \mathbf{p}$, it will follow that

$$C(\mathbf{y} - \mathbf{w}_C) = C\mathbf{y} - \mathbf{p}, \tag{7.2.22}$$

and this identity yields the inequality

$$\|\mathbf{y} - \mathbf{w}_C\| \leqq \|C^{-1}\| \|C\mathbf{y} - \mathbf{p}\|. \tag{7.2.23}$$

Here the matrix norm $\|C^{-1}\|$ is the standard matrix norm consistent with the vector norm (7.2.20). It is thus evident, that in order to get an accurate bound on $\|\mathbf{y} - \mathbf{w}_C\|$, we shall require accurate bounds on both $\|C\mathbf{y} - \mathbf{p}\|$ as

well as on $\|C^{-1}\|$. Such bounds are indeed possible, under our assumptions on μ and ν. Essentially, the same assumptions on μ and ν will then also enable us to bound $\|A^{-1}\|$ and $\|B^{-1}\|$, and they will thus enable us to get accurate bounds on $\|\mathbf{y} - \mathbf{w}_A\|$, and on $\|\mathbf{y} - \mathbf{w}_B\|$, with $\mathbf{w}_A$ and $\mathbf{w}_B$ the solutions of equations $A\mathbf{w}_A = \mathbf{p}$ and $B\mathbf{w}_B = \mathbf{p}$, respectively.
Next, since $y \in \mathbf{L}_\alpha(\mathcal{D})$, we can obtain an *a priori* bound on

$$\varepsilon_N = \sup_{x\in\Gamma} \left| y(x) - \sum_{j=-N}^{N} y(z_j) S(j,h) \circ \phi(x) \right|. \qquad (7.2.24)$$

We shall now be able to show that

$$\sup_{x\in\Gamma} \left| y(x) - \sum_{j=-N}^{N} w_{C,j} S(j,h) \circ \phi(x) \right| \leqq \varepsilon_N + \|\mathbf{y} - \mathbf{w}_C\|, \qquad (7.2.25)$$

and similarly in the case where the components $w_{C,j}$ of the vector $\mathbf{w}_C$ are replaced in (7.2.25) by the components of the vector $\mathbf{w}_A$, or of the vector $\mathbf{w}_B$.

We now proceed to establish convergence along the lines of the outline just given.

As pointed out above, while Assumption 7.2.1 suffices to bound $\|A\mathbf{y}-\mathbf{p}\|$, it is convenient to make the stronger Assumption 7.2.2 in order to bound $\|B\mathbf{y} - \mathbf{p}\|$ and $\|C\mathbf{y} - \mathbf{p}\|$.

Lemma 7.2.3 *Let Assumption 7.2.1 be satisfied.*
(a) Then there exists a constant c_A independent of N, such that

$$\|A\mathbf{y} - \mathbf{p}\| \leqq c_A N^{1/2} e^{-(\pi d \alpha N)^{1/2}}. \qquad (7.2.26)$$

(b) Let Assumption 7.2.2 be satisfied. Then there exists a constant c_B independent of N, such that

$$\|B\mathbf{y} - \mathbf{p}\| \leqq c_B N^{1/2} e^{-(\pi d \alpha N)^{1/2}}. \qquad (7.2.27)$$

Moreover, in this case,

$$\|C\mathbf{y} - \mathbf{p}\| \leqq \frac{1}{2}(c_A + c_B) N^{1/2} e^{-(\pi d \alpha N)^{1/2}}. \qquad (7.2.28)$$

Proof. Part (a): Define K_m and $\omega_{m,j}$ for $m = 0, 1, 2$ and $j = -N, \cdots, N$, by

$$\begin{aligned} K_m(x,z) &= \frac{1}{2\pi i [\phi'(x)]^m} \left(\frac{\partial}{\partial x}\right)^m \frac{\sin[\pi\phi(x)/h]}{\phi(z) - \phi(x)}, \\ \omega_{m,j}(x) &= \frac{1}{[\phi'(x)]^m} \left(\frac{d}{dx}\right)^m S(j,h) \circ \phi(x). \end{aligned} \qquad (7.2.29)$$

Since $y \in \mathbf{L}_\alpha(\mathcal{D})$, it follows that $\phi' y \in \mathbf{H}^1(\mathcal{D})$. Hence, by Theorems 4.2.1 and 4.2.2, we have the identity

$$\begin{aligned} y(x) &- \sum_{j=-\infty}^{\infty} \omega_{0,j}(x) y(z_j) \\ &= \int_{\partial\mathcal{D}} \frac{K_0(x,z)}{\sin[\pi\phi(z)/h]} \phi'(z) y(z) dz. \end{aligned} \tag{7.2.30}$$

Letting v_k denote the kth component of the vector $A\mathbf{y} - \mathbf{p}$, it follows that

$$v_k = v_k^{(1)} + v_k^{(2)}, \tag{7.2.31}$$

where

$$\begin{aligned} v_k^{(1)} &= \frac{h^2}{[\phi'(x)]^2} \left[y^{(2)}(z_k) + \mu(z_k) y'(z_k) \right] \\ &\quad -h^2 \sum_{j=-\infty}^{\infty} \left[\omega_{2,j}(z_k) + \frac{\mu(z_k)}{\phi'(z_k)} \omega_{1,j}(z_k) \right] y(z_j) \\ &= h^2 \int_{\partial\mathcal{D}} \frac{K_2(z_k,z) + \frac{\mu(z_k)}{\phi'(z_k)} K_1(z_k,z)}{\sin[\pi\phi(z)/h]} \phi'(z) y(z) dz, \\ v_k^{(2)} &= h^2 \sum_{|j|>N} \left[\omega_{2,j}(z_k) + \frac{\mu(z_k)}{\phi'(z_k)} \omega_{1,j}(z_k) \right] y(z_j). \end{aligned} \tag{7.2.32}$$

From the definition (7.2.29) one may readily deduce the expressions

$$\begin{aligned} K_0(z_k,z) &= 0, \\ K_1(z_k,z) &= \frac{(-1)^k}{2ih[\phi(z) - kh]}, \\ K_2(z_k,z) &= \frac{(-1)^k}{2ih} \left\{ -\frac{(1/\phi')'(z_k)}{\phi(z) - kh} + \frac{2}{[\phi(z) - kh]^2} \right\}. \end{aligned} \tag{7.2.33}$$

Also, setting

$$\begin{aligned} u(z) &= \Re\phi(z), \\ M_1 &= \sup_{x\in\Gamma} \left| \frac{\mu(x)}{\phi'(x)} \right|, \\ M_2 &= \sup_{x\in\Gamma} \left| \left(\frac{1}{\phi'} \right)'(x) \right|, \\ M_3 &= M_1 + M_2, \end{aligned} \tag{7.2.34}$$

recalling that $\Im\phi(z) = \pm d$ and that $|\phi(z) - kh| \geqq d$ if $z \in \partial\mathcal{D}$, we find that

$$h^2 \left| K_2(z_k, z) + \frac{\mu(z_k)}{\phi'(z_k)} K_1(z_k, z) \right| \leqq \frac{c\,h}{2[(u(z) - kh)^2 + d^2]^{1/2}}, \qquad (7.2.35)$$

where $c = M_3 + 2/d$. Therefore,

$$\begin{aligned} \|A\mathbf{y} - \mathbf{p}\| &= \left(\sum_{k=-N}^{N} |v_k|^2 \right)^{1/2} \\ &\leqq \left(\sum_{k \in \mathbf{Z}} |v_k|^2 \right)^{1/2} \\ &\leqq \left(\sum_{k \in \mathbf{Z}} |v_k^{(1)}|^2 \right)^{1/2} + \left(\sum_{k \in \mathbf{Z}} |v_k^{(2)}|^2 \right)^{1/2}, \end{aligned} \qquad (7.2.36)$$

where, by Problems 1.6.9 and 7.2.3

$$\begin{aligned} &\sum_{k \in \mathbf{Z}} |v_k^{(1)}|^2 \\ &\quad \leqq \sum_{k \in \mathbf{Z}} \left| \int_{\partial \mathcal{D}} \frac{c}{[(u(z) - kh)^2 + d^2]^{1/2}} \frac{|\phi'(z) y(z)|}{|\sin[\pi \phi(z)/h]|} |dz| \right|^2 \\ &\quad \leqq \sum_{k \in \mathbf{Z}} \frac{c}{k^2 h^2 + d^2} \left(\int_{\partial \mathcal{D}} \frac{|\phi'(z) y(z) dz|}{|\sin[\pi \phi(z)/h]|} \right)^2 \\ &\quad \leqq \frac{c_1\, h^{-1}}{[\sinh(\pi d/h)]^2}. \end{aligned} \qquad (7.2.37)$$

Also, since

$$\begin{aligned} &h^2 \left| \omega_{2,j}(z_k) + \frac{\mu(z_k)}{\phi'(z_k)} \omega_{1,j}(z_k) \right| \\ &\leqq \left| \delta_{k-j}^{(2)} \right| + h M_3 \left| \delta_{k-j}^{(1)} \right|, \\ &\leqq [1 + hM_3] \gamma_{k-j}, \end{aligned} \qquad (7.2.38)$$

M_3 defined as in (7.2.34), and $\gamma_k = \max\left\{ \left| \delta_k^{(2)} \right|, \left| \delta_k^{(1)} \right| \right\}$. Hence, it follows that $\sum_{k \in \mathbf{Z}} \gamma_k^2 = c_5^2 < \infty$. Moreover, since $y \in \mathbf{L}_\alpha(\mathcal{D})$, implies that $|y(z_j)| \leqq c e^{-\alpha |j| h}$, we find from (7.2.32), that

$$\begin{aligned}
&\sum_{k=-N}^{N} \left|v_k^{(2)}\right|^2 \\
&\quad = \sum_{k=-N}^{N} \left| \sum_{|j|>N} \left\{ \delta_{k-j}^{(2)} + h \left[\left(\frac{1}{\phi'}\right)'(z_k) - \frac{\mu(z_k)}{\phi'(z_k)} \right] \delta_{k-j}^{(1)} \right\} y(z_j) \right|^2 \\
&\quad \leqq (1+hM_3)^2 \sum_{k=-N}^{N} \left| \sum_{|j|>N} \gamma_{k-j}\, c\, e^{-\alpha|j|h} \right|^2 \qquad (7.2.39) \\
&\quad = (1+hM_3)^2 c^2 \sum_{|j|>N} \sum_{|\ell|>N} \sum_{k\in\mathbf{Z}} \gamma_{k-j}\gamma_{k-\ell} e^{-\alpha|j|h} e^{-\alpha|\ell|h} \\
&\quad \leqq c_3 h^{-2} e^{-2\alpha N h}.
\end{aligned}$$

We remark that the terms involving $(\nu y)(z_j)$ and $\sigma(z_j)$ do not appear in these inequalities since they are identically zero for the range of indecies in these sums. Hence, it thus follows that

$$\|A\mathbf{y} - \mathbf{p}\| = \left(\sum_{k=-N}^{N} |v_k|^2 \right)^{1/2} \leqq c_4 N^{1/2} e^{-(\pi d \alpha N)^{1/2}}. \qquad (7.2.40)$$

This completes Part (a) of the proof.

Part (b): Let us select an arbitrary integer in the range $[-N, N]$, and let us write S for $S(k,h)\circ\phi$. Then, by equations (7.2.14) and (7.2.15), we have

$$\begin{aligned}
0 &= h \int_\Gamma \frac{(Ly)(x) S(x)}{\phi'(x)} dx \\
&= h \int_\Gamma \tfrac{1}{\phi'}(x) \left[\left(\tfrac{S}{\phi'}\right)''(x) - \left(\tfrac{\mu S}{\phi'}\right)'(x) + \tfrac{\nu S}{\phi'}(x) - \tfrac{\sigma S}{\phi'}(x) \right] \phi'(x) y(x) dx \\
&= v_k^{(1)} + v_k^{(2)} + v_k^{(3)},
\end{aligned} \qquad (7.2.41)$$

where $v_k^{(1)}$ denotes the kth component of the vector $B\mathbf{y} - \mathbf{p}$,

$$v_k^{(2)} = \sum_{|j|>N} \left[\delta_{k-j}^{(2)} + h\delta_{k-j}^{(1)} \left(\left(\frac{1}{\phi'}\right)'(z_j) - \frac{\mu}{\phi'}(z_j) \right) \right] y(z_j), \qquad (7.2.42)$$

and where $v_k^{(3)}$ denotes the error of infinite-point quadrature, which we may express explicitly by means of the formula in (4.2.6). That is, from (4.2.2) in Theorem 4.2.1,

$$v_k^{(3)} = \frac{i}{2}\int_{\partial\mathcal{D}} \frac{\kappa(z,h)\left[\left(\frac{S}{\phi'}\right)'' - \left(\frac{\mu S}{\phi'}\right)' + \frac{\nu S}{\phi'} - \frac{\sigma S}{\phi'}\right]}{\phi'(z)\sin[\pi\phi(z)/h]}\phi'(z)y(z)dz, \quad (7.2.43)$$

with $\kappa(z,h) = \exp\{(i\pi\phi(z)/h)\,\mathrm{sgn}\Im\phi(z)\}$, so that for $z \in \partial\mathcal{D}$, $|\kappa(z,h)| = e^{-\pi d/h}$.

Now, under Assumption 7.2.2, each of the functions $(1/\phi')'$, $(1/\phi')''/\phi'$, μ/ϕ', $(\mu/\phi')'/\phi'$, and $\nu/[\phi']^2$ belong to $\mathbf{H}^\infty(\mathcal{D})$, and we also have $\sigma/[\phi']^2 \in \mathbf{L}_\alpha(\mathcal{D})$. We again set $u(z) = \Re\phi(z)$, and we again recall that if $z \in \partial\mathcal{D}$, then $|\phi(z) - kh| \geqq d$. It then follows, after explicitly expressing the numerator in (7.2.43), that there exists a constant c_5 which is independent of h, such that

$$\left|v_k^{(3)}\right| \leqq c_5 e^{-\pi d/h}\int_{\partial\mathcal{D}} \frac{|\phi'(z)y(z)dz|}{[(u(z)-kh)^2 + d^2]^{1/2}}. \quad (7.2.44)$$

Hence, by proceeding as for (7.2.37), we get the bound

$$\begin{aligned}\left(\sum_{k=-N}^{N}\left|v_k^{(3)}\right|^2\right)^{1/2} &\leqq \left(\sum_{k\in\mathbf{Z}}\left|v_k^{(3)}\right|^2\right)^{1/2} \\ &\leqq c_6 h^{-1/2}e^{-\pi d/h}.\end{aligned} \quad (7.2.45)$$

By proceeding as for (7.2.39), we also obtain the bound

$$\left(\sum_{k=-N}^{N}\left|v_k^{(2)}\right|^2\right)^{1/2} \leqq c_7 h^{-1}e^{-\alpha N h}, \quad (7.2.46)$$

and it thus follows that

$$\|B\mathbf{y} - \mathbf{p}\| = \left(\sum_{k=-N}^{N}|v_k|^2\right)^{1/2} \leqq c_8 N^{1/2}e^{-(\pi d\alpha N)^{1/2}}. \quad (7.2.47)$$

Finally, from (7.2.19), (7.2.40) and (7.2.47) it now follows that

$$\begin{aligned}\|C\mathbf{y} - \mathbf{p}\| &= \tfrac{1}{2}(\|A\mathbf{y} - \mathbf{p}\| + \|B\mathbf{y} - \mathbf{p}\|) \\ &\leqq c_9 N^{1/2}e^{-(\pi d\alpha N)^{1/2}}.\end{aligned} \quad (7.2.48)$$

This completes the proof of Lemma 7.2.3. ■

Next, we shall consider obtaining accurate bounds for $\|C^{-1}\|$, and we shall also consider bounds for $\|A^{-1}\|$ and $\|B^{-1}\|$. For sake of simplicity of argument, we shall consider only the special cases when μ/ϕ' is real-valued on Γ, which is the case, for example, when Γ is either $\mathbb{R}$, or an interval on $\mathbb{R}$, and when μ is real-valued on Γ. We thus first examine the eigenvalues of the square matrices $I^{(m)}$ of order $2N+1$, where $I^{(m)} = [\delta^{(m)}_{j-k}]$, and where $\delta^{(m)}_{j-k}$ denotes the (j,k)th element of the matrix $I^{(m)}$, and is defined as in (1.10.19). *Toeplitz matrices* are a convenient tool to enable us to achieve our aim. Let us therefore introduce the notion of a Toeplitz matrix. Let a sequence of numbers $\{c_k\}_{k=-\infty}^{\infty}$ denote the Fourier coefficients of an arbitrary function $f \in \mathbf{L}^1(-\pi,\pi)$, defined by (see Section 1.6)

$$c_k = \frac{1}{2\pi}\int_{-\pi}^{\pi} f(x)\, e^{-ikx}\, dx. \tag{7.2.49}$$

Every square matrix T whose (j,k)th entry is c_{j-k} is called a Toeplitz matrix.

Lemma 7.2.4 *Let $I^{(m)}$ be defined as in the above paragraph.*
(a) Then $I^{(m)}$ is a Toeplitz matrix, generated by the function $f(x) = (ix)^m$.
(b) If m is an odd positive integer, that is, if $m = 2n+1$, then $I^{(m)}$ is a skew-symmetric, singular matrix, with eigenvalues $i\lambda_j$, with

$$-\pi^{2n+1} \leqq \lambda_j \leqq \pi^{2n+1}. \tag{7.2.50}$$

If m is an even positive integer, that is, if $m = 2n$, then $I^{(m)}$ is a non-singular matrix, with eigenvalues $(-1)^n\lambda_j$, with $\lambda_j \in (0,\pi^{2n})$.
(c) If we denote the eigenvalues of $I^{(2)}$ by $-\lambda_j^{(2)}$, then

$$4\sin^2[\pi/(4N+4)] \leqq \lambda_j^{(2)} \leqq \pi^2. \tag{7.2.51}$$

Proof. Part (a): Substituting the definition (1.10.19) of $\delta^{(k)}_{m-n}$ into the Fourier integral representation (1.10.10) of $S(k,1)(x)$ and differentiating under the integral sign immediately yields a representation of the form (7.2.49), with $f(x) = (-ix)^m$, which proves part (a) of the lemma.
Part (b): For part (b), let us first consider the function f in the above definition of a Toeplitz matrix to be real-valued on $(-\pi,\pi)$. Let $\mathbf{u} = (u_{-N},\cdots,u_N)^T$ denote an arbitrary complex vector, such that $\|u\| = 1$. Then it is readily seen that

$$\mathbf{u}^* T \mathbf{u} = \frac{1}{2\pi}\int_{-\pi}^{\pi} f(x)\left|\sum_{k=-N}^{N} u_k e^{ikx}\right|^2 dx \tag{7.2.52}$$

It thus follows from (7.2.52) that all eigenvalues of T lie between the minimum and maximum values of f. It is therefore clear, that if m is odd, that is, if $m = 2n+1$, then all the eigenvalues of the Toeplitz matrix generated by the function $f(x) = x^{2n+1}$ lie on the interval $(-\pi^{2n+1},\pi^{2n+1})$. Moreover, $\det\{I^{(2n+1)}\} = 0$, as shown by taking $\mathbf{u}$ in (7.2.52) such that $u_{-k} = u_k$ for $k = 1,\cdots,N$, and such that at least one $u_k \neq 0$, which results in the sum $\sum_{k=-N}^{N} u_k e^{ikx}$ being an even function. If m is even, i.e., if $m = 2n$, then all the eigenvalues of the Toeplitz matrix generated by the function $f(x) = x^{2n}$ lie on the interval $(0, \pi^{2n})$. This establishes part (b) of Lemma 7.2.4.

Part (c): Indeed, the representation (7.2.49) enables us to further deduce that, if for $\ell = 1, 2$, we have Toeplitz matrices T_ℓ, each of the same order, generated by real-valued functions f_ℓ, such that for all $x \in (-\pi, \pi)$ we have $f_1(x) \leqq f_2(x)$, and if μ_ℓ denotes the maximum of the eigenvalues of T_ℓ, then $\mu_1 \leqq \mu_2$. This fact is readily demonstrated by letting $\mathbf{u}$ be an eigenvector of either T_1 or T_2, and substitution into (7.2.52). Now it turns out that the eigenvalues of the special tri-diagonal Toeplitz matrix of order $2N+1$ which has a "-2" on its diagonal, and a "1" on its sup and sub-diagonal are well-known to people using finite difference methods to solve boundary value problems, i.e., the eigenvalues of this matrix are given explicitly by $\lambda_j = -4/\sin^2[j\pi/(4N+4)]$, $j = 1, \cdots, 2N+1$. Furthermore, it is readily seen that this particular tri-diagonal Toeplitz matrix is generated by the function $f(x) = 2[\cos(x) - 1]$. Finally, we have $2[\cos(x) - 1] \geqq -x^2$, for all $x \in (-\pi, \pi)$. Part (c) of Lemma 7.2.4 follows.

This completes the proof of Lemma 7.2.4. ∎

Lemma 7.2.5 *Let A, B, and C be defined as in (7.2.11), (7.2.17), and (7.2.19) respectively. Let $N \geqq 1$, and let h be defined as in (7.2.7), and let Assumption 7.2.1 be satisfied.*

(a) There exists a constant c independent of N, such that

$$\begin{aligned} \|A\| &\leqq \pi^2 \left(1 + cN^{-1/2}\right), \\ \|B\| &\leqq \pi^2 \left(1 + cN^{-1/2}\right), \\ \|C\| &\leqq \pi^2 \left(1 + cN^{-1/2}\right). \end{aligned} \tag{7.2.53}$$

(b) If μ/ϕ' is real-valued on Γ, and if

$$\Re\left\{\left[\left(\frac{1}{\phi'}\right)''(x) - \left(\frac{\mu}{\phi'}\right)'(x) + \frac{2\nu}{\phi'}(x)\right]\frac{1}{\phi'}(x)\right\} \leqq 0, \quad x \in \Gamma, \tag{7.2.54}$$

then

$$\|C^{-1}\| \leqq \frac{4N^2}{\pi^2}\left(1 + cN^{-1}\right), \tag{7.2.55}$$

with c a constant independent of N. Moreover, in this case each of the inequalities

$$\|A^{-1}\| \leqq \frac{4N^2}{\pi^2}\left(1 + c_1 N^{-1}\right), \tag{7.2.56}$$

$$\|B^{-1}\| \leqq \frac{4N^2}{\pi^2}\left(1 + c_1 N^{-1}\right), \tag{7.2.57}$$

is satisfied for all N sufficiently large, and for some constant c_1 independent of N.

Proof. Part (a): Let us define numbers γ_1 and γ_2 by

$$\begin{aligned}\gamma_1 &= \sup_{x\in\Gamma} |[(1/\phi')' - \mu/\phi'](x)|, \\ \gamma_2 &= \sup_{x\in\Gamma} \left|\frac{\nu(x)}{[\phi'(x)]^2}\right|.\end{aligned} \tag{7.2.58}$$

By (7.2.51) we have $\|I^{(2)}\| \leqq \pi^2$. Hence, we have, using the definition of A given in (7.2.11), the definition of γ_j given in (7.2.58), the definition of h given in (7.2.7), and our assumption that $N \geqq 1$, that

$$\begin{aligned}\|A\| &\leq \pi^2 + h\gamma_1\pi + h^2\gamma_2 \\ &= \pi^2[1 + cN^{-1/2}],\end{aligned} \tag{7.2.59}$$

with c a constant that is independent of N.
Now (see Theorem 4.1.2), if Assumption 7.2.1 is satisfied on a domain $\mathcal{D} = \mathcal{D}(d)$, then, by Problem 4.1.8 it follows that Assumptions 7.2.2 will be satisfied on a domain $\mathcal{D}' = \mathcal{D}(d')$, where $0 < d' < d$. Since the entries in B and C are functions evaluated on Γ, and since $\Gamma \subset \mathcal{D}(d') \subset \mathcal{D}(d)$, we may assume that since Assumption 7.2.1 is satisfied, Assumption 7.2.2 is also satisfied as far as bounding $\|B\|$ and $\|C\|$ is concerned. Moreover, a bounds on $\|B\|$ and $\|C\|$ can be similarly obtained as our bound on $\|A\|$, and we therefore omit the details of obtaining these bounds.

This completes the proof of part (a) of the Lemma.

Part (b): Since the inequality (7.2.54) only involves values of ϕ, μ and ν or derivatives of these functions on Γ, we may, in fact assume, as far as this inequality is concerned, that Assumption 7.2.2 is satisfied.
Let us set $g = (1/\phi')' - \mu/\phi'$, $D_1 = D(g)$, $D_2 = (1/2)D(g'/\phi' + 2\nu/[\phi']^2) \equiv \operatorname{diag}(d_{2,-N}, \cdots, d_{2,N})$, where by (7.2.54), we have $d_{2,j} \leqq 0$ on Γ. It is evident that $E \equiv D_1 I^{(1)} + I^{(1)} D_1$ is a skew-symmetric matrix, and, hence, if $\mathbf{w} = (w_{-N}, \cdots, w_N)^T$ is an arbitrary complex vector with $\|\mathbf{w}\| = 1$, it follows that $\Re(\mathbf{w}^* E\mathbf{w}) = 0$. Therefore, by (7.2.51) we have

$$\begin{aligned}\Re(\mathbf{w}^* C\mathbf{w}) &\leqq -4\sin^2\left(\frac{\pi}{4N+4}\right) + \max_{j=-N,\cdots,N} d_{2,j} \\ &\leqq -4\sin^2\left(\frac{\pi}{4N+4}\right).\end{aligned} \tag{7.2.60}$$

Part (b) of Lemma 7.2.5 thus follows for the case of $\|C^{-1}\|$.
What about $\|A^{-1}\|$ and $\|B^{-1}\|$?
Recalling the definitions of A and B we may note that

$$A - B = h\left(D(g)I^{(1)} - I^{(1)}D(g)\right) - h^2 D(g'/\phi'). \tag{7.2.61}$$

Let the assumptions of part (b) of Lemma 7.2.5 be satisfied. It is consistent with the applications of this chapter to consider the matrix $A - B$ operating on a vector $\mathbf{y}$ belonging to the special $2N + 1$ dimensional vector space Y of vectors $\mathbf{y} = (y(z_{-N}), \cdots, y(z_N))^T$, with $y \in \mathbf{L}_\alpha(\mathcal{D})$, an arbitrary but fixed function independent of N. *While A^{-1} and B^{-1} may not exist for some bounded, small values of N, we shall now show that these inverses do, in fact, exist for all N sufficiently large in the space Y, and moreover, as $N \to \infty$, $\|A^{-1}\| \sim \|B^{-1}\| \sim \|C^{-1}\|$.*

Thus, let us take $y \in \mathbf{L}_\alpha(\mathcal{D})$, so that we have for some fixed constant $c > 0$,

$$|y(x)| \leqq c \frac{|\rho(x)|^\alpha}{[1 + |\rho(x)|]^{2\alpha}}. \tag{7.2.62}$$

Let us furthermore assume without loss of generality, that

$$\|y\|_\Gamma^2 \equiv \int_\Gamma |y(x)|^2 |\phi'(x)\, dx| = 1, \tag{7.2.63}$$

and let us note by Equation (4.2.13) that

$$\int_{\partial\mathcal{D}} c^2 \frac{|\rho(z)|^{2\alpha} |\phi'(z)\, dz|}{[1 + |\rho(z)|]^{4\alpha}} = 2\, c^2 \frac{[\Gamma(2\alpha)]^2}{\Gamma(4\alpha)} < 2\, c^2 \frac{2}{2\alpha}. \tag{7.2.64}$$

Taking $h = [\pi d/\alpha N)]^{1/2}$, setting $\mathbf{y} = (y(z_{-N}), \cdots, y(z_N))^T$, and taking

$$\|\mathbf{y}\|_h^2 \equiv h \sum_{j=-N}^{N} |y(z_j)|^2 = h\|\mathbf{y}\|^2, \tag{7.2.65}$$

with $\|\mathbf{y}\|$ defined as in (7.2.20), we have

$$h^{1/2}\|\mathbf{y}\| = (1 + \varepsilon_N)\|y\|_\Gamma = 1 + \varepsilon_N, \tag{7.2.66}$$

where, since $\int_{\partial\mathcal{D}} |y(x)^2\, \phi'(x)\, dx| \leqq c_6$, since (see (3.1.37))

$$h \sum_{|j|>N} |y(z_j)|^2 \leqq c_7 e^{-2(\pi d\alpha N)^{1/2}} \|\mathbf{y}\|,$$

with c_7 independent of N, we have, by Theorem 3.2.1, that

$$\begin{aligned} |\varepsilon_N| &\leqq \left[\frac{c_6 e^{-2(\pi d\alpha N)^{1/2}}}{1 - e^{-2(\pi d\alpha N)^{1/2}}} + c_7 e^{-2(\pi d\alpha N)^{1/2}} \right]^{1/2} \\ &\leqq c_8 e^{-(\pi d\alpha N)^{1/2}} \end{aligned} \tag{7.2.67}$$

with c_8 independent of h.

Now, it is readily seen upon setting $g(x) = (1/\phi'(x))' - \mu(x)/\phi'(x)$, that the kth component of the vector $(A - B)\mathbf{y}$ may be obtained by applying collocation to the identity

$$- hg(x) \left\{ \frac{hy'(x)}{\phi'(x)} \right\} + \left\{ \frac{h\,(g(x)y(x))'}{\phi'(x)} \right\} - \frac{h^2 g'(x) y(x)}{\phi'(x)} \equiv 0, \tag{7.2.68}$$

and then replacing the terms in curly brackets in Equation (7.2.68) by an approximation obtainable by proceeding as for (7.2.9). Thus, setting

$$M_1 = \|g\|_{\mathcal{D}}, \quad M_2 = \|g'/\phi'\|_{\mathcal{D}}, \tag{7.2.69}$$

and recalling by Assumption 7.2.2 that each of the constants M_j is finite, it follows, proceeding as in the proof of part (b) of Lemma 7.2.3, that

$$\|\frac{1}{2}(A - B)\mathbf{y}\| \leqq c_2 N^{1/2} e^{-(\pi d \alpha N)^{1/2}} \tag{7.2.70}$$

with c_2 a constant that depends on M_j, c, and α, but is independent of N. Hence the identity

$$A\mathbf{y} = C\mathbf{y} + \frac{1}{2}(A - B)\mathbf{y}, \tag{7.2.71}$$

(7.2.20), and (7.2.51) yield

$$\begin{aligned} \|A\mathbf{y}\| &\geqq \|C\mathbf{y}\| - \|\tfrac{1}{2}(A - B)\mathbf{y}\| \\ &\geqq 4\sin^2\left(\tfrac{\pi}{4N+4}\right)\|\mathbf{y}\| - c_2 N^{1/2} e^{-(\pi d \alpha N)^{1/2}} \|\mathbf{y}\|. \end{aligned} \tag{7.2.72}$$

The bound of $\|A^{-1}\|$ given in (7.2.57) thus follows, for all N sufficiently large from (7.2.72). The analysis for the proof of the bound for $\|B^{-1}\|$ given in (7.2.57) is similar, and we omit this part of the proof. ∎

Theorem 7.2.6 *Let Assumption 7.2.2 be satisfied. Let y denote the exact solution of (7.2.4), and let y_N, defined as in (7.2.5), denote the approximate solution, where the vector $\mathbf{w} = (w_{-N}, \cdots, w_N)^T$ denotes the exact solution of the system of equations (7.2.18). If the inequality (7.2.55) is satisfied, then there exists a constant c_2 which is independent of N, such that*

$$\sup_{x \in \Gamma} |y(x) - y_N(x)| \leqq c_2 N^{1/2} e^{-(\pi d \alpha N)^{1/2}}. \tag{7.2.73}$$

Proof. Letting η_N be defined by

$$\eta_N(x) \equiv \sum_{k=-N}^{N} y(z_k)\, S(k,h) \circ \phi(x). \tag{7.2.74}$$

it follows, by the triangle inequality, that

$$|y(x) - y_N(x)| \leqq |y(x) - \eta_N(x)| + |\eta_N(x) - y_N(x)|. \tag{7.2.75}$$

By assumption, $y \in \mathbf{L}_\alpha(\mathcal{D})$, and therefore, by Theorem 4.2.5, there exists a constant c_3 independent of N, such that

$$\sup_{x \in \Gamma} |y(x) - \eta_N(x)| \leqq c_3 N^{1/2} e^{-(\pi d \alpha N)^{1/2}}. \tag{7.2.76}$$

The term on the extreme right-hand side of (7.2.75) satisfies the relation

$$\begin{aligned}
&|\eta_N(x) - y_N(x)| \\
&= \left| \sum_{j=-N}^{N} [y(z_j) - w_j] S(j,h) \circ \phi(x) \right| \\
&\leqq \sum_{j=-N}^{N} |y(z_j) - w_j| |S(j,h) \circ \phi(x)| \\
&\leqq \left(\sum_{j=-N}^{N} |y(z_j) - w_j|^2 \right)^{1/2} \left(\sum_{k=-N}^{N} |S(j,h) \circ \phi(x)|^2 \right)^{1/2} \\
&\leqq \left(\sum_{j=-N}^{N} |y(z_j) - w_j|^2 \right)^{1/2} = \|\mathbf{y} - \mathbf{w}\|,
\end{aligned} \tag{7.2.77}$$

where, to obtain the last line in (7.2.77), we used the notation of (7.2.20) and (7.2.21), as well as the results of Problem (7.3.10), which tells us that if $x \in \Gamma$, then $\sum_{k \in \mathbf{Z}} |S(k,h) \circ \phi(x)|^2 = 1$.
Finally, from (7.2.18), (7.2.28), (7.2.56), and Lemma 7.2.3, we have

$$\begin{aligned}
\|\mathbf{y} - \mathbf{w}\| &= \|C^{-1} C(\mathbf{y} - \mathbf{w}\| \\
&= \|C^{-1}[C\mathbf{y} - \mathbf{p}\| \\
&\leqq \|C^{-1}\| \, \|C\mathbf{y} - \mathbf{p}\| \\
&\leqq c_3 N^{5/2} e^{-(\pi d \alpha N)^{1/2}},
\end{aligned} \tag{7.2.78}$$

with c_3 a constant independent of N.
The result (7.2.73) now follows from (7.2.75) to (7.2.78). ■

In view of Theorem 6.3.5 and Lemma 7.2.5 we immediately have

Lemma 7.2.7 *Let E denote either A, B, or C, as defined in (7.2.11), (7.2.17) or (7.2.18), respectively, and consider the equation*

$$E\mathbf{w} = \mathbf{p}, \tag{7.2.79}$$

with $\mathbf{p}$ *defined as in (7.2.11). Let* E_c *and* $\mathbf{p}_c$ *denote the entities* E *and* $\mathbf{p}$, *respectively, as stored in the computer. Let* $\mathbf{w}_e$ *denote the exact solution of (7.2.73), and let* $\mathbf{w}_c$ *denote the computed solution. If* $\|E_c - E\| \|E^{-1}\| < 1$, *then there exists a constant* c *independent of* N, *such that*

$$\|\mathbf{w}_c - \mathbf{w}_e\| \leqq \varepsilon_1. \tag{7.2.80}$$

with ε_1 *given by*

$$\varepsilon_1 \equiv 4N^2\{1 + c/N\} \left[\frac{\|E_c - E\|}{\|E\|} + \frac{\|\mathbf{p}_c - \mathbf{p}\|}{\|\mathbf{p}\|} \right]. \tag{7.2.81}$$

By Theorem 7.2.6 and Lemma 7.2.7, we immediately have

Corollary 7.2.8 *Let* ε_1 *be defined as in (7.2.81), and let* $y_N = y_{N,c}$ *be defined as in (7.2.5), with* $\mathbf{w} = \mathbf{w}_c = (w_{-N}, \cdots, w_N)^T$ *denoting the computed solution of the equation* $E_c\mathbf{w} = \mathbf{p}_c$, *where* E *and* $\mathbf{p}_c$ *are defined as in Theorem 7.2.6 and Lemma 7.2.7. If the conditions of Theorem 7.2.6 and Lemma 7.2.7 are satisfied, then there exists a constant* c_1 *independent of* N, *such that for all* N *sufficiently large, we have*

$$\sup_{x \in \Gamma} |y(x) - y_{N,c}(x)| \leqq \varepsilon_1 + c_1 N^{5/2} e^{-(\pi d \alpha N)^{1/2}}. \tag{7.2.82}$$

Let E denote any one of the matrices A, B, or C. We observe that due to the accuracy of Sinc approximation in the space $\mathbf{L}_\alpha(\mathcal{D})$, the condition number $\|E\| \|E^{-1}\| \sim 4N^2$ is not unduly large, and hence, in practice, the error term ε_1 will usually be negligible compared with the second error term in (7.2.82). We also note that the condition number $\|E\| \|E^{-1}\| \sim 4N^2$ yields an upper bound on the worst-case error of approximation. We expect that we will have $\|E\| \|E^{-1}\| \leqq cN$ for nearly all (except eigenvalue) problems encountered in applications. Indeed, by the analysis of Section 6.2.4, we can expect the error bound of Corollary 7.2.8 to apply even for the case of simple eigensolutions, while for the case of multiple eigensolutions we must take a suitable fractional root of the error, according to Corollary 6.2.10.

7.2.5 Non-Linear Equations

The above approximations (7.2.9) and (7.2.13) through (7.2.15) readily lend themselves to reduce a non-linear ordinary differential equation boundary value problem to a system of algebraic equations. Consider, for example, the problem

$$\begin{aligned} (Ly)(x) &\equiv y''(x) + G(x, y(x), y'(x)) = 0, \quad x \in \Gamma, \\ y(a) &= y(b) = 0; \end{aligned} \tag{7.2.83}$$

if this equation has a solution $y \in \mathbf{L}_\alpha(\mathcal{D})$, then we may approximate both $y''(z_k)/[\phi'(z_k)]^2$ as well as $y'(z_k)/[\phi'(z_k)]^2$, by proceeding as in Section 7.2.1. At this point we simply make the replacements indicated in (7.2.8), or (7.2.13) through (7.2.14), to get a system of $2N+1$ non-linear algebraic equations for $w_{-N}, \cdots, w_N$, the solution of which, while not necessarily easily obtained, yields an accurate approximate solution (7.2.5) of the boundary value problem (7.2.83).

7.2.6 Non-Homogeneous Boundary Conditions

Non-zero boundary conditions at endpoints require a slight modification of the form of $y_N(x)$ given in (7.2.2). For example, given the boundary value problem

$$\begin{aligned} (Lu)(x) &= u''(x) + F(x, u(x), u'(x)) = 0, \quad x \in \Gamma, \\ u(a) &= u_a, \quad u(b) = u_b, \end{aligned} \tag{7.2.84}$$

where at least one of u_a or u_b is non-zero, we could proceed in one of two ways:
(i) Make the substitution $u = y + [u_a + \rho u_b]/[1+\rho]$ in the differential equation (7.2.84) and thus derive a new boundary value problem of the form (7.2.83); or
(ii) Proceed directly to approximate the solution of (7.2.83) via a Galerkin approximation of the form $u(x) \approx [u_a + \rho(x) u_b]/[1+\rho(x)] + y_N(x)$, with y_N defined as in (7.2.5).
At this point the reader may prefer the choice (i), relying on the lemmas and theorems stated above, to guide him to a solution with which he can be confident. On the other hand, after he has acquired some experience in solving boundary value problems via Sinc methods, he may feel equally confident with both procedures, and he may, in fact, prefer procedure (ii).

Mixed conditions are also easily dealt with. Consider, for example, a differential equation $(Ly) = 0$ in the form given in either (7.2.4) or (7.2.83) over an interval (a, b) of finite length, subject to the boundary condition

$$y(a) = p, \quad cy(b) + dy'(b) = q \tag{7.2.85}$$

with $|c|^2 + |d|^2 > 0$. In this case we replace the approximation $y_N(x)$ in (7.2.5) with

$$\begin{aligned} y(x) \approx y_N(x) = p\frac{b-x}{b-a} &+ c_{N+1}(x-a) + B(x-a)^2 \\ &+ \sum_{k=-N}^{N} c_k (x-b) S(k,h) \circ \phi(x). \end{aligned} \tag{7.2.86}$$

Clearly, y_N satisfies (7.2.85), at $x = a$, since the sum $\sum_{k=-N}^{N}(x-b)S(k,h)\circ\phi(x)$ is zero at $x = a$ and at $x = b$. Also, y_N' exists on all of $\overline{\Gamma}$ except at $x = a$. In particular, the derivative of $\sum_{k=-N}^{N}(x-b)S(k,h)\circ\phi(x)$ is zero at $x = b$. Hence, substituting y_N for y in the differential equation $(Ly) = 0$ enables us to eliminate B in (7.2.86), to get

$$y_N(x) = p\frac{b-x}{b-a} + \frac{pd+(b-a)q}{c(b-a)^3+2d(b-a)^2}(x-a)^2$$
$$+ \sum_{k=-N}^{N} c_k(x-b)S(k,h)\circ\phi(x) + c_{N+1}\omega(x), \tag{7.2.87}$$

with

$$\omega(x) = \left[1 - \frac{\{c(b-a)+d\}(x-a)}{c(b-a)^2+2d(b-a)}\right](x-a), \quad \phi(x) = \log\left(\frac{x-a}{b-x}\right). \tag{7.2.88}$$

The approximation y_N in (7.2.87) now satisfies the boundary conditions (7.2.85). There are $2N+2$ unknowns in $y_N(x) : c_{-N}, \cdots, c_N, c_{N+1}$. The coefficients of these unknowns are the basis functions for purposes of approximation of $y(x)$ by $y_N(x)$ via a Galerkin scheme. Since y and y' may both be assumed to be bounded at $x = b$, but y' might not be bounded at a, it suffices to define our inner product by

$$(f,g) = \int_a^b f(x)\overline{g(x)}(x-a)\,dx. \tag{7.2.89}$$

The numbers $c_{-N}, \cdots, c_N, c_{N+1}$ are thus determined by approximate solution of the system of $2N+2$ equations

$$\int_\Gamma (Ly_N)(x)\{(x-b)S(k,h)\circ\phi(x)\}(x-a)dx = 0,$$
$$\int_\Gamma (Ly_N)(x)\{\omega(x)\}(x-a)dx = 0, \tag{7.2.90}$$

where these expressions may, for example, be reduced to algebraic equations by collocation, after replacing the continuous inner product (7.2.89) by the discrete one, in the form

$$(Ly_N, \omega_k)_D \equiv h\sum_{j=-N}^{N+1} \frac{Ly_N(z_j)\omega_k(z_j)(z_j-a)}{\phi'(z_j)} = 0, \quad k = -N, \cdots, N+1, \tag{7.2.91}$$

with ω_k denoting the coefficient of c_k in (7.2.87).

We end this subsection with a warning about a possible pitfall of computation. It is tempting to deal with the mixed condition (7.2.85) more simply by means of the approximation

$$y_N(x) = p\frac{b-x}{b-a} + \sum_{k=-N}^{N} c_k S(k,h) \circ \phi(x) + c_{N+1}\frac{x-a}{b-a}, \tag{7.2.92}$$

with $c_{-N}, \cdots, c_N, c_{N+1}$ unknown. Then, since the z_k "bunch up" near a and b, it is tempting to take care of the second condition in (7.2.85) via the approximation

$$y'(b) \approx \frac{y_N(b) - y_N(z_N)}{b - z_N} = \frac{c_{N+1} - p}{b-a} - \frac{c_N}{b - z_N}. \tag{7.2.93}$$

However, this does not work; although $y_N(x)$ given by (7.2.92) may be an accurate approximation to y on (a,b), $y_N'(x)$ may be a very poor approximation to $y'(x)$ near $x = a$ or $x = b$. That is, although the above approximation scheme (7.2.92) guarantees that $|y(z_N) - y(b)|$ is small,

$$\left| y'(b) - \frac{y_N(b) - y_N(z_N)}{b - z_N} \right| \tag{7.2.94}$$

may, in fact, be unacceptably large.

7.2.7 Symmetric Sinc–Galerkin Method

An important contribution to Sinc methods for solving boundary value problems is *Lund's Symmetric Sinc–Galerkin method.* The main advantage of this method is that it is particularly well suited to the solution of self-adjoint ODE–BVP. An excellent example of this is seen in Problem (7.2.4), with $\mu \equiv 0$, in the case when Γ is an interval, since it transforms a self-adjoint problem into a linear algebra problem in which the coefficient matrix is real and symmetric. Another possible advantage in the case when $1/\phi'$ is uniformly bounded on Γ is that the approximating solution that this method produces then converges in the energy norm $\left(\int_\Gamma \left\{|y(x)|^2 + |y'(x)|^2\right\} dx\right)^{1/2}$. A possibly theoretical disadvantage of the method is that its derivation requires more stringent conditions than the Sinc–Galerkin method of Sections 7.2.1 through 7.2.4. This disadvantage may be artificial, since the method has been shown to work on examples for which the methods of Section 7.2.1 through 7.2.4 work, i.e., even if the conditions for the derivation of the Symmetric Sinc–Galerkin method are not satisfied. More work is necessary, to determine if this is, in fact, always the case. The Symmetric Sinc–Galerkin method also requires a different selection of the parameter h than the standard Sinc–Galerkin method of Sections 7.2.1 through 7.2.4; it is convenient, however, to delay this part of the discussion until after the derivation of the method.

Let us now derive the Symmetric Sinc–Galerkin method.

We again return to approximating the solution to problem (7.2.4), but this time we approximate the solution y via the expression

$$Y_N(x) \equiv [\phi'(x)]^{-1/2} \sum_{j=-N}^{N} v_j S(j,h) \circ \phi(x). \tag{7.2.95}$$

In this approximating form the v_j are to be determined to approximate $[\phi'(z_j)]^{1/2} y(z_j)$. Upon substituting Y_N defined as in (7.2.95) into the differential equation (7.2.4) and evaluating the result at z_k for $k = -N, \cdots, N$, we obtain the system of equations

$$A_s \mathbf{v} = h^2 D\left(\frac{1}{[\phi']^{3/2}}\right) \mathbf{s}, \tag{7.2.96}$$

with $\mathbf{v} = (v_{-N}, \cdots, v_N)^T$, $\mathbf{s}$ is defined as in (7.2.11), and with

$$\begin{aligned} A_s &= I^{(2)} - hD\left(\tfrac{\mu}{\phi'}\right) I^{(1)} \\ &\quad + h^2 D\left(\tfrac{1}{2\phi'}\left(\tfrac{1}{\phi'}\right)'' - \tfrac{1}{4}\left[\left(\tfrac{1}{\phi'}\right)'\right]^2 + \tfrac{\mu}{2\phi'}\left(\tfrac{1}{\phi'}\right)' + \tfrac{\nu}{[\phi']^2}\right). \end{aligned} \tag{7.2.97}$$

As may be expected in view of Subsections 7.2.1 to 7.2.3, a similar result is obtained via the use of integration by parts. Instead of the inner product (7.2.6), we now use

$$< f, g >= \int_\Gamma \frac{f(x)\overline{g(x)}}{[\phi'(x)]^{1/2}} dx. \tag{7.2.98}$$

Here we select y_N, not as in (7.2.95), but as in (7.2.5), and then we evaluate the integrals

$$< L y_N, S(k,h) \circ \phi >= 0, \quad k = -N, \cdots, N \tag{7.2.99}$$

using the formula

$$< f, g >_D \equiv h \sum_{j=-N}^{N} \frac{f(z_j)\overline{g(z_j)}}{[\phi'(z_j)]^{3/2}}, \tag{7.2.100}$$

after integrating by parts to eliminate derivatives in y, and dropping the intermediate boundary terms, which can be shown to vanish under assumptions which we shall state below. We thus arrive at the system of equations

$$B_s' \mathbf{w} = h^2 D\left(\frac{1}{[\phi']^{3/2}}\right) \mathbf{s}, \tag{7.2.101}$$

with $\mathbf{w} = (w_{-N}, \cdots, w_N)^T$ as in (7.2.5), $\mathbf{s}$ as in (7.2.11), and

$$
\begin{aligned}
B'_s &= B_s D\left([\phi']^{1/2}\right) \\
B_s &= I^{(2)} - hI^{(1)} D\left(\tfrac{\mu}{\phi'}\right) \\
&\quad + h^2 D\left(\tfrac{1}{2\phi'}\left(\tfrac{1}{\phi'}\right)'' - \tfrac{1}{4}\left[\left(\tfrac{1}{\phi'}\right)'\right]^2 - \tfrac{\mu}{2\phi'}\left(\tfrac{1}{\phi'}\right)' - \tfrac{\mu'}{[\phi']^2} + \tfrac{\nu}{[\phi']^2}\right).
\end{aligned}
\tag{7.2.102}
$$

We now make the transformation

$$
\begin{aligned}
[\phi']^{1/2} y &= v \\
D\left([\phi']^{1/2}\right) \mathbf{w} &= \mathbf{v}
\end{aligned}
\tag{7.2.103}
$$

to arrive at the equation

$$
B_s \mathbf{v} = h^2 D\left(\frac{1}{[\phi']^{3/2}}\right) \mathbf{s}. \tag{7.2.104}
$$

We note here, that in the important case $\mu = 0$, the above matrices A_s and B_s become identical. Furthermore it can be seen in this case, by inspection of these matrices, that if $\mu \equiv 0$, then A_s and B_s are symmetric matrices.

As in Subsection 7.1.3, we may also consider a third system,

$$
C_s \mathbf{v} = h^2 D\left(\frac{1}{[\phi']^{3/2}}\right) \mathbf{s}, \tag{7.2.105}
$$

where $\mathbf{v}$ and $\mathbf{s}$ are the same as above, and where

$$
\begin{aligned}
C_s &= \tfrac{1}{2}\left(A_s + B_s\right) \\
&= I^{(2)} - \tfrac{h}{2}\left[I^{(1)} D\left(\tfrac{\mu}{\phi'}\right) + D\left(\tfrac{\mu}{\phi'}\right) I^{(1)}\right] \\
&\quad + h^2 D\left(\tfrac{1}{2\phi'}\left(\tfrac{1}{\phi'}\right)'' - \tfrac{1}{4}\left[\left(\tfrac{1}{\phi'}\right)'\right]^2 + \tfrac{2\nu - \mu'}{2[\phi']^2}\right).
\end{aligned}
\tag{7.2.106}
$$

The matrix C_s is convenient for purposes of proving the convergence of the scheme, since the part of C_s given by

$$
K \equiv -\frac{h}{2}\left[I^{(1)} D\left(\frac{\mu}{\phi'}\right) + D\left(\frac{\mu}{\phi'}\right) I^{(1)}\right] \tag{7.2.107}
$$

is skew-symmetric for μ/ϕ' real-valued on Γ.

Assumption 7.2.9 *(a) Let us assume for the differential equation (7.2.4), that all of the functions*

$$
\frac{\mu}{\phi'}, \quad \left(\frac{\mu}{\phi'}\right)' \frac{1}{\phi'}, \quad \frac{\nu}{[\phi']^2}, \quad \left(\frac{1}{\phi'}\right)', \quad \left(\frac{1}{\phi'}\right)'' \frac{1}{\phi'}
$$

belong to $\mathbf{H}^\infty(\mathcal{D})$. *Furthermore, we assume that the problem (7.2.4) has a solution* y *with* $[\phi']^{1/2} y \in \mathbf{L}_\alpha(\mathcal{D})$.
(b) Let part (a) of this assumption be satisfied, and let us assume furthermore, that μ/ϕ' *is real-valued on* Γ, *and that*

$$\Re\left\{\frac{1}{2\phi'}\left(\frac{1}{\phi'}\right)'' - \frac{1}{4}\left[\left(\frac{1}{\phi'}\right)'\right]^2 + \frac{2\nu - \mu'}{2[\phi']^2}\right\} \leqq 0 \qquad (7.2.108)$$

on Γ.
(c) Let parts (a) and (b) of this assumption be satisfied, and assume that $1/\phi'$ *is uniformly bounded in* $\mathcal{D}$.

We remark that the boundedness assumption of $1/\phi'$ in part (c) of Assumption 7.2.9 is satisfied for the important special maps of $\phi(x) = \log\{(x-a)/(b-x)\}$, $\phi(x) = \log\{\sinh(x)\}$, as well as for the identity map. It is not satisfied for all solutions of the problem (7.2.4) for the case of the map $\phi(x) = \log(x)$, or for the map $\phi(x) = \log\{x + \sqrt{1+x^2}\}$.

Given a vector $\mathbf{v} = (v_{-N}, \cdots, v_N)^T$, where $v_j \in \mathbb{C}$, we again define $\|\mathbf{v}\|$ by

$$\|\mathbf{v}\| = \left(\sum_{k=-N}^{N} |v_k|^2\right)^{1/2}, \qquad (7.2.109)$$

and given a matrix E of order $2N+1$, we define $\|E\|$ by

$$\|E\| = \sup_{\|\mathbf{v}\|=1} \|E\mathbf{v}\|. \qquad (7.2.110)$$

Lemma 7.2.10 *Let Assumption 7.2.9 - (c) be satisfied. Then:*
(a) $[\phi']^{1/2-k} y^{(k)} \in \mathbf{L}_\alpha(\mathcal{D})$, $k = 0, 1, 2, \cdots$;
(b) $[\phi']^{1-k} y^{(k)} \in \mathbf{H}^2(\mathcal{D})$, $k = 0, 1, 2, \cdots$;
(c) $y \in \mathbf{H}^2(\mathcal{D})$;
(d) y *belongs to the space* S *of all functions* w *defined on* Γ, *normed by*

$$\|w\|_S = \left(\int_\Gamma \left[|w(x)|^2 + |w'(x)|^2\right] dx\right)^{1/2} \qquad (7.2.111)$$

such that $\|y\|_S < \infty$; *and*
(e) $\int_\Gamma [y'(x)]^2 dx = -\int_\Gamma y(x) y''(x) dx$.

Proof. Let us first prove part (a) of the lemma, by induction. Under the assumption that $w \equiv [\phi']^{1/2-k} y^{(k)} \in \mathbf{L}_\alpha(\mathcal{D})$, it follows from Theorem 4.1.2 that

$$w'/\phi' = [\phi']^{1/2-k-1} y^{(k+1)} - (1/2 - k)(1/\phi')' w \in \mathbf{L}_\alpha(\mathcal{D}). \qquad (7.2.112)$$

Since, however, $(1/\phi')'$ is assumed to be uniformly bounded in $\mathcal{D}$, it follows from (7.2.111) that $[\phi']^{1/2-k-1}y^{(k+1)} \in \mathbf{L}_\alpha(\mathcal{D})$.
For part (b) note for example, that once part (a) of the lemma has been validated, it follows that $[\phi']^{1-2k}(y^{(k)})^2 \in \mathbf{L}_{2\alpha}(\mathcal{D})$, and, therefore, $[\phi']^{2-2k}(y^{(k)})^2 \in \mathbf{H}^1(\mathcal{D})$, so that $[\phi']^{1-k}y^{(k)} \in \mathbf{H}^2(\mathcal{D})$.
For part (c), note that

$$\int_{\partial\mathcal{D}} |y|^2 |dz| \leqq c \int_{\partial\mathcal{D}} |\phi' y|^2 |dz|, \tag{7.2.113}$$

with $c = \sup_{z\in\mathcal{D}} |\phi'(z)|^{-2}$ (see Problem 4.4.14).
The proof of part (d) is self-evident, using the established parts (b) and (c).
For part (e), note that since both $[\phi']^{1/2}F$ (by assumption) and $[\phi']^{-1/2}F'$ (by part (a) of the lemma) belong to $\mathbf{L}_\alpha(\mathcal{D})$, so does $F\,F'$, the product of these two functions. Hence $F\,F'$ vanishes at the endpoints of Γ. We thus obtain part (e) of the lemma, after integration by parts of $\int_\Gamma (F')^2 dz$, since the boundary term, $F\,F'|_\Gamma$ vanishes. ■

Theorem 7.2.11 *Let h be determined as in (7.2.7).*
(a) *Let Assumption 7.2.9-(a) be satisfied. If $y(x)$ denotes the solution of Problem (7.2.4), if $\mathbf{y} = (y(z_{-N}), \cdots, y(z_N))^T$, and if E_s denotes any one of the matrices A_s, B_s or C_s defined above, then there exists a constant c_1 independent of N, such that*

$$\left\| E_s D\left([\phi']^{1/2}\right)\mathbf{y} - h^2 D\left(\frac{1}{[\phi']^{3/2}}\right)\mathbf{s}\right\| \leqq c_1 N^{1/2} e^{-(\pi d \alpha N)^{1/2}}. \tag{7.2.114}$$

(b) *If Assumption 7.2.9-(b) is satisfied, then there exists a constant c_2 independent of N, such that*

$$\|C_s^{-1}\| \leqq \frac{4N^2}{\pi^2}\left[1 + \frac{c_2}{N}\right]. \tag{7.2.115}$$

In addition, if $\mathbf{v} = (v_{-N}, \cdots, v_N)^T$ denotes the solution to the problem (7.2.104), and if $y(x)$ denotes the solution to (7.2.4), then there exists a constant c_3 independent of N, such that

$$\sup_{x\in\Gamma}\left| [\phi'(x)]^{1/2} y(x) - \sum_{j=-N}^{N} v_j S(j,h)\circ\phi(x)\right| \leqq c_3 N^{5/2} e^{-(\pi d\alpha N)^{1/2}}. \tag{7.2.116}$$

Furthermore, A_s^{-1} and B_s^{-1} also exist, and their norms are bounded by the right-hand side of (7.2.115) for all N sufficiently large, and if the vector $\mathbf{v}$ is

determined by either (7.2.96) or by (7.2.104), then the inequality (7.2.116) again holds for all N sufficiently large.
(c) *Let* $\mathbf{v}$ *be determined as in (7.2.105), let* $\|y\|_S$ *be defined as in (7.2.111), and let* Y_N *be defined as in (7.2.95). If Assumption 7.2.9-(c) is satisfied, then there exists a constant* c_4 *independent of* N*, such that*

$$\|y - Y_N\|_S \leqq c_4 N^{5/2} e^{-\pi d \alpha N)^{1/2}}. \tag{7.2.117}$$

Moreover, in this case, if $\mathbf{v}$ *is determined by either (7.2.96) or by (7.2.104), then (7.2.117) again holds for all* N *sufficiently large.*

Proof. The proof is a straightforward repetition of some of the arguments used in Subsection 7.2.4, as well as an application of Lemma 7.2.10, and is left to the reader in Problem 7.2.16. ■

We emphasize that the choice of h made in Theorem 7.2.11 is optimal for the uniform Sinc approximation of $[\phi']^{1/2} y$ on Γ, and it is also the best choice for Sinc approximation of y over Γ with respect to the norm $\|y\|_S$ as defined in (7.2.110). This choice of h *is not* the best choice for the *uniform* approximation of y on Γ. For example, for the case of approximation over $[0,1]$, if $\phi' y \in \mathbf{L}_\alpha(\mathcal{D})$, then the best h for uniform approximation of y on Γ is given by $h = \{\pi d/[(\alpha + 1/2)N]\}^{1/2}$. The choice of h given by (7.2.7) yields an approximation which is more accurate near the mid-range of the interval $[0,1]$ and a poorer approximation near the endpoints of this interval.

In applications, the case $\mu \equiv 0$ and Γ an interval is a frequently occurring one, particularly when the coefficients ν and σ are real-valued on Γ. The boundary value problem for this case is

$$\begin{aligned} (Ly)(x) &= y''(x) + \nu(x) y(x) - \sigma(x) = 0, \quad x \in \Gamma, \\ y(a) &= 0, \quad y(b) = 0. \end{aligned} \tag{7.2.118}$$

This is the most frequently occurring form of *self-adjoint problems*. The eigenvalues of such differential equations are real, and it is desirable to have this feature be reflected in the resulting system of linear algebraic equations. This feature is *not* reflected by the Sinc–Galerkin method of the previous subsection, except in special circumstances, such as, when $\phi(x) = x$. For example, even in the case when Γ is a finite interval (a,b) and $\mu \equiv 0$, the eigenvalues of the coefficient matrices A, B, and C defined in Sections 7.2.1 to 7.2.4 have non-zero imaginary components. The symmetric Sinc–Galerkin procedure remedies this situation. In this case the system of equations to be solved becomes

$$B_s \mathbf{v} = h^2 D(1/[\phi']^{3/2}) \mathbf{s} \tag{7.2.119}$$

with

$$B_s = I^{(2)} + h^2 D\left(\frac{1}{2\phi'}\left(\frac{1}{\phi'}\right)'' - \frac{1}{4}\left[\left(\frac{1}{\phi'}\right)'\right]^2 + \frac{\nu}{[\phi']^2}\right). \tag{7.2.120}$$

7.2.8 More Examples

In this section we present some illustrative examples of applications of the above derived Sinc–Galerkin methods.

Example 7.2.12 Let us examine the simple problem,

$$\varepsilon^2 y'' - y = -1, \quad 0 < x < 1, \quad y(0) = y(1) = 0 \tag{7.2.121}$$

whose solution has a *boundary layer* in the neighborhoods of each of the points $x = 0$ and $x = 1$. Evidently, we see from the differential equation in (7.2.121) that the exact solution y(x) of the problem (7.2.121) is an entire function of x. In addition, we can deduce, via the use of integral equations (see Problem 7.2.9), that this solution $y(x)$ is uniformly bounded by 1 on the disc $\mathcal{D}(1/2; 1/2)$, which has center 1/2 and radius 1/2. Hence, upon taking $\phi(x) = \log\{(x/(1-x)\}$, it follows that for unspecified $h > 0$, we have

$$\sup_{0<x<1}\left|y(x) - \sum_{k=-\infty}^{\infty} y(z_k)S(k,h)\circ\phi(x)\right| \leqq ce^{-(\pi d\alpha N)^{1/2}} \tag{7.2.122}$$

with c a constant which is independent of h and also of ε. Now it is also evident using either a WKB approximation, or using the integral equation referred to above, that the boundary layers are on the intervals $0 < x < \varepsilon$ and $1-\varepsilon < x < 1$. For example, in the neighborhood of $x = 0$ we can expect the solution $y(x)$ to satisfy an inequality such as $|y(x)| < C_1[1-\exp(-x/\varepsilon)]$, with C_1 independent of ε, and similarly, for x near 1. Using this estimate, it is readily seen that for N sufficiently large such that $e^{-Nh}/\varepsilon < 1$, then

$$\left|\sum_{|k|>N} y(z_k)S(k,h)\circ\phi(x)\right| \leqq \frac{c_2\varepsilon h}{e^{-Nh}}, \tag{7.2.123}$$

with c_2 a constant independent of h, N, and also of ε. By equating $e^{-\pi^2/(2h)}$ and e^{-Nh}/ε, we get

$$h = \frac{\log(\varepsilon) + \sqrt{(\log(\varepsilon))^2 + 2N\pi^2}}{2N}. \tag{7.2.124}$$

This choice of h is thus the right choice for solving problem (7.2.121) via a method of Section 7.2.7. Upon substituting into (7.2.119), in which we take $\mu = 0$, and $\nu = \sigma = -1/\varepsilon^2$, we get

$$\sup_{0<x<1} \left| y(x) - \sum_{k=-N}^{N} v_k S(k,h) \circ \phi(x) \right| \leqq cN^2 \varepsilon^{-1/2} e^{-(1/2)\sqrt{(\log \varepsilon)^2 + 2N\pi^2}} \tag{7.2.125}$$

with c independent of N, and under the assumption that $e^{-Nh}/\varepsilon < 1$. These estimates are indeed reliable, as is readily verified by comparison with the exact solution,

$$y(x) = \frac{2\sinh(\frac{x}{2\varepsilon})\sinh(\frac{1-x}{2\varepsilon})}{\cosh(\frac{1}{2\varepsilon})}. \tag{7.2.126}$$

For example, with $\varepsilon = 0.3$, we get a maximal error which is less than 4.2×10^{-7} using $N = 24$.
The same values of N and h yield similar accuracy for the problem

$$\varepsilon^2 y'' - y = -1/\sqrt{x(1-x)}, \quad 0 < x < 1, \quad y(0) = y(1) = 0. \tag{7.2.127}$$

which has both a boundary layer as well as an algebraic-type singularity at $x = 0$ and $x = 1$.

Example 7.2.13 (Lundin): Let us consider the non-linear "spinor field" problem,

$$y'' - y + \frac{y^3}{x^2} = 0, \quad -\infty < x < \infty, \quad y(-\infty) = y(\infty) = 0. \tag{7.2.128}$$

The properties of the solution to this problem are exhibited in Problem 7.2.10. It may be shown that problem (7.2.128) has a countably infinite number of solutions. Moreover, all solutions of the problem (7.2.128) are odd functions of x, and in particular, it is known that the problem has a unique solution, which is positive on $(0, \infty)$. By Problem 7.2.10 (d), this solution belongs to the class $\mathbf{L}_\alpha(\mathcal{D}_d)$, with $\alpha = 1$, and $d = 1/\sqrt{1 + y'(0)^2}$, a result which we may get from a knowledge of the approximate value of $y'(0)$ gotten via the use of the classical "shooting" method for solving this problem. Let us define $\|f\|$ by

$$\|f\| = \left(\int_{\mathbb{R}} [|f(x)|^2 + |f'(x)|^2] \, dx \right)^{1/2}. \tag{7.2.129}$$

Hence, taking $N = 70$, selecting $h = [\pi d/(\alpha N)]^{1/2}$, making the approximation

$$Y_N(x) = \sum_{k=1}^{N} a_k \{S(k,h)(x) - S(-k,h)(x)\}. \tag{7.2.130}$$

and then solving the problem of minimization of $\|y_N\|$ subject to the condition that

$$\|Y_N\|^2 = \int_{\mathbf{R}} \frac{Y_N^4(x)}{x^2} dx, \tag{7.2.131}$$

the resulting answer we get is accurate to 5 significant figures on $(0, \infty)$. We remark here, that the solution of the non-linear system of equations (7.2.131) was carried out by Powell's method, which converges slowly for this particular problem.

A simpler method to solve this problem which has been tried and shown to be effective, is to use the initial value solution method of Section 7.1 in the form of the classical "shooting" method. To this end, we may note that the positive solution to the problem (7.2.128) belongs to the space $\mathbf{L}_\alpha(\mathcal{D}_d)$, with $\mathcal{D}$ defined as in (1.7.12), so that we may take $\phi(x) = \log(\sinh(x))$, $\alpha = 1$, and $d = 1/\{1 + y'(0)^2\}^{1/2}$. An approximate value of $y'(0)$ will suffice here; such an approximate value may be obtained using approximate solutions with a non-optimal h. We would seek an approximate solution of the form $Y_N(x) = \sum_{k=-N}^{N} a_k S(k, h) \circ \phi(x)$. We would thus simultaneously solve the initial value differential equation problem $y'' - y + y^3/x^2 = 0$, $y(0) = 0$, $y'(0) = c$, after first writing this differential equation in the form of a system of two first-order equations, along with the Fréchet derivative problem given in Problem 7.2.10, which we also first convert to a pair of first-order equations, and then determine the value of c iteratively, by solving the equation $a_N(c) = 0$ via Newton's method.

Example 7.2.14 (Jarratt): Let us illustrate the use of the Sinc–Galerkin method to determine the solution to an ordinary differential equation eigenvalue problem. The model problem which we shall consider takes the form

$$\begin{gathered} Ly(x) \equiv -y''(x) - \nu(x)y(x) = \lambda\rho(x)y(x), \quad a < x < b \\ y(a) = y(b) = 0. \end{gathered} \tag{7.2.132}$$

As usual for this model problem, we shall assume that the *potential function* $\nu \leqq 0$ on (a, b), and that the *weight function* ρ is positive on this interval. We shall use the method of Subsection 7.2.7 to derive an approximate solution to this problem. To this end, we shall assume that both $\nu/[\phi']^2$ and $\rho/[\phi']^2$ belong to $\mathbf{H}^\infty(\mathcal{D})$, with ϕ and $\mathcal{D}$ defined as in Example 1.7.5, and that $(1/\phi')''/\phi'$ is uniformly bounded in $\mathcal{D}$. In this case, letting N denote a positive integer, and selecting the matrix B_s by (see 7.2.120)

$$B_s = -I^{(2)} + h^2 D\left(\frac{([1/\phi']^{1/2})''[\phi']^{1/2} - \nu}{[\phi']^2}\right), \tag{7.2.133}$$

we arrive at the eigenvalue problem

$$B_s \mathbf{w} = \mu D \left(\frac{\rho}{[\phi']^2} \right) \mathbf{w} \tag{7.2.134}$$

for the eigenvalue μ approximating λ and for the vector $\mathbf{w}$ such that

$$y_N(x) \equiv [\phi'(x)]^{-1/2} \sum_{k=-N}^{N} w_k \, S(k,h) \circ \phi(x) \tag{7.2.135}$$

approximates the corresponding eigensolution $y(x)$ of (7.2.132). It follows by our assumptions on ν and ρ and by Theorem 7.2.11 that all solutions μ of the eigenvalue problem (7.2.134) are positive, and the set of corresponding eigenvectors w may be selected to be real, and orthonormal. Moreover, we may show, by proceeding as in Subsection 6.2.4, or by Problem 7.2.13, that, e.g., corresponding to the kth eigenvalue μ of B_s, (listed in increasing order of magnitude) and the associated vector $\mathbf{w}$ defined by (7.2.134), there corresponds an eigenvalue λ and an associated eigenfunction y of Problem 7.2.132, such that if $y_N(x)$ is defined as in (7.2.135), then

$$\begin{aligned} |\lambda - \mu| &\leqq c N^{5/2} e^{-(\pi d \alpha N)^{1/2}}, \\ \left(\int_a^b |y(x) - y_N(x)|^2 dx \right)^{1/2} &\leqq c_1 N^{5/2} e^{-(\pi d \alpha N)^{1/2}}, \end{aligned} \tag{7.2.136}$$

with c and c_1 constants which may depend on k but are independent of N. Consider, for example, the case of a harmonic oscillator, that is, when $\nu(x) = -x^2 - 2x^{-2}$ and $\rho(x) = 1$. One may, in this case, deduce that a solution $y(x)$ of the problem (7.2.132) has to satisfy the relation $y(x) \sim c_1 x^{3/2}$ as $x \to 0$, and also, by using the classical WKB method, that $y(x) \sim c_2 x^{-1/2} e^{-x^2/2}$, $x \to \infty$. By using these estimates, and by inspection of the differential equation, one may thus deduce that the solution $y \in \mathbf{L}_\alpha(\mathcal{D})$, with $\mathcal{D}$ defined as in (1.7.12), with $d = \pi/2$, and with $\alpha = 3/2$. By using the above procedure, taking $\phi(x) = \log(\sinh(x))$, and $N = 24$, one can approximate the first four eigenvalues $(\lambda_1, \cdots, \lambda_4)$ of this problem, to within errors bounded by 0.304×10^{-8}, 0.194×10^{-7}, 0.496×10^{-7}, and 0.331×10^{-5} respectively.

Example 7.2.15 Let us illustrate an application of Sinc–Galerkin methods to the solution of an *inverse problem*. Given the differential equation

$$y''(x) = q(x)\, y(x) + f(x), \quad 0 < x < 1, \tag{7.2.137}$$

we wish to determine the function q knowing that $y(0) = y'(0) = 0$, and knowing (or being able to measure) $y(1)$ for every given f. Let us recall some relevant properties of Equation (7.2.137) which will assist us to obtain an approximate solution to this problem.

It is easily shown that if U and V denote two arbitrary linearly independent solutions of the homogeneous equation

$$u''(x) - q(x)u(x) = 0, \tag{7.2.138}$$

then $U(x)V'(x) - U'(x)V(x) = c$, a non-zero constant, and hence if q is well defined, so that the problem (7.2.138) does in fact have a smooth solution on [0,1], the function

$$K(x,t) = c^{-1}\{U(t)V(x) - U(x)V(t)\} \tag{7.2.139}$$

is well-defined on $[0,1]\times[0,1]$. Moreover, this function satisfies the equations

$$\begin{aligned} K(x,x) &= 0, \quad x \in [0,1], \\ [K_x(x,t)]_{t=x} &= 1, \quad x \in [0,1], \\ \tfrac{d^2}{dt^2}K(1,t) - q(t)\,K(1,t) &= 0, \quad t \in [0,1]. \end{aligned} \tag{7.2.140}$$

Furthermore, if we define y by the expression

$$y(x) = \int_0^1 K(x,t)\, f(t)\, dt, \tag{7.2.141}$$

then y satisfies the equation (7.2.137) as well as the initial conditions $y(0) = y'(0) = 0$.

For purposes of accurate Sinc approximation, let us make the additional assumption that all solutions of the homogeneous equation (7.2.138) belong to the class $\mathbf{M}_\alpha(\mathcal{D})$, as defined in Section 4.1, where $0 < \alpha \leqq 1$, $0 < d < \pi$, and where $\mathcal{D} = \{z \in \mathbb{C} : |\arg[z/(1-z)]| < d\}$. It then follows by inspection of (7.2.138) that $u''/[\phi']^2 \in \mathbf{L}_\alpha(\mathcal{D}')$ for all functions u which satisfy (7.2.138), where $\mathcal{D}'$ is related to $\mathcal{D}$ as in Theorem 4.1.2. Also, it then follows that $q/[\phi']^2 \in \mathbf{L}_\alpha(\mathcal{D}')$. Notice also, that $K(1,\cdot) \in \mathbf{M}_\alpha(\mathcal{D})$, and since $K(1,1) = 0$, it follows by setting

$$\kappa(t) = K(1,t) - (1-t)K(1,0), \tag{7.2.142}$$

that $\kappa \in \mathbf{L}_\alpha(\mathcal{D})$. Hence by the third equation in (7.2.140), we have

$$q(t) = [K(1,t)]^{-1} \frac{d^2}{dt^2}\kappa(t). \tag{7.2.143}$$

Now, corresponding to some positive integers M and N, where $N < M$, we set $h = [\pi d/(\alpha N)]^{1/2}$, $z_k = e^{kh}/[1+e^{kh}]$, and we take $f(t) = h^{-1}S(k,h) \circ \phi(t)\,\phi'(t)$ in (7.2.141), to get the identity

$$\begin{aligned} y_k &\equiv \int_0^1 K(1,t)S(k,h) \circ \phi(t)\,\phi'(t)\,dt \\ &= hK(1,z_k) + \varepsilon_k, \end{aligned} \tag{7.2.144}$$

where by Problem 4.2.11, $\varepsilon_k = \mathcal{O}(e^{-\pi d/h})$. Hence, since $K(1, z_{-M}) = K(1,0) + \mathcal{O}(e^{-Mh})$, taking M sample values of $y_k(1)$ defined as in (7.2.144), a good approximation to $\kappa(t)$ at $t = z_k$ is

$$\kappa(z_k) = y_k - (1 - z_k) y_{-M}. \tag{7.2.145}$$

Also, by Section 7.2.1, we have the approximation

$$h^2 \frac{\kappa''(z_k)}{\phi'(z_k)^2} \approx \sum_{j=-M}^{M} \left[\delta_{k-j}^{(2)} + h \left(\frac{1}{\phi'} \right)' (z_k) \delta_{k-j}^{(1)} \right] \kappa(z_j), \tag{7.2.146}$$

where the error of this approximation can be accurately bounded by proceeding as in Section 7.2.4. Hence, we get the approximation

$$\frac{q(z_k)}{\phi'(z_k)^2} = \frac{\sum_{j=-M}^{M} \left[\delta_{k-j}^{(2)} + h \left(\frac{1}{\phi'} \right)' (z_k) \delta_{k-j}^{(1)} \right] \kappa(z_j)}{h y_k}, \tag{7.2.147}$$

this result being reliable for those values of k for which the relative error $\varepsilon_k / [h y_k]$ of the approximation in the denominator is tolerable. Since $K(1, t) \to 0$ as $t \to 1$, we restrict k, such that $k \in [-N, N]$, $(0 < N < M)$, where N is selected such that $e^{-\pi d/h + \alpha N h} = e^{-(\pi d \alpha N)^{1/2}}$.

We could obtain precise bounds on the error of the above approximations by proceeding as in Section 7.2.4. We could also consider the effects of noisy data, by proceeding as in Section 6.9.

As noted above, $q/[\phi']^2 \in \mathbf{L}_\alpha(\mathcal{D}')$, and thus the numbers defined in (7.2.147), can now be used to obtain an accurate Sinc interpolation of $q/[\phi']^2$ over $[0, 1]$, via the formula

$$\frac{q(x)}{\phi'(x)^2} \approx \sum_{k=-N}^{N} \frac{q(z_k)}{\phi'(z_k)^2} S(k, h) \circ \phi(x). \tag{7.2.148}$$

•

Problems for Section 7.2

7.2.1 Prove that the boundary terms in (7.2.13) vanish.

7.2.2 Verify (7.2.14) and (7.2.15).

7.2.3 Prove in detail the inequalities in (7.2.37).

7.2.4 Use the results of Lemma 7.2.4 to obtain upper bounds on the norms of the matrices A, B, and C which are defined as in Lemma 7.2.5.

7.2.5 Give a detailed proof of Lemma 7.2.6.

7.2.6 Give a detailed proof of Lemma 7.2.7.

7.2.7 Show that the boundary terms of the integrations by parts for deriving (7.2.101) through (7.2.103) appear in the form

$$
\begin{aligned}
&< \mu y', S(k,h)\circ\phi > \\
&\qquad = bt_1 + \int_\Gamma y(x)\left([\phi']^{-1/2}\mu S(k,h)\circ\phi\right)'(x)dx
\end{aligned}
$$

$$
\begin{aligned}
&< y'', S(k,h)\circ\phi > \\
&\qquad = bt_2 + \int_\Gamma y(x)\left([\phi']^{-1/2} S(k,h)\circ\phi\right)''(x)dx,
\end{aligned}
$$

and they are explicitly given by

$$
\begin{aligned}
bt_1 &= [\phi'(x)]^{-1/2} y(x)\mu(x) S(k,h)\circ\phi(x)\Big|_a^b, \\
bt_2 &= y'(x)[\phi'(x)]^{-1/2} S(k,h)\circ\phi(x) \\
&\quad - y(x)\{[\phi'(x)]^{1/2} S'(k,h)\circ\phi(x) \\
&\quad + \left([\phi'(x)]^{-1/2}\right)' S(k,h)\circ\phi(x)]\Big|_a^b .
\end{aligned}
$$

(b) Show that if $[\phi']^{1/2} y \in \mathbf{L}_\alpha(\mathcal{D})$, then each of the terms bt_1 and bt_2 vanishes.

7.2.8 Give a detailed proof of Theorem 7.2.11.

7.2.9 Prove that the solution to Problem (7.2.121) is uniformly bounded on the disc $D(1/2; 1/2)$, independently of $\varepsilon \in (0,1)$.

7.2.10 Let $y(x)$ denote a solution of the problem (7.2.128), and let $y'(0) = c$, $c > 0$.

(a) Prove that y is analytic and uniformly bounded in the strip $\mathcal{D}_d$, with $d = 6^{1/2}/(1+c)$.

(b) Prove that there exists a constant $C = C(c) > 0$ such that $|y(x)| < Ce^{-x}$ for all $x \in \mathcal{D}_d$.

(c) Prove that if $y(x) > 0$ on $(0,\infty)$, then the the Fréchet derivative problem $\eta'' + \eta - 3y(x)^2\eta/x^2 = 0$, $\eta(0) = \eta(\infty) = 0$ has the unique solution $\eta \equiv 0$.

(d) Use the result in (c) to prove that the positive solution of the problem (7.2.128) is unique.

(e) (Lundin, [Lu1]) Let $y(x)$ be defined as in part (c) of this problem. Define $\|f\|$ by

$$\|f\| = \left(\int_{\mathbb{R}} \{|f(x)|^2 + |f'(x)|^2\} \, dx \right)^{1/2},$$

and let f be such that $\|f\| < \infty$. Prove that there exists a constant $\kappa \in (0, 2)$ such that

$$6 \int_0^\infty y(x)^2 |f(x)|^2 x^{-2} dx < (2 - \kappa) \|f\|^2.$$

(f) (Lundin, [Lu1]) Let $y(x)$ be defined as in part (c), and let $\| \cdot \|$ be defined in part (e) of this problem. Let Y_N be defined by

$$Y_N(x) = \sum_{k=1}^{N} a_k \{S(k, h)(x) - S(-k, h)(x)\}.$$

Prove that the problem of minimization of $\|Y_N\|$ subject to the condition that

$$\|Y_N\|^2 = \int_{\mathbb{R}} \frac{Y_N^4(x)}{x^2} dx$$

has a unique solution for all such Y_N with all $a_k \geqq 0$ and at least one $a_k > 0$.

7.2.11 Use the results of the previous problem to solve the problem (7.2.128) via a shooting method, as outlined in Example 7.2.13.

7.2.12 Give a detailed proof of the inequalities in (7.2.136). [Hint: See Eggert, Jarratt and Lund [EgJaL1]].

7.2.13 Consider, at the outset, a special case of Problem 7.2.4, namely, when $\Gamma = (0, 1)$, and the functions μ, ν, and σ all belong to $\mathbf{M}_\alpha(\mathcal{D})$, with $\mathcal{D}$ defined as in (1.7.11), with $a = 0$ and $b = 1$. (a) Show that every solution y of equation (7.2.4) also belongs to $\mathbf{M}_\alpha(\mathcal{D})$. Show that in this case the functions y' and y'' also belong to $\mathbf{M}_\alpha(\mathcal{D})$.

(b) Use an approximation of the form

$$\begin{aligned} y(x) \approx y_N(x) &\equiv (Ty)(x) + x^2(1-x)^2 \sum_{k=-N}^{N} c_k S(k, h) \circ \phi(x) \\ (Ty)(x) &\equiv x(1-x)^2 c_{-N-2} + x^2(1-x)^3 c_{-N-1} \\ &\quad + x^3(1-x)^2 c_{N+1} + x^2(1-x) c_{N+2} \end{aligned}$$

in which the c_k are constants, and $\phi(x) = \log\{x/(1-x)\}$ to accurately approximate y, as well as y' and y'' on Γ.

(c) Show that the resulting approximation $y_N(x)$ given above may be used to accurately approximate y, y', and y'' uniformly on [0,1], provided that $h = [\pi d/(\alpha N)]^{1/2}$, $z_k = e^{kh}/(1+e^{kh})$, and where for $k = -N, \cdots, N$, the c_k approximate the values

$$c_k \approx z_k^{-2}(1-z_k)^{-2}\{y(z_k) - (Ty)(z_k)\}.$$

Upon substituting y_N for y in (7.2.4) and then approximating

$$\int_0^1 \{Ly_N(x)\}\omega_k(x)\,dx = 0$$
$$k = -N-2, -N-1, \cdots, N+2$$

by means of the quadrature formula (see Section 4.2)

$$\int_0^1 g(x)\,dx \approx h \sum_{k=-N-2}^{N+2} z_k(1-z_k)g(z_k)$$

with the ω_k the coefficients of c_k in (7.2.102), we get the linear system of equations

$$(Ly_N)(z_k) = 0, \quad k = -N-2, -N-1, \cdots, N+2$$

for the $2N+5$ coefficients c_k, $k = -N-2, \cdots, N+2$.

(d) Prove, using the results of Subsection 7.2.1, that there exists a constant C independent of N, such that

$$\begin{aligned}
\sup_{x\in\Gamma} |y(x) - y_N(x)| &\leqq CN^{5/2}e^{-(\pi d\alpha N)^{1/2}} \\
\sup_{x\in\Gamma} |y'(x) - y_N'(x)| &\leqq CN^{6/2}e^{-(\pi d\alpha N)^{1/2}} \\
\sup_{x\in\Gamma} |y''(x) - y_N''(x)| &\leqq CN^{7/2}e^{-(\pi d\alpha N)^{1/2}}.
\end{aligned}$$

7.3 Analytic Solutions of PDE

In this section we discuss some properties of analyticity of solutions to partial differential equations, as well as singularities which may occur due to the effects of corners in regions.

A feature about solving partial differential equations (PDE) is that in constructing algorithms for their solution, every efficient numerical algorithm which was designed to solve a class of specific problems from some other area of numerical approximation may be applied towards obtaining an approximate solution of a partial differential equation.

7.3.1 Analyticity of Solutions in All Variables

In this subsection we shall discuss analyticity requirements for approximating solutions of differential equations over rectangular regions, curvilinear regions, and regions with corners.

As for the case of ordinary differential equations, existence theorems are also known for partial differential equations. Let us simply state, without proof, one important result of this type, for linear differential equations. In the statement of the theorem which follows, β is a *multi-index* i.e., we use the notations $\beta = (\beta_1, \cdots, \beta_n)$, with $\beta_j \geqq 0$, $a^\beta = a(x_1, \cdots, x_n)^{\beta_1 \cdots \beta_n}$, $D_j = \partial/\partial x_j$, and $D^\beta = D_1^{\beta_1} \cdots D_n^{\beta_n}$, $|\beta| = \sum_1^n \beta_j$ is the *order* of the derivative D^β, and $x = (x_1, \cdots, x_n)$.

Theorem 7.3.1 (Cauchy–Kowalevsky Theorem:) *Let the differential equation*

$$\sum a^\beta D^\beta u = f \tag{7.3.1}$$

of order m be given, with the coefficients a^β and f analytic in a neighborhood of the origin, and with the coefficient of D_1^m is $\neq 0$ when $x = 0$. Then, for every function w, analytic in a neighborhood of the origin, there exists a unique solution u of (7.3.1) analytic in a neighborhood of the origin and satisfying the boundary conditions

$$D_1^j(u - w) = 0 \quad \text{when} \quad x^1 = 0 \quad \text{and} \quad j < m. \tag{7.3.2}$$

Proof. The somewhat lengthy proof of this theorem may be found in standard textbooks on differential equations, and is omitted here. ∎

We may similarly extend the proof of Theorem 7.1.8 to the case of partial differential equations. The details are left to Problem 7.3.1.

7.3.2 Analyticity for Sinc Approximation

We next discuss the type of analyticity which we require for solving partial differential equations by Sinc methods. As seen previously in Sections 6.5 and 6.6, we shall not require analyticity in all variables simultaneously, but rather, only in one variable at a time, while the other variables are fixed.

Based on a given partial differential equation problem, the class of regions V which we shall consider are a union of regions of the form

$$\begin{aligned} K_n &= \{(\xi^1, \cdots, \xi^n) \in \mathbb{R}^n : u^1 \leqq \xi^1 \leqq v^1, u^2(\xi^1) \leqq \xi^2 \leqq v^2(\xi^1), \\ &\quad \cdots, u^n(\xi^1, \cdots, \xi^{n-1}) \leqq \xi^n \leqq v^n(\xi^1, \cdots, \xi^{n-1})\} \end{aligned} \tag{7.3.3}$$

where it is assumed that the u^i and v^i are either identically infinite, or else they are bounded a.e. on their domain of definition.

It is perhaps important to emphasize here that the variables defining K_n in (7.3.3) need not be Cartesian coordinate variables; rather, these are the variables under which the differential equation is given. These variables might, for example be spherical, cylindrical, or any other coordinate variables.

In practice, methods that approximate a solution of a differential equation problem are most easily applied to regions which are cartesian products of one-dimensional regions. In particular, the Sinc approximations of the previous section readily lend themselves to approximations over rectangular regions of the form

$$V_n = \{(x^1, \cdots, x^n) \in \mathbb{R}^n : u^i \leqq x^i \leqq v^i, i = 1, \cdots, n\}, \tag{7.3.4}$$

with (u^i, v^i) any one of either a finite interval, $(0, \infty)$, $(-\infty, 0)$, or $(-\infty, \infty)$. Indeed, we can even consider the case when x_i ranges over a contour Γ_i.

Let us next consider a suitable function space on V_n, with V_n defined as in (7.3.4). We would like to consider functions $F(\xi^1, \cdots, \xi^n)$ which belong to the space $\mathbf{M}_\alpha(\mathcal{D}_i)$ with respect to the variable ξ^i, when the other variables are fixed. More precisely, let α and d be positive constants, such that $0 < \alpha \leqq 1$, and $0 < d < \pi$. Let ϕ_i be a conformal map of the region $\mathcal{D}_i \subset \mathbb{C}$ onto $\mathcal{D}_d$, with $\mathcal{D}_d$ defined as in (1.7.9), let $\psi_i = \phi_i^{-1}$ denote the inverse map, such that $\Gamma_i = [u_i, v_i] = \{\psi_i(w) : w \in \mathbb{R}\}$. Set $\rho^i = \exp(\phi^i)$, $h = [\pi d/(\alpha N)]^{1/2}$, and $z_k^i = \psi^i(kh)$.

Let $(x^1, \cdots, x^{i-1}, x^{i+1}, \cdots, x^n)$ be a given point in the region $\prod_{j \neq i} \Gamma_j$, and denote by $F^i = F^i(z)$ the function $F(x^1, \cdots, x^{i-1}, z, x^{i+1}, \cdots, x^n)$. Let $\mathbf{X}_\alpha(V_n)$ denote the family of all functions F defined on V_n such that $F^i \in \mathbf{M}_\alpha(\mathcal{D}_i)$ for $i = 1, \cdots, n$.

Define basis functions ω_k^i by means of the equations

$$\begin{aligned} \omega_k^i &= S(k,h) \circ \phi^i, \quad k = -N+1, \cdots, N-1, \\ \omega_{-N}^i &= \frac{1}{1+\rho^i} - \sum_{k=-N+1}^{N} \frac{1}{1+e^{kh}} S(k,h) \circ \phi^i, \\ \omega_N^i &= \frac{\rho^i}{1+\rho^i} - \sum_{k=-N}^{N-1} \frac{e^{kh}}{1+e^{kh}} S(k,h) \circ \phi^i. \end{aligned} \tag{7.3.5}$$

If $F \in \mathbf{X}_\alpha(V_n)$, then it is readily seen, by proceeding as in the proof of Theorem 6.5.2, that for $N > 1$, there exists a constant, C, independent of N, such that

$$
\begin{aligned}
\sup_{(x^1,\cdots,x^n)\in V_n} \Bigg| F(x^1,\cdots,x^n) - \sum_{k_1=-N}^{N} \cdots \\
\cdots \sum_{k_n=-N}^{N} F(z^1_{k_1},\cdots,z^n_{k_n})\omega^1_{k_1}(x^1)\cdots\omega^n_{k_n}(x^n)\Bigg| \\
\leqq C N^{1/2}\{\log(N)\}^{n-1} e^{-(\pi d\alpha N)^{1/2}}.
\end{aligned}
\tag{7.3.6}
$$

Although we have explicitly described Sinc approximation in the above paragraph, it may be more efficient or convenient to select a non-Sinc approximating basis corresponding to some of the variables x^i. For example, if corresponding to a particular integer $i \in [1, n]$, u^i and v^i are finite, and respectively equal to -1 and 1 for the sake of this discussion, and if x^i is independent of the variables x^j, $j \neq i$, the solution U of the differential equation defined on V_n is an analytic function with respect to x^i in the ellipse $\mathcal{E}_R$ defined in (1.7.6), with $R > 1$, then it is more efficient to use Chebyshev polynomials than Sinc functions for approximation in the variable x^i. Similarly, if $(u^i, v^i) = (-\pi, \pi)$, and if U is known to be a periodic and smooth function of x^i, then it is better to use Fourier polynomials than Sinc functions in the variable x^i. In these latter two cases it may be possible to achieve an $\mathcal{O}(e^{-cn})$ error with polynomial approximation of degree n, whereas we would only achieve a $\mathcal{O}(e^{-c'n^{1/2}})$ error if Sinc approximation is used. Aside from their simplicity of implementation, Sinc approximations handle singularities of the solution on the boundary well; they are quite efficient if the solution has a boundary layer, if the problem is stiff, or if the solution is defined on an infinite range.

Given a problem over the region K_n defined as in (7.3.3), we first transform the region K_n into the region V_n of the form (7.3.4). This can always be accomplished via simple "quasi-linear" transformations. For example, if u^i and v^i defined as in (7.3.3) are finite everywhere, we set

$$
\xi^i = u^i + (v^i - u^i)x^i \tag{7.3.7}
$$

so that if, for example, this is the case for $i = 1, \cdots, n$, the new region of the variables x^i is now the cube $\prod_{i=1}^n [0,1]$. If $u^i = -\infty$, $v^i = \infty$, we do not apply any transformation, unless some other advantages may be gained, depending on the problem. If u^i is a function, and $v^i = \infty$, we set

$$
\xi^i = u^i + x^i, \tag{7.3.8}
$$

and if $u^i = \infty$ and v^i is a function, we set

$$
\xi^i = u^i - x^i. \tag{7.3.9}
$$

It may be readily shown that if the functions u^i and v^i belong to $\mathbf{M}_\gamma(\mathcal{D}^i)$ in each of the variables x^{i-k}, $k = 1, \cdots, i-1$, with γ a positive constant, then the transformed function under (7.3.7), (7.3.8), or (7.3.9) in the variables x^i belongs to the space $\mathbf{X}_{\alpha\gamma}(V_n)$. (See Problem 7.3.2).

We recommend decomposing the original domain (say $\mathcal{B}$) of the solution of the partial differential equation into a finite number of subdomains, V_n, such that the solution of the partial differential equation belongs to the space $\mathbf{X}_{\alpha'}(V_n)$, for each V_n. We will then be able to construct a solution to within an accuracy given on the right-hand side of (7.3.6) (in which α is replaced by α') throughout the region $\mathcal{B}$.

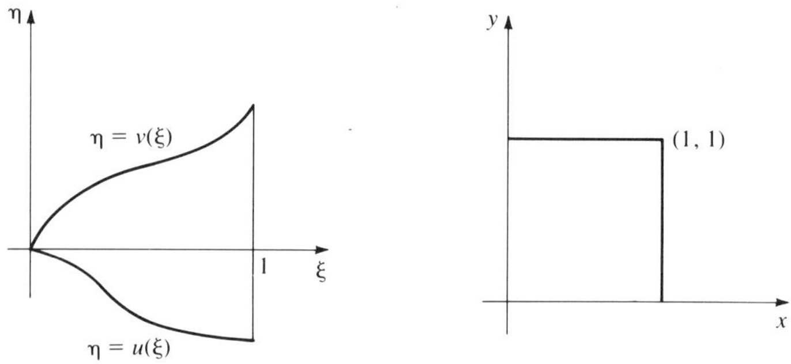

Figure 7.3.1: Transformation under (7.3.12).

This process of domain decomposition also provides the key to proper parallel computation. That is, after subdividing the original domain, $\mathcal{B}$, into a finite number of domains, V_n, as above, the majority of the computations can be carried out simultaneously, over each domain V_n. In this process of subdivision, and in setting up the the system of algebraic equations, we must of course take care to identify the common unknowns on the common boundaries of the regions, in a manner discussed previously in Section 6.5.6.

Example 7.3.2 Let us consider Laplace's equation

$$U_{\xi\xi} + U_{\eta\eta} = 0, \quad (\xi, \eta) \in K_2, \tag{7.3.10}$$

with K_2 the region

$$K_2 = \{(\xi, \eta) \in \mathbb{R}^2 : 0 < \xi < 1, u(\xi) < \eta < v(\xi)\}. \tag{7.3.11}$$

It is assumed that the functions u and v belong to the class $\mathbf{M}_\alpha(\mathcal{D})$, with $\mathcal{D}$ defined as in (1.7.11), and $(a, b) = (0, 1)$. Hence, according to (7.3.7), we apply the transformation

$$\xi = x, \quad \eta = u + (v - u)y, \tag{7.3.12}$$

and these functions enable us to deduce the transformation identities

$$\begin{aligned}\frac{\partial}{\partial \xi} &= \frac{\partial}{\partial x} - \frac{u' + (v' - u')y}{v - u}\frac{\partial}{\partial y} \\ \frac{\partial}{\partial \eta} &= \frac{1}{v - u}\frac{\partial}{\partial y}.\end{aligned} \tag{7.3.13}$$

By using this expression in (7.3.10), we readily transform Laplace's equation (7.3.10) into the equation

$$\begin{aligned} U_{xx} - 2\left(\frac{-u' - (v' - u')y}{v - u}\right)U_{xy} + \left(\frac{[u' + (v' - u')y]^2 + 1}{(v - u)^2}\right)U_{yy} \\ + \left\{\frac{u' + (v' - u')y}{v - u}\left[\frac{v' - u'}{v - u} + \frac{\partial}{\partial x}\left(\frac{u' + (v' - u')y}{v - u}\right)\right]\right\}U_y = 0. \end{aligned} \tag{7.3.14}$$

This last equation is to be solved on the region $V_2 = (0,1) \times (0,1)$ by use of Sinc approximation, or (if convenient) some other method of approximation. Notice, by inspection of the resulting equation (7.3.14), that if u and v have singularities at endpoints of their interval of definition, then the solution U will also have singularities on the boundary of V_2. Similarly, since $v - u$ appears in the denominator of the new equation, it follows that if $u(0) = v(0)$, then the solution U will have a singularity at the edge $x = 0$ of K_2.

The transformation procedure described above enables us to consider problems whose solution may have singularities at some or all points of the boundary of a region. In applications it is more frequently the case that singularities of solutions of differential equations are more isolated. Consider, for example, the case of a two-dimensional elliptic partial differential equation, defined over a "triangular" domain, K_2, which may be characterized in terms of polar coordinates, as follows:

$$K_2 = \{(r, \theta) : \theta_1 < \theta < \theta_2, 0 < r < f(\theta)\}, \tag{7.3.15}$$

with $f(\theta)$ an analytic function of θ in a domain containing the interval $[\theta_1, \theta_2]$. Let us assume, moreover, that corresponding to every fixed $(r_0, \theta_0) \in \overline{K_2}$, the solution $u = u(r, \theta)$ has the following properties:

(i) $u(\cdot, \theta_0)$ belongs to $\mathbf{M}_\alpha(\mathcal{D})$, with $\mathcal{D}$ defined as in (1.7.11), with $a = 0$, $b = f(\theta)$; and

(ii) $u(r_0, \cdot)$ is analytic and bounded in a domain containing the interval $[\theta_1, \theta_2]$.

It is readily seen, in this case, that upon replacing r by $t\,f(\theta)$, the resulting function $V(t, \theta) = U(tf(\theta), \theta)$ satisfies the properties that for all fixed $t \in [0,1]$ it is an analytic function of θ in a domain containing the interval

$[\theta_2, \theta_2]$, that for each $\theta \in [\theta_1, \theta_2]$, belongs to $\mathbf{M}_\alpha(\mathcal{D})$ as a function of t, with $\mathcal{D}$ defined as in (1.7.11), with $(a, b) = (0, 1)$. We may now carry out efficient approximation of $U(t, \theta)$ on the region $V_2 = \{(t, \theta) : 0 < t < 1, \theta_1 < \theta < \theta_2\}$ using a Sinc bases in the variable t, and polynomial bases in the variable θ.

7.3.3 Singularities Due to Corners and Edges

Solutions of differential equations may have singularities, even though the coefficients of the equations are smooth. Later, in Section 7.6 we shall exemplify a Navier–Stokes differential equation, which is non-linear, but whose solutions may have discontinuities, even though all coefficients of the equation are smooth, and even though the domain of existence of the solution is $\mathbb{R}^n$. This is not the case for linear differential equations, whose solutions are smooth in the interior of the domain of the solutions, wherever the coefficients of the equation are smooth. Corners in domains are frequently occurring phenomena in applications, and indeed, solutions of even linear differential equations cease to be analytic at such points, even though the coefficients of the equations are smooth there. The existence of corners drastically decreases the rate of convergence of the solution of differential equation by classical methods, such as by finite difference, finite element, or polynomial (e.g., Fourier polynomial) methods of solution, and it is therefore important to recognize the nature of such singularities for these methods. Current modifications of classical methods introduce special "boundary elements" which mimic the singular behavior of the solution, but, to be able to do this effectively, requires a knowledge of the exact nature of the singularity, which may at times be difficult to deduce. It is also important to know the locations of such singularities for Sinc methods, since we need to ensure that singularities occur at boundaries of domains, and hence these singularities will affect domain decomposition. On the other hand, we no longer need to know the exact nature of the singularities in order to achieve near optimal convergence. The presentation given here of the type of singularities that arise due to corners in the domain is due to Fichera; it is sufficiently precise to enable us to deduce an optimally accurate Sinc approximation for the solution. In addition, this presentation shows that it is far from trivial to determine the exact nature of the singularities in the solution of differential equations, as a result of corners in the domain of the solution.

Let B be a bounded domain in $\mathbb{R}^3$, and set $A = \mathbb{R}^3 \setminus B$. We denote $\sigma = \partial A$ to be the boundary of A, and we consider the problem

$$\begin{aligned} \nabla^2 u &= 0 \quad \text{in} \quad A \\ u &= b \quad \text{on} \quad \partial A, \end{aligned} \tag{7.3.16}$$

with $b \in \mathbf{C}^{(\infty)}(\mathbb{R}^3)$. We assume that B is a body with a finite number of corners and edges.

The solution u of (7.3.16) is representable in the form

$$u(P) = \frac{1}{2\pi} \int \int_{\partial A} \frac{\delta(Q)}{|P-Q|} d\sigma(Q), \quad P \in A. \tag{7.3.17}$$

Let O denote the origin in $\mathbb{R}^3$, and let O belong to ∂A. Let S' denote a small sphere with center O and radius r, set $S'' = A \cap S'$, and let S^* be defined by

$$S^* = \lim_{r \to 0} \left\{ \omega \in \mathbb{R}^3 : \omega = \frac{P-O}{|P-O|}, \quad P \in S'' \right\}. \tag{7.3.18}$$

Then S^* is a uniquely defined subset of the unit sphere $S \in \mathbb{R}^3$. Let H be defined by

$$H = \left\{ P \in \mathbb{R}^3 : 0 \leqq |P-O| \leqq 1, \quad \frac{P-O}{|P-O|} \in S^* \text{ for } P \neq O \right\}. \tag{7.3.19}$$

Let $\omega_1, \cdots, \omega_q$ be points of the unit sphere S which are vertices of S^*, denote by I_j the segments

$$I_j = \{P \in \mathbb{R}^3 : P - O = |P-O|\omega_j, \ 0 \leqq |P-O| \leqq 1\}, \tag{7.3.20}$$

and let μ_j denote the size of the interior angle of ∂S^* at ω_j (see Figure 7.3.2).

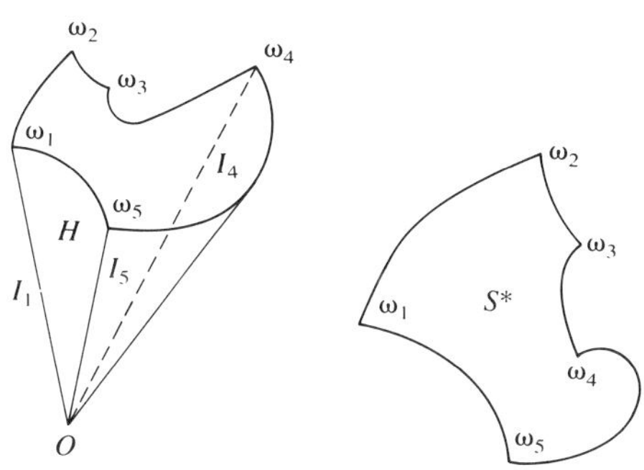

Figure 7.3.2 The Regions S^* and H of (7.3.18) and (7.3.19)

The *Laplace–Beltrami* operator in three dimensions is defined by

$$L_\omega \equiv \frac{\partial^2}{\partial \theta^2} + \cot(\theta)\frac{\partial}{\partial \theta} + \frac{1}{\sin^2(\theta)} \cdot \frac{\partial^2}{\partial \phi^2}. \tag{7.3.21}$$

We define the following eigenvalue problem on S^* in terms of L_ω:

$$\begin{aligned} L_\omega v + \lambda v &= 0 \quad \text{in} \quad S^*, \\ v &= 0 \quad \text{on} \quad \partial S^*. \end{aligned} \tag{7.3.22}$$

The problem (7.3.22) has solutions v_i in S^*, such that

$$\begin{aligned} L_\omega v_i + \lambda_i v_i &= 0 \quad \text{in} \quad S^*, \\ v_i &= 0 \quad \text{on} \quad \partial S^*, \\ i &= 1, 2, \cdots, \end{aligned} \tag{7.3.23}$$

where $0 < \lambda_1 < \lambda_2 \leqq \lambda_3 \leqq \cdots$, such that $\lambda_i \to \infty$ as $i \to \infty$, and where v_1 may be chosen to be positive on S^*. We then have

Theorem 7.3.3 (Fichera): *For $P \in H \setminus \{I_1 \cup I_2 \cup \cdots \cup I_q\}$ we have, as $P \to O$,*

$$\nabla[u - b](P) = \mathcal{O}\left(\rho^{\alpha-1} \prod_{j=1}^{q} \tau_j(\omega)\right), \tag{7.3.24}$$

with

$$\begin{aligned} \alpha &= \frac{\sqrt{1 + 4\lambda_1} - 1}{2}, \\ \rho &= |P - O|, \\ \omega &= \frac{P - O}{|P - O|}, \end{aligned} \tag{7.3.25}$$

and

$$\tau_j(\omega) = \begin{cases} |\omega - \omega_j|^{\frac{\pi}{\mu_j} - 1} & \text{if} \quad \mu_j > \frac{\pi}{2} \\ |\omega - \omega_j| \log \frac{3}{|\omega - \omega_j|} & \text{if} \quad \mu_j = \frac{\pi}{2} \\ |\omega - \omega_j| & \text{if} \quad \mu_j < \frac{\pi}{2}. \end{cases} \tag{7.3.26}$$

Moreover, these relations are sharp with regards to order.

One may note from Theorem 7.3.3 that the exact value of α required in Equation (7.3.24) is not too easily obtained, since it itself requires the solution of a partial differential equation. Fortunately, in the case of Sinc methods, we do not require the exact value of α to achieve a near-optimal rate of convergence of Sinc approximation of the solution. Also, the asymptotic forms given in (7.3.26) suggest that the solution u may not have a singularity at an edge where the interior angle $\mu < \pi/2$. That the solution u can in fact have a singularity in this case is illustrated in Problem 7.3.3.

Example 7.3.4 Let us consider the following Dirichlet problem over a region K_2 in the plane:

$$\begin{aligned} \nabla^2 u = u_{\xi\xi} + u_{\eta\eta} &= -1, \quad (\xi, \eta) \in K_2, \\ u &= 0, \quad (\xi, \eta) \in \partial K_2, \end{aligned} \tag{7.3.27}$$

with K_2 is defined by

$$K_2 = \{(\xi, \eta) \in \mathbb{R}^2 : 0 < \xi < 1,\ 0 < \eta < a\,\xi^\alpha\} \tag{7.3.28}$$

and with a and α positive numbers, and such that $0 < \alpha \leqq 1$. For this example, we know by Fichera's theorem, that the solution to the problem (7.3.27) belongs to the class $\mathbf{Lip}_1(K_2)$, and consequently, the solution $u(\xi, \eta)$ will belong to the class $L_\alpha(\mathcal{D})$, in each variable separately, the other variable being fixed. More precisely, corresponding to an arbitrary fixed point $(\xi, \eta) \in K_2$, $u(\xi, \cdot) \in \mathbf{L}_1(\mathcal{D})$ with

$$\mathcal{D} = \left\{\eta \in \mathbb{C} : \left|\arg \frac{\eta}{\xi^\alpha - \eta}\right| < d\right\}, \tag{7.3.29}$$

and $u(\cdot, \eta) \in \mathbf{L}_1(\mathcal{D})$ with

$$\mathcal{D} = \left\{\xi \in \mathbb{C} : \left|\arg \frac{\xi - (\eta/a)^{1/\alpha}}{1 - \xi}\right| < d\right\}. \tag{7.3.30}$$

Now, let us perform the transformation (7.3.12), in which we take $u = 0$, and $v = a\xi^\alpha$, to get

$$\begin{pmatrix} \xi \\ \eta \end{pmatrix} = \begin{pmatrix} x \\ a x^\alpha y \end{pmatrix}. \tag{7.3.31}$$

In this case, we find, using the procedure of Example 7.3.2, that

$$\begin{aligned} & u_{xx} - \tfrac{2\alpha y}{x} u_{xy} + \left(\tfrac{1}{a^2 x^{2\alpha}} + \tfrac{\alpha^2 y^2}{x^2}\right) u_{yy} + \tfrac{\alpha y}{x^2}(1+\alpha) u_y = -1,\ (x,y) \in V_2 \\ & u = 0,\ (x, y) \in \partial V_2, \end{aligned} \tag{7.3.32}$$

where $V_2 = (0,1) \times (0,1)$. Moreover, it is easily deduced, since $u \in \mathbf{L}_\alpha(B)$ as a function of (ξ, η), that in terms of (x, y), u now belongs to $\mathbf{Lip}_\alpha(0,1)$ as a function of x, and to $\mathbf{Lip}_1(0,1)$ as a function of y. It thus follows that as functions of (x, y), we have $u(\cdot, y) \in \mathbf{L}_\alpha(\mathcal{D})$, and $u(x, \cdot) \in \mathbf{L}_1(\mathcal{D})$, where in each of these cases, $\mathcal{D}$ denotes the region defined as in (1.7.11) for the case of $(a, b) = (0, 1)$.

•

Problems for Section 7.3

7.3.1 Consider a system of non-linear partial differential equations of the form

$$\frac{\partial w_i}{\partial x_1} = f_i, \quad i = 1, 2, \cdots, m.$$

Prove the following theorem, which guarantees the existence of analytic solutions of non-linear partial differential equations.

Theorem: In the above equation, let each f_i be a function of $x_1, \cdots, x_n$, of $w_1, \cdots, w_m$, and of all of the first-order partial derivatives of the form $\frac{\partial w_k}{\partial x_j}$, with $j = 1, \cdots, n$; $k = 1, \cdots, m$, $(j,k) \neq (1,i)$. Assume furthermore that each of these functions f_i is analytic in a neighborhood of a system of particular values

$$(x_i)_0,\ (w_k)_0, \left(\frac{\partial w_j}{\partial x_\ell}\right)_0$$

that are assigned to f_i. Then each of the m equations given at the outset of this problem has an integral which is analytic in a neighborhood of the point $((x_1)_0, \cdots, (x_n)_0)$, and such that the ith equation reduces for $x_1 = (x_1)_0$ to a given function w_i which is an analytic function of the $n-1$ variables $(x_2, \cdots, x_n)$, and such that the values of w_i and of $\left(\frac{\partial \phi_i}{\partial x_j}\right)_0$ are precisely $(w_i)_0$ and $\left(\frac{\partial w_i}{\partial x_j}\right)_0$, $i = 1, \cdots, m$; $j = 2, \cdots, n$.

7.3.2 Assume that $f \in \mathbf{Lip}_\alpha(K_n)$, with K_n defined as in (7.3.3). Now make the transformation (7.3.4), mapping V_n onto K_n, with the functions u^i and v^i belonging to $\mathbf{L}_{\gamma^i}(\mathcal{D}^i)$, as defined in Subsection 7.3.2, with γ_i a positive number. Prove that the transformed function belongs to $\mathbf{Lip}_{\alpha'}(V_n)$, with $\alpha' = \min_i\{\alpha\gamma^i\}$.

7.3.3 Given the Poisson problem

$$E_{xx} + E_{yy} = F(x+iy), \quad x+iy \in \mathcal{D},$$

let

$$w = u + iv = \phi(z) \quad \Leftrightarrow z = x + iy = \psi(w) \quad (\psi = \phi^{-1})$$

denote a conformal map from $\mathcal{D}$ onto $\mathcal{D}'$.

(a) Use the conformal map, ψ, to transform the above Poisson problem into the problem

$$|\psi'|^{-2}[E_{uu} + E_{vv}] = F \circ \psi, \quad u + iv \in \mathcal{D}'.$$

(b) Let $(x,y) = (0,0)$ be a vertex of $\mathcal{D}$ at which two analytic boundary arcs of $\mathcal{D}$ meet with an interior angle $\mu\pi$, $0 < \mu < 2$, and $1/\mu$ not an integer. Let F be defined as above, such that $F \in \mathbf{Hol}(\mathcal{D})$, and let $F(0) \neq 0$. Prove that if n is any integer, $n > [1/\mu]$, then there are non-negative integers j and k, such that $j + k = n$, and such that $\frac{\partial^n U}{\partial x^i \partial y^j}$ is unbounded at $(x,y) = (0,0)$.

7.3.4 Let u denote the solution to the Dirichlet problem

$$\begin{array}{rcl} u_{xx} + u_{yy} & = & 0 \quad \text{in} \quad B, \\ u & = & g \quad \text{on} \quad \partial\mathcal{B}, \end{array}$$

with B an bounded region in $\mathbb{R}^2$, and with $g \in \mathbf{C}^\infty(\mathbb{R}^2)$. Let ∂B, the boundary of B, be of class $\mathbf{C}^2$, except for a corner point, located at $z = x + iy = 0$, with the value of angle of the corner, measured from the interior of B being $\mu\pi$. Prove that as $z = 0$,

$$\nabla(u-g)(z) = \begin{cases} \mathcal{O}(|z|) & \text{if} \quad \mu < \pi, \\ \mathcal{O}(|z|\log(1/|z|) & \text{if} \quad \mu = \pi, \\ \mathcal{O}(|z|^{1/\mu - 1}) & \text{if} \quad \mu > \pi. \end{cases}$$

7.4 Elliptic Problems

In this section we shall give several types of examples of the solution of elliptic partial differential equations by Sinc methods.

Example 7.4.1 Burke [Bur1], Bowers and Lund [BoL1]. We consider the Poisson problem,

$$\begin{array}{rcl} \nabla^2 u = u_{xx} + u_{yy} & = & -f, \quad (x,y) \in S, \\ u & = & 0, \quad (x,y) \in \partial S, \end{array} \qquad (7.4.1)$$

with S the unit square, $S = (0,1) \times (0,1)$. Under the assumption that f is smooth on S, we start with the approximation

$$u(x,y) \approx u_N(x,y) \equiv \sum_{k=-N}^{N} \sum_{\ell=-N}^{N} u_{k,\ell} S(k,h) \circ \phi(x) S(\ell,h) \circ \phi(y), \qquad (7.4.2)$$

and we then have a choice of either using collocation, as described in Subsection 7.2.1, or using the integration by parts procedure of Subsection 7.2.2. In the latter case we would define the inner product

$$(f,g) = \int\int_S \frac{f(x,y)\overline{g(x,y)}}{\phi'(x)\phi'(y)} dx dy \qquad (7.4.3)$$

and eliminate the derivatives of u via integration by parts as in Section 7.2.2, to arrive at the equation

$$BUD(1/[\phi']^2) + D(1/[\phi']^2)UB^T = -h^2 D(1/[\phi']^2) F D(1/[\phi']^2) \qquad (7.4.4)$$

with U and A square matrices of order $2N+1$, B defined as in (7.2.17), $U = [u_{ij}]$, $F = [f(z_i, z_j)]$, and $D(v)$ a diagonal matrix defined as in the earlier sections of this chapter. Multiplying through by the inverses of $D(1/\phi')$ both on the left and on the right, setting

$$\begin{gathered} C = D(\phi')B = X\Lambda X^{-1}, \quad V = X^{-1}U(X^T)^{-1}, \\ E = X^{-1}F(X^T)^{-1}, \end{gathered} \tag{7.4.5}$$

with $\Lambda = \text{diag}(\lambda_{-N}, \cdots, \lambda_N)$ a diagonal matrix, we get the equation

$$\Lambda V + V\Lambda = -h^2 E. \tag{7.4.6}$$

Solving this equation for the (i,j)th element v_{ij} of V, we get

$$v_{ij} = -h^2 e_{ij}/(\lambda_i + \lambda_j), \tag{7.4.7}$$

with $E = [e_{ij}]$. It may be readily shown, proceeding along the lines of the proof of Lemma 7.2.5, that all of the λ_i have strictly negative real parts, and so Equation (7.4.6) has a unique solution V. Once V has been obtained, we can get U from the identity $U = XVX^T$. One can show using potential theory integral equation methods that if for each fixed $x \in [0,1]$ $f(x,\cdot) \in \mathbf{L}_\alpha(\mathcal{D})$, and if in addition, for each fixed $y \in [0,1]$, $f(\cdot,y) \in \mathbf{L}_\alpha(\mathcal{D})$, then the solution u of Problem (7.4.1) belongs to $\mathbf{L}_1(\mathcal{D})$, with $\mathcal{D}$ defined as in (1.7.11), with $(a,b) = (0,1)$. Let us suppose, furthermore, that we may take $d = \pi/2$ in the definition of $\mathcal{D}$, which is the case, for example if $f \equiv 1$. Then, with $N = 16$, $h = \pi/(2N)^{1/2}$ we get an approximation of the form (7.4.2) which is accurate to five decimals on S.

It is not always possible to use the compact matrix notation of Equation (7.4.4) when solving PDE–BVP. To this end, we state the following Lemma:

Lemma 7.4.2 *Let A, B, $U = [u_{ij}]$, and $C = [c_{ij}]$ be $m \times m$, $n \times n$, $m \times n$, and $m \times n$ matrices, respectively. Define* $\mathbf{u}$, $\mathbf{c}$ *and the Kronecker product* $A \otimes B$ *by*

$$\begin{aligned} \mathbf{u} &= (u_{11}, u_{21}, \cdots, u_{m1}, u_{12}, \cdots, u_{m2}, u_{13}, \cdots, u_{mn})^T, \\ \mathbf{c} &= (c_{11}, c_{21}, \cdots, c_{m1}, c_{12}, \cdots, c_{m2}, c_{13}, \cdots, c_{mn})^T, \\ A \otimes B &= \begin{bmatrix} a_{11}B & a_{12}B & \cdots & a_{1m}B \\ a_{21}B & a_{22}B & \cdots & a_{2m}B \\ \vdots & \vdots & \vdots & \vdots \\ a_{m1}B & a_{m2}B & \cdots & a_{mm}B \end{bmatrix} \end{aligned} \tag{7.4.8}$$

Then the system of equations

$$BUA^T = C \tag{7.4.9}$$

can also be written in the equivalent matrix-vector form:

$$(A \otimes B)\mathbf{u} = \mathbf{c}. \tag{7.4.10}$$

Moreover, if we denote the eigenvalues of A and B respectively by μ_i and λ_j, then the eigenvalues of $A \otimes B$ are $\mu_i \lambda_j$.

Proof. The proof is straightforward, and we omit it. ∎

We may observe, in particular, that a system of the type considered in Example 7.4.1, such as $AU + UB^T = C$, with A an $m \times m$ matrix, U an $m \times n$ matrix, B an $n \times n$ matrix, and $C = [c_{ij}]$ an $m \times n$ matrix, can also be written in the form

$$(I_n \otimes A + B \otimes I_m)\mathbf{u} = \mathbf{c}, \tag{7.4.11}$$

with I_m and I_n unit matrices of order m and n respectively, and $\mathbf{u}$ and $\mathbf{c}$ are concatenations of the columns of U and C respectively. Indeed, it is *always* possible to derive a large matrix-vector system of equations of the form of (7.4.11), whereas it is not always possible to carry out a solution as in (7.4.4) through (7.4.7). We may also remark that the matrix multiplying u in (7.4.11) is rather sparse, with blocks of the matrix A being on the diagonal, and where each off-diagonal block consists of the unit matrix multiplied by a single number, an entry of B.

Example 7.4.3 The present example serves to illustrate how Sinc methods relate to other methods for solving elliptic boundary value problems. We consider a Poisson problem in polar coordinates, namely,

$$\begin{aligned} \nabla^2 U &= U_{rr} + \frac{1}{r} U_r + \frac{1}{r^2} U_{\theta\theta} = -G, \ (r,\theta) \in K_2, \\ U &= 0 \text{ on } \partial K_2, \end{aligned} \tag{7.4.12}$$

with K_2 the region

$$K_2 = \{(r,\theta) : 0 < r < f(\theta), \ 0 < \theta < \tau\}, \tag{7.4.13}$$

and with G a function which is holomorphic in a domain K_2^*, such that $K_2 \subset K_2^*$. It may be shown, in this case, that the solution U of (7.4.12) is in fact an analytic function on the domain K_2.
We assume, in addition, that $f \in \mathbf{Hol}(\mathcal{D}(\tau))$, with $\mathcal{D}(\tau)$ a simply-connected domain containing the interval $(0,\tau)$, and that $f(\theta) \neq 0$ if $0 \leqq \theta \leqq \tau$.
It is easily established, proceeding as in the proof of Theorem 6.5.6, that for each fixed $\theta \in [0,\tau]$, the solution U of (7.4.12) is an analytic function of r in the domain $\mathcal{D} = \{r \in \mathbb{C} : |\arg[r/(f(\theta) - r)]| < d\}$, and moreover, d may be taken to be arbitrarily large. Let us assume for the moment, that $\overline{K_2} \subset K_2^*$. The singular nature of the solution at $r = 0$ and at $r = f(\theta)$ is then revealed in the solution of Problem 7.3.4. At $r = 0$, the boundary arcs of K_2 intersect at an interior angle of magnitude τ; if π/τ is not an integer, then all derivatives of U of order $[\pi/\tau]$ or less are uniformly bounded in K_2, while if $r > 0$, the derivatives of U with respect to r of order greater than

$[\pi/\tau]$ are unbounded at $r = 0$. The solution U will of course be an analytic function of $\theta \in \mathcal{D}(\tau)$. On the other hand, it may be less easy to deduce the singular nature of the solution U if some of the corner points of K_2 are also singularities of G, although for purposes of efficient Sinc approximation it suffices to deduce that U is dominated by a fractional power of distance, as one approaches a singularity in the solution.

Let us assume, henceforth, that $\pi < \tau < 2\pi$. If we also have $\overline{K_2} \subset K_2^*$, then $U(\cdot, \theta) \in \mathbf{Lip}_\alpha(\mathcal{D} \cap B_0)$, with B_0 a small neighborhood of $r = 0$, and with $\alpha = \pi/\tau < 1$. Also, $U(\cdot, \theta) \in \mathbf{Lip}_1(\mathcal{D} \cap B_\tau)$, with B_τ a small neighborhood of $r = f(\theta)$.

The transformation

$$\begin{aligned} \theta &= s, \\ r &= t\,f(\theta) = t\,f(s) \end{aligned} \tag{7.4.14}$$

transforms K_2 into $V_2 = \{(t, s) : 0 < t < 1, 0 < s < \tau\}$. We easily determine, for this transformation, that

$$\begin{aligned} \frac{\partial}{\partial \theta} &= \frac{\partial}{\partial s} - \frac{t f'(s)}{f(s)} \frac{\partial}{\partial t}, \\ \frac{\partial}{\partial r} &= \frac{1}{f(s)} \frac{\partial}{\partial t} \end{aligned} \tag{7.4.15}$$

and using these identities, we transform (7.4.12) into the problem of determining $V(t, s) = U(tf(s), s)$ such that

$$\begin{aligned} &\left[1 + \left(\tfrac{f'(s)}{f(s)}\right)^2\right] V_{tt} - \tfrac{2f'(s)}{tf(s)} V_{ts} + \tfrac{1}{t^2} V_{ss} \\ &+ \tfrac{1}{t}\left[1 + \left(\tfrac{f'(s)}{f(s)}\right)^2 - \tfrac{d}{ds}\left(\tfrac{f'(s)}{f(s)}\right)\right] V_t = -f^2(s) G, \ (t, s) \in V_2 \\ &V(t, s) = 0, \quad (t, s) \in \partial V_2. \end{aligned} \tag{7.4.16}$$

Evidently, it now follows by our assumptions on f, that for each fixed $t \in [0, 1]$, $V(t, s)$ is an analytic function of $s \in \mathcal{D}(\tau)$, with $\mathcal{D}(\tau)$ is some simply-connected domain containing the interval $[0, \tau]$. It also follows from our discussions above, that for each fixed $s \in [0, \tau]$, $V(s, \cdot) \in \mathbf{L}_\alpha(\mathcal{D})$, with $\alpha = \pi/\tau < 1$, $\mathcal{D} = \{t \in \mathbb{C} : |\arg[t/(1-t)]| < d\}$, and with $d > 0$, but otherwise arbitrary.

Let us next comment on the accuracies, advantages and pitfalls and amount of work (i.e., the complexity) of various methods for solving the problem (7.4.12), or the equivalent problem (7.4.16), whichever is more convenient for a particular method.

Method (a): The use of equi-spaced finite differences is a very popular method to solve elliptic partial differential equations. In this case, it makes no difference to our final condition and complexity estimates whether we apply this method to (7.4.12) or to (7.4.16)—hence, let us assume that we are solving (7.4.16). We propose to use standard first-order and second-order central difference approximations to approximate respectively the first and second derivatives of V, using N_t points in the t direction, and N_s points in the s direction. Thus the total number of points is $M = N_t N_s$. Due to the sparsity of the matrix, it may be shown that it then takes $\mathcal{O}(M)$ operations to solve the resulting system of linear algebraic equations. Using estimates of the eigenvalues of one-dimensional matrices given, e.g., in the proof of part (c) of Lemma 7.2.4, and using Lemma 7.4.2, we may conclude that the condition number of the coefficient matrix of the system of equations is $\mathcal{O}(M^2)$. Since V behaves like t^α near $t = 0$, we make an $\mathcal{O}(N_t^{-\alpha})$ error when approximating $N_t^{-2} V_t t$ in the neighborhood of $t = 0$ by its central difference approximation, with step-size N_t^{-1}. The corresponding errors in the approximation of the derivatives of V with respect to s are $\mathcal{O}(N_s^{-2})$. We may clearly expect errors of these orders of magnitude in the final solution. Setting each of these errors equal to ε, and thus eliminating N_s, we find that the *complexity*, i.e., the total amount of work required to achieve a solution accurate to within ε, is $\mathcal{O}(M) = \mathcal{O}(N_t^{(2+\alpha)/2}) = \mathcal{O}(\varepsilon^{-\frac{2+\alpha}{2\alpha}})$.

Method (b): A method of current research interest for circumventing the effect of the singularity at $r = 0$ of the above problem is to use finite differences, as above, with an equi-spaced mesh in the s variable, but with a graded mesh in the t variable. As we noted above, we can then achieve an $\varepsilon = \mathcal{O}(N_s^{-2})$ error by using N_s points in the s variable; furthermore, it may be shown (see Problem 2.1.3) that by choosing a suitable mesh, one can then also achieve an $\varepsilon = \mathcal{O}(N_t^{-2})$ error by choosing N_t points in the t variable. Upon equating each of these errors, and thus eliminating N_s, we find that the amount of work required to achieve an error of order ε is at least $\mathcal{O}(M) = \mathcal{O}(N_t^2) = \mathcal{O}(\varepsilon^{-1})$. The use of a variable mesh generally tends to worsen the condition number of a system of equations, although, if the matrix in the t variable is set up properly, (as, e.g., we have done using Sinc methods in Section 7.2.2), then this matrix can be written as a product of a badly conditioned diagonal matrix times a relatively well conditioned matrix (or as a sum of a relatively small number of such matrices), and in this latter case the bad computational effects of ill conditioning due to a variable mesh are circumvented. The well conditioned matrix in the t variable can then be considered to have a condition number which is of the order of N_t^2, and hence we may assume that the effective condition number of the resulting coefficient matrix of the final system of equations to be at least $\mathcal{O}(M^2)$.

Method (c): Finite element methods are also very popular for solving elliptic partial differential equations. Piecewise linear elements are the most frequently used among such methods. In the latter case, when the mesh points are located at the same points as those of a standard equi-spaced finite difference mesh stemming from a one-dimensional three-point finite difference approximation to a second derivative, the coefficient matrix of the system of linear equations to be solved is the same as that of the above equi-spaced finite difference matrix. Hence, in this case, the condition of the matrix and the complexity of finite element methods is the same as the condition and complexity of finite difference methods, regardless of whether one uses a fixed or a variable mesh. However, (as for Sinc methods), once the system of equations has been solved, the finite element approximation provides an explicit method to interpolate the solution throughout the domain. Special *boundary elements* are also used occasionally, to circumvent the bad effects of the singularity of the solution at the corner of the domain, and in the latter case it is possible to circumvent the use of a more complicated variable mesh. This boundary element procedure can be quite effective provided that the boundary element can be well selected. For example, if the function G in (7.4.12) is smooth in a neighborhood of $r = 0$, then the singularity of the solution is, in effect determined by the interior angle at $r = 0$ (see Problem 7.3.3). On the other hand, if G also has a singularity at $r = 0$, then it may not be so easy to select a good boundary element.

Method (d): Another popular procedure is to use a product Fourier polynomial basis, and to solve the resulting linear algebra problem using an efficient FFT (Fast Fourier Transform) algorithm, which is based on the formulas in Sections 2.2 and 3.3. Thus, using a Fourier polynomial in each variable, of degree N_t in the t variable, and of degree N_s in the s variable, it follows by Problem 2.2.11 that the error of Fourier polynomial approximation with respect to the t variable will be $\mathcal{O}(N_t^{-2\alpha})$. The error of Fourier polynomial approximation in the s variable will be $\mathcal{O}(N_s^{-2})$. By equating these errors, we get $N_s = N_t^{\alpha} = \mathcal{O}(\varepsilon^{-1/2})$. The total amount of work required to solve the resulting system of equations of order $M = N_s N_t = N_t^{1+\alpha}$ and thus achieve a solution accurate to within ε is $\mathcal{O}(M \log(M)) = \mathcal{O}(N_t^{1+\alpha} \log N_t) = \mathcal{O}(\varepsilon^{-\frac{1+\alpha}{2\alpha}} \log[1/\varepsilon])$. We know, from the results of Section 2.1, that if $f(\theta)$ is periodic in θ, of period τ, then a Fourier polynomial in the s variable can be constructed to have $\mathcal{O}(e^{-\gamma N_s})$ accuracy. In this latter case we find that $N_s = (2\alpha/\gamma)\log(N_t) = \mathcal{O}(|\log[\varepsilon]|)$, and hence in this case, the complexity is $M\log(M) = \mathcal{O}(N_t \log[N_t]) = \mathcal{O}(\varepsilon^{-\frac{1}{2\alpha}} |\log[\varepsilon]|^2)$. The condition number of the resulting coefficient matrix is not unduly large (see Problem 2.2.10).

Method (e): The use of Chebyshev polynomials provides yet another popular and frequently highly accurate method to approximate solutions of partial differential equations. If we use Chebyshev polynomials in each variable, of degree N_t in the t variable and of degree N_s in the s variable, we can again achieve $\mathcal{O}(N_t^{-2\alpha})$ accuracy in the t variable, and, in the case when $V(t,\cdot) \in \mathbf{Hol}(\mathcal{D}(\tau))$, with $\mathcal{D}(\tau) \supset [0,\tau]$, the error of approximation in the s variable will be $\mathcal{O}(e^{-\gamma N_s})$, with γ a positive number. Upon again equating these errors, and setting $M = N_t N_s$, we find that if the matrix is Toeplitz, then the total amount of work required to solve the system of equations is $\mathcal{O}(M \log[M])$, and thus, the total amount of work required to achieve an answer that is correct to within a ε error then is $\mathcal{O}(\varepsilon^{\frac{1}{2\alpha}} |\log(\varepsilon)|^2)$. By Problem (2.1.7), the condition number of the resulting coefficient matrix will be at least $\mathcal{O}(M)$. We may in fact use $\mathcal{O}(M^3)$ operations to solve the system of linear equations, in which case the complexity will be $\mathcal{O}(\varepsilon^{-\frac{3}{2\alpha}} |\log(\varepsilon)|^3)$.

In order to simplify our comparisons of methods for the remaining examples to follow, we shall assume that $V(\cdot, s)$ has the same analytic features in the t variable as t^α, for each fixed $s \in [0,1]$, and that $V(t,\cdot)$ is an analytic function of s in the disc $\mathcal{D}(1/2; R)$, for some $R > 1/2$.

Method (f): Let us now use standard Sinc approximation in both the t variable as well as in the s variable. For the case of Sinc approximation in the t variable, we can take $d_t = \pi$, $h_t = [\pi d_t/(\alpha N_t)]^{1/2}$, to get the error $\varepsilon_t = \mathcal{O}(N_t^{1/2} e^{-\pi(\alpha N_t)^{1/2}})$. In the case of approximation in the s variable, we can take $d_s = \pi/2$, and $\alpha_s = 1$. Corresponding to a positive integer N_s, we thus take $h_s = [\pi d_s/(\alpha_s N_s)^{1/2} = \pi/(2N_s)^{1/2}$, and we get an error of approximation $\varepsilon_s = \mathcal{O}(N_s^{1/2} e^{-\pi(N_s/2)^{1/2}})$. By equating these two errors, we get $N_s \sim 2\alpha N_t$. The matrix size is $M = (2N_t + 1)(2N_s + 1)$, and the amount of work required to solve the system of linear equations is $\mathcal{O}(M \log[M])$. Since both N_t and N_s are of the order of $|\log(\varepsilon)|^2$, we thus find that the complexity is $\mathcal{O}(|\log(\varepsilon)|^4 \log[|\log(\varepsilon)|]$. The condition number of the resulting coefficient matrix will be $\mathcal{O}(M^2)$ (see Lemma 7.2.5).

Method (g): Suppose that we next approximate $V(t,s)$ using variable spacing and variable degree splines in t, as described in Problem 2.1.4, and using Chebyshev polynomials in the s variable. Then, the approximation error in the t variable is $\varepsilon_t = \mathcal{O}(\exp\{-a(\alpha N_t)^{1/2}\})$, with $a = -2\log(2^{1/2} - 1) = 1.7627....$ Using Chebyshev polynomials of degree N_s in the s variable, we get an error $\varepsilon_s = \mathcal{O}(e^{-\gamma N_s})$. By equating these two errors, we get $|\log(\varepsilon)| = a(\alpha N_t)^{1/2} = \gamma N_s$. The size of the matrix is $M = N_t N_s = [a/(\alpha\gamma)](\alpha N_t)^{3/2}$. The amount of work required to solve the system of linear equations, i.e., the complexity, is

$$cM^3 = c\frac{1}{a^6}\frac{1}{\alpha^3\gamma^3}|\log(\varepsilon)|^9.$$

If non-orthogonal splines are used, then the condition number of the resulting coefficient matrix is so bad that only limited accuracy is possible via this procedure (see Problem 3.2.12). If the same (or equivalently accurate) approximating splines are represented via, for example, Legendre polynomials, then the condition number with respect to the t variable is a more reasonable $\mathcal{O}(N_t^2)$. Since the condition number with respect to the s variable is $\mathcal{O}(N_s)$, by Problem 2.1.7, the condition number of the final approximating matrix will be $\mathcal{O}(N_t^2 N_s) = \mathcal{O}(M^{5/3})$.

Method (h): We again use Chebyshev polynomial approximation in the s variable, as described above, but we use $2N_t + 1$-point Sinc approximation, with basis $S(k,h) \circ \phi$ in the t variable, taking $\phi(t) = \log[t/(1-t)]$. Taking $d = \pi$, so that $h_t = \pi/(\alpha N_t)^{1/2}$, we achieve an error $\varepsilon_t \sim \mathcal{O}(e^{-\pi(\alpha N)^{1/2}})$. The approximation error in the s variable is the same as that given in *Method (g)* above. We again equate these two erros, to get $|\log(\varepsilon)| = \pi(\alpha N_t)^{1/2} = \gamma N_s$. The size of the matrix is $M = (2N_t+1)N_s \sim [2\pi/(\alpha\gamma)](\alpha N_t)^{3/2}$. The amount of work required to solve the system of linear equations, i.e., the complexity, is

$$cM^3 = c\frac{8}{\pi^6}\frac{1}{\alpha^3\gamma^3}|\log(\varepsilon)|^9.$$

However, because of the different values of the constants in the methods (g) and (h), to get the same accuracy, the present method requires only about one quarter as much work as Method (g). In this case the condition number of the resulting coefficient matrix is again $\mathcal{O}(M^{5/3})$.

Method (i): Suppose that we again use Chebyshev polynomial approximation in the s variable, as described above, but we now use $2N_t + 1$-point Sinc approximation in the t variable, via the procedure described in Problem 4.2.6 to approximate x^α. Thus, the error in the t-approximation is now $\varepsilon_t = \mathcal{O}(\exp\{-N_t/\log(N_t)\})$, and the error with respect to the s variable is as that in Method (g) above. We emphasize here, that in order to use the transformation of Problem 4.2.6, and then to get this bound, we require the solution $V(\cdot, s)$ to be an analytic function of t on a Riemann surface, as described in Example 4.2.13, and while the function x^α does have this analyticity property, we expect that the solution to Equation (7.4.16) will only rarely have this property. By equating approximation errors in the s and t approximations, we get $|\log(\varepsilon)| = N_t^2/\log(N_t) = \gamma N_s$, and hence, setting $M = (2N_t + 1)N_s$, and assuming an $\mathcal{O}(M^3)$ amount of work to solve the linear system, we find that the complexity of this procedure is $|\log(\varepsilon)|^6[\log|\log(\varepsilon)|]^3$. The condition number of the resulting coefficient matrix is not known at this time, although, unfortunately, it may in fact be unmanageably large, as we have experienced when attempting to solve one-dimensional boundary value problems via the use of a double exponential transformation.

We remark, that the coefficient matrices of the procedures in Methods (g), (h), and (i) are rather sparse, and thus it may in fact be possible to obtain the solution of the resulting linear system of equations with fewer than $\mathcal{O}(M^3)$ operations. The most efficient methods to solve the linear system of equations for these methods have not yet been determined, due to the novelty of the methods. We have selected the same $\mathcal{O}(M^3)$ as the total amount of work required in each of these methods for sake of comparison of the complexities of the methods; also, the solution of the linear systems of equations of each of the methods can clearly be achieved with this amount of work.

Examples 7.4.1 or 7.4.3 illustrate two of several ways of approximating the solution of Laplace's equation over a region in the plane. Another, different though quite well known procedure which we shall not present here is to use the Cauchy singular equation methods of Sections 1.3 and 6.7 to reduce the problem to the solution of an integral equation problem over the boundary of the region. The next example illustrates an unusual, though quite efficient procedure to obtain an approximate solution of Laplace's equation over a region in the plane.

Example 7.4.4 Let Ω denote a simply connected and bounded domain in $\mathbb{C}$, and let

$$\begin{aligned} \nabla^2 u &= 0, \qquad z \in \Omega \\ u &= g, \qquad z \in \partial\Omega \end{aligned} \tag{7.4.17}$$

with

$$\partial\Omega = \bigcup_{i=1}^{L} \Gamma_i. \tag{7.4.18}$$

We also assume that for each $i = 1, 2, \cdots, L$, there exists a function $\phi_i : \mathcal{D}_i \to \mathcal{D}_d$, with ϕ_i and $\mathcal{D}_i$ satisfying Definition 4.1, if in Definition 4.1 we replace ϕ by ϕ_i, and $\mathcal{D}$ by $\mathcal{D}_i$, and with $\mathcal{D}_d$ defined as in (1.7.9). We assume, in addition, that $\psi_i = \phi_i^{-1} : \mathbb{R} \to \Gamma_i$, and furthermore, that the endpoints a_i and b_i of Γ_i satisfy the equations $a_{i+1} = b_i$, $i = 1, 2, \cdots, L$, and $a_{L+1} = a_1$. For $h > 0$, and $k \in \mathbb{Z}$, set $z_{i,k} = \psi_i(kh)$. We shall also assume that if $\mathcal{D}_i^+$ denotes that part of $\mathcal{D}_i$ which lies to the left of Γ_i as Γ_i is traversed from a_i to b_i, then $\Omega \subset \mathcal{D}_i$ for each i. Let $T(k,h) \circ \phi$ be defined as in Problem 4.3.12. Then (see Problems 1.8.17 and 4.3.12) given an arbitrary set of numbers $\{v_{i,k}\}_{i=1}^{L}\,{}_{k=-N}^{N}$, the sum

$$v_N(z) \equiv \sum_{i=1}^{N} \sum_{k=-N}^{N} v_{i,k} T(k,h) \circ \phi_i(z) \tag{7.4.19}$$

is harmonic in Ω, and satisfies the relations

$$\lim_{z \to z_{j,\ell}} v_N(z) = v_{j,\ell} + \sum_{i=1,\cdots,L; i \neq j} \sum_{k=-N}^{N} v_{i,k} T(k,h) \circ \phi_i(z_{j,\ell}). \tag{7.4.20}$$

These properties of $v_N(z)$ suggest to attempt to find an approximate solution to the problem (7.4.17) by means of the approximation

$$u(z) \approx u_N(z) \equiv p_{L-1}(z) + \sum_{i=1}^{N} \sum_{k=-N}^{N} v_{i,k} T(k,h) \circ \phi_i(z), \tag{7.4.21}$$

with p_{L-1} a polynomial of degree $L-1$.

Equating $u_N(z) = g(z)$ for $z = z_{j,\ell}$, using (7.4.20), and then eliminating p_{L-1} using, the values of g at the corner points, a_i, and via, e.g., the Lagrange polynomial representation of p_{L-1}, we get a linear system of equations of the form

$$\begin{bmatrix} I & B_{12} & \cdots & B_{1L} \\ B_{21} & I & \cdots & B_{2L} \\ \vdots & \vdots & \ddots & \vdots \\ B_{L1} & B_{L2} & \cdots & I \end{bmatrix} \begin{bmatrix} \mathbf{v}_1 \\ \mathbf{v}_2 \\ \vdots \\ \mathbf{v}_L \end{bmatrix} = \begin{bmatrix} \mathbf{g}_1^* \\ \mathbf{g}_2^* \\ \vdots \\ \mathbf{g}_L^* \end{bmatrix}, \tag{7.4.22}$$

with $\mathbf{v}_i = (v_{i,-N}, \cdots, v_{i,N})^T$, with I is the unit matrix of order $2N+1$, and with the entries of the matrix $B_{i,j}$ linear combinations of the values of the bases $T(k,h) \circ \phi_j$ evaluated at the points a_j and also at the points $z_{i,k}$.

Our experience with this procedure has shown that the system of equations in (7.4.22) is both non-singular and well-conditioned; indeed, we have found that the matrix in this system is frequently diagonally dominant.

Let us now assume that the arcs Γ_{i-1} and Γ_i intersect at an interior angle $\mu_i \pi$ at a_i. Assuming that $g \in \mathbf{L}_{\alpha'}(\mathcal{D}_i)$ for $i = 1, 2, \cdots, L$, we set $\mu = \max_i\{\mu_i\}$, and

$$\alpha = \min\{\alpha' \ if \ \mu \leqq 1, \ \alpha'/\mu \ if \ \mu > 1\},$$

and we then select $h = [\pi d/(\alpha N)]^{1/2}$. It then follows (see Problem 7.3.3 and Theorem 7.3.3) that u_N approximates g on $\partial\Omega$ to within an error bounded by $C N^{1/2} e^{-(\pi d \alpha N)^{1/2}}$. Furthermore, it then follows by the maximum principle, that u_N then approximates the solution u of Problem (7.4.17) to the same accuracy throughout Ω.

•

Problems for Section 7.4

7.4.1 Prove that the condition number of the matrix for solving the problem of Example 7.4.3 by *Method (a)* is at least $\mathcal{O}(M^2)$. [Hint: See the proof of Lemma 7.2.5, and use the result based on the fact that if matrices A and B have eigenvalues λ_m and μ_n, then the eigenvalues of the Kronecker product matrix $A \otimes B$ are $\lambda_m \mu_n$.]

7.4.2 Describe explicitly, the system of linear algebraic equations for solving the problem of Example 7.4.1, but when the differential equation Problem (7.4.1) is replaced by

$$\begin{aligned} u_{xx} + g(x,y)u_{yy} &= -f(x,y), \quad (x,y) \in S \\ u &= 0, \quad (x,y) \in \partial S. \end{aligned}$$

7.4.3 Derive the Sinc–Galerkin equations for (7.4.16) using the collocation approximation procedure of Section 7.2.1.

7.4.4 Compare the complexities of methods (g) and (h) in Example 7.4.3 in the case when both methods require $M \log(M)$ amount of work.

7.5 Hyperbolic Problems

In this section we illustrate the application of some Sinc procedures to obtain approximate solutions of some simple hyperbolic partial differential equations.

Example 7.5.1 It is easily verified that the solution to the wave equation problem

$$\begin{aligned} \frac{1}{c^2} u_{tt} &= u_{xx}, \quad x \in \mathbb{R},\ t > 0, \\ u(x,0) &= f(x), \\ u_t(x,0) &= g(x), \end{aligned} \tag{7.5.1}$$

is

$$u(x,t) = \frac{f(x+ct) + f(x-ct)}{2} + \frac{1}{2c}\int_{x-ct}^{x+ct} g(\xi)\, d\xi. \tag{7.5.2}$$

This is the form of the solution given in textbooks on partial differential equations. In applications we are still faced with the problem of approximating the integral in (7.5.2). To this end, we assume that both functions f and g belong to $\mathbf{L}_\alpha(\mathcal{D}_d)$, with $0 < \alpha \leqq 1$, and $0 < d < \pi$. Then $G \in \mathbf{M}_\alpha(\mathcal{D}_d)$, with

$$G(x) = \int_{-\infty}^{x} g(\xi)\, d\xi, \tag{7.5.3}$$

and we therefore have

$$u(x,t) = \frac{f(x+ct)+f(x-ct)}{2} + \frac{G(x+ct)-G(x-ct)}{2c}. \tag{7.5.4}$$

Let us now take $h = [\pi d/(\alpha N)]^{1/2}$, let us set $f_k = f(kh)$, $\mathbf{f} = (f_{-N}, \cdots, f_N)^T$, $g_k = g(kh)$, $\mathbf{g} = (g_{-N}, \cdots, g_N)^T$, and then by proceeding according to Section 7.1, let us define the vector $\mathbf{G}$ by $\mathbf{G} = hI^{(-1)}\mathbf{g}$. We can then set

$$\begin{aligned} f_h(x) &= \sum_{k=-N}^{N} f_k S(k,h)(x), \\ G_h(x) &= \frac{G_N e^x}{1+e^x} + \sum_{k=-N}^{N} \left[G_k - \frac{G_N e^{kh}}{1+e^{kh}} \right] S(k,h)(x). \end{aligned} \tag{7.5.5}$$

It then follows that there exist constants, C_1 and C_2, independent of N, such that

$$\begin{aligned} \sup_{x \in \mathbf{R}} |f(x) - f_h(x)| &\leqq C_1 N^{1/2} e^{-(\pi d \alpha N)^{1/2}} \\ \sup_{x \in \mathbf{R}} |G(x) - G_h(x)| &\leqq C_2 N^{1/2} e^{-(\pi d \alpha N)^{1/2}}. \end{aligned} \tag{7.5.6}$$

Hence, it follows that there exists a constant, C, independent of N, such that

$$\begin{aligned} &\sup_{x \in \mathbf{R}, t>0} \left| u(x,t) - \frac{f_h(x+ct)+f_h(x-ct)}{2} - \frac{G_h(x+ct)-G_h(x-ct)}{2c} \right| \\ &\leqq C N^{1/2} e^{-(\pi d \alpha N)^{1/2}}. \end{aligned} \tag{7.5.7}$$

Example 7.5.2 A problem, such as

$$\begin{aligned} \frac{1}{c^2} u_{tt} &= u_{xx} + f(x,t), \quad 0 < x < 1, \ t > 0, \\ u(x,0) &= 0, \quad u_t(x,0) = 0, \\ u(0,t) &= 0, \quad u(1,t) = 0. \end{aligned} \tag{7.5.8}$$

is usually solved via *the method of lines*, i.e., we can use a method of Section 7.2 to replace the term u_{xx} in (7.5.8), and then we can use an initial value method of Section 7.2.1 to solve the resulting ODE–IVP. To this end, we take $\phi(x) = \log[x/(1-x)]$, and $\chi(t) = \log[\sinh(t)]$, this latter map being preferable, over $\chi(t) = \log(t)$ for solution of initial value problems on the interval $(0, \infty)$, due to its broader range of convergence. Then, if the function

G defined by $G(x,t) = f(x,t)/[\phi'(x)^2 \chi'(t)^2]$ satisfies the conditions that $G(\cdot,t) \in \mathbf{L}_\alpha(\mathcal{D}_x)$ for all $t \in [0,\infty]$, with $\mathcal{D}_x = \{x \in \mathbb{C} : |\arg[x/(1-x)]| < d\}$, and also, if $G(x,\cdot) \in \mathbf{L}_\alpha(\mathcal{D}_t)$ for all $x \in [0,1]$, then we may expect the differential equation problem (7.5.8) together with the *boundary conditions* $u(x,0) = u(x,\infty) = u(0,t) = u(1,t) = 0$ to have a solution u which satisfies the same conditions as G. Let us select $h = [\pi d/(\alpha N)]^{1/2}$, $x_m = \phi^{-1}(mh)$, and let us use the method of Subsection 7.2.2 to replace the derivative term u_{xx} in the differential equation (7.5.8). To this end, we take $A = I^{(2)} + hD(1/\phi')I^{(1)}$, we set $\mathbf{u}(t) = (u_{-N}(t), \cdots, u_N(t))^T$, with $u_j(t)$ an approximation of $u(x_j,t)$, and we set $\mathbf{f}(t) = (f(x_{-N},t), \cdots, f(x_N,t))^T$, and we thus replace the differential equation (7.5.8) by the system of ordinary differential equations:

$$h^2 D\left(\frac{1}{[\phi']^2}\right)\frac{d^2\mathbf{u}(t)}{dt^2} = A\mathbf{u}(t) + h^2 D\left(\frac{1}{[\phi']^2}\right)\mathbf{f}(t). \tag{7.5.9}$$

Indeed, setting

$$\begin{aligned} D\left(\tfrac{1}{[\phi']^2}\right)\mathbf{u} &= \mathbf{v} \\ \mathcal{A} &= h^{-2}AD([\phi']^2), \\ \frac{d\mathbf{v}}{dt} &= \mathbf{w} \end{aligned} \tag{7.5.10}$$

we get the system of first-order equations

$$\frac{d}{dt}\begin{pmatrix}\mathbf{v}\\ \mathbf{w}\end{pmatrix} = \begin{pmatrix}0 & I\\ \mathcal{A} & 0\end{pmatrix}\begin{pmatrix}\mathbf{v}\\ \mathbf{w}\end{pmatrix} + \begin{pmatrix}0\\ D\left(\frac{1}{[\phi']^2}\right)\mathbf{f}\end{pmatrix}. \tag{7.5.11}$$

The system (7.5.11) can now be solved by methods of Section 7.1.

We mention here that it has already been proved that the above problem (7.5.8) can also be solved by treating it as a boundary value problem, and then solving this problem by means of the symmetric Sinc approximation procedure of Section 7.2.7. We leave this procedure to Problem 7.5.2 at the end of this section.

•

Problems for Section 7.5

7.5.1 * Describe a Sinc procedure which uses *the method of characteristics* to solve a hyperbolic problem of the form

$$au_{xx} + 2bu_{xy} + cu_{yy} = e$$

where a, b, c, and e may be functions of x, y, u, u_x, and u_y, and where u is given on an analytic arc Γ which is not a characteristic. The characteristics are defined by the equations

$$\frac{dy}{dx} = \frac{-b \pm \sqrt{b^2 - ac}}{a}.$$

7.5.2 (a) Reduce the problem of Equation (7.5.8) to a system of linear algebraic equations via the use of the symmetric Sinc–Galerkin procedure of Section 7.2.7, using a Sinc approximation of the form

$$\begin{aligned} &[\phi'(x)\chi'(t)]^{1/2} u(x,t) \\ &\approx \sum_{m=-M_x}^{N_x} \sum_{n=-M_t}^{N_t} w_{mn} S(m,h) \circ \phi(x) S(n,k) \circ \chi(t). \end{aligned}$$

(b) Now solve the resulting problem for

$$f(x,t) = \frac{e^{-t}}{\sqrt{t}} \frac{1}{x} \left\{ t^2 - \left[\frac{3}{4} - 3t + t^2 \right] x^2 \log(x) \right\}$$

for which the true solution is $u(x,t) = t^{3/2} e^{-t} x \log(x)$. Take $h = k = \pi/(M_x)^{1/2}$, and specifically, take $M_x = N_x = 32$, $M_t = 16$, $N_t = 4$, and then solve the system of linear equations, to get an approximate solution for which the maximum error at the Sinc points is 0.767×10^{-4} (See McArthur, Bowers, and Lund: [M1, M1BoL1,M1BoL2]).

7.6 Parabolic Problems

In applications, parabolic problems can be simple problems, such as the heat equation—or more sophisticated conservation law problems, which are used to describe "shocks", or "explosions" in real life. Let us illustrate an example of each of these.

Example 7.6.1 A heat equation problem which is simple to describe is to determine the temperature $u = u(x,t)$ which satisfies the equations

$$\begin{aligned} u_t &= u_{xx}, \quad x \in \mathbb{R}, \quad t > 0, \\ u(x,0) &= f(x). \end{aligned} \tag{7.6.1}$$

The "solution" to this equation, which can be found in all elementary textbooks on partial differential equations and which is valid for a large class of functions f, is

$$u(x,t) = \frac{1}{2\sqrt{\pi t}} \int_{\mathbb{R}} \exp\left\{ -\frac{(x-\xi)^2}{4t} \right\} f(\xi)\, d\xi. \tag{7.6.2}$$

It is easily shown that if $t > 0$, then the solution u is an entire function of x, for all bounded functions f defined on $\mathbb{R}$. For smooth functions f, e.g., if

$f \in \mathbf{L}_\alpha(\mathcal{D})$, we can use the trapezoidal rule (i.e., one of the Sinc methods) to approximate the integral in (7.6.2). Taking $h = [2\pi d/(\alpha N)]^{1/2}$, we get the approximation

$$u(x,t) \approx u_h(x,t) \equiv \frac{1}{2\sqrt{\pi t}} \sum_{k=-N}^{N} \exp\left\{-\frac{(x-kh)^2}{4t}\right\} f(kh), \tag{7.6.3}$$

which is accurate even for small values of t (see Section 3.2, and Theorem 5.8.1).

Example 7.6.2 Let us next consider a similar problem, namely to find the temperature $u(x,t)$ of a slab of length 1 located on the interval $(0,1)$ on the x-axis, where it has an initial temperature distribution $f(x)$, and such that the ends of the rod are kept at temperature *zero* for all times. The differential equation for u now takes the form

$$\begin{aligned} u_t &= u_{xx}, & x \in (0,1),\ t > 0, \\ u(x,0) &= f(x), & x \in (0,1), \\ u(0,t) &= u(1,t) = 0, & t > 0. \end{aligned} \tag{7.6.4}$$

While it has already been demonstrated that this problem can be effectively solved by treating it as a boundary value problem, we prefer to illustrate the solution of it here by a combination of a Sinc procedure and the *method of lines*. To this end, we assume that the function f belongs to $\mathbf{L}_\alpha(\mathcal{D})$. Taking $h = [\pi d/(\alpha N)]^{1/2}$, Sinc collocation with respect to x then results in a system of Volterra integral equations

$$\mathbf{u}(t) = \int_0^t h^{-2} D([\phi']^2) A \mathbf{u}(\tau)\, d\tau + \mathbf{f}, \tag{7.6.5}$$

with $\phi(x) = \log[x/(1-x)]$, with $u(t) = (u_{-N}(t), \cdots, u_N(t))^T$, $\mathbf{f} = (f(z_{-N}), \cdots, f(z_N))^T$ and with the square matrix A of order N defined with respect to this mapping ϕ as in (7.2.11). We next collocate with respect to the t variable via the use of the indefinite integration formula (4.5.9), in which we take the mapping $\chi(t) = \log\{\sinh(t)\}$. Thus, defining a matrix E by $E = kI^{(-1)}D(1/\chi')$, with $t_j = \chi^{(-1)}(jk)$, $j = -M, \cdots, M$, we can determine a rectangular $(2N+1) \times (2M+1)$ matrix $U = [u_{i,j}]$ as the solution to the matrix problem

$$U = h^{-2} D([\phi']^2) A U E^T + F \tag{7.6.6}$$

in which the vector F has the same dimensions as the vector U, and every column of F consists of the same vector $\mathbf{f}$. The problem (7.6.6) may now be solved by simultaneous diagonalization of the matrices $h^{-2}D([\phi']^2)A$ and E^T. Presently, the exact relationships of the spacings h and k which guarantee the stability of the procedure is not known. It is also unclear, at this time, if a symmetric Sinc–Galerkin discretization with respect to x is preferable. For example, if $f(x) = \sin(x)$, we have found $h = 2k$ to be satisfactory, and to yield 4 decimals accuracy, with $N = 16$.

Example 7.6.3 A problem which frequently arises in applications is a scalar conservation law problem of the type

$$\begin{aligned} &u_t + \sum_{i=1}^{n} f_i(u, \mathbf{x}, t)_{x^i} + f_0(u, \mathbf{x}, t) = \varepsilon \nabla^2 u, \\ &\mathbf{x} = (x^1, \cdots, x^n) \in \mathbb{R}^n, \quad t \in \mathbb{R}_+ = (0, \infty) \\ &u(\mathbf{x}, 0) = u_0(\mathbf{x}). \end{aligned} \tag{7.6.7}$$

The functions f_i in (7.6.7) are typically holomorphic functions, and if this is also the case for u_0, then it may be shown that the solution to (7.6.7) is also holomorphic, provided that $\varepsilon > 0$. On the other hand, as $\varepsilon \to 0$, the solution to (7.6.7) can have *shock discontinuities*, and this feature makes the approximate solution to (7.6.7) both interesting and challenging.
It is advantageous to deduce the existence and properties of the solutions to (7.6.7), as well as for constructing approximate solutions, to transform the problem (7.6.7) into the integral equation problem

$$u = v + \mathcal{F} u. \tag{7.6.8}$$

The notation of (7.6.8) may be described as follows. We let $K(\mathbf{x}, t)$ denote the fundamental solution to the problem

$$u_t - \varepsilon \nabla^2 u = 0 \tag{7.6.9}$$

i.e.,

$$K(\mathbf{x}, t) = (4\pi\varepsilon t)^{-n/2} e^{-|\mathbf{x}|^2/(4\varepsilon t)}. \tag{7.6.10}$$

Then

$$\begin{aligned} v(\mathbf{x}, t) &= \int_{\mathbb{R}^n} K(\mathbf{x} - \xi, t)\, u_0(\xi)\, d\xi \\ (\mathcal{F}u)(x, t) &= -\int_0^t \int_{\mathbb{R}^n} K_i(\mathbf{x} - \xi, t - \tau) f_i(u(\xi, \tau), \xi, \tau) d\xi d\tau \\ K_0(\mathbf{x}, t) &= K(\mathbf{x}, t) \\ K_i(\mathbf{x}, t) &= \frac{\partial}{\partial x^i} K(\mathbf{x}, t), \quad i = 1, \cdots, n. \end{aligned} \tag{7.6.11}$$

In studying the existence and uniqueness of the solution of (7.6.8), researchers have found that under suitable conditions, and restriction of the range of t, the right-hand side of (7.6.8) is a contraction. For example, if

$$F(u, \mathbf{x}, t) \equiv f_0(u, \mathbf{x}, t) + \sum_{i=0}^{n} f_{i,x^i}(u, \mathbf{x}, t) \tag{7.6.12}$$

is bounded on bounded sets of u, and $-F_u$ is bounded above for real arguments of u and $(\mathbf{x}, t) \in \mathbb{R}^n \times (0, t)$, if $|f_0(0, \mathbf{x}, t)| \leqq c_0$, $|u_0| \leqq M$, and $-F_u \leqq c_1$, then it can be shown that the solution of (7.6.8) is bounded by $(M + c_0 T)e^{c_1 T}$, for $(\mathbf{x}, t) \in \mathbb{R}^n \times (0, T)$. Under the same, or similar conditions, it is possible to show that the solution of (7.6.8) does in fact belong to the class $\mathbf{M}_{\alpha_i}(\mathcal{D}_i)$ with respect to the x^i variable, (with every x_j $(j \neq i)$ and t fixed and real) and also to $\mathbf{M}_{\alpha_t}(\lceil_t)$ with respect to the t variable, $t \in \mathcal{D}_t$, a sector in the complex t-plane, $\alpha_t = (1/2)\min_i\{\alpha_i\}$, with $x_1, \cdots, x_n$ fixed and real, provided only that $u_0(x_1, \cdots, x_{i-1}, \cdot, x_{i+1}, \cdots, x_n)$ belongs to $\mathbf{M}_{\alpha_i}(\mathcal{D}_i)$, for all points $x_1, \cdots, x_{i-1}, x_{i+1}, \cdots, x_n \in \mathbb{R}^{n-1}$. To this end, we remark that a non-smooth or non-analytic $u_0(\mathbf{x})$ can be accurately approximated by an analytic function, via the method of Equation (5.8.6). The method (5.8.6) is particularly convenient, since the function v in (7.6.8) can then be explicitly expressed by approximation of the first convolution integral in (7.6.11).

A tried and tested method is to approximate the solution u by a linear combination of the form (7.3.6), i.e.,

$$u(\mathbf{x}, t) \approx u_N(\mathbf{x}, t) \equiv \sum_{j_1=-N}^{N} \cdots \sum_{j_n=-N}^{N} u_{j_1, \cdots, j_n}(t)\omega_{j_1}^1(x^1) \cdots \omega_{j_n}^n(x^n). \tag{7.6.13}$$

After substituting (7.6.13) into (7.6.8), and performing explicit approximations of the spatial integrals via the formula (3.2.10), we can then perform successive approximations,

$$\mathbf{u}^{m+1}(t) = \mathbf{v}(t) + \int_0^t \Phi(t - \tau)\mathbf{u}^m(\tau)d\tau, \tag{7.6.14}$$

where we have exemplified only the discretization with respect to the space variable $\mathbf{x}$, in order to emphasize the fact that we now have a system of Volterra integral equations of the type (6.4.2), which can be solved as described in Section 6.4. In (7.6.14) the vector $\mathbf{u}$ is a vector of order $M = \prod_{i=1}^n (2N_i + 1)$. The integrations with respect to τ involve integrals of the form

$$w(t) = \int_0^t (t - \tau)^{-1-n/2} \exp\left\{-\frac{|\mathbf{x} - \xi|^2}{4\varepsilon(t - \tau)}\right\} f(\tau)d\tau, \tag{7.6.15}$$

which are required for $t = t_{-N_{n+1}}, \cdots, t_{N_{n+1}}$, and these are tedious, if done as described in the proof of Theorem 6.4.3. Although not yet tried, we recommend here the use of the convolution method of Section 4.6 as an efficient means of approximating (7.6.15). Such an approach is possible, since in the case of n odd, the Laplace transform of $t^{-1-n/2}\exp\{-|\mathbf{x}|^2/(4\varepsilon t)\}$ is explicitly known, i.e.,

$$g(s) = \int_0^\infty t^{-1-n/2} \exp\left\{-\frac{|\mathbf{x}|^2}{4\varepsilon t} - st\right\} dt = 2s^{-1/2} K_{-n/2}\left[\frac{|\mathbf{x}|s^{1/2}}{\varepsilon^{1/2}}\right] \tag{7.6.16}$$

with $K_{-n/2}$ the standard Hankel function. If n is an odd integer, the right-hand side of (7.6.16) can be explicitly expressed; if n is an even integer, we can use the Sinc approximation (6.9.22) to approximate the integral (7.6.16). Thus setting $\hat{g}(s) = g(1/s)$, and then, in the notation of Section 4.5, forming the matrix

$$G = \hat{g}\left(hI^{(-1)}D(1/\phi')\right), \tag{7.6.17}$$

we can replace the τ-integrations in (7.6.15) by the collocation

$$(w_{-N_{n+1}}, \cdots, w_{N_{n+1}}) = (f_{-N_{n+1}}, \cdots, f_{N_{n+1}})G^T. \tag{7.6.18}$$

That is, this collocation enables us to replace (7.6.14) by the scheme

$$\hat{\mathbf{u}}^{(m+1)} = \hat{\mathbf{v}} + \hat{\Phi}\hat{\mathbf{u}}^{(m)}G^T, \tag{7.6.19}$$

where the vector $\mathbf{u}$ of order M in (7.6.14) has now become a rectangular matrix $\hat{\mathbf{u}}$ on the left-hand side of (7.6.19), of order $M \times (2N_{n+1}+1)$, and the right-hand term in (7.6.19) must be interpreted to be a rectangular system of non-linear equations. Instead of (7.6.19), Newton's method is usually a better method to solve the system of non-linear equations, especially if equations (7.6.19) are stiff, as it is frequently the case in applications, since by Newton's method (see the derivation of Equation (7.1.46)) we can usually select a much larger interval $(0, T)$.

The above scheme has already been tried and found to be an effective method of solving Burger's equation, i.e., the equation

$$u_t + uu_x = \varepsilon u_{xx} \tag{7.6.20}$$

as well as some two dimensional Navier–Stokes problems. Sinc methods have thus proven to be both accurate and efficient for solving such problems, in the case of $\varepsilon > 0$.

The $\varepsilon = 0$ solution may at times be desired in applications. We remark, to this end, that it may be shown, if the solution $u(\mathbf{x}, t, \varepsilon)$ has the property that $u(\mathbf{x}, t, 0) \in \mathbf{Lip}_\alpha$ in a neighborhood of a point $(\mathbf{x}, t)$, then $u(\mathbf{x}, t, \varepsilon) \in \mathbf{Lip}_{\alpha/2}$ as a function of ε, in some sector

$$\{\varepsilon \in \mathbb{C} : |\arg(\varepsilon)| < d\}.$$

Hence, using Sinc methods, we can efficiently compute some solutions for a few different values of $\varepsilon > 0$ and by Subsection 5.6.5 we can use these solutions in the Thiele algorithm to accurately extrapolate the solution $u(\mathbf{x}, t, 0)$.

•

Problems for Section 7.6

7.6.1 Describe an iterative scheme, based on Newton's method, to solve (7.6.19).

7.6.2 * Prove that if the solution $u(\mathbf{x}, t, \varepsilon)$ has the property that $u(\mathbf{x}, t, 0) \in \mathbf{Lip}_\alpha$ in a neighborhood of a point $(\mathbf{x}, t)$, then $u(\mathbf{x}, t, \varepsilon) \in \mathbf{Lip}_{\alpha/2}$ as a function of ε, in some sector

$$\{\varepsilon \in \mathbb{C} : |\arg(\varepsilon)| < d\}.$$

7.6.3 (a) Show that if g is bounded in $\mathbb{R} \times (0, \infty)$, then the problem

$$\begin{aligned} u_t - u_{xx} &= g(x,t), \quad x \in \mathbb{R}, \quad t > 0, \\ u(x,0) &= 0, \quad x \in \mathbb{R}. \end{aligned}$$

has a "solution", given by

$$u(x,t) = \int_0^t \frac{1}{2\sqrt{\pi(t-\tau)}} \int_{\mathbb{R}} \exp\left\{-\frac{(x-\xi)^2}{4(t-\tau)}\right\} g(\xi, \tau)\, d\xi.$$

(b) Let ϕ be defined as in (1.7.16), assume that $g(x, \cdot)$ is holomorphic in the region $\mathcal{D}$ defined in (1.7.12) for all $t \in (0, \infty)$, and that $g(\cdot, t)$ is holomorphic in the region $\mathcal{D}_d$, defined in (1.7.9) for all $x \in \mathbb{R}$. Assume also that $u(x, \cdot) \in \mathbf{M}_\alpha(\mathcal{D})$, for each fixed $x \in \overline{\mathbb{R}}$, and that $u(\cdot, t) \in \mathbf{M}_\alpha(\mathcal{D}_d)$, for each fixed $t \in [0, \infty]$. Take $h = [\pi d/(\alpha N)]^{1/2}$, and use the transforms of Problems 1.5.1 (c) and 1.5.1 (b) in the method from Section 4.6 to get an accurate approximation of u on $\mathbb{R} \times (0, \infty)$.

Historical Remarks on Chapter 7

This chapter is a mixture of both known and new results. Section 7.1 summarizes some known analyticity properties of solutions of ordinary differential equations, and derives some new ones for the spaces $\mathbf{L}_\alpha(\mathcal{D})$ and $\mathbf{M}_\alpha(\mathcal{D})$ in which Sinc approximations excel.

The results of Section 7.2.1 are new, although a brief summary of these results appeared in *Stenger et al.*, [SKyOrPk1]. An algorithm of the computer implementation of the methods of this section has been completed, and will be published elsewhere.

The collocation method of Section 7.2.1 is new. The historically earliest Sinc–Galerkin method, the integration-by-parts procedure of Subsection

7.2.2 was published in [S23]. The method of Subsection 7.2.3, combining the collocation procedure of Subsection 7.2.1 and the integration by parts procedure of Section 7.2.2 is new. The proof of convergence in Subsection 7.2.4, of the procedures of Subsections 7.2.1 to 7.2.3 is new. Subsections 7.2.5 and 7.2.6 are, in essence, a summary of results presented previously in [S24]. The important symmetric Sinc–Galerkin procedure of Section 7.2.7 was originally due to *Lund* [L2]. It is perhaps appropriate to mention here that *Lund*, and his colleagues and students at Montana State University have done a great deal of important work in the computer implementation of Sinc–Galerkin methods for the solution of ordinary and partial differential equations. In [Bi2,Bi6], *Bialecki* made the important observation that Lund Symmetric Sinc–Galerkin method for the equation (7.2.4), with $\mu = 0$ could also be obtained by collocation, starting with the approximation (7.2.94), and he proved (for some situations) the convergence of the procedure with respect to the $\|w\|_S$ norm defined as in (7.2.111). The equivalence presented in Subsection 7.2.7, of Lund's integration by parts procedure, and *Bialecki's* approximation (7.2.94) in the case of μ in (7.2.4) not being zero, as well as the general convergence of these methods are new. The boundary layer example, 7.2.12, was originally programmed by *Burke* in [Bur1]. The short Example 7.2.13 of the Klein–Gordon equation does not do justice to *Lundin's* [Lu1,Lu2] beautiful algorithmic and analytic work on this problem. Example 7.2.14 illustrates some of the important work of *Jarratt* [Ja1,Ja2,JaLBo1], who demonstrated that Sinc–Galerkin methods are an effective tool to obtain accurate approximations to eigenvalues and eigensolutions of Sturm–Liouville problems whose coefficients may have singularities, or over semi-infinite and infinite intervals. Example 7.2.15 illustrates the inversion of a second order ordinary differential equation by Sinc methods. The author is grateful to *J. Muldowney* for discussion of this procedure.

The material of Section 7.3 is a summary and extension of known results on analyticity properties of solutions to partial differential equations. The Cauchy–Kowalewsky theorem [Tu1] is well known. Section 7.3.2 consists of results presented first in [S26,S27] as well as an extension of some of these results. The detailed discussion of the elementary mappings which make the Sinc solution of boundary value problems over curvilinear regions as easy as over rectangular regions is new. *Fichera's* theorem [Fr1] is important, in that it gives sufficient insight into the nature of singularities of solutions of elliptic problems due to corners of regions to be enable us to solve such problems via an optimal Sinc procedure.

The solution in Section 7.4 of the model Poisson's problem (7.4.1) was done by *Burke* [Bur1] and by *Bowers and Lund* [BoL1]. Example 7.4.3 is new; it estimates the accuracy, condition number of the matrix, and complexity of various popular methods of solving a Poisson problem over a triangular-like region in the plane. Example 7.4.4, a novel efficient Sinc-related method of solving Laplace's equation over a region bounded by a finite number of analytic arcs in the plane is new.

The elementary treatment in Example 7.5.1 of the solution of the wave equation over $(x, t) \in \mathbb{R} \times (0, \infty)$ is new. The initial value approach in Example 7.5.2, to the solution of the wave equation over $(x, t) \in (0, 1) \times (0, \infty)$ is also new. *McArthur* [M1] has previously shown that a boundary value approach is also an effective method to solve this problem. *McArthur's* work [M1,M1BoL1,M1BoL2] provides an important contribution, in that it exhibits the structure and sparse nature of the matrices encountered in the Sinc solution of partial differential equations.

A computer algorithm for the Sinc solution of the elementary heat equation problem viewed as a boundary value problem in Example 7.6.1 of Section 7.6 was first attempted by *Burke* [Bur1], and later, by *Lund and Bowers*, in [BoL1]. In [Lk1], *Lippke* recently presented an excellent example, showing how effective Sinc methods are for solving boundary value problems. Our brief discussion of the conservation law problem of Example 7.6.2 does not do justice to the beautiful and extensive work by *Stromberg* [St1] on this subject. We have presented *Stromberg's* procedure, with the addition of some new ways of solving the time part of this problem, taking account of new Sinc formulas which have been developed since *Stromberg* did his work. We have also omitted the presentation of practically useful work of *Yin* [Y2] on the parallel computation of Navier–Stokes problems over planar regions, via the use of Sinc methods, and where the domain decomposition is based on analyticity, since this work was incomplete during the writing of this text.

References

N.H. Abel

[Ab1] *Solution de quelques problèmes à l'aide d'integrales définies*, Œuvres complètes D'Abel, v. 1, 11–27.

N.I. Achieser

[Ac1] *Vorlesungen Über Approximationstheorie*, Akademie–Verlag, Berlin (1953).

G.S. Ammar and W.B. Gragg

[AGg1] *The Generalized Schur Algorithm for the Superfast Solution of Toeplitz Systems*, pp. 1–22 of "Rational Approximation and Its Applications in Mathematics and Physics," edited by J. Gilewitz, M. Pindor and W. Siemaszko, Springer Lecture Notes in Mathematics # 1237, Springer-Verlag, Heidelberg (1985).

[AGg2] *The Implementation and Use of the Generalized Schur Algorithm*, pp. 265–279 of "Computational and Combinatorial Methods in Systems Theory", Edited by C.I. Byrnes and A. Lindquist, North–Holland, Elsevier, New York (1986).

G.S. Ammar, W.B. Gragg, and L. Reichel

[AGgRc1] *Determination of Pisarenko Frequency Estimates as Eigenvalues of an Orthogonal Matrix*, pp.143–145 of Advances in Algorithms for Signal Processing II, Editor. F.D. Luk, SPIE , v. 826 (1987).

E.G. Anastasselou and N.I. Ioakimidis

[AuIk1] *A New Method for Obtaining Exact Analytical Formulæ for the Roots of transcendental Functions*, Lett. Math. Phys., v. 8 (1984) 135–143.

J.–E. Andersson

[Ad1] *Optimal Quadrature of* $\mathbf{H}^p$ *Functions*, Math. Z., v. 172 (1980) 55–62.

J.–E. Andersson, and B.D. Bojanov

[AdBo1] *A Note on the Optimal Quadrature in* $\mathbf{H}^p$, Numer. Math., v. 44 (1984) 301–308.

D.D. Ang, J.R. Lund, and F. Stenger

[AnLS1] *Complex Variable and Regularization Methods of Inversion of the Laplace Transform*, Math. Comp., v. 83 (1989) 589–608.

P.M. Anselone

[Al1] *Collectively Compact Operator Approximation Theory and Application to Integral Equations*, Prentice–Hall, Englewood Cliffs, N.J. (1971).

R. Askey

[As1] *Orthogonal Polynomials and Special Functions*, SIAM (1975).

K.E. Atkinson

[At1] *Survey of Numerical Methods for the Solution of Fredholm Integral Equations of the Second Kind*, SIAM (1976).

G.A. Baker, Jr. and P.G. Graves–Morris

[BGm1] *Padé Approximants, Part I: Basic Theory*, Encyl. of Math. and Its Applications, v. 13, Addison-Wesley, Reading (1981).

[BGm2] *Padé Approximants, Part II: Extensions and Applications*, Encyl. of Math. and Its Applications, v. 14, Addison-Wesley, Reading (1981).

R.W. Barnard, W.T. Ford, and Hsing Y. Wang

[BaFoWa1] *On Zeros of Interpolating Polynomials*, SIAM J. Math. Anal., v. 17 (1986) 734–744.

S. Beignton and B. Noble

[BnNo1] *An Error Estimate for Stenger's Quadrature Formula*, Math. Comp., v. 38 (1982) 539–545.

M.J. Berggren, S. Johnson, C. Wilcox, and F. Stenger

[BeJWS] *Rational Function Frequency Extrapolation in Ultrasonic Tomography*, pp. 19–34 of "Wave Phenomena, Modern Theory and Applications," C. Rogers and T.B. Moodie, eds. Elsevier Science Publishers (B.B), North–Holland (1984).

A. Ben–Israel

[B–I1] *An Iterative Method for Computing the Generalized Inverse of an Arbitrary Matrix*, Math. Comp. v. 19 (1965) 452–455.

B. Bialecki

[Bi1] *Sinc Method of Solution of CSIE and Inverse Problems*, Ph.D. Thesis, University of Utah (1987).

[Bi2] *Sinc-type Approximation in* $\mathbf{H}^1$*-Norm, with Application to Boundary Value Problems*, Journ. Comp. and Appl. Math., v. 25 (1989) 289–303.

[Bi3] *A Modified Sinc Quadrature Rule for Functions with Poles Near the Arc of Integration*, BIT, v. 29 (1989) 464–476.

[Bi4] *A Sinc Quadrature Rule for Hadamard Finite Part Integrals*, Numer. Math., v. 57 (1990) 263–269.

[Bi5] *Sinc–Hunter Quadrature Rule for Cauchy Principal Value Integrals*, Math. Comp., v. 55 (1990) 665–681.

[Bi6] *Sinc-Collocation Methods for Two–Point Boundary Value Problems*, IMA J. Numer. Anal., v. 11 (1991) 357–375.

[Bi7] *Sinc–Nyström Method for Numerical Solution of a Dominant System of Cauchy Singular Integral Equations Given on a Piecewise Smooth Contour*, Siam J. Numer. Anal. v. 26 (1989) 1194–1211.

B. Bialecki and F. Stenger

[BiS1] *Sinc–Nyström Method for Numerical Solution of One-Dimensional Cauchy Singular Integral Equations Given on a Smooth Arc in the Complex Plane*, Math. Comp., v. 51 (1988) 133–165.

R.P. Boas, Jr.

[Boa1] *Entire Functions*, Academic Press, New York (1954).

[Boa2] *Summation Formulas and Band Limited Signals*, Tôhuko Math. J., v. 24 (1972) 121–125.

B.D. Bojanov

[Boj1] *The Optimal Method for the Integration of a a Certain Class of Analytic Functions*, Proc. Second Spring Conference of the Bulgarian Math. Society, Vidin, 1973, Izdat Bulgar. Akad. Nauk, Sofia, 1974, pp. 75-86 (In Bulgarian); English summary, Math. and Math. Education.

[Boj2] *Best Quadrature Formula for a Certain Class of Analytic Functions*, Zastosowania Matematyki Appl. Mat. XIV(1974) 441–447.

E. Borel

[Bor1] *Sur l'interpolation*, C.R. Acad. Sci. Paris, v. 124 (1897) 673–676.

[Bor2] *Mémoire sur les Séries Divergente*, Ann. École Norm. Sup., v. 16 (1899) 9–131.

K.L. Bowers and J. Lund

[BowL1] *Numerical Solution of Singular Poisson Solvers, via the Sinc Galerkin Method*, SIAM J. Numer. Anal., v. 24 (1987) 36–51.

Claude Brezinski

[Bre1] *Convergence Acceleration Methods: The Past Decade*, J. Comput. Appl. Math., v. 12 and 13 (1985) 19–36.

Vincent Broman

[Bro1] *Initial Value Methods for Integral Equations*, M.Sc. Thesis, University of Utah, (1979).

H.G. Burchard and K. Höllig

[BuH1] *N-Width and Entropy of $\mathbf{H}^p$ Classes in $\mathbf{L}^q(-1,1)$*, SIAM J. Math. Anal., v. 16 (1985) 405–421.

R. Burke

[Bur1] *Application of the Sinc–Galerkin Method to the Solution of Boundary Value Problems*, M. Sc. Thesis, University of Utah, 1977.

P.L. Butzer

[Bz1] *A Survey of the Whittaker–Shannon Expansion Theorem and Some of Its Extensions*, J. Math. Res. Exposition, v. 3 (1983) 185–212.

P.L. Butzer and W. Engels

[BzEn1] *Dyadic Calculus and Sampling Theorems for Functions with Multidimensional Domains, I II*, Inform. and Control, v. 52 (1982) 333–363.

P.L. Butzer and R. Nessel

[BzN1] *Fourier Analysis and Approximation, Vol. 1*, Academic Press, London (1971).

P.L. Butzer and R. Stens

[BzSt1] *The Euler–MacLaurin Summation Formula, the Sampling Theorem, and Approximate Integration over the Real Axis*, Linear Algebra Appl., vols. 52/53 (1983) 141–155.

C. Carathéodory

[Ca1] *Theory of Functions*, vols. 1 and 2, Chelsea, New York (1958).

A. Cauchy

[Cu1] *Mémoire sur les Développements des Fonctions en Séries Périodiques*, Mémoires de l'Institut, v. 6 (1826) 603–613.

W. Cauer

[C1] *Bemerkung über eine Extremalaufgabe, von E. Zolotareff*, Z. Angew. Math. Mech., v.20 (1940) 358.

M.M. Chawla and V. Kaul

[ChKu1] *Optimal Rules for Numerical Integration Round the Unit Circle*, BIT, v.13 (1973) 145–152.

Y.H. Chiu

[Ci1] *Integral Equation Solution of* $\nabla^2 u + k^2 u = 0$ *in the Plane*, Ph.D. Thesis, University of Utah, 1977.

P.G. Ciarlet, M.H. Schultz and R.S. Varga

[Cr1] *Numerical Methods of High Order Accuracy for Nonlinear Boundary Value Problems, I. One Dimensional Problem*, Numer. Math., v. 2 (1967) 394–430.

D. Cope

[Co1] *Convergence of Piessen's Method for Numerical Inversion of the Laplace Transform on the Real Line*, SIAM J. Numer. Anal., (to appear).

P. J. Davis

[Da1] *Errors of Numerical Approximation for Analytic Functions*, J. Rational Mech. Anal., v. 2 (1953) 303–313.

[Da2] *On the Numerical Integration of Periodic Analytic Functions*, pp. 303–313 of "On Numerical Approximation," R. Langer, ed., University of Wisconsin Press, Madison. (1953).

[Da3] *The Schwarz Function and Its Applications*, Carus Mathematical Monographs # 17 MAA (1974).

P.J. Davis and P. Rabinowitz

[DaRz1] *Methods of Numerical Integration*, Second Edition, Academic Press, Orlando. (1984).

C. de Boor

[dB1] *A Practical Guide to Splines*, Springer–Verlag, New York (1978).

R. deVore and K. Scherer

[dVSr1] *Variable Knot, Variable Degree Spline Approximation to* x^β, in *Quantum Approximation*, Academic Press (1980) 121–131.

F. de Hoog

[dH] *A New Algorithm for Solving Toeplitz Systems of Equations*, Lin. Algebra Appl., vols. 88/89 (1987) 122–133.

J.D. Donaldson and D. Elliott

[DnE1] *Quadrature II: The Estimation of Remainders in Certain Quadrature Rules*, Tech. Rep. 24, Dept. of Mathematics, Univ. of Tasmania (1970).

E. de Donker and R. Piessens

[DrPs1] *Automatic Computation of Integrals with Singular Integrand over a Finite or Infinite Range*, Report TW22, Applied Mathematics and Programming Division, Katholike Universiteit, Leuven (1975).

DeeAnn Dorman

[Do1] *Methods for Accelerating Convergence*, M.Sc. Thesis, University of Utah, 1983.

D.P. Dryanov, Q.I. Rahman and G. Schmeisser

[DrRaSr1] *Converse Theorems in the Theory of Approximate Integration*, Constr. Approx., v. 6 (1990) 321–334.

U. Eckhardt

[Ec1] *Einige Eigenschaften Wilfscher Quadraturformeln*, Numer. Math., v. 12 (1968) 1–7.

A. Edrei, E.B. Saff and R.S. Varga

[EdSaV1] *Zeros of Sections of Power Series*, Springer Lecture Notes in Mathematics # 1002 Springer-Verlag, New York (1983).

N. Eggert, M. Jarratt, and J. Lund

[EgJaL1] *Sinc Function Computation of the Eigenvalues of Sturm–Liouville Problems*, J. Comput. Phys., v. 69 (1987) 209–229.

N. Eggert and J. Lund

[EgL1] *The Trapezoidal Rule for Analytic Functions of Rapid Decrease*, J. Computational and Appl. Math., v. 27 (1989) 389–406.

M. Eiermann, X. Li, and R.S. Varga

[EiLiV] *On Hybrid Semi-Iterative Methods*, SIAM J. Numer. Anal. v. 26 (1989) 152–168.

M. Eiermann, R.S. Varga, and W. Niethammer

[EiVNi1] *Iterationsverfahren für Nichtsymmetrische Gleichungssysteme und Approximationsmethoden in Komplexen*, Jber. d. Df. Math.–Verein., v. 89 (1987) 1–32.

S.W. Elacott and M.H. Gutknecht

[ElGu1] *The Carathéodory–Fejér Extension of a Finite Geometric Series*, IMA Journal of Numerical Analysis, v.3 (1983) 221–227.

D. Elliott

[E1] *Some Observations on the Convergence Theories of Anselone, Noble, and Linz*, University of Tasmania Mathematics Department, Technical Report No. 122 (1979).

[E2] *A Note on the Convergence of Operator Equations*, University of Tasmania Mathematics Department, Technical Report No. 124 (1979).

D. Elliott and F. Stenger

[ES1] *Sinc Method of Solution of Singular Integral Equations*, pp. 155–166 of "IMACS Conference on CSIE", Philadelphia, P.A. (1984).

H. Engels

[En1] *Numerical Quadrature and Cubature*, Academic Press, London. (1980).

H.W. Engl and C.W. Groetsch

[Wi1] *Inverse and Ill-Posed Problems*, Academic Press, Inc. (1987).

W. Feller

[Fe1] *An Introduction to Probability Theory and its Applications*, v. 2, John Wiley and Sons, New York (1971).

H.E. Fettis

[Fi1] *Numerical Calculation of Certain Definite Integrals by Poisson's Summation Formula*, M.T.A.C., v. 9 (1955) 85–92.

G. Fichera

[Fr1] *Asymptotic Behavior of the Electric Field Near the Singular Points of the Conductor Surface*, Academi Nazionale dei Lincei, v. 60 (1976) 1–8.

W. Forst

[Fs1] *Zur Optimalität interpolatorischer Quadraturformeln periodischer Funktionen*, Numer. Math., v. 25 (1975) 15–21.

B.G. Gabdulhaev

[Gb1] *A General Quadrature Process and its Application to the Approximate Solution of Singular Integral Equations*, Soviet Math., Dokl., v. 9 (1968) 386–389.

[Gb2] *Approximate Solution of Singular Integral Equations by the Method of Mechanical Quadratures*, Soviet Math., Dokl., v. 9 (1968) 329–332.

F.D. Gakhov

[Gk1] *Boundary Value Problems*, translated from Russian by I.N. Sneddon, Pergamon Press, Oxford. (1966).

D.G. Gardner, J.C. Gardner, G. Laush, and W.W. Meinke

[GdGdLuMi1] *Method for the Analysis of Multicomponent Exponential Decay Curves*, J. Chem. Phys., v. 12, #1 (1959) 978–986.

W. Gautschi

[Ga1] *Norm Estimates for Inverses of Vandermonde Matrices*, Numer. Math., v. 23 (1975) 337–347.

[Ga2] *How (Un)stable are Vandermonde Systems?*, pp. 193–210 of Asymptotics and Computational Analysis, v. 124 of Lecture Notes in Pure and Applied Mathematics, edited by R. Wong, Marcel Dekker, Inc., New York (1990).

[Ga3] *A Survey of Gauss–Christoffel Quadrature Formulæ*, pp. 72–147 of "E.B. Christoffel: the Influence of his Work in Mathematics and the Physical Sciences," Edited by P.L. Butzer and F. Fehér, Birkhäuser, Basel. (1981).

[Ga4] *Computational Methods in Special Functions—A Survey*, pp. 1–98 of "Theory and Applications of Special Functions," Academic Press, New York (1975).

W. Gautschi and R.S. Varga

[GaV1] *Error bounds for Gaussian Quadrature of Analytic Functions*, SIAM J. Numer. Anal. v. 20 (1983) 1170–1186.

W.B. Gearhart and F. Stenger

[GeS1] *An Approximate Convolution Equation of a Given Response*, pp. 168–196 of "Optimal Control Theory and Its Applications" Springer–Verlag Lecture Notes on Economics and Mathematical Systems, v. 106, Springer-Verlag, New York (1974).

I.M. Gel'fand and B.M. Levitan

[GfLv1] *On the Determination of a Differential Equation from its Spectral Function*, AMS Translations, v.2 (1955) 253–304.

D. Gilbarg and N.S. Trudinger

[GiTu1] *Elliptic Partial Differential Equations of Second Order*, (Second Edittion) Springer-Verlag, New York (1983).

D.S. Gilliam, J.R. Schulenberger and J.L. Lund

[GmScL1] *Spectral Representation of the Laplace and Stieltjes Transforms*, Mat. Aplic. Comp., v. 7 (1988) 101–107.

M. Golomb

[Gol] *Lectures on Theory of Approximation*, Argonne National Laboratory (1962).

G.M. Golusin

[Gl1] *Geometric Theory of Functions of a Complex Variable*, Translations of Mathematical Monographs, AMS (1969).

R.F. Goodrich and F. Stenger

[GcS1] *Movable Singularities and Quadrature*, Math. Comp., v. 24 (1970) 283–300.

E.T. Goodwin

[Goo1] *The Evaluation of Integrals of the Form* $\int_{-\infty}^{\infty} a(x)\, e^{-x^2}\, dx$, Proc. Camb. Philos. Soc., v. 45 (1949) 241–245.

I. S. Gradshteyn and I. M. Ryzhik

[GdR1] *Tables of Integrals, Series and Products*, 4th Edition, Academic Press, New York (1965).

W.B. Gragg

[Gg1] *The Padé Table and Its Relation to Certain Algorithms of Numerical Analysis*, SIAM Rev., v. 14 (1972) 1–62.

[Gg2] *Truncation Error Bounds for g–Fractions*, Numer. Math., v. 11 (1968) 370–379.

W.B. Gragg and D.D. Warner

[GgWa1] *Two Constructive Results in Continued Fractions*, SIAM J. Numer. Anal., v.20 (1983) 1187–1197.

P.R. Graves–Morris

[Gm1] *Efficient Reliable Interpolation*, pp. 28–63 of "Padé Approximation," Editors M. de Bruijn and H., von Rossum, Springer Lecture Notes in Math., v. 888 Springer-Verlag, Heidelberg (1981).

P.R. Graves–Morris and E.B. Saff

[GmSa1] *Row Convergence Theorems for Generalized Inverse Vector-Valued Padé Approximants*, Inst. for Constructive Mathematics, University of South Florida, Report ICM # 87–012.

U. Grenander and G. Szegö

[GrSz1] *Toeplitz Forms and Their Applications*, University of California Press, Berkeley (1958).

B. Guo

[Go1] *The $h-p$ Version of Finite Element Method in Two Dimensions*, Ph.D. Thesis, University of Maryland (1985).

B. Guo and I. Babuska

[GoBa1] *The $h-p$ Version of the Finite Element Method. Part I. The Basic Approximation Results. Part II. General Results and Applications*, Computational Mechanics (1986) Part I, 24–41, Part II, (to appear).

S.-Å. Gustafson

[Gs1] *Convergence Acceleration on a General Class of Power Series*, Computing, v. 21 (1978) 53–69.

[Gs2] *Two Computer Codes for Convergence Acceleration*, Computing, v. 21 (1978) 87–91.

M.H. Gutknecht

[Gu1] *Continued Fractions Associated with the Newton–Padé Table*, ETH–Zentrum, IPS Research Report No. 89–01 (1989).

S. Haber

[H1] *The Error in the Numerical Integration of Analytic Functions*, Quart. Appl. Math., v. 29 (1971) 411–420.

[H2] *The Tanh Rule for Numerical Integration*, SIAM J. Numer. Anal., v. 14 (1977) 668–685.

[H3] *Two Formulas for Numerical Indefinite Integration*, Math. Comp., v. 60 (1993) 279–296.

M. Hagmann and F. Stenger

[HaS1] *Unique Advantages of Sinc Function Basis in the Method of Moments*, in Proceedings of the Conference on Electromagnetic Insonification, IEEE , v. 54 (1980) 35–37.

R.V.L. Hartley

[Ht1] *The Transmission of Information*, Bell System Tech. J., v. 7 (1928) 535–560.

E. Hayashi, L.N. Trefethen, and M.H. Gutknecht

[HyTrGu1] *The CF Table*, Department of Mathematics, MIT Numerical Analysis Report (1987) 87–3.

P. Henrici

[Hn1] *Applied Computational and Complex Analysis*, v. 1 (1974); v. 2 (1977); v. 3 (1986) John Wiley and Sons, New York.

J.R. Higgins

[Hs1] *Five Short Stories about the Cardinal Series*, Bull. A.M.S., v.12 (1985) 45–89.

Y. Ikebe

[I1] *The Galerkin Method for Numerical Solution of Fredholm Integral Equations of the Second Kind*, SIAM Rev., v. 14 (1972) 465–491.

Y. Ikebe, M. Kowalski, and F. Stenger

[IKS1] *Rational Approximation of the Step, the Filter, and the Impulse Function*, pp. 441–454 of "Asymptotic and Computational Analysis," Edited by R. Wong, Marcel Dekker, New York (1990).

Y. Ikebe, T.Y. Li, and F. Stenger

[ILiS1] *Numerical Solution of Hilbert's Problem*, in "Theory of Approximation with Applications," eds. A.G. Law and B. Sahney, Academic Press, New York (1976) 338–358.

E.L. Ince

[In1] *Ordinary Differential Equations*, Dover, New York (1960).

M.E.H. Ismail

[I1] *Contiguous Relations, Basic Hypergeometric Functions, and Orthogonal Polynomials*, Inst. for Constructive Mathematics, ICM # 88–103 University of South Florida (1988).

S. Järner

[J1] *On Weighted Zolotarev Problems*, Department of Mathematics, Göteborg, (1982).

M. Jarratt

[Ja1] *Approximation of Eigenvalues of Sturm–Liouville Differential Equations by the Sinc-Collocation Method*, Ph.D. Thesis, Montana State University (1987).

[Ja2] *Eigenvalue Approximations on the Entire Real Line*, pp. 133-144 of "Computation and Control," Edited by K.L. Bowers and J. Lund, Birkhäuser, Basel (1989).

M. Jarratt, J. Lund, and K.L. Bowers

[JaLBo1] *Galerkin Schemes and the Sinc–Galerkin Method for Singular Sturm–Liouville Problems*, J. Comput. Phys., v. 89 (1990) 41–62.

A. Jerri

[Je1] *The Shannon Sampling Theorem—Its Various Extensions and Applications: a Tutorial Review*, Proc. IEEE, v. 65 (1977) 1565–1596.

S.A. Johnson, Y. Zhou, M.L. Tracy, M.J. Berggren, and F. Stenger

[JZTBS1] *Inverse Scattering Solutions by Sinc Basis, Multiple Source Moment Method,—Part III: Fast Algorithms*, Ultrasonic Imaging, v. 6 (1984) 103–116.

T. Kato

[La1] *Perturbation Theory for Linear Operators*, Springer-Verlag, New York (1969).

R.B. Kearfott

[K1] *A Sinc Approximation for the Indefinite Integral*, Math. Comp. , v. 41 (1983) 559–572.

R.B. Kearfott, K. Sikorski, and F. Stenger

[KSiS1] *A Sinc Adaptive Algorithm for Solving Elliptic Partial Differential Equations*, in manuscript.

F. Keinert

[Ke1] *Uniform Approximation to $|x|^\beta$ by Sinc Functions*, J. Approx. Th., v. 66 (1991) 44–52.

W. Knauff

[Kn1] *Fehlernormen zur Quadratur Analytischer Funktionen*, Computing, v. 17 (1977) 309–322.

[Kn2] *Gewichtsoptimale Quadraturformeln bei Analytischer Funktionen*, Computing, v. 18 (1977) 59–66.

W. Knauff and R. Kress

[KnKs1] *Optimale Approximation Linearer Funktionale auf Periodischen Funktionen*, Numer. Math., v. 22 (1974) 187–205.

H. Kober

[Ko1] *A Dictionary of Conformal Representation*, Dover, New York (1957).

V.A. Kotel'nikov

[Kt1] *On the Carrying Capacity of the 'Ether' and Wire in Telecommunications*, "Material for the First All–Union Conference on Questions of Communication," Izd. Red. Upr. Svyazi RKKA, Moscow (1933).

M. Kowalski, K. Sikorski, and F. Stenger

[KwSiS1] *Special Topics in Approximation Theory*, textbook, to be published by Oxford, New York.

H.P. Kramer

[Kra1] *A Generalized Sampling Theorem*, J. Math. Phys. v. 38 (1959) 68–72.

L. Kratz

[Kz1] *The Reduction of Double Integrals to Boundary-Type Single Integrals for Solutions of Partial Differential Equations*, Idaho State University Department of Mathematics manuscript.

[Kz2] *The Integral of Solutions of Hyperbolic Equations*, Idaho State University Department of Mathematics manuscript.

[Kz3] *Reducing Dimensionality in Integrating Solutions of Differential Equations*, Ph.D. Thesis, University of Utah (1975).

R. Kress

[Ks1] *Interpolation auf einem Unendlichen Intervall*, Computing, v. 6 (1970) 274–288.

[Ks2] *Über die numerische Berechnung Konjugierter Funktionen*, Computing, v. 10 (1972) 177–187.

[Ks3] *Ein Ableitungsfrei Restglied für die trigonometrische Interpolation Periodischer Analytischer Funktionen*, Numer. Math. v. 16 (1971) 389–396.

[Ks4] *Zur Numerischen Integration Periodischer Funktionen nach der Rechteckregel*, Numer. Math., v. 20 (1972) 87–92.

[Ks5] *On General Hermite Trigonometric Interpolation*, Numer. Math., v. 20 (1972) 125–138.

[Ks6] *On Error Norms of the Trapezoidal Rule*, SIAM J. Numer. Anal., v. 15 (1978) 433–443.

[Ks7] *Zur Quadratur Unendlicher Integrale bei Analytischen Funktionen*, Computing, v. 13 (1974) 267–277.

[Ks8] *Linear Integral Equations*, Springer, Applied Mathematical Sciences, v. 82, Springer-Verlag, New York (1982).

M. Lachance, E.B. Saff, and R.S. Varga

[LcSV] *Inequalities for Polynomials with a Prescribed Zero*, Math. Z., v. 168 (1979) 105–116.

P. Lancaster

[La1] *Theory of Matrices*, Academic Press, New York (1969).

F.M. Larkin

[Lk1] *Optimal Approximation in Hilbert Spaces with Reproducing Kernel Functions*, Math. Comp., v. 24 (1970) 911–921.

B.Ja. Levin

[Ln1] *Distribution of Zeros of Entire Functions*, Translations of Mathematical Monographs, AMS, Providence, R.I. (1964).

B.M. Levitan and M.G. Gasynov

[LvGv1] *Determination of a Differential Equation by Two of Its Spectra*, Russian Math. Surveys, v. 19 (1964) 1–63.

D.L. Lewis

[Le1] *A Fully Galerkin Method for Parabolic Problems*, Ph.D. Thesis, Montana State University (1989).

D.L. Lewis and J. Lund

[LeL] *The Space–Time Sinc–Galerkin Method for Parabolic Problems*, Int. J. Numer. Methods Eng., v. 24 (1987) 1629–1644.

D.L. Lewis, J. Lund, and K.L. Bowers

[LeLBo1] *The Space–Time Sinc–Galerkin Method for Parabolic Problems*, Int. J. Numer. Methods Eng., v. 24 (1987) 1629–1644.

E. Lindelöf

[Ln1] *Le Calcul des Résidus*, Chelsea, New York (1947).

P. Linz

[Lz1] *A General Theory for the Approximate Solution of Operator Equations of the Second Kind*, SIAM J. Numer. Anal., v. 14 (1977) 543–554.

P. Lipow and F. Stenger

[LpS1] *How Slowly Can Quadrature Formulas Converge?*, Math. Comp., v. 26 (1972) 917–922.

A. Lippke

[Lk1] *Analytic Solution and Sinc Function Approximation in Thermal Conduction with Nonlinear Heat Generation*, J. Heat Transfer (Transactions of the ASME), v. 113 (1991) 5–11.

H.L. Loeb and H. Werner

[LoW1] *Optimal Numerical Quadratures in* $\mathbf{H}^p$ *Spaces*, Math. Z., v. 138 (1974) 111–117.

G.G. Lorentz

[Lr1] *Approximation of Functions*, Holt, Rinehart, and Winston, New York (1966).

J. Lund

[L1] *Numerical Evaluation of Integral Transforms*, Ph.D. Thesis, University of Utah, (1978).

[L2] *Symmetrization of the Sinc–Galerkin Method for Boundary Value Problems*, Math. Comp., v. 47 (1986) 571–588.

J. Lund and K.L. Bowers

[LBow2] *Sinc Methods for Quadrature and Differential Equations*, textbook, SIAM (1992).

J. Lund, K.L. Bowers, and K. McArthur

[LBoM1] *Symmetrization of the Sinc–Galerkin Method with Block Techniques for Elliptic Equations*, IMA J. Numer. Anal., v. 9 (1989) 29–46.

J. Lund and B. Riley

[LRl1] *A Sinc-Collocation Method for the Computation of the Eigenvalues of the Radial Schrödinger Equation*, IMA J. Numer. Anal., v. 4 (1984) 83–98.

L. Lundin

[Lu1] *A Cardinal Method of Solution of the Klein–Gordon Equation*, Ph.D. Thesis, Univ. of Utah (1975).

[Lu2] *A Cardinal Function Method of Solution of the Equation* $\Delta u = u - u^3$, Math. Comp., v. 35 (1980) 747–756.

L. Lundin and F. Stenger

[LuS1] *Cardinal–Type Approximation of a Function and Its Derivatives*, SIAM J. Numer. Anal., v. 10 (1979) 139–160.

R.J. Marks, II

[M1] *Introduction to Shannon Sampling and Interpolation*, Springer Series in Electrical Engineering, Springer-Verlag, New York (1990).

E. Martensen

[Ma1] *Zur Numerischen Auswertung Unendlicher Integrale*, ZAMM, v. 48 (1968) T83–T85.

K.M. McArthur

[M1] *Sinc–Galerkin Solution of Second–Order Hyperbolic Problems in Multiple Space Dimension*, Ph.D. Thesis, Montana State University (1987).

K.M. McArthur, K.L. Bowers, and J. Lund

[MBoL1] *Numerical Implementation of the Sinc –Galerkin Method for Second Order Hyperbolic Equations*, Numer. Methods Partial Diff. Eq., v. 3 (1987) 169–185.

[MBoL2] *The Sinc Method in Multiple Space Dimensions: Model Problems*, Numer. Math., v. 56 (1990) 789–816.

J. McNamee

[Mc1] *Error–Bounds for the Evaluation of Integrals by the Euler–MacLaurin Formula and by Repeated Gauss–Type Formulæ*, Math. Comp., v. 18 (1964) 368–381.

J. McNamee, F. Stenger, and E.L. Whitney

[McSW1] *Whittaker's Cardinal Function in Retrospect*, Math. Comp. , v. 25 (1971) 141–154.

G. Meinardus

[Me1] *Approximation von Funktionen und Ihre Numerische Behandlung*, Springer Tracts in Natural Philosophy, v. 4, Springer-Verlag, Berlin (1964).

P.A.P. Moran

[Mn1] *Approximate Relation Between Series and Integrals*, Math. Comp., v. 12 (1958) 34–37.

[Mn2] *Numerical Integration in the Presence of Singularities*, Acta Mathematica Scienta, v. 1 (1981) 83–85.

M. Mori

[Mo1] *An IMT–Type Double Exponential Formula for Numerical Integration*, Publ. RIMS, Kyoto Univ., v. 14 (1978) 713–729.

[Mo2] *Quadrature Formulas Obtained by Variable Transformation and the DE Rule*, Journ. Comput. Appl. Math., v. 12 13 (1985) 119–130.

[Mo3] *Developments in the Double Exponential Formulas for Numerical Integration*, Department of Applied Physics, Univ. of Tokyo (preprint).

D. Morley

[Mr1] *On the Convergence of a Newton–like Method for Closed Surfaces*, Ph.D. Thesis, University of Utah, (1979).

K. Murota and M. Iri

[MuIr1] *Parameter Tuning and Repeated Application of the IMT–Type Transformation in Numerical Quadrature*, Numer. Math., v. 38 (1982) 327–363.

N.I. Muskhelishvili

[Mv1] *Singular Integral Equations*, Nordhoff, Gröningen (1953).

National Bureau of Standards

[Nbs1] *Handbook of Mathematical Functions*, National Bureau of Standards Applied Math. Series (1964).

Z. Nehari

[Ne1] *Conformal Mapping*, McGraw–Hill, New York (1952).

P. Nevai

[N1] *Orthogonal Polynomials*, Memoirs Amer. Math. Soc., v. 213 (1970).

D.J. Newman

[Nw1] *Rational Approximation to* $|x|$, Michigan Math. J., v. 11 (1964) 11–14.

[Nw2] *Quadrature Formulas for* $\mathbf{H}^p$ *Functions*, Math. Z., v. 166 (1979) 111–115.

B. Noble

[No1] *Error Analysis of Collocation Methods for Solving Fredholm Integral Equations*, "Topics in Numerical Analysis," J.J.H. Miller (ed.), Academic Press, New York (1973) 211–232.

H. Nyquist

[Ny1] *Certain Topics in Telegraph Transmission Theory*, Trans. Amer. Inst. Elect. Engrg., v. 47 (1928) 617–644.

F.W.J. Olver

[O1] *Asymptotics and Special Functions*, Academic Press, New York (1974).

M. Oreilly

[Or1] *Sinc Function Solution of the Inverse Helmholtz Equation*, Ph. D. Thesis, University of Utah (1991).

H. Padé

[P1] *Sur la Représentation Approchée d'une Fonction par des Fractions Rationelles*, Ann. Fac. Sci. École Norm. Sup., v. 9 (1982) 1–93.

D.W. Peaceman and H.H. Rachford, Jr.

[PaR1] *The Numerical Solution of Parabolic and Elliptic Partial Differential Equations*, J. Soc. Indust. Appl. Math., v. 3 (1955) 28–41.

P.P. Petrushev and V.A. Popov

[PePo1] *Rational Approximation of Real Functions*, Cambridge University Press, Cambridge (1987).

S.K. Pichorides

[Pi1] *On the Best Values of the Constants in the Theorems of Riesz, Zygmund, and Kolmogorov*, Studia Mathematica, T.XLIV (1972) 165–179.

G. Plana

[Pl1] *Sur une nouvelle expression analytique des nombres Bernoulliens*, Academia di Torino, v. 25 (1820) 403–418.

G. Polyà amd G. Latta

[PyLa] *Complex Variables*, John Wiley and Sons, New York (1974).

M.J.D. Powell

[Pw1] *Approximation Theory and Methods*, Cambridge University Press, Cambridge (1981).

Q.I. Rahman and G. Schmeisser

[RaSr1] *Characterization of the Speed of Convergence of the Trapezoidal Rule*, Numer. Math. v. 57 (1990) 123–138.

L. Reichel

[Rc1] *A Matrix Problem with Application to Rapid Solution of Integral Equations*, Institute for Constructive Mathematics, University of South Florida, Report ICM # 86–009.

J.R. Rice

[Re1] *The Approximation of Functions*, Addison-Wesley, Reading, Mass. (1964).

N. Richter–Dyn

[Ri1] *Minimal Interpolation and Approximation in Hilbert Spaces*, SIAM J. Numer. Anal., v. 8 (1971) 583–597.

B. Riley

[Rl1] *A Sinc Collocation Method for Weakly Singular Integral Equations*, pp. 263–275 of "Computation and Control," Edited by K.L. Bowers and J. Lund, Birkhäuser, Basel (1989).

T.J. Rivlin

[Rv1] *An Introduction to the Approximation of Functions*, Blaisdell, Waltham, Mass. (1969).

E.B. Saff

[Sa1] *Polynomial and Rational Approximation in the Complex Domain*, Institute for Constructive Mathematics, University of South Florida, Report ICM # 86–006.

E.B. Saff and J.C. Snader

[SaSn1] *The Error for Quadrature Methods: A Complex Variable Approach*, Institute for Constructive Mathematics, University of South Florida, Report ICM # 86–002.

E.B. Saff and A.D. Snider

[SaSi] *Fundamentals of Complex Analysis*, Prentice–Hall, Englewood Cliffs, N.J. (1976).

E.B. Saff and V. Totik

[SaTo1] *Limitations of the Carathéodory–Fejèr Method for Polynomial Approximation*, Institute for Constructive Mathematics, University of South Florida, Report ICM # 88–007.

T.W. Sag and G. Szekeres

[SgSz1] *Numerical Evaluation of High Dimensional Integrals*, Math. Comp., v. 18 (1964) 245–253.

S. Schaffer and F. Stenger

[SrS1] *Multigrid–Sinc Methods*, Appli. Math. and Comp. J. , v. 19 (1986) 35–47.

A. Schönhage

[Sh1] *Zur Quadratur Holomorpher Periodischer Funktionen*, J. Approximation Theory, v. 13 (1975) 341–347.

C. Schwartz

[Sz1] *Numerical Integration of Analytic Functions*, J. Comput. Physics, v. 4 (1969) 19–29.

[Sz2] *High Accuracy Approximation Techniques for Analytic Functions*, J. Comput. Physics, v. 26 (1985) 411–415.

J. Schwing

[Sw1] *Numerical Solution of Integral Equations in Potential Theory Problems*, Ph.D. Thesis, University of Utah, (1976).

C.E. Shannon

[Sn1] *A Mathematical Theory of Communication*, Bell Systems Tech. J., v. 27 (1948) 379–423 and 623–656.

A. Sharma, H.P. Dikshit, V. Singh, and F. Stenger

[ShDSnS1] *Overconvergence in Chebyshev Polynomial Approximation*, J. Approx. Theory, v. 52 (1988) 339–349.

K. Sikorski

[Si1] *Optimal Quadrature Algorithms in* $\mathbf{H}^p$ *Spaces*, Numer. Math., v. 39 (1982) 405–410.

K. Sikorski, J. Schwing, and F. Stenger

[SiSwS1] *ALGORITHM 614. A FORTRAN Subroutine for Numerical Integration in* $\mathbf{H}^p$, ACM TOMS, v. 10 (1984) 152–160.

K. Sikorski and F. Stenger

[SiS1] *Optimal Quadratures in* $\mathbf{H}^p$ *Spaces*, ACM–TOMS, v. 10 (1984) 140–151.

R.C. Smith, K.L. Bowers, and J. Lund

[SmBoL1] *Efficient Numerical Solution of Fourth–Order Problems in the Modeling of Flexible Structures*, pp. 283–297 of "Computation and Control", Birkhäuser, Basel (1989).

[SmBoL2] *A Fully Sinc-Galerkin Method for Euler–Bernoulli Beam Models*, Numer. Methods Partial Differential Equations, v. 8 (1992) 171–202.

K. Smith

[Sh1] *Power Series From a Computational Point of View*, Springer–Verlag, New York (1987).

W. Splettstösser, R.L. Stens, and G. Wilmes

[SpStWm1] *On Approximation by the Interpolating Series of G. Valiron*, Functiones et Approximatio Commentarii Mathematici, v. XI (1981) 39–56.

W. Squire

[Sq1] *Numerical Evaluation of Integrals Using Moran Transformations*, Aerospace Engrg. Report No. Tr–14 (1969).

F. Stenger

[S1] *Bounds on the Error of Gauss–type Quadratures*, Num. Math., v. 8 (1966) 150–160.

[S2] *Error Bounds for the Evaluation of Integrals by Repeated Gauss–type Formulas*, Num. Math., v. 9 (1966) 200–213.

[S3] *Error Bounds for Asymptotic Solution of Differential Equations: I. The Distinct Eigenvalue Case*, J. Res. NBS–B. Math. and Math. Phys., v. 708 (1966) 167–186.

[S4] *Error Bounds for Asymptotic Solutions of Differential Equations, II. The General Case*, J. Res. NBS–B. Math. and Math. Phys., v. 708 (1966) 187–210.

[S5] *On the Asymptotic Solution of the Riccati Differential Equation*, (21 typed pages), in manuscript.

[S6] *Kronecker Product Extension of Linear Operators*, SIAM J. Numer. Anal., v. 5 (1968) 422–435.

[S7] *The Asymptotic Approximation of Certain Integrals*, J. Math. Anal., v. 1 (1970) 392–404.

[S8] *Constructive Proofs for Approximation by Inner Functions*, J. Approx. Theory, v. 4 (1971) 372–386.

[S9] *The Reduction of Two–Dimensional Integrals into a Finite Number of One–Dimensional Integrals*, Aeq. Math., v. 6 (1971) 278–287.

[S10] *Transform Methods of Obtaining Asymptotic Expansions of Definite Integrals*, SIAM J. Math. Anal., v. 3 (1972) 20–30.

[S11] *The Approximate Solution of Wiener–Hopf Integral Equations*, J. Math. Anal. Appl., v. 37 (1972) 687–724.

[S12] *The Approximate Solution of Convolution–type Integral Equations*, SIAM J. Math. Anal., v. 4 (1973) 536–555.

[S13] *An Algorithm for Solving Wiener–Hopf Integral Equations*, COMM ACM, v. 16 (1973) 708–710.

[S14] *Integration Formulas Based on the Trapezoidal Formula*, J. Inst. Maths Applics, v. 12 (1973) 103–114.

[S15] *On the Convergence and Error of the Bubnov–Galerkin Method*, pp. 448–464 of "SIAM Proceedings of the Conference on the Numerical Solution of Ordinary Differential Equations," Springer–Verlag Lecture Notes in Mathematics, v. 362, Springer-Verlag, New York (1973).

[S16] *An Analytic Function which is an Approximate Characteristic Function*, SIAM J. Numer. Anal., v. 12 (1975) 239–254.

[S17] *Connection Between a Cauchy System Representation of Kalaba and Fourier Transforms*, App. Math. and Comput., v. 1 (1975) 83–91.

[S18] *Computing the Topological Degree of a Mapping in* $\mathbb{R}^n$, Numer. Math., v. 25 (1975) 23–38.

[S19] *Approximations, via the Whittaker Cardinal Function*, J. Approx. Theory, v. 17 (1976) 222–240.

[S20] *Remarks on Integration Formulas Based on the Trapezoidal Formula*, J. Inst. Maths Applics, v. 19 (1977) 145–147.

[S21] *Upper and Lower Estimates on the Rate of Convergence of Approximations in* $\mathbf{H}^p$, Bull. AMS, v. 84 (1978) 145–148.

[S22] *Optimum Convergence of Minimum Norm Approximations in* $\mathbf{H}^p$, Numer. Math., v. 29 (1978) 345–362.

[S23] *A Sinc–Galerkin Method of Solution of Boundary Value Problems*, Math. Comp., v. 33 (1979) 85–109.

[S24] *Numerical Methods Based on the Whittaker Cardinal, or Sinc Functions*, SIAM Rev., v. 23 (1981) 165–224.

[S25] *Polynomial, Sinc, and Rational Function Methods for Approximating Analytic Functions*, pp. 244–251 of "Proceedings of the 1983 Tampa Conference on Rational Approximation and Interpolation," Springer–Verlag Lecture Notes in Math., # 1105, Springer -Verlag, New York (1984).

[S26] *Sinc Methods for Solving Partial Differential Equations*, in "Contributions of Mathematical Analysis to the Approximate Solution of Partial Differential Equations," A. Miller, Ed., Proceedings of the Centre for Mathematical Analysis, Australian National University, v. 7 (1984) 40–64.

[S27] *Sinc Methods of Approximate Solution of Partial Differential Equations*, pp. 244-251 of "IMACS Proceedings on Advances in Computer Methods for Partial Differential Equations," Philadelphia, P.A.(1984).

[S28] *Explicit, Nearly Optimal, Linear Rational Approximations with Preassigned Poles*, Math. Comp., v. 47 (1986) 225–252.

[S29] *Explicit Approximate Methods for Computational Control Theory*, pp. 299–316 of "Computation and Control," Edited by K.L. Bowers and J. Lund, Birkhäuser, Basel (1989).

[S30] *The Scattered Field Due to an Axially–Symmetric Body with Corners and Edges*, in manuscript.

[S31] *Asymptotic Ultrasonic Inversion Based on Using More than One Frequency*, Acoustical Imaging 11 (1982) 425–444.

F. Stenger, M. Hagmann, and J. Schwing

[SHSw1] *Algorithm for the Electromagnetic Scattering Due to an Axially Symmetric Body with an Impedance Boundary Condition*, J. Math. Anal. Appl., v. 78 (1980) 531–573.

F. Stenger, B. Keys, M. O'Reilly, and K. Parker

[SOrPk1] *Sinc Indefinite Integration and Initial Value Problems*, pp. 281–282 of "*Numerical Integration*, Recent Developments, Software and Applications", edited by T.O. Espelid and A. Genz, NATO ASI Series, Series C: Mathematical and Physical Sciences, v. 357, Kluwer Academic Publishers, Dordrecht (1992).

[SOrPk2] *Sinc Computer Package for Initial Value Problems*, (in preparation).

F. Stenger, K. McArthur, and J. McNamee

[SMMc1] *Analysis of Numerical Methods*, textbook, (to appear).

F. Stenger, M.J. Berggren, Y. Li, and S.A. Johnson

[SBLjJ1] *An Adaptive, Noise Tolerant, Frequency Extrapolation Algorithm for Diffraction Connected Ultrasonic Tomography*, IEEE Ultrasonics Symposium Proceedings (1983) 726–731.

T.J. Stieltjes

[Si1] *Recherches sur les Fractions Continues*, Ann. Fac. Sci. Toulouse, v. 8 (1894) 1–22; (1895) 1–47.

M. Stromberg

[St1] *Solution of Shock Problems by Methods Using Sinc Functions*, Ph.D. Thesis, University of Utah (1988).

[St2] *Approximate Solution of Quasilinear Equations of Conservation Law Type*, pp. 316–331 of "Computation and Control," Edited by K.L. Bowers and J. Lund, Birkhäuser, Basel (1989).

[St3] *Simplificial Sinc Methods for Approximation and Quadrature on Polyhedra, with Applications*, Dept. of Math., Texas Tech Univ. (preprint).

M. Sugihara

[Su1] *Multidimensional Numerical Integration Using Good Lattice Points*, Ph.D. Thesis, University of Tokyo (1982).

[Su2] *Methods of Numerical Integration of Oscillatory Functions by the DE-Formula with the Richardson Extrapolation*, Jour. Comp. Appl. Math., v. 17 (1987) 47–68.

[Su3] *A Class of Functions for which the Trapezoidal Rule Gives the Exact Value of Integral over the Infinite Integral*, Jour. Comp. Appl. Math., v. 20 (1987) 1–6.

J. Szabados

[Ss1] *Uniform Approximation of Continuous Functions by Rational Functions*, Ph.D. Thesis, Hungarian Academy of Sciences (1968).

O. Szasz

[Sz1] *Generalization of S. Bernstein's Polynomials to the Infinite Interval*, J. Res. NBS, v. 45 (1950) 239–245.

G. Szegö

[Sze1] *Orthogonal Polynomials*, AMS Colloquium Publications, v. XXIII, (1939).

H. Takahasi and M. Mori

[TaMo1] *Quadrature Formulas Obtained by Variable Transformation*, Numer. Math., v. 21 (1973) 206–219.

[TaMo2] *Double Exponential Formulas for Numerical Integration*, Publ. RIMS, Kyoto Univ., v. 9 (1974) 721–741.

T.N. Thiele

[Th1] *Differences Reciproques*, Overs. danske Vids. Selsk. Forhandl. (Sitzber., Akad. Kopenhagen) (1906).

A.N. Tikhonov and V.Y. Arsenin

[Th1] *Solution of Ill-Posed Problems*, Winston, Washington, D.C. (1977).

E.C. Titchmarsh

[Tt1] *Eigenfunction Expansions Associated with Second-Order Differential Equations*, Part I, Clarendon Press, Oxford (1962).

J. Todd

[Td1] *Inequalities of Chebyshev, Zolotareff, Cauer, and W.B. Jordan*, in "Inequalities," Edited by O. Shisha, Academic Press, New York (1967) 321–329.

[Td2] *A Legacy from E.I. Zolotarev (1847–1878)*, The Mathematical Intelligencer, v.10 (1988) 50–53.

H. Toda and H. Ono

[ToO1] *Some Remarks for Efficient Usage of the Double Exponential Formulas*, Res. Inst. Math. Sci. Kyoto Univ., No. 339 (1978) 74–109.

L.N. Trefethen

[Tr1] *Paraxial Approximation—Numerical Chebyshev–Padé Approximations to* $\sqrt{1-x^2}$, preprint, Courant Institute (1984).

[Tr2] *MATLAB Programs for CF Approximation*, Dept. of Math., MIT preprint.

L.N. Trefethen and M.H. Gutknecht

[TrGu1] *On Convergence and Degeneracy in Rational Padé and Chebyshev Approximation*, SIAM J. Math. Anal., v. 16 (1985) 198–210.

H.L. Turrittin

[Tt1] *Convergent Solutions of Ordinary Differential Equations in the Neighborhood of an Irregular Singular Point*, Acta. Math., v. 93 (1955) 27–66.

R.S. Varga

[V1] *Topics in Polynomial and Rational Interpolation and Approximation*, Les Presses de l'Université de Montréal no. 55 (1982).

E. Venturino

[Ve1] *On Solving Singular Integral Equations, via a Hyperbolic Tangent Quadrature Rule*, Math. Comp., v. 47 (1986) 159–167.

H. Wallin

[W1] *On the Convergence Theory of Padé Approximants*, pp. 461–469 of "Linear Operators and Approximations," ISNA v. 20, AMS, Providence, R.I. (1960).

J.L. Walsh

[Wa1] *Interpolation and Approximation by Rational Functions in the Complex Domain*, 3rd Edition, AMS Colloquium Publications, v. 20 (1960).

E.T. Whittaker

[We1] *On the Functions which are Represented by Expansion of the Interpolation Theory*, Proc. Roy. Soc. Edinburgh, v. 35 (1915) 181–194.

J.M. Whittaker

[Wh1] *Interpolatory Function Theory*, Cambridge University Press, Cambridge (1935).

D.V. Widder

[Wi1] *The Laplace Transform*, Princeton University Press, Princeton, N.J. (1971).

K. Wilderotter

[Wi1] *n–Widths of $\mathbf{H}^p$–Spaces in $\mathbf{L}_q(-1,1)$*, Jour. Complexity, v.8 (1992) 324–335.

H. Wilf

[Wl1] *Exactness Condition in Numerical Quadrature*, Numer. Math., v. 6 (1964) 315–319.

G. Yin

[Y1] *Sinc Collocation Method with Orthonormalization for Singular Poisson–like Problems*, to appear in Math. Comp.

[Y2] *Sinc Method for Incompressible Navier–Stokes Equations*, Ph.D. Thesis, Univ. of Utah (1991).

A.I. Zayed

[Z1] *Sampling Expansion for the Continuous Bessel Transform*, Appl. Anal., v. 27 (1988) 57–65.

A.I. Zayed, G. Hinsen, and P.L. Butzer

[ZHiBu1] *On Lagrange Interpolation and Kramer–Type Sampling Theorems Associated with Sturm–Liouville Problems*, Aachen Rheinish Westfälische Technische Hochschule Publication (1984).

Index

SYMBOLS

SPRINGER SERIES IN COMPUTATIONAL MATHEMATICS

Volume 8
E. Hairer, S.P. Nørsett, G. Wanner
Solving Ordinary Differential Equations I
Nonstiff Problems
1987.
Second edition in preparation.

Volume 9
Z. Ditzian, V. Totik
Moduli of Smoothness
1987. IX, 227 pp.
Hardcover ISBN 3-540-96536-X

Volume 10
Yu. Ermoliev. R.J.-B. Wets (Eds.)
Numerical Techniques for Stochastic Optimization
1988. XV, 571 pp. 62 figs.
Hardcover ISBN 3-540-18677-8

Volume 11
J-P. Delahaye
Sequence Transformations
With an Introduction by C. Brezinski
1988. XXI, 252 pp. 164 figs.
Hardcover ISBN 3-540-15283-0

Volume 12
C. Brezinski
History of Continued Fractions and Pade Approximants
1990. VIII, 561 pp. 6 figs.
Hardcover ISBN 3-540-15286-5

Volume 13
E.L. Allgower, K. Georg
Numerical Continuation Methods
An Introduction
1990. XIV, 388 pp. 37 figs.
Hardcover ISBN 3-540-12760-7

Volume 14
E. Hairer, G. Wanner
Solving Ordinary Differential Equations II
Stiff and Differential-Algebraic Problems
1991. XV, 601 pp. 150 figs.
Hardcover ISBN 3-540-53775-9

Volume 15
F. Brezzi, M. Fortin
Mixed and Hybrid Finite Element Methods
1991. X, 360 pp. , 65 figs.
Hardcover ISBN 0-387-97582- 9

Springer-Verlag
New York
Berlin
Heidelberg
London
Paris
Tokyo
Hong Kong
Barcelona
Budapest